EYE MOVEMENTS:
A WINDOW ON MIND AND BRAIN

EYE MOVEMENTS: A WINDOW ON MIND AND BRAIN

Edited by

ROGER P. G. VAN GOMPEL
University of Dundee

MARTIN H. FISCHER
University of Dundee

WAYNE S. MURRAY
University of Dundee

and

ROBIN L. HILL
University of Edinburgh

ELSEVIER

Amsterdam • Boston • Heidelberg • London • New York • Oxford • Paris
San Diego • San Francisco • Singapore • Sydney • Tokyo

Elsevier
The Boulevard, Langford Lane, Kidlington, Oxford OX5 1GB, UK
Radarweg 29, PO Box 211, 1000 AE Amsterdam, The Netherlands

First edition 2007

British Library Cataloguing in Publication Data
A catalogue record for this book is available from the British Library

Library of Congress Cataloging-in-Publication Data
A catalog record for this book is available from the Library of Congress

ISBN-13: 978-0-08-044980-7
ISBN-10: 0-08-044980-8

For information on all Elsevier publications
visit our website at books.elsevier.com

Printed and bound in the United Kingdom
Transferred to Digital Print 2009

To Alan Kennedy

CONTENTS

PREFACE

This book documents much of the state of the art in current eye-movement research by bringing together work from a wide variety of disciplines. The book grew out of ECEM 12, the very successful 12th European Conference on Eye Movements that was held at the University of Dundee in August 2003. This biennial conference attracts attendants with very diverse backgrounds who present the latest findings in eye-movement research, and it is probably unique in the way that it fosters interactions between specialists from quite different backgrounds, but all of whom work with eye movements in one form or another. The format of this book is similar in that it aims to provide an overview of the latest research in a wide variety of eye-movement disciplines and intends to encourage cross-fertilisation between these different areas. Thus, this book is intended to be an authoritative reference for everyone with an interest in eye-movement research.

The book is not merely a collection of research presented at ECEM 12. Because there is always a necessary delay between the time that findings are presented at a conference and subsequent publication, we asked conference attendees to report their latest research. In addition, we asked a number of authors who did not attend ECEM 12 to write a chapter related to aspects of eye-movement research that were not in fact presented at ECEM 12. The chapters were carefully selected from abstracts that the authors sent us and each chapter was peer-reviewed.

We would like to thank the reviewers of the chapters, who greatly contributed to the quality of this book. A list of all reviewers is included on p. xxxv. We would also like to thank the people at Elsevier Science, who worked hard to make this book possible. Finally, we are very grateful to our colleagues in the School of Psychology at the University of Dundee for their support. In particular, Nick Wade for his helpful advice and suggestion for the title of this book; Ben Tatler, who reviewed several chapters at short notice and even offered to stand-in in case one of the chapters dropped out (and did not complain when that didn't happen); and last, but certainly not least, Alan Kennedy, without whom this book would not exist. It is because of Alan's long-standing contributions to the field and many organisational efforts that ECEM 12 was held in Dundee and was such a great success. Furthermore, Alan shaped the eye-movement research group in the School of Psychology at the University of Dundee and made it an attractive and pleasant environment in which to work. Without him, we would probably not be in bonny Dundee and we may not have conducted eye-movement research with anything like the enthusiasm we now have. We therefore dedicate this book to Alan.

LIST OF CONTRIBUTORS

Irene Armstrong
Queen's University
Centre for Neuroscience Studies, Queen's University, Kingston, Ontario, Canada, K7L 3N6

Karl G. D. Bailey
Andrews University
Behavioral Sciences Department, Andrews University, 100 Old US 31, Berrien Springs, MI, USA, 49104-0030

Raymond Bertram
University of Turku
Department of Psychology, University of Turku, FIN-20014 Turku, Finland

Lo J. Bour
University of Amsterdam
Department of Neurology/Clinical Neurophysiology, Academic Medical Centre, University of Amsterdam, Meibergdreef 9, 1105 AZ Amsterdam, The Netherlands

James R. Brockmole
University of Edinburgh
Department of Psychology, University of Edinburgh, 7 George Square, Edinburgh EH8 9JZ, Scotland, United Kingdom

Matthew R. G. Brown
University of Western Ontario
University of Western Ontario, Neuroscience Program, 1151 Richmond St, SDRI rm. 216, London, Ontario, Canada, N6G 2V4

Monica S. Castelhano
University of Massachusetts, Amherst
Department of Psychology, University of Massachusetts, 423 Tobin Hall, Amherst, MA 01003, USA

Charles Clifton, Jr.
University of Massachusetts, Amherst
Department of Psychology, University of Massachusetts, Amherst, MA 01003, USA

Brian Coe
Queen's University
Centre for Neuroscience Studies, Queen's University, Kingston, Ontario,
Canada, K7L 3N6

Eleanor Cross
University of Nottingham
School of Psychology, University of Nottingham, Nottingham NG7 2RD, England,
United Kingdom

Delphine Dahan
University of Pennsylvania
Department of Psychology, University of Pennsylvania, 3401 Walnut street, Philadelphia,
PA 19104, USA

Peter De Graef
University of Leuven
Laboratory of Experimental Psychology, University of Leuven, Tiensestraat 102,
B-3000 Leuven, Belgium

Heiner Deubel
Ludwig-Maximilians-Universität
Department Psychologie, Ludwig-Maximilians-Universität, Leopoldstr.13,
81369 München, Germany

Jason Droll
University of California
Psychology Department, University of California, Santa Barbara, CA 93106-9660, USA

Ava Elahipanah
University of Toronto, Mississauga
Department of Psychology, University of Toronto at Mississauga, 3359 Mississauga Road
N. RM 2037B, Mississauga, Ontario, Canada L5L 1C6

Ralf Engbert
University of Potsdam
Computational Neuroscience Lab, Department of Psychology, University of Potsdam,
Karl-Liebknecht-Str. 24–25, 14476 Potsdam, Germany

Stefan Everling
University of Western Ontario
Robarts Research Institute, 100 Perth Drive, London, Ontario, Canada, N6A 5K8

Fernanda Ferreira
University of Edinburgh
Department of Psychology, University of Edinburgh, 7 George Square,
University of Edinburgh, Edinburgh, EH8 9JZ, Scotland, United Kingdom

John M. Findlay
University of Durham
Department of Psychology, University of Durham, South Road, Durham, DH1 3LE,
England, United Kingdom

Martin H. Fischer
University of Dundee
Department of Psychology, University of Dundee, Dundee DD1 4HN, Scotland,
United Kingdom

Kristen A. Ford
University of Western Ontario
University of Western Ontario, Neuroscience Program, 1151 Richmond St,
SDRI rm. 216, London, Ontario, Canada, N6G 2V4

Lynn Gareze
University of Durham
Department of Psychology, University of Durham, South Road, Durham, DH1 3LE,
England, United Kingdom

Michael Greig
University of Waterloo
Gait & Posture Lab, Department of Kinesiology, University of Waterloo,
200 University Ave. West, Waterloo, ON, N2L 3G1, Canada

Mary M. Hayhoe
University of Rochester
Center for Visual Science, University of Rochester, Rochester, NY 14627, USA

John M. Henderson
University of Edinburgh
Department of Psychology, University of Edinburgh, 7 George Square,
Edinburgh EH8 9JZ, Scotland, United Kingdom

Robin L. Hill
University of Edinburgh
Human Communication Research Centre, School of Informatics, University of Edinburgh, 2 Buccleuch Place, Edinburgh, EH8 9LW, Scotland, United Kingdom

Ignace Th. C. Hooge
Utrecht University
Psychonomics, Helmholtz Instituut, Utrecht University, Heidelberglaan 2, NL-3584 CS Utrecht, The Netherlands

Louise Humphreys
University of Nottingham
School of Psychology, University of Nottingham, Nottingham NG7 2RD, England, United Kingdom

Jukka Hyönä
University of Turku
Department of Psychology, University of Turku, FIN-20014 Turku, Finland

Albrecht Inhoff
State University of New York, Binghamton
Binghamton University, State University of New York, Psychology Department, P.O. Box 6000, Binghamton, NY 13902-6000, USA

Rebecca L. Johnson
University of Massachusetts, Amherst
Department of Psychology, Tobin Hall, Amherst, MA 01003, USA

Barbara J. Juhasz
Wesleyan University
Department of Psychology, Wesleyan University, 207 High Street, Middletown, CT 06459-0408, USA

Reinhold Kliegl
University of Potsdam
Cognitive Psychology, Department of Psychology, University of Potsdam, Karl-Liebnecht-Str. 24–25, 14476 Potsdam, Germany

Pia Knoeferle
Saarland University
FR 4.7, Computational Linguistics, Building C71, Saarland University, Postbox 151150, 66041 Saarbrücken, Germany

Carmen Koch
Ludwig-Maximilians-Universität
Department Psychologie, Ludwig-Maximilians-Universität, Leopoldstr.13,
81369 München, Germany

Gustav Kuhn
University of Durham
Department of Psychology, University of Durham, South Road, Durham, DH1 3LE,
England, United Kingdom

Michael F. Land
University of Sussex
Department of Biology and Environmental Science, University of Sussex, Falmer,
Brighton BN1 9QG, England, United Kingdom

Michael Mack
Vanderbilt University
411 Wilson Hall, Vanderbilt University, Nashville, TN 37212, USA

David Melcher
Oxford Brookes University
Department of Psychology, Oxford Brookes University, Gipsy Lane, Oxford OX3 0BP,
England, United Kingdom and University of Trento, Italy.

Neil Mennie
University of Nottingham
School of Psychology, University of Nottingham, University Park, Nottingham, NG7
2RD, United Kingdom

Antje S. Meyer
University of Birmingham
School of Psychology, University of Birmingham, Edgbaston, Birmingham, B15 2TT,
England, United Kingdom

Concetta Morrone
Università Vita-Salute San Raffaele
Facoltà di Psicologia, Dipartimento di Neuroscienze, Università Vita-Salute San Raffaele,
Via Olgettina 58, Milano, Italy

Douglas P. Munoz
Queen's University
Centre for Neuroscience Studies, Queen's University, Kingston, Ontario,
Canada, K7L 3N6

Wayne S. Murray
University of Dundee
School of Psychology, University of Dundee, Dundee DD1 4HN, Scotland, United Kingdom

Alison Novak
Queen's University
Motor Performance Laboratory, Centre for Neuroscience Studies, Queen's University, Kingston, ON, K7L 3N6, Canada

Antje Nuthmann
University of Potsdam
Cognitive Psychology, Department of Psychology, University of Potsdam, Karl-Liebknecht-Str. 24–25, 14476 Potsdam, Germany

Eelco A. B. Over
Utrecht University
Physics of Man, Helmholtz Instituut, Utrecht University, Princetonplein 5, NL-3584 CC Utrecht, The Netherlands

Aftab E. Patla
University of Waterloo
Gait & Posture Lab, Department of Kinesiology, University of Waterloo, 200 University Ave. West, Waterloo, ON, N2L 3G1, Canada

Jeff B. Pelz
Rochester Institute of Technology
Center for Imaging Science, Rochester Institute of Technology, 54 Lomb Memorial Drive, Rochester, NY 14623, USA

Alexander Pollatsek
University of Massachusetts, Amherst
Department of Psychology, University of Massachusetts, Tobin Hall, Amherst, MA 01993-7710 USA

Ralph Radach
Florida State University
Department of Psychology, Florida State University, Tallahassee, FL 32306-1270, USA

Keith Rayner
University of Massachusetts, Amherst
Department of Psychology, University of Massachusetts, Amherst, MA 01003, USA

Erick D. Reichle
University of Pittsburgh
University of Pittsburgh, 635 LRDC, 3939 O'Hara Street, Pittsburgh, PA 15260, USA

Ronan Reilly
National University of Ireland, Maynooth
Department of Computer Science, Callan Building, National University of Ireland, Maynooth, Maynooth, Co. Kildare, Ireland

Eyal M. Reingold
University of Toronto, Mississauga
Department of Psychology, University of Toronto at Mississauga, 3359 Mississauga Road N. RM 2037B, Mississauga, Ontario, Canada, L5L 1C6

Constantin Rothkopf
University of Rochester
Center for Visual Science, Brain and Cognitive Sciences Department, University of Rochester, Meliora Hall, Rochester, NY 14627-0270, USA

Anne Pier Salverda
University of Rochester
Department of Brain and Cognitive Sciences, University of Rochester, Meliora Hall, Rochester, NY 14627, USA

Jiye Shen
University of Toronto, Mississauga
Department of Psychology, University of Toronto at Mississauga, 3359 Mississauga Road N. RM 2037B, Mississauga, Ontario, Canada, L5L 1C6

Adrian Staub
University of Massachusetts, Amherst
Department of Psychology, University of Massachusetts, Amherst, MA 01003, USA

Michael K. Tanenhaus
University of Rochester
Department of Brain and Cognitive Sciences, University of Rochester, Meliora Hall, Rochester, NY 14627, USA

Benjamin W. Tatler
University of Dundee
School of Psychology, University of Dundee, Dundee, DD1 4HN, Scotland, United Kingdom

S. Sebastian Tomescu
Queen's University
Faculty of Medicine, Undergraduate Medical Education Office, 68 Barrie Street, Kingston, ON, K7L 3N6, Canada

Geoffrey Underwood
University of Nottingham
School of Psychology, University of Nottingham, Nottingham NG7 2RD, England, United Kingdom

Johan N. van der Meer
University of Amsterdam
Department of Neurology/Clinical Neurophysiology, Academic Medical Centre, University of Amsterdam, Meibergdreef 9, 1105 AZ, Amsterdam, The Netherlands

Femke van der Meulen
University of Birmingham

Roger P. G. van Gompel
University of Dundee
School of Psychology, University of Dundee, Dundee DD1 4HN, Scotland, United Kingdom

Anke M. van Mourik
Nederlands Kanker Instituut
Nederlands Kanker Instituut, Plesmanlaan 121, 1066 CX Amsterdam, The Netherlands

Françoise Vitu
CNRS, Université de Provence
Laboratoire de Psychologie Cognitive, CNRS, Université de Provence, Centre St Charles, Bâtiment 9, Case D, 3 place Victor Hugo, 13331 Marseille Cedex 3, France

Björn N. S. Vlaskamp
Utrecht University
Psychonomics, Helmholtz Instituut, Utrecht University, Heidelberglaan 2, NL-3584 CS Utrecht, The Netherlands

Nicholas J. Wade
University of Dundee
School of Psychology, University of Dundee, Dundee DD1 4HN, Scotland, United Kingdom

Gerald Westheimer
University of California
Division of Neurobiology, University of California, 144 Life Sciences Addition, Berkeley, CA 94720-3200, USA

Linda R. Wheeldon
University of Birmingham
School of Psychology, University of Birmingham, Edgbaston, Birmingham B15 2TT, England, United Kingdom

Sarah J. White
University of Leicester
School of Psychology, Henry Wellcome Building, University of Leicester, Lancaster Road, Leicester, LE1 9HN, United Kingdom

Shun-nan Yang
Smith-Kettlewell Eye Research Institute
Smith-Kettlewell Eye Research Institute, 2318 Fillmore Street, San Francisco, CA 94115, USA

REVIEWERS

We would like to thank the reviewers of the chapters:

Karl Bailey
Raymond Bertram
Marc Brysbaert
Martin Corley
Heiner Deubel
John Findlay
Ken Forster
Adam Galpin
Stefan Glasauer
Harold Greene
Zenzi Griffin
Thomas Haslwanter
Mary Hayhoe
Jukka Hyönä
Juhani Järvikivi
Rebecca Johnson
Barbara Juhasz
Alan Kennedy
Reinhold Kliegl
Pia Knoeferle
Gustav Kuhn
Simon Liversedge
Mike Land
Antje Meyer
Robin Morris
Aftab Patla
Jeff Pelz
Martin Pickering
Joel Pynte
Ralph Radach
Keith Rayner
Eyal Reingold
Christoph Scheepers
Ben Tatler
Geoff Underwood
Boris Velichkovsky
Gerald Westheimer
Sarah White

Chapter 1

EYE-MOVEMENT RESEARCH: AN OVERVIEW OF CURRENT AND PAST DEVELOPMENTS

ROGER P. G. VAN GOMPEL, MARTIN H. FISCHER AND WAYNE S. MURRAY
University of Dundee, UK

ROBIN L. HILL
University of Edinburgh, UK

Eye Movements: A Window on Mind and Brain
Edited by R. P. G. van Gompel, M. H. Fischer, W. S. Murray and R. L. Hill

Abstract

This opening chapter begins with an overview of the issues and questions addressed in the remainder of this book, the contents of which reflect the wide diversity of eye-movement research. We then provide the reader with an up-to-date impression of the most significant developments in the area, based on findings from a survey of eye-movement researchers and database searches. We find that the most heavily cited publications are not necessarily those rated as most influential. It is clear that eye-movement research is published across a wide variety of journals and the number of articles is increasing, but the relative proportion of eye-movement articles has remained almost constant. The United States produces the largest number of these articles, but other countries actually produce more per capita. Finally, we see that computational modelling, new eye-tracking technologies and anatomical and physiological mapping of the visual-oculomotor system are viewed as the most important recent developments.

> Orpheus called the eyes, the looking glasses of nature: Hesichius, the doores for the Sunne to enter in by: Alexander the Peripatecian, the windowes of the mind, because that by the eyes we doe cleerely see what is in the same, we pearce into the deepe thoughts thereof, and enter into the privities of his secret chamber. (Du Laurens, 1596, translated by Surphlet, 1599, p. 19)

When Du Laurens (1596), a sixteenth-century French anatomist and medical scientist, referred to the eyes as *windowes of the mind,* he reflected a view both ancient and modern. And it seems clear today that people's eye movements do indeed reveal much about the workings of both mind and brain. In recent years, the study of eye movements has become one of the most important and productive ways for investigating aspects of mind and brain across a wide variety of topic areas. This book contains a collection of chapters that reflect that breadth and, we believe, the sort of productivity that stems from this type of research.

The 12th European Conference on Eye Movements, ECEM12, held in Dundee, Scotland, provided the foundation for this collection. This biennial conference brings together a large number of eye-movement researchers drawn from a wide variety of different disciplines, such as clinical researchers, neurophysiologists, cognitive psychologists, vision scientists and ergonomists, to mention but a few, and this book contains a similarly wide range of research topics. The research is grouped into eight parts, each of which presents a different area central to current eye-movement research. Each section begins with an overview chapter that provides a critical review of current research in the area, followed by a series of chapters reporting new findings and addressing new topics. Of course, the organisation into sections does not imply that there are no links between the chapters in the different sections. Although eye-movement research can be divided into several fairly well-defined research areas, much research draws on findings from more than a single area, and research frequently crosses the boundaries between different disciplines – perhaps one reason why the interdisciplinary nature of ECEM has always been so successful.

The chapters in this book present an overview of current developments in eye-movement research. However, eye-movement research has become a vast area, so naturally, some research fields are better represented than others. Included in this book are areas that have recently developed as well as established areas with a much longer tradition in eye-movement research. Recent developments include the history of eye-movement research (Part 1), eye movements as a method for investigating spoken language processing (Part 6), and eye movements in natural environments (Part 8). The modelling of eye movements (Part 4), although not really new, has also seen a great deal of recent change and development. Areas that have been longer established but continue to produce important new findings are physiology and clinical studies of eye movements (Part 2), transsaccadic integration (Part 3), eye-movements and reading (Part 5), and eye movements as a method for investigating attention and scene perception (Part 7).

In this opening chapter, we aim to highlight the most important developments in eye-movement research. We begin by providing an overview of the chapters in the different sections of this book. This should provide the reader with a thorough subjective impression of the most important current developments in eye-movement research. We then follow this with a more objective form of assessment. We consider the state of eye-movement research (both in the present and the past) by reporting results from a survey sent to participants of ECEM12 and a journal database search. Based on this combination, our aim is to provide the reader with a synoptic overview of current and past developments across the field in general.

1. Overview of the parts in this book

1.1. History of eye-movement research

Looking backwards is the perfect way to begin a book on eye-movement research. Knowing the genesis and evolution of any subject is critical to understanding the motivation and driving forces (both psychological and technological) that lie behind scientific progress in the area. Our first part therefore examines the historical foundations of eye-movement research; something which has permeated through an extremely rich and diverse range of multidisciplinary empirical topics.

The opening chapter by Wade places the birth of the modern scientific study of eye movements firmly in the eighteenth century, before charting its historical progress via the thoughts and works of many generations of eminent scientists. Central to this was the growth in physiological experimentation, and the examination/exploitation of visual vertigo, torsion, and afterimages in particular. Wade also describes how through the ingenious use of technology, techniques for acquiring eye-movement records and studying scan paths were developed to facilitate reading research and the study of visual perception.

The immediate post-war period is the focus of Westheimer's historical account. He documents how multidisciplinary scientific work forged under conditions of war later sparked a mini-renaissance in instrumentation and neurophysiological techniques. This naturally fed through into a greater understanding of the oculomotor system and helped establish the investigation of eye movements as a major tool in psychological research. Following Westheimer, Land's exposition begins in the 1950s but rapidly brings us up to date. Again, he relates how the development of modern technology, choosing head-mounted mobile eye-trackers as his specific example, has led to advances across a range of eye-movement research areas. The flexibility afforded by such devices frees us from many constraints (often literally in the case of head-vices and bite-bars) and opens up the "real" world to greater investigation. Together, these three chapters make it clear that the act of looking backwards results in our forward vision becoming sharper and brighter.

1.2. Physiology and clinical studies of eye movements

The second part of the book deals with physiological and clinical studies of eye movements. First, Munoz, Armstrong, and Coe provide an authoritative overview on how to use simple eye-movement tasks to diagnose both the developmental state and the possible dysfunctions of the cortical and subcortical oculomotor system. Their accumulator model allows an understanding of performance profiles across a wide range of deficits, such as Parkinson's disease, Tourette syndrome, and Attention Deficit Hyperactivity Disorder.

In the second Chapter in this part, Ford, Brown, and Everling report an fMRI study which shows that anti-saccade task performance depends on specific brain activation prior to stimulus presentation. Their work describes the relationship between frontal brain activity and the accuracy of a subsequent saccadic decision. It also illustrates how eye-tracking can be combined with high-tech brain imaging methods that have become popular over the last two decades.

The final chapter in this part is by Bour, Van der Meer, and Van Mourik. These authors describe commutative eye rotations in congenital nystagmus (a fixation instability resulting from visual or oculomotor deficits) and a computational model that predicts anatomical aberrations in a patient with congenital nystagmus. Together, the contributions in this part illustrate how a multifaceted clinical and neurophysiological approach, spanning behavioural and brain imaging studies as well as computer simulations and patient studies, can help us understand the functioning of eye-movement control structures in the brain.

1.3. Transsaccadic integration

Part 3 focuses on a topic that has intrigued investigators for many years. As De Graef points out in the first chapter in this part, there has been an ongoing mystery associated with the question of how we "integrate" information across a temporally and spatially discontinuous train of "snapshots". This mystery has been sufficiently deep to even encourage some to conclude that the question is a mistaken one, and in various important senses we do not in fact integrate information across successive fixations at all. But the chapters in this part suggest that the pendulum is now swinging back. De Graef argues that the bases behind that view are flawed and that there is good reason to consider the mechanisms involved in the integration of information across fixations during the process of scene and object recognition, especially as these relate to the role played by parafoveal preview.

In the second Chapter in this part, Koch and Deubel address the question of how we achieve a stable representation of the location of objects in the world across fixations. That is, how the visual system maps and re-maps the continually changing retinal input into a consistent and stable perceptual frame. They suggest that this is based on the use of "landmarks" in the visual field and demonstrate in a series of experiments that relatively low-level configurational properties of the landmarks drive the relocation process, with the visual system apparently operating on the (perfectly reasonable) assumption that the world is unlikely to have changed significantly while the eyes are being moved.

In the final chapter, Melcher and Morrone address questions concerning the integration of information related to both the location and the form of objects across fixations. Their experiments provide evidence that visually detailed but temporally inexact information is retained across saccades, and they suggest that our subjective impression of a stable visual world is derived from a combination of a fast-acting saccadic re-mapping, reliant, again, on the assumption of invariance over time, and a more durable scene-based memory.

1.4. Modelling of eye movements

Modelling of eye movements has been one of the clearest growth areas in eye-movement research since the late 1990s (see section 2.3). It seems clear that great advances have been made in this area, at least in so far as this relates to reading, but this is certainly not an area without controversy. The nature of these controversies and some important issues arising from them are well documented in the introductory chapter by Radach, Reilly, and Inhoff. These authors provide a very useful overview and classification of model types and then proceed to highlight some of the important questions that models of eye-movement control in reading need to address. One of these is the role played by attention, but as Radach et al. point out, this concept itself is frequently underspecified and models need to be more explicit about the characteristics of attention that are assumed. They conclude the chapter with some proposals regarding model testing and some challenges for the modelling enterprise.

In the following chapter, Reichle, Rayner, and Pollatsek explore aspects of the E-Z Reader model in the context of lexical ambiguity. This serves two primary purposes: It allows additional investigation of the distinction proposed in the model between the "familiarity check" (L1) and the completion of lexical processing (L2) and it uses lexical ambiguity as a tool to investigate the way in which the model handles the effects of higher-level sentence context on word identification. The authors incorporate changes to E-Z Reader that closely resemble the assumptions made by the reordered access model of lexical ambiguity resolution, and thus this represents a step further along the path that they have noted for some time as a necessary extension to this model.

Yang and Vitu then concentrate on the modelling of saccade targeting procedures during reading. In contrast to the many models which propose a uniform, generally word-based, oculomotor targeting process, they show how well a model postulating a combination of targeting strategies performs. They suggest that strategy-based saccades of a relatively constant length are common at early time intervals during a fixation, while visually guided (word-based) saccades only become frequent after longer latencies. They conclude that eye guidance in reading is the result of a dynamic coding of saccade length rather than the sort of cognitively based aiming strategies related to the requirements of lexical identification and the shift of attention that are assumed by many of the current models.

In the final chapter of this part, Engbert, Nuthmann, and Kliegl discuss a model that does assume word-based targeting strategies – SWIFT – but they focus on the question of when those strategies do not produce the desired outcome and result in "mislocated fixations". They use simulations with the SWIFT model as a way of checking an algorithm that iteratively decomposes observed landing position distributions into mislocated and well-located cases, and conclude that "mislocated fixations" occur remarkably frequently, especially when the intended target is a short word. They conclude by discussing the link between mislocated fixations and the inverted optimal viewing position (IOVP) effect.

1.5. Eye movements and reading

One of the most productive areas in eye-movement research is the investigation of reading processes. This part begins with a comprehensive review of the research into human sentence processing that has arisen from studying eye movements whilst people read. Clifton, Staub, and Rayner have compiled a list of 100 key articles in this area of language processing. Beginning with word recognition, they proceed to summarise the results of studies investigating the effects of syntactic, semantic, and pragmatic factors on sentence comprehension.

Word recognition, or rather decomposition, is the focus of the chapter by Juhasz. She examines the role of semantic transparency in English compound words, embedded in neutral-context sentences. Her results suggest that there is an early pre-semantic decomposition of compound words during reading and that decomposition occurs for both transparent and opaque compounds. In their chapter, Bertram and Hyönä also investigate the reading of compound words, but in Finnish sentences. Using an eye-contingent paradigm (altering what is displayed on a computer screen depending on where the eye is looking or when it is moving), they manipulated the letter information available in the parafovea for words with long or short first constituents and conclude that there is no morphological preview benefit: morphological structure is only determined in the fovea.

White re-examines the hypothesis that foveal load (i.e. processing difficulty) modulates the amount of parafoveal (pre)processing that takes place. She finds that manipulating localised foveal load, in the form of orthographic regularity and misspellings, does not influence the probability of the next word being skipped. However, there is still an effect on reading times and so she concludes that this may be evidence for different, independent mechanisms influencing eye-movement control. The final chapter also exploits misspellings, but this time in the parafovea itself. Johnson explores non-adjacent transposed-letter effects through the use of an eye-contingent boundary change of display. She shows that transposed-letter effects continue to occur even when the manipulated letter positions are non-adjacent and that they arise similarly across vowels and consonants.

1.6. Eye movements as a method for investigating spoken language processing

Part 6 concerns a relatively new area of eye-movement research; the use of eye-tracking as a tool for investigating spoken language processing. Research in this area took off following the seminal article by Tanenhaus, Spivey-Knowlton, Eberhard, and Sedivy (1995). What has come to be known as the *visual world method* – recording eye movements directed to objects and scenes while the participant listens to spoken text is becoming increasingly popular. In his overview chapter, Tanenhaus, a pioneer in this field, provides an overview of the most important developments since the publication of their article. He also discusses data analysis techniques and assumptions underlying the link between fixation behaviour and language processing models. Finally, he compares the results from reading studies with studies on spoken language comprehension using the visual world method, and thereby also provides a link with Part 5.

The remaining chapters in Part 6 all report research using the visual world method. The results reported by Dahan, Tanenhaus, and Salverda provide important evidence against the view that participants in visual-world experiments covertly name the pictures prior to word onset. Furthermore, they show that fixations made to pictures of words that are auditorily presented are affected both by the amount of time that people can inspect the picture before the word is presented and by the position of the pictures in the display. Bailey and Ferreira provide an excellent example of how the visual world method can be used to investigate issues that are specific to spoken (rather than written) language comprehension by investigating the effects of disfluencies (uhh) on sentence comprehension. They show that disfluencies had no effect on both online and offline parsing preferences. However, there was evidence that listeners identified ambiguities during disfluencies.

Recently, researchers have also started to use the visual world method to investigate speech production. Wheeldon, Meyer, and Van der Meulen use this method to investigate speech planning when speakers make anticipation errors (e.g., they say *whistle* instead of *ladder, whistle*). They show that the order of fixating the objects is the same, regardless of whether the words are named in the correct order or whether an anticipation error occurred. This suggests that anticipation errors are not due to premature shifts in visual attention. In the last chapter of this part, Knoeferle employs the visual world method to investigate how information provided by the scene interacts with linguistic information provided in the sentence. Her experiment shows that sentence interpretation in German word order ambiguities is closely time-locked to the point in time when the combination of both linguistic and visual information disambiguates the sentence. Hence both linguistic and visual information crucially affect sentence comprehension.

1.7. Eye movements as a method for investigating attention and scene perception

The study of eye movements has been an extremely valuable tool for investigating attention and scene perception, and this part illustrates some of the recent research in this area. The part starts with a critique of the currently fashionable concept of "visual saliency" by Henderson, Brockmole, Castelhano, and Mack. They show that the pattern of eye movements during visual search in real-world scenes cannot be accounted for by visual saliency. This work is in conflict with the currently popular approach of predicting eye behaviour on the basis of image statistics, an approach that is illustrated in the following chapter by Underwood, Humphreys, and Cross. They report effects of semantic congruency between target and background, visual saliency, and the gist of a scene on the inspection of objects in natural scenes. From these chapters it seems clear that the relative contribution of low-level visual factors and high-level cognitive factors to eye-movement control remains a topic of hot debate.

The following two chapters in this part examine the classical visual search task where observers look through a set of items in a display to determine whether a predefined target is present or absent. The two visual search studies clearly illustrate the benefits of adding eye-movement recording to this established line of research into covert attention and visual cognition. Hooge, Vlaskamp, and Over recorded fixation durations as a function of the difficulty of both the previous and the current item, as well as the expected difficulty of the search task, while the chapter by Shen, Elahipanah, and Reingold documents an important distinction between local context effects and higher-level congruency effects on eye-movement control in visual search. Both chapters resonate with the familiar theme that both top-down and bottom-up factors have an impact on eye behaviour. In the concluding chapter, Gareze and Findlay report that scene context had no effect on object detection and eye gaze capture in their study, thus making the case for a predominance of lower-level factors. This message conflicts with the results reported by Underwood et al., and the authors suggest that factors such as the eccentricity of objects from fixation and their size are important experimental variables that need to be explored in future research on eye-movement control during scene perception.

1.8. Eye movements in natural environments

With the development of new eye-trackers that allow participants to freely move around in their environment, one area that has seen a recent upsurge in interest is the investigation of eye movements in natural environments. In their chapter, Hayhoe, Droll, and Mennie provide an overview of these developments. They argue that fixation patterns are learnt behaviour: People have to learn the structure and dynamic properties of the world, the order in which a task is carried out, learn where to find the relevant information in the scene, and learn where to look to balance competing demands of the task. Therefore, as their literature review indicates, the nature of the task critically influences fixation behaviour. Pelz and Rothkopf provide one demonstration of this in their chapter. In their experiment, the extent to which people fixate on the path in front of them depends on whether they walk on pavement in a man-made environment or on a dirt path in a wooded environment. Furthermore, when standing still, fixation durations were longer in the wooded than in the man-made environment, although there was no difference during walking. Patla, Tomescu, Greig, and Novak also investigated fixation behaviour during walking. Their data suggest that there is a tight link between fixation behaviour and route planning. They present a model of route planning that is consistent with fixation behaviour. Finally, Tatler and Kuhn used eye-tracking to investigate how magicians misdirect observers. Their results show that observers follow the magician's gaze during the trick, so their fixations are directed away from the disappearing object. Fixation behaviour was also affected by whether observers had previously seen the trick. But, interestingly, even when observers watched the trick a second time and spotted it, they usually did not fixate on the disappearing object, suggesting that detection is linked to covert visual attention.

2. Questionnaire study and journal database search

The above overview gives an impression of current developments in eye-movement research, while the chapters in the remainder of this book describe those developments in more detail and refer to the latest research in these areas. However, any book is likely to underrepresent some areas of research while overrepresenting others. In order to gain a more objective overview across the area in general, in terms of both the current state of eye-movement research and the past developments, we carried out a questionnaire study and searched a database of journal articles.

For the questionnaire study, we sent out an e-mail to all attendants of ECEM 12 asking for their comments and impressions in response to a number of questions. This questionnaire was sent between March and April 2006. The attendants were asked the following four questions:

1. In what area (or areas) of eye-movement research do you work? (They could choose one of the parts in this book or specify a different area.)
2. List up to 3 eye-movement papers (in order) that you consider to have been most influential in the development of the field.
3. What do you consider to be the best journal(s) for keeping up to date with developments in eye-movement research?
4. What do you consider to be the most important development(s) in eye-movement research in the past 5 years (please use 1 or 2 sentences)?

Forty-four attendants (out of the approximately 200 who received the questionnaire) responded. Twenty-one of the respondents were authors of one of the chapters in this book.

For our database search, we used ISI Web of Knowledge (http://wos.mimas.ac.uk/), which contains articles from over 8600 journals published since 1970, of which 1790 journals are in the social sciences. This allowed us to search for journal articles with ‘eye(-)movement(s)’ in the title or keywords and for citation counts for selected eye-movement articles. Web of Knowledge also provides citation counts for selected publications that are not journal articles (e.g., books and book chapters). However, it does not output any publications other than journal articles when it is searched for ‘eye(-)movement(s)’ in the title and keywords. The searches were run on 12 and 15 May 2006.

Table 1 shows the number of respondents working in eye-movement research areas that correspond to the parts in the book. The majority of respondents were involved in research in either reading (21) or attention and scene perception (18). Both areas have traditionally been very productive in eye-movement research, and this is reflected here and in the chapters (Parts 5 and 7) in this book. There were also many respondents (15) who indicated that they were involved in the modelling of eye movements, an area which

Table 1
Number of respondents to the questionnaire working in the different eye-movement research areas

Research area	Number of researchers
1. History of eye-movement research	2
2. Physiology and clinical studies of eye movements	10
3. Transsaccadic integration	5
4. Modelling of eye movements	15
5. Eye-movements and reading	21
6. Eye-movements as a method for investigating spoken language processing	9
7. Eye-movements as a method for investigating attention and scene perception	18
8. Eye movements in natural environments	5
9. Other	7

Note: Number of researchers: Number of respondents to the questionnaire who specified that they worked in each of the eye-movement research areas ($N = 44$).

has seen important developments in recent years, as can be seen in Part 4 of this book. There was also reasonable representation in the areas of physiology and clinical studies of eye movements (10) and spoken language processing (9). This latter area has grown dramatically since the mid-1990s, and the relatively high number of respondents working in the area reflects that recent growth. Physiology and clinical studies of eye movements has traditionally been a very strong area within eye-movement research in terms of number and impact of publications, and is therefore possibly somewhat underrepresented here. Fewer people indicated that they worked on transsaccadic integration (5), eye movements in natural environments (5), and the history of eye-movement research (2). This seems a fair representation of the number of people working in these areas, although of course that does not necessarily reflect on the importance of the research. Seven respondents indicated that they conducted a different type of research; eye-movement behaviour during driving was mentioned twice. A high number of respondents (24) indicated that they worked in more than one area, showing the highly cross-disciplinary nature of eye-movement research (and perhaps this bias in those who attend ECEM). Overall, these data suggest that the questionnaire results are likely to be fairly representative of the views of a broad cross section of researchers working in the area of eye-movement research.

2.1. Most influential and most cited publications

Table 2 shows the eye-movement publications that our respondents considered most influential in the development of the field. The eye-movement model of reading proposed by Reichle, Pollatsek, Fisher, and Rayner (1998) received most points, highlighting its importance for both reading and modelling research. However, it was not the paper that was most often mentioned as most influential. Both Findlay and Walker's (1999) model of saccade generation and Rayner's (1998) overview on eye movements during reading

Table 2
Eye-movement publications considered to have been most influential in the development of the field by respondents to the questionnaire

Publication	Points	Times mentioned as most influential	Citation count	Area
1. Reichle, Pollatsek, Fisher, & Rayner (1998)	27	3	155	4, 5
2. Findlay & Walker (1999)	16	4	155	7
3. Rayner (1998)	15	4	346	5, 7
4. Tanenhaus, Spivey-Knowlton, Eberhard, & Sedivy (1995)	15	3	167	6
5. McConkie & Rayner (1975)	11	2	255	5
6. Yarbus (1967)	10	2	1049	8
7. Deubel & Schneider (1996)	9	0	212	7
8. Becker & Jürgens (1979)	8	2	335	4
9. Frazier & Rayner (1982)	6	2	362	5
10. Rayner (1975)	6	0	236	5

Note: *Points*: For each respondent ($N = 44$), the most influential paper counted as 3 points, the second most influential paper as 2 points, and the third most influential paper as 1 point.

Times mentioned as most influential: Number of respondents who indicated that this publication has been most influential (i.e. it received 3 points).

Citation count: Number of citations (including incorrect citations) in all databases in Web of Knowledge, searched on 12 May 2006.

Area: Area according to the classification in Table 1.

and information processing were mentioned more frequently as the most influential paper. Tanenhaus, Spivey-Knowlton, Eberhard, and Sedivy (1995) was equal to Reichle et al. (1998) in terms of this measure of importance. Most of the publications in Table 2 are concerned with reading research (5) or attention and scene perception (3); perhaps not surprisingly, given that many respondents worked in these areas. In contrast, the history of eye movements, physiology and clinical studies, and transsaccadic integration are not represented in the table, and this may also be related to the background of the respondents.

It is interesting to see whether those publications considered influential by specialists in the field of eye movements are also the ones most often cited. Or to put it another way, are the most-cited publications also considered to be most influential? Citation counts are often used as a means of quantifying the impact of publications, so we should expect to see a good correlation between these two measures.

In order to determine whether the publications that were considered to be most influential also had high citation counts, we conducted a search in Web of Knowledge, using all available databases: science, social sciences, and arts and humanities. In this citation count, we included incorrect citations where, for example, the exact title of the book was wrong, as long as the authors' surnames and year of publication were correct. As shown in Table 2, all publications in the top 10 received a fairly high number of citations, but there is no clear correlation between the citation count and the rankings.

Table 3
Most cited eye-movement publications out of all mentioned by the respondents to the questionnaire

Publication	Points	Times mentioned as most influential	Citation count	Area
1. Robinson (1963)	3	1	1303	9
2. Yarbus (1967)	10	2	1049	8
3. Just & Carpenter (1980)	2	0	646	5
4. Robinson (1975)	3	1	567	2
5. Guitton, Buchtel, & Douglas (1985)	1	0	529	2
6. Rayner & Pollatsek, (1989)	2	0	480	5
7. Duhamel, Colby, & Goldberg (1992)	1	0	460	2,3
8. Frazier & Rayner (1982)	6	2	362	5
9. Rayner (1998)	15	4	346	5,7
10. Becker & Jürgens (1979)	8	2	335	4

Note: *Points:* For each respondent (N = 44), the most influential paper counted as 3 points, the second most influential paper as 2 points, and the third most influential paper as 1 point.

Times mentioned as most influential: Number of respondents who indicated that this publication has been most influential (i.e. it received 3 points).

Citation count: Number of citations (including incorrect citations) in all databases in Web of Knowledge, searched on 12 May 2006.

Area: Area according to the classification in Table 1.

Of the 10 publications in Table 2, the two that were considered most influential have the fewest citations, while Yarbus's (1967) book, which was ranked sixth, had by far the most citations (1049).

To further explore this question, we have listed in Table 3 the 10 most cited publications from all those mentioned by at least one of the respondents to the questionnaire. Four of the publications that were in the top 10 in Table 2 are also in Table 3, suggesting at least some correspondence between citation counts and publications that specialists considered influential. However, the other six publications were only mentioned by one respondent, including Robinson's (1963) article on using a scleral search coil for measuring eye movements, which received by far the highest number of citations. Furthermore, the two most influential articles in Table 2 (Findlay & Walker, 1999; Reichle et al., 1998) do not even reach the top 10 in Table 3.

Table 4 confirms that the most cited publications are not always seen as most influential. In this table, we have listed the 10 most cited journal articles (according to Web of Knowledge) that have 'eye(-)movement(s)' in their title or keywords. Of course, this search is somewhat limited, since some eye-movement articles do not have 'eye(-)movement(s)' in their title or keywords. However, it does enable us to determine whether articles that are well cited are considered to be influential by specialists in the area. Clearly, this is not the case: of the 10 most cited articles, only one – by Duhamel, Colby, and Goldberg (1992) – was mentioned, and this article was only mentioned as the third most influential by a single respondent.

Table 4
Most cited articles with 'eye(-)movement(s)' in title or keywords in any of the databases in Web of Knowledge

Publication	Points	Times mentioned as most influential	Citation count	Area
1. Corbetta, Miezin, Dobmeyer, et al. (1991)	0	0	852	2,7
2. Corbetta, Miezin, Shulman, et al. (1993)	0	0	758	2,7
3. Dawes, Liggins, Leduc, et al. (1972)	0	0	626	9
4. Robinson (1972)	0	0	554	2
5. Duhamel, Colby, & Goldberg (1992)	1	0	458	2,3
6. Zee, Yamazaki, Butler, et al. (1981)	0	0	448	2
7. Keller (1974)	0	0	438	2
8. Park & Holzman (1992)	0	0	429	2
9. Goldman-Rakic (1995)	0	0	419	2
10. Benca, Obermeyer, Thisted, et al. (1992)	0	0	417	9

Note: *Points:* For each respondent ($N = 44$), the most influential paper counted as 3 points, the second most influential paper as 2 points, and the third most influential paper as 1 point.

Times mentioned as most influential: Number of respondents who indicated that this publication has been most influential (i.e. it received 3 points).

Citation count: Number of citations (including incorrect citations) in all databases in Web of Knowledge, searched on 12 May 2006.

Area: Area according to the classification in Table 1.

The discrepancy between the citation counts and responses to the questionnaire might have several causes. First, these citation counts do not include any publications other than journal articles (and occasionally, conference proceedings), explaining why Yarbus (1967), Rayner and Pollatsek (1989), and Robinson (1975) do not appear in Table 4. Clearly, this is a limitation in the use of this database (and many others). However, it should be noted that most of the publications mentioned by our respondents were indeed journal articles, and this consequently cannot account for all of the discrepancy. Second, citation counts may be more biased towards early publications than the opinions of specialists, since earlier publications have had more chance to accumulate citations. But there is little evidence for this. The mean year of publication in Table 4 is 1985, which is comparable to 1986 in Table 2 and later than 1981 in Table 3. Thus, the dates of the publications mentioned by the respondents and those which have most citations are not that dissimilar, though clearly, we would wish to argue that our survey data is likely to be more "up to date" in terms of current impact. Third, it is possible that the type of research that was most common amongst our group of correspondents differed from the type of research that receives most citations. Most of the articles in Table 4 are concerned with physiology and clinical studies, but as we mentioned before, researchers from these areas were probably underrepresented in our sample. It may be that articles on physiology and clinical studies of eye movements receive a large number of citations because many people work in these fields. Similarly, there were no articles on reading in Table 4, while reading researchers were the largest group among questionnaire respondents.

Since most (though certainly not all) of our respondents were social scientists (rather than clinicians and medical researchers), we might expect to see a clearer correspondence between citation counts and the respondents' responses if we consider social science publications only. In order to check this, we ran a search in Web of Knowledge using the social sciences database only. The 10 most cited journal articles are shown in Table 5. We now see two publications, Frazier and Rayner (1982) and Rayner (1998), that were frequently mentioned as influential by the respondents to the questionnaire. This suggests that some of the discrepancy between the citation counts and the opinions of our respondents resulted from a "social science bias" on the part of our respondents. However, it nonetheless does not account for all of the discrepancy: Table 5 shows that many of the most cited social science articles were not mentioned by our respondents either.

It seems that none of the above explanations can completely account for the discrepancy between citation counts and responses to our questionnaire. It appears that citation counts may therefore be a rather poor reflection of the actual significance of particular publications. One reason for this may be that some well-cited publications have 'eye(-)movement(s)' in their title or keywords, but do not have this as their main focus of their investigation. For example, many researchers would agree that MacDonald, Pearlmutter, and Seidenberg (1994) has been influential in the area of sentence processing but their paper has not been influential for the field of eye-movement research. Other

Table 5
Most cited articles with 'eye(-)movement(s)' in title or keywords in the social sciences database in Web of Knowledge

Publication	Points	Times mentioned as most influential	Citation count	Area
1. Corbetta, Miezin, Dobmeyer, et al. (1991)	0	0	852	2, 7
2. Corbetta, Miezin, Shulman, et al. (1993)	0	0	758	2, 7
3. Park & Holzman (1992)	0	0	429	2
4. Goldman-Rakic (1995)	0	0	419	2
5. Benca, Obermeyer, Thisted, et al. (1992)	0	0	417	9
6. MacDonald, Pearlmutter, & Seidenberg (1994)	0	0	386	5
7. Frazier & Rayner (1982)	6	2	362	5
8. Rensink, O'Regan, & Clark (1997)	0	0	345	7
9. Rayner (1998)	15	4	346	5, 7
10. Corbetta, Akbudak, Conturo, et al. (1998)	0	0	330	2, 7

Note: *Points:* For each respondent ($N = 44$), the most influential paper counted as 3 points, the second most influential paper as 2 points, and the third most influential paper as 1 point.

Times mentioned as most influential: Number of respondents who indicated that this publication has been most influential (i.e. it received 3 points).

Citation count: Number of citations (including incorrect citations) in all databases in Web of Knowledge, searched on 12 May 2006.

Area: Area according to the classification in Table 1.

publications may receive many citations because they provide a representative example of a particular type of research, method, or theory, without being the original, groundbreaking work that influenced subsequent research. Finally, of course, there is also the possibility that some publications are often cited because they provide a good example of a particularly outrageous claim, a theory that is falsified or research that is flawed (readers will have to come up with their own examples falling into this category!).

In conclusion, it is clear that while there is some relationship between citations and influence, the most heavily cited journal articles are not necessarily the most influential publications. This suggests that we should be cautious when interpreting citation counts and that citation counts are not necessarily the optimal way of quantifying how influential particular publications may be. Of course, questionnaire studies also have limitations. It is difficult to obtain a sufficiently large and representative sample of respondents. Furthermore, the respondents may not always be entirely objective and may be biased towards their own publications, those from within their research group or area, or those that are part of the theoretical framework in which they are working. When considering the impact of publications, both methods have their value: most researchers would probably agree that the majority of publications listed in Tables 2–5 provide a representative sample of influential publications in the field of eye-movement research.

2.2. Journals

We also asked our respondents to list the journals that they considered best for keeping up to date with developments in eye-movement research. Table 6 shows the top 10 most mentioned journals. It is clear that *Vision Research* is the favourite journal of most respondents, followed at a distance by two of the Journals of Experimental Psychology (*HPP & LMC*). We see that *Vision Research* also has many articles with 'eye(-)movement(s)' in the title (in the period: 2001–2005), whereas both Journals of Experimental Psychology have few. All journals in Table 6 have a relatively high impact factor (see Tables 6–8 for the definition), mostly higher than 2.

Table 7 ranks all journals in Web of Knowledge according to the number of articles that have 'eye(-)movement(s)' in their title in the period 2001–2005. *Investigative Ophthalmology & Visual Science*, which has a special section on eye movements, strabismus, amblyopia, and neuro-ophthalmology, contains the most articles. It is closely followed by *Perception*, with 79. Interestingly, neither journal was mentioned very often by respondents to the questionnaire; in fact, *Investigative Ophthalmology & Visual Science* was never mentioned. This may be because the majority of our respondents were social scientists, but this does not explain why none of the nine respondents who said they worked on physiology and clinical studies of eye movements mentioned it. *Perception* has a fairly low impact factor, so this may explain why this journal was seldom mentioned by our respondents.

Table 8 ranks all those journals that are in the social sciences database (but not in the science or arts & humanities databases). All eye-movement articles in *Perception* were in the social sciences, so this journal now ranks highest. By contrast, none of the eye-movement articles in *Investigative Ophthalmology & Visual Science* are social

Table 6
Journals considered to be the best for keeping up to date with developments in eye-movement research by respondents to the questionnaire

Journal	Times mentioned	'Eye(-)movement(s)' in title	Impact factor
1. *Vision Research*	27	45	2.03
2. *Journal of Experimental Psychology: Human Perception and Performance*	13	6	2.88
3. *Journal of Experimental Psychology: Learning, Memory, and Cognition*	12	5	2.81
4. *Journal of Neurophysiology*	9	50	3.81
5. *Perception & Psychophysics*	8	10	1.73
6. *Journal of Memory and Language*	7	9	2.82
6. *Psychological Review*	7	0	7.99
8. *Experimental Brain Research*	6	49	2.12
9. *Cognition*	5	5	3.78
10. *Quarterly Journal of Experimental Psychology*	4	4	1.77

Note: *Times mentioned:* Number of respondents ($N = 44$) who mentioned the journal.

'Eye(-)movement(s)' in title: Number of articles with 'eye(-)movement(s)' in title in all databases in Web of Knowledge for the period 2001–2005, searched on 15 May 2006.

$$\textit{Impact factor} = \frac{\text{Number of times that articles published in the journal in 2003–2004 were cited during 2005}}{\text{Number of articles published in the journal in 2003–2004}}$$

Table 7
Journals with most articles that have 'eye(-)movement(s)' in the title during 2001–2005 in all Web of Knowledge databases

Journal	Times mentioned	'Eye(-)movement(s)' in title	Impact factor
1. *Investigative Ophthalmology & Visual Science*	0	80	3.64
2. *Perception*	3	79	1.39
3. *Sleep*	0	51	4.95
4. *Journal of Neurophysiology*	9	50	3.85
5. *Experimental Brain Research*	6	49	2.12
6. *Vision Research*	27	45	2.03
7. *Journal of Neuroscience*	1	23	7.51
8. *International Journal of Psychology*	0	23	0.65
9. *Schizophrenia Research*	0	20	4.23
10. *Biological Psychiatry*	0	19	6.78

Note: *Times mentioned:* Number of respondents ($N = 44$) who mentioned the journal.

'Eye(-)movement(s)' in title: Number of articles with 'eye(-)movement(s)' in title in all databases in Web of Knowledge for the period 2001–2005, searched on 15 May 2006.

$$\textit{Impact factor} = \frac{\text{Number of times that articles published in the journal in 2003–2004 were cited during 2005}}{\text{Number of articles published in the journal in 2003–2004}}$$

Table 8
Journals with most articles that have 'eye(-)movement(s)' in the title during 2001–2005 in the social sciences database in Web of Knowledge

Journal	Times mentioned	'Eye(-)movement(s)' in title	Impact factor
1. *Perception*	3	79	1.39
2. *International Journal of Psychology*	0	23	0.65
3. *Schizophrenia Research*	0	20	4.23
4. *Vision Research*	27	17	2.03
5. *Behavioral and Brain Sciences*	1	12	9.89
6. *Journal of Cognitive Neuroscience*	0	12	4.53
7. *Memory & Cognition*	1	11	1.57
8. *Perception & Psychophysics*	8	10	1.73
9. *Journal of Memory and Language*	7	9	2.82
10. *Journal of Psychophysiology*	0	9	0.97

Note: *Times mentioned:* Number of respondents ($N = 44$) who mentioned the journal.

'Eye(-)movement(s)' in title: Number of articles with 'eye(-)movement(s)' in title in all databases in Web of Knowledge for the period 2001–2005, searched on 15 May 2006.

$$\textit{Impact factor} = \frac{\text{Number of times that articles published in the journal in 2003–2004 were cited during 2005}}{\text{Number of articles published in the journal in 2003–2004}}$$

science, so this journal is absent from Table 8. In Table 8 we now see some correspondence between the number of articles with 'eye(-)movement(s)' in their title and number of times the journal was mentioned by our respondents, although the correspondence is still not great. One possible reason for this is that the impact factor of some social science journals that contain many eye-movement articles (especially the *International Journal of Psychology* and, to a lesser extent, *Perception*) is relatively low. Consequently, the quality of the eye-movement articles published in these journals may not be as high as those in other journals. Table 9 suggests another possible reason. For three 10-year periods, it lists the number of articles with 'eye(-)movement(s)' in their title together with the number of articles with eye(-)movement(s) per 1000 articles published for seven journals. We see that the per mille of eye-movement articles is relatively high for the five journals that were most often mentioned by our respondents (see Table 6: *Vision Research*, *Journal of Experimental Psychology*: *HPP*, *Journal of Experimental Psychology*: *LMC*, *Journal of Neurophysiology*, *Perception & Psychophysics*). The exception is the *Journal of Experimental Psychology*: *LMC* for the period 1976–1985, but this is partly because it did not exist under this name until 1982. By contrast, this measure is quite low for *Investigative Ophthalmology & Visual Science* and the *International Journal of Psychology*. This suggests that our respondents assigned more importance to the relative number of eye-movement articles in a journal rather than the absolute number. Certain journals are therefore seen to have a greater affinity for eye-movement research.

It is clear from the above that eye-movement articles are published across a wide variety of journals. This is of course related to the fact that eye-movement research is a

Table 9
Number of articles with ‘eye(-)movement(s)’ in title for different journals for 10-year periods from 1976 onwards.

	1976–1985		1986–1995		1996–2005	
	Number	‰	Number	‰	Number	‰
Vision Research	65	29.0	60	20.4	83	21.3
Journal of Experimental Psychology: Human Perception and Performance	8	11.6	9	12.4	13	13.3
Journal of Experimental Psychology: Learning, Memory, and Cognition	0	0	9	9.7	9	9.3
Journal of Neurophysiology	37	24.6	76	24.6	95	14.8
Perception & Psychophysics	20	12.5	17	12.3	17	15.4
Investigative Ophthalmology & Visual Science	19	6.4	103	4.4	193	3.7
International Journal of Psychology	0	0	12	3.0	27	1.8

Note: *Number:* Number of articles with ‘eye(-)movement(s)’ in title in all databases in Web of Knowledge for the period 2001–2005, searched on 15 May 2006.

‰: Number of articles with ‘eye(-)movement(s)’ in title per 1000 articles published during each period.

very diverse field which encompasses many different disciplines, with many researchers using eye-movement methodology as a research tool for investigating various different kinds of cognitive and perceptual processes, rather than necessarily focussing on the characteristics of eye movements per se. There is no single journal that includes all these different disciplines. *Vision Research*, which is the favourite journal for most of the respondents to the questionnaire, contains articles on physiology and clinical studies, attention, scene perception, modelling, and low-level reading processes, and so it includes a fairly substantial part of the full range of eye-movement research. However, it normally does not have any articles that use eye-movements to investigate higher levels of language processing (syntax, semantics, and discourse processing). Of the other journals that were often mentioned by our respondents, both *Journals of Experimental Psychology* include a wide variety of eye-movement research in experimental psychology, but no physiological and clinical studies involving eye movements. The *Journal of Neurophysiology* includes physiological studies, but few behavioural studies, while *Perception & Psychophysics* focuses on particular behavioural areas, but not others.

Until very recently, there was no specialised eye-movement journal that contained research from all the different eye-movement disciplines, but this has changed with the introduction of the online *Journal of Eye Movement Research* (http://www.jemr.org). One might argue that such a specialised journal is not needed, because eye-movement research is too diverse and the different disciplines are too unrelated to bring them together in a single journal. Many researchers only read articles from their own eye-movement area, and hence articles outside this area may be irrelevant. However, evidence suggests that this assumption is incorrect. First, it is well known that many research areas develop as a consequence of cross-fertilisation between different areas. This is also the case for eye-movement research. For example, researchers using eye-tracking to investigate spoken language processing have been very interested in eye-movement research investigating attention and scene perception (Part 6 in this book). Eye-movement behaviour during reading and scene perception has interesting commonalities and differences, and is therefore often compared (e.g., Rayner, 1998). Furthermore, eye movements have been modelled for both reading and scene perception, and findings in these two areas can inform one another (see, for example, the comment in the chapter by De Graef, Chapter 7). Finally, many disciplines in eye-movement research have benefited from findings from physiological and clinical studies, and, in turn, physiological and clinical studies have drawn on findings from behavioural studies. In sum, there is significant overlap between the different disciplines in eye-movement research and cross-fertilisation has often been very fruitful. However, because eye-movement research is distributed across different journals and many researchers mainly read journals in their own specialised field, there is a risk that many eye-movement researchers are unaware of much of the relevant research and findings outside their own field. This is likely to inhibit the integration of research across the different disciplines and therefore hold back progress. Perhaps an even greater concern derives from the increased availability of eye-movement technology to researchers with little general background in the area and the consequent risk of their failing to appreciate some of the technical questions and considerations that need to be addressed when planning and analysing their experiments. A specialised eye-movement journal that brings together research from different disciplines would be a place where they could find answers to their questions about the various eye-tracking technologies, experimental methods, and data analysis techniques.

Second, the fact that ECEM conferences have been so successful strongly suggests that bringing together eye-movement researchers from different disciplines is worthwhile. One of the reasons why many researchers go to these conferences is because it enables them to learn about eye-movement research outside their own specialised area. Third, the fact that most ECEM conferences result in an edited volume that contains eye-movement research from all disciplines and the fact that these volumes are read by many people in the field indicate that there is a need to bring together eye-movement research from the different disciplines. A specialised eye-movement journal serves this function. There are, of course, many factors that determine the success of a journal, such as the quality of the publications, speed of the review and publication process, the editorial board, readership, its availability in libraries and on the web, advertising, publicity, inclusion in search databases, and the confidence that researchers have in it. But we hope the *Journal of Eye Movement Research* will facilitate integration between different eye-movement disciplines and advance eye-movement research in general.

2.3. Development of the area over time

The chapters by Wade, Land, and Westheimer in this book discuss historical developments in eye-movement research in a qualitative way. Using quantitative data from our database searches and questionnaire, we attempted to gain an insight into the development of eye-movement research as a scientific field of interest in the period from 1970 onwards. The 1970s are often seen as the start of the modern area of eye-movement research, so it seems appropriate to take a closer look at developments in the past 35 years or so.

The first question we addressed is whether the number of journal articles on eye movements has grown during that time period. This would give us an idea of whether research in the area is on the increase (as most of us would like to think) or decline. As mentioned in the previous section, Table 9 shows the number of eye-movement articles for seven representative journals for three 10-year periods. The number of articles with 'eye(-) movement(s)' in the title went up in all journals except *Perception & Psychophysics*, for which it stayed roughly the same. Overall, there appears to be a steady increase across the seven journals. However, when we look at the per mille measure, then there is no clear increase in the proportion of eye-movement articles. Together, this indicates that more and more eye-movement articles are being published in the selected journals, but this appears to be simply a reflection of the fact that these journals accept more and more articles, regardless of whether they involve eye movements.

Of course, a few representative journals do not tell the whole story. The picture may be different for other journals, and over the years, new journals publishing eye-movement research may appear, with a resultant increase in eye-movement articles. Table 10 presents the number and per mille articles with 'eye(-)movement(s)' in the title for 5-year periods from 1971 onwards. We have listed the results from all databases in Web of Knowledge (social science, science, and arts & humanities) plus the results from the social science database only. The results from all databases combined show a clear and steady increase in the number of published eye-movement articles. However, again when we look at the per mille of all articles, there is no obvious increase. In fact, the per mille eye-movement

Table 10
Number of articles with 'eye(-)movement(s)' in title for 5-year period from 1971 onwards

	All databases		Social sciences database	
Period	Nr. eye-movement articles	‰ of all articles	Nr. eye-movement articles	‰ of all articles
1971–1975	536	0.23	174	0.45
1976–1980	613	0.18	290	0.47
1981–1985	722	0.17	267	0.43
1986–1990	759	0.17	228	0.37
1991–1995	908	0.19	344	0.53
1996–2000	1021	0.18	388	0.54
2001–2005	1259	0.20	498	0.67

Note: *Nr. eye-movement articles:* Number of articles with 'eye(-)movement(s)' in title.

‰ of all articles: Number of articles with eye(-)movement(s) in title per 1000 articles published during each period.

articles was highest between 1971 and 1975, and there seems to be a slight dip in the per mille eye-movement articles in the 1980s. We conclude that the number of eye-movement articles is going up, but this increase is proportional to the total number of articles being published. There is no evidence that eye-movement research has become more dominant or productive relative to other areas. However, this consistent pattern does reinforce the notion that eye-movement research is a stable and reliable form of investigation rather than a passing scientific or technological fad.

The picture is slightly different when we consider journal articles in the social sciences only. As before, we see an increase in the number of eye-movement articles since 1971 (though again, there is a slight dip in the 1980s). But we also see that the per mille eye-movement articles in the social sciences has gradually increased since the late 1980s. In particular, the number of eye-movement articles between 2001 and 2005 is much larger than in any previous period. This clearly suggests that in recent years, eye-movement research has become more productive or more popular in the social sciences. The fact that many social scientists responded to our questionnaire and that they are well represented in the chapters in this book may be a further reflection of the gaining popularity of eye-movement research in this area.

We also tried to gain an impression of whether the countries where eye-movement research is conducted have changed over time. Table 11 ranks countries that produced most journal articles with ‘eye(-)movement(s)’ in the title for the period 1976–1980 in all databases in Web of Science. The United States produced by far the largest number

Table 11
Countries producing most articles with ‘eye(-)movement(s)’ in the title. Web of Science, all databases, 1976–1980

Country	Nr articles	% of all articles	Nr articles/million population
1. USA	292	47.6	1.44
2. West Germany	40	6.5	0.64
3. England	39	6.4	0.85
4. Canada	33	5.4	1.53
5. France	28	4.6	0.54
6. Japan	27	4.4	0.25
7. Czechoslovakia	16	2.6	1.10
8. Italy	12	2.0	0.22
9. Switzerland	12	2.0	1.88
10. Australia	11	1.8	0.83

Note: 7.3% of articles did not specify the country of origin.

Nr articles: Number of articles with ‘eye(-)movement(s)’ in title in all databases in Web of Knowledge for the period 1976–1980, searched on 15 May 2006.

% of all articles: Percentage of articles with ‘eye(-)movement(s)’ in title out of all eye-movement articles from all countries.

Nr articles/million population: Number of articles with ‘eye(-)movement(s)’ in title per million population. Population data from Bradfield, Keillor, and Pragnell (1976).

of eye-movement articles during this period: 47.6% of all relevant articles. It is followed at a long distance by West Germany and England. Of course, one might argue that this is not a fair comparison, as the United States has a much larger population than the other countries in the table. If we consider the number of eye-movement articles per million population for all countries in Table 11, both Switzerland and Canada rank higher than the United States.

Table 12 shows the same data for the period 2001–2005. We see that there are not many striking differences with that of the period 1976–1980. The position of the United States has become slightly less dominant, as indicated by the percentages relative to all eye-movement articles (40.2% vs. 47.6% of eye-movement articles in 2001–2005 vs. 1976–1980 respectively), while the percentage of eye-movement articles produced in Germany, Japan, and England has clearly increased. Czechoslovakia and Switzerland have disappeared from the top 10 (Czechoslovakia has, of course, ceased to exist as a country), whereas The Netherlands and Scotland are now included. Interestingly, the number of eye-movement articles per million population is very high for The Netherlands and even more so for Scotland.

Table 13 presents the results for the period 1976–1980 from the social sciences database only. The United States produces over half of all eye-movement articles in the social sciences during this period, followed at a long distance by England and Canada. It is noteworthy that Czechoslovakia ranks fourth in this table. Israel has a high number of eye-movement articles per million population, but it should be noted that its total number

Table 12
Countries producing most articles with 'eye(-)movement(s)' in the title. Web of Science, all databases, 2001–2005

Country	Nr articles	% of all articles	Nr articles/million population
1. USA	506	40.2	1.72
2. Germany	141	11.2	1.71
3. Japan	139	11.0	1.09
4. England	138	11.0	2.79
5. Canada	68	5.4	2.13
6. France	57	4.5	0.95
7. Netherlands	49	3.9	3.01
8. Italy	40	3.2	0.69
9. Australia	31	2.5	1.54
10. Scotland	29	2.3	5.80

Note: 4.2% of articles did not specify the country of origin.

Nr articles: Number of articles with 'eye(-)movement(s)' in title in all databases in Web of Knowledge for the period 2001–2005, searched on 15 May 2006.

% of all articles: Percentage of articles with 'eye(-)movement(s)' in title out of all eye-movement articles from all countries.

Nr articles/million population: Nr of articles with 'eye(-)movement(s)' in title per million population. Population data from Bruinsma, Koedam, Dilworth, and Stuart-Jones (2005).

Table 13
Countries producing most articles with 'eye(-)movement(s)' in the title. Web of Science, social sciences, 1976–1980

Country	Nr articles	% of all articles	Nr articles/million population
1. USA	156	53.8	0.77
2. England	22	7.6	0.48
3. Canada	20	6.9	0.93
4. Czechoslovakia	11	3.8	0.75
5. West Germany	11	3.8	0.18
6. France	10	3.4	0.19
7. Japan	10	3.4	0.09
8. Israel	5	1.7	1.62
9. Bulgaria	4	1.4	0.47
10. Switzerland	4	1.4	0.63

Note: 5.9% of articles did not specify the country of origin.

Nr articles: Number of articles with 'eye(-)movement(s)' in title in the social sciences database in Web of Knowledge for the period 1976–1980, searched on 15 May 2006.

% of all articles: Percentage of articles with 'eye(-)movement(s)' in title out of all eye-movement articles from all countries.

Nr articles/million population: Number of articles with 'eye(-)movement(s)' in title per million population. Population data from Bradfield, Keillor, and Pragnell (1976).

of eye-movement articles was only five. In general, there were few eye-movement articles in the social sciences during this period produced in any country apart from the United States. Interestingly, however, the foundation for international conferences focusing on eye movements was forged in Europe, with the first ECEM in Bern, Switzerland, in 1981.

When we look at Table 14, which ranks the countries that produced most social science eye-movement articles for the period 2001–2005, we see that the percentage of eye-movement articles produced in the United States is quite a bit smaller than in the 1976–1980 period. By contrast, the percentages for other countries such as England, Germany, and Japan have gone up. In fact, the United States is the only country that occurs in both tables for which the number of eye-movement articles per million population is lower in 2001–2005 than in 1976–1980. This suggests that within the social sciences, the United States is losing some of its dominance in eye-movement research, although it remains the case that most eye-movement articles continue to be produced by US-based researchers. Further comparison of Table 14 with Table 13 shows that Czechoslovakia, Israel, Bulgaria, and Switzerland have disappeared from the top 10 and have been replaced by Scotland, The Netherlands, Australia, and Spain. Finally, the number of eye-movement articles per million population shows that Scotland produces by far the most eye-movement articles in the social sciences, while England and The Netherlands also do well.

It is difficult to gauge the most important developments in eye-movement research from database searches, but very easy to ask practitioners in the area for their views.

Table 14
Countries producing most articles with 'eye(-)movement(s)' in the title. Web of Science, social sciences, 2001–2005

Country	Nr articles	% of all articles	Nr articles/million population
1. USA	185	37.1	0.63
2. England	82	16.5	1.66
3. Germany	56	11.2	0.68
4. Japan	33	6.6	0.26
5. Canada	31	6.2	0.97
6. Scotland	27	5.4	5.40
7. Netherlands	26	5.2	1.60
8. France	22	4.4	0.37
9. Australia	14	2.8	0.69
10. Spain	11	2.2	0.25

Note: 2.6% of articles did not specify the country of origin.

Nr articles: Number of articles with 'eye(-)movement(s)' in title in the social sciences database in Web of Knowledge for the period 2001–2005, searched on 15 May 2006.

% of all articles: Percentage of articles with 'eye(-)movement(s)' in title out of all eye-movement articles from all countries.

Nr articles/million population: Number of articles with 'eye(-)movement(s)' in title per million population. Population data from Bruinsma, Koedam, Dilworth, and Stuart-Jones (2005).

We therefore asked the respondents of our questionnaire what they thought the most important developments in eye-movement research have been over the past 5 years. As shown in Table 15, most respondents mentioned computational modelling (especially in the areas of reading and scene perception), and this is reflected in this book

Table 15
Most important developments in eye-movement research in the past 5 years according to respondents to the questionnaire

	Number of times mentioned
1. Computational modelling (mainly in reading and scene perception)	19
2. Developments in eye-tracking technology (inexpensive, easy to use trackers, free-view trackers)	14
3. Developments in the anatomical and physiological mapping of the visual-oculomotor system	6
4. Eye movements as a research tool for investigating spoken language processing	5
5. The use of eye movement methodology in combination with fMRI	3
6. Eye movements in naturalistic settings and tasks	3

Note: $N = 44$.

(Part 4). Many also mentioned recent developments in eye-tracking technology such as the introduction of inexpensive, easy-to-use eye-trackers and free-view trackers. Although advances in eye-tracking technology are not really new developments in research per se, new technologies frequently trigger new types of research. Many chapters in this book demonstrate ways in which new eye-tracking technologies have been used to advance new types of research. Developments in the anatomical and physiological mapping of the visual-oculomotor system was also regularly mentioned, despite the fact that relatively few researchers working on physiology and clinical studies responded. The chapters in Part 2 show some of these developments. Eye movements as a tool for investigating spoken language processing also figures prominently in this book (Part 6), eye-movement methodology in combination with fMRI is used in Chapter 6 by Ford, Brown, and Everling, and Part 8 is on eye movements in naturalistic settings and tasks.

3. Conclusions

As is clear from the discussion of the different parts in this book, eye-movement research is an extremely diverse field that brings together researchers from many different sub-disciplines. Several new areas have developed recently, such as eye movements for investigating spoken language processing, historical eye-movement research, and eye movements in natural environments. Other, long established areas have also seen many important new advances.

The analyses of the responses to a questionnaire sent out to attendants of ECEM and a citation count from Web of Knowledge gave us an impression of the most influential publications in eye-movement research. Interestingly, the results from the questionnaire and the citation counts proved more disparate than expected. It appears that the publications which are considered most influential by eye-movement specialists are not necessarily the ones most frequently cited. Our analyses of both the questionnaire responses and articles in Web of Knowledge indicated that eye-movement articles are published in a wide variety of different journals. Only very recently has a specialised eye-movement journal (the *Journal of Eye Movement Research*) been established. Given the cross-disciplinary nature of eye-movement research and the success of ECEM and ECEM edited volumes, this journal could make an important contribution to the field of eye-movement research.

Looking at the number of journal articles published since 1970, we see a steady increase in the number of eye-movement articles. However, the number of articles relative to those published in all other areas remains fairly constant, except in the social sciences where the relative number of eye-movement articles has gone up in recent years. The countries where most eye-movement research is produced have also remained similar, with the United States dominating eye-movement output across our period of investigation. However, smaller countries such as Scotland, The Netherlands, and England produce a larger proportion of eye-movement articles relative to their populations. And we do, of course, take great delight seeing the pre-eminent position of Scotland in this respect!

Finally, the respondents to our questionnaire indicated that the most important developments in eye-movement research since 2001 involved computational modelling, new eye-tracking technologies, anatomical and physiological mapping of the visual–oculomotor system, eye movements during spoken language processing, the use of eye-movement methodology in combination with fMRI, and eye movements in naturalistic settings and tasks. We expect that in the near future, these areas of research will continue to gain in importance, while at the same time, well-established areas of eye-movement research investigating reading, attention, and scene perception will continue to flourish. We hope that you, the reader, agree that ECEM12, and this book which grew out of it, have provided a good reflection of the importance and development of this area, and we foresee future ECEMs maintaining their position as a leading nexus of this knowledge and serving as a prime source for the dissemination of work related to all aspects of eye-movement research.

References

Becker, W., & Jürgens, R. (1979). Analysis of the saccadic system by means of double step stimuli. *Vision Research, 19*, 967–983.

Benca, R. M., Obermeyer, W. H., Thisted, R. A., & Gillin, J. C. (1992). Sleep and psychiatric-disorders: A metaanalysis. *Archives of General Psychiatry, 49*, 651–668.

Bradfield, R. M., Keillor, J.-P., & Pragnell, M. O. (1976). *The international year book and statemen's who's who.* Kingston upon Thames: Kelly's Directories Limited.

Bruinsma, S., Koedam, A., Dilworth, J., & Stuart-Jones, M. (2005). *The international year book and statemen's who's who.* Leiden: Martinus Nijhoff.

Corbetta, M., Akbudak, E., Conturo, T. E., Snyder, A. Z., Ollinger, J. M., Drury, H. A., et al. (1998). A common network of functional areas for attention and eye movements. *Neuron, 21*, 761–773.

Corbetta, M., Miezin, F. M., Dobmeyer, S., Shulman, G. L., & Petersen, S. E. (1991). Selective and divided attention during visual discriminations of shape, color, and speed: Functional-anatomy by positron emission tomography. *Journal of Neuroscience, 11*, 2383–2402.

Corbetta, M., Miezin, F. M., Shulman, G. L., & Petersen, S. E. (1993). A pet study of visuospatial attention. *Journal of Neuroscience, 13*, 1202–1226.

Dawes, G. S., Fox, H. E., Leduc, B. M., Liggins, G. C., & Richards, R. T. (1972). Respiratory movements and rapid eye-movement sleep in the foetal lamb. *Journal of Physiology, 220*, 119–143.

Deubel, H., & Schneider, W. X. (1996). Saccade target selection and object recognition: Evidence for a common attentional mechanism. *Vision Research, 36*, 1827–1837.

Du Laurens, A. (1596/1599/1938). *A discourse of the preservation of the melancholic diseases; of rheumes, and of old age* (Trans. R. Surphlet). London: The Shakespeare Association.

Duhamel, J. R., Colby, C. L., & Goldberg, M. E. (1992). The updating of the representation of visual space in parietal cortex by intended eye-movements. *Science, 255*, 90–92.

Findlay, J. M., & Walker, R. (1999). A model of saccade generation based on parallel processing and competitive inhibition. *Behavioral and Brain Sciences, 22*, 661–674.

Frazier, L., & Rayner, K. (1982). Making and correcting errors during sentence comprehension: Eye movements in the analysis of structurally ambiguous sentences. *Cognitive Psychology, 14*, 178–210.

Goldman-Rakic, P. S. (1995). Cellular basis of working-memory. *Neuron, 14*, 477–485.

Guitton, D., Buchtel, H. A., & Douglas, R. M. (1985). Frontal-lobe lesions in man cause difficulties in suppressing reflexive glances and in generating goal-directed saccades. *Experimental Brain Research, 58*, 455–472.

Just, M. A., & Carpenter, P. A. (1980). A theory of reading: From eye fixations to comprehension. *Psychological Review, 87,* 329–354.

Keller, E. L. (1974). Participation of medial pontine reticular-formation in eye-movement generation in monkey. *Journal of Neurophysiology, 37*, 316–332.

MacDonald, M. C., Pearlmutter, N. J., & Seidenberg, M. S. (1994). The lexical nature of syntactic ambiguity resolution. *Psychological Review, 101*, 676–703.

McConkie, G. W., & Rayner, K. (1975). The span of the effective stimulus during a fixation in reading. *Perception and Psychophysics, 17*, 578–586.

Park, S., & Holzman, P. S. (1992). Schizophrenics show spatial working memory deficits. *Archives of General Psychiatry, 49*, 975–982.

Rayner, K. (1975). The perceptual span and peripheral cues in reading. *Cognitive Psychology, 7,* 65–81.

Rayner, K. (1998). Eye movements in reading and information processing: 20 years of research. *Psychological Bulletin, 124*, 372–422.

Rayner, K., & Pollatsek, A. (1989). *The psychology of reading*. Englewood Cliffs, NJ, US: Prentice-Hall, Inc.

Reichle, E. D., Pollatsek, A., Fisher, D. L., & Rayner, K. (1998). Toward a model of eye movement control in reading. *Psychological Review, 105*, 125–157.

Rensink, R. A., O'Regan, J. K., & Clark, J. J. (1997). To see or not to see: The need for attention to perceive changes in scenes. *Psychological Science, 8*, 368–373.

Robinson, D. A. (1963). A method of measuring eye movement using a scleral search coil in a magnetic field. *IEEE Transactions on Biomedical Engineering, BM10*, 137–145.

Robinson, D. A. (1972). Eye-movements evoked by collicular stimulation in alert monkey. *Vision Research, 12*, 1795–1808.

Robinson, D. A. (1975). Oculomotor control signals. In G. Lennerstrand, & P. Bach-y-Rita (Eds.), *Basic mechanisms of ocular motility and their clinical implications* (pp. 337–374). Oxford: Pergamon Press.

Tanenhaus, M. K., Spivey Knowlton, M. J., Eberhard, K. M., & Sedivy, J. C. (1995). Integration of visual and linguistic information in spoken language comprehension. *Science, 268*, 1632–1634.

Yarbus, A. L. (1967). *Eye movements and vision*. New York: Plenum Press.

Zee, D. S., Yamazaki, A., Butler, P. H., & Gucer, G. (1981). Effects of ablation of flocculus and paraflocculus on eye-movements in primate. *Journal of Neurophysiology, 46*, 878–899.

PART 1

HISTORY OF EYE-MOVEMENT RESEARCH

Edited by

ROBIN L. HILL

Chapter 2

SCANNING THE SEEN: VISION AND THE ORIGINS OF EYE-MOVEMENT RESEARCH

NICHOLAS J. WADE
University of Dundee, UK

Eye Movements: A Window on Mind and Brain
Edited by R. P. G. van Gompel, M. H. Fischer, W. S. Murray and R. L. Hill

Abstract

Interest in recording eye movements has been informed by studies of vision. The importance of scanning a scene was recognised theoretically before it was examined experimentally. The contrast between the restricted range of distinct vision and the experience of a uniform and clear visual field focused attention on rapid eye movements. However, they were reported initially in the context of visual vertigo following body rotation; characteristics of the slow and fast phases of nystagmus were recorded by comparing the apparent motion of an afterimage (formed before rotation) with that of a real image. Afterimages were next employed to record torsion when the eye was in tertiary positions. Saccadic eye movements when reading or viewing pictures were measured with a variety of devices from the late nineteenth century. Researchers were genuinely surprised by what they found: contrary to our experience of a stable world and of smooth transitions between objects fixated, the eyes moved rapidly and discontinuously. Vision was generally restricted to the short periods of fixation between saccades.

The history of research on eye movements is fascinating, but it has not received the attention associated with many other aspects of vision (see Wade & Tatler, 2005). Scanning has been neglected relative to the seen. It is, however, the seen (vision) that has provided the stimulus for examining the details of scanning (eye movements). The restriction of distinct vision to a small region around the visual axis led to considerations of how a full impression of the visual surroundings could be formed. However, the precise pattern of scanning was to elude students of seeing for centuries. Knowledge about the limited region of clear vision was known to the ancients, but speculations regarding the manner in which it could be overcome were not advanced until the eighteenth century – before sophisticated devices were available for recording eye movements.

We tend to associate experimental studies of eye movements with the more detailed examinations of reading that were undertaken at the end of the nineteenth century and the beginning of the twentieth. We also generally acknowledge that the link between eye movements and image processing was established experimentally by an American. However, many might be surprised by the identity of the American scientist, who worked over a century before Raymond Dodge (1871–1942). I will argue that William Charles Wells (1757–1817) should be accorded the credit for initiating the experimental study of eye movements. He might not be well known, but the ingenuity of his experiments was impressive. Wells was born to Scottish parents in Charlestown, South Carolina, educated in Scotland, and he practised medicine in London. He wrote a monograph on binocular vision in 1792 after which he was elected Fellow of the Royal Society of London. His experiments on and explanation of the formation of dew (Wells, 1814) resulted in the award of the Rumford Medal and in his election as Fellow of the Royal Society of Edinburgh. He also provided an account of natural selection in 1813, as Charles Darwin (1809–1882) later acknowledged. His initial researches were concerned principally with binocular vision, but he also conducted experimental studies of eye movements, visual resolution, visual motion, visual persistence, accommodation and the effects of belladonna on it. His medical practice was not a thriving one although his small circle of distinguished friends held him in high regard. He wrote a memoir of his life in his last year (see Wells, 1818); this, together with his monograph on vision and related articles, is reprinted in Wade (2003a). Wells examined aspects of vertigo and eye movements as they related to visual motion and it was in this context that he crossed swords with Erasmus Darwin (1731–1802).

Darwin's major work on physiology and psychology was *Zoonomia*, the first volume of which was published in 1794 and the second in 1796. The first chapter in the book was concerned with motion, after which is one on eye movements, although its title is "The motions of the retina demonstrated by experiments". Afterimages (or ocular spectra as he called them) were the tool used to examine the effects of eye movements, and he drew heavily on an article ostensibly written by his son, Robert Waring Darwin (1766–1848), some years earlier (R. Darwin, 1786). In fact, Robert's son, Charles Darwin (1887), suggested that the article had been written by Erasmus! The situation remains one of uncertainty regarding the authorship of the article (see Wade, 2002). One of the features described and demonstrated by Darwin was that of the instability of the eyes during steady fixation: "When we look long and attentively at any object, the eye cannot always be kept intirely motionless; hence, on inspecting a circular area of red silk placed on white paper, a lucid crescent or edge is seen to librate on one side or the other of the red circle" (R. Darwin, 1786, p. 341). Erasmus Darwin repeated the observation in his *Zoonomia*, as well as other demonstrations utilising ocular spectra. Thus, afterimages provided the means for demonstrating the involuntary motions of the eye.

Erasmus Darwin's work conflates two traditions regarding the study of eye movements – the philosophical and the medical. Empiricist philosophers have used muscular movements as an explanatory tool for educating vision about three-dimensional space. Paradoxically, the concerns of physicians have been more empirical, that is more practical. They have observed movements of the eyes, particularly when they have been abnormal. This usually applied to squint, or strabismus, which was one of the first topics Darwin addressed. Darwin was an empiricist in the mould of Bishop Berkeley, whose new theory of vision was published in 1709: the perception of space was considered to be a consequence of motions within it. Darwin incorporated motion as a fundamental part of perception: "The word *perception* includes both the action of the organ of sense in consequence of the impact of external objects, and our attention to that action; that is, it expresses both the motion of the organ of sense, or idea, and the pain or pleasure that succeeds or accompanies it" (1794, p. 12, original italics).

Darwin embraced the long philosophical tradition that held eye movements close to its theoretical heart. Eye movements were of value to the empiricists, but prior to Darwin they were rarely examined empirically. That is, the ways in which the eyes moved were not investigated. Rather, eye movements provided a useful theoretical weapon with which empiricists could attack nativist ideas about the perception of space, and it provided a vehicle for explaining perceptual learning. Muscular sensations could be associated with visual stimulation, and the third dimension of space could slowly be learned. Not all students of the senses accepted Berkeley's theory. Prominent among these was William Porterfield (ca. 1696–1771), a Scottish physician who wrote extensively about eye movements. He rejected Berkeley's theory on logical grounds: Porterfield argued that touch is as arbitrary in its representation of space as is vision, and therefore cannot teach vision external dimensions.

Porterfield wrote two long articles on eye movements in 1737 (Figure 1, left) and 1738; one was on external and the other was on internal motions of the eye. In the course of the latter, Porterfield coined the term "accommodation" for the changes in focus of the eye for different distances. He also examined an aphakic patient, in whom the lens in one eye had been removed, and demonstrated that the lens is involved in accommodation. However, it is his analysis of eye movements during scanning a scene and reading that are of greater interest here. His ideas and observations were repeated in his *Treatise* (Figure 1, right), published over twenty years later. Porterfield's studies started, as had earlier ones, from an understanding that distinct vision was limited to a small region of the visual field – around the point of fixation. This did

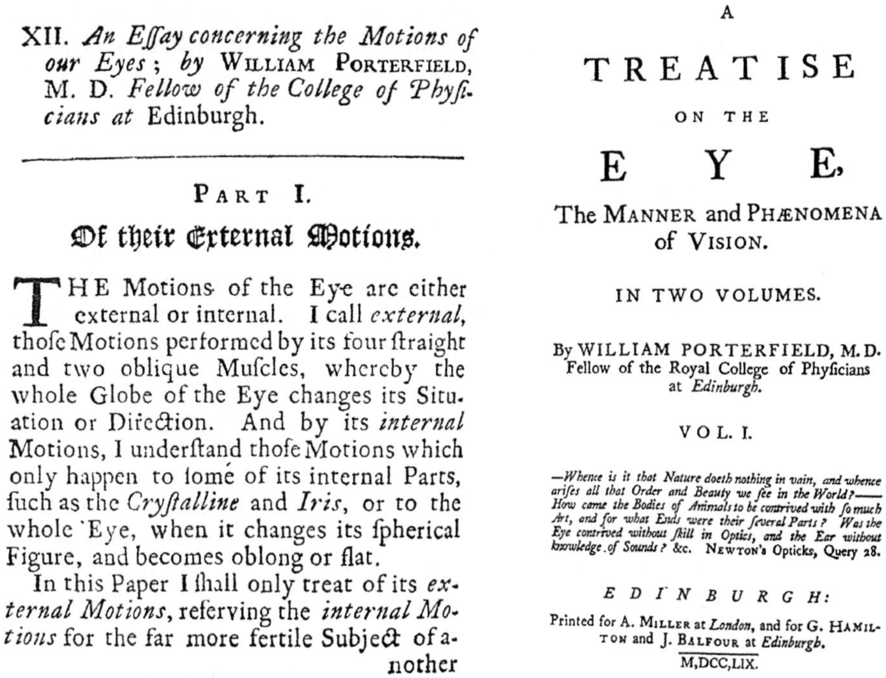

XII. *An Eſſay concerning the Motions of our Eyes*; *by* WILLIAM PORTERFIELD, M. D. *Fellow of the College of Phyſicians at* Edinburgh.

PART I.

Of their External Motions.

THE Motions of the Eye are either external or internal. I call *external*, thoſe Motions performed by its four ſtraight and two oblique Muſcles, whereby the whole Globe of the Eye changes its Situation or Direction. And by its *internal* Motions, I underſtand thoſe Motions which only happen to ſome of its internal Parts, ſuch as the *Cryſtalline* and *Iris*, or to the whole Eye, when it changes its ſpherical Figure, and becomes oblong or flat.

In this Paper I ſhall only treat of its *external Motions*, reſerving the *internal Motions* for the far more fertile Subject of another

A

TREATISE

ON THE

EYE,

The MANNER and PHÆNOMENA of VISION.

IN TWO VOLUMES.

By WILLIAM PORTERFIELD, M.D. Fellow of the Royal College of Phyſicians at *Edinburgh*.

VOL. I.

—*Whence is it that Nature doeth nothing in vain, and whence ariſes all that Order and Beauty we ſee in the World?—— How came the Bodies of Animals to be contrived with ſo much Art, and for what Ends were their ſeveral Parts? Was the Eye contrived without ſkill in Optics, and the Ear without knowledge of Sounds?* &c. NEWTON's Opticks, Query 28.

EDINBURGH:

Printed for A. MILLER at *London*, and for G. HAMILTON and J. BALFOUR at *Edinburgh*.

M,DCC,LIX.

Figure 1. The title pages of Porterfield's (1737) first essay on eye movements (left) and his *Treatise* on vision (right).

not correspond with our visual experience of the scene, and he described this paradox eloquently:

> Now, though it is certain that only a very small Part of any Object can at once be clearly and distinctly seen, namely, that whose Image on the *Retina* is in the *Axis* of the Eye; and that the other Parts of the Object, which have their Images painted at some Distance from this same *Axis*, are but faintly and obscurely perceived, and yet we are seldom sensible of this Defect; and, in viewing any large Body, we are ready to imagine that we see at the same Time all its Parts equally distinct and clear: But this is a vulgar Error, and we are led into it from the quick and almost continual Motion of the Eye, whereby it is successively directed towards all the Parts of the Object in an Instant of Time. (Porterfield, 1737, pp. 185–186, original italics)

Porterfield applied this understanding of the requirement for rapid eye movements to reading itself, although his analysis was logical rather than psychological:

> Thus in viewing any Word, such as MEDICINE, if the Eye be directed to the first Letter M, and keep itself fixed thereon for observing it accurately, the other Letters will not appear clear or distinct. . . . Hence it is that to view any Object, and thence to receive the strongest and most lively Impressions, it is always necessary we turn our

> Eyes directly towards it, that its Picture may fall precisely upon this most delicate and sensible Part of the Organ, which is naturally in the *Axis* of the Eye. (Porterfield, 1737, pp. 184–185, original capitals and italics)

Thus, Porterfield did not provide empirical support for the ideas he developed. Like Erasmus Darwin, Porterfield was part of the medical tradition of examining eye movements. Unlike Darwin, he had an acute awareness of optics and its integration with vision and its defects. He also appreciated the historical background in which his researchers were placed. Moreover, he applied his understanding of eye movements to a wide range of phenomena, including visual vertigo. It was from vertigo that the first signs of discontinuous eye movements derived: the fast and slow phases of nystagmus were demonstrated with the aid of afterimages.

1. Visual vertigo

The visual motion of the world following body rotation was clearly described in antiquity (see Wade, 1998, 2000), but Porterfield (1759) added an eye-movement dimension to it. In fact he denied the existence of eye movements following rotation because he was not aware of feeling his eyes moving. That is, the index of eye movement he used was the conscious experience of it:

> If a Person turns swiftly round, without changing his Place, all Objects about will seem to move in a Circle to the contrary Way, and the Deception continues, not only when the Person himself moves round, but, which is more surprising, it also continues for some time after he stops moving, when the Eye, as well as the Objects, are at absolute Rest. Why, when the Eye turns round, Objects appear to move round the contrary Way, is not so difficult to explain; for, tho', properly speaking, Motion is not seen, as not being in itself the immediate Object of Sight, yet, by the Sight, we easily know when the Image changes its Place on the *Retina*, and thence conclude, that either the Object, the Eye, or both are moved . . . But the great Difficulty still remains, *viz*. Why, after the Eye ceases to move, Objects should for some Time still appear to continue in Motion, tho' their Pictures on the *Retina* be truly at rest, and do not at all change their Place. This, I imagine, proceeds from a Mistake we are in, with respect to the Eye; which, tho' it be absolutely at rest, we nevertheless conceive it as moving the contrary way to that in which it moved before: From which Mistake with respect to the Motion of the Eye, the Objects at rest will appear to move in the same way, which the Eye is imagined to move in, and consequently will seem to continue their Motion for some Time after the Eye is at rest. (Porterfield, 1759, pp. 425–426)

Porterfield sought to accommodate this visual vertigo within his broad analysis of visual motion. In modern terminology he was suggesting that it was the signals for eye movements, rather than the eye movements themselves, that generated the visual motion following body rotation.

Porterfield's description stimulated others to examine vertigo and to provide interpretations of it, some of which involved eye movements. Robert Darwin gave an afterimage interpretation of vertigo: "When any one turns round rapidly on one foot, till he becomes dizzy and falls upon the ground, the spectra [afterimages] of the ambient objects continue to present themselves in rotation, or appear to librate, and he seems to behold them for some time still in motion" (1786, p. 315). That is, afterimages were formed in peripheral regions of the retina during body rotation and they would induce eye movements to place them in central vision when the body rotation ceased. Since the afterimages were stabilised on the retina, eye movements would never change their retinal location and so they would appear to move. Erasmus Darwin also examined vertigo in his chapter on eye movements (the opening page of which is shown in Figure 2, left) and he reinforced the peripheral afterimage theory:

> After revolving with your eyes open till you become vertiginous, as soon as you cease to revolve, not only the circum-ambient objects appear to circulate around

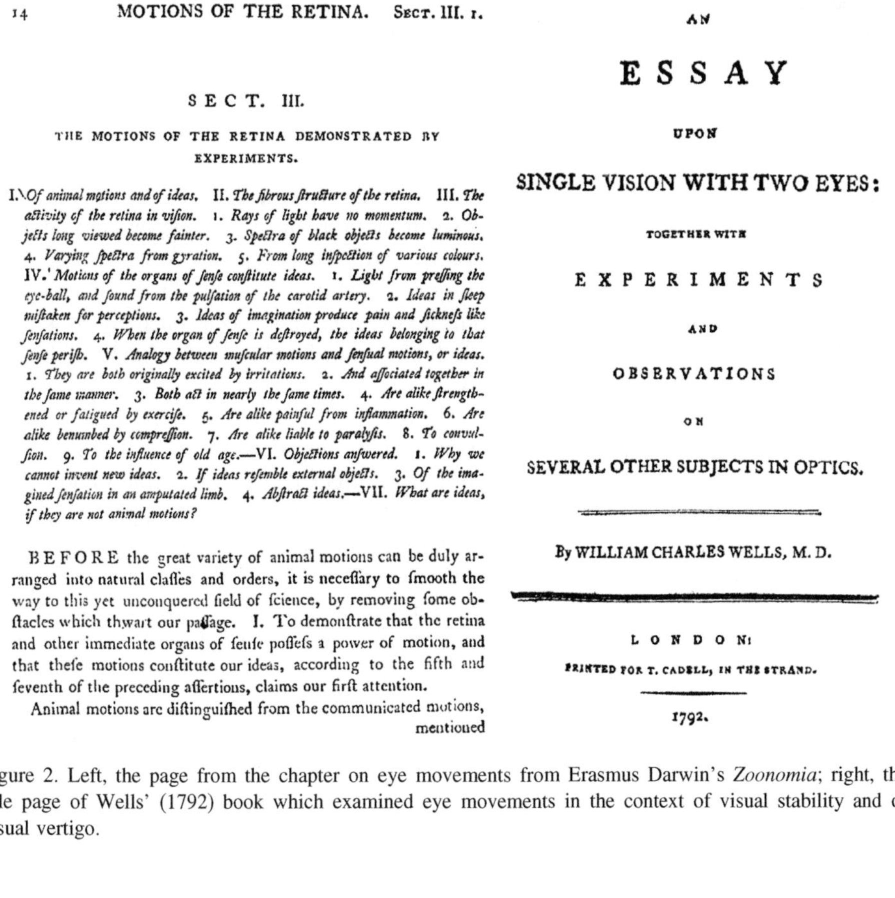

14 MOTIONS OF THE RETINA. SECT. III. 1.

SECT. III.

THE MOTIONS OF THE RETINA DEMONSTRATED BY EXPERIMENTS.

I. *Of animal motions and of ideas.* II. *The fibrous ſtructure of the retina.* III. *The activity of the retina in viſion.* 1. *Rays of light have no momentum.* 2. *Objects long viewed become fainter.* 3. *Spectra of black objects become luminous.* 4. *Varying ſpectra from gyration.* 5. *From long inſpection of various colours.* IV. *Motions of the organs of ſenſe conſtitute ideas.* 1. *Light from preſſing the eye-ball, and ſound from the pulſation of the carotid artery.* 2. *Ideas in ſleep miſtaken for perceptions.* 3. *Ideas of imagination produce pain and ſickneſs like ſenſations.* 4. *When the organ of ſenſe is deſtroyed, the ideas belonging to that ſenſe periſh.* V. *Analogy between muſcular motions and ſenſual motions, or ideas.* 1. *They are both originally excited by irritations.* 2. *And aſſociated together in the ſame manner.* 3. *Both act in nearly the ſame times.* 4. *Are alike ſtrengthened or fatigued by exerciſe.* 5. *Are alike painful from inflammation.* 6. *Are alike benumbed by compreſſion.* 7. *Are alike liable to paralyſis.* 8. *To convulſion.* 9. *To the influence of old age.*—VI. *Objections anſwered.* 1. *Why we cannot invent new ideas.* 2. *If ideas reſemble external objects.* 3. *Of the imagined ſenſation in an amputated limb.* 4. *Abſtract ideas.*—VII. *What are ideas, if they are not animal motions?*

BEFORE the great variety of animal motions can be duly arranged into natural claſſes and orders, it is neceſſary to ſmooth the way to this yet unconquered field of ſcience, by removing ſome obſtacles which thwart our paſſage. I. To demonſtrate that the retina and other immediate organs of ſenſe poſſeſs a power of motion, and that theſe motions conſtitute our ideas, according to the fifth and ſeventh of the preceding aſſertions, claims our firſt attention.

Animal motions are diſtinguiſhed from the communicated motions, mentioned

AN

ESSAY

UPON

SINGLE VISION WITH TWO EYES:

TOGETHER WITH

EXPERIMENTS

AND

OBSERVATIONS

ON

SEVERAL OTHER SUBJECTS IN OPTICS.

By WILLIAM CHARLES WELLS, M. D.

LONDON:

PRINTED FOR T. CADELL, IN THE STRAND.

1792.

Figure 2. Left, the page from the chapter on eye movements from Erasmus Darwin's *Zoonomia*; right, the title page of Wells' (1792) book which examined eye movements in the context of visual stability and of visual vertigo.

> you in a direction contrary to that, in which you have been turning, but you are liable to roll your eyes forwards and backwards; as is well observed, and ingeniously demonstrated by Dr. Wells in a late publication on vision. (p. 571)

Here we get the first reference to the experiments that Wells had conducted on visual vertigo. These were reported in his monograph concerned with binocular single vision; the title page is shown in Figure 2, right. The text of the book has been reprinted in Wade (2003a), and his experiments and observations on eye movements and vertigo are described in one of the "other subjects in optics". The characteristics of eye movements following rotation were clearly described. He formed an afterimage (which acted as a stabilised image) before rotation so that its apparent motion could be compared to that of an unstabilised image when rotation ceased. The direction of the consequent slow separation of the two images and their rapid return (nystagmus) was dependent on the orientation of the head and the direction of body rotation. In the course of a few pages Wells laid the foundations for understanding both eye movements and visual vertigo, which he referred to as giddiness:

> If the eye be at rest, we judge an object to be in motion when its picture falls in succeeding times upon different parts of the retina; and if the eye be in motion, we judge an object to be at rest, as long as the change in the place of its picture upon the retina, holds a certain correspondence with the change of the eye's position. Let us now suppose the eye to be in motion, while, from some disorder in the system of sensation, we are either without those feelings, which indicate the various positions of the eye, or are not able to attend to them. It is evident, that, in such a state of things, an object at rest must appear to be in motion, since it sends in succeeding times its picture to different parts of the retina. And this seems to be what happens in giddiness. I was first led to think so from observing, that, during a slight fit of giddiness I was accidentally seized with, a coloured spot, occasioned by looking steadily at a luminous body, and upon which I happened at that moment to be making an experiment, was moved in a manner altogether independent of the positions I conceived my eyes to possess. To determine this point, I again produced the spot, by looking some time at the flame of a candle; then turning myself round till I became giddy, I suddenly discontinued this motion, and directed my eyes to the middle of a sheet of paper, fixed upon the wall of my chamber. The spot now appeared upon the paper, but only for a moment; for it immediately after seemed to move to one side, and the paper to the other, notwithstanding I conceived the position of my eyes to be in the mean while unchanged. To go on with the experiment, when the paper and spot had proceeded to a certain distance from each other, they suddenly came together again; and this separation and conjunction were alternately repeated a number of times; the limits of the separation gradually becoming less, till, at length, the paper and spot both appeared to be at rest, and the latter to be projected upon the middle of the former. I found also, upon repeating and varying the experiment a little, that when I had turned myself from left to right, the paper moved from right to left, and the spot consequently the contrary way; but that when I had turned

> from right to left, the paper would then move from left to right. These were the appearances observed while I stood erect. When I inclined, however, my head in such a manner, as to bring the side of my face parallel to the horizon, the spot and paper would then move from each other, one upward and the other downward. But all these phenomena demonstrate, that there was a real motion in my eyes at the time I imagined them to be at rest; for the apparent situation of the spot, with respect to the paper, could not possibly have been altered, without a real change of the position of those organs. To have the same thing proved in another way, I desired a person to turn quickly round, till he became very giddy; then to stop himself and look stedfastly at me. He did so, and I could plainly see, that, although he thought his eyes were fixed, they were in reality moving in their sockets, first toward one side, and then toward the other. (Wells, 1792, pp. 94–97)

Thus, Wells used afterimages to provide an index of how the eyes move by comparing them with real images. He confirmed his observations by looking at the eyes of another person who had rotated. By these means he cast doubt on evidence derived from subjective impressions of how the eyes were moving. Darwin (1794) responded negatively to this conclusion because it contradicted his own theory. He provided what he considered to be crucial evidence that eye and visual motions were not related. He denied the occurrence of ocular torsion or rolling of the eyes and so he conducted an ingenious experiment:

> . . . in which the rolling of the eyes does not take place at all after revolving, and yet the vertigo is more distressing than in the situations above mentioned. If any one looks steadily at a spot in the ceiling over his head, or indeed at his own finger held up high over his head, and in that situation turns round till he becomes giddy; and then stops, and looks horizontally; he now finds, that the apparent rotation of objects is from above downwards, or from below upwards; that is, the apparent circulation of objects is now vertical instead of horizontal, making part of a circle round the axis of the eye; and this without any rolling of his eyeballs. The reason of there being no rolling of the eyeballs perceived after this experiment, is, because the images of objects are formed in rotation round the axis of the eye, and not from one side to the other of the axis of it; so that, as the eyeball has not the power to turn in its socket round its own axis, it cannot follow the apparent motions of these evanescent spectra, either before or after the body is at rest. (Darwin, 1794, p. 572)

Thus, Darwin considered that if ocular torsion was not possible then rotary motion could not be associated with it! Accordingly, he dismissed the correlation between eye movements and visual motion for all forms of post-rotational vertigo. This contradiction provided a signal service to the understanding of eye movements in vertigo because it stimulated Wells (1794a, 1794b; reprinted in Wade, 2003a) to conduct experiments to refute it. Not only was each of Darwin's criticisms countered, but Wells also added several additional facts about eye movements and vertigo. First, post-rotational eye movements involve slow rotation in one direction and rapid rotations in the reverse direction; vision is suppressed during the rapid return. Secondly, the eyes can undergo torsional nystagmus. Thirdly, nystagmus (and the apparent rotation that accompanies it) can be suppressed by directed attention. Fourthly, the extent of torsional nystagmus is less than that of horizontal nystagmus. Wells' work was ignored or neglected, perhaps as a consequence of Erasmus Darwin's hostile reaction to it. Darwin (1801) did, however, invent the device that could be used to study the effects of rotation – the human centrifuge. Initially, the device was used as a treatment for the insane (Wade, 2005; Wade, Norsell, & Presly, 2005) and only later was it applied to the study of vertigo. One such rotating device was used by Jan Evangelista Purkinje (1787–1869) in his studies of vertigo.

Purkinje essentially repeated Wells' experiments, but was ignorant of them. Indeed, Purkinje's experiments were inferior to those by Wells, but both adopted interpretations of visual vertigo in terms of eye movements. Purkinje (1820, 1825) added a novel method for studying vertigo and eye movements – galvanic or electrical stimulation of the ears. Stimulating the sense organs with electricity from a voltaic pile was widely applied in the nineteenth century (see Wade, 2003b). The technique was amplified by Eduard Hitzig (1838–1907). In 1871, he examined eye and head movements during galvanic stimulation of the ears and he provided a delightful description of nystagmus: it was like a fisherman's float drifting slowly downstream and then being snatched back. The 1870s was the decade of added interest in eye-movement research because of its assistance in determining semicircular canal function. Post-rotational eye movements were measured and related to the hydrodynamic theory, which was proposed independently by Ernst Mach (1838–1916), Josef Breuer (1842–1925), and Alexander Crum Brown (1838–1922).

Breuer (1874) provided a similar description of post-rotational nystagmus to Wells, but he was able to relate the pattern of eye movements to the function of the semicircular canals. Breuer argued that during rotation the eyes lag behind the head in order to maintain a steady retinal image; then they make rapid jerky motions in the direction of head rotation. The eye movements reduce in amplitude and can stop with rotation at constant angular velocity. When the body rotation ceases the eyes rotate in the same direction as prior head rotation, and the visual world appears to move in the opposite direction interspersed with rapid returns. He also stated, like Wells, that there is no visual awareness during these rapid returns. This is another clear reference to saccadic suppression, although he did not use the term "saccade".

Afterimages were also employed by Mach (1873, 1875), who rediscovered Wells' method for examining post-rotational eye movements:

> I observed nystagmic movements described by Breuer in the following manner. I produced a strong and lasting afterimage with a piece of flaming magnesium wire. If I now rotate actively about the body axis in a clockwise direction (as seen from above), I notice how the afterimage sticks on an object and then suddenly catches up. The eyes rotate slowly counterclockwise and jerk back clockwise. This movement gradually weakens until finally the afterimage rotates with the observer. After stopping, the afterimage moves slowly across the objects clockwise, interrupted by jerklike movements in the opposite direction. The objects then move counterclockwise. (Young, Henn, & Scherberger, 2001, p. 84)

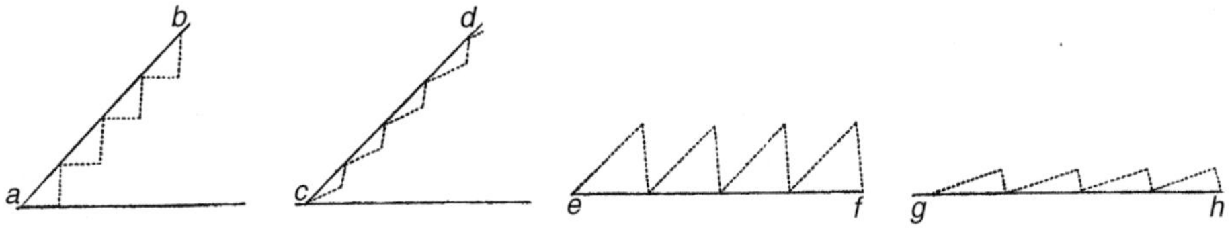

Figure 3. Schematic diagrams by Crum Brown (1878) of eye movements during and after body rotation: "When a real rotation of the body takes place the eyes do not at first perfectly follow the movement of the head. While the head moves uniformly the eyes move by jerks. Thus, in the diagram, Fig. (i), where the abscissæ indicate time and the ordinate the angle described, the straight line *ab* represents the continuous rotatory motion of the head and the dotted line the continuous motion of the eye. Here it will be seen that the eye looks in a fixed direction for a short time, represented by one of the horizontal portions of the dotted line *ab*, and then very quickly follows the motion of the head, remains fixed for a short time, and so on. After the rotation has continued for some time the motion of the eye gradually changes to that represented by the dotted line *cd* in Fig. (ii). The eye now never remains fixed, but moves for a short time more slowly than the head, then quickly makes up to it, then falls behind, and so on. At last the discontinuity of the motion of the eye disappears, and the eye and the head move together. If now the rotation of the head be stopped (of course the body stops also) the discontinuous movements of the eyeballs recommence. They may now be represented by the dotted line in Fig. (iii). The intermittent motion of the eyes gradually becomes less, passing through a condition such as that shown by the dotted line in Fig. (iv), and at last ceases" (Crum Brown, 1878, p. 658)

In addition to observing an afterimage, he applied the time-honoured technique of placing a finger by the side of the eye, and also using pressure figures as stabilised retinal images. However, perhaps the clearest descriptions of eye movements during and following body rotation were given by Crum Brown (1878), who provided diagrams of the steady head and jerky eye movements (Figure 3). Wells' account of the dynamics of eye movements following rotation was beautifully refined by Crum Brown, although no reference was made to Wells. Like most other historians of the vestibular system, Crum Brown championed Purkinje as the founder of experimental research linking eye movements to vestibular stimulation (see Wade & Brožek, 2001; Wade, 2003b).

In the early twentieth century, two aspects of eye movements and vertigo attracted attention. The first was the use of post-rotational nystagmus as a clinical index of vestibular function. These characteristics of nystagmus were defined more precisely by Robert Bárány (1876–1936; 1906, 1913), who was awarded the Nobel Prize in 1914 for his vestibular researches. Indeed, the rotating chair is now called the Bárány chair. He also refined the method of stimulating the semicircular canals with warm and cold water so that the eye movements they induce could be easily observed. The second aspect was the use of post-rotational eye movements as a screening test for aviators.

Aircraft flight placed demands on the vestibular receptors that were beyond the normal range. Only the human centrifuge had subjected the human frame to similar forces. It had been devised by Erasmus Darwin as a treatment for insanity, it was adopted as an instrument for generating vertigo, and now it was applied to simulating the pressures of aircraft flight. Griffith (1920) examined eye movements with aircraft pilots following body rotation. Initially aviators were selected on the basis of their vestibular sensitivity, as determined by tests on a Bárány chair. However, seasoned pilots were not so susceptible to vertigo, and he argued that they had habituated to the repeated rotations to which they had been exposed. In order to examine habituation more rigorously, Griffith tested students on repetitive rotations in a modified Bárány chair. They were exposed to 10 rotations of 20 s, alternating in direction, per session and they were tested over many days. Measures of the duration of apparent motion and the number and duration of nystagmic eye movements were recorded after the body was stopped:

> We have found that, as turning is repeated from day to day, the duration of after-nystagmus, the number of ocular movements made, and the duration of the apparent movement rapidly decrease. The major part of this decrease occurs within the first few days. The decrease takes place not only from day to day but also within a period of trials on any single day (Griffith, 1920, p. 46).

The topic of vestibular habituation attracted Dodge (1923) and he sought to determine how the eyes moved during and after rotation. The problem of adaptation to rotation is a general one, and it is relevant to the relative immunity to motion sickness of those, like seafarers, who are regularly exposed to the conditions which can induce it. As he remarked, "The very existence of habituation to rotation was vigorously denied during the war by those who were responsible for the revolving chair tests for prospective aviators" (Dodge, 1923, p. 15). Dodge had previously measured eye movements during reading and was noted for the ingenuity of the recording instruments he made. Recording eye movements during rotation provided a particular challenge:

> Just before the War I became particularly interested in recording the *reflex compensatory* eye-movements that occur when the semi-circular canals are stimulated by rotation of the subject. The instrumental problem was peculiarly difficult and intriguing. Since a moving field is more or less in evidence if the eyes are open during rotation of the subject, reflex compensatory movements tend to be complicated by pursuit movements during direct observation, even when a strong convex lens is used to prevent clear vision. To study the reflex in pure form, the first requirement was to record the movements of the closed eyes. (Dodge, 1930, pp. 111–112, original italics)

In examining the possibilities he had noticed that the convexity of the cornea was visible as a moving bulge beneath the eyelid. Dodge (1921) mounted a mirror over the closed eyelid and was able to record eye movements by the reflections from it. With it he was able to confirm the results of Griffith: without the possibility of visual fixation, after-nystagmus declines with repeated rotations. In section 3 it will become evident that Dodge was a pioneer of recording eye movements generally, and during reading in particular. It is of interest to note that following his developments in these novel areas he engaged in examining eye movements following body rotation – the problem tackled by previous pioneers over a century earlier. Indeed, Dodge shared with Purkinje a willingness to engage in heroic experiments. When Purkinje gained access to a rotating chair he noted the effects of being rotated for one hour in it. Dodge similarly subjected himself to a gruelling regime: "The experiment consisted of a six-day training period during which the subject (myself) was rotated in the same direction one hundred and fourteen times each day at as nearly uniform speed as possible" (Dodge, 1923, p. 16). The amplitude of post-rotational nystagmus decreased form day to day throughout the experiment. Any feelings of dizziness also disappeared with prolonged practice, and the experience was said to be soothing.

As a footnote to this brief survey of visual vertigo, Dodge was also ignorant of Wells' work when he used his photographic technique to study the effects of body rotation on eye movements in 1923.

2. Torsion

Visual vertigo and nystagmus are usually examined following rotation of the body around a vertical axis, whereas torsion results from lateral rotation around the roll axis. The extent of torsion is much smaller than that of lateral nystagmus, and in the eighteenth and nineteenth centuries there was much debate about whether it occurred at all. Galen had described the oblique muscles that could produce eye rolling or ocular torsion, but evidence of its occurrence was to wait many centuries. The problem related to reconciling the anatomy of the eye muscles with the difficulty of observing small rotations of the eye around the optic axis.

Torsion was easier to observe as a consequence of inclining the head, and it is in this context that Scheiner (1619) hinted at its occurrence: "in an eye movement in which the middle part of the eye remains stationary, it is because it moves by a corresponding head rotation" (p. 245). A more precise description was provided by John Hunter (1728–1793) who outlined the function that the oblique muscles could serve; when the head was tilted to one side, they could rotate the eyes in their sockets in the opposite direction:

> Thus when we look at an object, and at the same time move our heads to either shoulder, it is moving in the arch of a circle whose centre is the neck, and of course the eyes would have the same quantity of motion on this axis, if the oblique muscles did not fix them upon the object. When the head is moved towards the right-shoulder, the superior oblique muscle of the right-side acts and keeps the right-eye fixed on the object, and a similar effect is produced upon the left-eye by the action of the inferior oblique muscle; when the head moves in the contrary direction, the other oblique muscles produce the same effect. This motion of the head may, however, be to a greater extent than can be counteracted by the action of the oblique muscles. (Hunter, 1786, p. 212)

Hunter's analysis of eye movements was on the basis of oculomotor anatomy rather than direct observation or measurement. He was concerned with the fact that the eyes could retain a stable position with respect to their surroundings even when the head and body moved, and it was such considerations that led him to examine torsion. He described three types of eye movements and related them to the operation of the four straight and two oblique eye muscles. The oblique muscles were said to be involved in torsion and the straight muscles moved the eyes from one stationary object to another or pursued a moving object. He did not, however, report the manner in which the eyes moved under these three circumstances.

> The eye being an organ of sense, which is to receive impressions from without, it was necessary it should be able to give its motions that kind of direction from one body to another, as would permit its being impressed by the various surrounding objects; and it was also necessary, that there should be a power capable of keeping the eye fixed, when our body or head was in motion.... From all of which we find these three modes of action produced; first, the eye moving from one fixed object to another; then the eye moving along with an object in motion; and last, the eye keeping its axis to an object, although the whole eye, and the head, of which it makes a part, are in motion.... These two first modes of action are performed by the straight muscles. (Hunter, 1786, pp. 209–210)

It is clear from the dispute between Wells and Darwin that Darwin did not accept that torsion occurred at all, and he was by no means alone in this view. It was Wells' (1794b) demonstration of torsional nystagmus that should have resolved the dispute. The method he used – forming an afterimage of a horizontal line and then comparing it with a real line – should have opened the way to detailed measurements of eye position. Instead, it was forgotten, only to be rediscovered by Christian Ruete (1810–1867; 1845), and applied extensively thereafter.

Charles Bell (1774–1842) argued on the grounds of anatomy and experiment "that the oblique muscles are antagonists to each other, and that they roll the eye in opposite directions, the superior oblique directing the pupil downwards and outwards, and the inferior oblique directing it upwards and inwards" (1823, p. 174). The experiments he conducted on rabbits and monkeys consisted of either noting changes in tension of the oblique muscles during eye movements or sectioning an oblique muscle in one eye and noting the consequences. Three years later, Johannes Müller (1801–1858; 1826) tried to measure torsion by marking the conjunctiva and noting any rotations of the marks. He was unable to see any and concluded that the eyes did not undergo torsion. Much of the remaining part of his book on eye movements was concerned with convergence and binocular single vision.

In contrast to Müller's conclusion, observational studies by Alexander Friedrich Hueck (1802–1842) lent support to the occurrence of torsion. Hueck (1838) examined rotation of the eyes in the opposite direction to head tilt in more detail and tried to measure its magnitude. He stated that the eyes remain in the same relation to gravity for head tilts up to 28°; that is, there was perfect compensation for head inclination:

> The observation of actual eye torsion shows that vertical diameter of the eyeball remains vertical if the head is inclined sideways by up to 28°, and therefore during this movement the image maintains its location unchanged. If we incline the head to

> the side by more than 38°, the eyeball can no longer retain its position with respect to the head rotation (Hueck, 1838, p. 31).

Subsequent experiments have not supported Hueck's claim: while rotation of the eyes does occur, its magnitude is only a fraction of the head tilt. This would now be called ocular countertorsion, and it has a gain of around 0.1 (Howard, 1982). Thus, when the head is tilted laterally by 30° the eyes rotate in the opposite direction by about 3°. It was because the magnitude of countertorsion is small that it was difficult to measure with any degree of precision and there was much debate concerning its occurrence at all. A review of the nineteenth-century debates concerning countertorsion was provided by Nagel (1896), who was able to resolve the dispute regarding its occurrence in animals but not in humans. Howard and Templeton (1966) and Howard (1982, 2002) describe the techniques that have been applied to measure countertorsion in the twentieth century.

In the mid-nineteenth century, ocular torsion assumed considerable theoretical importance. Concern with the geometry of eye rotations and with the specification of their constraints led to the formulations of laws that could only be verified by means of measuring its magnitude. Although many were involved in this endeavour, the work was summarised and integrated by Hermann Helmholtz (1821–1894). A number of preliminary issues required resolving before the geometry of eye movements could be described with accuracy. The first concerned the centre of rotation of the eye. Despite the fact that small translations of the eye occurred, the centre of rotation was determined to be 13.6 mm behind the vertex of the cornea (Helmholtz, 1867, 1924a, 1924b, 1925, 2000). The orbit of the eye can take a variety of locations as the head moves, and so a further question related to the position the head should adopt for experiments on eye movements. Helmholtz defined the planes of the head with regard to the symmetry between the two sides and the line joining the two eyes. The former was called the median plane and the latter the transverse plane. The head was said to be upright when the median plane was vertical and the transverse plane horizontal.

With the head in this erect position some reference position for the eyes was required as was some system of coordinates relative to which rotations could be described. The primary position was defined as the position of the eyes when they are horizontal and perpendicular to the axis between their centres of rotation (the interocular axis). The adoption of fixed axes around which the eye can rotate is arbitrary. Helmholtz described two systems: the Fick axis system in which the vertical axis is assumed to be fixed to the skull and the Helmholtz system in which the horizontal axis is so fixed. When the eye moves from the primary position along a horizontal or a vertical axis it assumes a secondary position; movements to any other (oblique) location involve the eye in a tertiary position.

Frans Cornelis Donders (1818–1889; 1847) measured the eye position in secondary and tertiary locations and provided a general relationship that Helmholtz (1867) elevated to the status of a law:

> The law, given first by Donders and confirmed by all subsequent investigations, is that when the position of the line of fixation is given with respect to the head, the

> angle of torsion will invariably have a perfectly definite value for that particular adjustment; which is independent not only of the volition of the observer but of the way in which the line of fixation arrived in the position in question. (Helmholtz, 1925/2000, p. 44)

Thus, no matter what path the eye pursues to reach a given position, the degree of torsion will be the same. As Helmholtz remarked,

> The best way of verifying these facts is by using after-images, as was originally suggested by Ruete. The way to do it is to stand opposite a wall covered with paper on which horizontal and vertical lines can be distinguished, the pattern, however, not being so sharply outlined as to make it difficult to see after-images on it. The best background is one of a smooth pale grey colour. Directly opposite the observer's eye, and on the same level with it, a black or coloured ribbon is stretched horizontally, two or three feet long; standing out in sharp contrast to the colour of the wall-paper . . . Now let the observer look intently for a little while at the middle of the band, and then, without moving his head, turn his eyes suddenly to another place on the wall. An after-image of the band will appear there, and by comparing it with the horizontal lines of the wall-paper, the observer can see whether the after-image is horizontal or not. The after-image itself is developed on those points of the retina belonging to the retinal horizon; and during the motions of the eyes it indicates those parts of the visual field on which the retinal horizon is projected. (Helmholtz, 1925/2000, p. 45)

This is essentially the technique Wells (1794b) had used in order to determine the existence of torsional nystagmus. In Helmholtz's case the degree of deviation of the afterimage from the rectangular grid could be determined and the diagram he provided to show the degrees of torsion in secondary and tertiary positions is shown in Figure 4, left. The afterimages remain vertical or horizontal in secondary but not in tertiary positions.

Donders' law was formulated for movements of the eye from the primary position. Another lawful relationship was formulated by Johann Benedict Listing (1808–1882). Whereas Donders' law stated that the eye always assumed the same orientation at a particular location, no matter how it had been reached, Listing's law specified the orientation that was adopted in a particular position. He did not publish an account of it himself but it was initially described in Ruete's (1853) book on ophthalmology. Again it was Helmholtz who named it Listing's law; it was concerned with describing the axis about which the eye rotates from the primary position. Helmholtz described it in the following way:

> Accordingly, for the case of a pair of emmetropic eyes with parallel lines of fixation, the law of their motion may be stated as follows: *When the line of fixation is brought from its primary position into any other position, the torsional rotation of the eyeball in this second position will be the same as if the eye had been turned around a fixed axis perpendicular to the initial and final directions of the line of fixation.* This is known as Listing's law of ocular movements, because it was first stated by him in this form. (Helmholtz, 1925/2000, p. 48, original italics)

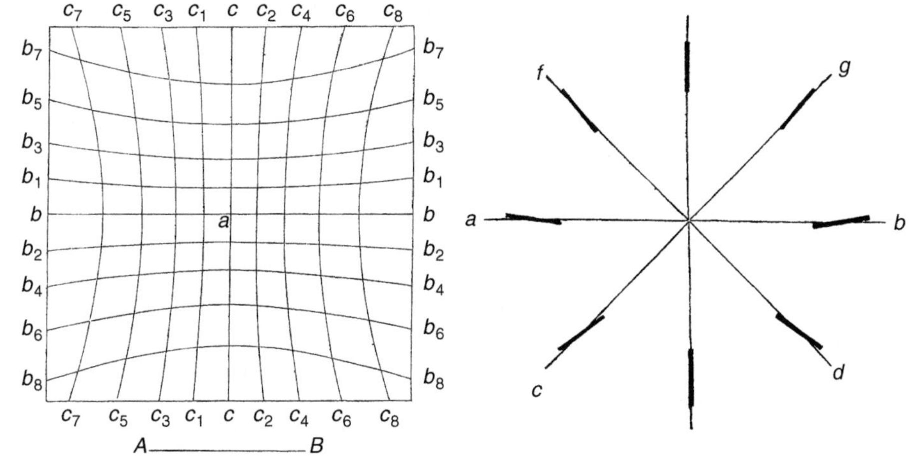

Figure 4. Left, eye orientation in primary, secondary and tertiary positions; right, deviations from Listing's law with convergence. Both images from Helmholtz (1867).

Both Donders' and Listing's laws apply for eye movements from the primary position; that is, with the eyes parallel and directed to the horizon. Empirical studies generally supported the laws, and Helmholtz reported that eye rotations could be measured to within half a degree using the afterimage method. However, most eye movements are directed to objects that are nearer than optical infinity and so will involve some degree of convergence. Alfred Wilhelm Volkmann (1800–1875; 1846) demonstrated that departures from the laws occur with convergence (Figure 4, right).

One of the great advantages that Helmholtz bestowed on eye-movement research was the introduction of the bite bar for controlling head position. His arch-rival Ewald Hering (1834–1918) made many contributions to the study of eye movements amongst which was an improved version of Helmholtz's bite bar. Hering (1879a, 1942) was much more concerned with binocular than monocular eye movements. He made the distinction between eye movements in distant vision (with the axes parallel and horizontal) and near vision (with convergence and depression or elevation of the eyes). He did, however, conduct experiments, like those described by Helmholtz, in which the orientations of line afterimages were determined in a variety of eye positions in distant vision. A vertical line afterimage was formed in the primary position of the eye and then its orientation was determined in secondary and tertiary positions. Hering found, as Volkmann had before him, that this situation did not apply to near vision:

> *In symmetric convergence postures, the mid-vertical sections are inclined at the upper end, more outward or less inward than Listing's law stipulates, and that this deviation grows on the one hand with an increase of the angle of convergence, on the other with the lowering of the visual plane, and at the most to about 5°. Furthermore with asymmetric convergence, the relative position of both retinas to*

> *one another, and therefore the divergence of mid-vertical or horizontal sections does not differ markedly from that with equally symmetric convergence. (Hering, 1942, p. 133, original italics)*

For Hering, the two eyes moved as a single unit and this union was not dependent upon learning or association, as Helmholtz (1873) maintained. Hering stated this point repeatedly and he referred to the yoked actions of the two sets of eye muscles. His diagrams reflected this, too; the motions of each eye were referred to those of an imaginary eye between them. In the context of visual direction the locus from which objects are seen was called the cyclopean eye. For binocular eye movements he referred to the directions relative to this 'imaginary single eye' as the binocular visual line. Fixations occur between movements of the eyes and Hering showed a clear understanding of the functions that such movements served. That is, his concerns were not only with where the eye alights but how it reaches that point. Technical constraints tended to restrict experimental studies to determining the orientation of the eye rather than how it moved. Nonetheless, the purpose of such movements was described as well as the range over which they can occur:

> In ordinary seeing the visual point constantly changes its position; for, in order to recognize the outer objects, we must examine one after another, all their separate parts and endeavor to perceive them with utmost acuity. This can be accomplished only by bringing their images upon the two retinal centers. With small objects the mere turning of the eyeballs in their sockets suffices. But if the object of regard is of large dimensions, we resort to turning the head, and if necessary, the upper body, and finally even changing our location. Rarely do we devote our attention to one spot, even for the duration of a second. Rather does our glance spring from point to point, and in its wanderings it is followed slavishly by the visual point, in whose movements we have greater agility than in any other movement, for no organ is so continuously used as the visual organ. (Hering, 1942, p. 83)

The nature of such movements was studied by Hering (1879b) in a short article published in the same year as the long handbook chapter (Hering, 1879a). Hering employed after-images again in these studies, but they were not the principal source of evidence regarding the ways in which the eyes moved. Rather, they provided a means of calibrating another technique that he adopted for assessing rapid eye movements and fixations.

3. Eye movements during reading

Hering's work on binocular eye movements seems to have diverted interest from his brief but insightful experiments on eye movements during reading (Hering, 1879b). He employed afterimages to study eye movements, but in a radically different way from that applied by Mach (1875) and Breuer (1874) in the context of vertigo. Hering accepted that the movements between real images and afterimages reflected the movements of the eyes, and used this as a source of evidence to support a novel method of measurement.

He used two rubber tubes, rather like a miniature stethoscope, placed on the eyelids to listen to the sounds of the ocular muscles. With this technique he observed, "Throughout one's observations, one hears quite short, dull clapping sounds, which follow each other at irregular intervals" (1879b, p. 145). He found that these 'clapping' sounds were evident when observers read lines of text but disappeared if they were instructed to fixate a stationary target. The sounds were attributed to contractions of the oculomotor muscles accompanying eye movements. Hering provided evidence for this conclusion by comparing the occurrences of clapping sounds with the movements of an afterimage: "every clapping sound corresponds to a displacement of the afterimage" (p. 145). Thus, Hering described the discontinuity of eye movements and recognised the class of rotations that we now refer to as saccadic. He was amongst the first to offer a description of the discontinuity of eye movements outside the context of vestibulo-ocular reflexes.

Hering's experiment is significant not only because he compared afterimage movement to the sounds of muscular movements, but also because he applied the technique to reading: "One can observe the clapping sounds very clearly during reading. Although the eyes appear to glide steadily along the line, the clapping sounds disclose the jerky movement of the eyeball" (p. 146). Hering's report of his experiments was published in the same year as Louis-Émile Javal (1839–1909) gave a brief description of "saccades" during reading. Javal is generally considered to have instigated research in eye movements during reading. Statements to this effect can be found in Huey (1908), Vernon (1931), and Woodworth (1938).

In fact, Javal said virtually nothing about eye movements in his eight essays on the physiology of reading (see Wade & Tatler, 2005), and saccades were only mentioned in passing in the final article. Moreover, he was not referring to his own work: "Following the research of M. Lamare in our laboratory, the eye makes several saccades during the passage over each line, about one for every 15–18 letters of text. It is probable that in myopes the eye reacts with a rapid change in accommodation with every one of these saccades" (Javal, 1879, p. 252). Javal tried, unsuccessfully, to attach a pointer to the eye so that eye movements could be recorded on a smoked drum. He also tried, with a similar lack of success, to measure the deflections of light from a mirror attached to the eye. Throughout, his concern was with distinguishing between horizontal and vertical eye movements. It was Lamare who observed and recorded the jerky or saccadic movements during reading in 1879. However, Lamare did not describe his experiments until 13 years later. He tried various methods, including observing the eyes of another person, but

> The method that gives the best results is one by which the movements are heard via a drum with an ebonite membrane in the centre and to which a small tube is attached; the tube is in contact with the conjunctiva or eyelid and is connected to both ears by rubber tubes... The apparatus yields distinctive sounds which an assistant can count and add, and note for each line. The return movement of the eyes to the beginning of a line gives a longer and louder noise that is easy to recognise; one counts the number of *saccades* from the start of the line to be able to *note* the number of divisions that occur in a line. (Lamare, 1892, p. 357, original italics)

Lamare's intricate technique for recording eye movements was clearly described by Javal's successor as director of the ophthalmological laboratory at the Sorbonne, Marius Hans Erik Tscherning (1854–1939). Tscherning did not mention Javal in the context of his brief description of saccadic eye movements in his *Optique physiologique* (1898). He noted,

> The eyes are, therefore, in perpetual motion which is made by jerks: they fix a point, make a movement, fix another point, and so forth. While reading, the eyes move also by jerks, four or five for each line of an ordinary book. *Lamare* constructed a small instrument, formed by a point which is supported on the eye across the upper eyelid, and which is fastened to the ears of the observer by rubber tubes. With this instrument each movement of the eye causes a sound to be heard. We hear four or five sounds during the reading of one line, and a louder sound when we begin to read a new line. (1900, p. 299)

Lamare's technique bears a remarkable similarity to that employed by Hering in 1879, and they should both be accorded the credit for demonstrating the discontinuity of eye movements in reading. It is of interest to note that Tscherning's (1898) book in French used the word "saccades" which was translated as "jerks" in the English edition (Tscherning, 1900). The term "saccades" was incorporated into English by Dodge (1916): "German and Scandinavian writers are commonly using the descriptive class term 'saccadic' to denote the rapid eye movements for which we have only the arbitrary name of 'type I'. I am not sure with whom the term originated, but it seems worth adopting" (pp. 422–423).

As the nineteenth century ended and the twentieth began there was a proliferation of eye-tracking technologies. Huey (1898, 1900) and Delabarre (1898) developed eye-trackers that used a lever attached to an eye-cup to record movements of the eyes on the surface of a smoked drum. Lever devices were limited by their mechanics; they applied additional force to the eye and their inertia resulted in considerable overshoots in the eye-movement traces recorded (Figure 5). Alternative devices were developed rapidly in which direct attachment between eye and recording surface was not required. Orschansky (1899) attempted to record light reflected from a mirror attached to the surface of an eye-cup and variants of this principle were used throughout the twentieth century. An alternative to recorded light reflected from an eye-cup, or other attachment on the surface of the eye, was to record light reflected directly from the surface of the eye itself. The key figure in the development of such devices was Dodge (Dodge, 1903, 1904; Dodge & Cline, 1901). Photographic devices that required no attachment on the eye were more comfortable for the participants and many of the modern eye-trackers are based on this principle.

In 1900, Dodge presented a very early exploration of the possibility of perception within eye movements. This paper built on earlier observations made during his time with Benno Erdmann (1851–1921) in which they conducted experiments concerning vision during reading (Erdmann & Dodge, 1898). They had been using mirrors to observe subjects' eye movements whilst reading text. When looking into the mirrors themselves Erdmann and Dodge both found it impossible to see their own rapid eye movements.

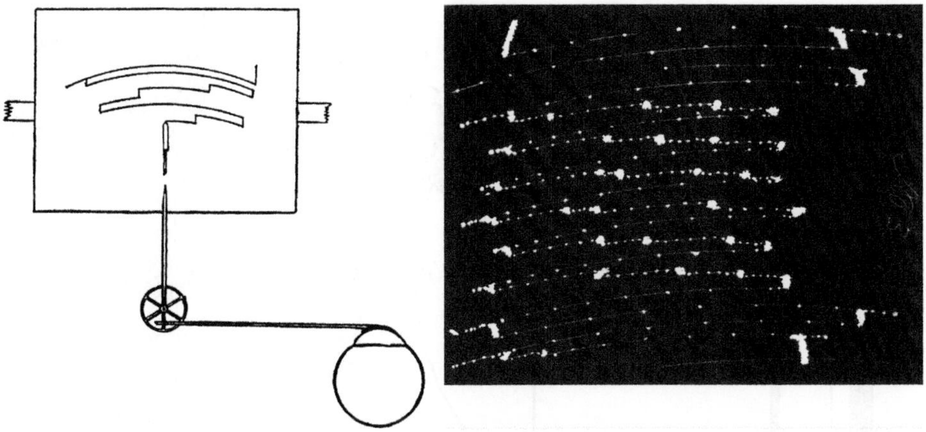

Figure 5. Left, Huey's (1898) technique for recording eye movement during reading. A lever was attached to a plaster eye-cup and moved a pen over a rotating smoked drum. The initial records shown are for calibration across the lines, followed by jerks and pauses during reading. Right, a record of eye movements during reading using the lever device (from Huey, 1900). The tracing on the smoked drum was photographed and then engraved. "The curve shows the movements of the eye in reading six lines, preceded and followed by two free movements of the eye each way, in which it was swept from one end of the line to the other, the beginning and end alone being fixated" (Huey, 1900, p. 290).

Until these reports by Erdmann and Dodge, it seems that the general consensus in the field was that eye movements were themselves an integral part of the processes of visual perception. It was believed that perception continued during eye movements and that the continuous movement of gaze over an object would be sufficient for its perception. Dodge, however, recognised that this was not the case. Critically, Dodge appreciated the errors and pitfalls of self-observation when describing eye movements and perception, in the same way that Wells had distrusted Porterfield's recourse to subjective experience over 100 years earlier. Consequently, Dodge employed the use of an assistant to observe his eye movements, or to be observed. He also developed a method for photographically recording eye movements (Dodge & Cline, 1901), which is shown in Figure 6.

Dodge's observations and studies of visual perception during and between eye movements have been of central importance to understanding visual processing of text. However, Wells (1794b) offered a very early report of the possible lack of visual perception during eye movements in the context of post-rotational vision. He wrote,

> For I mentioned that, if, while giddy, and in possession of the spectrum [afterimage] of a small luminous body, I direct my eyes to a sheet of white paper, fixed to a wall, a spot immediately appears upon the paper; that the spot and paper afterwards separate from each other to a certain distance, the latter seemingly moving from left to right, if I had turned from right to left; but from right to left if I had turned the contrary way; and that they suddenly come together again. My conclusion from this experiment is, that, although the eye during it moves forwards and backwards, still

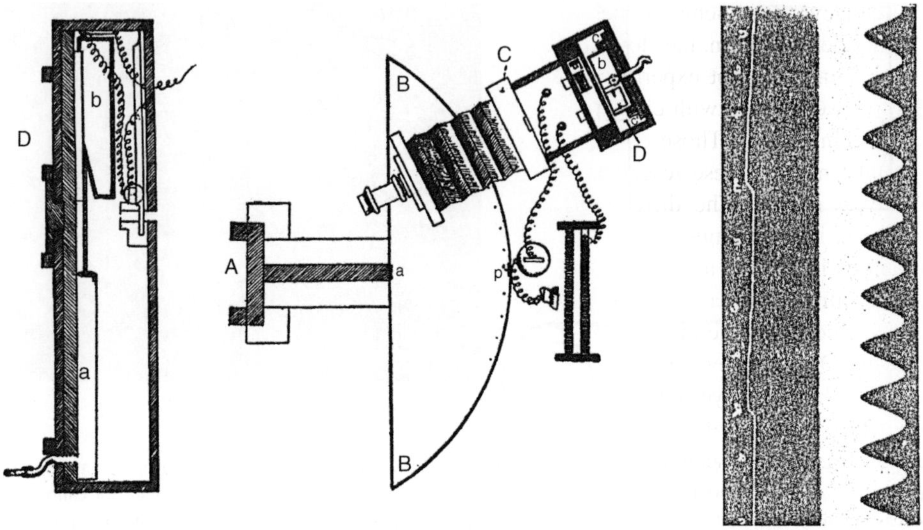

Figure 6. Left, the photographic apparatus used by Dodge and Cline (1901) to record eye movements. Right, their first published trace of an eye movement (from Dodge & Cline, 1901).

> the two motions are not exactly similar, but that in one the picture travels slowly enough over the retina to allow me to attend to the progression of the paper; while in the latter the passage of the picture is so rapid, that no succession of the paper's apparent places can be observed. (pp. 905–906)

During fixations the image on the retina is kept largely stationary. When gaze is relocated, there is a brief period of blindness followed by the reception of another relatively static view of the new fixation target. Consequently the input that our eyes supply to the visual system is in the form of a series of relatively static glimpses of the world separated by brief periods of blindness. This is very different from the smooth perception that we experience. Crum Brown appreciated the significance of saccades and fixations, and described them eloquently:

> We fancy that we can move our eyes uniformly, that by a continuous motion like that of a telescope we can move our eyes along the sky-line in the landscape or the cornice of a room, but we are wrong in this. However determinedly we try to do so, what actually happens is, that our eyes move like the seconds hand of a watch, a jerk and a little pause, another jerk and so on; only our eyes are not so regular, the jerks are sometimes of greater, sometimes of less, angular amount, and the pauses vary in duration, although, unless we make an effort, they are always short. During the jerks we practically do not see at all, so that we have before us not a moving panorama, but a series of fixed pictures of the same fixed things, which succeed one another rapidly. (Crum Brown, 1895, pp. 4–5)

Ballistic eye movements and periods of fixation are a feature of post-rotational nystagmus. During body rotation, the slow phase kept an image relatively stable on the retina, and the rapid return was not experienced visually. These aspects of involuntary eye-movement control were stated with clarity and economy first by Wells, and later by Mach, Breuer, and Crum Brown. Those concerned with voluntary eye movements, as in reading, did not take note of these researches, and painstakingly rediscovered the manner in which the eyes moved. The differentiating feature between the two endeavours is that the saccades in nystagmus are in one direction, opposite to the slow phase. Saccades and fixations in reading are progressive; saccades are smaller and in the same direction until the termination of a line. The importance of Wells' work in the development of our understanding of eye movements has been completely overlooked.

After Dodge's development of the photographic eye-tracker there followed something of a revolution in eye-movement research and a proliferation of new experiments in this field (see Taylor, 1937). Other researchers developed similar convenient and effective eye-trackers and research extended beyond the domain of reading. Researchers began to consider whether the newly described saccade-and-fixate strategy applied to tasks other than reading, and it was soon realised that this was the case.

4. Eye movements over patterns

George Malcolm Stratton (1865–1957) employed a photographic technique to examine eye movements when viewing patterns. This was an important new direction for eye-movement research and served to highlight the importance of saccades outside the context of reading. Like his contemporaries, Stratton was surprised by the discontinuity of eye movements: "The eye darts from point to point, interrupting its rapid motion by instants of rest. And the path by which the eye passes from one to another of these of these resting places does not seem to depend very nicely upon the exact form of the line observed" (Stratton, 1902, p. 343). This quotation highlights a central aspect of Stratton's work: he demonstrated that the path taken by the eye during saccades did not appear to relate to the form of the pattern viewed (Figure 7). Stratton (1906) returned to this issue of the relationship between patterns and eye movements in a later article, which opened by saying that: "our pleasure in graceful curves could not be due to the ease and smoothness of the eye's own motion in viewing these curves. For the ocular movement itself, when photographically recorded, is found to be interrupted and jerky and most unlike the figures we enjoy" (Stratton, 1906, p. 82). The article went on to consider whether eye movements can account for simple visual illusions, and whether eye movements can explain the aesthetics of symmetry. Stratton found no evidence for eye-movement explanations of the Müller-Lyer, Poggendorff, or Zöllner illusions:

> If some special form of the eye's movement is the cause of a particular illusion, such a movement must be present on every occasion when the illusion is felt. But in the actual records we find plenty of instances where the movements which have been

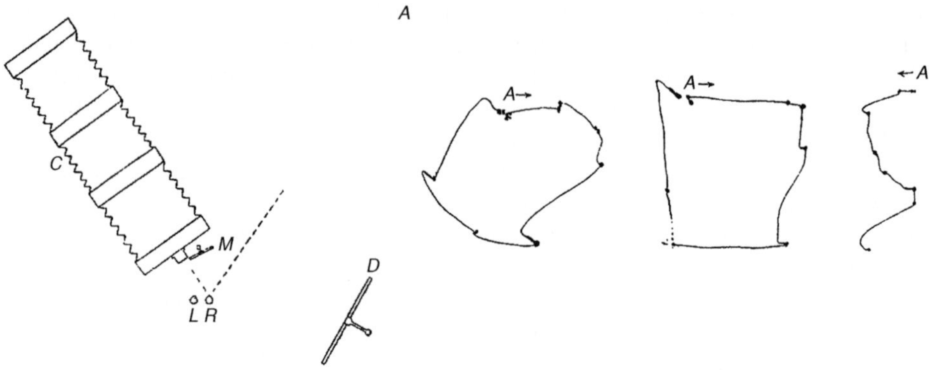

Figure 7. Left, Stratton's camera for recording eye movements, and right, records for tracking a circle, square and an S shape (both from Stratton, 1902).

> supposed to produce a given illusion are present when the illusion is absent, and absent when the illusion is present. One can hardly believe, then, that the supposed causal connection really exists. (Stratton, 1906, p. 93)

In his investigation of the eye movements of observers while looking at a variety of symmetrical and non-symmetrical figures, Stratton was again surprised by what he found when he examined the photographic negatives recorded: "one is struck by the almost grotesque unlikeness between the outline observed and the action of the eye in observing it. For the most part the eye moves irregularly over the figure, seeking certain points of vantage from which the best view of important features may be obtained. And these positions are marked by the eye's momentary resting there" (1906, p. 94). Again, the path taken by the eye during movements did not relate to the form of the figure being viewed, but Stratton also observed that the positions fixated did not show any clear and consistent relationship with the figure:

> Now these points of rest are evidently of more consequence to the observer than the path by which the eye reaches them; indeed the form of any single path between two stops usually bears no observable resemblance to the outline which the subject was taking in, and which in many cases he believes his eye to be accurately following. But even these points of rest are not so arranged as to supply of themselves a rough sense of the form perceived, after the manner of an outline pricked disconnectedly in paper. The points of the eye's rest in the records are usually too few and too inexact to give any such clear and connected perception of the form as the observer regularly and readily obtains. (Stratton, 1906, p. 94)

The phrasing of the article suggests some degree of despair in Stratton's reports of the lack of correspondence between the eye movements and the figure being observed. The degree of surprise and disbelief evident in this article highlights the fact that the saccade and fixate behaviour of the eye was still very much a new aspect of vision research.

This continued to surprise and intrigue researchers even nearly 30 years after Javal's report of Lamare's work and Hering's article describing saccadic eye movements. Not only was there no clear relationship between the eye movements and form of the figure viewed, but there was also no clear relationship between the symmetry of the figure viewed and the symmetry of the eye movements made when viewing it.

Stratton's work is significant because it attempts to bridge the gap between visual phenomena (illusions), cognition (aesthetic judgements), and the underlying mechanisms (eye movements). Thus, both Stratton and Dodge defined new directions for eye-movement research, highlighting the discrepancy between cognition and perception in vision. Stratton explicitly states the inadequacy of eye movements as an explanation for aesthetic judgements: "The sources of our enjoyment of symmetry, therefore, are not to be discovered in the form of the eye's behaviour. A figure which has for us a satisfying balance may be brought to the mind by most unbalanced ocular motions" (1906, p. 95). It is clear from Stratton's closing remarks in his article on eye movements and symmetry that he appreciated the gulf between perception and cognition and that he was aware that his work was defining new questions in vision that would be addressed in the future.

Charles Hubbard Judd (1873–1946) worked initially on geometrical illusions and was dismissive of interpretations based on angle expansion and perspective. Rather he favoured what he called "the movement hypothesis": "The more intense the sensation of movement, the greater will be the estimation of the distance; conversely, the less the intensity of the sensations of movement, the shorter the estimated distance" (Judd, 1899, p. 253). He did not equate the hypothesis with actual movement nor did he, at this stage, provide any measurements of the extent of eye movements. Judd was impressed by the work of Dodge and Stratton, but felt that the eye-trackers used by both were somewhat limited in the range of tasks to which they could be applied: Dodge's eye-tracker was designed for recording movements only along a straight line (as occurs in reading) and Stratton's lacked temporal resolution and required a dark room in which the photographic plate was exposed during experimental recording. Judd developed a "kinetoscopic" eye-tracker in which a small fleck of Chinese white, which had been affixed to the cornea, was photographed; eye movements could thus be discerned in two dimensions, and did not require a dark room (Judd, McAllister & Steele, 1905). While Judd's eye-tracker offered specific advantages over previous photographic devices, it was still somewhat limited in its temporal resolution, typically operating at about 8–9 frames per second (in comparison to Dodge's 100 Hz Photochronograph). One impressive feature of Judd's eye-tracker was that it allowed binocular recordings to be taken.

Having developed this new photographic eye-tracker, Judd and colleagues went on to use it to investigate eye movements between simple targets (McAllister, 1905) and the role of eye movements in the Müller-Lyer illusion (Judd, 1905a), the Poggendorff illusion (Cameron and Steele, 1905), and the Zöllner illusion (Judd & Courten, 1905). In the first of these reports (McAllister, 1905) observers were asked to alternate fixation between two targets. The nature of these targets was varied comprising different arrangements of dots and lines. In his investigation of the Müller-Lyer illusion, Judd (1905a) found that there was some evidence for there being differences in the eye movements when looking at the overestimated and underestimated regions of the illusions, and also that practice influenced eye movements:

> The positive outcome of our researches consists, therefore, in the exhibition of a consistent tendency on the part of five subjects to show restricted movements in looking across the underestimated Müller-Lyer figure and freer movements in looking across the overestimated figure. One of these subjects who performed a series of adjustments with sufficient frequency to overcome the illusion showed a marked change in the character of the movement. (Judd, 1905a, p. 78)

In the case of the Poggendorff illusion, eye movements again correlated with the experience of the illusion (Cameron and Steele, 1905). However, this was not the case for the Zöllner illusion: "The usual fact . . . is that the eye movement, and consequently the sensations from these movements, are in the wrong direction to account for the illusion" (Judd & Courten, 1905, p. 137). It is interesting to note that Stratton and Judd appeared to disagree in their investigations of the role of eye movements in visual illusions. While Judd offered eye-movement explanations for the Müller-Lyer and Poggendorff illusions, Stratton, as we have seen, proposed that no such explanations were possible.

Judd's interest in interpreting the results of the studies on illusions was in addressing the relationship between movement and perception (Judd, 1905b). Judd sought to use his results to dismiss the notion that perceptions might arise directly from movements themselves; rather he stressed the importance of visual information from the retina both in forming perceptions and in coordinating the movements:

> When the eye moves toward a point and the movement does not at first suffice to bring the point in question on the fovea, the retinal sensations which record the failure to reach the desired goal will be a much more powerful stimulus to new action than will any possible muscle sensation. (Judd, 1905b, p. 218)

While much of Judd's discussion of his eye-movement studies focused upon the discussion of their relation to theories of movement sensation, he did also notice that the pattern of eye movements was likely to be influenced by the instructions given to the observers during the experiments. The recognition that eye movements were not entirely directed by the stimulus being viewed echoed the opinion expressed by Stratton at the same time, but it was not investigated in any great depth by either Judd or Stratton.

A more thorough treatment of this question was not conducted until some years later when Judd's former doctoral student, Guy Thomas Buswell (1891–1994), conducted an impressive series of eye movement studies. While Buswell's work on reading is probably his most renowned work, it was from his study of eye movements when viewing pictures that one of his major contributions to eye-movement research arises. The latter work was published in his impressive monograph *How people look at pictures* (Buswell, 1935). In the monograph, Buswell reports eye-movement data recorded from 200 participants each viewing multiple pictures, such that his data comprised almost 2000 eye-movement records each containing a large number of fixations. This volume of eye-movement data is impressive by modern standards, but particularly so given the technology at the time and the need to transform manually the horizontal and vertical position of the eye indicated in the eye-movement records into precise locations on the pictures viewed. This work was the first to explore in a systematic way eye movements of observers while viewing complex pictures, rather than text or simple patters, and represented somewhat of a revolution in eye-movement research. Using complex images and scenes has become a central aspect of modern eye-movement research and is an important part of understanding eye movements in everyday vision.

Buswell's monograph explores a wide range of issues regarding the eye movements made while viewing pictures, including some surprisingly modern concerns: he looked at the overall distribution of fixations on pictures; he compared the first few fixations on a painting to the last few; he compared the durations of fixations made early in viewing to those made near the end of viewing; he looked at how fixation duration changed with viewing time; he compared the consistency between different observers when viewing the same picture; and he looked at the influence of instructions given to observers upon their eye movements when viewing a picture.

Buswell made density plots of where all participants fixated when viewing pictures and showed that not all locations and objects in pictures are fixated, with particular "centers of interest" where fixations are concentrated (Figure 8). He also appreciated that

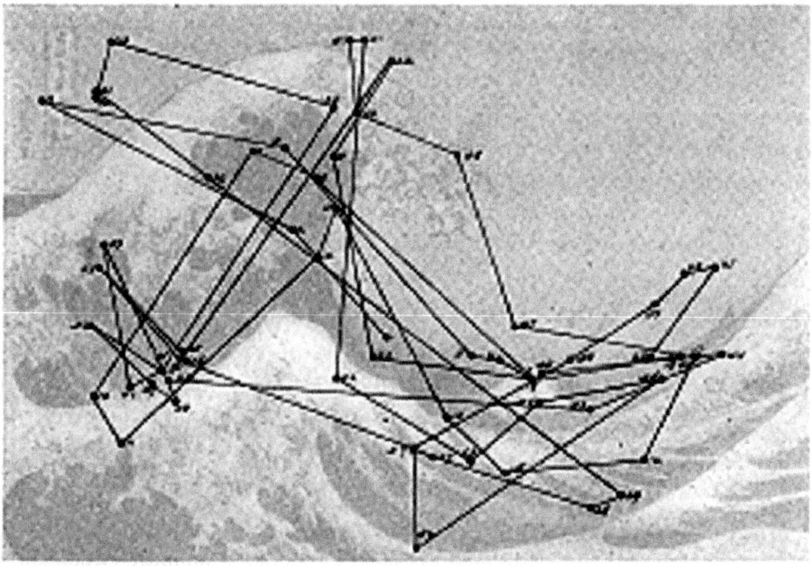

Figure 8. Eye-movement record of one subject viewing "The Wave" painted by Hokusai. Each black circle represents a fixation made by the observer. The lines indicate the saccades that moved the eye from one fixation to the next. Each fixation point is numbered in sequence from the start of viewing (after Buswell, 1935).

there could be quite large individual differences in where people fixate when viewing pictures:

> The positions of the fixations indicate clearly that for certain pictures the center or centers of attention are much more limited than in other pictures. The fact that the density plots represent composite diagrams from a great many different subjects obviously results in a wider distribution of fixations on account of the varied interests of different individuals in looking at the same picture. However, the density plots do give a rather clear indication as to what parts of a given picture are likely to prove most interesting to a random selection of subjects. (Buswell, 1935, p. 24)

The issue of individual differences in the patterns of eye movements was explored in detail in a section at the end of the second chapter. The differences that were present in the locations fixated by individuals when viewing each image were also reflected in the durations of the fixations, with a large variation between observers in their average fixation duration on each picture. Buswell's investigation of individual differences extended to exploring differences between artistically trained individuals and those without training; between children and adults; and between Western and Oriental participants. In all cases, differences between the groups were small: "The average differences between the groups were so much less than the individual differences within each group that the results cannot be considered significant" (Buswell, 1935, p. 131). Differences were found in the eye-movement data that emerged over the course of viewing a picture for some time. The regions fixated in the picture were more consistent between observers for the first few fixations than for the last few on each picture. Buswell also found that fixation duration increased over the course of viewing a picture for some time.

Buswell devoted a chapter of the book to looking at the influence of the characteristics of the picture upon where is fixated. This work is very reminiscent of that conducted some years earlier by Stratton although he did not cite Stratton's work. In places Buswell appears to argue that eye movements do tend to follow lines in pictures. This is contrary to Stratton's suggestion that eye movements do not appear to be particularly influenced by the form of the figure being viewed. However, other aspects of Buswell's data suggest less concordance between eye movements and the characteristics of the picture. When Buswell showed participants more basic designs and patterns he found that:

> The effect of different types of design in carrying the eye swiftly from one place to another is apparently much less than is assumed in the literature of art. . . . The writer should emphasize that the data from eye movements are not to be considered as evidence either positively or negatively for any type of artistic interpretation. (Buswell, 1935, p. 115).

Like Stratton, Buswell felt that the pattern of eye movements was insufficient to explain our visual experience and so highlighted the need to appeal to cognitive explanations of vision.

Perhaps the most often overlooked aspect of Buswell's work was his investigation of how different instructions given to observers prior to viewing an image can influence the eye movements made during viewing. For example, when presented with a picture of Chicago's Tribune Tower, eye movements were first recorded while the participant viewed the picture "without any special directions being given. After that record was secured, the subject was told to look at the picture again to see if he could find a person looking out of one of the windows of the tower" (Buswell, 1935, p. 136). Very different eye-movement records were obtained in these two situations, demonstrating that cognitive factors such as the viewer's task can have a strong effect upon how a picture is inspected. Such descriptions of influences played by high-level factors upon eye movements are not typically associated with Buswell, but rather it is Yarbus who is generally regarded to be the first to have offered such an observation (Yarbus, 1967).

Buswell's impressive monograph illustrates the rapid changes that were taking place in eye-movement research in the first half of the twentieth century. Understanding of saccadic eye movements was rapidly increasing, as was the technology with which they could be measured; with these changes came both new questions about vision and the ability to address them in increasingly realistic viewing conditions and with increasing flexibility and precision. Buswell's book discussed a wide range of issues regarding the relationship between eye movements and visual experience and these questions have been reflected in much of the eye-movement research that had followed.

5. Conclusion

In an historical context, investigations of eye movements pose several paradoxes. On the one hand, they are a fundamental feature of our exploration of the world. We are aware that our own eyes move and we can readily detect movements in the eyes of those we observe. Indeed, eye movements are potent cues in social intercourse – we might see "eye to eye" with someone, we might determine honesty when we "look someone straight in the eye" or we might denigrate someone because they have a "shifty look". On the other hand, throughout the long descriptive history of studies of eye movements a vital characteristic of them remained hidden from view, both in ourselves and in our observations of others. The rapid discontinuous nature of eye movements is a relatively recent discovery, as are the small involuntary movements that accompany fixation. For most of recorded history, the eyes were thought to glide over scenes to alight on objects of interest, which they would fix with unmoved accuracy.

Another paradox is that initial knowledge about eye movements derived from generating stimuli that did not move with respect to the eye when the eye moved. The first of these was the afterimage which, since the late eighteenth century, has been applied to determining the ways the eyes moved. More complex photographic recording devices appeared a century later, and the assistance of computers was incorporated three quarters of a century later still. Nonetheless, the insights derived from the skilful use of afterimages have tended to be overlooked in the histories of eye movements. One of the reasons that less attention has been paid to studies using afterimages is that they became tainted with other "subjective" measures of eye movements as opposed to the "objective" recording methods of the twentieth century. A second factor was that the early studies were concerned with vertigo, which resulted in involuntary movements of the eyes. Moreover, histories of vertigo have been similarly blind to the early studies of it employing afterimages.

These points apply particularly to the experiments conducted by Wells (1792, 1794a, 1794b), both in the context of voluntary and involuntary eye movements. In the case of voluntary eye movements, Wells demonstrated that afterimages appeared to move with active eye movements whereas they appeared stationary when the eye was moved passively (by finger pressure); this procedure became one of the critical tests of afference vs efference, of inflow vs outflow. Thus, the essential aspects of visual stability during eye movements were examined using afterimages, and the procedure was repeated throughout the nineteenth century. There was nothing novel about afterimages themselves at that time. Wells produced them by the time-honoured technique of viewing a bright stimulus (like a candle flame) for many seconds; the afterimage remained visible for long enough for him to examine its apparent movement under a range of conditions. One of these was vertigo generated by rotating the body around a vertical axis.

Wells' work was forgotten, but the techniques he pioneered were rediscovered by those scientists puzzling over the mechanisms of vestibular function. Mach, Breuer, and Crum Brown all employed afterimages to measure the movements of the eyes after body rotation. They were able to link vertigo with their hydrodynamic theory of semicircular canal activity. Crum Brown forged the link between eye movements during vertigo and normal scanning of scenes; again afterimages were the tool used to demonstrate how scanning proceeds via a sequence of rapid movements interspersed with brief stationary pauses.

Afterimages had been enlisted to attempt to examine eye movements during reading by Javal but the contrast between the printed letters and paper made the afterimage difficult to discern. Novel methods were required and he was instrumental in trying some, but with little success. Javal's student, Lamare (1892), developed an indirect method that was more successful and yielded interesting and unexpected results. Lamare's technique was to use a tube attached to the eyelid or conjunctiva, from which sounds could be heard by the observer whenever the eyes moved. Using this technique, Lamare observed that, contrary to expectation and introspection, eye movements during reading were discontinuous. A very similar acoustic technique had been employed by Hering (1879b) to observe eye movements. However, Hering's technique had the added advantage of comparing the sounds thought to arise from eye movements to the movement of afterimages. Thus, Hering was able to confirm that the sounds coincided with movements of the afterimage and were therefore likely to be associated with eye movements.

The objective eye-trackers developed in the late nineteenth and early twentieth centuries allowed crucial new insights into the true nature of eye movements (see Venezky, 1977). Researchers were unanimously surprised by what they found; eye movements were not as smooth and continuous as they subjectively appeared. With these new devices for measuring eye movements so began a proliferation of interest in the nature of eye movements. The technological advancements allowed new questions to be addressed and indeed identified new and unexpected questions in the psychology and physiology of eye movements and their relation to cognition and visual experience. Eye-tracking technology continues to evolve, and with it so do the range of questions that can be addressed. Increasingly, the eyes are measured when inspecting objects in three-dimensional space – the scene is being scanned!

References

Bárány, R. (1906). Untersuchungen über den vom Vestibularapparat des Ohres reflektorisch ausgelösten rhythmischen Nystagmus und seine Begleiterscheinungen. *Monatsschrift für Ohrenheilkunde, 40*, 193–297.

Bárány, R. (1913). Der Schwindel und seine Beziehungen zum Bogengangapparat des inneren Ohres. Bogengangapparat und Kleinhirn. (Historische Darstellung. Eigene Untersuchungen.) *Naturwissenschaften, 1*, 396–401.

Bell, C. (1823). On the motions of the eye, in illustration of the uses of the muscles and of the orbit. *Philosophical Transactions of the Royal Society, 113*, 166–186.

Berkeley, G. (1709). *An essay towards a new theory of vision*. Dublin: Pepyat.

Breuer, J. (1874). Über die Function der Bogengänge des Ohrlabyrinthes. *Wiener medizinisches Jahrbuch, 4*, 72–124.

Buswell, G. T. (1935). *How people look at pictures: A study of the psychology of perception in art*. Chicago: University of Chicago Press.

Cameron, E. H., & Steele, W. M. (1905). The Poggendorff illusion. *Psychological Monographs, 7*, 83–111.

Crum Brown, A. (1878). Cyon's researches on the ear. II. *Nature, 18*, 657–659.

Crum Brown, A. (1895). *The relation between the movements of the eyes and the movements of the head*. London: Henry Frowde.

Darwin, C. (1887). *The life of Erasmus Darwin* (2nd ed.) London: Murray.

Darwin, E. (1794). *Zoonomia; or, the laws of organic life* (Vol. 1). London: Johnson.

Darwin, E. (1796). *Zoonomia; or, the laws of organic life* (2nd ed., Vol. 1). London: Johnson.

Darwin, E. (1801). *Zoonomia; or, the laws of organic life* (3rd ed., Vol. 4) London: Johnson.

Darwin, R. W. (1786). New experiments on the ocular spectra of light and colours. *Philosophical Transactions of the Royal Society, 76*, 313–348.

Delabarre, E. B. (1898). A method of recording eye movements. *American Journal of Psychology, 9*, 572–574.

Dodge, R. (1900). Visual perception during eye movement. *Psychological Review, 7*, 454–465.

Dodge, R. (1903). Five types of eye movement in the horizontal meridian plane of the field of regard. *American Journal of Physiology, 8*, 307–329.

Dodge, R. (1904). The participation of eye movements and the visual perception of motion. *Psychological Review, 11*, 1–14.

Dodge, R. (1916). Visual motor functions. *Psychological Bulletin, 13*, 421–427.

Dodge, R. (1921). A mirror-recorder for photographing the compensatory movements of the closed eyes. *Journal of Experimental Psychology, 4*, 165–174.

Dodge, R. (1923). Habituation to rotation. *Journal of Experimental Psychology, 6*, 1–35.

Dodge, R. (1930). Raymond Dodge. In C. Murchison (Ed.), *A history of psychology in autobiography* (Vol. 1, pp. 99–121). Worcester, MA: Clark University Press.

Dodge, R., & Cline, T. S. (1901). The angle velocity of eye movements. *Psychological Review, 8*, 145–157.

Donders, F. C. (1847). Beitrag zur Lehre von den Bewegungen des menschlichen Auges. *Höllandischen Beiträge zu den anatomischen und physiologischen Wissenschaften, 1*, 104–145.

Erdmann, B., & Dodge, R. (1898). *Psychologische Untersuchungen über das Lesen auf experimenteller Grundlage*. Halle: Niemeyer.

Griffith, C. R. (1920). The organic effects of repeated bodily rotation. *Journal of experimental Psychology, 3*, 15–47.

Helmholtz, H. (1867). *Handbuch der physiologischen Optik.* In G. Karsten (Ed.), *Allgemeine Encyklopädie der* (Vol. 9). Leipzig: Voss.
Helmholtz, H. (1873). *Popular lectures on scientific subjects* (Trans. E. Atkinson). London: Longmans, Green.
Helmholtz, H. (1924a). *Helmholtz's treatise on physiological optics* (Vol. 1, Trans. J. P. C. Southall). New York: Optical Society of America.
Helmholtz, H. (1924b). *Helmholtz's treatise on physiological optics.* (Vol. 2, Trans. J. P. C. Southall). New York: Optical Society of America.
Helmholtz, H. (1925). *Helmholtz's treatise on physiological optics* (Vol. 3, Trans. J. P. C. Southall). New York: Optical Society of America.
Helmholtz, H. (2000). *Helmholtz's treatise on physiological optics* (3 vols., Trans. J. P. C. Southall). Bristol: Thoemmes.
Hering, E. (1879a). Der Raumsinn und die Bewegungen des Auges. In L. Hermann (Ed.), *Handbuch der Physiologie, Vol. 3, Part 1: Physiologie des Gesichtssinnes* (pp. 343–601). Leipzig: Vogel.
Hering, E. (1879b). Über Muskelgeräusche des Auges. *Sitzberichte der kaiserlichen Akademie der Wissenschaften in Wien. Mathematisch-naturwissenschaftliche Klasse, 79*, 137–154.
Hering, E. (1942). *Spatial sense and movements of the eye* (Trans. C. A. Radde). Baltimore: The American Academy of Optometry.
Hitzig, E. (1871). Ueber die beim Galvanisiren des Kopfes entstehenden Störungen der Muskelinnervation und der Vorstellung vom Verhalten im Raume. *Archiv für Anatomie, Physiologie und wissenschaftliche Medicin*, 716–771.
Howard, I. P. (1982). *Human visual orientation.* New York: Wiley.
Howard, I. P. (2002). *Seeing in depth: Vol. 1. Basic mechanisms.* Toronto: Porteous.
Howard, I. P., & Templeton, W. B. (1966). *Human spatial orientation.* London: Wiley.
Hueck, A. (1838). *Die Achsendrehung des Auges.* Dorpat: Kluge.
Huey, E. B. (1898). Preliminary experiments in the physiology and psychology of reading. *American Journal of Psychology, 9*, 575–586.
Huey, E. B. (1900). On the psychology and physiology of reading. I. *American Journal of Psychology, 11*, 283–302.
Huey, E. B. (1908). *The psychology and pedagogy of reading.* New York: Macmillan.
Hunter, J. (1786). *Observations on certain parts of the animal œconomy.* London: Published by the author.
Javal, L. É. (1879). Essai sur la physiologie de la lecture. *Annales d'Oculistique, 82*, 242–253.
Judd, C. H. (1899). A study of geometrical illusions. *Psychological Review, 6*, 241–261.
Judd, C. H. (1905a). The Müller-Lyer illusion. *Psychological Monographs, 7*, 55–81.
Judd, C. H. (1905b). Movement and consciousness. *Psychological Monographs, 7*, 199–226.
Judd, C. H., & Courten, H. C. (1905). The Zöllner illusion. *Psychological Monographs, 7*, 112–139.
Judd, C. H., McAllister, C. N., & Steele, W. M. (1905). General introduction to a series of studies of eye movements by means of kinetoscopic photographs. *Psychological Monographs, 7*, 1–16.
Lamare, M. (1892). Des mouvements des yeux dans la lecture. *Bulletins et Mémoires de la Société Française d'Ophthalmologie, 10*, 354–364.
Mach, E. (1873). Physikalische Versuche über den Gleichgewichtssinn des Menschen. *Sitzungsberichte der Wiener Akademie der Wissenschaften, 68*, 124–140.
Mach, E. (1875). *Grundlinien der Lehre von den Bewegungsempfindungen.* Leipzig: Engelmann.
McAllister, C. N. (1905). The fixation of points in the visual field. *Psychological Monographs, 7*, 17–53.
Müller, J. (1826). *Zur vergleichenden Physiologie des Gesichtssinnes des Menschen und der Thiere, nebst einen Versuch über die Bewegung der Augen und über den menschlichen Blick.* Leipzig: Cnobloch.
Nagel, W. A. (1896). Über kompensatorische Raddrehungen der Augen. *Zeitschrift für Psychologie und Physiologie der Sinnesorgane, 12*, 331–354.
Orschansky, J. (1899). Eine Methode die Augenbewegungen direct zu untersuchen. *Centralblatt für Physiologie, 12*, 785–790.
Porterfield, W. (1737). An essay concerning the motions of our eyes. Part I. Of their external motions. *Edinburgh Medical Essays and Observations, 3*, 160–263.

Porterfield, W. (1738). An essay concerning the motions of our eyes. Part II. Of their internal motions. *Edinburgh Medical Essays and Observations, 4*, 124–294.

Porterfield, W. (1759). *A treatise on the eye, the manner and phœnomena of vision* (Vol. 2). Edinburgh: Hamilton and Balfour.

Purkinje, J. (1820). Beyträge zur näheren Kenntniss des Schwindels aus heautognostischen Daten. *Medicinische Jahrbücher des kaiserlich-königlichen öesterreichischen Staates, 6*, 79–125.

Purkinje, J. (1825). *Beobachtungen und Versuche zur Physiologie der Sinne. Neue Beiträge zur Kenntniss des Sehens in subjectiver Hinsicht.* Berlin: Reimer.

Ruete, C. G. T. (1845). *Lehrbuch der Ophthalmologie für Aerzte und Studirende.* Braunschweig: Vieweg.

Ruete, C. G. T. (1853). *Lehrbuch der Ophthalmologie* (2nd ed.). Leipzig: Vieweg.

Scheiner, C. (1619). *Oculus, hoc est fundamentum opticum.* Innsbruck: Agricola.

Stratton, G. M. (1902). Eye-movements and the aesthetics of visual form. *Philosophische Studien, 20*, 336–359.

Stratton, G. M. (1906). Symmetry, linear illusions and the movements of the eye. *Psychological Review, 13*, 82–96.

Taylor, E. A. (1937). *Controlled reading.* Chicago: University of Chicago Press.

Tscherning, M. (1898). *Optique Physiologique.* Paris: Carré and Naud.

Tscherning, M. (1900). *Physiologic optics* (Trans. C. Weiland). Philadelphia: Keystone.

Venezky, R. L. (1977). Research on reading processes: A historical perspective. *American Psychologist, 32*, 339–345.

Vernon, M. D. (1931). *The experimental study of reading.* Cambridge: Cambridge University Press.

Volkmann, A. W. (1846). Sehen. In R. Wagner (Ed.), *Handwörterbuch der Physiologie* (Vol. 3, Pt. 1, pp. 265–351). Braunschweig: Vieweg.

Wade, N. J. (1998). *A natural history of vision.* Cambridge, MA: MIT Press.

Wade, N. J. (2000). William Charles Wells (1757–1817) and vestibular research before Purkinje and Flourens. *Journal of Vestibular Research, 10*, 127–137.

Wade, N. J. (2002). Erasmus Darwin (1731–1802). *Perception, 31*, 643–650.

Wade, N. J. (2003a). *Destined for distinguished oblivion: The scientific vision of William Charles Wells (1757–1817).* New York: Kluwer/Plenum.

Wade, N. J. (2003b). The search for a sixth sense: The cases for vestibular, muscle, and temperature senses. *Journal of the History of the Neurosciences, 12*, 175–202.

Wade, N. J. (2005). The original spin doctors – the meeting of perception and insanity. *Perception, 34*, 253–260.

Wade, N. J., & Brožek, J. (2001). *Purkinje's vision. The dawning of neuroscience.* Mahwah, NJ: Lawrence Erlbaum Associates.

Wade, N. J., Norrsell, U., & Presly, A. (2005). Cox's chair: "a moral and a medical mean in the treatment of maniacs". *History of psychiatry, 16*, 73–88.

Wade, N. J., & Tatler, B. W. (2005). *The moving tablet of the eye: The origins of modern eye movement research.* Oxford: Oxford University Press.

Wells, W. C. (1792). *An essay upon single vision with two eyes: Together with experiments and observations on several other subjects in optics.* London: Cadell.

Wells, W. C. (1794a). Reply to Dr. Darwin on vision. *The Gentleman's Magazine and Historical Chronicle, 64*, 794–797.

Wells, W. C. (1794b). Reply to Dr. Darwin on vision. *The Gentleman's Magazine and Historical Chronicle, 64*, 905–907.

Wells, W. C. (1814). *An essay on dew and several appearances connected with it.* London: Taylor and Hessey.

Wells, W. C. (1818). *Two essays: One upon single vision with two eyes; the other on dew.* London: Constable.

Woodworth, R. S. (1938). *Experimental psychology.* New York: Holt.

Yarbus, A. L. (1967). *Eye movements and vision.* New York: Plenum Press.

Young, L. R., Henn, V., & Scherberger, H. (Trans. & Ed.). (2001). *Fundamentals of the theory of movement perception by Dr. Ernst Mach.* New York: Kluwer/Plenum.

Chapter 3

EYE MOVEMENT RESEARCH IN THE 1950s

GERALD WESTHEIMER

University of California, USA

Eye Movements: A Window on Mind and Brain
Edited by R. P. G. van Gompel, M. H. Fischer, W. S. Murray and R. L. Hill

Abstract

The decade and a half following the Second World War was a particularly interesting period in human eye-movement research. The enlistment of scientists and engineers in technical projects for the war effort showed many of them the way towards practical applications. This included not only electronic devices and computers, but also instrumentation and procedures for solving problems in biology and medicine. Accompanying this development was a change in attitude in the direction of applying rigorous modes of analysis. With instrumentation derived from war surplus material and under the banner of "systems theory," eye movements, the pupil and accommodation responses were recorded to step, pulse, ramp and sinusoidal stimuli. Transfer functions were calculated and open- and closed-loop behaviour investigated. Because this was the first round, researchers did not see, or because of their physicalist biases were even unprepared and unable to see, possible mismatches between their biological systems and their analytical probe.

Though the study of human eye movements has a venerable history, a constellation of circumstances made the 1950s a particularly fertile period of advances. Up to then, progress had been sporadic. In psychology and education, eye movements were being recorded during reading and the results used in attempts to measure learning deficits (Carmichael & Dearborn, 1947). The American Optical Company marketed a moving film camera device for that purpose. There was some interest in ocular motility among neuro-ophthalmologists (Adler, 1950; Cogan, 1956) but almost none within the ranks of neuromuscular physiologists, whose focus then was on lower motor reflexes that could be studied in spinal preparations.

1. Change in mindset

The Second World War brought about a fundamental change. Large numbers of physical scientists and engineers enlisted in research efforts that may best be described as applied science. Specific examples are radar, submarine detection, control of guns, bombs, vehicles and aircraft. The projects had deterministic end points and the whole exercise was to achieve a defined goal by whatever means that could be marshalled. The war had wide popular support and almost everyone was engaged in it without reservation. As a consequence, the personnel tackling the scientific and technical projects were of the highest calibre; a whole generation of the intellectual elite became effective problem solvers who learned what was needed to achieve practical success.

In the years following the war this attitude and the hardware that resulted from the war effort began to be channelled in new directions: electronics and computers are the most prominent of these, but instrumentation and procedures for biological research also were beneficiaries. Until then, the armamentarium of neurophysiology – electrodes, stimulators, amplifiers – was one of a kind. Good academic departments had workshops in which skilled technicians laboriously hand-designed and crafted research apparatus. Reliable oscilloscopes, invaluable in this work, were just becoming more readily available. And so did microscopes, optical and electron.

Perhaps even more important was the change in the mindset that these wartime activities generated. An example will illustrate. In the 1940s mathematics programme of Sydney University, as taught by Cambridge wranglers, we used as a textbook Piaggio's *Treatise on Differential Equations* (Piaggio, 1944). Here are some quotations from the Preface: "The theory of differential equations, said Sophus Lie, is the most important branch of modern mathematics . . . If we travel along the purely analytical path, we are soon led to discuss infinite series, existence theorems and the theory of functions." As a graduate student in 1951, I came across a textbook covering much the same ground as Piaggio's. But Trimmer's book (Trimmer, 1950) was entitled *The response of physical systems* and here are excerpts from *its* preface: "The book is based on material used for lectures and laboratory sessions of a course 'Instrumentation' ". Trimmer goes on to write about "the emergence of a discipline called 'system response' " and about "the relation of instrumentation and system response to the newly defined '*Cybernetics*' " quoting Norbert Wiener's book (Wiener, 1948). In addition to Wiener's Cybernetics, he might have mentioned an even more seminal contemporary, Shannon's information theory (Shannon & Weaver, 1949). Within a span of a few years, the lecture courses in differential equations changed their emphasis from existence theorems to instrumentation.

Eye movement had been a subject dear to the hearts of earlier generations of psychologists. The single most influential paper in the whole subject, clearly distinguishing between the saccades and pursuit movements, was written by R. Dodge, a Yale psychologist (Dodge, 1903) and the topic has had a presence in experimental psychology since Wundt's time. In the 1950s one strand of it concerned itself with the steadiness of the eyes during normal fixation.

2. Micronystagmus

Some years earlier Adler and Fliegelmann had shown in inventive experiments that the eyes were never still even during intersaccadic intervals (Adler & Fliegelman, 1934). Movements, called micronystagmus, were of sufficient magnitude – up to several minutes of arc – to raise concern about their role in visual acuity. Some people theorized that they actually sharpened vision (Marshall & Talbot, 1942) while others felt that their effect would be detrimental. The general problem was seen to be of such import that considerable ingenuity was employed by several laboratories to record the micronystagmus. Electro-oculography had nowhere near the required spatial resolution (Marg, 1951), and standard photography could not reliably distinguish between rotational eye movements, which would displace images on the retina, and translational movements of the whole eyeball, which would not. The procedure of choice, developed in the laboratory of both Riggs at Brown University (Riggs & Ratliff, 1950), and Ditchburn at Reading (Ditchburn & Ginsborg, 1953), involved reflecting light from a mirror attached firmly to the eye by means of tight-fitting contact lenses. Other techniques were occasionally employed, such as a drop of mercury applied to the limbus (Barlow, 1952) or even more imaginative ones utilizing light reflected from retinal structures (Cornsweet, 1958).

Floyd Ratliff, later to gain prominence as Hartline's collaborator, devoted his PhD thesis, working under Lorrin Riggs, to recording micronystagmus during short time intervals and relating it to the measured visual acuity in these epochs (Ratliff, 1952). His conclusion was that the two were negatively correlated. The opposing view, namely that eye-movements aided acuity, was never articulated in sufficient detail to allow its empirical validation. Altogether, the topic of visual resolution was not advanced materially by these exercises, because it was shown 20 years later (Westheimer & McKee, 1975) that target movements in the velocity range of micronystagmus do not impair acuity.

Something else about micronystagmus intrigued researchers at the time, namely the possibility that it played a role in keeping the visual process active. As a result of Hartline's work, and later Kuffler's, on single retinal ganglion cells, it had become apparent that most neural activity occurred at the onset and offset of light stimuli, and little during steady illumination. This insight, together with the observation that afterimages fade and that entoptic visualization of Haidinger's brushes (Ratliff, 1958) or the Purkinje vascular tree requires temporal transients, made people wonder what vision would be like when the optical image on the retina eye was totally stabilized. Nulling out micronystagmus requires considerable experimental expertise, involving not only contact lens mirrors but also multi-stage optical paths (Ditchburn & Fender, 1955; Ditchburn & Ginsborg, 1952). In the end it was demonstrated that when there was no jiggling of the optical image on the retina, vision does indeed fade. Nowadays we would call this a manifestation of the temporal bandpass characteristics of retinal processing; transients are necessary to activate the optic nerve impulses, though the phototransductive stages of vision have good DC output. Description of experiences with stabilized vision became quite a cottage industry, especially when it was extended into the realm of colour.

3. Systems theory

A much more direct expression of the Zeitgeist of 1950s' science and technology was the application of "systems theory" to oculomotorics, sparked predominantly by engineers and fellow-travelling physicians, biologists and psychologists. Systems theory here was not just the – still very popular – fitting of engineering models to biological data; rather it was the enveloping concept that such models capture its essence, in other words that the prime approach is to examine biological systems as if they were black boxes whose properties and the interaction between them can be satisfactorily described and analyzed by equations. A *bon mot* of the times is revealing: What do you find when you open the black box? Other black boxes.

At the outset, the research framework was uncomplicated: examine the eye-movement responses of a human observer as if they were manifestations of a mechanical, or perhaps opto-electronic, control system. That is, present an unambiguous stimulus in the domain of visual target position (the input) and record and analyze the resultant eye movement (the output). Adequate photomechanical and later electro-optical instrumentation could be built, utilizing the experimentalists' major supply line: war surplus mechanical, electrical and optical components. The shelves of laboratories then were groaning with elaborate and originally very expensive lenses, mirrors, filters, gear trains, electric motors and circuit boards, which had been intended for airplanes, submarines, radar installations and such.

The first order of business was to identify closed-loop responses to step, pulse, ramp, sinusoidal stimuli and so on. In this round, a great deal of clarity was obtained: there is a reaction time (delay) between onset of stimulus and onset of response of the order of 120 ms. Of the two classes of movement that are utilized, saccades are discretely deployed ballistiform with a step-like forcing function, while pursuit movements can be smoothly modified. An important step in systems theory is to understand the role of feedback, where the difference between the actual and the needed response is detected and utilized for correction. It was usually supposed that the intent is to foveate a target appearing (or attracting attention) in the peripheral visual field and hence, in this formulation, the error is angular distance of the target from the fovea. A popular engineering model at the time was the sampled-data system, in which the error was sampled at discrete time intervals (Jury, 1958). Needless to say, this can cause complicated, oscillatory behaviour. This kind of thinking leads to the next step: open the loop, that is, record the system's response to an error signal that is maintained, regardless of the resultant response. If the input is sinusoidal in a good range of temporal frequencies, such an approach can be particularly revealing, because now one obtains a Bode plot of gain and phase changes. The descriptor is called transfer function and if the input and the output remain expressed in the form of sinusoidal functions, the behaviour of the whole system in any other situation can be predicted by mere multiplication. In an equivalent approach, the elemental function is a brief pulse and the computational procedure is only marginally more complicated.

Engineering-minded people were thrilled with this development, because it gave them the chance to transfer to high-level biology many of the techniques and insights that they had so successfully employed in the design of aircraft and war materiel. The name of the discipline, servoanalysis, speaks volumes. Its origin in the world of inanimate things does not mean that it is entirely mechanistic. Thresholds, noise, dead-space in the realm of error detection, these are just a few of the concepts that had also to be contended with, and that were regarded by the engineering mind of the researchers merely as further challenges to their ingenuity. It was felt that when these were mastered, the human oculomotor system would be characterized by diagrams with boxes connected by arrows, each element representing an expression in terms of mathematically definable symbols. In preparation for this goal, many laboratories were conducting experiments to measure open- and closed-loop transfer functions. Unsurprisingly, major players in this arena were the talented members of the MIT electronics laboratory, an outgrowth of the wartime Radiation Laboratory, and publisher of a 28-volume series of textbooks with such titles as *Waveforms*, *Thresholds Signals* and *Theory of Servomechanisms* (James, Nichols, & Phillips, 1947). For reasons which may have to do with a particular reading of dialectical materialism, many laboratories in the then "eastern bloc" had the term "cybernetics" in their name. A particularly successful venture in Germany, the Max-Planck Institute for Biocybernetics in Tuebingen, housed a group formed by Werner Reichardt, where this kind of thinking was applied most productively to insect flight.

The investigation of the vestibular and vestibulo-ocular apparatus, which is very amenable to this kind of analysis, followed a somewhat separate track at the time, presumably because, for aeronautical and astronautical considerations, it was the focus of concentration in military centres. Only in the 1960s, when it became widely apparent just how intertwined the vestibular and oculomotor systems were, did the two research streams merge.

Incidentally, while we are here talking about oculo-motility, a parallel development took place in the domain of visual sensory functions. In the mid-1950s two seminal papers by engineers were published: in 1954, DeLange subjected the temporal aspect of the human light sense to an analysis by sinusoidal flashes (DeLange, 1954) and Otto Schade in 1956 did the same for its spatial aspects using sinusoidal gratings (Schade, 1956). A significant difference between these studies and the oculomotor ones should be mentioned.

In the latter, one measured a true transfer function, obtaining the ratio of measured input and output amplitudes. In the psychophysical studies one uses the observer's subjective detection threshold as a fixed measure of output and plots the stimulus amplitude needed to generate it.

These modes of investigation perfectly mirrored the prevailing approach to science. They were, of course, pervaded by the analytical and mechanistic mindframe immanent in the science of the time. I would argue that they were necessary stages, productive, centred on the then current state of knowledge, confirming the discipline's standing midway between neurophysiology and cognition. Criticism of lack of a holistic predisposition are out of place, because there were no precedents for success in such a direction, *vide* the floundering of Gestalt psychology as a heuristic movement.

4. Limitations to the approach

The problem lies in an entirely different direction. If the use of the engineers' system theory in eye-movement research suffered from a fatal flaw, it was that deep down it is inapplicable, and not on philosophical grounds of reductionism, but because the two are a poor match. This came home to me in two wallops in the summer of 1952 when working on my PhD thesis entitled "The Response of the human oculomotor system to visual stimuli in the horizontal plane". I had fitted a second-order differential equation to saccades, regarding them as responses to a step forcing function. That worked well, but the simple corollary, viz. that the shape of the responses should be the same for saccades of all amplitudes, was not borne out: saturation of peak velocity started already with about 20-deg saccades (Westheimer, 1954b).Unfortunately, non-linearities enter the biological systems research as soon as you open the door. Not long after that, while recording responses to periodic square-wave and sinusoidal stimuli, I realized that, lo and behold, after just a handful of cycles, there no longer was a reaction time. Responses were synchronous or even preceded the stimulus (Westheimer, 1954a). Prediction was taking place and one could no longer rightly talk of the response to a stimulus in a systems-theoretical way.

Right at the beginning of a systems-theoretical analysis of a reasonably tractable biological apparatus – one that could be given straightforward unambiguous stimuli, that had easily measurable responses, that had no load and did not suffer from parametric feedback via muscle spindles – the limitations of the approach had to be confronted head-on.

Books such as Trimmer's in the very first chapter warned about possible non-linearities and the consequent intractability, but in their treatment of systems such books sailed right on as if in real life non-linearities could be ignored. So one proceeded to deal with oculomotor, accommodation and pupil responses staying entirely within the mode of linear analysis. Of course, one has to start somewhere and in the application of the formulations to electrical and mechanical systems one can indeed go a long linear distance before one is pulled up. Most practitioners of oculomotor research came from physics or engineering or, even if their original training was in medicine, had strong biases in that direction. However, the first thing any clinician or practicing biologist learns is that in their domain nothing is straightforward or linear and, when there is a poor fit between the biological preparation and the analytical probe, to proceed with caution, or to look for other avenues.

This underlying problem did not entirely disappear when the next stage of oculomotor research took over in the 1960s. A more nuanced approach was followed in the exploration of eye-movement responses, taking account of the perceptual and cognitive dimensions by examining the various strategies that human observers employ. When confronted with predictive responses, engineers would just insert an operator in their equations that shifted the response backward in time, as if that constituted a step in the scientific understanding of the system. Experimental psychologists were more aware of the kind of explanations and constructs that would have standing in such a research area.

In addition, a clearer picture began to emerge of the neural pathways through which oculomotor responses operate. Control of the intra-ocular musculature is largely confined to brainstem neurons and responses to some stimuli do not involve the cortex. Hence it is quite appropriate to talk of the "pupillary reflexes" in the traditional, Sherringtonian parlance and expect a servoanalytic approach to be not entirely out of place at the outset (Stark & Sherman, 1957). But this kind thinking begins to fail in the case of ciliary muscle and accommodation responses. Fincham, the most significant contributor to our understanding of the accommodation mechanism, still thought that focusing was a reflex akin to pupillary constriction (Fincham, 1951). But it soon became clear that the error signal, the crucial element in any servomechanism, could not be handled by unsophisticated midbrain structures (Campbell & Westheimer, 1959). And once the cerebral cortex becomes involved, with its vast interrelation and plasticity, engineering formulations, no matter how clever and elegant, become inadequate.

As modern neurophysiological exploration with microelectrodes gained ground in the 1950s, especially in the hands of Eccles and other members of the revived Sherrington motor-research school, it started also to be applied to the oculomotor system. Whereas the spinal reflexes could be studied in the encéphale isolé preparation, studies of the neural paths of oculomotor system needed the experimental animal to remain alert and behaving. Once the technical problems of recording neuronal firing when a visual stimulus elicited an eye movement were solved, an enormous and very productive enterprise developed. At the outset there still was much servo-analytical modelling, but soon the realities of the intricate neural apparatus began to assert themselves and the researchers in this discipline settled down to the long haul of unravelling the intricate midbrain and cerebellar circuitry involved in translating to the eye muscles the animal's intention to place a particular retinal target on the fovea. This research is continuing apace.

In retrospect, one aspect of the 1950s eye-movement enterprise constituted its major limitation. Even if open-loop transfer functions could be established with some semblance of linearity, generality and validity, they are still intended to describe only the performance in a servomechanistic setting: keeping a target on the fovea as it experiences displacements. But in a richly textured visual world, a conscious observer's oculomotor behaviour depends on many more influences. The current movement towards "natural scenes" is trying to encompass them. Will the problems in such multi-layered research environment make the next-generation of scientist look back with nostalgia to a time when their predecessors' aim was merely to succeed with operators and transfer functions from engineering handbooks?

References

Adler, F. H. (1950). *Physiology of the eye*. St. Louis: Mosby.

Adler, F. H., & Fliegelman, M. (1934). Influence of fixation on the visual acuity. *A.M.A. Archives of Ophthalmology, 12*, 475–483.

Barlow, H. B. (1952). Eye movements during fixation. *The Journal of Physiology, 116*, 290–306.

Campbell, F. W., & Westheimer, G. (1959). Factors influencing accommodation responses of the human eye. *Journal of the Optical Society of America, 49*, 568–571.

Carmichael, L., & Dearborn, W. F. (1947). *Reading and visual fatigue*. Boston: Houghton Mifflin.

Cogan, D. G. (1956). *Neurology of the ocular muscles*. Springfield, IL: Thomas.

Cornsweet, T. N. (1958). A new method for the measurement of small eye movements. *Journal of the Optical Society of America, 48*, 921–928.

DeLange, H. (1954). Relationship between the critical flicker frequency and a set of low-frequency characteristics of the eye. *Journal of the Optical Society of America, 44*, 380–389.

Ditchburn, R. W., & Fender, D. H. (1955). The stabilized retinal image. *Optica Acta, 2*, 128–133.

Ditchburn, R. W., & Ginsborg, B. L. (1952). Vision with stabilized retinal images. *Nature, 170*, 36–37.

Ditchburn, R. W., & Ginsborg, B. L. (1953). Involuntary eye movements during fixation. *The Journal of Physiology, 119*, 1–17.

Dodge, R. (1903). Five types of eye movement in the horizontal meridian plane of the field of regard. *American Journal of Physiology, 8*, 307–329.

Fincham, E. F. (1951). The accommodation reflex and its stimulus. *British Journal of Ophthalmology, 35*, 381–393.

James, H. M., Nichols, N. B., & Phillips, R. S. (Eds.). (1947). *Theory of servomechanisms*. New York: McGraw-Hill.

Jury, E. I. (1958). *Sampled data control systems*. New York: Wiley.

Marg, E. (1951). Development of electro-oculography. *A.M.A. Archives of Ophthalmology, 45*, 169–185.

Marshall, W. H., & Talbot, S. A. (1942). Recent evidence for neural mechanisms in vision leading to a general theory of sensory acuity. *Biological Symposia, 7*, 117–164.

Piaggio, H. T. H. (1944). *An elementary treatise on differential equations*. London: Bell.

Ratliff, F. (1952). The role of physiological nystagmus in monocular visual acuity. *Journal of Experimental Psychology, 43*, 163–172.

Ratliff, F. (1958). Stationary retinal image requiring no attachments to the eye. *Journal of the Optical Society of America, 48*, 274–275.

Riggs, L. A., & Ratliff, F. (1950). Involuntary motion of the eye during monocular fixation. *Journal of Experimental Psychology, 40*, 687–701.

Schade, O. H. (1956). Optical and photoelectric analog of the eye. *Journal of the Optical Society of America, 46*, 721–739.

Shannon, C. E., & Weaver, W. (1949). *The mathematical theory of communication*. Urbana: U. Illinois Press.

Stark, L., & Sherman, P. M. (1957). A servo-analytic study of consensual pupil reflex to light. *Journal of Neurophysiology, 20*, 17–26.

Trimmer, J. D. (1950). *The response of physical systems*. New York: John Wiley.

Westheimer, G. (1954a). Eye movement responses to horizontally moving stimuli. *A.M.A. Archives of Ophthalmology, 52*, 932–941.

Westheimer, G. (1954b). Mechanism of saccadic eye movements. *A.M.A. Archives of Ophthalmology, 52*, 710–724.

Westheimer, G., & McKee, S. P. (1975). Visual acuity in the presence of retinal-image motion. *Journal of the Optical Society of America, 65*(7), 847–850.

Wiener, N. (1948). *Cybernetics: or, Control and communication in the animal and the machine*. Cambridge, MA: Technology Press.

Chapter 4

FIXATION STRATEGIES DURING ACTIVE BEHAVIOUR: A BRIEF HISTORY

MICHAEL F. LAND

University of Sussex, UK

Eye Movements: A Window on Mind and Brain
Edited by R. P. G. van Gompel, M. H. Fischer, W. S. Murray and R. L. Hill

Abstract

The study of the relationships of fixation sequences to the conduct of everyday activities had its origins in the 1950s but only started to flourish in the 1990s as head-mounted eye-trackers became readily available. The main conclusions from a decade of study are as follows: (i) that eye movements are not driven by the intrinsic salience of objects, but by their relevance to the task in hand; (ii) appropriate fixations typically lead manipulations by up to a second, and the eye often leaves the fixated object before manipulation is complete; (iii) many fixations have identifiable and often surprising roles in providing information for locating, guiding and checking activities. The overall conclusion is that, in contrast to free viewing, the oculomotor system is under tight top-down control, and eye movements and actions are closely linked.

1. Introduction: before 1990

Eye-movement recordings have been made for over a century (Wade & Tatler, 2005), but until comparatively recently such recordings were confined to tasks in which the head could be held stationary in a laboratory setting. These early studies included tasks such as reading aloud (Buswell, 1920), typing (Butch, 1932), and even playing the piano (Weaver, 1943). An important outcome of all these investigations was that eye movements typically lead actions, usually by about a second. Vision is thus in the vanguard of action, and not just summoned up as specific information is required.

The studies just quoted all used bench-mounted devices of various kinds to record the eye movements of restrained subjects. For tasks involving active movements of participants a different system is required, and all the devices that have been used successfully for such recordings over the last 50 years have been head-mounted. The first of these mobile eye-trackers was made by Norman Mackworth in the 1950s (Thomas, 1968). His device consisted of a head-mounted ciné camera which made a film of the view from the subject's head, onto which was projected, via a periscope, an image of the corneal reflex which corresponded to the direction of the visual axis. Mackworth and Thomas (1962) used a TV camera version of the device to study the eye movements of drivers. However, the development of more user-friendly devices had to wait until the 1980s, when video cameras became smaller and computers could be enlisted in the analysis of the image from the eye. A review of the methodologies currently employed is given by Duchowski (2003).

The general question that mobile eye-trackers could begin to answer was the nature of the relation between patterns of fixations and the actions that they are associated with. In this chapter we will mainly be concerned with gaze direction, that is the points to which the fovea is directed. Shifts of gaze usually involve movements of the eyes and head, and often of the whole body. To record the whole pattern of movement that results in a gaze shift requires the separate measurement of the eye, head and body contributions. The interrelations of the different components are quite complex. For example, head movements augment eye movements during gaze shifts, but during fixations the effects of head movement are cancelled out by the vestibulo-ocular reflex, which rotates the eye with a velocity equal and opposite to the head movement (Guitton & Volle, 1987). A similar reciprocity exists for head and trunk rotations, mediated by the vestibulo-collic reflex (Land, 2004). Mounting the eye-tracker on the head obviates the need to record head and body movements separately, and only eye-in-head movements have to be measured in order to retrieve gaze direction on the image from the (also head-mounted) scene camera (see Figures 6 and 8).

Such eye-trackers allow us to address a variety of questions. Which points in the scene are selected, given that the fovea can only view one point at a time? What kinds of information do each fixation supply? And what are the time relations between eye movements and actions?

Although not concerned with active tasks, a key figure in the development of current ideas about active vision was Alfred Yarbus. In his book *Eye movements and vision*

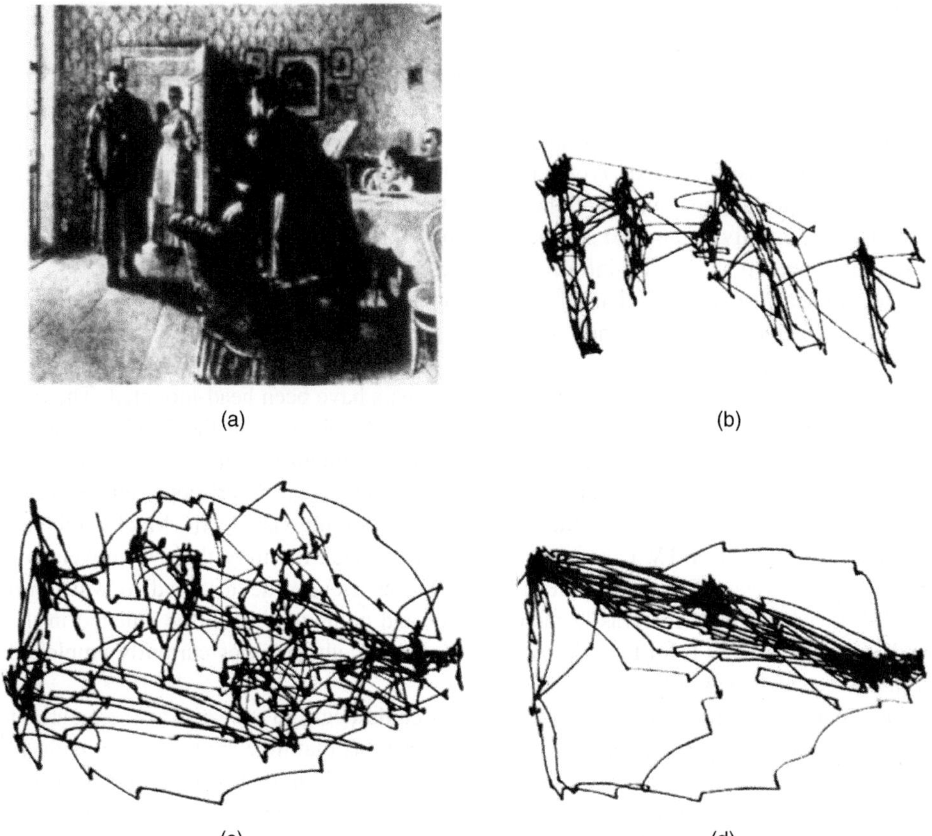

Figure 1. Eye movements made by subjects while examining I. E. Repin's painting "An Unexpected Visitor", with different questions in mind (adapted from Yarbus, 1967). (a) The picture. (b) "Remember the clothes worn by the people." (c) "Remember the positions of the people and objects in the room." (d) "Estimate how long the 'unexpected visitor' had been away from the family." Saccades are the thin lines; fixations are the knot-like interruptions.

(Yarbus, 1967) he demonstrated convincingly that the kinds of eye movements people make when viewing a scene depend on what information they are trying to get from it, and not just on the eye-catching power ("intrinsic salience") of the objects in that scene (Figure 1). He provided an account of eye movements in which central control, related to the task in hand, was seen as being more important than reflex-like responses to stimulus objects. In fact Guy Buswell had somewhat anticipated Yarbus' approach in his book *How People Look at Pictures*, which also demonstrated that different patterns of eye movement result when different questions were asked (Buswell, 1935).

During manipulative activities of various kinds, we require different kinds of visual information from the world at different times ("Where's the hammer?" "Is the kettle boiling?"). We may thus expect that, like Yarbus' subjects, our eye-movement patterns reflect the information needs of the tasks we are engaged in.

2. The block copying task of Dana Ballard: Two useful maxims

One of the first detailed studies of eye movements in relation to manipulative activity was by Ballard, Hayhoe, Li, and Whitehead (1992). They used a task in which a model consisting of coloured blocks had to be copied using blocks from a separate pool. Thus the task involved a repeated sequence of looking at the model, selecting a block, moving it to the copy and setting it down in the right place (Figure 2). The most important finding

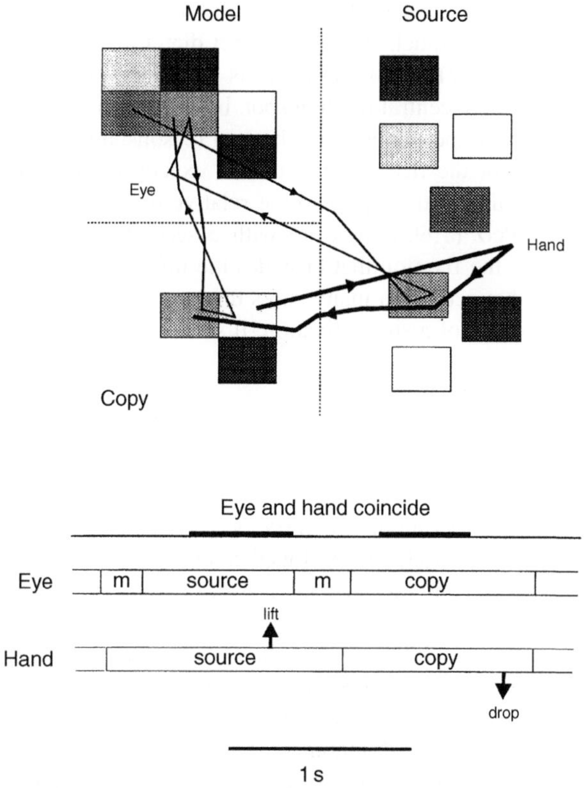

Figure 2. The block copying task of Ballard et al. (1992). A Copy (bottom left) of the Model is assembled from randomly positioned blocks in the Source area. Typical movements of hand and eye are shown, together with their timing in a typical cycle. The eyes not only direct the hands, but also perform checks on the Model to determine the colour and location of the block being copied.

was that the operation proceeds in a series of elementary acts involving eye and hand, with minimal use of memory. Thus a typical repeat unit would be as follows: Fixate (block in model area); remember (its colour); fixate (a block in source area of the same colour); pickup (fixated block); fixate (same block in model area); remember (its relative location); fixate (corresponding location in model area); move block; drop block. The eyes have two quite different functions in this sequence: to guide the hand in lifting and dropping the block, and, alternating with this, to gather the information required for copying (the avoidance of memory use is shown by the fact that separate glances are used to determine the colour and location of the model block). The only times that gaze and hand coincide are during the periods of about half a second before picking up and setting down the block (as with other tasks the eyes have usually moved on before the pickup or drop are complete).

The main conclusion from this study is that the eyes look directly at the objects they are engaged with, which in a task of this complexity means that a great many eye movements are required. Given the relatively small angular size of the task arena, why do the eyes need to move so much? Could they not direct activity from a single central location? Ballard et al. (1992) found that subjects could complete the task successfully when holding their gaze on a central fixation spot, but it took three times as long as when normal eye movements were permitted. For whatever reasons, this strategy of "do it where I'm looking" is crucial for the fast and economical execution of the task. This strategy seems to apply universally. With respect to the relative timing of fixations and actions, Ballard, Hayhoe, and Pelz (1995) came up with a second maxim: the "just in time" strategy. In other words the fixation that provides the information for a particular action immediately precedes that action; in many cases the act itself may occur, or certainly be initiated, within the lifetime of a single fixation. It seems that memory is used as little as possible.

3. Everyday life tasks: making tea and sandwiches

Activities such as food preparation, carpentry or gardening typically involve a series of different actions, rather loosely strung together by a flexible "script". They provide examples of the use of tools and utensils, and it is of obvious interest to find out how the eyes assist in the performance of these tasks.

Land, Mennie, and Rusted (1999) studied the eye movements of subjects whilst they made cups of tea. When made with a teapot, this common task involves about 45 separate acts (defined as "simple actions that transform the state or place of an entity through manual manipulation"; Schwartz, Montgomery, Palmer, & Mayer, 1991). Figure 3 shows the 26 fixations made during the first 10 s of the task. The subject first examines the kettle (11 fixations), picks it up and looks towards the sink (3 fixations), walks to the sink whilst removing the lid from the kettle (inset: 4 fixations), places the kettle in the sink and turns on the tap (3 fixations), then watches the water as it fills the kettle (4 fixations). There is only one fixation that is not directly relevant to the task (to the sink tidy on the right).

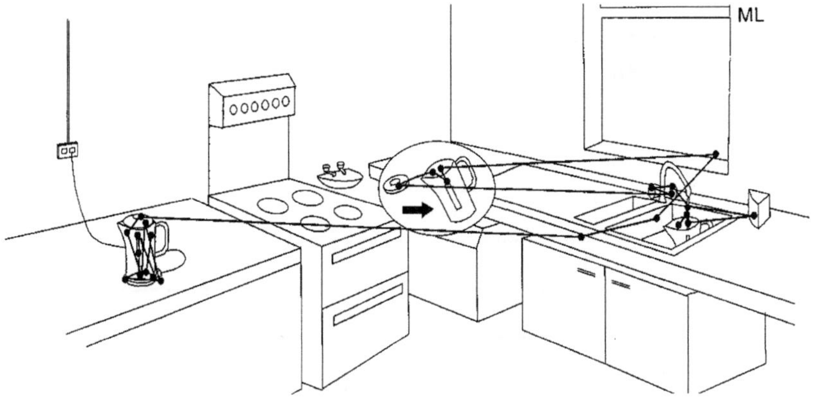

Figure 3. Fixations and saccades made during the first 10 s of the task of making a cup of tea (lifting the kettle and starting to fill it). Note that fixations are made on the objects that are relevant at the time (kettle, sink, lid, taps, water stream) and that only one fixation is irrelevant to the task (the sink tidy on the right). Two other subjects showed remarkably similar fixation patterns (from Land et al., 1999).

Two other subjects showed remarkably similar numbers of fixations when performing the same sequence. The principal conclusions from this sequence are as follows:

1. Saccades are made almost exclusively to objects involved in the task, even though there are plenty of other objects around to grab the eye.
2. The eyes deal with one object at a time. This corresponds roughly to the duration of the manipulation of that object, and may involve a number of fixations on different parts of the object.

There is usually a clear "defining moment" when the eyes leave one object and move on to the next, typically with a combined head and eye saccade. These saccades can be used to "chunk" the task as a whole into separate "object-related actions", and they can act as time markers to relate the eye movements to movements of the body and manipulations by the hands. In this way the different acts in the task can be pooled, to get an idea of the sequence of events in a "typical" act. The results of this are shown in Figure 4. Perhaps surprisingly, it is the body as a whole that makes the first movement in an object-related action. Often the next object in the sequence is on a different work surface, and this may necessitate a turn or a few steps before it can be viewed and manipulated. About half a second later the first saccade is made to the object, and half a second later still the first indications of manipulation occur. The eyes thus lead the hands. Interestingly, at the end of each action the eyes move on to the next object about half a second before manipulation is complete. Presumably the information that they have supplied remains in a buffer until the motor system requires it.

Almost identical results were obtained by Mary Hayhoe (Hayhoe, 2000; Hayhoe, Srivastava, Mruczec, & Pelz, 2003) in a study of students making peanut butter and jelly sandwiches. She found the same attachment of gaze to task-related objects and the same absence of saccades to irrelevant objects as with the tea-making gaze led manipulation,

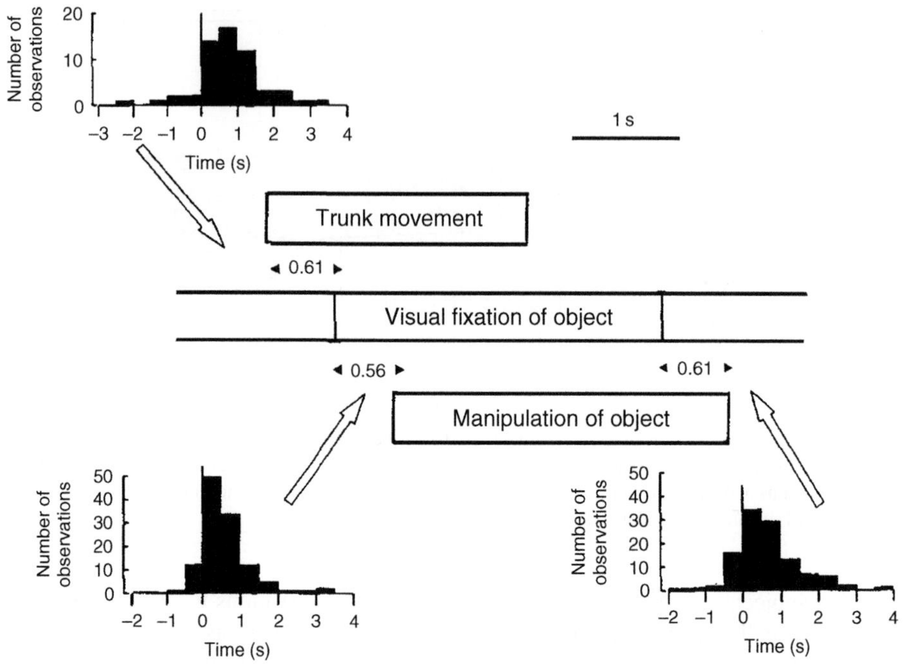

Figure 4. The average sequence of events during the 40 or so "object related actions" that comprise the tea-making task (3 subjects). When a trunk movement is required (for example from one work surface to another) this precedes the first fixation on the next object by about half a second. Similarly the first fixation precedes the first movement of the hands by about the same amount. At the end of the action, the eyes have already moved to the next object in the sequence about half a second before manipulation is complete, implying that information is retained in a buffer.

although by a somewhat shorter interval. This difference is probably attributable to the fact that the sandwich-making was a sit-down task only involving movements of the arms. Two other differences that may have the same cause are the existence of more short duration (<120 ms) fixations than in the tea-making study and the presence of more "unguided" reaching movements (13%) mostly concerned with the setting down of objects. There was a clear distinction in both studies between "within object" saccades which had mean amplitudes of about 8° in both, and "between object" saccades which were much larger, up to 30° in the sandwich-making on a restricted table top, and 90° in tea-making in the less restricted kitchen (Land & Hayhoe, 2001).

From their tea-making study Land et al. (1999) concluded that individual fixations had four main functions: locating (an object for future use), directing (hand to object), guiding (one object with respect to another, e.g., lid to pan), and checking (that some condition is met). The last three (directing, guiding, checking) comply with the maxims (*do it where I'm looking* and *just in time*) of Ballard et al. (1992, 1995), but the first (locating) does not. There is no action at the time, and information is stored for future use. In a study of hand washing, Pelz & Canosa (2001) found a small number of similar "look-ahead fixations" to objects to be contacted a few seconds later, as did Hayhoe et al. (2003) during sandwich-making. These fixations show that, in contrast to the apparent outcome of some "change blindness" studies, positional information is sometimes retained across time intervals corresponding to many fixations (see Tatler, 2002; Tatler, Gilchrist, & Land, 2005).

4. Ball games

Some ball sports are so fast that there is barely time for the player to use his normal oculomotor machinery. Within less than half a second (in baseball or cricket) the batter has to judge the trajectory of the ball and formulate a properly aimed and timed stroke. The accuracy required is a few cm in space and a few ms in time (Regan, 1992). Half a second gives time for one or at the most two saccades, and the speeds involved preclude smooth pursuit for much of the ball's flight. How do practitioners of these sports use their eyes to get the information they need?

Part of the answer is anticipation. Ripoll, Fleurance, and Caseneuve (1987) found that international table-tennis players anticipated the bounce and made a pre-emptive saccade to a point close to the bounce point. Land and Furneaux (1997) confirmed this (with more ordinary players). They found that shortly after the opposing player had hit the ball the receiver made a saccade down to a point a few degrees above the bounce point, anticipating the bounce by about 0.2 s. At other times the ball was tracked around the table in a normal non-anticipatory way: tracking in this case was almost always performed by means of saccades rather than smooth pursuit. The reason why players anticipate the bounce is that the location and timing of the bounce are crucial in the formulation of the return shot. Up until the bounce, the trajectory of the ball as seen by the receiver is ambiguous. Viewed monocularly, the same retinal pattern in space and time would arise from a fast ball on a long trajectory or a slow ball on a short one (Figure 5). (Whether either stereopsis or looming information is fast enough to provide a useful depth signal is still a matter of debate). This ambiguity is removed the instant the timing and position of the bounce are established. Therefore the strategy of the player is to get gaze close to the bounce point (this need not be exact) before the ball does, and lie in wait. The saccade that effects this is interesting in that it is not driven by a "stimulus", but by the player's estimate of the location of something that has yet to happen.

In cricket, where – unlike baseball – the ball bounces before reaching the batsman, Land and McLeod (2000) found much the same thing as in table tennis. With fast balls the batsmen watched the delivery and then made a saccade down to the bounce point, the eye arriving 0.1 s or more before the ball (Figure 6). They showed that with a knowledge of the time and place of the bounce the batsman has the information he needs to judge where and when the ball will reach his bat. Slower balls involved more smooth pursuit. With good batsmen this initial saccade had a latency of only 0.14 s, whereas poor or non-batsmen had more typical latencies of 0.2 s or more.

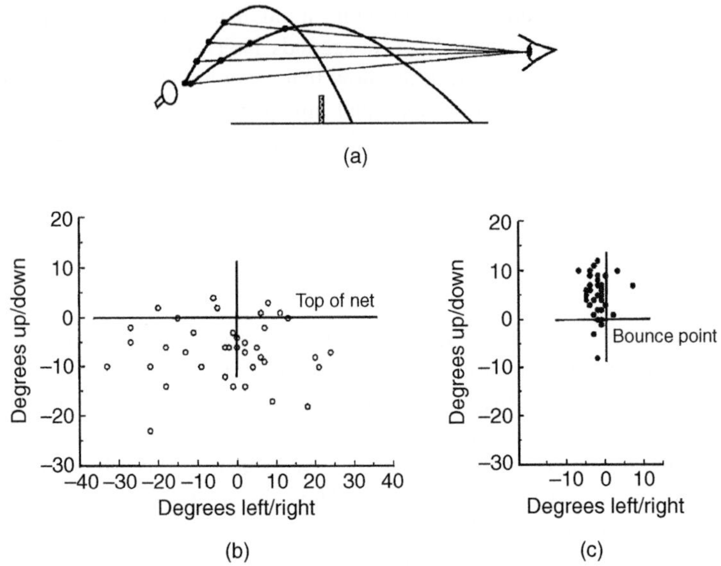

Figure 5. (a) The ambiguity problem in ball games. Seen from the receiver's viewpoint the trajectory of the ball is ambiguous. A slow short ball will have the same retinal velocity trajectory as a faster long ball. The ambiguity is removed by determining the bounce point, after which the time and location of arrival of the ball become predictable. (b) and (c) Landing points of the first saccade made by the receiver after the ball has been struck by the opponent. In (b) these are shown relative to the table surface, in (c) relative to the bounce point. Typically the saccades are aimed a few degrees above the point where the ball will bounce 1–200 ms later (based on Land & Furneaux, 1997).

In baseball the ball does not bounce, and so that source of timing information is not available. Bahill and LaRitz (1984) examined the horizontal head and eye movements of batters facing a simulated fastball. Subjects used smooth pursuit involving both head and eye to track the ball to a point about 9 feet from them, after which the angular motion of the ball became too fast to track (a professional tracked it to 5.5 feet in front: he had exceptional smooth pursuit capabilities). Sometimes batters watched the ball onto the bat by making an anticipatory saccade to the estimated contact point part way through the ball's flight. This may have little immediate value in directing the bat, because, according to Peter McLeod, the stroke is committed as much as 0.2 s before contact (McLeod, 1987), but may be useful in learning to predict the ball's location when it reaches the bat, especially as the ball often "breaks" (changes trajectory) shortly before reaching the batter. According to Bahill and LaRitz (1984) "The success of good players is due to faster smooth-pursuit eye movements, a good ability to suppress the vestibulo-ocular reflex, and the occasional use of an anticipatory saccade".

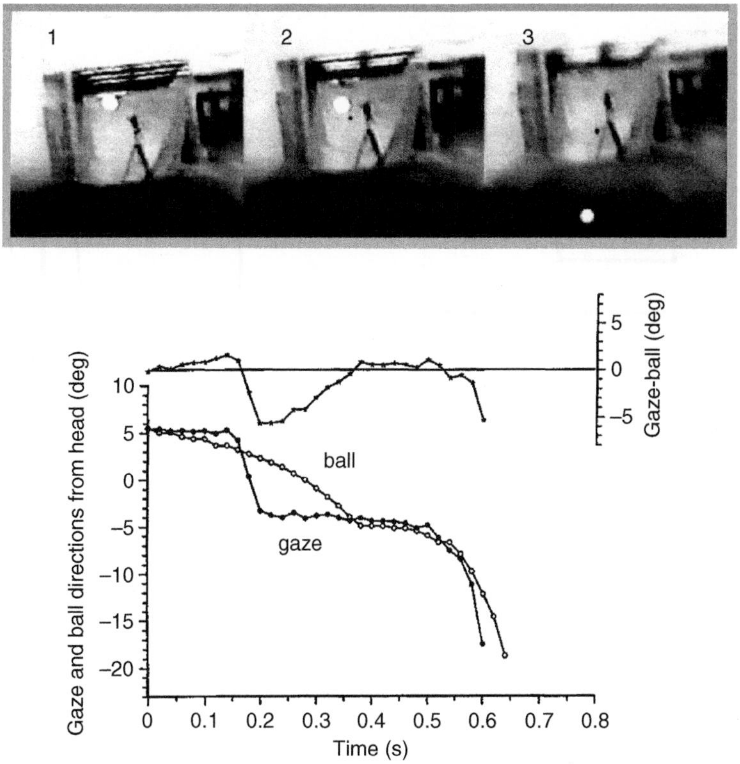

Figure 6. The "pre-emptive" saccade made by a batsman facing a medium-paced ball delivered by a bowling machine. Upper part shows 3 frames from the eye-movement camera on the batsman's head. (1) shows batsman's gaze (white spot) on the bowling machine; (2) about 0.13 safter the delivery showing the ball (small black dot) descending slightly ahead of gaze; (3) about 0.1 s later, showing the end-point of the saccade which lands very close to the point at which the ball will bounce. The ball is still only slightly below its position in (2). Lower graph shows the time course of gaze direction and ball position as seen from the batsman's head. Upper inset shows the difference between gaze and ball positions, and the period in which the batsman takes his eye off the ball by as much as 5° (from Land & McLeod, 2000).

5. Driving

Driving is a complex skill that involves dealing with the road itself (steering, speed control), other road users (vehicles, cyclists, moving and stationary pedestrians) and attention to road signs and other relevant sources of information. It is thus a complex task, and one would expect a range of eye movement strategies to be employed. I will first consider steering, as this is a prerequisite for all other aspects of driving.

When steering a car on a winding road, vision has to supply the arms and hands with the information they need to turn the steering wheel the right amount. What is this control signal, and how is it obtained? As pointed out by Edmund Donges as early as

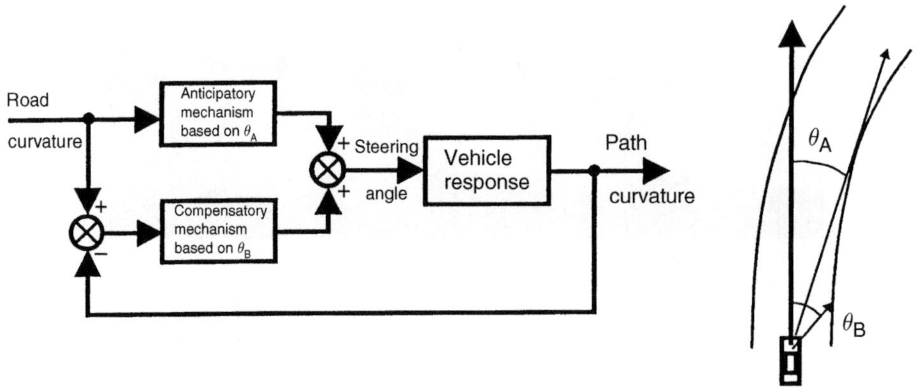

Figure 7. Two-component model of steering, modified from Donges (1978). Road curvature affects path curvature by two routes: feed-forward information is derived from distant regions of the road, such as the tangent points of bends (θ_A), and feedback information from the relative movements of the much closer lane edge (θ_B).

1978, there are basically two sorts of signal available to drivers: feedback signals (lateral and angular deviation from the road centre-line, differences between the road curvature and the vehicle's path curvature), and feed-forward or anticipatory signals obtained from more distant regions of the road up to 2 s ahead in time (corresponding to 90 feet or 27 m at 30 mph). Donges (1978) used a driving simulator to demonstrate that both of these kinds of signal were indeed used in steering, although he did not discuss how they might be obtained visually (Figure 7).

Eye movement studies both on real roads (Land & Lee, 1994) and on a simulator (Land & Horwood, 1995) have confirmed Donges' two-level model of driving, and gone some way to establishing how the eyes find the appropriate information. Earlier studies, mainly on US roads that had predominantly low curvatures, had found only a weak relationship between gaze direction and steering (e.g., Zwahlen, 1993). However, on a winding road in Scotland, where continuous visual control was essential, a much more precise relationship was seen. Land and Lee (1994) found that drivers spent much of their time looking at the tangent point on the up-coming bend (Figure 8). This is the moving point on the inside of the bend where the driver's line of sight is tangential to the road edge; it is the point that protrudes into the road, and is the only point in the flow-field that is clearly defined visually (see also Underwood, Chapman, Crundall, Cooper, & Wallén, 1999). The angular location of this point relative to the vehicle's line of travel (effectively the driver's trunk axis if he is belted in) predicts the curvature of the bend: larger angles indicate steeper curvatures. Potentially this angle is a signal that can provide the feed-forward signal required by Donges' analysis, and Figure 9 does indeed show that curves of gaze direction and steering wheel angle are almost identical. The implication is that this angle, which is equal to the eye-in-head plus the head-in-body angle when the driver is looking at the tangent point, is translated more or less directly into the motor control signal for

Figure 8. Four views of Queen's drive, Edinburgh, from an eye-movement camera, showing representative records of gaze direction (white dot). Upper frames show gaze on the tangent points of right and left bends, lower left on a straight road, and lower right on a jogger. The lower third of each frame shows an inverted image of the eye, from which the gaze direction is derived.

the arms. Cross-correlating the two curves in Figure 9 shows that gaze direction precedes steering wheel angle by about 0.8 s. This provides the driver with a reasonable "comfort margin", but the delay is also necessary to prevent steering taking place before the bend has been reached. (In fact a delay of 0.8 s is typical of many eye-action tasks involving a buffer: about 1 s is observed in reading aloud, typing and sight reading music; a slightly shorter delay of 0.6 s was seen during tea-making actions.)

Simulator studies showed that feed-forward information from the distant part of the road was not on its own sufficient to give good steering (Land & Horwood, 1995). When the near region of the simulated road was removed from view, curvature matching was still accurate, but position-in-lane control was very poor. To maintain good lane position required a view of the road only a few metres ahead, and this region provided much of the feedback information identified in the Donges model. Interestingly this part of the road was rarely fixated compared with the more distant tangent point region, but it was certainly seen and used; it is typically about 5° obliquely below the tangent point. Mourant and Rockwell (1970) had already concluded that lane position is monitored with peripheral vision. They also argue that learner drivers first use foveal vision for lane keeping, then increasingly move foveal gaze to more distant road regions, and learn to

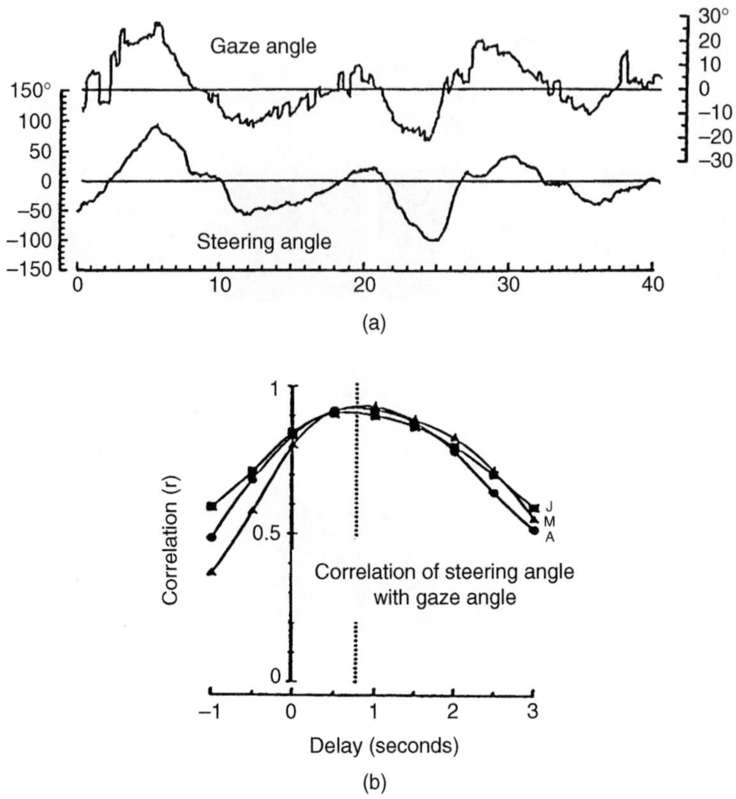

Figure 9. (a) The similarity between gaze angle relative to the line of travel (θ_A on Figure 7) and steering wheel angle over a 40-s period. The curves are almost identical, apart from small saccadic excursions to left and right, implying that gaze angle may be the control variable for steering wheel angle. (b) Correlation between gaze and steering wheel angles for different lead times for 3 drivers. The maximum correlation occurs when gaze leads steering by about 0.8 s, implying that this is the delay in the anticipatory pathway in Figure 7.

use their peripheral vision to stay in lane. Summala, Nieminen, and Punto (1996) reached similar conclusions. The principal conclusion from these studies is that neither the far-road feed-forward input nor the near-road feedback input are sufficient on their own, but the combination of the two allows fast accurate driving (Land, 1998).

A feature of Figure 9 and similar records is that the eyes are not absolutely glued to the tangent point, but can take time out to look at other things. These excursions are accomplished by gaze saccades and typically last between 0.5 and 1 s. The probability of these off-road glances occurring varies with the stage of the bend that the vehicle has reached, and they are least likely to occur around the time of entry into a new bend. At this point drivers fixated the tangent point 80% of the time. It seems that special attention is required at this time, presumably to get the initial estimate of the bend's

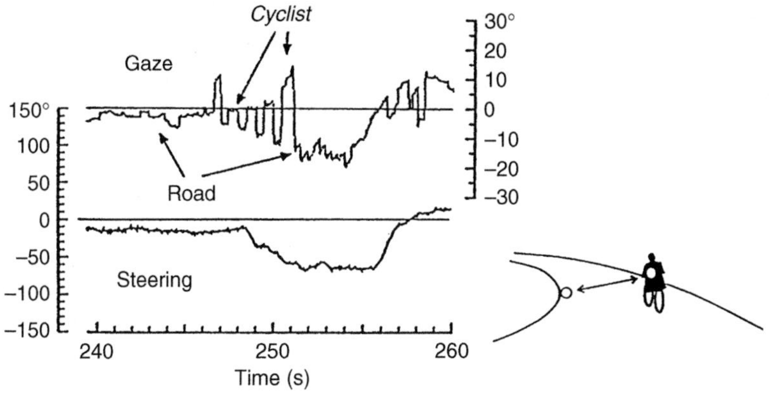

Figure 10. An example of time-sharing during driving. The driver's gaze alternates between the cyclist and tangent point at half-second intervals, but the steering is only related to the direction of the tangent point.

curvature correct. A confirmation of this came from Yilmaz and Nakayama (1995), who used reaction times to a vocal probe to show that attention was diverted to the road just before simulated bends, and that sharper curves demanded more attention than shallower ones. The fewer and shallower the bends in the road, the more time can be spent looking off the road, and this probably accounts for the lack of a close relation between gaze direction and steering on studies of driving on freeways and other major roads where attentional demands related to steering are low.

Sometimes the eye must be used for two different functions at the same time, and as there is only one fovea and off-axis vision is poor, the visual system has to resort to time-sharing. A good example of this is shown in Figure 10, in which the driver is negotiating a bend and so needs to look at the tangent point, while passing a cyclist who needs to be checked on repeatedly. The record shows that the driver alternates gaze between tangent point and cyclist several times, spending half a second on each. The lower record shows that he steers by the road edge, which means that the coupling between eye and hand has to be turned off when he views the cyclist (who would otherwise be run over!). Thus not only does gaze switch between tasks, so does the whole visuo-motor control system. Presumably, whilst looking at the cyclist, the information from the tangent point is kept "on hold" at its previous value.

The last example has shown how the visual system is able to divide its time between different activities. In urban driving this is even more important as each traffic situation and road sign competes for attention. To my knowledge there has been no systematic study of where drivers look in traffic, but from our own observations it is clear that drivers foveate the places from which they need to obtain information: the car in front, the outer edges of obstacles, pedestrians and cyclists, road signs and traffic lights and so on. In general, speeds of 30 mph or less only require peripheral lane-edge (feedback) information for adequate steering. Thus the necessity to use distant tangent points is much reduced, freeing up the eyes for the multiple demands of dealing with other road users. Just as with open-road steering both foveal and peripheral vision are involved. Toshiaki Miura has shown that as the demands of traffic situations increase, peripheral vision is sacrificed to provide greater attentional resources for information uptake by the fovea (Miura, 1987).

6. Conclusions

The various studies described here all tend to confirm the correctness of the conclusions from the first "block copying" study by Ballard et al. (1992), that fixations are made to the place where information is needed for the component of a task that is about to be undertaken. Further, the time between the first saccade to that point and the beginning of the action itself is typically a second or less. With the exception of occasional "look ahead" saccades, in which objects to be used a few seconds in the future are briefly fixated, there is very little evidence that the visual system stores up information for future use beyond that 1 s limit. It seems that in general the visuo-motor system can only do one thing at a time, and it does it immediately.

6.1. What the oculomotor system needs to know

In endeavouring to supply the information the motor system needs for its tasks, the oculomotor system also needs information. Specifically it needs to know where to go to and what to look for. I found it extraordinary that in a kitchen with which subjects had only about 30 s familiarity, there was hardly any sign that they had difficulty locating key objects: obvious search behaviour was very rare. Even when the next object in the task – a mug, say – was behind them, it rarely took more than two saccades to locate it. The implication is that, as each new object in the task is called for by the script, the visual system is provided with both a search image with which to identify the object and also information about its location, stored either during the preliminary look around the kitchen or during "look ahead" saccades made during the execution of the task itself. In addition to locating appropriate objects, the eyes are involved in "checking" operations. Again, the visual system needs to know where to look and what to look for in order to bring a particular act to a close – for example, the level of milk in the cup or whether the kettle is boiling. All this argues strongly that the program the oculomotor system executes, although distinct from the program for the manipulation itself, is inseparable from it. For every motor act there is a loosely corresponding set of eye movements, and just as each motor manipulation is different, so are the eye movements that go with it. All this, of course, has to be learnt. Presumably most of this learning takes place in the early childhood years during which motor abilities of various kinds develop. About the early development of oculomotor competences we know next to nothing.

6.2. How close do we need to look?

In all the studies discussed in this chapter the eyes look at points that are particularly informative for the ongoing action: in food preparation it is the object being manipulated at the time, in steering it is the tangent point, in ball games the bounce point and so on. How accurately these points are targeted by gaze depends on the spatial scale of the task. Thus in reading each fixation takes in about 7 letter spaces, which with standard print at 40 cm means that saccades are about 1.33° long, and so the maximum angular distance from any letter in the line being read is half this, 0.67°. At the other extreme, the average size of "within object" saccades in both tea-making and sandwich-making was about 8°, implying that the centre of the viewed target is rarely more than about 4° from the foveal direction (Land & Hayhoe, 2001). Thus in food preparation visual targeting is 6 times less precise than in reading, presumably because the large size of culinary objects requires correspondingly less accuracy.

Most other estimates are between these extreme values. For example, Johansson, Westling, Bäckström, and Flanagan (2001) used a high-precision eye-tracker to study performance in a task in which a bar was grasped, and lifted to make contact with a target switch, avoiding a projecting obstacle on the way (Figure 11). They found that – as in other tasks – the eyes always fixated certain distinct landmarks (the grasp site, the target, and the surface to which the bar returned) and frequently but not always fixated the obstacle and the tip of the bar. They estimated the precision of fixation by determining the diameters of circles that enclosed 90% of the fixation points for all 9 subjects: these were 3.3° for the tip of the bar and 5.6° for the obstacle. For the target itself most fixations were within a 3° circle, and they regard 3° as the diameter of the "functional fovea" for this task. This implies a maximum target eccentricity of 1.5°.

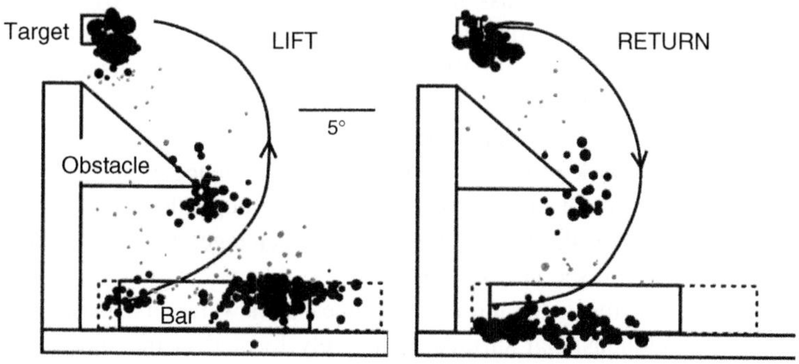

Figure 11. Accuracy of fixation during a task in which a bar is lifted past an obstacle to make contact with a target and then set down again. All fixations of nine subjects are shown. Five landmarks are consistently viewed: on the LIFT the grasp site, left tip of the bar, target, and tip of the obstacle. On the RETURN the support surface is viewed rather than the grasp site. Black circles are fixations within 3° of one of these landmarks (more than 90%), and grey circles are fixations outside these regions. Areas of circles indicate fixation durations (modified from Johansson et al., 2001).

How far from the fovea can useful information be obtained? Again this is likely to depend on the scale of the task. Shioiri and Ikeda (1989) studied the extraction of information from pictures, using a window which was contingent on eye position. They found that the maximum area over which high-resolution pictorial information could be extracted was about 10° across: larger windows provided no extra information. This implies that no further "useful resolution" is available outside about 5° from the fixation point. However, this cannot be universally true. Land et al. (1999) found that subjects could make accurate single eye-and-head saccades to appropriate objects that were up to 50° from the current fixation point. Even allowing a role for object position memory, some information that permitted object identification must have been available even in what would usually be considered the far periphery.

6.3. Intrinsic salience and top-down control

In both tea-making and sandwich-making (Land & Hayhoe, 2001) we were particularly impressed by the way gaze moved from one task-relevant object to the next, ignoring all other objects that were not involved in the activity. The proportion of task-irrelevant objects viewed (other than during periods of waiting – for the kettle to boil, for example) was under 5% in both studies (see Figure 3). Just as Yarbus (1967) had concluded from his study (see Figure 1), it appears that – in real tasks – the eyes are driven much more by top-down information from the script of the activity, rather than by the "intrinsic salience" of objects in the scene. In one sandwich-making experiment involving four subjects 50% of the objects on the table were irrelevant to the task (pliers, scotch tape, forks etc.). In the interval before the task commenced, while the eyes were scanning the table, the proportion of irrelevant objects fixated was 52%. When the task started, this reduced to 18%. Presumably this represented a shift from target selection based on intrinsic salience to one based on task instructions. Shinoda et al. (2001) reached similar conclusions with a virtual reality driving task that required the detection of STOP signs; they found that detection was heavily affected by both task instructions and local context (they are rarely detected mid-block, compared with at intersections). Triesch, Ballard, Hayhoe, and Sullivan (2003) also demonstrated the importance of the immediate context in another virtual reality task in which the sizes of bricks changed at different stages in the procedure. Whether or not the changes were noticed depended crucially on whether brick size was important for the task at the moment of change. This led them to propose another maxim: "What you see is what you need".

Most recent ideas on the generation of saccades to new targets involve a "salience map". This is a two-dimensional surface, tentatively located in the superior colliculus, in which peaks of excitation correspond to objects in the image. These peaks compete with each other in a winner-takes-all manner to reach a threshold that triggers a saccade. Some versions of this salience map concentrate exclusively on bottom-up properties of the image, such as orientation, intensity and colour (e.g. Itti & Koch, 2000), whilst others allow a degree of top-down control to influence the state of the map (Findlay & Walker, 1999). The studies reported here emphasize this top-down influence, since it is clear that eye movements are very closely coupled to the script of the action as a whole. As Johansson et al. (2001) put it: "The salience of gaze targets arises from the functional sensorimotor requirements of the task". (The only problem here is that the word "salience" rather loses its original meaning of conspicuity, and its definition becomes almost becomes circular – an object is salient if it gets looked at it, for whatever reason).

6.4. What next?

In a thoughtful and influential article, Viviani (1990) was sceptical that much could be learned about the cognitive organization underlying eye movements from the study of visual search under free viewing conditions. Even in the dramatic case of Yarbus' "An Unexpected Visitor" (see Figure 1), Viviani comments, "what Yarbus meant is that something about an observer's interest can be inferred from the scanpath. But the scanpath, by itself, does not show how visual details are apprehended or used to plan subsequent movements". Since the work of Ballard et al. (1992), and the other more naturalistic studies outlined here, it seems that in active vision the patterns of eye movements are more constrained than in free viewing, and the relationship of individual fixations to cognitive processes involved in the task is more explicit. The *do it where I'm looking* strategy implies that each type of visually controlled action has an appropriate assemblage of fixations associated with it, reflecting the moment by moment information requirements of the task. However, although there is often quite impressive agreement both within and between subjects as to what is fixated and to some extent when, it nevertheless remains impossible to predict with certainty the timing and destination of each saccade in a sequence. One task for the future is to try to sort out the relationship between the task-related predictability of fixation patterns, and the various kinds of temporal and spatial unpredictability that accompany these patterns. At present it is unclear whether this unpredictability is simply due to "noise" in the oculomotor system, or whether there is a structure to it which is yet to be discovered.

A second direction for future research relates to the nature of the learning process that matches fixation patterns to actions. Eye-movement patterns, whether in reading, playing music, driving, ball sports or domestic activity, are not learned by instruction but by implicit processes about which little is known. The task of tracking eye-movement patterns through childhood, as some skill such as piano playing or table tennis is acquired, seems daunting because of the time scale involved. There are, however, some tasks that are learned as an adult over a reasonably short period of time – driving is an obvious example – where a detailed analysis of the learning process might provide answers to questions about the way that eye-movement strategies develop, and eye movements and motor actions become efficiently coupled in time and space. Very recently (September 2005) a study has been published that addresses this very problem. Sailer, Flanagan, and Johansson (2005) have devised a task in which a somewhat difficult-to-use mouse-like tool is employed to move a cursor on a screen, to locate (and remove) targets that come up in sequence. The main finding was that learning to use the tool occurred in three stages.

Initially the subjects found it hard to control vertical and horizontal movements at the same time. Little apparent progress was made, and it took as much as 20 s for each "hit" to be made. Interestingly, gaze tracked the cursor with a lag, implying that vision was being used to provide information to the motor system about its progress. In the next phase learning was rapid and, although diagonal moves were still difficult, the time for each hit came down to about 2 s. Gaze was now level with or even led the cursor. In the final phase diagonal moves were made confidently, and gaze, instead of tracking the cursor, went straight to the next target. All this makes excellent sense. The role of vision changes, during the 20 min it takes to learn the task, from providing feedback to an inexpert motor system to giving feed forward information to a system that is now competent to perform a targeted action.

References

Bahill, A. T., & LaRitz, T. (1984). Why can't batters keep their eyes on the ball? *American Scientist, 72*, 249–253.

Ballard, D. H., Hayhoe, M. M., Li, F., & Whitehead, S. D. (1992). Hand-eye coordination during sequential tasks. *Philosophical Transactions of the Royal Society of London, B 337*, 331–339.

Ballard, D., Hayhoe, M., & Pelz, J. (1995). Memory representations in natural tasks. *Cognitive Neuroscience, 7*, 66–80.

Buswell, G. T. (1920). An experimental study of the eye-voice span in reading. *Supplementary Educational Monographs, no. 17*. Chicago: Chicago University Press.

Buswell, G. T. (1935). *How people look at pictures: A study of the psychology of perception in art.* Chicago: Chicago University Press.

Butch, R. L. C. (1932). Eye movements and the eye-hand span in typewriting. *Journal of Educational Psychology, 23*, 104–121.

Donges, E. (1978). A two-level model of driver steering behavior. *Human Factors, 20*, 691–707.

Duchowski, A. T. (2003). *Eye tracking methodology: Theory and practice.* London: Springer-Verlag.

Findlay, J. M., & Walker, R.(1999). A model of saccade generation based on parallel processing and competitive inhibition. *Behavioral & Brain Sciences, 22*, 661–721.

Guitton, D., & Volle, M. (1987). Gaze control in humans: Eye-head coordination during orienting movements to targets within and beyond the oculomotor range. *Journal of Neurophysiology, 58*, 427–459.

Hayhoe, M. (2000). Vision using routines: A functional account of vision. *Visual Cognition, 7*, 43–64.

Hayhoe, M. M., Srivastava, A. A., Mruczec, R., & Pelz, J. B. (2003). Visual memory and motor planning in a natural task. *Journal of Vision, 3*, 49–63.

Itti, L., & Koch, C. (2000). A saliency-based search mechanism for overt and covert shifts of visual attention. *Vision Research, 40*, 1489–1506.

Johansson, R. S., Westling, G., Bäckström, A., & Flanagan, J. R. (2001). Eye-hand coordination in object manipulation. *Journal of Neuroscience, 21*, 6917–6932.

Land, M. F. (1998). The visual control of steering. In L. R. Harris, & M. Jenkin (Eds.), *Vision and action* (pp. 163–180). Cambridge University Press.

Land, M. F. (2004). The coordination of rotations of the eyes, head and trunk in saccadic turns produced in natural situations. *Experimental Brain Research, 159*, 151–160.

Land, M. F., & Furneaux, S. (1997). The knowledge base of the oculomotor system. *Philosophical Transactions of the Royal Society of London, B 352*, 1231–1239.

Land, M. F., & Hayhoe, M. (2001). In what ways do eye movements contribute to everyday activities. *Vision Research, 41*, 3559–3565.

Land, M. F., & Horwood, J. (1995). Which parts of the road guide steering? *Nature, 377*, 339–340.

Land, M. F., & Lee, D. N. (1994). Where we look when we steer. *Nature, 369*, 742–744.

Land, M. F., & McLeod, P. (2000). From eye movements to actions: How batsmen hit the ball. *Nature Neuroscience, 3*, 1340–1345.

Land, M. F., Mennie, N., & Rusted, J. (1999). The roles of vision and eye movements in the control of activities of daily living. *Perception, 28*, 1311–1328.

McLeod, P. (1987). Visual reaction time and high-speed ball games. *Perception, 16*, 49–59.

Mackworth, N. H., & Thomas, E. L. (1962). Head-mounted eye-movement camera. *Journal of the Optical Society of America, 52*, 713–716.

Miura, T. (1987). Behavior oriented vision: Functional field of view and processing resources. In J. K. O'Regan, & A. Lévy-Schoen (Eds.), *Eye movements: From physiology to cognition* (pp. 563–572). Amsterdam: North-Holland.

Mourant, R. R., & Rockwell, T. H. (1970). Mapping eye-movement patterns to the visual scene in driving: An exploratory study. *Human Factors, 12*, 81–87.

Pelz, J. B., & Canoza, R (2001). Oculomotor behavior and perceptual categories in complex tasks. *Vision Research, 41*, 3587–3596.

Regan, D. (1992). Visual judgments and misjudgments in cricket, and the art of flight. *Perception, 21*, 91–115.

Ripoll, H., Fleurance, P., & Caseneuve D. (1987). Analysis of visual patterns of table tennis players. In J. K. O'Regan & A. Levy-Schoen (Eds.), *Eye movements: From physiology to cognition* (pp. 616–617). Amsterdam: North-Holland.

Sailer, U., Flanagan, J. R., & Johansson, R. S. (2005). Eye-hand coordination during learning of a novel visuomotor task. *Journal of Neuroscience, 25*, 8833–8842.

Schwartz, M. F., Montgomery, M. W., Palmer, C., & Mayer, N. H. (1991). The quantitative description of action disorganization after brain damage: A case study. *Cognitive Neuropsychology, 8*, 381–414.

Shinoda, H., Hayhoe, M. M., & Shrivastava, A. (2001). What controls attention in natural environments? *Vision Research, 41*, 3535–3545.

Shioiri, S., & Ikeda, M. (1989). Useful resolution for picture perception as a function of eccentricity. *Perception, 18*, 347–361.

Summala, H., Nieminen, T., & Punto, M. (1996). Maintaining lane position with peripheral vision during in-vehicle tasks. *Human Factors, 38*, 442–451.

Tatler, B. W. (2002). What information survives saccades in the real world? In J. Hyönä, D. Munoz, W. Heide, & R. Radach (Eds.), *The brain's eye: Neurobiological and clinical aspects of oculomotor research* (pp. 149–163). Amsterdam: Elsevier.

Tatler, B. W., Gilchrist, I. D., & Land, M. F. (2005). Visual memory for objects in natural scenes: From fixations to object files. *Quarterly Journal of Experimental Psychology, 58*, 931–960.

Thomas, E. L. (1968). Movements of the eye. *Scientific American, 219*(2), 88–95.

Triesch, J., Ballard, D. H., Hayhoe, M. M., Sullivan, B. T. (2003). What you see is what you need. *Journal of Vision, 3*, 86–94.

Underwood, G., Chapman, P., Crundall, D., Cooper, S., & Wallén, R. (1999). The visual control of steering and driving: where do drivers look when negotiating curves. In A. G. Gale et al. (Eds.), *Vision in Vehicles VII.* (pp. 245–252). Amsterdam: Elsevier.

Viviani, P. (1990). Eye movements and visual search: cognitive, perceptiual and motor aspects. In E. Kowler (Ed.), *Eye movements and their role in cognitive processes* (pp. 353–393). Amsterdam: Elsevier.

Wade, N. J., & Tatler, B. W. (2005). *The moving tablet of the eye: The origins of modern eye movement research.* Oxford: Oxford University Press.

Weaver, H. E. (1943). A study of visual processes in reading differently constructed musical selections. *Psychological Monographs, 55*, 1–30.

Yarbus, A. (1967). *Eye movements and vision.* New York: Plenum Press.

Yilmaz, E. R., & Nakayama, K. (1995). Fluctuation of attention levels during driving. *Investigative Ophthalmology and Visual Science, 36*, S940.

Zwahlen, H. T. (1993). Eye scanning rules for drivers: how do they compare with actual observed eye-scanning behavior? *Transportation Research Record, 1403*, 14–22.

PART 2

PHYSIOLOGY AND CLINICAL STUDIES OF EYE MOVEMENTS

Edited by

MARTIN H. FISCHER

Chapter 5

USING EYE MOVEMENTS TO PROBE DEVELOPMENT AND DYSFUNCTION

DOUGLAS P. MUNOZ, IRENE ARMSTRONG and BRIAN COE

Queen's University, Canada

Eye Movements: A Window on Mind and Brain
Edited by R. P. G. van Gompel, M. H. Fischer, W. S. Murray and R. L. Hill

Abstract

Recording of saccadic eye movements has proved to be a valuable tool for investigation of brain function and dysfunction. Recent neurophysiological studies have revealed that the time from target appearance to saccade initiation can be modeled as an accumulator function in which both baseline and rate of rise of saccade-related activity contribute toward achieving threshold for movement initiation. In this chapter, we review recent saccadic eye-movement studies designed to track abilities across development and in disorders of frontal cortex and basal ganglia. Studies can be designed to probe the ability to initiate automatic vs voluntary saccades or to suppress saccades. The accumulator model can be used to explain normal developmental changes in voluntary saccade control that are present in normal development as well as in attention deficit hyperactivity disorder (ADHD), Parkinson's disease (PD), and Tourette syndrome (TS).

1. Introduction

One of the most important functions of the central nervous system is the generation of movement in response to sensory stimulation. The visual guidance of saccadic eye movements represents one form of sensory-to-motor transformation that has provided significant insight into our understanding of motor control and sensorimotor processing. The eyes have a simple and well-defined repertoire of movements, and the neural circuitry regulating the production of saccadic eye movements is now understood at a level that is sufficient to now link activation in cortical and subcortical areas with behavior and dysfunction. As a result, deficits in eye-movement control of various patient groups can now provide greater insight into the neural substrate underlying the pathophysiology. To properly interpret these insights one must have an understanding of the entire visuomotor loop involved in saccade control, from the visual input on the eye, through the many cortical and subcortical regions of the brain, to the motor output of the brain stem on the muscles that move the eyes, creating behaviors we can measure.

The primate retina has a specialized region in its center, the fovea, which serves the central 1° of the visual field and provides the greatest visual acuity (Perry and Cowey, 1985). In most visual areas of the primate brain, the fovea has the greatest representation, emphasizing the importance of foveal vision in many aspects of visual processing and visually guided behavior (Dow, Snyder, Vautin, & Bauer, 1981; Van Essen, Newsome, & Maunsell, 1984). To maximize the efficiency of foveal vision, we must have the ability to align the fovea rapidly upon objects in the visual world and then keep the fovea aligned upon these objects for a sufficient period of time for the visual system to perform a detailed analysis of the image. Saccadic eye movements are used to redirect the fovea from one point of interest to another and a fixation mechanism is used to keep the fovea aligned on the target during subsequent image analysis. This alternating behavior of saccade–fixation is repeated several hundred thousand times a day and is critical for complex acts such as reading or driving an automobile.

Saccades can be triggered by the appearance of a visual stimulus in the periphery (e.g., the sudden appearance or motion of a novel visual stimulus), or initiated voluntarily, in the absence of any overt sensory stimuli, motivated by the goals of the individual. They can also be suppressed during periods of visual fixation. In special situations where visually guided saccade plans and internally motivated saccade plans compete, the brain must inhibit the automatic response and instead promote the internally motivated saccade in order to perform the desired behavior. Several experimental paradigms have been devised to investigate the control of saccades in these different behavioral situations (see below).

Our understanding of the neural circuitry controlling saccades has increased dramatically in the past 30 years as a result of human behavioral, imaging, and clinical studies as well as animal behavioral, physiological, anatomical, and pharmacological studies. Figure 1 highlights some of the important brain areas that have been identified. Critical nodes in the network include regions of the parietal and frontal cortices, basal ganglia, thalamus, superior colliculus (SC), cerebellum, and brainstem reticular formation (see Hikosaka, Nakamura, & Nakahara, 2000; Leigh and Zee, 1999; Moschovakis,

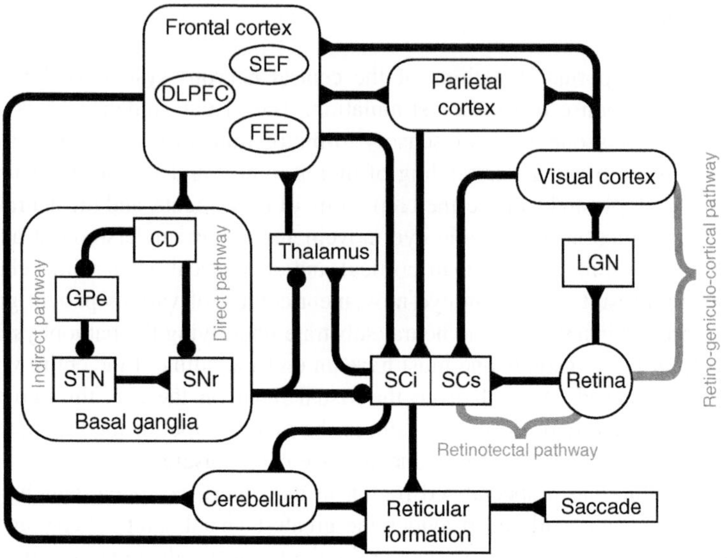

Figure 1. Schematic of brain areas involved in the control of visual fixation and saccadic eye movements. See text for details. See Appendix 1 for the list of abbreviations.

Scudder, & Highstein, 1996; Munoz, Dorris, Paré, & Everling, 2000; Munoz & Everling, 2004; Munoz & Fecteau, 2002; Munoz & Schall, 2003; Schall and Thompson, 1999; Scudder, Kaneko, & Fuchs, 2002; Sparks, 2002; Wurtz and Goldberg, 1989 for detailed review of aspects of the circuitry). Because these brain areas span much of the central nervous system, neurological immaturity, degeneration, or malfunction may influence the ability of a subject to maintain visual fixation and generate fast and accurate saccades. Consequently, many neurological and psychiatric disorders are accompanied by disturbances in eye movements and visual fixation which can now be used in the identification of the affected brain regions.

2. Overview of brain areas involved in saccade control

Eye movements are controlled by the synergistic action of the six extraocular muscles. The extraocular muscle motoneurons (MN) discharge a burst of action potentials to move the eyes and a tonic discharge to keep the eyes at a fixed position (see Leigh & Zee, 1999 for review). The burst discharge of the MN is generated by the brainstem premotor circuitry located in the mesencephalic, pontine, and medullary regions of the brainstem reticular formation (see Moschovakis et al., 1996; Scudder et al., 2002; Sparks, 2002 for detailed reviews). Excitatory and inhibitory burst neurons (EBN and IBN), which innervate the MN directly, are silent during fixation and discharge bursts of action potentials for saccades in a specific direction. Other neurons located in the brainstem reticular formation control the discharge of EBN and IBN. Long-lead burst neurons (LLBN) discharge a high-frequency burst of action potentials for saccades and they also have a low-frequency buildup of activity before the burst. It is believed that LLBN project to the EBN and IBN to provide the burst input. The EBN and IBN are subject to potent inhibition from omnipause neurons (OPN) which discharge tonically during all periods of fixation and pause for saccades in any direction. Thus, in order to generate a saccade, the OPN must be silenced and then the LLBN activate the appropriate pools of EBN and IBN to produce the saccade command that is sent to the MN. Following completion of the saccade, the OPN reactivate and inhibit the EBN and IBN, thus preventing the eyes from moving any further. The tonic activity of the OPN ensures that any early, presaccadic activity among other premotor elements cannot lead to spurious activity among EBN and IBN, which would disrupt fixation.

Inputs to the brainstem premotor circuitry arise from several structures including the frontal cortex, SC, and cerebellum. Although our understanding of how these inputs are coordinated to control the actions of the brainstem premotor circuit precisely are incomplete, significant progress has been made in recent years.

The SC plays a critical role in the control of visual fixation and saccadic eye movements. The superficial layers of the SC (SCs) contain neurons that receive direct retinal inputs as well as inputs from other visual areas (Robinson & McClurkin, 1989). These visual neurons are organized into a visual map of the contralateral visual hemifield.

The intermediate layers of the SC (SCi) contain neurons whose discharges are modulated by saccadic eye movements and visual fixation (see Munoz et al., 2000; Munoz & Fecteau, 2002 for review). These neurons are organized into a retinotopically coded motor map specifying saccades into the contralateral visual field. Neurons that increase their discharge before and during saccades, referred to as saccade neurons, are distributed throughout the SCi. Neurons that are tonically active during visual fixation and pause during saccades, referred to as fixation neurons, are located in the rostrolateral pole of the SC where the fovea is represented. These saccade and fixation neurons in the SC project directly to the brainstem premotor circuitry in the reticular formation to influence behavior.

The (SCi) receives inputs from posterior parietal cortices, frontal cortices, and basal ganglia which all play a role in the voluntary selection of potential saccadic targets to ultimately influence behavior. Visual inputs that are crucial for maintaining visual fixation or generating saccades are directed from visual cortex, through the parietal lobe, to the (SCi). One area in particular that lies at the sensory-motor interface is the lateral intraparietal area (LIP). Projections from LIP to the (SCi) are involved in sensory-motor transformations and attentional processing (see Andersen et al., 1997; Colby & Goldberg, 1999; Glimcher, 2001 for detailed reviews).

The frontal cortex receives direct projections from the visual cortex but areas like the frontal eye fields (FEF) are strongly interconnected with parietal visual areas (see Schall, 1997; Schall & Thompson, 1999 for reviews). This is a vital point as FEF may act as a central hub connecting several frontal areas such as supplementary eye fields (SEF), and the dorsolateral prefrontal cortex (DLPFC) with the parietal cortex, the (SCi) and also the basal ganglia. The SEF and the DLPFC are known to play a role in working memory and decision making. The numerous connections between FEF, parietal lobe, and SCi makes this an excellent system to combine stimulus-driven saccade signals and internally driven voluntary saccade signals. Indeed, it is not uncommon to find neurons with similar firing patterns in all three areas (Munoz & Schall, 2003; Paré & Wurtz, 2001; Wurtz, Sommer, Paré, & Ferraina, 2001). This is not to say that all information must go through the FEF, because the FEF, SEF, and DLPFC all project to the SCi. In addition FEF and SEF also project to the cerebellum and brainstem reticular formation directly.

The frontal cortex also connects through the basal ganglia (see Figure 1) to participate in presaccadic processing (for detailed review, see Hikosaka et al., 2000, 2006). These pathways through the basal ganglia allow for the integration of motivation and reward information with saccade planning. There is a *direct pathway* in which the frontal areas project to the caudate nucleus (CD) to excite GABAergic neurons which in turn project directly to the substantia nigra pars reticulata (SNr). The neurons in the SNr form the major output of the basal ganglia. They are GABAergic and they project to the SCi and the thalamus. The thalamus then projects back to frontal and parietal cortices. Via this direct pathway through the basal ganglia, activation of cortical inputs will lead to disinhibition of the SC and thalamus because the signals pass through two inhibitory synapses.

There is also an *indirect pathway* through the basal ganglia in which a separate set of GABAergic neurons in the CD project to the external segment of the globus pallidus (Gpe). Neurons in the Gpe are GABAergic and project to the subthalamic nucleus (STN). Neurons in the STN then send excitatory projections to the SNr, which in turn projects to the SCi and thalamus. Thus, the indirect pathway travels through three inhibitory synapses and activation of cortical input will serve to inhibit the SCi and thalamus.

3. Saccadic eye-movement tasks

Figure 2 illustrates some of the saccadic eye-movement tasks used to probe brain function and dysfunction. The pro-saccade task (Figure 2a) is used to probe the ability of subjects to initiate automatic visually triggered saccades. The anti-saccade task (Figure 2b) is used to probe the ability of subjects to suppress the automatic visually triggered saccade and instead initiate a voluntary response in the opposite direction. It is believed that these tasks probe very different mechanisms. Visually triggered saccades can be triggered by visual inputs to the saccade-generating circuit, while anti-saccades require both saccadic suppression and voluntary saccade execution (see Munoz and Everling, 2004 for review).

The pro- and anti-saccade tasks can be used in combination with a variety of additional conditions. For example, fixation state can be manipulated by the presence or absence of the fixation point at the time of target appearance. There are both exogenous and endogenous components of fixation control (Paré and Munoz, 1996; Reuter-Lorenz, Oonk, Barnes, & Hughes, 1995). The endogenous component of fixation is required to maintain steady fixation independent of whether there is a visual stimulus on the fovea, while the exogenous component is mediated by the presence of a visible stimulus on the fovea.

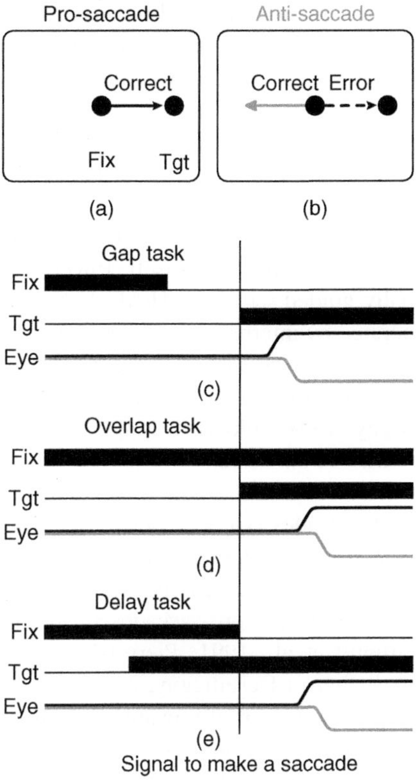

Figure 2. Some of the paradigms used to investigate saccadic function and dysfunction. (a) In the pro-saccade task, the participant is instructed to look (solid black arrow) from the center position (Fix) to the peripheral target (Tgt). (b) In the anti-saccade task, the subject is instructed to suppress the automatic pro-saccade (dotted black arrow) and instead generate a voluntary saccade in the opposite direction (solid gray arrow). Direction errors are defined as saccades triggered toward the target. (c) In the gap condition, the central fixation point disappears 200 ms prior to target appearance. (d) In the overlap condition, the fixation point remains illuminated during target presentation. (e) In the delayed saccade task, the participant is instructed to maintain fixation upon the central fixation point until it disappears.

The presence or absence of the exogenous component influences performance in pro- and anti-saccade tasks (Fischer & Weber, 1992; Munoz, Broughton, Goldring, & Armstrong, 1998). In the gap saccade task (Figure 2c), the initial fixation point disappears and the subject is in complete darkness for some period of time prior to target appearance. The prior disappearance of the fixation point removes exogenous fixation at the time of target appearance and only endogenous fixation signals remain to suppress saccade initiation. In this condition, reaction times (RTs) are reduced and the frequency of express saccades is facilitated (Fischer & Boch, 1983; Fischer & Weber, 1993; Paré & Munoz, 1996). In the anti-saccade task, the insertion of the gap period prior to target appearance leads to increases in the percentage of direction errors (Fischer & Weber, 1992; Munoz et al., 1998). In the overlap condition (Figure 2d), the fixation point remains illuminated at the time of target appearance leading to increased saccadic reaction time (SRT) and reduced error rates.

Saccades are triggered via parallel descending pathways from the cerebral cortex to the (SCi) and brainstem reticular formation. Visually triggered saccades are initiated by the sudden appearance of a visual stimulus and are mediated by the (SCi), with important inputs from the visual and posterior parietal cortices (Guitton, Buchtel, & Douglas, 1985; Hanes & Wurtz, 2001; Schiller, Sandell, & Maunsell, 1987). Lesions of posterior parietal cortex increase RT of visually guided saccades (Heide & Kompf, 1998).

Volitional saccades, generated by internal goals, sometimes in the absence of any overt triggering stimulus, rely upon circuitry that includes higher brain centers such as the frontal cortex and the basal ganglia (Dias & Segraves, 1999; Gaymard, Ploner, Rivaud-Pechoix, & Pierrot-Deseilligny, 1998; Hikosaka et al., 2000; Pierrot-Deseilligny, Israel, Berthoz, Rivaud, & Gaymard, 1991). Lesions of the FEF have only a modest effect on visually guided saccades, but they produce significant impairment in the generation of voluntary or memory-guided saccades (Dias & Segraves, 1999; Gaymard et al., 1998; Gaymard, Ploner, Rivaud, Vermersch, & Pierrot-Deseilligny, 1999). These movements have increased RTs and reduced saccadic velocities. Lesions of the DLPFC reduce the ability of subjects to suppress reflexive pro-saccades in the anti-saccade task (Guitton et al., 1985; Pierrot-Deseilligny et al., 1991; Pierrot-Deseilligny et al., 2003; Ploner, Gaymard, Rivaud-Pechoux, & Pierrot-Deseillegny, 2005). Thus, a critical function of the DLPFC may be the voluntary suppression of unwanted or visually triggered saccades.

Single-cell recording in non-human primates has identified the neurophysiological correlates of saccadic suppression, preparation, and execution that are used to interpret behavioral performance. Fixation neurons in the (SCi) and FEF are tonically active during visual fixation and pause for saccades, while saccade neurons are silent during visual fixation and burst during saccade production (see Munoz & Schall, 2003; Munoz & Everling, 2004 for review). The drop in fixation activity that occurs during the gap period in the gap saccade task correlates with reduced SRT and represents the neural correlate of the gap effect and fixation disengagement (Dorris & Munoz, 1995; Dorris, Paré, & Munoz, 1997). This drop in fixation activity leads to disinhibition of the saccadic system and a reciprocal increase in low-frequency pre-target activity among subsets of saccade neurons in the brainstem reticular formation (i.e., LLBN), SCi, and FEF (Everling & Munoz, 2000; Dorris et al., 1997; Munoz et al., 2000). This increase in activity prior to target appearance (i.e., pre-target activity) represents the neural correlate for saccadic preparation because its intensity is correlated negatively to SRT (Everling & Munoz, 2000; Dorris et al., 1997; Dorris & Munoz, 1998; Munoz et al., 2000).

Correct performance on anti-saccade trials (i.e., successful suppression of visually triggered pro-saccades) requires that saccade neurons in the FEF and (SCi) be inhibited before target appearance (see Munoz & Everling, 2004 for review). Fixation neurons in the FEF and SC appear to carry this saccadic suppression signal because they discharge at a higher frequency during this instructed fixation period on correct anti-saccade trials.

This task-dependent modulation of neuronal excitability in the FEF and (SCi) is adaptive and essential for successful performance. On anti-saccade trials, target appearance elicits a phasic visual response among saccade neurons in the (SCi) and FEF contralateral to the stimulus that could serve to trigger a direction error. This activity must be suppressed and saccade neurons in the opposite (SCi) and FEF activated to generate the correct anti-saccade. We have hypothesized that DLPFC and/or SNr provide the essential saccadic suppression signals on anti-saccade trials that are required to inhibit the saccade neurons prior to target appearance (Munoz & Everling, 2004). Immaturity or dysfunction of prefrontal cortex and/or basal ganglia will influence the ability to selectively recruit these saccadic suppression signals making it harder to inhibit unwanted or reflexive saccades.

4. Accumulator models describe reaction times

Models have been developed to explain the stochastic variability of RT (Luce, 1986). The accumulator model has been particularly useful at interpreting neurophysiological and behavioral data related to the initiation of saccadic eye movements. This type of model supposes that in response to a stimulus, a signal grows until it reaches a threshold thereby triggering a movement in response to the stimulus (Figure 3). Models of this sort include three sources for the stochastic variability evident in RTs: variable baseline (e.g., Trappenberg, Dorris, Munoz, & Klein, 2001); variable threshold (e.g., Grice, Nullmeyer, & Spiker, 1982; Nazir & Jacobs, 1991); and variable rate of rise from baseline to threshold (e.g., Carpenter, 1988; Ratcliff, 1978). How are aspects of these models instantiated at the level of the single cell, a single brain area, and the entire saccadic generating circuitry? The pattern of movement-related activity recorded from saccade neurons in the FEF and SC of monkeys performing various saccade tasks has been analyzed to evaluate these alternative models of RT (Dorris et al., 1997; Dorris & Munoz, 1998; Everling et al., 1998; Everling & Munoz, 2000; Hanes & Schall, 1996; Paré & Hanes, 2003). Both pre-target (i.e., variable baseline) and post-target (i.e., rate of rise) information processing can contribute to the accumulation of activity toward the threshold.

Figure 3 illustrates how an accumulator model can be implemented to describe the behavior of subjects performing pro- and anti-saccade tasks. In the pro-saccade task (Figure 3a), many saccade neurons (i.e., visuomotor neurons) in the (SCi) and FEF contralateral to the target will discharge both a phasic visual response following the appearance of the visual target (boxed in area in Figure 3a) and a motor-related burst that is time-locked to the saccade. If the level of pre-target activity is high, then the visual response will add to it thus exceeding the saccadic threshold and an express saccade will be triggered (solid thick lines in Figure 3a). If pre-target activity is low, then the visual response will not reach threshold and the system will have to wait for the subsequent accumulation of activity to threshold to trigger a regular latency saccade. In the gap condition, the early disappearance of the fixation point leads to disinhibition of saccade neurons in the (SCi) and FEF which elevates pre-target activity, making it easier to trigger express saccades.

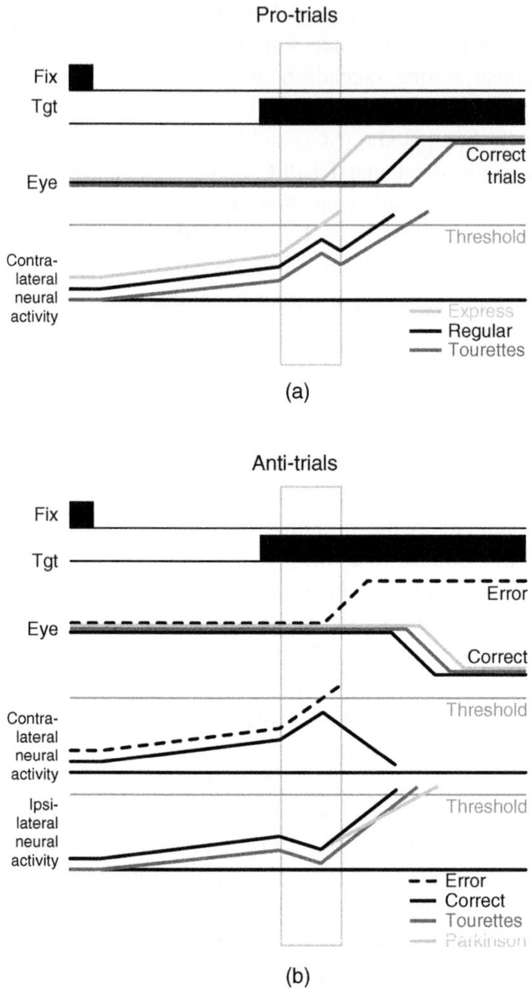

Figure 3. Race model to describe behavior. Activity among saccade neurons in the SC and FEF accumulates toward threshold to initiate a saccade. (a) In the pro-saccade task, pre-target and post-target factors contribute to accumulation of activity. Immediately following target appearance, many saccade neurons discharge a phasic visual response (boxed in area). With sufficient pre-target activity, the phasic visual response can lead to the immediate threshold crossing, triggering an express saccade (light gray trace), whereas reduced pre-target activity can lead to increased RTs (dark gray trace). (b) In the anti-saccade task, pre-target and post-target factors contribute to threshold crossing. The phasic visual response (boxed in area) is registered on the contralateral side of the brain, however, to initiate a correct anti-saccade, activity must cross threshold on the ipsilateral side of the brain. If pre-target activity is too high, the phasic visual response will exceed saccade threshold and a direction error will be triggered (dashed trace).

In the anti-saccade task (Figure 3b), there are two processes racing toward threshold: a process initiated on the contralateral side of the brain by the appearance of the target which serves to initiate the automatic prepotent response; and another process initiated on the ipsilateral side of the brain by the inversion of the stimulus vector to initiate a voluntary anti-saccade. To perform the task correctly, the process related to the initiation of the automatic pro-saccade (i.e., direction error) must be cancelled or suppressed, to allow time for the voluntary response (i.e., correct saccade) to grow toward threshold. Neurophysiological studies have revealed that a critical step in the completion of the anti-saccade task is the reduction of excitability of saccade neurons in the Sc_i and FEF before the target appears (Everling et al., 1998, 1999; Everling & Munoz, 2000). If pre-target activity is too high, then the visual response will sum with the elevated pre-target activity to trigger a direction error (dashed line in Figure 3b). If pre-target activity is suppressed, then the visual response will not exceed threshold and instead activity can accumulate on the side ipsilateral to the target so that a correct anti-saccade can be triggered (solid traces in Figure 3b).

5. Normal Development

Performance on saccadic eye-movement tasks varies dramatically across the life span (Abel, Troost, & Dell'Osso, 1983; Biscaldi, Fischer, & Stuhr, 1996; Bono et al., 1996; Fischer, Biscaldi, & Gezeck, 1997; Moschner & Baloh, 1994; Munoz et al., 1998; Olincy, Ross, Young, & Freedman, 1997; Pratt, Abrams, & Chasteen, 1997; Sharpe & Zackon, 1987; Spooner, Sakala, & Baloh, 1980; Warabi, Kase, & Kato, 1984; Wilson, Glue, Ball, & Nutt, 1993). We have now investigated saccadic eye-movement performance in over 300 normal participants between the ages of 4 and 85 years (Munoz et al., 1998, 2003). All of these participants performed separate blocks of pro- and anti-saccade trials in which target location (left or right) and fixation condition (gap or overlap) were randomly interleaved within each block. Figures 4a and b illustrates the systematic variations in SRT that occur between the ages of 4 and 85 years in the pro- and anti-saccade tasks, respectively. The "U-shaped" pattern is present in both pro- and anti-saccade tasks across both gap and overlap conditions. Mean SRTs tend to be greatest for the youngest and oldest subjects and are at a minimum for subjects around 20 years of age. Note that the gap effect, the difference between gap and overlap conditions, was constant across subject age.

The amount of intra-subject variability in SRT, expressed as the coefficient of variation (CV) (standard deviation/mean * 100), also varied systematically across subject age (Figures 4c, d). CV was greatest for the youngest and oldest subjects in our sample and was minimal for adult subjects between the ages of about 20–70 years.

Express saccades are the shortest latency visually triggered saccades (Fischer & Ramsberger, 1984; Fischer et al., 1993; Paré & Munoz, 1996) that humans can make and, in our lab (Munoz et al., 1998, 2003), they are initiated between 90 and 140 ms after target appearance. Figure 4e shows the percentage of express saccades elicited in the pro-saccade task with gap and overlap conditions. Express saccades were most prevalent

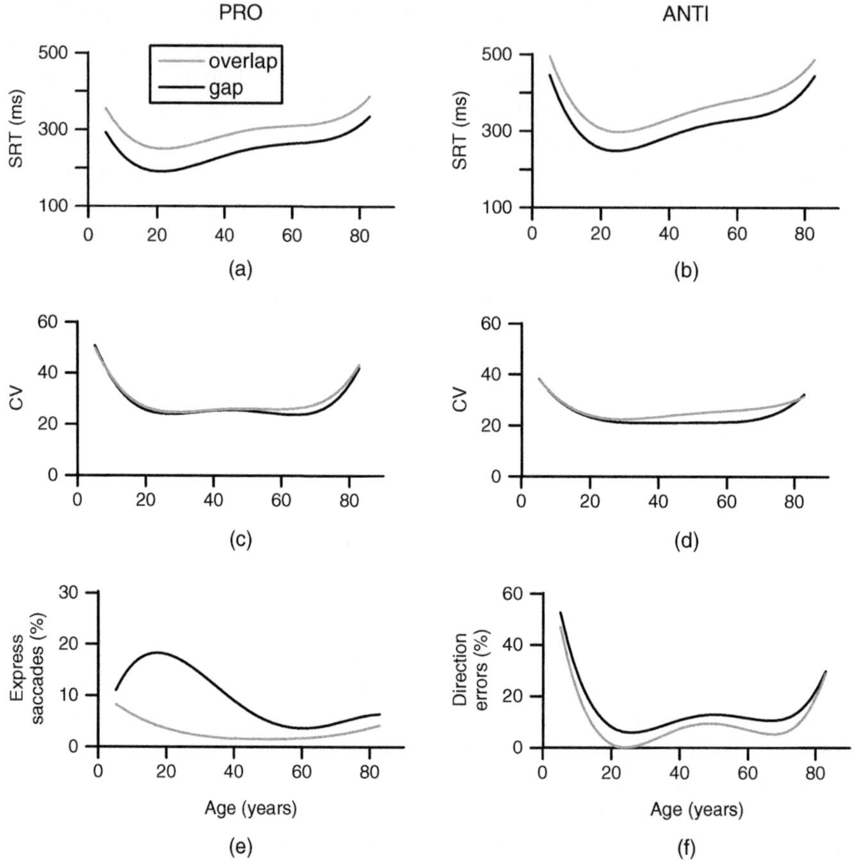

Figure 4. Behavioral performance from 300 control subjects ranging in age from 4–85 years performing the immediate pro-saccade (a, c, e) and anti-saccade tasks (b, d, f) in the gap (gray lines) and overlap (black lines) conditions. (a, b): Mean saccadic reaction time; (c, d): Intra-subject variability in RT expressed as the coefficient of variation (CV = standard deviation/mean * 100); (e): Percent of express saccades elicited in the pro-saccade task; (f): % of direction errors (automatic saccades made to the target) in the anti-saccade task. Subjects aged 4–15 years are represented in bin widths of one year, subjects aged 16–30 years are represented in bins of three years and subjects older than 30 years are represented in bin widths of five years width. The mean of each bin was plotted (centered on the bin) and a 4th-degree polynomial was fit to the data.

in the gap condition. More importantly, the percentage of express saccades diminished with subject age and reached a minimum among the elderly.

Figure 4f shows the percentage of direction errors triggered in the anti-saccade task. The curve is U-shaped. There is a rapid improvement in saccadic suppression attained throughout 5–20 years. Then, saccadic suppression abilities begin to falter among subjects greater than 70 years.

The normative data illustrated in Figure 4 can be accounted for with the race model (see Figure 3). We hypothesize that young children lack strong voluntary control over saccade-generating circuits because they have more variable RTs, initiate more express saccades in the pro-overlap condition, and generate a greater percentage of direction errors in the anti-saccade task. As a consequence, pre-target activity among (SCi) and FEF saccade neurons presumably varies considerably from trial to trial (i.e., variable baseline). The result is more variability in RTs, including higher percentages of automatic express saccades, and increased percentage of direction errors in the anti-saccade task. The time course of the improvement in saccade control between the ages of 5–20 is consistent with several changes in frontal lobe function and connectivity (see Paus, 2005 for review). Such maturation of frontal functional connectivity could produce the changes in performance in the saccadic tasks.

6. Eye-Movement Abnormalities in clinical studies

There are multitudes of studies investigating eye-movement dysfunction in a variety of neurological and psychiatric disorders (*Attention Deficit/Hyperactivity Disorder*: Feifel et al., 2004; Huang-Pollock, & Nigg, 2003; Klein, Raschke, & Brandenbusch, 2003; Munoz et al., 2003; *Tourette syndrome*: Farber, Swerdlow, & Clementz, 1999; LeVasseur, Flanagan, Riopelle, & Munoz, 2001; *Autism*: Goldberg et al., 2002; Minshew, Luna, & Sweeney, 1999; van der Geest, Kemner, Camfferman, Verbaten, & van Engeland, 2001; *Huntington's Disease*: Blekher et al., 2004; Fawcett, Moro, Lang, Lozano, & Hutchison, 2005; Winograd-Gurvich et al., 2003; *Parkinson's Disease*: Chan, Armstrong, Pari, Riopelle, & Munoz, 2005; Crawford, Henderson, & Kennard, 1989; Kimmig, Haussmann Mergner, & Lucking, 2002; Vidailhet et al., 1994; *Schizophrenia*: Avila, Hong, & Thaker, 2002; Broerse, Crawford, & den Boer, 2001; Clementz, 1996; Obayashi, Matsushima, Ando, H., Ando, K., & Kojima, 2003; see also Everling & Fischer, 1998; Sweeney, Takarae, Macmillan, Luna, & Minshew, 2004 for reviews) and the normative data illustrated in Figure 4 can be used as the backdrop for comparison. Here, we focus on the eye-movement abnormalities identified in three specific disorders – ADHD, PD, and TS. This comparison illustrates how eye-movement recording can be used to gain insight into pathophysiology.

7. Attention Deficit Hyperactivity Disorder

Attention deficit hyperactivity disorder (ADHD), a neurobehavioral disorder estimated to affect approximately 5% of children is characterized by the symptoms of impulsiveness, hyperactivity, and inattention that often persist into adulthood (Barkley, 1997). Response inhibition may be an important component of the disability because ADHD subjects have difficulty suppressing inappropriate behavioral responses (Mostofsky et al., 2001; Shue & Douglas, 1992). At present, the etiology of ADHD remains poorly defined.

Several observations support a hypothesis of a frontostriatal deficit, possibly involving dysfunction in dopamine transmission, which may produce the symptoms of ADHD (Castellanos et al., 2002; Feifel, Farber, Clementz, Perry, & Anllo-Vento, 2004; Kates et al., 2002; Seidman, Valera, & Bush, 2004; Willis & Weiler, 2005; see Castellanos, 2001; Castellanos & Tannock 2002 for reviews).

We hypothesized that children and adults diagnosed with ADHD may have specific difficulties in oculomotor tasks requiring the suppression of automatic or unwanted saccadic eye movements. To test this hypothesis, we compared the performance of 114 ADHD and 180 control participants ranging in age from 6 to 59 years (Munoz et al., 2003). In the pro-saccade task, mean SRT was elevated modestly but significantly in ADHD, relative to age-matched controls (Figure 5a). This increase was present in both the gap and the overlap conditions. Perhaps more dramatically, intra-subject variability was also increased significantly in ADHD (Figure 5c). In other words, response latencies among ADHD subjects were more variable. Although ADHD subjects tended to generate more express saccades than control subjects, this difference did not reach significance (Figure 5e).

In the anti-saccade task, mean SRT (Figure 5b) and CV (Figure 5d) were also elevated for children and adults with ADHD. Most importantly, ADHD subjects also initiated many more direction errors in the anti-saccade task (Figure 5f). Thus, ADHD participants had considerable difficulty exerting voluntary control over saccade generation. These findings are consistent with fronto-striatal pathophysiology. We hypothesize that this pathophysiology results in poor voluntary control over the saccade-generating circuitry. This can be modeled with the race model (see Figure 3). Figure 3a shows how variable levels of pre-target activity among saccade neurons in the (SCi) and FEF can lead to saccades with variable RTs. Poor control over pre-target activity will lead to increased variability in RTs and, on anti-saccade trials, when pre-target activity is too high, the phasic visual response registered on saccade neurons contralateral to the stimulus will trigger anti-saccade errors (see Figure 3b). Thus, we conclude that pathophysiology in ADHD specifically leads to poor control of excitability of saccade neurons prior to target appearance.

8. Parkinson's disease

The motor impairments of Parkinson's disease (PD), including muscle rigidity and slowness of movement (Taylor, Saint-Cyr, & Lang, 1986; Owen et al., 1993; Berry, Nicoloson, Foster, Behrmann & Sagar, 1999; Lezak, 1995), result from degeneration of dopaminergic neurons in the substantia nigra pars compacta (Leenders & Oertel, 2001; Bergman & Deuschl, 2002). In addition to their slowed movements, individuals with PD are often impaired in their ability to suppress automatic behavioral responses (Henik, Dronkers, Knight, & Osimani, 1993; Owen et al., 1993).

We investigated saccade control in 18 PD patients and compared their performance to age- and sex-matched control participants (Chan et al., 2005). In the pro-saccade task,

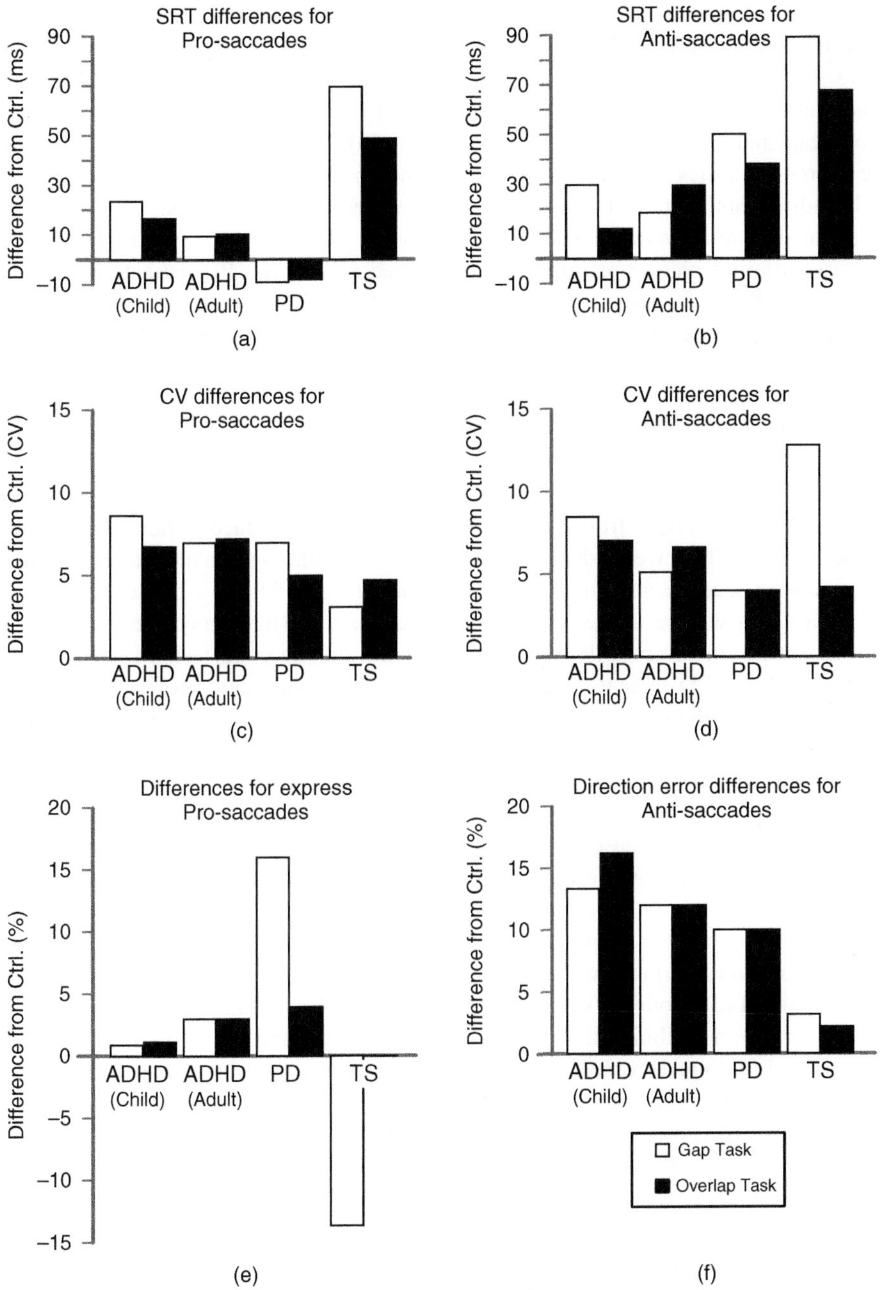

Figure 5. Behavioral performance of ADHD (Child and Adult), PD, and TS patients contrasted to age- and sex-matched control subjects in the pro-saccade (a, c, e) and anti-saccade (b, d, f) tasks. Values on ordinate are expressed as differences from control values (see Figure 4).

PD patients had shorter mean SRT that just failed to reach significance (see Figure 5a). Recall that among normal elderly, mean SRT was increased, compared to young adults (Figure 4a). Thus, the PD patients were performing like younger adults in the pro-saccade task. They had faster SRTs (although non-significantly different from control) in both gap and overlap conditions (see Figure 5a), and they generated significantly more express saccades (Figure 5e). This latter finding was particularly surprising because elderly individuals tend not to make express saccades. However, unlike young controls, PD patients had more variable SRT, expressed as an increase in CV (see Figure 5c).

A very different picture of impairment emerged among the PD patients in the anti-saccade task. SRTs for correct anti-saccades were significantly slower (see Figure 5b) and more variable (see Figure 5d), compared to age-matched controls. In addition, PD patients generated a significantly greater proportion of direction errors in both the gap and overlap conditions (see Figure 5f).

The deficit in automatic saccade suppression and increased variability in SRT that we observed among PD patients is consistent with a disorder of the prefrontal-basal ganglia circuit. Impairment of this pathway may lead to disinhibition or release of the automatic saccade system from top-down inhibition and produce deficits in volitional saccade control. In the race model (see Figure 3), this manifests as an increase in pre-target activity among saccade neurons in the (SCi) and FEF in PD. As a consequence, more express saccades will be triggered in the pro-saccade task (light gray line in Figure 3a) and more direction errors will be triggered in the anti-saccade task (dashed lines in Figure 3b). When PD patients are able to suppress the automatic pro-saccade on anti-saccade trials, then their SRTs are exaggerated (Figure 5b). Thus, we hypothesize that the rise to threshold that takes place among saccade neurons in the (SCi) and FEF ipsilateral to the stimulus is slowed or abnormal (light gray traces in Figure 3b).

Previous investigations on motor and cognitive control in PD have identified similar deficits to what we have described. Cognitive processes, such as attention control, are also impaired in PD (Brown & Marsden, 1990). Individuals with PD were faster than controls on a reflexive visual-orienting task (Briand, Hening, Poizner, & Sereno, 2001) and showed impairment in suppression of visuomotor activation (Praamstra & Plat, 2001). In the Stroop task, participants are presented with color or neutral words in various colors and asked to ignore the word and name its color. Individuals with PD demonstrate greater difficulty in inhibiting the reflexive response to read the word (Henik et al., 1993). These parallel findings across various cognitive and oculomotor tasks suggest a common mechanism underlying a general deficit in automatic response suppression in PD.

It has long been known that the slow, hypokinetic movements of PD can be improved through the provision of external cues (Cunnington et al., 1995; Jahanshahi et al., 1995; Morris, Iansek, Matyas, & Summers, 1996). For example, stride length can be increased with external visual cues (Morris et al., 1996) and reaching movement speeds can be improved during visually cued conditions (Majsak, Kaminski, Gentile, & Flanagan, 1998); thus, automatic motor actions (i.e., movements to an external visual cue) appear to be spared in PD, unlike movements which are volitional. Even more intriguing is the observation that long-latency reflexes are also altered in PD. Tatton and colleagues (Tatton & Lee, 1975; Tatton,

Eastman, Bedingham, Verrier, & Bruce, 1984) demonstrated that long-loop reflexes, which are presumed to include transcortical pathways, are exaggerated in PD (see also Mortimer & Webster, 1979; Rothwell, Obeso, Traub, & Marsden, 1983). These exaggerated "M2" responses in PD have been attributed to reduced inhibition onto cortical motor output neurons. Thus, it appears that PD patients are hyper excitable to sensory stimuli, and automatic responses to external stimuli are enhanced or exaggerated.

9. Tourette Syndrome

Tourette Syndrome (TS) is an inherited condition characterized by the presence of motor and phonic tics which can be worsened by anxiety or fatigue (Singer, 1997) and improved by concentration (Jankovic, 1997). Although the physiological basis for tics and TS remains unknown, a substantial amount of evidence suggests a disorder of frontal-striatal circuits (Kates et al., 2002; Singer, 1997). It has been suggested that TS may result from overactivity of the direct pathway through the basal ganglia (Hallet, 1993). TS patients may therefore experience abnormal control of voluntary saccadic eye movements.

Previous studies examining saccades in TS patients have reported conflicting results. Pro-saccade RTs in TS patients have been reported as normal or only slightly elevated (Farber et al., 1999; Straube, Mennicken, Reidel, Eggert, & Muller, 1997), but saccade durations may be reduced (Farber et al., 1999). Anti-saccades have greater RT (Farber et al., 1999; Straube et al., 1997), and peak velocities may be reduced (Straube et al., 1997). The frequency of direction errors among TS subjects performing the anti-saccade task has been reported as normal (Straube et al., 1997) or abnormally high (Farber et al., 1999; Narita, Shawkat, Lask, Taylor, & Harris, 1997).

We initially hypothesized that because of overactivity in the direct pathway, TS subjects would have faster RTs, more express saccades in the pro-saccade task and more direction errors in the anti-saccade task. To test these hypotheses, we investigated saccade control in 10 TS patients and compared performance to age-matched control participants (LeVasseur et al., 2001).

Contrary to our initial hypothesis, TS patients were significantly *slower* than control subjects in saccade intiation (Figures 5a–b). In addition, TS patients did not initiate more direction errors in the anti-saccade task (Figure 5f). These results suggest that the ability to inhibit automatic visually triggered saccades was not impaired in TS. Instead, it suggests that pre-target activity in TS was below that of control subjects leading to prolonged mean SRT (dark gray lines in Figures 3a and b).

10. Delayed saccade task

We were initially surprised when TS patients did not initiate more direction errors in the anti-saccade task. However, a previous study (Flanagan, Jakobson, & Munhall, 1999) noted that TS patients initiated tics most frequently in a task with a long and variable

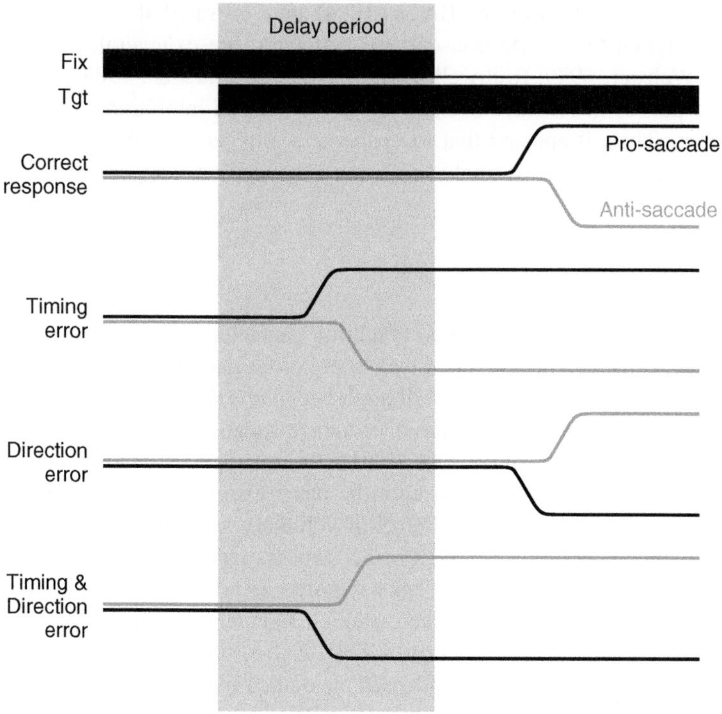

Figure 6. Schematic representation of types of responses possible in the interleaved pro-saccade (black traces)/ anti-saccade(gray traces) saccade delay task. See text for details.

delay period. Therefore we also measured eye movements in the delayed saccade task with a random delay period interleaving pro- and anti-saccade trials randomly (Figure 6). Subjects could therefore initiate several types of errors: (1) timing errors, correct saccade direction but initiated during the variable delay period; (2) direction errors, incorrect saccade direction initiated at the correct time, after the delay period; and (3) combined timing and direction errors.

In this task, TS, PD, and ADHD subjects made more errors relative to age-matched control subjects (Figure 7). The distribution of the different types of errors varied between tasks and disorders. Here we focus on the distribution of only timing errors (Figure 7a) – those saccadic errors triggered in the correct direction but which occurred prematurely, during the variable delay epoch. ADHD, PD, and TS patients all generated more timing errors when compared to age-matched control subjects. However, the time when these errors were generated varied across the different patient groups. Among ADHD and PD patients, timing errors were initiated early in the delay epoch – the occurrence of timing errors did not increase for longer delay intervals (Figure 8). In sharp contrast, TS patients made few timing errors during short delay epochs and instead generated more timing errors later in the delay epoch. This is evident by the fact that the occurrence of

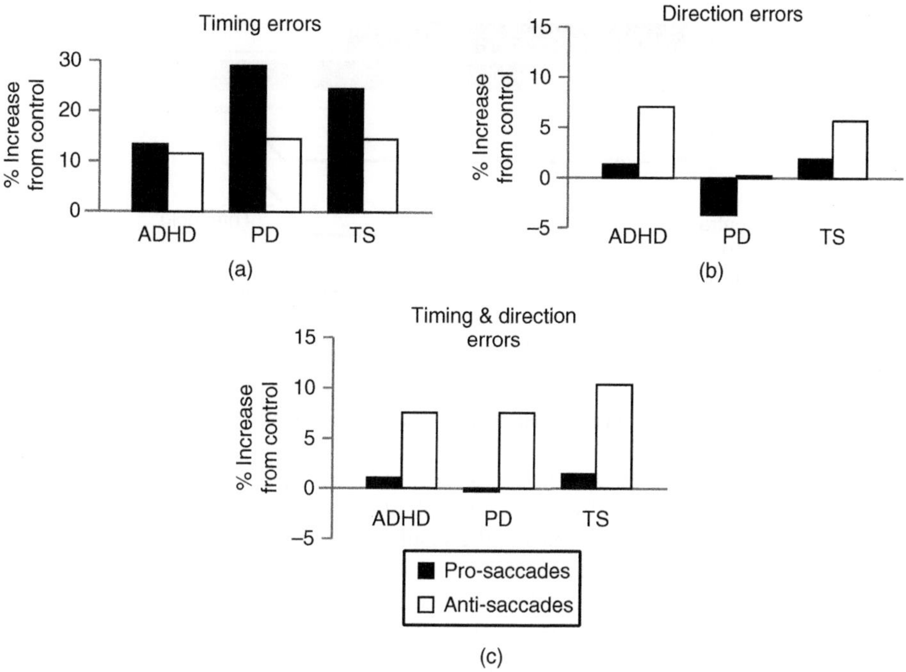

Figure 7. Performance of ADHD, PD, and TS subjects relative to age-matched controls in the delay saccade task.

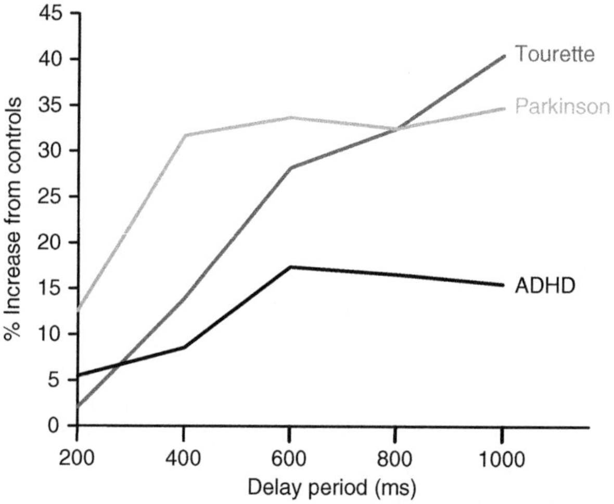

Figure 8. Timing errors as a function of delay duration in the delay saccade task. Data collapsed across pro- and anti-saccade tasks. Timing errors were constant for ADHD and PD across delay interval, whereas in TS, timing errors increased with increasing delay interval.

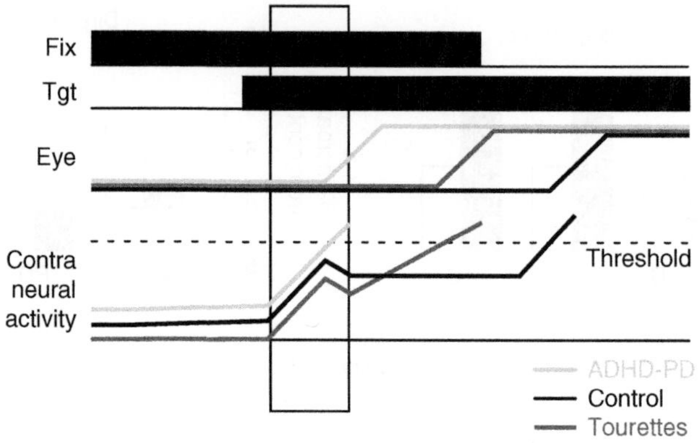

Figure 9. Race model predictions of performance in the delay saccade task. See text for details.

timing errors increased for longer and longer delay periods (see Figure 8), suggestive of disruption of a different type of saccadic suppression mechanism.

Figure 9 provides an explanation for the results we obtained in the delayed saccade task in ADHD, PD, and TS. As described above, we believe that in ADHD and PD, there is poor control over excitability in the saccadic generating circuitry. As a result, SRTs are variable and it is difficult for these subjects to suppress automatic visually triggered saccades (e.g., light gray lines in Figure 9). In the delayed saccade task this results in timing errors, saccades that are triggered shortly after target appearance. However, in TS, the timing errors occurred later in the delay period (see Figure 8). The prolonged SRT in the immediate pro- and anti-saccade tasks (see Figures 5a, b), suggests that in TS patients, saccade neurons in the (SCi) and FEF are at a lower level of excitability (i.e., reduced pre-target activity). However, during the delay epoch, pre-saccadic activity drifts toward threshold triggering premature responses later in the delay epoch (Figure 9).

11. Conclusions

Fixation and saccadic signals are distributed across a network of brain areas that extends from the parietal and frontal cortices, through the basal ganglia and thalamus, to the SC, cerebellum, and brainstem reticular formation. Evidence is accumulating to show that these competing signals may interact at multiple levels of the neuraxis. Thus, it is likely that specific functions are not localized to only one brain area. Rather, they may be distributed across multiple areas. Activity among saccade neurons in some of these areas (e.g., SCi and FEF) accumulates toward threshold to trigger saccadic eye movements. This chapter summarizes how knowledge of the circuit can be used to tease apart deficits in different patient groups. The race model can be used to interpret these deficits and make specific predictions about how brain pathology can influence excitability in the saccade-generating circuitry in the brain.

Appendix A

A.1. List of abbreviations

ADHD: attention deficit hyperactivity disorder
CD: caudate nucleus
CV: coefficient of variation
DLPFC: dorsolateral prefrontal cortex
EBN: excitatory burst neuron
FEF: frontal eye field
GABA: γ-amino-butyric acid
GPe: external segment of the globus pallidus
IBN: inhibitory burst neuron
LGN: lateral geniculate nucleus
LIP: lateral intraparietal area
LLBN: long-lead burst neuron
MN: motoneuron
OPN: omnipause neuron
PD: Parkinson's disease
RT: reaction time
SC: superior colliculus
SCi: intermediate layers of the superior colliculus
SCs: superficial layers of the superior colliculus
SEF: supplementary eye field
SNr: substantia nigra pars reticulata
SRT: saccadic reaction time
STN: subthalamic nucleus
TS: tourette syndrome

References

Abel, L. A., Troost, B. T., & Dell'Osso, L. F. (1983). The effects of age on normal saccadic characteristics and their variability. *Vision Research, 23*, 33–37.

Andersen, R.A., Snyder, L. H., Bradley, D. C., & Xing, J. (1997). Multimodal representation o space in the posterior parietal cortex and its use in planning movements. *Annual Review of Neuroscience*, *20*, 303–330.

Avila, M. T., Hong, E. & Thaker, G. K. (2002). Current progress in schizophrenia research – Eye movement abnormalities in schizophrenia: What is the nature of the deficit? *Journal of Nervous and Mental Disease*, *190*, 479–480.

Barkley, R. A. (1997). *ADHD and the nature of self-control*. New York: Guildford.

Bergman, H. & Deuschl, G. (2002). Pathophysiology of Parkinson's disease: From clinical neurology to basic neuroscience and back. *Movement disorders, 17*, S28–S40.

Berry, E. L., Nicoloson, R. I., Foster, J. K., Behrmann, M., & Sagar, H. J. (1999). Slowing of reaction time in Parkinson's disease: the involvement of the frontal lobes. *Neuropsychologia, 37*, 787–795.

Biscaldi, M., Fischer, B., & Stuhr, V. (1996). Human express-saccade makers are impaired at suppressing visually evoked saccades. *Journal of Neurophysiology, 76*, 199–214.

Blekher, T. M., Yee, R. D., Kirkwood, S. C., Hake, A. M., Stout, J. C., Weaver, M. R., et al. (2004). Oculomotor control in asymptomatic and recently diagnosed individuals with the genetic marker for Huntington's disease. *Vision Research, 44*, 2729–2736.

Bono, F., Oliveri, R. L., Zappia, M., Aguglia U., Puccio, G., & Quattrone, A. (1996). Computerized analysis of eye movements as a function of age. *Archives in Gerontology and Geriatrics, 22*, 261–269.

Briand, K. A., Hening, W., Poizner, H., & Sereno, A. B. (2001). Automatic orienting of visuospatial attention in Parkinson's disease. *Neuropsychologia, 39*, 1240–1249.

Broerse, A., Crawford, T. J. & den Boer, J. (2001). Parsing cognition in schizophrenia using saccadic eye movements: A selective review. *Neuropsychologia, 39*, 742–756.

Brown, R. G., & Marsden, C. D. (1990). Cognitive function in Parkinson's disease: From description to theory. *Trends in Neurosciences, 13*, 21–29.

Carpenter, R. H. S., (1988). *Movements of the eyes*, London: Pion.

Castellanos, F. X. (2001). Neural substrates of attention-deficit hyperactivity disorder. *Advances in Neurology, 85*, 197–206.

Castellanos, F. X., Lee, P. P., Sharp, W., Jeffries, N. O., Greenstein, D. K., Clasen, L. S. et al. (2002). Developmental trajectories of brain volume abnormalities in children and adolescents with attention-deficit/hyperactivty disorder. *Journal of the American Medical Association, 288*, 1740–1748.

Castellanos, F. X., & Tannock, R. (2002). Neuroscience of attention-deficit/hyperactivity disorder: The search for endophenotypes. *Nature Reviews Neuroscience, 3*, 617–628.

Chan, F., Armstrong, I. T., Pari, G., Riopelle, R. J., Munoz, D. P. (2005). Deficits in saccadic eye-movement control in Parkinson's disease. *Neuropsychologia, 43*, 784–796.

Clementz, B. A. (1996). The ability to produce express saccades as a function of gap interval among schizophrenia patients. *Experimental Brain Research, 111*, 121–130.

Colby, C. L., & Goldberg, M. E. (1999). Space and attention in parietal cortex *Annual Review of Neuroscience, 22*, 319–349.

Crawford, T. J., Henderson, L., & Kennard, C. (1989). Abnormalities of nonvisually guided eye movements in Parkinson's disease. *Brain, 112*, 1573–1586.

Cunnington, R., Iansek, R., Bradshaw, J. L., & Phillips, J. G. (1995). Movement-related potentials in Parkinson's disease: Presence and predictability of temporal and spatial cues. *Brain, 118*, 935–950.

Dias, E. C., & Segraves, M. A. (1999). Muscimol-induced inactivation of monkey frontal eye field: Effect on visually and memory-guided saccades. *Journal of Neurophysiology, 81*, 2191–2214.

Dorris, M. C. & Munoz, D. P. (1995). A neural correlate for the gap effect on saccadic reaction times in monkey. *Journal of Neurophysiology*, *73*, 2558–2562.

Dorris, M. C., & Munoz, D. P. (1998). Saccadic probability influences motor preparation signals and time to saccadic initiation. *Journal of Neruoscience, 18*, 7015–7026.

Dorris, M. C., Paré, M., & Munoz, D. P. (1997). Neuronal activity in monkey superior colliculus related to the initiation of saccadic eye movements. *Journal of Neuroscience, 17*, 8566–8579.

Dow, B. M., Snyder, A. Z., Vautin, R. G., & Bauer, R. (1981). Magnification factor and receptive-field size in foveal striate cortex of the monkey. *Experimental Brain Research, 44*, 213–228.

Everling, S., Dorris, M. C., & Munoz, D. P. (1998). Reflex suppression in the anti-saccade task is dependent on prestimulus neural processes. *Journal of Neurophysiology, 80*, 1584–1589.

Everling, S., Dorris, M. C., Klein, R. M. & Munoz, D. P. (1999). Role of primate superior colliculus in preparation and execution of anti- and pro-saccades. *Journal of Neuroscience*, *19*, 2740–2754.

Everling, S., & Fischer, B. (1998). The antisaccade: A review of basic research and clinical studies. *Neuropsychologia, 36*, 885–889.

Everling, S., & Munoz, D. P. (2000). Neuronal correlates for preparatory set associated with pro-saccades and anti-saccades in the primate frontal eye field. *Journal of Neuroscience, 20*, 387–400.

Farber, R. H., Swerdlow, N. R., & Clementz, B. A. (1999). Saccadic performance characteristics and the behavioural neurology of Tourette's syndrome. *Journal of Neurology Neurosurgery and Psychiatry, 66*, 305–312.

Fawcett, A. P., Dostrovsky, J. O., Lozano, A. M., & Hutchison, W. D. (2005). Eye movement-related responses of neurons in human subthalamic nucleus. *Experimental Brain Research, 162*, 357–365.

Fawcett, A. P., Moro, E., Lang, A. E., Lozano, A. M., & Hutchison, W. D. (2005). Pallidal deep brain stimulation influences both reflexive and voluntary saccades in Huntington's disease. *Movement Disorder, 20*, 371–377.

Feifel, D., Farber, R. H., Clementz, B. A., Perry, W., & Anllo-Vento, L. (2004). Inhibitory deficits in ocular motor behavior in adults with attention-deficit/hyperactivity disorder. *Biological Psychiatry, 56*, 333–339.

Fischer, B., Biscaldi, M., & Gezeck, S. (1997). On the development of voluntary and reflexive components in human saccade generation. *Brain Research, 754*, 285–297.

Fischer, B. & Boch, R., (1983). Saccadic eye movements after extremely short reaction times in the monkey. *Brain Research, 260*, 21.

Fischer, B., & Ramsberger, E. (1984). Human express saccades: extremely short reaction times of goal directed eye movements. *Experimental Brain Research, 57*, 191–195.

Fischer, B., & Weber, H. (1992). Characteristics of "anti" saccades in man. *Experimental Brain Research, 89*, 415–424.

Fischer, B. & Weber, H. (1993). Express saccades and visual attention, *Behavioural and Brain Sciences*, 16, 533.

Fischer, B., Weber, H., Biscaldi, M., Aiple, F., Otto, P., & Stuhr, V. (1993). Separate populations of visually guided saccades in humans: reaction times and amplitudes. *Experimental Brain Research, 92*, 528–541.

Flanagan, J. R., Jakobson, L.S., & Munhall, K. G. (1999). Anticipatory grip adjustments are observed in both goal-directed movements and movement tics in an individual with Tourette's syndrome. *Experimental Brain Research, 128*, 69–75.

Gaymard, B., Ploner, C. J., Rivaud-Pechoix, S., & Pierrot-Deseilligny, C. (1999). The frontal eye field is involved in spatial short-term memory but no in reflexive saccade inhibition. *Experimental Brain Research, 129*, 288–301.

Gaymard, B., Ploner, C. J., Rivaud, S., Vermersch, A. I., & Pierrot-Deseilligny, C. (1998). Cortical control of saccades. *Experimental Brain Research, 123*, 159–163.

Glimcher, P. W. (2001). Making choices: the neurophysiology of visual-saccadic decision making. *Trends in Neurosciences, 24*, 654–659.

Goldberg, M. C., Lasker, A. G., Zee, D. S., Garth, E., Tien, A., & Landa, R. J. (2002). Deficits in the initiation of eye movements in the absence of a visual target in adolescents with high functioning autism. *Neuropsychologia, 40*, 2039–2049.

Grice, G. R., Nullmeyer, R., Spiker, V. A. (1982). Human reaction time: Toward a general theory. *Journal of Experimental Psychology General, 111*, 135.

Guitton, D., Buchtel, H. A., & Douglas, R., M. (1985). Frontal lobe lesions in man cause difficulties in suppressing reflexive glances and in generating goal-directed saccades. *Experimental Brain Research, 58*, 455–472.

Hallet, M. (1993). Physiology of basal ganglia disorders: an overview. *Canadian Journal of Neurological Science, 20*, 177–183.

Hanes, D. P., & Schall, J. D. (1996). Neural control of voluntary movement initiation *Science, 274*, 427–430.

Hanes, D. P., & Wurtz, R. H. (2001). Interaction of the frontal eye field and superior colliculus for saccade generation. *Journal of Neurophysiology, 85*, 804–815.

Heide, W., & Kompf, D. (1998). Combined deficits of saccades and visuo-spatial orientation after cortical lesions *Experimental Brain Research, 123*,164–171.

Henik, A., Dronkers, N. F., Knight, R. T., & Osimani, A. (1993). Differential-effects of semantic and identity priming in patients with left and right-hemisphere lesions. *Journal of Cognitive Neuroscience, 5*, 45–55.

Hikosaka, O., Nakamura, K., Nakahara, H. (2006). Basal ganglia orient eyes to reward. *Journal of Neurophysiology*, *95*, 567–584.

Hikosaka, O., Takikawa, Y., & Kawagoe, R. (2000). Role of the basal ganglia in the control of purposive saccadic eye movements. *Physiological Review, 80*, 953–978.

Huang-Pollock, C. L., & Nigg, J. T. (2003). Searching for the attention deficit in attention deficit hyperactivity disorder: The case of visuospatial orienting. *Clinical Psychology Review, 802*, 801–830.

Jahanshahi, M., Jenkins, I. H., Brown, R. G., Marsden, C. D., Passingham, R. D., & Brooks, D. J. (1995). Self-initiated versus externally triggered movements. I. An investigation using measurement of regional cerebral blood flow with PET and movement-related potentials in normal and Parkinson's disease participants. *Brain, 118*, 913–933.

Jankovic, J. 1997. Tourette syndrome: Phenomenology and classification of tics. *Neurologic Clinics, 15*, 267–275.

Kates, W. R., Frederikse, M., Mostofsky, S. H., Folley, B. S., Cooper, K., Mazur-Hopkins, P. et al. (2002). MRI parcellation of the frontal lobe in boys with attention deficit hyperactivity disorder or Tourette syndrome. *Psychiatry Research Neuroimaging, 116*, 63–81.

Kimmig, H., Haussmann, K., Mergner, T., & Lucking, C. H. (2002). What is pathological with gaze shift fragmentation in Parkinson's disease? *Journal of Neurology, 249*, 683–692.

Klein, C., Raschke, A., & Brandenbusch, A. (2003). Development of pro- and antisaccades in children with attention-deficit hyperactivity disorder (ADHD) and healthy controls. *Psychophysiology, 40*, 17–28.

Leenders, K. L., & Oertel, W. H. (2001). Parkinson's disease: Clinical signs and symptoms, neural mechanisms, positron emission tomography and therapeutic interventions. *Neural Plasticity, 8*, 99–110.

Leigh, R. J., & Zee, D. S. (1999). *The neurology of eye movements*. Philadelphia: Davis.

LeVasseur, A. L., Flanagan, J. R., Riopelle, R. J., & Munoz, D. P. (2001). Control of volitional and reflexive saccades in Tourette's syndrome. *Brain, 124*, 2045–2058.

Lezak, M. D. (1995). *Neuropsychological assessment 3rd Edition*. Oxford University Press: New York.

Luce, R.D. (1986). *Response times: Their role in inferring elementary mental organization*. Oxford: Oxford University Press.

Majsak, M. J., Kaminski, T., Gentile, A. M., & Flanagan, J. R. (1998). The reaching movements of patients with Parkinson's disease under self-determined maximal speed and visually cued conditions. *Brain, 121*, 755–766.

Minshew, N. J., Luna, B., & Sweeney, J. A. (1999). Oculomotor evidence for neocortical systems but not cerebellar dysfunction in autism. *Neurology, 52*, 917–922.

Morris, M. E., Iansek, R., Matyas, T. A., & Summers, J. J. (1996). Stride length regulation in Parkinson's disease: Normalization strategies and underlying mechanisms. *Brain, 119*, 551–568.

Mortimer, J. A., & Webster, D. D. (1979). Evidence for a quantitative association between emg stretch responses and Parkinsonian rigidity. *Brain Research, 162*, 169–173.

Moschner, C., & Baloh, R. W. (1994). Age-related changes in visual tracking. *Journal of Gerontology, 49*, M235–M238.

Moschovakis, A. K., Scudder, C. A., & Highstein, S. M. (1996). The microscopic anatomy and physiology of the mammalian saccadic system. *Progress in Neurology, 50*, 133–245.

Mostofsky, S. H., Lasker, A. G., Cutting, L. E., Denckla, M. B., & Zee D. S. (2001). Oculomotor abnormalities in attention deficit hyperacitivty disorder: A preliminary study. *Neurology, 57*, 423–430.

Munoz, D. P., Armstrong, I. T., Hampton, K. A., & Moore, K. D. (2003). Altered control of visual fixation and saccadic eye movements in attention-deficit hyperactivity disorder. *Journal of Neurophysiology, 90*, 503–514.

Munoz, D. P., Broughton, J. R., Goldring, J. E., & Armstrong, I. T., (1998). Age-related performance of human subjects on saccadic eye movement tasks. *Experimental Brain Research, 121*, 391–400.

Munoz, D. P., Dorris, M. C., Paré, M., & Everling, S. (2000). On your mark, get set: Brainstem circuitry underlying saccadic initiation. *Canadian Journal of Physiology and Pharmacology, 78*, 934–944.

Munoz, D. P., & Everling, S. (2004). Look away: The anti-saccade task and the voluntary control of eye movement. *Nature Reviews Neuroscience, 5*, 218–228.

Munoz, D. P., & Fecteau, J.H. (2002). Vying for dominance: dynamic interactions control visual fixation and saccadic initiation in the superior colliculus. In J. Hyona, D. P. Munoz, W. Heide, & R. Radach (Eds.) *The brain's eye: Neurobiological and clinical aspects of oculomotor research*, Amsterdam: Elsevier, 3–19.

Munoz, D. P., & Schall, J. D. (2003). Concurrent, distributed control of saccade initiation in the frontal eye field and superior colliculus. In W. D. Hall, & M. K. Moschovakis (Eds.) *The oculomotor system: New approaches for studying sensorimotor integration*. Boca Raton, FL: CRC.

Narita, A. S., Shawkat, F. S., Lask, B., Taylor, D. S. I., & Harris, C. M. (1997). Eye movement abnormalities in a case of Tourette syndrome. *Developmental Medicine and Child Neurology, 39*, 270–273.

Nazir, T. A., & Jacobs, A. M. (1991). The effects of target discriminability and retinal eccentricity on saccade latencies – An analysis in terms of variable-criterion theory. *Psychological Research – Psychologische Forschung, 53*, 281–289.

Obayashi, R., Matsushima, E., Ando, H., Ando, K., & Kojima, T. (2003). Exploratory eye movements during the Benton Visual Retention Test: Characteristics of visual behavior in schizophrenia. *Psychiatry and Clinical Neurosciences, 57*, 409–415.

Olincy, A., Ross, R. G., Young, D. A., & Freedman, R. (1997). Age diminishes performance on an antisaccade eye movement task. N*eurobiology of Aging, 18*, 483–489.

Owen, A. M., James, M., Leigh, P. N., Summers, B. A., Quinn, N, & Marsden, C. D. (1993). Fronto-striatal cognitive deficits at different stages of Parkinson's disease. *Brain, 115*, 1727–1751.

Owen, A. M., Roberts, A. C., Hodges, J. R., Summers, B. A., Polkey, C. E., & Robbins, T. W. (1993). Contrasting mechanisms of impaired attentional set-shifting in patients with frontal-lobe damage or Parkinson's disease. *Brain, 116*, 1159–1175.

Paré, M., & Hanes, D.P. (2003). Controlled movement processing: superior colliculus activity associated with countermanding saccades. *Journal of Neuroscience*, 23: 6480–6489.

Paré, M. & Munoz, D. P. (1996). Saccadic reaction times in the monkey: advanced preparation of oculomotro programs is primarily responsible for express saccade occurrence. *Journal of Neurophysiology, 76*, 3666–3681.

Paré, M., & Wurtz, R. H. (2001). Progression in neuronal processing for saccadic eye movements from parietal cortex area LIP to superior colliculus. *Journal of Neurophysiology, 85*, 2545–2562.

Paus, T. (2005). Mapping brain maturation and cognitive development during adolescence. *Trends in Cognitive Sciences, 9*, 60–68.

Perry, V. H., & Cowey, A. (1985). The ganglion cell and cone distributions in the monkey's retina: Implications for central magnification factors. *Vision Research, 25*, 1795–1810.

Pierrot-Deseilligny, C., Israel, I., Berthoz, A., Rivaud, S., & Gaymard, B. (1993). Role of the different frontal-lobe areas in the control of the horizontal component of memory-guided saccades in man. *Experimental Brain Research, 95*, 166–171.

Pierrot-Deseilligny, C., Rivaud, S., Gaymard, B., & Agid, Y. (1991). Cortical control of reflexive visually-guided saccades. *Brain, 114*, 1473–1485.

Pierrot-Deseilligny, C., Muri, R. M., Ploner, C. J., Gaymard, B., Demeret, S., & Rivaud-Pechoux, S. (2003). Decisional role of the dorsolateral prefrontal cortex in ocular motor behavior. *Brain 126*, 1460-1473.

Ploner, C. J., Gaymard, B. M., Rivaud-Pechoux, S., & Pierrot-Deseilligny, C. (2005). The prefrontal substrate of reflexive saccade inhibition in humans. *Biological Psychiatry, 57*, 1159–1165.

Praamstra, P., & Plat, F. M. (2001). Failed suppressions of direct visuomotor activation in Parkinson's disease. *Journal of Cognitive Neuroscience, 13*, 31–43.

Pratt, J., Abrams, R. A., & Chasteen, A. L. (1997). Initiation and inhibition of saccadic eye movements in younger and older adults: An analysis of the gap effect. *Journal of Gerontology, 52*, 103–107.

Pratt, J., Abrams, R. A., & Chasteen, A. L. (1997). Initiation and inhibition of saccadic eye movements in younger and older adults: an analysis of the gap effect. *Journal of Gerontology, 52*, 103–107.

Ratcliff, R. (1978). A theory of memory retrieval, *Psychological Review, 85*, 59–108.

Reuter-Lorenz, P. A., Oonk, H., M., Barnes, L. L., & Hughes H. D. (1995). Effects of warning signals and fixation point offsets on the latencies of pro- versus antisaccades: Implications for an interpretation of the gap effect. *Experimental Brain Research, 103*, 287–293.

Robinson, D. L., & McClurkin, J. W. (1989). The visual superior colliculus and pulvinar. *Review of Oculomotor Research, 3*, 337.

Rothwell, J. D. Obeso, J. A., Traub, M. M. & Marsden, C. D. (1983). The behavior of the long latency stretch reflex in patients with Parkinson's disease. *Journal of Neurology Neurosurgery and Psychiatry, 46*, 35–44.

Schall, J. D. (1997). Visuomotor areas of the frontal lobe. *Cerebellum Cortex, 12*, 527–638.

Schall, J. D., & Thompson, K. G. (1999). Neural selection and control of visually guided eye movements. *Annual Review of Neuroscience, 22*, 241–259.

Schiller, P. H., Sandell, J. H., & Maunsell, J. H. R. (1987). The effect of frontal eye field and superior colliculus lesions on saccadic latencies in the rhesus-monkey. *Journal of Neurophysiology, 57*, 1033–1049.

Scudder, C. A., Kaneko, C. S, & Fuchs, A. F. (2002). The brain stem burst generator for saccadic eye movements: a modern synthesis. *Experimental Brain Research, 142*, 439–462.

Seidman, L. J., Valera, E. M., & Bush, G. (2004). Brain function and structure in adults with attention-deficit/hyperactivity disorder. *Psychiatric Clinics of North America, 27*, 323.

Sharpe, J. A., & Zackon, D. H. (1987). Senescent saccades: Effects oaging on their accuracy, latency and velocity. *Acta Otolaryngol, 104*, 422–428.

Shue, K. L., & Douglas, V. (1992). Attention deficit hyperactivity disorder and the frontal lobe syndrome. *Brain and Cognition, 20*, 104–124.

Singer, H. S. (1997). Neurobiology of Tourette syndrome. *Neurologic Clinics, 15*, 357.

Sparks, D. L. (2002). The brainstem control of saccadic eye movements. *Nature Reviews Neuroscience, 3*, 952–964.

Spooner, J. W., Sakala, S. M., & Baloh, R. W. (1980). Effect of aging on eye tracking *Archives of Neurology, 37*, 575–576.

Straube, A., Mennicken, J. B., Riedel, M., Eggert, T., & Muller, N (1997). Saccades in Gilles de la Tourette's syndrome.*Movement Disorders, 12*, 536–546.

Sweeney, J. A., Takarae, Y., Macmillan, C., Luna, B., & Minshew, N. J. (2004). Eye movements in neurodevelopmental disorders. *Current Opinion in Neurology, 17*, 37–42.

Tatton, W. G., & Lee, R. G. (1975). Evidence for abnormal long-loop reflexes in rigid Parkinsonian patients. *Brain Research, 100*, 671–676.

Tatton, W. G., Eastman, M. J., Bedingham, W., Verrier, M. C., & Bruce, I. C. (1984). Defective utilization of sensory input as the basis for bradykinesia, rigidity and decreased movement repertoire in Parkinson's disease: A hypothesis. *Canadian Journal of Neurological Sciences, 11*, 136–143.

Taylor, A. E., Saint-Cyr, J. A., & Lang, A. E. (1986). Frontal lobe dysfunction in Parkinson's disease. *Brain, 109*, 845–883.

Trappenberg, T. P., Dorris, M. C., Munoz, D. P., & Klein, R. M. (2001). A model of saccade initiation based on the competitive integration of exogenous and endogenous signals in the superior colliculus. *Journal of Cognitive Neuroscience, 15*, 256–271.

van der Geest, J. N., Kemner, C., Camfferman, G., Verbaten, M. N., & van Engeland, H. (2001). Eye movements, visual attention, and autism: A saccadic reaction time study using the gap and overlap paradigm. *Biological Psychiatry, 50*, 614–619.

Van Essen, D. C., Newsome, W. T., & Maunsell, J. H. R. (1984). The visual-field representation in striate cortex of the macaque monkey – asymmetries, anisotropies, and individual variability. *Vision Research, 24*, 429–448.

Vidailhet, M., Rivaud, S., Gouider-Khouja, N., Pillon, B., Bonnet, A. M., Gaymard, B et al. (1994). Eye movements in Parkinsonian syndromes. *Annals of Neurology, 35*, 420–426.

Warabi, T., Kase, M., & Kato, T. (1984). Effect of aging on the accuracy of viasually guided saccadic eye movements. *Annals of Neurology, 16*, 449–454.

Willis, W. G. and Weiler, M. D. (2005). Neural substrates of childhood attention-deficit/hyperactivity disorder: Electroencephalographic and magnetic resonance imaging evidence. *Developmental Neuropsychology, 27*, 135–182.

Wilson, S. J., Glue, P., Ball, D., & Nutt, D. J. (1993). Saccadic eye movement parameters in normal subjects. *Electroencephalography in Clinical Neurophysiology, 86*, 69–74.

Winograd-Gurvich, C. T., Georgiou-Karistianis, N., Evans, A., Millist, L., Bradshaw, J. L., Churchyard, A. et al. (2003). Hypometric primary saccades and increased variability in visually-guided saccades in Huntington's disease. *Neuropsychologia, 41*, 1683–1692.

Wurtz, R. H., & Goldberg, M. E. (1989). *The neurobiology of saccadic eye movements*. Amsterdam: Elsevier.

Wurtz, R. H., Sommer, M. A., Paré, M., & Ferraina, S. (2001). Signal transformations from cerebral cortex to superior colliculus for the generation of saccades. *Vision Research, 41*, 3399–3412.

Chapter 6

ANTI-SACCADE TASK PERFORMANCE IS DEPENDENT UPON BOLD ACTIVATION PRIOR TO STIMULUS PRESENTATION: AN FMRI STUDY IN HUMAN SUBJECTS

KRISTEN A. FORD, MATTHEW R. G. BROWN, AND STEFAN EVERLING

University of Western Ontario, Canada

Eye Movements: A Window on Mind and Brain
Edited by R. P. G. van Gompel, M. H. Fischer, W. S. Murray and R. L. Hill

Abstract

We used a blocked design experiment to localize functional brain areas as measured by fMRI. Subjects were instructed before stimulus onset either to look at the stimulus (pro-saccade) or to look away from the stimulus (anti-saccade). The cortical areas localized in the blocked design experiment comparing anti- and pro-saccades were then used to examine activation during widely spaced event-related trial presentations. Eye movements were recorded and BOLD signal activation during the event-related experiment was grouped into correct anti-saccades and errors (saccades towards the stimulus on anti-saccade trials). Correct anti-saccades were associated with significantly more activity in the right dorsolateral prefrontal cortex, the anterior cingulate cortex, and the supplementary eye fields compared with error anti-saccades, during the instruction period before stimulus appearance. Our findings suggest that rapidly presented trials in blocks elicit similar patterns of BOLD activation to widely spaced event-related trials, and confirm that activation of localized frontal cortical areas prior to stimulus presentation is associated with subjects' performance in the anti-saccade task.

The cognitive control of action requires the ability to suppress automatic responses that may not be suitable in a given context and instead generate voluntary behaviours. The prefrontal cortex has been proposed to play a role in this process allowing human and animals to exercise behavioural flexibility and pursue long-term goals (Duncan, 2001; Miller, 2000; Miller & Cohen, 2001).

This flexible control of movement has been investigated using the anti-saccade task (Hallett, 1978) in which correct task performance requires subjects to look away from a flashed visual stimulus and to generate a saccade to the mirror opposite location (Everling and Fischer, 1998; Munoz and Everling, 2004). Correct anti-saccade task performance requires both the suppression of an automatic pro-saccade towards the stimulus and the generation of the anti-saccade away from the stimulus to an empty location in the visual field. Subjects sometimes fail to suppress a reflexive saccade towards the stimulus when instructed to generate an anti-saccade. These errors are especially frequent in young children and patients with certain neurological or psychiatric disorders that involve the frontal cortex and/or the basal ganglia (Everling and Fischer, 1998; Munoz and Everling, 2004). Therefore, it has been proposed that these areas provide top-down suppression signals during the preparation to execute an anti-saccade.

Differences between pro- and anti-saccade trials have been investigated using event-related functional magnetic resonance imaging (fMRI) studies in humans. These studies have demonstrated that a number of frontal areas including the dorsolateral prefrontal cortex (DLPFC) (Connolly, Goodale, Menon, & Munoz, 2002; Curtis and D'Esposito 2003; Desouza, Menon, & Everling, 2003; Ford, Goltz, Brown, & Everling, 2005), supplementary eye fields (SEF) (Curtis and D'Esposito, 2003; Ford et al., 2005), and frontal eye fields (FEF) (Connolly et al., 2002; Desouza et al., 2003; Ford et al., 2005) exhibit differences between pro- and anti-saccade trials during the preparatory period, lending support to the hypothesis that these areas provide top-down suppression signals during the preparation to execute an anti-saccade.

However, to more fully explore the role of these brain regions as potential sources of top-down suppression signals during the preparation to execute an anti-saccade, a direct comparison between correct and error anti-saccade trials has been investigated. If in fact the suppression of saccade neurons in FEF and superior colliculus is necessary for correct anti-saccade task performance as has been suggested by single neuron recordings in the monkey (Everling & Munoz, 2000; Everling, Dorris, & Munoz, 1998) and these frontal areas play a role in providing top-down suppression signals as has been suggested by single neuron recordings in SEF (Amador, Schlag-Rey, & Schlag, 2003; Schlag-Rey, Amador, Sanchez, & Schlag, 1997) and other studies of frontal area function (Gaymard, Ploner, Rivaud, Vermersch, & Pierrot-Deseilligny, 1998; Guitton, Buchtel, & Douglas, 1985; Pierrot-Deseilligny et al., 2003), then it can be hypothesized that frontal brain areas should show increased activation on trials in which the automatic pro-saccade is successfully suppressed (correct anti-saccade trial) as compared to trials in which the automatic pro-saccade is not suppressed (error anti-saccade trial).

Several recent fMRI studies have examined differences in preparatory activation between correct and error anti-saccades. One study by Curtis and D'Esposito (2003) utilized a region of interest (ROI) analysis. However, previous research indicates that a large network of brain areas are differentially modulated during the execution of pro- and anti-saccades, and this limited ROI analysis may not have captured all potentially modulated areas. A recent study from our group examined differences between pro-saccades, correct anti-saccades and error anti-saccades during the preparatory and saccade periods utilizing an event-related trial design and whole brain general linear model (GLM) analysis (Ford, Goltz, Brown, & Everling, 2005). Results from this study indicated that a number of areas exhibited increased activation during the preparatory period for correct anti-saccades as compared with errors, including right DLPFC, anterior cingulate cortex (ACC) and pre-SEF. In addition, one area in the superior frontal sulcus (SFS) showed the opposite pattern of activation, with increased activation for errors compared to correct anti-saccades. However, this analysis left several questions unanswered. First, we wished to directly compare cortical activation patterns elicited by rapid blocked trial presentations to those elicited by slow event-related trial presentations. A number of recent fMRI studies have utilized long preparatory periods in an effort to dissociate temporal processing within individual trials (Connolly et al., 2002; Curtis and D'Esposito, 2003; Desouza et al., 2003; Ford et al., 2005). By combining our new blocked experimental data with our event-related data previously analyzed, we investigated the validity of attributing differences exhibited in blocked design experiments to differences demonstrated in slow event-related design experiments. Second, by localizing functional areas using our new blocked design experiment, we were able to perform a new analysis of our event-related data. This new analysis makes no assumptions about the particular temporal waveform shape of the BOLD signal response, therefore allowing for greater sensitivity in our analysis and allowing us to further examine this data for potentially significant differences.

To address these questions subjects performed blocks of pro- and anti-saccade trials, which were analyzed using a whole brain analysis. We then compared these activations to those we previously localized using our slow event-related design, and examined the brain areas shown to be differently modulated using event-related trials with long preparatory periods.

1. Methods

1.1. Subjects

Ten subjects (Seven male, three female, mean age of 28 years) provided informed consent and participated in this study. All were right handed, could see clearly at arms length without glasses and reported no history of head injury, epilepsy, neurological or psychiatric disorder. The experiments were approved by the University of Western Ontario Review

Board for Health Sciences Research Involving Human Subjects and are in accordance with the 1964 Declaration of Helsinki.

1.2. Experimental task

1.2.1. Blocked design experiment

Blocks of pro- and anti-saccades were alternated with blocks of fixation, with a fixation control block at the beginning of each scan (Figure 1a). A change in the colour of the white fixation cross-conveyed the instruction to generate a pro- or anti-saccade upon stimulus presentation. The colours (red, green, or blue) used as the pro- or anti-saccade instruction were kept constant across functional scans within subjects and were randomized across subjects. Each pro- or anti-saccade block (see Figure 1a) began with the fixation cross changing color from white to either a red, green or blue. After 1 s, the central fixation cross-disappeared, and 0.2 s of darkness (gap period) were introduced before a peripheral stimulus (white circle, 2°) was flashed for 0.5 s, either 10° to the left or 10° to the right of the initial fixation cross. Subjects were required to maintain fixation on the central cross during the fixation blocks, the instruction period and the gap period. Subjects were instructed to look towards the stimulus on pro-saccade trials and to look away from the stimulus to its mirror position in the opposite hemifield on anti-saccade trials. The white central fixation cross then reappeared for 0.5 s before the start of the next trial, and subjects were instructed to return gaze to center and fixate the white cross after executing each saccade. Eight saccades (either pro- or anti-saccades) were executed within each 20 s block. Each block was repeated 8 times (four blocks each of pro- and anti-saccades) in each functional scan. Each subject performed three- or four-blocked design functional scans. Each scan was 5.33 min. long.

1.2.2. Event-related Experiment

Figure 1b shows the timing of event-related trials. For a detailed description of the event-related trial design refer to our previously published work which also examined this functional data (Ford et al., 2005).

1.2.3. Visual display and eye-tracking

Visual stimuli were generated using SuperLab Pro 2.0 software (Cedrus, San Padro, CA) and presented using fiber optics housed in dual stalks that are placed in front of a subject's eyes, allowing binocular presentation of visual stimuli (SMI iView-fMRI Eye™ tracking (SensoMotoric Instruments, Needham/Boston, MA) and Silent Vision™ SV-40 21 (Avotec, Stuart, FL)). Eye-tracking was identical for event-related and blocked experiments. For further details of eye-tracking refer to previously published work by Ford and colleagues (2005).

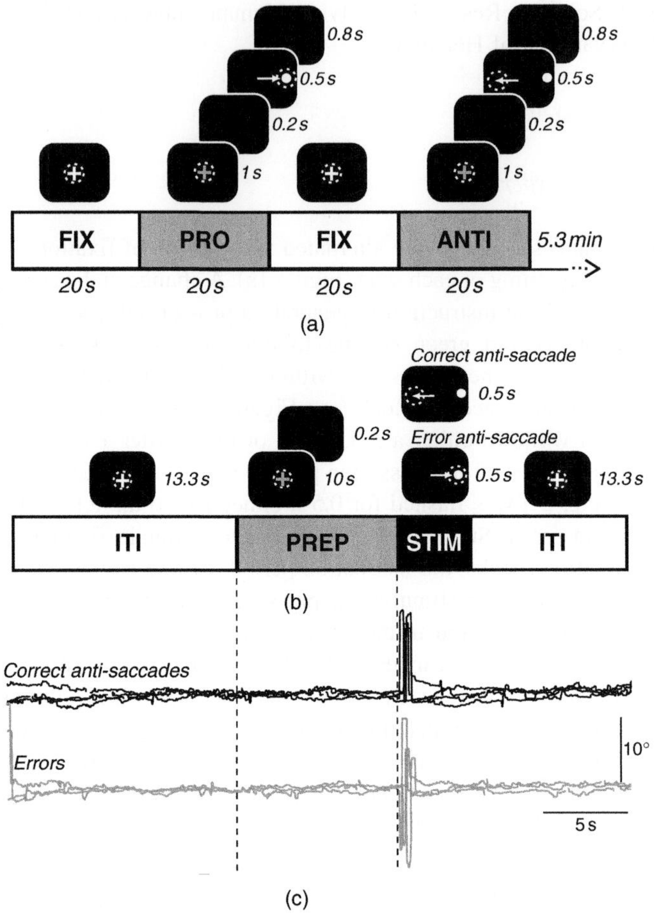

Figure 1. (a) Visual presentation sequence of blocked design experiment. Each experimental run consisted of 8 saccade blocks (4 pro-saccade, 4 anti-saccade) separated by 20 s blocks of fixation. Eight saccades were executed within each 20-s saccade block. *FIX*, fixation block; *PRO*, pro-saccade block; *ANTI*, anti-saccade block; (b) Visual presentation sequence of the event-related experiment for anti-saccade trials. Each experimental run consisted of 8 anti-saccade trials and 4 pro-saccade trials in a pseudo-random order. *ITI*, inter-trial interval; *PREP*, 10s preparatory period; *STIM*, stimulus presentation; (c) Horizontal eye-position traces from a single subject's scanning session during functional data acquisition, showing correct anti-saccade eye position traces (*black*) and error anti-saccades (*grey*). Note that the subject maintained fixation during the inter-trial period and preparatory period and anti-saccade errors were corrected.

1.2.4. Imaging and data analysis

All imaging data were acquired on a 4-Tesla whole body MRI system (Varian, Palo Alto, CA; Siemens, Erlangen, Germany). Imaging parameters were identical for event-related and blocked experiments and have been previously described (Ford et al., 2005). Analyses were conducted using BrainVoyager 2000 version 4.8 (Brain Innovation, Maastricht, The Netherlands). Functional runs were scaled to the Talairach standard (Talairach and Tournoux, 1988) and superimposed on anatomical scans for each subject. All functional images underwent motion correction, temporal filtering (linear trend removal, high-pass filter in Fourier domain with cut-off of 6 cycles/run for the event-related experiments and 4 cycles/run for the blocked design experiments, Gaussian filter in time domain with full width at half-maximum (FWHM) of 2.8 s), and spatial filtering (Gaussian filter in spatial domain with FWHM of 4.00 mm). For the blocked design experiments all eye movements were included in the analysis. For the event-related experiment, each trial was analyzed offline using the time-locked eye position traces recorded in the MR scanner, as previously described (Ford et al., 2005).

1.2.5. Statistical analysis of blocked experiment

Functional data from the blocked design experiment were statistically analyzed using the GLM framework with separate boxcar predictor functions for blocks of pro-saccades and anti-saccades. We then convolved these two predictors with the hemodynamic response function as implemented by the BrainVoyager software package. This convolution involves modelling the hemodynamic response function as a gamma curve after Boynton, Engel, Glover, & Heeger (1996) using the parameters tau = 1.25 and delta = 2.5. Data from left and right saccadic eye movements were combined. The GLM analysis resulted in statistical contrast activation maps for the comparison of pro-saccade blocks and anti-saccade blocks (Figure 2) ($p < 0.01$, corrected for multiple comparisons; and a cluster threshold size of >50 voxels). The functionally mapped brain regions which were shown to exhibit statistically significant differences in BOLD signal intensity between blocks of pro- and anti-saccades were then used to directly examine the BOLD signal differences between the preparatory periods of correct anti-saccades and error anti-saccades from each subject's event-related experiments.

1.2.6. New statistical analysis of event-related experiment

Mean BOLD signal time courses for each region identified in the blocked design experiment were then computed for each subject (Figure 3a, C). Raw data from each trial were transformed into % signal change values ((signal – baseline) $*$ 100/baseline), where baseline was defined as the average signal over the first 2 s of the trial on a trial-by-trial basis. Ten "within subject" mean activation curves were computed for each of correct anti-saccade and error anti-saccade tasks. Then, "between subjects" mean curves were computed by taking the mean across all 10 subjects for each of the two trial types. To examine the preparatory period we shifted our time series for statistical analysis 2 s forward in time and excluded the first data point to accommodate the hemodynamic response function. Our time series for statistical comparison of the 10 s preparatory period was

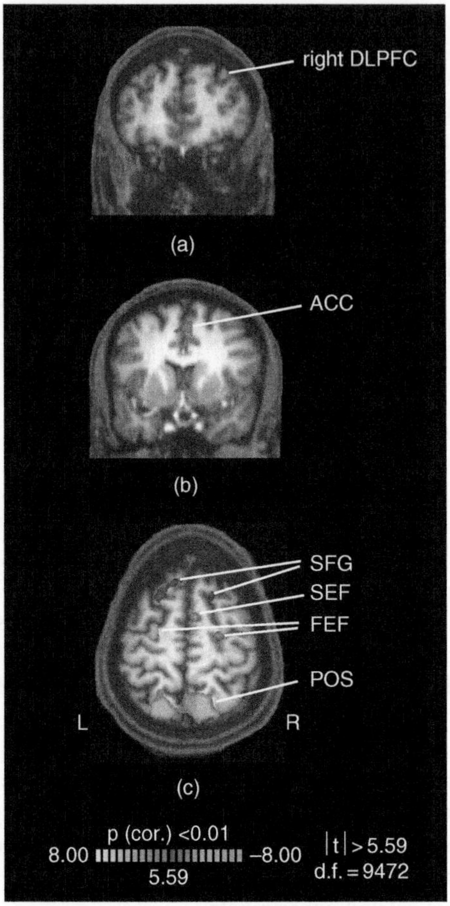

Figure 2. Group statistical activation maps generated from the general linear model (GLM) contrast comparing block of anti-saccades to blocks of pro-saccades from 10 subjects. *Red* and *yellow* regions exhibited significantly more BOLD activation for anti-saccades than for pro-saccades. *Blue* and *green* regions exhibited significantly more BOLD activation for pro-saccades than for anti-saccades. Bonferroni-corrected $p < 0.01$. L and R denote left and right. Maps obey neurological conventions. (a) *DLPFC*, dorsolateral prefrontal cortex; (b) *ACC*, anterior cingulate cortex; (c) *SFG*, superior frontal gyrus; *SEF*, supplementary eye fields; *FEF*, frontal eye fields; *POS*, parieto-occipital sulcus. (*See Color Plate 1.*)

therefore from 3 to 12 s after the onset of the instruction cue for a total of 10 s. The mean of these 10 data points was calculated separately for each subject, and then across subjects to formulate a group mean (Figure 3b, d, Figure 4). The Student's paired *t*-test was used to assess the statistical significance of differences during the preparatory period between correct anti-saccade and error anti-saccade BOLD signal intensities across subjects.

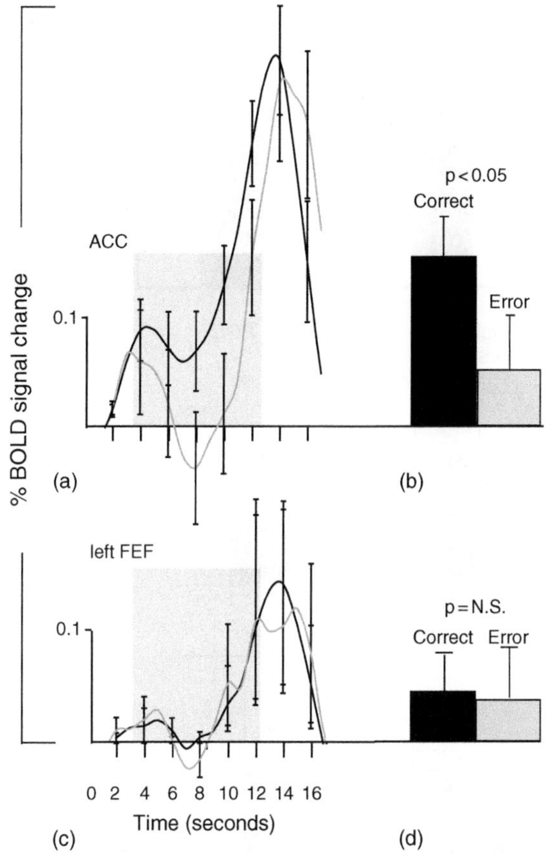

Figure 3. Preparatory period comparison between correct anti-saccades and error anti-saccades for event-related experiment. (a) BOLD signal time courses (*black* traces correct anti-saccade, *grey* traces error anti-saccade) averaged across 10 subjects for area ACC. Percent BOLD signal is plotted by time, where time point 1 is the onset of the instruction cue. The grey shaded area shows the data points included in the mean preparatory period comparison. Error bars indicate standard error of the mean (SE) across subjects at each time point; (b) Mean of preparatory period activation (*black* bars correct anti-saccade, *grey* bars error anti-saccade) averaged across 10 subjects for area ACC. Error bars represent SE across subjects; (c) Same as (A) but for left FEF; (d) Same as (B) but for left FEF. *ACC*, anterior cingulate cortex; *FEF*, frontal eye fields.

2. Results

2.1. Behaviour

During the performance of the event-related experiment all subjects generated errors on a number of the anti-saccade trials, that is they initially looked towards the peripheral stimulus before they looked away from it. Figure 1c shows horizontal eye position

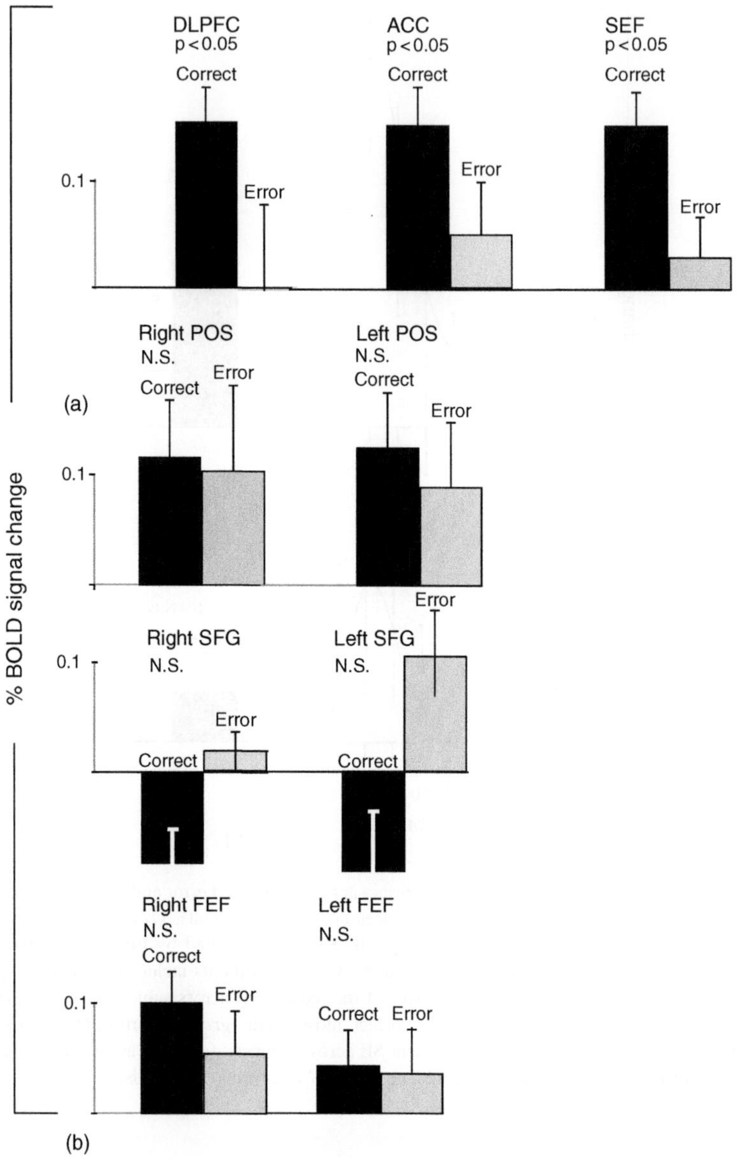

Figure 4. Mean of preparatory period activation (*black* bars correct anti-saccade, *grey* bars error anti-saccade) averaged across 10 subjects for all areas localized in the blocked GLM comparison. Error bars represent SE across subjects. (a) Functionally localized areas which show significantly different BOLD signal activation during the preparatory period between correct anti-saccades and error anti-saccades; (b) Functionally localized areas which failed to show significantly different BOLD signal activation during the preparatory period between correct anti-saccades and error anti-saccades. *DLPFC*, dorsolateral prefrontal cortex; *ACC*, anterior cingulate cortex; *SEF*, supplementary eye fields; *POS*, parieto-occipital sulcus; *SFG*, superior frontal gyrus; *FEF*, frontal eye fields.

traces from one subject performing anti-saccades in the MR scanner during the event-related experiment. The error rates of individual subjects ranged from 4 to 35% (16% ± 3.4% [mean ± standard error of the mean], median 11%). Subjects did not show significant differences in the percentage of errors between leftward (48.1%) and rightward (51.9%) stimulus presentations (paired t-test, $p = n.s$). Black traces show correct trials in which the subject looked away from a stimulus whereas grey traces show error trials in which the subject initially looked towards the stimulus before generating an anti-saccade.

As previously described (Ford et al., 2005) the mean SRT differed significantly between correct anti-saccades (355 ms) and error anti-saccades (326 ms). The SRT of anti-saccade errors indicates they were not express saccades. Correct anti-saccades had significantly longer mean SRTs than anti-saccade errors (paired t-test, $p < 0.05$).

2.2. *Functional imaging data*

2.2.1. *Blocked design experiment*

BOLD signal intensities were compared for anti-saccade blocks and pro-saccade blocks utilizing the GLM across data collected from 10 subjects ($p < 0.01$, corrected for multiple comparisons; see Section "Methods" for details). As shown in the group statistical activation map in Figure 2, significantly greater activation for anti-saccade blocks than pro-saccade blocks was exhibited bilaterally in areas corresponding to the FEF, the SEF, the ACC, the right DLPFC and bilaterally in an area in the parieto-occipital sulcus (POS). In contrast, pro-saccade blocks elicited greater BOLD signal activation than anti-saccade blocks bilaterally in an area in the superior frontal gyrus (SFG).

2.2.2. *Event-related experiment*

The regions functionally mapped in the blocked experiment demonstrated differential BOLD signal activations between blocks of pro- and anti-saccades. These functionally localized regions of cortex from the blocked experiment were then further examined by analyzing BOLD activations from correct and error anti-saccades in the event-related experiment.

Analysis of event-related trials during the mean of the 10s preparatory period before stimulus presentation and saccade execution showed no areas with significantly greater activation for error anti-saccades compared with correct anti-saccades; however, one area in the SFG located in Brodmann's area (BA) 9 did approach significance. There were, however, three areas localized in the blocked experiment which showed significantly greater activation for correct anti-saccades compared with errors. Figure 3a shows the group average BOLD signal time courses for correct anti-saccades (black line) and error anti-saccade (grey lines) for ACC which showed significantly higher BOLD activation for correct anti-saccades than error anti-saccades during the instruction period (paired t-test, <0.05). The grey shaded area shows the data points included in the mean (see Figure 3B).

As shown in Figure 4a, significantly higher BOLD signal activation was also found for the correct anti-saccade instruction period compared with the error anti-saccade instruction period in the right DLPFC and the SEF (paired t-test, $p < 0.05$). In contrast, Figure 3(c, d) shows the group average BOLD signal time courses for left FEF which failed to show significant differences between correct anti-saccades and error anti-saccades during the instruction period. All other areas localized in the block experiment failed to show any significant differences in mean preparatory BOLD activation between correct anti-saccades and errors (see Figure 4b).

3. Discussion

The present study had two main goals. First, to compare BOLD signal activations elicited by rapid blocked trial presentations of pro- and anti-saccades to those elicited by slow event-related trial presentations. It is important to recognize the potential differences between rapid trial presentations frequently used in single-neuron recording experiments conducted with monkeys and the event-related design used here to examine preparatory processes within a given trial. An event-related design utilizing a long preparatory period (in this case 10 s) and a 12 s inter-trial interval may not be directly comparable to electrophysiological experiments in which a trial may take place in the order of 1–2 s. It is possible that the preparation to make an eye movement that takes place in the order of 200 ms (Everling and Munoz 2000) is not reflective of the same process we have measured in this paradigm with our long preparatory period. Here, by demonstrating that we can localize similar cortical regions using rapid trials presentations in our blocked experiment to those localized using our event-related design (see Ford et al., 2005), we have further supported this event-related technique to dissociate processes taking place within a trial.

The second goal was to compare directly the prestimulus activations between correct anti-saccades and errors in cortical areas localized in a rapidly presented blocked experimental paradigm, utilizing a new analysis that makes no assumptions about the particular temporal waveform shape of the BOLD signal response during each event-related trial. Our previous analysis of this event-related data utilized multiple GLM predictor curves to model the BOLD signal response during preparatory and saccade periods (Ford et al., 2005). Although fruitful, we had hoped that by conducting this new analysis and examining the BOLD signal activation in a variety of areas independently of a model that assumes a hypothesized waveform shape, we could increase the sensitivity of our analysis. We have shown in this study that a number of cortical areas that exhibit differences in BOLD activation levels between blocks of pro- and anti-saccades demonstrate differences during the preparatory period of event-related trials between correct anti-saccades and errors. The new blocked design experiment yielded similar results to those found using only the event-related data. Our previous analysis found increased preparatory period activation for correct anti-saccades vs errors in the right DLPFC, ACC and pre-SEF. Here, we supported these results demonstrating that right DLPFC, ACC and a similar area in SEF (see Figures 3 and 4) showed the same pattern of BOLD signal activation even when localized independently using a new, rapidly presented blocked experimental paradigm. However, we failed to find significant differences in additional areas.

One novel finding of particular interest that has not yet been discussed is an area detected bilaterally in the SFG which showed a unique pattern of activation, with higher activation for blocks of pro-saccades as compared with blocks of anti-saccades. This area was both functionally and anatomically distinct from the cortical area we localized in the prefrontal cortex we refer to as DLPFC. Our previous analysis examining only data from slow event-related trials also localized a bilateral area in the SFS which showed an increase in the BOLD signal preceding anti-saccade errors compared with correct anti-saccades. The current analysis indicated that indeed this area did show a similar pattern of activation (higher activation for errors than correct anti-saccades during the preparatory period) but this difference was not found to be significant, although this comparison did approach significance. This activation appears to be located in BA 9 based on Petrides and Pandya (2001), and based on the combined, robust findings of our current and previous analysis deserves further examination. Previous research combining magnetoencephalography and positron emission tomography has suggested that BA 9 may be the source of the human motor readiness potential (Pedersen et al., 1998). It may be the case that our finding of increased activation in BA 9 on pro-saccade trials reflects the increased motor preparation involved in the performance of voluntary visually guided saccades, as compared with anti-saccade trials in which the automatic pro-saccade must be inhibited. This increase in motor preparation may also explain the trend towards an increase in activation in this area on error anti-saccade trials, in which the automatic pro-saccade to the visual stimulus was not inhibited.

Overall, our new comparison between blocks of pro- and anti-saccades found differences in areas corresponding to the FEF, SEF, ACC, DLPFC and POS and SFG (see Figure 2 and Table 1). The comparison between event-related correct and error

Table 1

Peak Activations for Comparisons

Region	*x*	*y*	*z*	Total volume of area mm^3
Anti-saccade blocks > Pro-saccade blocks				
left frontal eye field (FEF)	−23	−10	55	220
right frontal eye field (FEF)	26	−7	52	1026
right dorsolateral prefrontal cortex (DLPFC)	31	43	38	187
supplementary eye field (SEF)	7	4	52	399
anterior cingulate cortex (ACC)	7	6	44	373
left parieto-occipital sulcus (POS)	−9	−67	49	7378
right parieto-occipital sulcus (POS)	24	−62	51	6472
Pro-saccade blocks > Anti-saccade blocks				
left superior frontal gyrus (SFG)	−20	19	50	231
Right superior frontal gyrus (SFG)	14	25	46	253

anti-saccades during the preparatory period revealed significant differences in only three areas, the right DLPFC, ACC and SEF (see Figures 3 and 4) but not in the FEF, POS, or SFG. Although our new analysis did not detect differences in cortical areas beyond those demonstrated to show differences in our previous event-related analysis (Ford et al., 2005), we have succeeded in providing a valuable comparison between a rapidly presented blocked trial design and our slow event-related design utilizing long preparatory periods.

Our findings indicate that we can localize similar cortical regions using rapid trials presentations in a blocked design experiment to those localized using a event-related design with a long preparatory period and long inter-trial intervals (see Ford et al., 2005). In doing so, we have further supported our findings indicating a large network of frontal and posterior areas is modulated during the performance of anti-saccade compared with pro-saccade trials; however, it is the activation of more localized frontal cortical areas prior to stimulus presentation, which is associated with subjects' performance in the anti-saccade task.

References

Amador, N., Schlag-Rey, M., & Schlag, J. (2003). Primate antisaccade II: Supplementary eye field neuronal activity predicts correct performance. *Journal of Neurophysiology, 91*, 1672–1689.

Boynton, G. M., Engel, S. A., Glover, G. H., & Heeger, D. J. (1996). Linear systems analysis of functional magnetic resonance imaging in human V1. *Journal of Neuroscience, 16*, 4207–4221.

Connolly, J. D., Goodale, M. A., Menon, R. S., & Munoz, D. P. (2002). Human fMRI evidence for the neural correlates of preparatory set. *Nature Neuroscience, 5*, 1345–1352.

Curtis, C. E., & D'Esposito, M. (2003). Success and failure suppressing reflexive behavior. *Journal of Cognitive Neuroscience, 15*, 409–418.

Desouza, J. F., Menon, R. S., & Everling, S. (2003). Preparatory set associated with pro-saccades and anti-saccades in humans investigated with event-related FMRI. *Journal of Neurophysiology, 89*, 1016–1023.

Duncan, J. (2001). An adaptive coding model of neural function in prefrontal cortex. *Nature Neuroscience Review, 2*, 820–829.

Everling, S., Dorris, M. C., & Munoz, D. P. (1998). Reflex suppression in the anti-saccade task is dependent on prestimulus neural processes. *Journal of Neurophysiology, 80*, 1584–1589.

Everling, S. & Fischer, B. (1998). The antisaccade: A review of basic research and clinical studies. *Neuropsychologia, 36*, 885–899.

Everling, S. & Munoz, D. P. (2000). Neuronal correlates for preparatory set associated with pro-saccades and anti-saccades in the primate frontal eye field. *Journal of Neuroscience, 20*, 387–400.

Ford, K. A., Goltz, H. C., Brown, M. R., & Everling, S. (2005). Neural Processes associated with antisaccade task performance investigated with event-related fMRI. *Journal of Neurophysiology, 94*, 429–440.

Gaymard, B., Ploner, C. J., Rivaud, S., Vermersch, A. I., & Pierrot-Deseilligny, C. (1998). Cortical control of saccades. *Experimental Brain Research, 123*, 159–163.

Guitton, D., Buchtel, H. A., & Douglas, R. M. (1985). Frontal lobe lesions in man cause difficulties in suppressing reflexive glances and in generating goal-directed saccades. *Experimental Brain Research, 58*, 455–472.

Hallett, P. E. (1978). Primary and secondary saccades to goals defined by instructions. *Vision Research, 18*, 1279–1296.

Miller, E. K. (2000). The prefrontal cortex: No simple matter. *Neuroimage, 11*, 447–450.

Miller, E. K. & Cohen, J. D. (2001). An integrative theory of prefrontal cortex function. *Annual Review of Neuroscience, 24*, 167–202.

Munoz, D. P. & Everling, S. (2004). Look away: the anti-saccade task and the voluntary control of eye movement. *Nature Reviews Neuroscience, 5*, 218–228.

Pedersen, J. R., Johannsen, P., Bak, C. K., Kofoed, B., Saermark, K., & Gjedde, A. (1998). Origin of human motor readiness field linked to left middle frontal gyrus by MEG and PET. *Neuroimage, 8*, 214–220.

Petrides, M. & Pandya, D. N. (2001). Comparative cytoarchitectonic analysis of the human and the macaque ventrolateral prefrontal cortex and corticocortical connection patterns in the monkey. *European Journal of Neuroscience, 16*, 291–310.

Pierrot-Deseilligny, C., Muri, R. M., Ploner, C. J., Gaymard, B., Demeret, S., & Rivaud-Pechoux, S. (2003). Decisional role of the dorsolateral prefrontal cortex in ocular motor behaviour. *Brain, 126*, 1460–1473.

Schlag-Rey, M., Amador, N., Sanchez, H., & Schlag, J. (1997). Antisaccade performance predicted by neuronal activity in the supplementary eye field. *Nature, 390*, 398–401.

Talairach, J. & Tournoux, P. (1988). *Co-Planar stereotaxic atlas of the human brain: 3-dimensional proportional system: An approach to cerebral imaging*. Stuttgart: Thieme Medical Publishers.

Chapter 7

COMMUTATIVE EYE ROTATIONS IN CONGENITAL NYSTAGMUS

LO J. BOUR, JOHAN N. VAN DER MEER AND ANKE M. VAN MOURIK

University of Amsterdam, The Netherlands

Eye Movements: A Window on Mind and Brain
Edited by R. P. G. van Gompel, M. H. Fischer, W. S. Murray and R. L. Hill

Abstract

Listing's law can be achieved in two ways. First, neural activation of extraocular muscles could be encoded by the central nervous system. However, this requires complex calculations of the central nervous system. Secondly, Listing's law could be implemented by the anatomical structure of the ocular motor plant and then neural commands have to be encoded only in two dimensions.

Simulations with a dynamical model show that commutative eye movements can be generated solely by the anatomical structure of the eye globe. A necessary condition then is that during eye rotations to a specific location the extraocular muscles are kept in the same plane. In patients with congenital nystagmus (CN), rotations of the eye do not obey Listing's law. Simulations with the dynamical model demonstrate that this could be explained an aberrant anatomical structure, resulting in cross-talk between the horizontal, vertical and torsional planes.

1. Listing's law

Eye rotation in three dimensions (3D) should obey Listing's law, that is, all rotation axes describing the orientation of the eye lie in a single plane, called Listing's plane (Donders, 1876; Von Helmholtz, 1924). In this manner commutative eye rotations are ensured, which means that the sequence of consecutive rotations does not influence the final eye position and there will be no build-up of torsion. Mathematically, eye movements then are described with rotation vectors. Rotation vectors define eye orientation as a single rotation, with an angle φ around an axis $\mathbf{n}$, using the 'right-hand rule', from a reference orientation to the current orientation. The axis and angle are used to define the rotation vector $\mathbf{r}$ as:

$$\mathbf{r} = \tan(\varphi/2)\mathbf{n} \tag{1}$$

The components r_x, r_y and r_z indicate clockwise, vertical and horizontal rotation, respectively. The reference orientation is chosen as looking straight ahead with a zero torsional component: $\mathbf{r}_{\text{ref}} = (0, 0, 0)$.

In healthy subjects previous experimental findings of smooth pursuit and saccades show that under normal conditions within an accuracy of about 1–2° rotation axes describing the orientation of the eye are restricted to one single plane (Ferman, Collewijn, & Van den Berg, 1987c; Haslwanter et al., 1995; Straumann, Zee, Solomon & Kramer, 1996; Tweed and Vilis, 1987), that is Listing's plane. Exceptions to this rule, that is violations of Listing's law, occur for vestibularly generated eye movements (Tweed et al., 1999) and under special or pathological conditions (Nakayama and Balliet, 1977; Raphan, 1998). Furthermore, with convergence the primary position of both eyes generally changes. This implies that Listing's plane rotates with respect to the fronto-parallel plane (Bruno and Van den Berg, 1997; Mok, 1992; Van Rijn & Van den Berg, 1993).

There are two possibilities for the ocular motor system to obey Listing's law. First, neural activation of extraocular muscles could be specifically encoded by the central nervous system such that it compensates for non-commutativity of ocular rotations. A model has been developed (Tweed & Vilis, 1987, 1990; Tweed, Cadera, & Vilis, 1990; Tweed et al., 1999) that uses a non-commutative, rotational operator (quaternion), to generate innervation of the extraocular muscles. Since in the latter model complex neuronal computations are required, Schnabolk and Raphan (1994) developed an alternative model based on muscle torque. In their model, neural commands do not necessarily have to be commutative to control eye movements. The authors argue that neural innervation encodes muscle torque, which is represented by a vector and, therefore, commutes. It is assumed that the direction of pull of the muscles does not depend on eye position (corresponding with inter-muscular coupling only). It is also assumed that muscle force of a muscle pair depends linearly on innervation and is not influenced by muscle length (Robinson, 1975).

Under these assumptions the model holds for relatively small eye rotations ($<15°$). In addition, although the model generates realistic saccadic eye movements in 3D, it also generates post-saccadic drift. To avoid this pulse-step mismatch, Raphan (1997) has incorporated musculo-orbital coupling in the torque model. Muscle pulleys, which have previously been demonstrated histologically (Clark, Miller, & Demer, 1997, 1998, 2000), should account for this musculo-orbital coupling.

2. Muscle pulleys

Previous studies with CT and MRI scans have demonstrated that during eye rotations the bellies of the horizontal extraocular muscles (between insertion and origin) approximately stay in place with secondary positions of 30° in the vertical plane (Demer, Miller, Poukens, Vinters & Glasgow, 1995; Miller, 1989; Simonsz, Harting, De Waal, & Verbeeten, 1985). The same holds for the vertical rectus muscle bellies with secondary positions in the horizontal plane. Thus, the muscle bellies are kept in place by some structures in the surrounding tissue, which implies that the effective pulling direction of the rectus muscle with respect to the globe remains the same also for eye rotations larger than 15°.

Demer et al. (1995), Demer, Miller, & Poukens (1996) and Clark et al. (1997, 1998, 2000) have argued that previously identified anatomical structures, called muscle pulleys (Simonsz, 2003), act like sleeves around the extraocular muscles and serve as their functional origins. The position of the pulleys with respect to the eye globe determines their effect on the direction of pull of the muscle. In the most extreme situation, this direction of pull does not change at all regardless of eye position (Figure 1). It is suggested that the surrounding tissue contains elastin, collagen and smooth muscle, especially in the area where the pulleys should stabilize the muscle bellies. The smooth muscle tissue receives rich innervation. The neurotransmitters involved suggest both excitatory and inhibitory control, which indicates that the pulleys actively influence the paths of the extraocular muscles. The location of these pulleys and their displacement also were determined by high-resolution MRI. The results suggest that the pulleys are musculo-orbital tissue connections. On the other hand, the paths of the extraocular muscles posterior to the pulleys undergo small displacements during changes of gaze. These displacements are more consistent with inter-muscular mechanical coupling. Histological research results are not conclusive in this matter. There are arguments both for musculo-orbital and inter-muscular coupling.

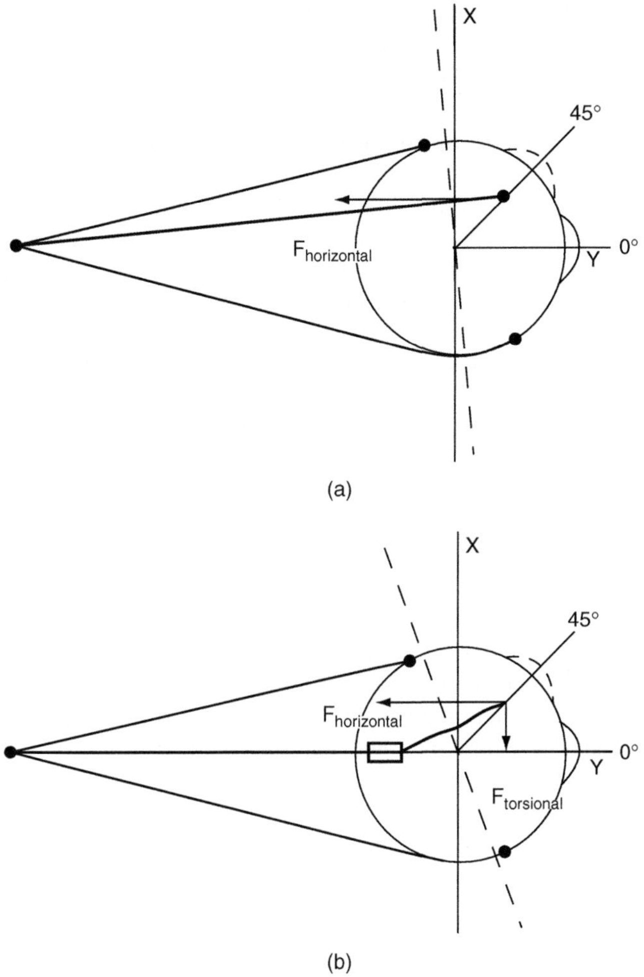

Figure 1. (a) Schematic diagram depicting the muscle forces acting on the eye globe without pulleys. A minor redirection of the horizontal muscle force in upgaze is neglected, so horizontal muscle force is independent of eye orientation. (b) Schematic diagram depicting the muscle forces acting on the eye globe with pulleys. Now the horizontal muscle force is redirected, such that it depends on eye orientation. The horizontal muscles now pull horizontally as well as torsionally.

Returning to the issue of commutativity, it has been suggested that rectus muscle pulleys ensure commutative rotations of the eye globe (Clark, 2000; Raphan, 1998). Hence, with rectus surgery, the existence of muscle pulleys should be taken into account (Demer, 1996) and their anatomical structure is of importance. However, in understanding fundamental principles underlying eye movement, the effect of the pulleys is of primary importance, rather than their anatomical form. Quaia & Optican (1998, 2003) have demonstrated with a dynamical model of the ocular motor plant that a correct position of orbital pulleys is in fact sufficient to obtain a perfect pulse-step match for any saccade in 3D.

3. Commutative eye movements and ocular motor instabilities

In patients with ocular motor instabilities and/or misalignments, including patients with strabismus or CN, 3D recordings demonstrate that rotation axes describing the orientation of the eye orientation typically are not restricted to a single horizontal plane (Dell'Osso and Daroff, 1975), but also have components in the vertical and torsional planes (Apkarian, Bour, van der Steen, & Faber, 1996 and Apkarian, Bour, Van der Steen, & Collewijn, 1999; Averbuch-Heller, Dell'Osso, Leigh, Jacobs, & Stahl, 2002; Korff et al., 2003; Ukwade, Bedell, & White, 2002). This may imply that in CN patients eye movements do not obey Listing's law. The question arises whether these two extra components of CN have their origin in the central nervous system or whether these components are the result of the anatomical structure of the eye-globe with its surrounding tissue (Averbuch-Heller et al., 2002). In the latter case, it is of interest whether aberrant anatomical structures of the eye-globe play a role in the generation of the three components of CN and/or the occurrence of misalignments.

A way to investigate this issue is to measure in 3D the amplitude and direction of particularly the fast phase of CN during various viewing conditions, including different gaze directions, binocular and monocular viewing. Eye-movement recordings are then compared with computer simulations of the dynamical model (Fetter, Haslwanter, Misslisch, & Tweed, 1997; Raphan, 1998; Schnabolk and Raphan, 1994), which predicts that dynamic variations in the amplitude and direction of CN are related to gaze in a systematic way.

By means of simulation of eye movements, the current study shows that fast components in the vertical and/or torsional plane may be explained by a displacement of musculo-orbital coupling or inter-muscular mechanical coupling with respect to the optimal position (no pulse-step mismatch). It is concluded that the observed torsional/vertical components of nystagmus in CN can be considered as the result of deviant plant characteristics only in the horizontal plane and their origin not necessarily has to be found in the central nervous system. In conclusion, non-commutative eye movements in CN patients with ocular motor abnormalities could be attributed to aberrant anatomical structures including a displacement of musculo-orbital and/or intermuscular coupling.

4. Methods

4.1. Subject

In the horizontal planes the patient displayed classic congenital nystagmus (Abadi and Dickinson, 1986; Apkarian et al., 1996; Dell'Osso and Daroff, 1975) and interocular amplitude disconjugacy. A variable esotropia and hypermetropia was present. The left eye was generally the preferred eye of fixation particularly in primary position. Ophthalmic evaluation revealed albinism with foveal and macular hypoplasia, peripheral fundus hypopigmentation and iris diaphany. Being an albino, the patient had blue irises, light brown hair colour and does not tan when exposed to sunlight. Optical refraction yielded spherical corrections of OD (right eye): S = +10; OS (left eye): S = +11.25. Corrected Snellen acuity at 6 m viewing distance was 0.1 (20/200) OD viewing, 0.15 (20/300) OS viewing. At near viewing (20 cm) OU (both eyes) acuity was 0.4 (20/50). The albino patient showed no evidence of stereopsis; color vision was normal. Visual fields as tested with static perimetry were normal.

4.2. Three dimensional recordings

Measurement of ocular rotation in 3D for the patient and five healthy control subjects was accomplished with the dual scleral induction coil technique (Bartl et al., 1995; Ferman et al., 1987a; Jansen, Ferman, Collewijn, & Van den Berg, 1987a,b). Raw data were digitized (12 bit) with a sample frequency of 500 Hz (Bour, 1984). Data were collected from ten channels; two channels for horizontal and vertical target position, four channels for OD and four channels for OS. For each eye, two channels are derived from the direction coil that picks up the eye rotation signals in the horizontal and vertical plane. The other two channels are derived from the torsion coil that yields the signals of clockwise and anticlockwise rotation of the eye in the torsional plane. In all cases, the eye orientations are calculated as rotation vectors from the signals, using Mathematica 5.0. (Wolfram, 1999) The various offsets were taken into account (Hess, Van Opstal, Straumann, & Hepp, 1992). Calibration consisted of an *in vitro* and an *in vivo* protocol.

For the *in vitro* calibration, the following procedure was implemented. The straight-ahead position (0°) was used as the reference target. The three gain components of the direction and the torsion coil (coil vectors) were obtained by directing a gimbal system to symmetrically positioned targets with respect to the reference target. Via a rotation matrix (Collewijn, 1975; Robinson, 1963), sine and cosine components of direction and torsion coils were converted to 3D Fick coordinates. After Haustein (1989), Fick coordinates were converted to rotation vectors (Apkarian et al., 1999).

For *in vivo* calibration, cooperation of the patient was required. The subject fixated a reference target positioned at straight ahead. To obtain the corresponding 'zero fixation' position for each eye, monocular measurements were implemented. Following the experimental protocol, all successive and successful "zero position fixations" within a given test session were marked for further offline calibration and analysis. To compensate for long-term drift and/or sudden shifts of eye-movement signals (particularly those from the torsion coil) repeated fixation of the zero fixation reference target was required. A third-order polynomial was fit through successive "zero-fixations" that extended over a given fixation measure. In addition, zero-fixations were parsed into foveation periods (Dell'Osso and Daroff, 1975).

Subjects had to look at a translucent screen at a distance of 114 cm which subtended a viewing angle of approximately 24 by 24°. The target consisted of a laser spot (0.1° diameter) that was back-projected on the screen and the position of the spot was changed by a servo-controlled mirror system. The laser spot jumped from the reference position (0, 0) to an eccentric position and back to the reference position. For the normal control subjects, 48 positions were located on 6 meridians at 0, 30, 60, 90, 120 and 150° with an eccentricity of 5, 10, 15 and 20°. The duration of each fixation position was 1 s. For the CN patients the stimulus grid consisted of 28 targets presented in horizontal, vertical and oblique meridians from primary position to ±30 eccentricity.

4.3. Analysis

Listing plane is determined by fitting a plane through the data points using a least-square fitting procedure. The equation of this plane is given by

$$\mathbf{r} \cdot \mathbf{n}_{\mathrm{lp}} = \mathrm{d} \tag{2}$$

where $\mathbf{n}_{\mathrm{lp}}$ is the "normal" vector of Listing's plane, and d is the distance of the plane to origin (0, 0, 0). With $\mathbf{n}_{\mathrm{lp}}$ and d, the orientation of this Listing's plane is determined. The deviation from Listing's plane is determined by calculating the standard deviation (SD) of the distance of the data points with respect to the plane. The x, y and z directions of the eye do not necessarily coincide with the x, y and z directions defined by the experimental setup. This has the effect that when Listing's plane is measured, it usually is skewed by some angle with respect to the fronto-parallel plane defined by y and z directions of the magnetic field setup. The rotation vector data points can be adjusted such that Listing's plane is equivalent to the fronto parallel plane. With the adjustment, the deviation from Listing's plane at each eye orientation is given by only the component r_{x}. The SD of the deviation from Listing's plane for all eye orientations N is

$$SD = \sqrt{\frac{1}{N-1}\sum_{i}^{N}(r_{x,i})^2} \tag{3}$$

The 'reference orientation' is the orientation when the eye is looking straight-ahead (along the x axis, defined by the magnetic field setup). The 'primary orientation' is the orientation when the eye is looking along the 'normal vector' of Listing's plane, given by $\mathbf{n}$. If the eye is rotating from this primary orientation to any other orientation, the component r_{x} (torsion) equals 0.

4.4. Three dimensional model simulations

The most important aspect of the dynamical eye-movement model of Schnabolk and Raphan (1994) is that the eye muscles generate torque to rotate the eye. When the restoring torque of the orbital tissues counterbalances the torque applied by the extra-ocular muscles, a unique equilibrium point is reached. The relationship between restoring torque and eye orientation yields the unique torque-orientation relationship. With this approach the signals from the central nervous system can be treated as vectors. These signals are converted to muscle torques, which are also vectors and thus commute. Therefore, the ocular motor plant must convert torque (vector) into orientation (non-vector). According to this model the problem of non-commutativity of eye orientation is then solved by the ocular motor plant and not by the central nervous system. The central nervous system only has to generate a two dimensional (2D) (horizontally and vertically) ocular motor command.

The diagram shows that the neural pulse-step generator, instead of the one-dimensional (1D) operator in the Robinson model (Robinson, 1975), uses three operators to transform the neural input signal $\mathbf{r}_\omega$ into a motoneuron signal $\mathbf{m_n}$. $\mathbf{H_p}$ is the system matrix for the velocity-to-position integrator, $\mathbf{G_p}$ is the coupling from the neural signal $\mathbf{r}_\omega$ to the integrator and $\mathbf{C_p}$ is a matrix that transforms the output of the integrator $\mathbf{x_p}$ to a motoneuron step signal $\mathbf{x'_p}$. For the direct premotor-to-motor coupling and the generation of the pulse from the velocity command there is a matrix $\mathbf{D}$. The sum of pulse ($\mathbf{D}\,\mathbf{r}_\omega$) and step ($\mathbf{C_p x_p}$) commands (Robinson, 1975) yields the total motoneuron signal $\mathbf{m_n}$, which is transformed by the muscle-matrix $\mathbf{M}$ into a torque-related signal $\mathbf{m}$. Thus, the transformation of neural input signal $\mathbf{r}_\omega$ into a torque-related signal $\mathbf{m}$ can be described by the following equations:

$$\begin{aligned} \frac{d\mathbf{x_p}}{dt} &= \mathbf{H_p x_p} + \mathbf{G_p r}_\omega \\ \mathbf{m_n} &= \mathbf{C_p x_p} + \mathbf{D r}_\omega \\ \mathbf{m} &= \mathbf{M m_n} \end{aligned} \tag{4}$$

A schematic diagram of the 3D model of eye-movement control is shown in Figure 2.

The final position of the eye is determined by the balance between the muscle force and the force exerted by the connective tissue. To obtain an accurate model, the description of the torque-orientation transformation has to account for both of these factors. The dynamical system (see Equation 5) describes the balance between the active torque $\mathbf{m}$ exerted by the extraocular muscles and the passive torque, where k is the coefficient of elasticity, $\mathbf{J}$ is the moment of inertia tensor, $\mathbf{B}$ is the matrix representing viscous damping, and ω and ω_h are the angular velocity vectors of the eye relative to space and the head relative to space, respectively. The resulting eye position in rotation vectors is defined by the angle φ and the vector $\mathbf{n}$. The actual derivation of the equations of this dynamical system that describes eye orientation as a function of the applied torque is too elaborate to discuss in this context (but see Schnabolk and Raphan, 1994).

$$\begin{aligned} \frac{d\omega}{dt} &= -\mathbf{J}^{-1}(\mathbf{B}\omega + k\varphi\mathbf{n}) - \frac{d\omega_h}{dt} + \mathbf{J}^{-1}\mathbf{m} \\ \frac{d\varphi}{dt} &= \langle \omega, \mathbf{n} \rangle \\ \frac{d\mathbf{n}}{dt} &= \frac{\omega \times \mathbf{n}}{2} + \frac{\mathbf{n} \times (\omega \times \mathbf{n})}{2} \cot\left(\frac{\varphi}{2}\right) \end{aligned} \tag{5}$$

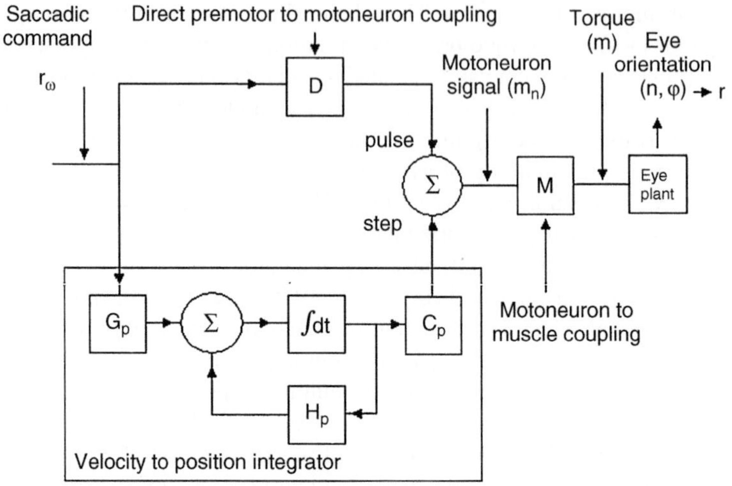

Figure 2. Schematic diagram of 3D model of eye-movement control. According to the Robinson model the internal saccade velocity command is split up in a direct pulse and a velocity-to-position integrated step. The pulse step signal of the motoneuron innervates the muscle. The 3D coupling from neural innervation to muscle force is given by matrix **M**. Finally, the active muscle torque counterbalances the passive visco-elastic forces of the eye globe with surrounding tissue and fat.

In this system Listing's law is automatically obeyed as long as the velocity commands that drive the pulse-step generator only have pitch and yaw components and no roll components. The model is physiologically quite accurate and generates realistic saccades and smooth pursuit eye movements. Raphan (1997) made adjustment to the original model of Schnabolk and Raphan (1994) and introduced a modifiable muscle-matrix **M** in such a way that it was dependent of eye orientation ($\mathbf{M}_p$). The original constant diagonal matrix used by Schnabolk and Raphan (1994) was multiplied by a rotation matrix $\mathbf{R}_M$, to incorporate the pulley-effect (Figure 3).

$$\mathbf{R}_M = \begin{pmatrix} n_1^2(1-\cos\delta)+\cos\delta & n_1n_2(1-\cos\delta)-n_3\sin\delta & n_1n_3(1-\cos\delta)+n_2\sin\delta \\ n_2n_1(1-\cos\delta)+n_3\sin\delta & n_2^2(1-\cos\delta)+\cos\delta & n_2n_3(1-\cos\delta)-n_1\sin\delta \\ n_3n_1(1-\cos\delta)-n_2\sin\delta & n_3n_2(1-\cos\delta)+n_1\sin\delta & n_3^2(1-\cos\delta)+\cos\delta \end{pmatrix}$$
$$\mathbf{M}_p = \mathbf{R}_M \cdot \mathbf{M} \tag{6}$$
$$\delta = k_\varphi \varphi$$

where $\mathbf{M}_P$ is the muscle matrix with pulleys, **n** is the axis of eye and muscle torque rotation, δ is the rotation angle of the muscle torque, φ is the rotation angle of the eye and k_φ is a constant ratio between the latter two that indicates the level of pulley-effect.

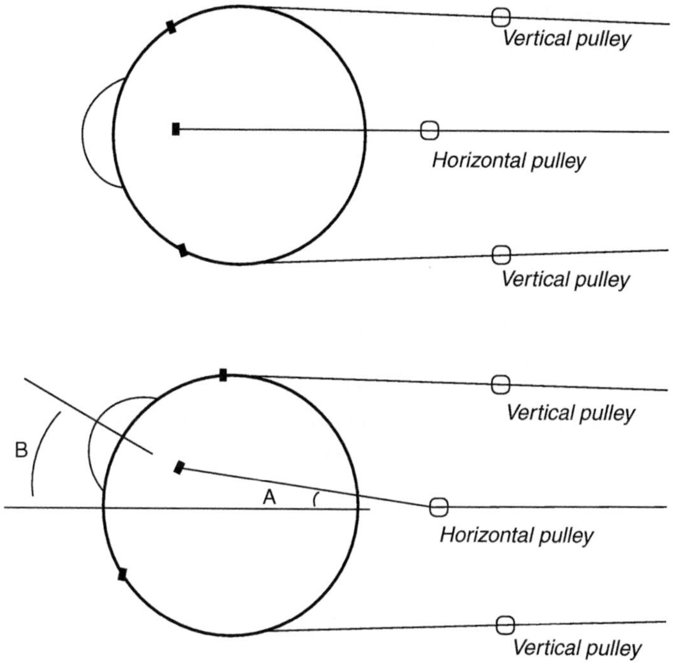

Figure 3. Instead of one pulley constant for all three muscle pairs, three pulley constants for each muscle pair can be used. The k_y and k_z value for the horizontal and vertical muscle pair are defined by the quotient of the angles: A/B.

4.5. Different pulleys for each muscle pair

For $k_\varphi = 0$, there is no pulley-effect whatsoever. The direction of muscle pull remains constant in the orbital frame and $\mathbf{R}_\mathrm{M}$ reduces to the identity matrix that was used in the original model. For $k_\varphi = 1$ the pulley effect is such that the direction of pull of the muscles always remains constant with respect to the eye regardless of the eye position. Simulations of Raphan (Fetter, 1997; Raphan, 1997) show that with $k_\varphi = 0.48$ the smallest transient torsional components are found during a saccade. This corresponds with the value $k_\varphi \approx 0.5$ that Quaia & Optican (1998) found for a negligible pulse-step mismatch.

A restriction of this approach is that there is no separate k_φ for the horizontal, vertical and torsional plane. Therefore, $\mathbf{R}_\mathrm{M}$ was replaced by the product of the pulley effect in three planes of rotation:

$$\vec{F}_{\text{active}} = \mathbf{R}_{k_x,k_y,k_z}(\vec{n}, \varphi) \cdot m\vec{m}_n \tag{7}$$

It must be noted that when the neural innervation is 2D, there is no "torsional" innervation, only horizontal and vertical innervation. The effect is that since $r_{\omega,\mathrm{x}} = 0$, $\mathrm{m}_{\mathrm{m,x}} = 0$, the value of k_x has no significance. The matrix $\mathbf{R}$ becomes essentially a 3 by 2 matrix, equal to Equation 6 with the left column removed. The pulley matrix converts neural input, which is two-dimensional, to muscle torque, which is three-dimensional:

$$\vec{F}_{\text{active}} = \mathbf{R}' \cdot \vec{m}\vec{m}_n$$

$$= m \begin{pmatrix} n_1 n_2(1-\cos k_y\varphi) - n_3 \sin k_y\varphi & n_1 n_3(1-\cos k_z\varphi) + n_2 \sin k_z\varphi \\ n_2^2(1-\cos k_y\varphi) + \cos k_y\varphi & n_2 n_3(1-\cos k_z\varphi) - n_1 \sin k_z\varphi \\ n_2 n_3(1-\cos k_y\varphi) + n_1 \sin k_y\varphi & n_3^2(1-\cos k_z\varphi) + \cos k_z\varphi \end{pmatrix} \begin{pmatrix} m_{n,y} \\ m_{n,z} \end{pmatrix} \quad (8)$$

Thus now the pull-direction of the extra ocular muscles can be adjusted for horizontal (k_z) and vertical (k_y) planes separately. Since the velocity command that drives the pulse-step generator only has pitch and yaw components the value of k_x is irrelevant. The dynamical model of Raphan with the addition of a 2D to 3D pulley-matrix $\mathbf{R}'$ was implemented in a computer-program that solved the differential equations numerically (Mathematica 5.0; Wolfram, 1999). The constants were the same as in the original model:

H = −0.03333
G = 0.3333
C = 2.944
D = 0.1389
$J = 5 * 10^{-7}\,\text{kg}\,\text{m}^2$
$m = 2.493 * 10^{-7}(\text{kg}\,\text{m}^2/\text{s}^2)(\text{spikes/s})$
$B = 7.476 * 10^{-5}\,\text{kg}\,\text{m}^2/\text{s/rad}$
$k = 4.762 * 10^{-4}\,\text{kg}(\text{m/s})^2/\text{rad}$
k_z = between 0 and 1
k_y = between 0 and 1
The angular head velocity ω_h was taken as zero.

5. Results

5.1. Normal controls

To determine Listing's plane, and the deviation from this plane in two normal control subjects, a 'calibration circle' paradigm has been used. With a least square error method Listing's plane was fitted through the 3D eye positions recordings during saccades and fixations. Subsequently, all data were rotated into the frontal parallel plane. The SD of the rotation vector data from this plane is then determined using Equation 2. The SD from Listing's plane for all normal control subjects did not exceed 1.3°.

5.2. Patient recordings

Eye-movement recordings of a patient with CN showed a component in the torsional plane that exceeded those of healthy normal controls. Figure 4 shows a recording obtained from a patient with CN. The left column shows the 3D recording when the patient makes

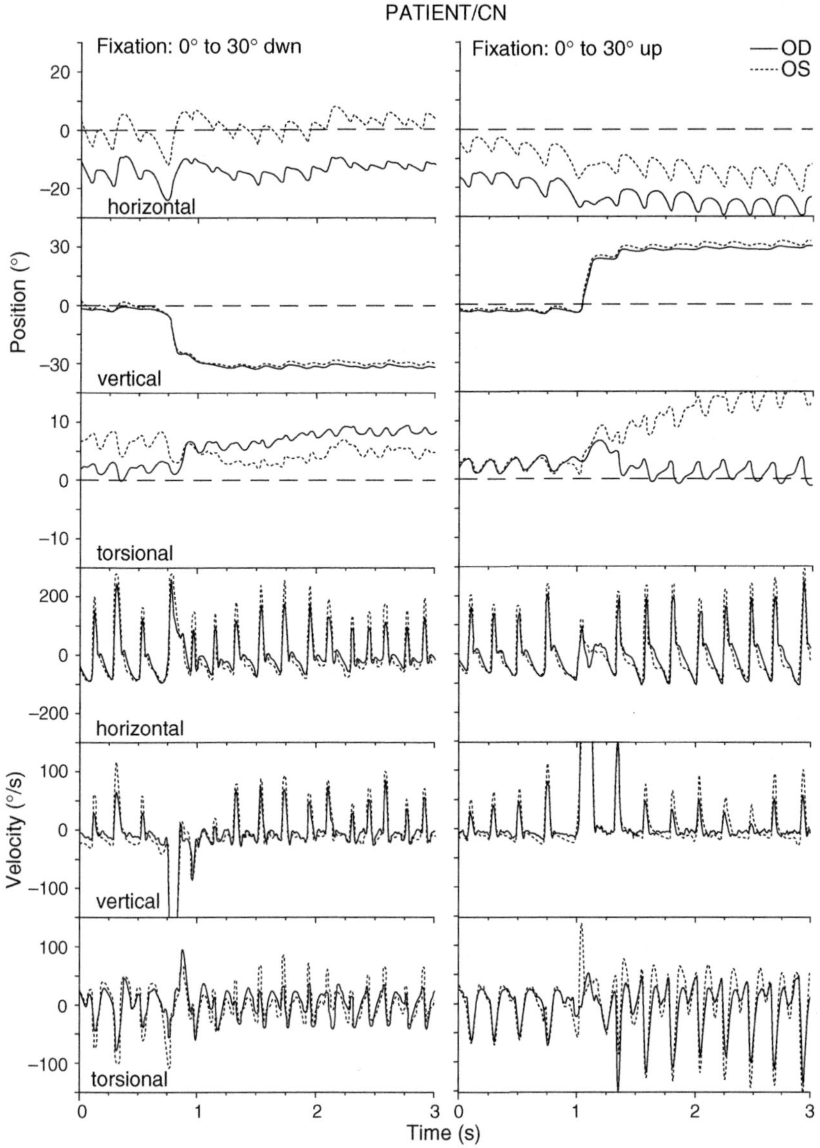

Figure 4. Three dimensional search-coil recording of a patient with congenital nystagmus. From top to bottom the position in the horizontal, vertical and torsional planes (rotation vectors) and the velocity in the horizontal vertical and torsional planes is shown. OD refers to the right eye and OS refers to the left eye. The horizontal CN clearly shows components in both the vertical and torsional planes. Note that a downward saccade for OD is associated with a clockwise torsion and a decrease in jerk amplitude in the torsional plane, whereas an upward saccade for OD is associated with a counterclockwise torsion and an increase of the jerk amplitude in the torsional plane.

a downward saccade and the right column shows the 3D recording when the patient makes an upward saccade.

The patient fixated targets at different eccentricities, which are shown on the left of Figure 5. Mean 3D eye position during fixation to all targets is summarized at the right of Figure 5. The bars drawn to the black dots represent the mean torsion at that particular eccentric position. The fitted Listing's plane in 3D is also shown by the projections on the top and at the right. The orientation of Listing's plane is listed by the angle ϕ and the vector **n**. Also shown in the Figure 5 is the value of the SD of eye movements representing the thickness of Listing's plane. In this case for the right eye this thickness is about 3.5°. This is much larger than has been measured in the normal control subjects. These results imply that CN patient's eye movements do not obey Listing's law.

To explain the non-Listing behaviour by anatomical factors in patients with CN, the dynamical model of Raphan was used. Only a signal in the horizontal direction was put into the model to generate the CN. The neural innervation r_ω (Figure 6) of this input signal was a superposition of a sinusoid and a repetitive phase-locked saccade. To avoid drift, the total integral of this signal was 0. Comparison of the CN waveform in the horizontal plane between the actual recording of the patient shown in Figure 4 and the simulated shown in Figure 7 demonstrates that there is a close resemblance.

To simulate a pattern of CN containing also a vertical and torsional component as has been observed in the actual recordings (Figure 4) with the same input signal, two important adjustments had to be made in the model. First, Listing's plane had to be displaced with respect to the fronto-parallel plane, and secondly, the pulley structures for the horizontal recti had to be displaced with respect to the pulley structures for the vertical recti.

The left panel of Figure 7 shows the result of the simulation with only a displaced pulley for the horizontal recti. A value of $k_y = 0.5$ means that there is optimal position of the pulleys for the vertical pulleys. A value of $k_z = 0.3$ means that the horizontal pulleys are located more backwards than the vertical pulleys and they do not fully compensate for non-commutativity. The right panel of Figure 7 depicts the result of the simulation when Listing's plane is also displaced from the fronto-parallel plane. The angle φ and amplitude vector **n** listed on the top of the panel represent the rotation vector associated with this displacement. The resemblance of this simulation shown in the right panel of Figure 7 with the actual recording of the right eye shown in Figure 4 is evident. After the vertical saccade a shift in the torsional plane can be seen which is clockwise with a downward saccade and anti-clockwise with an upward saccade. In addition, the decrease of the jerk left amplitude in the torsional plane with the downward saccade and the increase of the jerk left amplitude in the torsional plane with the upward saccade is both observed in the actual recording as in the simulation.

6. Discussion

As has been demonstrated by other studies (Straumann et al., 1996; Tweed et al., 1990), Listing's law holds for saccadic eye movements and the thickness of this plane does not

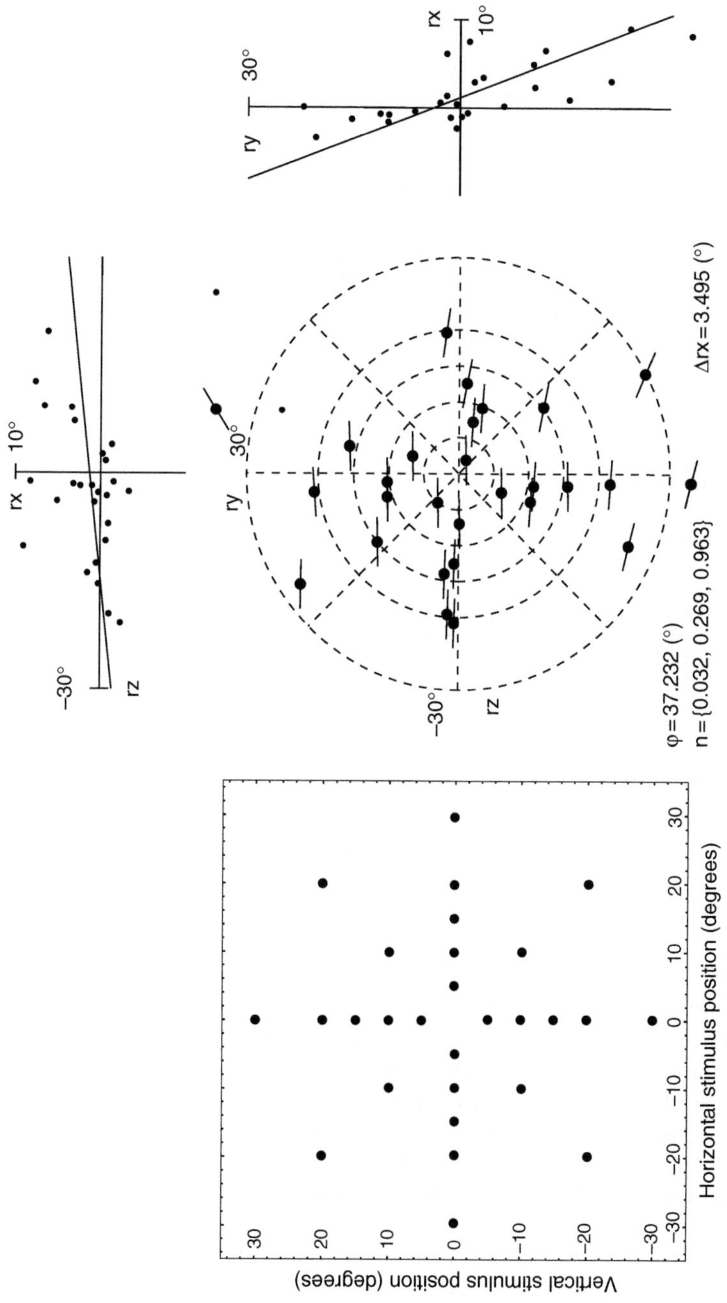

Figure 5. Determination of Listing's plane for the right eye of the CN patient.

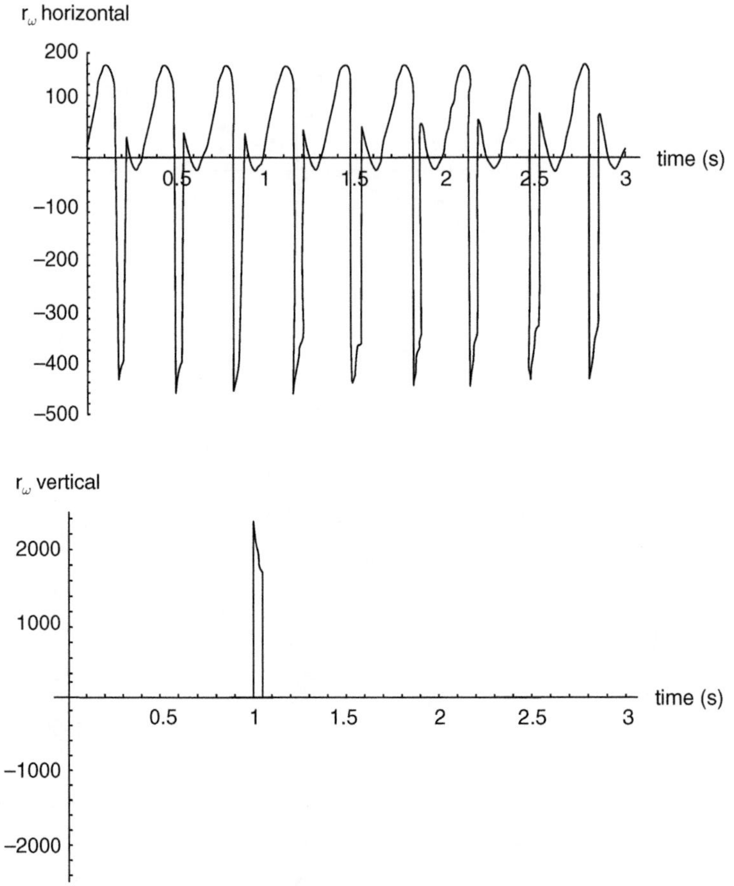

Figure 6. Neural velocity commands to simulate CN in the horizontal plane, that is, a pendular movement and a repetitive jerk left, and a saccade in the vertical plane, that is, a velocity pulse.

exceed an SD of approximately 1°. This is also found in our normal control subjects. However, Listing's plane does not necessarily have to lie parallel to the fronto-parallel plane and may be rotated even when a distant target is fixated (Bruno and Van den Berg, 1997; Mok, Cadera, Crawford, & Vilis, 1992; Van Rijn and Van den Berg, 1993). The analysis of recordings of 3D eye movements in a patient with CN showed a remarkably large SD from Listing's plane, namely three to four degrees. Moreover, on average there was a larger deviation from the fronto-parallel plane than in the control subjects. By means of a dynamical model (Fetter, 1997; Raphan, 1997; Schanbolk and Raphan, 1994) it was examined whether both deviations originated from aberrations from the optimal anatomy of the ocular motor plant, including the eye-globe muscles and surrounding tissue. Particular attention was paid to the function of the so-called "muscle pulleys",

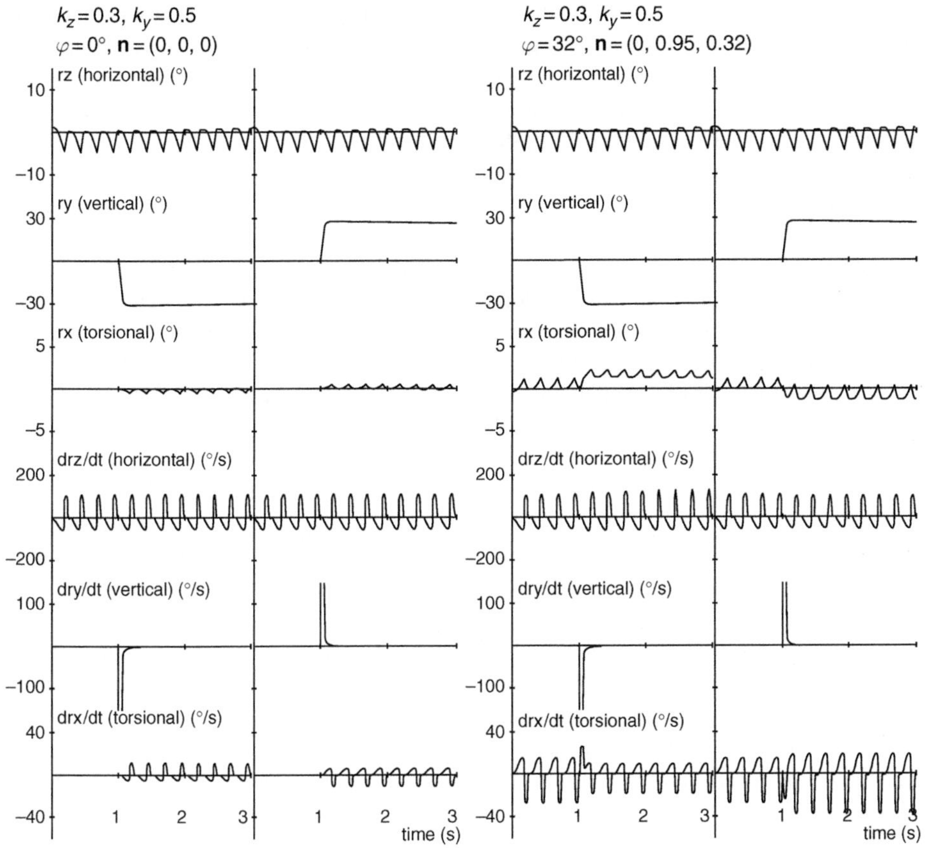

Figure 7. Simulation of a congenital nystagmus with a horizontal, vertical and torsional component. At the 1-second time either a downward or upward saccade is generated. The three upper tracings show the three components of the rotation vector, the lower three tracings show the derivative (velocity) in three directions. The left panel is the result of a simulation with Listing's plane parallel to the fronto-parallel plane, the right panel displays the results of the simulation with Listing's plane tilted. The values of the rotation vector associated with the displacement plane are indicated above. Both simulations have backward displaced pulleys.

which are thought to keep the muscle bellies in place during rotation of the eye (Demer et al., 1995). A neural drive to simulate the CN was constructed only in the horizontal direction. In the vertical direction, a 'standard' saccade signal to go up or down was used as an input to the model. Compared to previous studies (Quaia and Optican, 1998, 2003; Raphan, 1997), the dynamical model was refined with respect to an independent placement of the pulleys of the horizontal and vertical muscle pairs. Simulations showed a thickening of Listing's plane if one set of pulleys was misplaced. It can be concluded from the simulations that in case of only a horizontal neural drive of the CN, particularly in the torsional plane, a component of CN occurs that is consistent with an aberrant anatomy of the eye. The amplitude and direction of the fast phase of CN in the torsional plane appeared to be dependent on vertical gaze. This was also found in the experimental data recorded in the CN patient. Torsional components of CN are not uncommon (Averbuch-Heller, 2002); however, they are thought to originate in the central nervous system. Our study shows that there is no need to postulate a neural torsional generator to account for CN. Torsional components of CN can be explained by keeping the effects of specific anatomical structures into account. Patients with CN, particularly albinos, very often have strabismus, which may cause an imbalance in the horizontal recti and, therefore, the pulley effect may have changed in the horizontal plane (Oh et al., 2002). Moreover, albinos do have impaired binocular vision (Apkarian and Reits, 1989), which may lead to an insufficient neural drive to the extra-ocular muscles. It has also been demonstrated that patients with strabismus may have misplaced pulleys (Clark et al., 1997, 1998, 2000).

Some remarks have to be made about the model used for simulations. For instance, in the model the muscle torque is considered as a result of three muscle pairs. As has been pointed out by Schnabolk and Raphan (1994), the gain of the muscle pairs is not equal for each pair and the differences should be incorporated in the model. According to the Robinson model (1975), Quaia and Optican (2003) implemented these unequal forces of the six extraocular muscles into their model. This has not been implemented in our simulations, however, considering the effect of this adjustment to the model on the simulations in the paper of Quaia and Optican (2003), the effect on our type of eye-movement simulation should only be minor and it is not expected to influence our main conclusions. The presence of a vertical component of CN in the data, which is absent in the model simulations, probably finds its origin in the misalignment between the horizontal plane of the experimental set-up and the horizontal plane of the pulling direction of the horizontal recti: it does not change essentially in amplitude and direction as a function of gaze.

Furthermore, the role of the connective tissue still is not fully clear. Histological data show that there is evidence for both intermuscular and musculo-orbital coupling. The original model contains only intermuscular coupling, which yields approximately correct simulations for small angles. To sort out how large the pulley effect really is, *in vivo* experiment have been performed to determine the functional position of the pulleys (Demer et al., 2005). However, there is still a debate how the pulley effect is accomplished and whether the anatomical pulley structures really have a functional purpose (Van den Bedem et al., 2005).

As a general remark, it has to be stated that the impact of simulation results with this dynamical model should not be overestimated. Mathematically there are still several uncertain factors. For instance, for $\varphi = 0$ there is a singularity and the model becomes unstable. Therefore, φ has to be given a starting value that is unequal to zero.

Despite the aforementioned shortcomings, adjustment of the model to overcome these inaccuracies would not yield qualitatively different results. Hence, the effects found in the simulations can be compared to *in vivo* experimental data and the current dynamical eye-movement model may be used as a predictor for eye movements. Our simulations have demonstrated that the dynamical model gives a qualitative description of the observed eye movements, despite the uncertainties about its exact dynamical behaviour. Certainly, simulations with a model of the dynamics of eye movements can be of practical clinical use and may help in a better understanding of strabismus and CN.

Acknowledgement

The authors thank Dr. P. Apkarian and Dr. H van der Steen for data acquisition of the CN patient and for many helpful discussions of the experimental results.

References

Abadi, R. V. & Dickinson, C. M. (1986). Waveform characteristics in congenital nystagmus. *Documenta Ophthalmologica, 64*, 153–167.

Apkarian, P., Bour, L. J., Van der Steen, J. & Collewijn, H. (1999). Ocular motor disorders associated with inborn chiasmal crossing defects. In Becker et al. (Ed.), *Current oculomotor research* (pp. 403–13), New York: Plenum Press.

Apkarian, P., Bour, L. J., van der Steen, J. & de Faber, J. T. H. N. (1996). Chiasmal crossing defects in disorders of ocular motor function: three dimensional eye movement recordings in albinism and non- decussating retinal fugal fibre syndrome. *Investigative Ophthalmology & Visual Science, 37 (suppl.)*, 228.

Apkarian, P., & Reits, D. (1989). Global stereopsis in human albinos. *Vision Research, 29*, 1359–1370.

Averbuch-Heller, L., Dell'Osso, L. F., Leigh, R. J., Jacobs, J. B. & Stahl, J. S. (2002). The torsional component of "horizontal" congenital nystagmus. *Journal of Neuroophthalmol, 22*, 22–32.

Bartl, K., Siebold, C., Glasaucer, S., Helmchen, C. & Buttner, U. (1995). A simplified calibration method for three-dimensional eye movement recordings using search coils. *Vision Research, 36*, 997–1006.

Bour, L. J., van Gisbergen, J. A. M., Bruins, J., & Ottes, F. (1984). The double magnetic induction method for measuring eye movement – Results in monkey and man. *IEEE Transactions on Biomed Engineering, 31*, 419–427.

Bruno, P., & Van den Berg, A. V. (1997). Relative orientation of primary positions of the two eyes. *Vision Research, 37*, 935–947.

Collewijn, H., Van der Mark, F. & Jansen, T. C. (1975). Precise recording of human eye movement. *Vision Research, 15*, 447–450.

Clark, R. A., Miller, J. M. & Demer, J. L. (1997). Location and stability of rectus muscle pulleys. *Investigative Ophthalmology & Visual Science, 38*, 227–240.

Clark, R. A., Miller, J. M. & Demer, J. L. (1998). Displacement of the medial rectus pulley in superior oblique palsy. *Investigative Ophthalmology & Visual Science, 39*, 207–212.

Clark, R. A., Miller, J. M. & Demer, J. L. (2000). Three-dimensional location of human rectus pulleys by path inflections in secondary gaze positions *Investigative Ophthalmology & Visual Science, 41*, 3787–3797.

Dell'Osso, L. F., Daroff, R. B. (1975). Congenital nystagmus waveforms and foveation strategy. *Documenta Ophthalmologica, 39*,155–182.

Demer, J. L., Miller, J. M. & Poukens, V. (1996). Surgical implications of the rectus extraocular muscle pulleys. *Journal of Pediatric Ophthalmology Strabismus, 33*, 208–218.

Demer, J. L., Miller, J. M., Poukens, V., Vinters H. V. & Glasgow B. J. (1995). Evidence for fibromuscular pulleys of the recti extraocular muscles. *Investigative Opthalmology & Visual Science, 36*, 1125–1136.

Demer, J. L., Clark, R. A. (2005). Magnetic resonance imaging of human extraocular muscles during static ocular counter-rolling. *Journal of Neurophysiology, 94*, 3292–3302.

Donders, F. C. (1876). Versuch einer genetischen erklrung der augen-bewegungen. *Archiv fur die gesammte Physiologie des Menschen und der Thiere, 13*, 373–421.

Ferman, L., Collewijn H. & Van den Berg A. V. (1987a). A direct test of listing's law I: human ocular torsion measured in static tertiary positions. *Vision Research, 27,* 929–938.

Ferman, L., Collewijn H. & Van den Berg A. V. (1987b). A direct test of Listing's law II: Human ocular torsion measured under dynamic conditions. *Vision Research, 27,* 929–938.

Ferman, L., Collewijn, H., Jansen, T. C. & Van den Berg A. V. (1987c). Human gaze stability in horizontal, vertical and torsional direction during voluntary head movements, evaluated with a three-dimensional scleral induction coil technique. *Vision Research, 27,* 811–828.

Fetter, M., Haslwanter, T.,Misslisch, H.,Tweed.,D. (1997). *Three-Dimensional Kinematics of Eye, Head and Limb Movements*, Amsterdam: Harwood Academic Publishers.

Haslwanter, T. (1995). Mathematics of three-dimensional eye rotations. *Vision Research, 35,*1727–1739.

Haustein., W. (1989). Considerations on Listing's law and the primary position by means of a 3D matrix description of eye position control. *Biological Cybernetica, 60*, 411–420.

Hess, B. J. M., Van Opstal, A. J., Straumann, D. & Hepp, K. (1992). Calibration of three-dimensional eye position using search coil signals in the rhesus monkey. *Vision Research, 32*, 1647–1654.

Jansen, T. C., Ferman, L.,Collewijn, H. & Van den Berg, A. V. (1987). Human gaze stability in the horizontal, vertical and torsional direction during voluntary head movements, evaluated with a three-dimensional scleral induction coil technique. *Vision Research, 27*, 811–828.

Korff C. M., Apkarian P., Bour L. J., Meuli R., Verrey J. D. & Roulet Perez E. (2003). Isolated absence of optic chiasm revealed by congenital nystagmus, MRI and VEPs. *Neuropediatrics, 34,* 219–223.

Miller, J. M. (1989). Functional anatomy of normal human rectus muscles. *Vision Research, 29*, 223–240.

Mok, D., Ro, A., Cadera, W., Crawford, J. D. & Vilis, T. (1992). Rotation of Listing's plane during vergence. *Vision Research, 32,* 2055–2064.

Nakayama, K. & Balliet, R. (1977). Listing's law, eye position sense and perception of the vertical, *Vision Research, 17*, 453–457.

Oh, S. Y., Clark, R. A., Velez, F., Rosenbaum, A. L. & Demer, J. L. (2002). Incomitant strabismus associated with instability of rectus pulleys. *Investigative Ophthalmology & Visual Science, 43*, 2169–2178.

Quaia, C. & Optican, L. (1998). Commutative saccadic generator is sufficient to control a 3-D ocular plant with pulleys. *Journal of Neurophysiology, 79*, 3197–3215.

Quaia, C. & Optican, L. (2003). Dynamic eye plant models and the control of eye movements, *Strabismus, 11*, 17–31.

Raphan, T. (1998). Modeling control of eye orientation in three dimensions. I. Role of muscle pulleys in determining saccadic trajectory, *Journal of Neurophysiology, 79*, 2653–2667.

Robinson, D. A. (1963). A method of measuring eye movement using a scleral coil in a magnetic field. *IEEE Transactions on Biomedical Electronics BME-10*, 137–145.

Robinson, D. A. (1975). A quantitative analysis of extraocular muscle co-operation and squint. *Investigative Opthalmology, 14*, 801–825.

Schnabolk, C. & Raphan, T. (1994). Modeling 3D velocity-to-position transformation in oculomotor control. *Journal of Neurophysiology, 71*, 623–638.

Simonsz, H. J., Harting, F., De Waal, B. J. & Verbeeten, B. W. (1985). Sideways displacement and curved path of recti eye muscles, *Archives of Ophthalmology, 103*, 124–128.

Simonsz, H. J., (2003). First description of eye muscle 'poulies' by Tenon in 1805. *Strabismus, 11*, 59–62.

Straumann, D., Zee, D. S., Solomon, D. & Kramer, P. D. (1996). Validity of Listing's law during fixations, saccades, smooth pursuit eye movements, and blinks. *Experimental Brain Research, 112,* 135–146.

Tweed, D., Cadera, W. & Vilis, T. (1990). Computing three-dimensional eye position quaternions and eye velocity from search coil signals. *Vision Research, 30*, 97–110.

Tweed, D. & Vilis, T. (1987). Implications of rotational kinematics for the oculomotor system in three dimensions. *Journal of Neurophysiology, 56,* 832–849.

Tweed, D. & Vilis, T. (1990). Geometric relationships of eye position and velocity vectors during saccades, *Vision Research, 30,* 111–127.

Tweed, DB, Haslwanter TP, Happe V, Fetter M (1999). Non-commutativity in the brain, *Nature, 399*, 261–263.

Ukwade, M. T., Bedell, H. E., White, J. M. (2002). Orientation discrimination and variability of torsional eye position in congenital nystagmus. *Vision Research, 42*, 2395–2407.

Van den Bedem, S. P. M., Schutte, S.,Van der Helm, F. C. T., Simonsz, H. J. (2005). Mechanical properties and functional importance of pulley bands or *'faisseaux tendineux' Vision Research, 45*, 2710–2714.

Van Rijn, L. J. & Van de Berg, A. V. (1993). Binocular eye orientation during fixations: Listing's law extended to include eye vergence, *Vision Research, 33*, 691–708.

Von Helmholtz, H. (1924). *Treatise on physiological optics*, Southall, J. P. C. (Ed.). Rochester, New York: Optical Society of America.

Wolfram, S. (1999). *The mathematica book*, Cambridge University Press, Wolfram Media.

PART 3

TRANSSACCADIC INTEGRATION

Edited by

WAYNE S. MURRAY

Chapter 8

TRANSSACCADIC RECOGNITION IN SCENE EXPLORATION

PETER DE GRAEF

University of Leuven, Belgium

Eye Movements: A Window on Mind and Brain
Edited by R. P. G. van Gompel, M. H. Fischer, W. S. Murray and R. L. Hill

Abstract

Contrary to what is the case in theories of reading, theories of object and scene recognition have shown a remarkable blind spot to the fact that visual perception typically involves integration of information across a number of spatiotemporally discrete fixations. I present two main reasons which I think are responsible for this neglect: the apparent power of single-shot perception and the contested nature of transsaccadic integration. Both reasons are claimed to be flawed, as argued on the basis of a review of existing evidence and the presentation of some new data on transsaccadic preview benefits. It is concluded that a transsaccadic theory of object and scene recognition should be put on the agenda, inspired by new insights on the interaction of feed-forward and re-entrant visual processing streams.

1. Introduction

Visual acuity is not uniform across the retina and not all aspects of a visual stimulus are equally salient or equally pertinent to a viewer's current task. Inevitably, this leads the visual system to constantly sample different aspects and locations in its environment, which produces a continuous series of fixations and saccades that bring new stimulus aspects or locations into the processing focus. This implies that any visual entity is typically viewed in a series of discrete samples collected at different times and at different retinal and/or spatial coordinates resulting in qualitative differences in the informational content of each of the samples. The logical question therefore is whether and how these samples are integrated to construct a coherent percept.

In reading research, the inherent multi-sample nature of visual processing is at the core of an ever-increasing number of studies. The functional characteristics of eye movements in reading have been studied for well over a century (see Rayner, 1998, for a comprehensive review). With the development of new theoretical frameworks of language processing and technological advances in eye-tracking equipment and software in the 1970s (see Wade & Tatler, 2005, for a history of eye movement measurement), this has culminated in detailed studies of how overt eye movements are related to the ongoing covert processes of text comprehension. An important part of this work has focussed on the question of whether and how text samples acquired on successive fixations are integrated. Integration has been studied across a single fixation–saccade–fixation cycle to determine whether and how a peripheral preview of the word that is about to be fixated could be integrated with the subsequent foveal view of that word on the next fixation and thus facilitate word recognition (e.g., Balota, Pollatsek, & Rayner, 1985; Hyöna, Niemi, & Underwood, 1989; McConkie & Zola, 1979; Pollatsek, Lesch, Morris, & Rayner, 1992; Rayner & Well, 1996). Processes of integration have also been studied across multiple fixation–saccade–fixation cycles to understand how individual words are added into a developing syntactic and semantic representation of the sentence (e.g., Altmann, van Nice, Garnham, & Henstra, 1998; Brysbaert & Mitchell, 1996; Van Gompel, Pickering, Traxler, 2001) or, more globally, into a coherent comprehension of the discourse contained in the text (e.g., Cook & Myers, 2004; Kambe, Rayner, & Duffy, 2001).

Somewhat surprisingly, the great interest of reading research in the integration of information samples across eye movements has not been matched in other flourishing domains of vision science such as object and scene perception. In a sense, most prominent models of object and scene recognition have adopted what could be called an *in vitro* approach. They outline in great computational detail how object and scene structure can be recovered from the image, but the image is implicitly assumed to be a single, indefinitely sustained, uniform-acuity projection of scene and objects in the optic array (e.g., Biederman, 1987; Edelman, 1999; Oliva & Torralba, 2001; Tarr & Bülthoff, 1998). It has been tacitly ignored that in real-life perception any scene component or object has generally been glimpsed in peripheral vision before it is actually fixated, and that specific scene parts or objects often receive multiple distinct fixations during the first visual inspection of the scene or object. In both cases, the visual system will have at its disposal multiple distinct samples of the same stimulus, which can all be used in the recognition process.

While sample integration has not been incorporated in the dominant theoretical frameworks in object and scene *recognition* it would be false to claim that it has not been studied in object and scene *perception*. Recently, a whole new line of work is emerging aimed at identifying the buildup and stability of scene and object memory representations that are developed over the course of multiple fixation–saccade–fixation cycles (e.g., Hollingworth, 2004; Hollingworth & Henderson, 2002; Melcher, 2001, 2006; Tatler, Gilchrist, & Land, 2005; Tatler, Gilchrist & Rusted, 2003). In addition, for the past 25 years there has been a steady output of papers posing the question whether and how the visual contents that are acquired during a single fixation–saccade–fixation cycle are integrated into one percept.[1] More recently, the notion of transsaccadic perceptual integration is gaining increased acceptance as the neurophysiological underpinnings of this mechanism are starting to become clear (e.g., Khayat, Spekreijse, & Roelfsema, 2004; Melcher & Morrone, 2003; Prime, Niemeier, & Crawford, 2006).

However, in spite of this growing interest in multi-sample integration in visual perception, the concept has not (yet) made it into the mainstream of models accounting for the basic level identification of an object or a scene. In my opinion, there are two main reasons why this is so. First, there is increasingly compelling evidence that single-shot perception is very powerful, that is, that object and scene recognition can be achieved within a single fixation of modal duration (around 220 ms according to an overview presented by Henderson & Hollingworth, 1998). Second, in recent years there has been a strong line of empirical and theoretical work arguing the proposition that transsaccadic integration does not exist. In what follows, I will discuss these two reasons and I will attempt to demonstrate that they are insufficient grounds to further delay the development of a genuine transsaccadic theory of object and scene recognition.

2. The power of single-shot perception

2.1. Recognition at a glance

In the past, it has been demonstrated repeatedly that picture presentations well below the modal fixation duration are sufficient to recognize the general gist of a scene (e.g., Biederman, Mezzanotte, & Rabinowitz, 1982; Intraub, 1981; Potter, 1976) or to identify a depicted object (e.g., Biederman & Ju, 1988). Even when foveal masking dynamically curtailed stimulus presentation times on every fixation during extended scene exploration, viewers only seemed to require 50–70 ms of unmasked fixational content to allow for normal scene

[1] It is not my intention to extensively review this research, several excellent reviews are already available (e.g., O'Regan, 1992; Pollatsek & Rayner, 1992; Bridgeman, Van der Heijden, & Velichkovsky, 1994; Irwin, 1996; McConkie & Currie, 1996; Verfaillie et al., 2001; Deubel, Schneider & Bridgeman, 2002; Germeys, De Graef & Verfaillie, 2002; Henderson & Hollingworth, 2003a).

exploration (van Diepen, De Graef, & d'Ydewalle, 1995). In recent years, this impressive power of single-shot perception has been explored to its limits in a number of studies trying to detail the chronometry of visual object and scene categorization. In a seminal paper, Thorpe, Fize, and Marlot (1996) presented their participants with a central 20 ms exposure of a natural scene and required participants to release a button if they saw an animal (i.e., a go/no-go task). Event-related potentials (ERPs) were recorded throughout the task and showed a strong frontal negativity about 150 ms after the onset of stimuli for which participants correctly responded that there was no animal (i.e., on no-go trials). Thorpe et al. (1996) interpreted this as showing that 150 ms of processing time was sufficient to analyse the entire stimulus to such an extent that the presence of an animal could be ruled out.

However, one could also argue that 150 ms is a conservative estimate of the speed with which a stimulus can be processed because faster ERP effects have been documented (e.g., Delorme, Rousselet, Macé, & Fabre-Thorpe, 2004) for tasks involving specific image recognition (i.e., of a learned face) rather than overall image categorization (i.e., in animal/non-animal categories). Similarly, Keysers, Xiao, Földiak, and Perrett (2001) demonstrated that 14 ms of undistorted stimulus availability was enough to record target-selective activity in macaque temporal cortex, and that human target-detection performance was above-chance at a presentation rate of one image per 14 ms. Additional converging evidence for very fast stimulus processing comes from MEG studies indicating that 20 ms of natural scene processing is sufficient for above-chance discrimination from a distractor scene (Rieger, Braun, Bülthoff, & Gegenfurtner, 2005).

2.2. But is it recognition?

In view of these benchmarks of the visual system, it appears justified to conclude that all the visual processing that needs to be done on the input encoded during a single fixation can easily be completed within the course of that fixation, which raises the question why object and scene recognition would even need transsaccadic processing of multiple fixation samples? The answer to this question lies in the interpretation of the processes that are reflected in the fast physiological and behavioral effects that were cited above. Most of the tasks for which these effects were demonstrated involve the discrimination of specific target images from visually different distractor images. Therefore, one may wonder whether the results reveal true object and scene recognition or rather a sensitivity to low-level visual dissimilarities between target and distractor. VanRullen and Thorpe (2001) have argued that estimates of object recognition time can only be inferred from effects which reflect a task-dependent categorization of a stimulus as a target rather than a distractor. To obtain such estimates they employed the go/no-go categorization task as used by Thorpe et al. (1996, see above), and presented participants with blocks of trials in which the same two categories of images (animals and vehicles) alternately served as targets and distractors for the categorization task. ERPs recorded in this design showed differential activity for animals and vehicles after 75–80 ms following stimulus onset, regardless of which image category served as target. In addition, a target–distractor difference was observed at 150 ms poststimulus regardless of which image category served as target. Based on these data, VanRullen and Thorpe (2001) proposed that the early effect reflects a response to visual differences between the image categories, while the later effect provides an upper estimate of image recognition time.

Recently, however, Johnson and Olshausen (2003) argued that even the target–distractor effect at 150 ms poststimulus could be attributed to low-level visual dissimilarities. To ensure that ERP effects truly reflected target or distractor categorization and not a pre-categorical sensitivity to systematic visual differences between target and distractor, they ran the go/no-go categorization task with a different target on every trial. Across trials, specific images were equally often target as they were distractor, allowing for an assessment of the categorization response without the confound of visual target–distractor differences. Under these conditions, a categorization-related ERP effect emerged with a variable latency between 150 and 300 ms poststimulus. When compared to ERPs which Johnson and Olshausen (2003) recorded for a replication of the original animal vs no-animal categorization task, the same late effect showed, along with a faster fixed-latency target–distractor difference at 150 ms poststimulus. Taking into account that a spatial frequency analysis of animal and distractor images showed consistent differences in the power spectrum of the two image categories, Johnson and Olshausen concluded that visual category differences can be picked up within 150 ms, but recognition-based effects take between 150 and 300 ms to surface. Based on a study of priming effects on ERPs recorded in an object-naming task, Schendan and Kutas (2003) derived a similar estimate of the time required to access a stored structural representation of the object that is being viewed, that is 150–250 ms. Large, Kiss, and McMullen (2004) even extended these intervals by showing that differences in the type of object categorization that was required from participants only had ERP correlates after about 300 ms from stimulus onset. Specifically, at that point ERP responses to targets were stronger for superordinate categorizations (e.g., animal vs vehicle) than for basic level (e.g., dog vs cat) and for subordinate categorizations (e.g., beagle vs collie). Large et al. proposed that this shows a gross to detailed progression in object processing, with longer processing times required as a more specific identification is attempted.

2.3. *The need for a second shot*

Initial studies of the time course of object and scene recognition seemed to clearly show that the time of one modal duration fixation in scene perception (i.e., about 220 ms) is all that is needed to complete recognition of the fixated object or scene. However, more controlled follow-up work has now indicated that this time interval may not be sufficient to achieve object or scene recognition. When recognition processes are studied in the context of realistic scene exploration, this discrepancy between recognition and fixation time creates a complex puzzle. Consider what happens when a viewer freely explores an everyday scene. Eye recording in such a situation typically reveals a scanpath that consists of an alternating sequence of one or more within-object fixations followed by a between-object saccade (De Graef, Christiaens, & d'Ydewalle, 1990; Henderson, 2003;

Land, Mennie, & Rusted, 1999). Logically, this implies that recognition of an object can be achieved in three different ways. First, the gaze may remain on object n until it is identified, resulting in multiple within-object refixations. Second, visual information encoding may be completed during fixation of object n, while identification is completed during the subsequent between-object saccade to object $n+1$, or even during the first fixation on $n+1$. Third, processing of object n may start during fixation of object $n+1$, resulting in a presaccadic peripheral preview of n which is integrated with its subsequent postsaccadic foveal view, allowing for identification of n during its first fixation.

In reading research, these three mechanisms for transsaccadic (word) identification have all received ample attention in the study of within-word refixations (e.g., Reichle, Rayner, & Pollatsek, 1999), intrasaccadic and/or between-word spill-over effects (e.g., Irwin, 1998), and transsaccadic preview benefits (e.g., Rayner, Liversedge, & White, 2006) including their extreme manifestation in word skipping (e.g., Brysbaert, Drieghe, & Vitu, 2005). Moreover, quantitative models of eye-movement control in reading have provided an integrated account of the interplay between the three transsaccadic processes (Reichle, Rayner, & Pollatsek, 2003). In contrast, in scene perception research, treatment of the various possible types of transsaccadic identification has been much less extensive, primarily because the very notion of transsaccadic information integration has been severely challenged.

3. Transsaccadic information integration in scene exploration: *The Pit and the Pendulum*

3.1. The Pit and the Pendulum

The study of transsaccadic integration in object and scene perception is perhaps best described as an ongoing scientific mystery novel involving the two main props of Edgar Allan Poe's 1842 story *The Pit and the Pendulum*. The mystery is how to solve the discrepancy between the fact that the input to the visual system is a temporally and spatially discontinuous train of "snapshots", and the phenomenological experience of a stable and continuous visual world. Attempts to solve the mystery can be viewed as pendulum swings across a pit. On the one hand we have the *pit:* the saccadic period of 30–40 ms between fixations, during which visual information intake is suppressed (e.g., Matin, E., 1974; Matin, L., 1986; Ross, Morrone, Goldberg, & Burr, 2001). And on the other hand, we have the *pendulum*, which in the present context symbolizes scientific opinion swinging back and forth across the pit, with each new swing changing the view on whether or not there is actually something at the bottom of the pit.

3.2. A mirror at the bottom or no bottom at all?

A first swing across the pit launched the notion of an integrative visual buffer (McConkie & Rayner, 1976) or spatiotopic fusion (Feldman, 1985; Henderson &

Hollingworth, 2003a; Jonides, Irwin, & Yantis, 1982) according to which viewers could spatiotopically superimpose a presaccadic, peripheral stimulus and a postsaccadic, foveal stimulus. According to this scheme, transsaccadic integration of pre- and postsaccadic information samples was complete and visually detailed: the pit contained a faithful reflection of presaccadic fixation contents which are merged on a pixel-by-pixel basis with the postsaccadic fixation contents.

On the second swing across the pit, spatiotopic fusion was disproved in a number of studies (e.g., Bridgeman & Mayer, 1983; Henderson, 1997; Irwin, 1991; Pollatsek & Rayner, 1992). The notion of an intrasaccadic void was further reinforced by a flurry of *change blindness* studies (for an overview, see Simons, 2000; also see Rensink, 2002). In all these studies, the central proposition was that viewers are blind to change in their visual environment as long as the local transient associated with the change is somehow masked. Such masking can be achieved by simultaneously introducing attention-capturing local transients (O'Regan, Rensink, & Clark, 1999) or a global transient such as a grey-out of the image (Rensink, O'Regan, & Clark, 1997), a blink (O'Regan, Deubel, Clark, & Rensink, 2000), or a saccade (Grimes, 1996; Henderson & Hollingworth, 1999). Thus, change only seemed to be detectable on the basis of a within-fixation transient, and not on the basis of a comparison between pre- and postsaccadic images, which appeared to rule out a putative process of transsaccadic integration: the pit was bottomless. In fact, in two influential papers, O'Regan (1992; O'Regan & Noë, 2001) argued that there simply was no pit. Specifically, O'Regan proposed that there is no functional need to internally represent the outside world across a saccade, because that world is always there to be sampled in a fixation whenever needed. As soon as attentive sampling stops, however, viewers are afflicted with inattentional amnesia, that is, the failure to represent components of the visual world which they are no longer attending to (Rensink, 2000; Wolfe, Klempen, & Dahlen, 2000). This does not mean that we have no recollection of the state of the outside world from one fixation to the next: Whenever we have identified a setting or an object it can be stored in an evolving episodic memory trace of the scene we are currently viewing. O'Regan & Noë's (2001) point, however, is that this representation is a completely post-categorical and verbal summary, making abstraction of visual detail.

3.3. Construction at the bottom of the pit

On the third and currently ongoing swing across the pit, there appears to be a growing consensus that information gathered on multiple fixations during scene exploration is actively and selectively used to construct a task-relevant representation of the scene and objects that are viewed. This construction process is being studied at two levels.

First, there is the level of the overall scene for which, across multiple fixation-saccade-fixation cycles, an internal, episodic model of the contents and spatial layout of that particular scene is developed (Friedman, 1979; De Graef, 1992; Rayner & Pollatsek, 1992; Chun & Nakayama, 2000; Hollingworth & Henderson, 2002; Melcher, 2006; Tatler et al., 2003). Initially, this representation was assumed to make complete abstraction of visual detail and to be sparse, primarily constrained by the capacity limits of visual short-term memory (Alvarez & Cavanagh, 2004; Irwin & Andrews, 1996; Irwin & Zelinsky, 2002; Luck & Vogel, 1997). More recently, however, several authors have argued that every fixation on a scene leaves a visually detailed sensory trace which the visual system samples to construct a large-capacity online scene representation which contains as much visual detail as is required to perform the task at hand (Hollingworth, 2004; Melcher, 2006; Tatler, 2002; Tatler et al., 2005). Indeed, information selection and integration in an overall scene representation has been found to reflect the goals and stages of execution of even very complex tasks such as block copying (Hayhoe, Bensinger, & Ballard, 1998; Hayhoe, Karn, Magnuson, & Mruczek, 2001), driving in the real world (Crundall, 2005) or in a simulator (Shinoda, Hayhoe, & Shrivastava, 2001), playing chess (Reingold & Charnes, 2005), sandwich- and tea-making (Land & Hayhoe, 2001), or playing ping-pong (Land & Furneaux, 1997).

Second, there is the level of the individual object, identified across a single fixation–saccade–fixation cycle during which object information is sampled and integrated from a sequence of peripheral and foveal glimpses. As was the case for scenes, the dominant view on transsaccadic object perception was that it consisted of postcategorical, conceptual and non-visual priming in the object lexicon between the presaccadically activated node for the extrafoveally previewed saccade target and the postsaccadically activated node for that target once it is foveated (e.g., Gordon & Irwin, 2000). This view was supported by numerous demonstrations of failures to notice or react to intrasaccadic changes of visual object detail such as visible object contours (Henderson, 1997), object position (Pollatsek, Rayner, & Henderson, 1990), object size (Pollatsek, Rayner, & Collins, 1984), object in-depth orientation (Henderson & Hollingworth, 1999), individual object parts (Carlson, 1999), or even object exemplar (Archambault, O'Donnell, & Schyns, 1999). Countering this view, three lines of research can be distinguished which advocate the idea that transsaccadic object perception is not merely a matter of postcategorical priming by a peripherally attended saccade target, but rather involves the mandatory integration of specific visual object features across the saccade, even for objects that were not peripherally attended.

3.4. Previous research indicating transsaccadic integration of visual object detail

In a first line, change-blindness studies are challenged as having very limited relevance for understanding representation as it develops across consecutive fixation–saccade cycles (De Graef et al., 2001). First, change blindness disappears entirely when one is warned in advance about the location and type of change that will occur. In contrast, the failure to note certain intrasaccadic changes is resistant to such advance warning, indicating that very different mechanisms are at work (e.g., Verfaillie, 1997). Second, the finding that a viewer cannot overtly report a change does not mean that the changed information was not represented (e.g., Fernandez-Duque, Grossi, Thornton, & Neville, 2003; Fernandez-Duque & Thornton, 2000). This was recently confirmed for scene exploration by showing that when objects were intrasaccadically changed in a scene and when that change was not detected, fixation times on the changed object were still elevated relative to a no-change control (Henderson & Hollingworth, 2003b). In addition, a variety of change-detection studies using temporary postsaccadic blanking of the visual stimulus have revealed a richly detailed but maskable and rapidly decaying transsaccadic representation which is formed after every fixation (De Graef & Verfaillie, 2002; Deubel, Schneider, & Bridgeman, 2002; Gysen, Verfaillie, & De Graef, 2002). Recently, this earlier criticism of change blindness as an inappropriate basis for understanding transsaccadic representation has received eloquent support from some of the authors that launched the systematic study of change blindness (Simons & Ambinder, 2005; Simons & Rensink, 2005).

A second line of work, which counters the claim that visual object detail is not mandatorily represented across saccades, addresses the proposal that we have no on-line representation of those objects which we did not selectively attend to (e.g., Hollingworth, 2004). Contrary to this view, Germeys, De Graef, and Verfaillie (2002) demonstrated location-specific transsaccadic preview benefits on gaze durations for contextual or bystander objects: that is, objects that were present before, during, and after the saccade to another object. Although it is clear that, prior to the saccade to the target object, these bystander objects were abandoned by attention (if they ever were attended in the first place), Germeys et al. (2002) found that the bystanders were easier to identify than a new object at the same location. Importantly, this transsaccadic preview benefit only occurred when the bystander retained its location throughout the fixation–saccade–fixation cycle. This rules out an explanation in terms of location-independent postcategorical priming between stored object representations in a long-term object lexicon and firmly places the effect at the level of an on-line pre-attentive representation of the current visual stimulus.

The third line of work defending the notion of mandatory transsaccadic object perception involves transsaccadic object preview studies, aimed at revealing whether postsaccadic processing of an object on fixation n is modulated by presaccadic processing of that object on fixation $n-1$. To achieve this, the relation between pre- and postsaccadic object appearance is systematically manipulated in order to identify the object properties that are integrated across saccades. Specifically, during the critical saccade towards the target object, the presaccadic object image is replaced by the postsaccadic object image thus selectively altering or preserving particular features of the presaccadic image. Importantly, the transient associated with the change itself is imperceptible due to saccadic suppression, and participants are not asked to monitor and detect changes. Instead, measures of subsequent ease of identification of the postsaccadic object (e.g., naming latency or gaze durations) are collected to determine whether the (violation of) transsaccadic correspondence between pre- and postsaccadic object images has any effect.[2]

[2] The results of transsaccadic object preview studies have been corroborated by transsaccadic object change detection studies in which viewers are explicitly instructed to detect intrasaccadic changes by comparing pre- and postsaccadic object images (e.g.,Verfaillie & De Graef, 2000; De Graef, Verfaillie, & Lamote, 2001; Henderson & Hollingworth, 2003b). However, while transsaccadic integration can be inferred from transsaccadic change detection, it is also possible that the explicit nature of the task activates a comparison process which is not a mandatory part of transsaccadic perception.

For instance, Henderson and Siefert (1999; 2001) found increased transsaccadic preview benefits on object-naming latencies and gaze durations when the presaccadic left-right orientation of an object was preserved across a saccade. As mentioned above, Germeys et al. (2002) found shorter object gaze durations for all objects (i.e., saccade targets and saccade bystanders) that maintained their exact spatial position across a saccade. Earlier, the transsaccadic preservation of visual features such as global object shape had already been noted by Pollatsek et al. (1984). Specifically, these authors observed transsaccadic preview benefits on naming latencies for a fixated object that had been preceded by another, visually similar object peripherally previewed at the same location (e.g., a carrot preceding a baseball bat).

Based on these studies, several authors have proposed that transsaccadic object perception in scenes should be regarded as a basic, functional routine that has evolved in our everyday visuo-motor interaction with the world (Hayhoe, 2000; Verfaillie et al., 2001). The purpose of this routine is to take advantage of the fact that generally we are allowed more than one fixation on a given scene. For the objects in that scene this provides us with the opportunity to integrate object information sampled from a sequence of initially peripheral and ultimately foveal glimpses. Thus, foveal object identification can be jump-started by preliminary peripheral processing. In other words, while a single fixation may in principle be sufficient to identify an object, the preferred *modus operandi* is to increase speed and reliability of object identification by transsaccadic integration of foveal and extrafoveal evidence from multiple fixations. In the next section, I will report some recent evidence in support of this claim.

4. Some new data: Transsaccadic object recognition in scenes

To the best of my knowledge, mandatory transsaccadic object preview effects have never been examined in the context of full, realistic scenes, but have always been established with isolated objects or sparse arrays of isolated objects.[3] This is a potentially important problem if one wants to claim that transsaccadic object perception is the default mode for object recognition in scenes. Indeed, because objects in scenes are camouflaged and/or laterally masked by the background and nearby companion objects, the information extracted from a presaccadic, peripheral preview may be of such a low quality that it provides no constraints whatsoever on the subsequent, postsaccadic foveal identification process.

Thus, what was needed was a study in which viewers explore a complex scene in which their attention and subsequent gaze shift are guided towards a predesignated target location. Before they actually make the required saccade, the target location should be occupied by the postsaccadic target object (the *identical preview* condition), or by a different object (*different preview* condition), or by no object at all (*no preview* condition).

[3] Henderson and Hollingworth (2003b) did record gaze durations on intrasaccadically changed objects during full scene exploration, but this was in the context of an explicit change detection task making it difficult to decide whether the observed effects are mandatory or due to the requested comparison process.

During the saccade, the preview should be replaced by the same target object in all preview conditions. Upon fixation of the target, fixation parameters can be recorded in order to infer ease of target processing.

4.1. Recording object-fixation parameters in a non-object search task

To fill in this lacuna, a study was designed in which participants were presented with realistic black-on-white line drawings of real-world scenes from a pool made available by van Diepen and De Graef (1994) and extended by Hollingworth and Henderson (1998). Participants had to scan the scene in order to count the number of non-objects (a task first used in De Graef, Christiaens, & d'Ydewalle, 1990). The notion *non-object* was borrowed from Kroll and Potter (1984) and refers to a form that resembles real objects in terms of size range and the presence of a clear and closed part-structure but is completely meaningless and nameless (see Figure 3 for an example). By including a variable number of these non-objects in a scene populated with real objects, viewers are encouraged to scan the scene and saccade from one object-like form to the next in search of non-objects. In doing so, they also fixate real objects at which point they only have to determine whether the pictured object matches an object representation in long-term memory and then move on to the next object-like form. This object decision does not require memorization of object characteristics, nor does it require cognitive enhancement of degraded visual object information. In that sense the required object processing approximates what we usually do with most objects in our visual environment, namely register their presence and familiarity without explicitly retrieving their name, identity, or semantics.

During the non-object search task, eye-movement patterns were recorded, allowing the use of object-fixation parameters as a measure of ease of object identification for the real objects, non-object fixation times are never used in the analyses. Of course, there is no consensus on a single, "ideal" oculomotor measure of object perceptibility. In reading research, Inhoff and Radach (1998) found no less than 14 different oculomotor measures which were all claimed to reflect some type of perceptual or cognitive processing. In research on scene perception a similar variety of measures has been used, mostly in an exploratory fashion to determine whether they showed any effects at all (e.g., Antes & Penland, 1981; Henderson, Weeks, & Hollingworth, 1999). For this study, I selected three oculomotor measures based on a simple qualitative model of how ease of object identification could be reflected in the pattern of fixations on that object.

Specifically, it was assumed that when an object is fixated, perceptual encoding increases activation in the object lexicon until one entry reaches a criterion level and triggers object identification. In this view, objects are easier to process when their identification thresholds are reached more quickly due to a higher rate of activation accumulation and/or a lower identification threshold. There is a good deal of evidence that activation rate is directly reflected in fixation duration. Temporarily masking or degrading the stimulus in foveal vision increases fixation durations (Loftus, Kaufman, Nishimoto, & Ruthruff, 1992; Sanders & van Duren, 1998; van Diepen, Ruelens, & d'Ydewalle, 1999;

van Duren & Sanders, 1992). Similarly, higher identification thresholds affect fixation durations. For instance, words with a low frequency in the language (Rayner & Duffy, 1986; Rayner, Sereno, & Raney, 1996) or a low level of predictability from the sentence context (Rayner & Well, 1996; Sereno, 1995) were found to receive longer fixations. Hence, it seems a reasonable hypothesis that object-fixation time reflects the ease with which the fixated object can be encoded and identified.

However, objects generally are fixated more than once during scene exploration so the question arises *which* object-fixation time reflects ease of object identification. The first restriction is that only *first-pass* fixation times will be considered, that is, fixations during the first visit to the object. When the eye re-enters an object it has left before, this can be incidental in which case object processing will be atypically fast because the object is still active in the lexicon, or because processing is terminated as soon as the system detects it was already there. Alternatively, a re-entry can be purposeful to check for change or to resolve inconsistencies between the initial object interpretation and the contextual information gathered from other parts of the scene. In this case, fixation times are bound to be atypically slow and reflect a great deal of post-perceptual integration and comparison.

Even when only first object passes are considered, refixations are quite common and a variety of processing measures can and have been defined: single-fixation durations (i.e., discard all multiple-fixation passes), first-fixation duration, second-fixation or refixation duration, first-gaze or summed-fixation duration, average fixation duration, and number of first-pass fixations. From these measures, I selected first-fixation duration, first-gaze duration, and number of first-pass fixations, henceforth *first-pass density*. The rationale behind this selection is that on every first pass of an object, the perceptual system enters a fixation decision space which is schematically represented in Figure 1. As activation in the object lexicon starts to accumulate, a continuous evaluation is made of Δa, the amount of added activation per unit of time. As long as this activation rate stays above a threshold A, the ongoing fixation is maintained until eventually the identification threshold ID is reached and the fixation can be terminated. Any subsequent refixations are assumed to reflect post-identification processing. If, however, Δa does not reach A or drops below it before ID is reached, the system decides that the ongoing fixation is suboptimal and will refixate in order to achieve identification. Importantly, the decision to refixate is not governed exclusively by the low-level detection of a deviation from some optimal viewing position (as, for instance, stated by the strategy-tactics theory of eye-movement control in reading, O'Regan, 1992b). Instead, the probability of refixation is a function of activation rate which can be influenced by both high-level factors such as object familiarity or object predictability, and low-level factors such as stimulus contrast or perhaps deviation from the object's center of gravity (Henderson, 1993; Melcher & Kowler, 1999).

Given this model, the most reliable index of ease of object identification would in principle be single-fixation duration because all measures involving refixations may include post-identification components, while first-of-multiple fixations may be terminated prior to identification. However, discarding all multiple-fixation cases would greatly reduce the power of eye-movement studies and eliminate observations that may be just as valid as the single-fixation cases: In a direct comparison of word frequency effects on word-fixation parameters, Rayner et al. (1996) found the same effects on single-fixation durations, first-of-multiple fixations and first-gaze durations. I therefore opted to use first-fixation duration as a lower-limit estimate of ease of object identification, first-gaze duration as an upper-limit estimate and first-pass density as an index of duration-refixation tradeoffs: Shorter first-fixation durations only indicate easier object identification when they are not combined with higher first-pass densities. In the latter case, they in fact suggest a more difficult object identification.

One might argue that first-pass density is in fact a superfluous measure because refixations will always increase first-gaze duration. Thus, shorter first fixations combined with longer first gazes would indicate that the first fixations were prematurely interrupted because of a sub-criterion activation rate (i.e., an activation rate located in the grey area in Figure 1). However, Henderson (1993) pointed out that object refixations that follow short first fixations often are so short that they sum to gaze durations which are shorter than single-fixation durations. This suggests that there is a redundancy gain in refixating, possibly due to a higher processing rate of visual information during the initial stages of a fixation (Rayner, Inhoff, Morrison, Slowiaczek, & Bertera, 1981; van Diepen, De Graef, & d'Ydewalle, 1995). Therefore, a premature first-fixation cutoff does not always produce longer gaze durations but it does always lead to an increase in first-pass density.

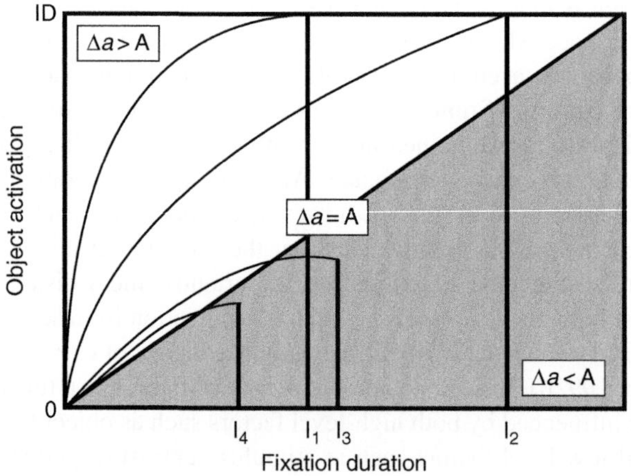

Figure 1. Fixation decision space relating activation in the object lexicon to fixation duration and refixation probability. Activation curves are plotted for 4 objects ranked for identifiability from easy (I_1) to difficult (I_4). Fixations are maintained until identification threshold (ID) is reached unless activation rate Δa drops below A (shaded area). In that case a refixation is planned. Note that lower identifiability can lead to longer single fixations ($I_2 > I_1$), or to a long first-of-multiple fixations ($I_3 > I_1$), or to a short first-of-multiple fixations ($I_4 < I_1$).

4.2. Measuring transsaccadic preview benefits in the non-object search task

In the present study, participants were told that they would take part in one of a series of experiments on how good people are at detecting various kinds of information in images of varying complexity. In this particular experiment, they would see line drawings depicting real-world scenes containing both real objects and non-objects. Each scene would have to be explored, starting at a peripheral position marked by a red rectangle (average eccentricity was 5°). After each trial, participants would have to answer (a) whether the object at the marked location was a non-object, and (b) whether any additional non-objects were present in the scene. All participants were informed that their accuracy in detecting the non-objects would be evaluated in two ways. First, following each stimulus they would have to use response keys to answer the two specified questions (one about the marked object, the other about the rest of the scene). Second, their eye movements would be registered during the entire scene exposure to determine whether they had indeed localized the non-objects in the display or had just guessed. Eye movements were recorded with a Generation 5.5 dual-Purkinje-image eye-tracker with a 1000 Hz sampling frequency and a spatial accuracy of 1 min of arc (Crane & Steele, 1985).

The course of a trial in this study is illustrated in Figure 2. First, the probe question appeared in the center of the screen (i.e., "*any non-objects?*"). When the participant had read this question, a button press replaced the question with a central fixation cross. If this was properly fixated for at least 200 ms, one of the preview displays appeared. Participants were instructed to always make their first eye movement to the center of the red rectangle in the preview display. As soon as the first eye movement was detected a display change command was initiated (typically 4–5 ms after saccade onset) and the preview display was replaced by a target display. The display change was started in mid-scan and was completed in one refresh (20 ms). Thus, the time required to intrasaccadically switch from preview to target display was about 25 ms, well below the duration of the viewers' saccades from the display center to the marked location (i.e., 45 ms). Following the display change, the eye-tracker monitored the position of the red rectangle for a period of 100 ms. If no fixation was detected inside the rectangle within that period, the trial was interrupted and participants received a text message instructing them to start scene exploration at the red rectangle. If a proper fixation was detected, then the target display remained on the screen until 4 s had elapsed since the onset of the preview display. Then, the target display was replaced by the first response screen with the question "*In the red rectangle?*" and the response alternatives "*left = yes, right = no*". Thus, participants were reminded that they had to decide whether the object in the rectangle was a non-object. Following a response, the second response screen appeared with the follow-up question "*In the rest of the scene?*" and the response alternatives "*left = yes, right = no*". Following the second response, a new trial was initiated.

To study transsaccadic preview benefits, the correspondence between pre- and post-saccadic contents of the location marked by the red rectangle was manipulated as shown in Figure 3. Presaccadically, the rectangle either contained no object (no preview), or it already contained the postsaccadic object (target preview), or it contained an object that

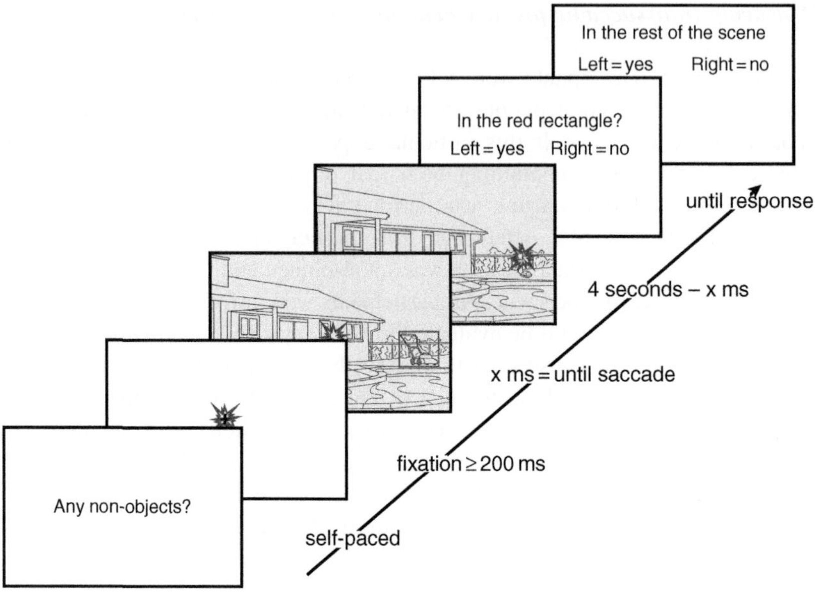

Figure 2. Course of a trial in the transsaccadic preview paradigm. After having read a probe question, participants initiate presentation of a central fixation cross (starburst symbolizes fixation position). Following a fixation of at least 200 ms, the preview display is presented (illustrated is the consistent target preview lawnmower) containing a red rectangle. Participants are instructed to make their first saccade to the rectangle. During the saccade, the scene is changed to the target display (illustrated is the inconsistent target vacuum cleaner) which remains on the screen for four seconds counting from scene onset. Subsequently, two response screens are presented to answer the probe question for the framed object and for the rest of the scene.

was semantically consistent with the scene and the postsaccadic target object (consistent companion preview), or it contained an object that was semantically inconsistent with the scene and the postsaccadic target object (inconsistent companion preview). Postsaccadically, the marked location always contained the target object, or a non-object while the rectangle had disappeared. Target-fixation parameters were recorded to infer ease of target object processing and to study whether that was affected by the type of preview. Specifically, if having a target preview helped postsaccadic target identification then target fixations should be shorter and fewer in the preview condition than in the no-preview condition. The consistent and inconsistent companion previews were used to examine whether transsaccadic object perception in scenes is (partly) mediated by episodic priming. Henderson and Anes (1994) and Germeys et al. (2002) demonstrated that transsaccadic preview benefits for isolated objects are partly based on pre-saccadic access to the preview's long-term representation in the object lexicon. Henderson (1992) argued for an automatic spreading of activation between episodically related individual object representations in the object lexicon, a claim recently supported by Bar (2004). The logical combination of these lines of research is that transsaccadic preview benefits in scene exploration are at least partly based on access to the preview's episodic

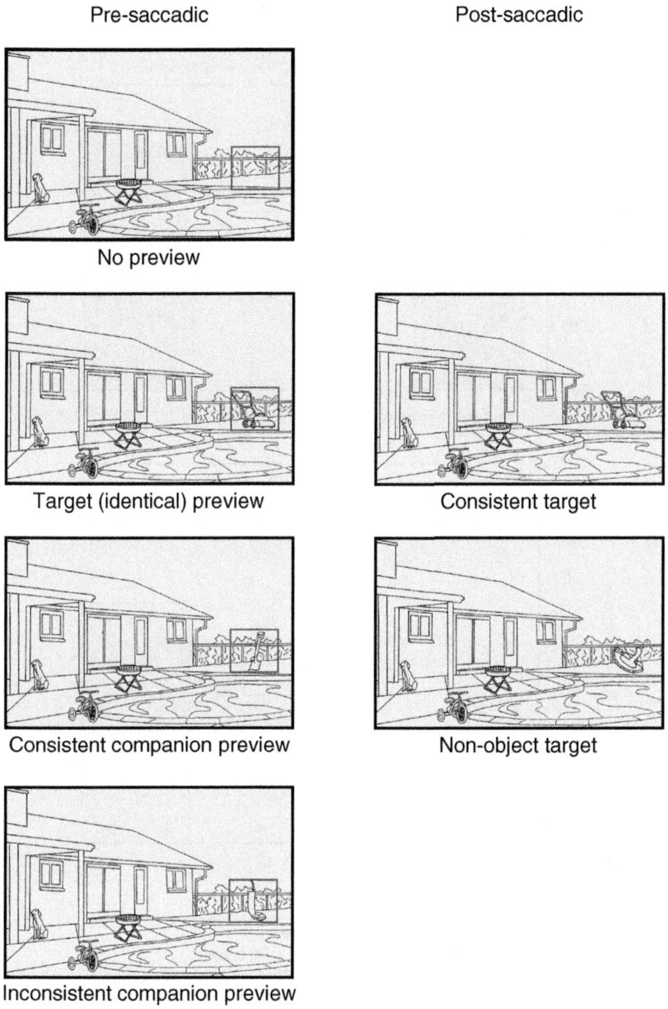

Figure 3. Four types of pre-saccadic preview display are orthogonally combined with two types of post-saccadic target display. Red rectangles appeared in the actual stimulus and indicate the target location for the participant's first saccade.

membership, that is, on knowledge about the preview's plausible companion objects in real-world settings. By including consistent and inconsistent companion previews, it could be determined whether a consistent preview would facilitate subsequent target processing relative to an inconsistent preview.

As can be seen in Figure 4, clear differences between preview conditions emerged, with fastest processing for the target (identical) preview and slowest for the inconsistent companion preview. A reliable transsaccadic preview benefit was found in the comparison

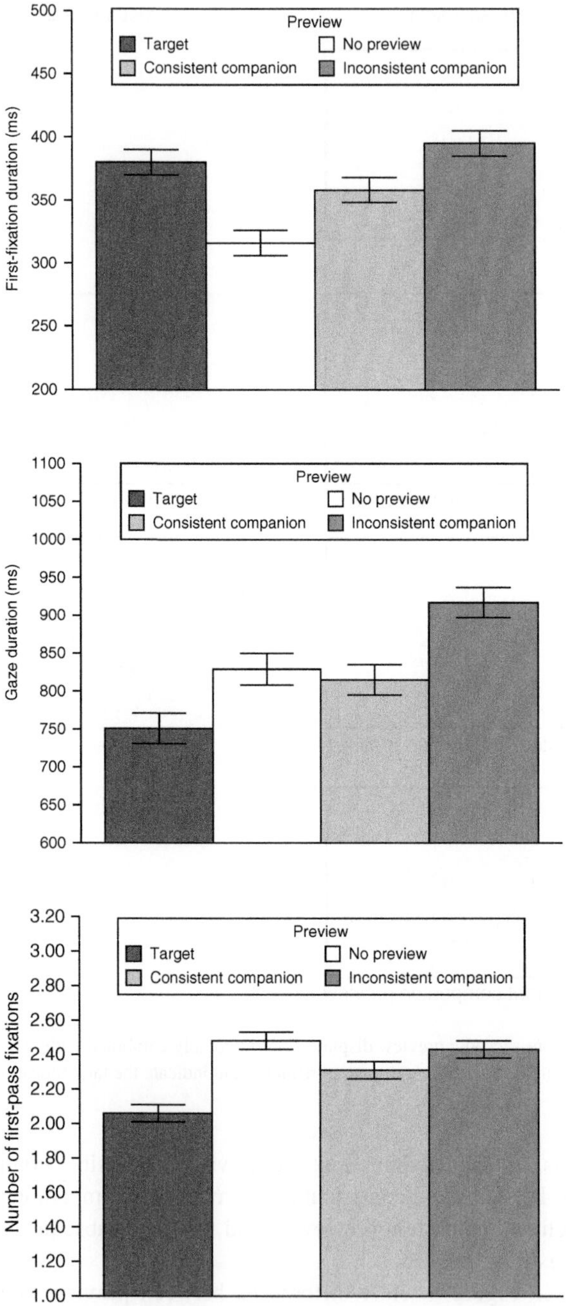

Figure 4. Target object fixation parameters. Means and standard errors for first-fixation duration (top), gaze duration (middle) and first-pass density (bottom) as a function of preview type.

between target preview and no preview, with a longer first fixation ($p < 0.0001$), a shorter gaze duration ($p < 0.02$) and fewer first-pass refixations ($p < 0.0001$) when the target was preceded by an identical preview. The pattern of a sustained first fixation followed by shorter and fewer additional fixations is compatible with a higher rate of information extraction as argued above (see Figure 1).

Of course, one could argue that the pattern is perhaps an artifact produced by the unavailability of a proper saccade target in the no-preview condition. Specifically, in that condition participants are saccading toward an empty rectangle rather than toward an object and therefore the postsaccadic deviation between object center and landing position may be systematically higher in the no-preview condition, causing a short first fixation followed by a quick corrective saccade and a refixation. Because postsaccadic deviation from the target's center did indeed prove to be higher in the no-preview than in the identical preview condition (0.93°) than in the other conditions (0.68°), this alternative explanation deserves further consideration. A first argument against it is that the same pattern of a longer first fixation, shorter gaze and fewer refixations is evident in the comparison between target preview and inconsistent companion preview. In this comparison, both preview objects were presaccadically presented at exactly the same location and no differences in postsaccadic deviation resulted, yet the same difference in fixation parameters was present. A second argument follows from an additional analysis: To correct for possible effects of postsaccadic deviation form the object's center, this variable was included as a covariate. Although greater postsaccadic deviation reliably decreased first-fixation duration ($p < 0.0001$), decreased gaze duration ($p < 0.009$), and increased first-pass density ($p < 0.04$), the pattern and reliability of preview effects were not changed.

In addition to a transsaccadic preview benefit, the data for consistent targets also showed transsaccadic priming: Targets preceded by an episodically related consistent companion preview were fixated shorter ($p < 0.0001$ for first-fixation duration and $p < 0.02$ for gaze) than targets preceded by an unrelated inconsistent companion preview. First-pass density showed the same pattern but non-reliable ($p = 0.14$).

4.3. Transsaccadic object perception in scenes: Conclusions

The test for transsaccadic preview benefits in non-object search yielded two main results. First, object perception in scenes is inherently transsaccadic. When a viewer is given an extrafoveal preview of an object, subsequent foveal processing of that object is faster than when no preview was available. This finding is new because it shows that even when peripheral figure-ground segregation of objects is complicated by the presence of a realistic scene background, preview information is sufficient to mandatorily influence subsequent foveal processing. In other words, the findings in full scenes have now been shown to be in line with earlier work demonstrating transsaccadic preview benefits for both words (e.g., Rayner & Morris, 1992) and isolated objects (e.g., Pollatsek et al., 1984) and provides yet another argument against the view that perception starts anew on every new fixation (O'Regan, 1992).

Second, transsaccadic facilitation of postsaccadic, foveal object processing by a presaccadic, extrafoveal object preview is partly mediated by inter-object connections in long-term memory that code episodic relatedness. Specifically, having a presaccadic preview of an object that is likely to appear in the same real-world settings as the postsaccadic target object facilitates target recognition relative to a situation where the presaccadic object does not belong to the target's episodic category. This transsaccadic priming effect is at odds with earlier failures to find transsaccadic object priming between semantically related, isolated objects (Henderson, 1992; Pollatsek et al., 1984). The present study suggests that for these episodic inter-object priming effects to operate the presence of a coherent scene setting is required. Bar (2003, 2004) has recently outlined a possible neural substrate for such a mechanism. Specifically, a coarse analysis of background and objects during the very first glance at a scene is assumed to project in parallel to the parahippocampal cortex (PHC) where it activates scene schemas, and to the prefrontal cortex where it activates a set of "*initial guesses*" about the possible identity of objects present in the scene. Both the activated schema and the set of possible objects project to inferior temporal (IT) cortex where they modulate the activation of stored object representations and thus facilitate the object recognition process. Because the episodic associations between objects are not stored in the object lexion itself (IT) but in the contextual associations stored in PHC, activation of those associations by the global setting information contained in a full scene is required to observe episodic priming effects.

5. Conclusion: Time to put a transsaccadic theory of recognition on the agenda

Earlier in this chapter, I have attempted to clarify why, in my opinion, current prominent theories of object and scene recognition provide no account of how recognition is achieved on the basis of information gathered from multiple distinct samples of the viewed stimulus, which are collected on spatially and temporally disparate fixations. Two reasons were identified why transsaccadic information integration is neglected in models of visual recognition.

First, there is the widespread conviction that the human visual system is so powerful that one short glance, well below the modal fixation duration, is more than enough to recognize a scene or an object. In a brief review of the existing evidence for that claim, I have tried to show that from a strictly chronometric point of view, single-fixation perception is sufficient to make low-level discriminations between object and scene categories, but may quite often fall short of true recognition. This would necessitate a second fixation if recognition is what the visual system tries to achieve before it moves on the next stimulus or stimulus location. Naturally, identification may not always be the goal of scene exploration (e.g., in single feature search, only the feature value is important not the spatio-temporal entity it is bound to). However, when identification is task-relevant then taking a second shot is likely to yield more reliable results than single-shot perception.

Second, for many years the very notion of transsaccadic information integration has been severely challenged by studies demonstrating (transsaccadic) change blindness and research disproving transsaccadic spatiotopic fusion. This has culminated in the prominent theoretical position that transsaccadic integration is a non-functional concept, because a visual on-line representation of the outside world is superfluous, given that the world itself can continuously be sampled anew on every fixation. In reply to this line of reasoning, I have reviewed evidence that shows that transsaccadic change blindness can easily be undone to reveal the presence of a visual on-line representation. In addition, I have reported new evidence indicating that during scene exploration, viewers routinely use the object information provided in a presaccadic, extrafoveal preview to speed up subsequent processing of that object during its postsaccadic foveation.

In conclusion, I hope to have shown that there are both chronometric and functional arguments to defend the claim that theories of object and scene recognition should make it their business to find out how the visual system exploits the advantage of having multiple extrafoveal and foveal samples of a stimulus in order to speed up and enhance the reliability of stimulus recognition. How exactly presaccadic identity hypotheses may influence postsaccadic foveal processing is outside the present scope, but answers to this question are bound to emerge from a rapidly growing set of studies on reverse hierarchy theory (Ahissar & Hochstein, 1997; Hochstein & Ahissar, 2002), the distinction between feed-forward and recurrent processing modes in vision (Lamme & Roelfsema, 2000; Ullman, 1996), and the notion of re-entrant visual processes (DiLollo et al., 2000). Despite some variations in naming, all these theories are centered around the claim that bottom-up and top-down processing streams can be distinguished behaviorally and neurophysiologically but are not functionally segregated. Specifically, all these theories propose that visual perception involves two streams of processing: A rapid, automatic, and pre-attentive feed-forward sweep of activation through the informational and cortical hiearchy which activates high-level, categorical stimulus representations; and a slower, attention-modulated, re-entrant or recurrent stream of processing originating in the higher-level representations and with a modulating effect on ongoing bottom-up processing of visual input.

The strength of this framework is twofold. First, it is sufficiently precise to outline testable mechanisms of how high-level categorical representations are integrated with lower-level, precategorical representations of the visual input, and this both at the computational level (e.g., Di Lollo et al., 2000; Di Lollo, Enns, & Rensink, 2002) and at the level of specific cortical mechanisms (e.g., Bar, 2003). Second, it is based on both psychophysical effects such as object substitution masking (e.g., Enns & DiLollo, 1997, 2001; Jiang & Chun, 2001; Lleras & Moore, 2003) and on neurophysiological data such as the existence of massive backprojections in the cortical hierarchy (Salin & Bullier, 1995) or the finding that over the course of their response, V1 neurons change their tuning from simple to more complex stimuli (Lamme, Zipser, & Spekreijse, 2002). By applying this framework to our eye-movement paradigms which were developed to study perception within the spatial and temporal information-integration constraints imposed by the oculomotor system, we should be able to develop a detailed account of the mechanisms involved in transsaccadic recognition of objects and scenes.

Acknowledgement

The writing of this chapter was supported by research grant GOA 2005/03 of the Research Fund K. U. Leuven and conventions G.0583.05 and 7.005.05 of the Fund for Scientific Research-Flanders.

References

Ahissar, M., & Hochstein, S. (1997). Task difficulty and learning specificity: Reverse hierarchies in sensory processing and perceptual learning. *Nature, 387*, 401–406.

Altmann, G., van Nice, K., Garnham, A., & Henstra, J. A. (1998). Late closure in context. *Journal of Memory and Language, 38*, 459–484.

Alvarez, G. A., & Cavanagh, P. (2004). The capacity of visual short-term memory is set both by visual information load and by number of objects. *Psychological Science, 15*, 106–111.

Antes, J. R., & Penland, J. G. (1981). Picture context effects on eye movement patterns. In D. F. Fisher, R. A. Monty, & J. W. Senders (Eds.), *Eye Movements: Cognition and Visual Perception* (pp. 157–170). Hillsdale, NJ: Erlbaum.

Archambault, A., O'Donnell, C., & Schyns, P. G. (1999). Blind to object changes: When learning the same object at different levels of categorization modifies its perception. *Psychological Science, 10*, 249–255.

Balota, D. A., Pollatsek, A., & Rayner, K. (1985). The interaction of contextual constraints and parafoveal visual information in reading. *Cognitive Psychology, 17*, 364–390.

Bar, M. (2003). A cortical mechanism for triggering top-down facilitation in visual object recognition. *Journal of Cognitive Neuroscience, 15*, 600–609.

Bar, M. (2004). Visual objects in context. *Nature Reviews Neuroscience, 5*, 617–629.

Biederman, I. (1987). Recognition-by-components: A theory of human image understanding. *Psychological Review, 94*, 115–148.

Biederman, I., & Ju, G. (1988). Surface vs edge-based determinants of visual recognition. *Cognitive Psychology, 20*, 38–64.

Biederman, I., Mezzanotte, R. J., & Rabinowitz, J. C. (1982). Scene perception: Detecting and judging objects undergoing relational violations. *Cognitive Psychology, 14*, 143–177.

Bridgeman, B., & Mayer, M. (1983). Failure to integrate visual information from successive fixations. *Bulletin of the Psychonomic Society, 21*, 285–286.

Bridgeman, B., Van der Heijden, A. H. C., & Velichkovsky, B. M. (1994). A theory of visual stability across saccadic eye movements. *Behavioral and Brain Sciences, 17*, 247–292.

Brysbaert, M., Drieghe, D., & Vitu, F. (2005). Word skipping: Implications for theories of eye movement control in reading. In G. Underwood (Ed.), *Cognitive Processes in Eye guidance* (pp. 53–77). Oxford: University Press.

Brysbaert, M., & Mitchell, D. C. (1996). Modifier attachment in sentence parsing: Evidence from Dutch. *Quarterly Journal of Experimental Psychology, 49A*, 664–695.

Carlson, L. A. (1999). Memory for relational information across eye movements. *Perception & Psychophysics, 61*, 919–934.

Chun, M. M., & Nakayama, K. (2000). On the functional role of implicit visual memory for the adaptive deployment of attention across scenes. *Visual Cognition, 7*, 65–81.

Cook, A. E., & Myers, J. L. (2004). Processing discourse roles in scripted narratives: The influences of context and world knowledge. *Journal of Memory and Language, 50*, 268–288.

Crane, H. D., & Steele, C. M. (1985). Generation-V dual-Purkinje-image eyetracker. *Applied optics, 24*, 527–537.

Crundall, D. (2005). The integration of top-down and bottom-up factors in visual search while driving. In G. Underwood (Ed.), *Cognitive Processes in Eye Guidance*(pp. 283–302). Oxford: Oxford University Press.

De Graef, P. (1992). Scene-context effects and models of real-world perception. In K. Rayner (Ed.), *Eye movements and visual cognition: scene perception and reading* (pp. 243–259). New York: Springer.

De Graef, P., Christiaens, D., & d'Ydewalle, G. (1990). Perceptual effects of scene context on object identification. *Psychological Research, 52*, 317–329.

De Graef, P., & Verfaillie, K. (2002). Transsaccadic memory for visual object detail. *Progress in Brain Research, 140*, 181–196.

De Graef, P., Verfaillie, K., Germeys, F., Gysen, V., & Van Eccelpoel, C. (2001). Trans-saccadic representation makes your Porsche go places. *Behavioral and Brain Sciences, 24*, 981–982.

De Graef, P., Verfaillie, K., & Lamote, C. (2001). Transsaccadic coding of object position: Effects of saccadic status and allocentric reference frame. *Psychologica Belgica, 41*, 29–54.

Delorme, A., Rousselet, G. A., Macé, M. J.-M., & Fabre-Thorpe, M. (2004). Interaction of top-down and bottom-up processing in the fast visual analysis of natural scenes. *Cognitive Brain Research, 19*, 103–113.

Deubel, H., Schneider, W. X., & Bridgeman, B. (2002). Transsaccadic memory of position and form. *Progress in Brain Research, 140*, 165–180.

Di Lollo, V., Enns, J. T., & Rensink, R. A. (2000). Competition for consciousness among visual events: The psychophysics of reentrant visual processes. *Journal of Experimental Psychology: General, 129*, 481–507.

Di Lollo, V., Enns, J. T., & Rensink, R. A. (2002). Object substitution without reentry? *Journal of Experimental Psychology: General, 131*, 594–596.

Edelman, S. (1999). *Representation and recognition in vision*. Cambridge, MA: MIT Press.

Enns, J. T., & Di Lollo, V. (1997). Object substitution: A new form of masking in unattended visual locations. *Psychological Science, 8*, 135–138.

Enns, J. T., & Di Lollo, V. (2001). An object substitution theory of visual masking. In T. F. Shipley, & P. J. Kellman (Eds.), *From fragments to objects: Segmentation and grouping in vision*(pp. 121–143). New York: Elsevier Science.

Feldman, J. A. (1987). A functional model of vision and space. In M. Arbib, & A. Hanson (Eds.), *Vision, Brain, and Cooperative Computation* (pp. 531–562). Cambridge: MIT Press.

Fernandez-Duque, D., Grossi, G., Thornton, I. M., & Neville, H. J. (2003). Representation of change: Separate electrophysiological markers of attention, awareness, and implicit processing. *Journal of Cognitive Neuroscience, 15*, 491–507.

Fernandez-Duque, D., & Thornton, I. M. (2000). Change detection without awareness: Do explicit reports underestimate the representation of change in the visual system? *Visual Cognition, 7*, 323–344.

Friedman, A. (1979). Framing pictures: The role of knowledge in automatized encoding and memory for gist. *Journal of Experimental Psychology: General, 108*, 316–355.

Germeys, F., De Graef, P., & Verfaillie, K. (2002). Transsaccadic identification of saccade target and flanker objects. *Journal of Experimental Psychology: Human Perception and Performance, 28*, 868–883.

Gordon, R. D., & Irwin, D. E. (2000). The role of physical and conceptual properties in preserving object continuity. *Journal of Experimental Psychology: Learning, Memory, and Cognition, 26*, 136–150.

Grimes, J. (1996). On the failure to detect changes in scenes across saccades. In K. Akins (Ed.), *Perception* (Vancouver Studies in Cognitive Science, Vol. 5, (pp. 89–110). New York: Oxford University Press.

Gysen, V., Verfaillie, K., & De Graef, P. (2002). The effect of stimulus blanking on the detection of intrasaccadic displacements of translating objects. *Vision Research, 42*, 2021–2030.

Hayhoe, M. M. (2000). Vision using routines: A functional account of vision. *Visual Cognition, 7*, 43–64.

Hayhoe, M. M., Bensinger, D. G., & Ballard, D. H. (1998). Task constraints in visual working memory. *Vision Research, 38*, 125–137.

Hayhoe, M. M., Karn, K., Magnuson, J., & Mruczek, R. (2001). Spatial representations across fixations for saccadic targeting. *Psychologica Belgica, 41*, 55–74.

Henderson, J. M. (1992). Identifying objects across saccades: Effects of extrafoveal preview and flanker object context. *Journal of Experimental Psychology: Learning, Memory, and Cognition, 18*, 521–530.

Henderson, J. M. (1993). Eye movement control during visual object processing: Effects of initial fixation position and semantic constraint. *Canadian Journal of Experimental Psychology, 47*, 79–98.

Henderson, J. M. (1997). Transsaccadic memory and integration during real-world object perception. *Psychological Science, 8*, 51–55.

Henderson, J. M. (2003). Human gaze control during real-world scene perception. *Trends in Cognitive Sciences, 7*, 498–504.

Henderson, J. M., & Anes, M. D. (1994). Roles of object-file review and type priming in visual identification within and across eye fixations. *Journal of Experimental Psychology: Human Perception and Performance, 20*, 826–839.

Henderson, J. M., & Hollingworth, A. (1998). Eye movements during scene viewing: An overview. In G. Underwood (Ed.), *Eye guidance in reading and scene perception* (pp. 269–298). Oxford: Elsevier.

Henderson, J. M., & Hollingworth, A. (1999). The role of fixation position in detecting scene changes across saccades. *Psychological Science, 10*, 438–443.

Henderson, J. M., & Hollingworth, A. (2003). Eye movements and visual memory: Detecting changes to saccade targets in scenes. *Perception & Psychophysics, 65*, 58–71.

Henderson, J. M., & Hollingworth, A. (2003). Global transsaccadic change blindness during scene perception. *Psychological Science, 14*, 493–497.

Henderson, J. M., Weeks, P. A. Jr., & Hollingworth, A. (1999). The effects of semantic consistency on eye movements during complex scene viewing. *Journal of Experimental Psychology: Human Perception and Performance, 25*, 210–228.

Hochstein, S., & Ahissar, M. (2002). View from the top: Hierarchies and reverse hierarchies in the visual system. *Neuron, 36*, 791–804.

Hollingworth, A. (2004). Constructing visual representations of natural scenes: The roles of short- and long-term visual memory. *Journal of Experimental Psychology: Human Perception and Performance, 30*, 519–537.

Hollingworth, A., & Henderson, J. M. (1998). Does consistent scene context facilitate object perception? *Journal of Experimental Psychology: General, 127*, 398–415.

Hollingworth, A., & Henderson, J. M. (2002). Accurate visual memory for previously attended objects in natural scenes. *Journal of Experimental Psychology: Human Perception and Performance, 28*, 113–136.

Hyöna, J., Niemi, P., & Underwood, G. (1989). Reading long words embedded in sentences: Informativeness of word parts affects eye movements. *Journal of Experimental Psychology: Human Perception and Performance, 15*, 142–152.

Intraub, H. (1981). Identification and processing of briefly glimpsed visual scenes. In D. F. Fisher, R. A. Monty, & J. W. Senders (Eds.), *Eye movements: Cognition and Visual Perception* (pp. 181–190). Hillsdale: Erlbaum.

Irwin, D. E. (1991). Information integration across saccadic eye movements. *Cognitive Psychology, 23*, 420–456.

Irwin, D. E. (1996). Integrating information across saccadic eye movements. *Current Directions in Psychological Science, 5*, 94–100.

Irwin, D. E. (1998). Lexical processing during saccadic eye movements. *Cognitive Psychology, 36*, 1–27.

Irwin, D. E., & Andrews, R. V. (1996). Integration and accumulation of information across saccadic eye movements. In T. Inui, & J. L. McClelland (Eds.), *Attention and performance XVI: Information integration in perception and communication* (pp. 125–155). Cambridge, MA: Bradford.

Irwin, D. E., & Zelinsky, G. J. (2002). Eye movements and scene perception: Memory for things observed. *Perception & Psychophysics, 64*, 882–895.

Jiang, Y., & Chun, M. M. (2001). Asymmetric object substitution masking. *Journal of Experimental Psychology: General, 27*, 895–918.

Johnson, J. S., & Olshausen, B. A. (2003). Timecourse of neural signatures of object recognition. *Journal of Vision, 3*, 499–512.

Kambe, G., Rayner, K., & Duffy, S. A. (2001). Global context effects on processing lexically ambiguous words: Evidence from eye fixations. *Memory & Cognition, 29*, 363–372.

Keysers, C., Xiao, D.-K., Földiák, P., & Perrett, D. I. (2001). The speed of sight. *Journal of Cognitive Neuroscience, 13*, 90–101.

Khayat, P. S., Spekreijse, H., & Roelfsema, P. R. (2004). Visual information transfer across eye movements in the monkey. *Vision Research, 44*, 2901–2917.

Kroll, J. F., & Potter, M. C. (1984). Recognizing words, pictures, and concepts: A comparison of lexical, object, and reality decisions. *Journal of Verbal Learning and Verbal Behavior, 23*, 39–66.

Lamme, V. A. F., & Roelfsema, P. R. (2000). The distinct modes of vision offered by feedforward and recurrent processing. *Trends in Neurosciences, 23*, 571–579.

Lamme, V. A. F., Zipser, K., & Spekreijse, H. (2002). Masking interrupts figure-ground signals in V1. *Journal of Cognitive Neuroscience, 14*, 1044–1053.

Land, M. F., & Furneaux, S. (1997). The knowledge base of the oculomotor system. *Philosophical Transactions of the Royal Society of London, SeriesB: Biological Sciences, 352*, 1231–1239.

Land, M. F., & Hayhoe, M. M. (2001). In what ways do eye movements contribute to everyday activities? *Vision Research, 41*, 3559–3565.

Land, M. F., Mennie, N., & Rusted, J. (1999). The roles of vision and eye movements in the control of activities in daily living. *Perception, 28*, 1311–1328.

Large, M.-E., Kiss, I., & McMullen, P. A. (2004). Electrophysiological correlates of object categorization: back to basics. *Cognitive Brain Research, 20.*

Lleras, A., & Moore, C. M. (2003). When the target becomes the mask: Using apparent motion to isolate the object-level component of object substitution masking. *Journal of Experimental Psychology: Human Perception and Performance, 29*, 106–120.

Loftus, G. R., Kaufman, L., Nishimoto, T., & Ruthruff, E. (1992). Effects of visual degradation on eye fixation durations, perceptual processing, and long-term visual memory. In K. Rayner (Ed.), *Eye movements and visual cognition: Scene perception and reading* (pp. 203–226). New York: Springer Verlag.

Luck, S. J., & Vogel, E. K. (1997). The capacity of visual working memory for features and conjunctions. *Nature, 390*(6657), 279–281.

Matin, E. (1974). Saccadic suppression: A review and analysis. *Psychological Bulletin, 81*, 899–917.

Matin, L. (1986). Visual localization and eye movements. In K. R. Boff, L. Kaufman, & J. P. Thomas (Eds.), *Handbook of perception and human performance*(Vol. 1, pp. 20.1–20.45). New York: Wiley.

McConkie, G. W., & Currie, C. B. (1996). Visual stability across saccades while viewing complex pictures. *Journal of Experimental Psychology: Human Perception and Performance, 22*, 563–581.

McConkie, G. W., & Rayner, K. (1976). Identifying the span of the effective stimulus in reading: Literature review and theories of reading. In H. Singer, & R. B. Ruddell (Eds.), *Theoretical models and processes in reading*. Newark, DE: International Reading Association.

McConkie, G. W., & Zola, D. (1979). Is visual information integrated across succesive fixations in reading? *Perception & Psychophysics, 25*, 221–224.

Melcher, D. (2001). Persistence of visual memory for scenes. *Nature, 412*, 401.

Melcher, D. (2006). Accumulation and persistence of memory for natural scenes. *Journal of Vision*, 8–17.

Melcher, D., & Kowler, E. (1999). Shape, surfaces and saccades. *Vision Research, 39*, 2929–2946.

Melcher, D., & Morrone, M. C. (2003). Spatiotopic temporal integration of visual motion across saccadic eye movements. *Nature Neuroscience, 6*, 877–881.

O'Regan, J. K. (1992). Optimal viewing position in words and the strategy-tactics theory of eye movements in reading. In K. Rayner (Ed.), *Eye Movements and visual cognition: Scene perception and reading* (pp. 333–354). New York: Springer Verlag.

O'Regan, J. K. (1992). Solving the "real" mysteries of visual perception: The world as an outside memory. *Canadian Journal of Psychology, 46*, 461–488.

O'Regan, J. K., Deubel, H., Clark, J. J., & Rensink, R. A. (2000). Picture changes during blinks: Looking without seeing and seeing without looking. *Visual Cognition, 7*, 191–211.

O'Regan, J. K., & Nöe, A. (2001). A sensorimotor account of vision and visual consciousness. *Behavioral and Brain Sciences, 24*, 939–1031.

O'Regan, J. K., Rensink, R. A., & Clark, J. J. (1999). Change-blindness as a result of "mudsplashes". *Nature, 398*, 34.

Oliva, A., & Torralba, A. (2001). Modeling the shape of the scene: A holistic representation of the spatial envelope. *International Journal of Computer Vision, 42*, 145–175.

Pollatsek, A., Lesch, M., Morris, R. K., & Rayner, K. (1992). Phonological codes are used in integrating information across saccades in word identification and reading. *Journal of Experimental Psychology: Human Perception and Performance, 18*, 148–162.

Pollatsek, A., & Rayner, K. (1992). What is integrated across fixations? In K. Rayner (Ed.), *Eye movements and visual cognition: Scene perception and reading* (pp. 166–191). New York: Springer.

Pollatsek, A., Rayner, K., & Collins, W. E. (1984). Integrating pictorial information across eye movements. *Journal of Experimental Psychology: General, 113*, 426–442.

Pollatsek, A., Rayner, K., & Henderson, J. M. (1990). Role of spatial location in integration of pictorial information across saccades. *Journal of Experimental Psychology: Human Perception and Performance, 16*, 199–210.

Potter, M. C. (1976). Short-term conceptual memory for pictures. *Journal of Experimental Psychology: Human Learning and Memory, 2*, 509–522.

Prime, S. L., Niemeier, M., & Crawford, J. D. (2006). Transsaccadic integration of visual features in a line intersection task. *Experimental Brain Research, 169*, 532–548.

Rayner, K. (1998). Eye movements in reading and information processing: 20 years of research. *Psychological Bulletin, 124*, 372–422.

Rayner, K., & Duffy, S. A. (1986). Lexical complexity and fixation times in reading: Effects of word frequency, verb complexity, and lexical ambiguity. *Memory & Cognition, 14*, 191–201.

Rayner, K., Inhoff, A. W., Morrison, R. E., Slowiaczek, M. L., & Bertera, J. H. (1981). Masking of foveal and parafoveal vision during eye fixations in reading. *Journal of Experimental Psychology: Human Perception and Performance, 7*, 167–179.

Rayner, K., Liversedge, S. P., & White, S. J. (2006). Eye movements when reading disappearing text: The importance of the word to the right of fixation. *Vision Research, 46*, 310–323.

Rayner, K., & Morris, R. K. (1992). Eye movement control in reading: Evidence against semantic preprocessing. *Journal of Experimental Psychology: Human Perception and Performance, 18*, 163–172.

Rayner, K., & Pollatsek, A. (1992). Eye movements and scene perception. *Canadian Journal of Psychology, 46*, 342–376.

Rayner, K., Sereno, S. C., & Raney, G. E. (1996). Eye movement control in reading: A comparison of two types of models. *Journal of Experimental Psychology: Human Perception and Performance, 22*, 1188–1200.

Rayner, K., & Well, A. D. (1996). Effects of contextual constraint on eye movements in reading. *Psychonomic Bulletin & Review, 3*, 504–509.

Reichle, E. D., Rayner, K., & Pollatsek, A. (1999). Eye movement control in reading: Accounting for initial fixation locations and refixations within the E-Z reader model. *Vision Research, 39*, 4403–4411.

Reichle, E. D., Rayner, K., & Pollatsek, A. (2003). The E-Z Reader model of eye movement control in reading: Comparisons to other models. *Behavioral and Brain Sciences, 26*, 445–526.

Reingold, E. M., & Charness, N. (2005). Perception in chess: Evidence from eye movements. In G. Underwood (Ed.), *Cognitive processes in eye guidance*(pp. 325–354). Oxford: Oxford University Press.

Rensink, R. A. (2000). The dynamic representation of scenes. *Visual Cognition, 7*, 17–42.

Rensink, R. A. (2002). Change detection. *Annual Review of Psychology, 53*, 245–277.

Rensink, R. A., O'Regan, J. K., & Clark, J. J. (2000). On the failure to detect changes in scenes across brief interruptions. *Visual Cognition, 7*, 127–145.

Rieger, J. W., Braun, C., Bülthoff, H. H., & Gegenfurtner, K. R. (2005). The dynamics of visual pattern masking in natural scene processing. *Journal of Vision, 5*, 275–286.

Ross, J., Morrone, M. C., Goldberg, M. E., & Burr, D. (2001). Changes in visual perception at the time of saccades. *Trends in Neurosciences, 24*, 113–121.

Salin, P., & Bullier, J. (1995). Corticocortical connections in the visual system: structure and function. *Physiological Reviews, 75*, 107–154.

Sanders, A. F., & van Duren, L. L. (1998). Stimulus control of visual fixation duration in a single saccade paradigm. *Acta Psychologica, 99*, 163–176.

Schendan, H. E., & Kutas, M. (2003). Time course of processes and representations supporting visual object identification and memory. *Journal of Cognitive Neuroscience, 15*, 111–135.

Sereno, S. C. (1995). Resolution of lexical ambiguitiy: Evidence from an eye movement paradigm. *Journal of Experimental Psychology: Learning, Memory, and Cognition, 21*, 582–595.

Shinoda, H., Hayhoe, M. M., & Shrivastava, A. (2001). What controls attention in natural environments? *Vision Research, 41*, 3535–3545.

Simons, D. J. (2000). Current approaches to change blindness. *Visual Cognition, 7*, 1–15.

Simons, D. J., & Ambinder, M. S. (2005). Change blindness: theory and consequences. *Current Directions in Psychological Science, 14*, 44–48.

Simons, D. J., & Rensink, R. A. (2005). Change blindness: Past, present, and future. *Trends in Cognitive Sciences, 9*, 16–20.

Tatler, B. (2002). What information survives saccades in the real world? *Progress in Brain Research, 140*, 149–163.

Tatler, B., Gilchrist, I. D., & Rusted, J. (2003). The time course of abstract visual representation. *Perception, 32*, 579–592.

Tatler, B. W., Gilchrist, I. D., & Land, M. F. (2005). Visual memory for objects in natural scenes: From fixations to object files. *Quarterly Journal of Experimental Psychology, 58A*, 931–960.

Tarr, M. J., & Bülthoff, H. H. (1998). Image-based object recognition in man, monkey and machine. *Cognition, 67*, 1–20.

Thorpe, S., Fize, D., & Marlot, C. (1996). Speed of processing in the human visual system. *Nature, 381*, 520–522.

Ullman, S. (1996). *High-level vision: Object recognition and visual cognition*. Cambridge, MA: MIT Press.

van Diepen, P. M. J., & De Graef, P. (1994). *Line-drawing library and software toolbox* (Psyc. Rep. No.165). Leuven, Belgium: University of Leuven, Laboratory of Experimental Psychology.

van Diepen, P. M. J., De Graef, P., & d'Ydewalle, G. (1995). Chronometry of foveal information extraction during scene perception. In J. M. Findlay, R. Walker, & R. W. Kentridge (Eds.), *Eye movement research: Mechanisms, processes and applications* (pp. 349–362). Amsterdam: North-Holland.

van Diepen, P. M. J., Ruelens, L., & d'Ydewalle, G. (1999). Brief foveal masking during scene perception. *Acta Psychologica, 101*, 91–103.

van Duren, L. L., & Sanders, A. F. (1992). The output code of a visual fixation. *Bulletin of the Psychonomic Society, 30*, 305–308.

Van Gompel, R. P. G., Pickering, M. J., & Traxler. M., J. (2001). Reanalysis in sentence processing: Evidence against current constraint-based and two-stage models. *Journal of Memory and Language, 454*, 225–258.

VanRullen, R., & Thorpe, S. (2001). The time course of visual processing: From early perception to decision-making. *Journal of Cognitive Neuroscience, 13*, 454–461.

Verfaillie, K. (1997). Transsaccadic memory for the egocentric and allocentric position of a biological-motion walker. *Journal of Experimental Psychology: Learning, Memory, and Cognition, 23*, 739–760.

Verfaillie, K., & De Graef, P. (2000). Transsaccadic memory for position and orientation of saccade source and target. *Journal of Experimental Psychology: Human Perception and Performance, 26*, 1243–1259.

Verfaillie, K., De Graef, P., Germeys, F., Gysen, V., & Van Eccelpoel, C. (2001). Selective transsaccadic coding of object and event-diagnostic information. *Psychologica Belgica, 41*, 89–114.

Wade, N., & Tatler, B. W. (2005). *The moving tablet of the eye: The origins of modern eye movement research*. New York: Oxford University Press.

Wolfe, J. M., Klempen, N., & Dahlen, K. (2000). Postattentive vision. *Journal of Experimental Psychology: Human Perception and Performance, 26*, 693–716.

Chapter 9

HOW POSTSACCADIC VISUAL STRUCTURE AFFECTS THE DETECTION OF INTRASACCADIC TARGET DISPLACEMENTS

CARMEN KOCH and HEINER DEUBEL

Ludwig-Maximilians-Universität, Germany

Eye Movements: A Window on Mind and Brain

Edited by R. P. G. van Gompel, M. H. Fischer, W. S. Murray and R. L. Hill

Abstract

Objects found at the end of a saccade serve as spatial landmarks. Here we investigate the landmark-based relocalisation process in four experiments by varying the spatial characteristics of the landmark object. The results of the first experiment show that the efficiency of a landmark is independent of transsaccadic displacements of the landmark object over a wide range of displacements. In an ambiguous mapping of the pre- and postsaccadic information, the visual system uses the centre of gravity of a stimulus configuration for postsaccadic localisation (Experiment 2 and 3). The final experiment demonstrates that the relational spatial information of the presaccadic layout of target and distractors is considered in the relocalisation process after the saccade. The results suggest that the mapping of pre- and postsaccadic information is based on spatially crude, low spatial frequency information.

Because of the anatomy of the human retina and the neural processing in later stages of the visual system, only a small part of the visual scene impinging on the retina can be seen with high resolution. In order to circumvent this shortcoming, humans shift their gaze about every 300 ms, with saccadic eye movements, to the new object of interest. These reorienting saccades are so habitual and automatic, that we are often not even aware of the continuous eye-movement activity. The failure to note the effects of eye movements is quite surprising, however, given the dramatic changes of the retinal visual information each saccade induces. Due to the high angular velocity of the eyes during the saccades there occurs a smearing of the retinal image during the saccade. Moreover, due to the continuously changing orientation of the eyes, the visual system has the task to re-map the changing retinal input into a consistent and stable perceptual frame and into an egocentric representation, which is needed for goal-directed action. In the work presented here we study the role of continuously present information on this relocalisation process, across saccadic eye movements.

Various studies have provided evidence for a suppression of perceptual processing during saccadic eye movements. More specifically, it has been demonstrated that saccadic suppression is most pronounced in the magnocellular pathways of the visual system, that is the pathway that is also responsible for the processing of motion signals (e.g., Burr, Morrone, & Ross, 1994; Ross, Morrone, Goldberg, & Burr, 2001). Recent fMRI studies support this finding. So, Kleiser, Seitz, and Krekelberg (2004) found evidence for a reduction of BOLD-signals immediately before saccade onset in brain areas that receive magnocellular input. This implies that saccade-induced motion signals are suppressed by a specific, saccade-related mechanism.

A consequence of the reduction of motion signals during saccades is that exogenous retinal movements due to motion of an object in the world are also prevented from being processed and registered. While detection of a small motion during fixation may be perfect, the sensitivity to movements during saccades is reduced by about four log units (Macknik, Fisher, & Bridgeman, 1991). In a classical paper, Bridgeman, Hendry, and Stark (1975) showed that an object displacement of up to about 30% of the saccade size goes unnoticed, provided that the displacement occurs during or close to the saccade. This finding is in contrast to theories that assume a cancellation of the retinal motion by extraretinal signals for perception (e.g., von Holst & Mittelstaedt, 1950; Sperry, 1950); without further assumptions these theories would predict that even a small transsaccadic displacement should be perceivable. This has lead some authors to propose that no or only very crude positional information about the presaccadic display is stored across the saccade (Bridgeman, van der Heijden, & Velichkovsky, 1994). Further findings also support the assumption that only high-level, abstract visual information is retained in transsaccadic memory. So, transsaccadic changes of the size of objects and transsaccadic changes from small to capital letters did not affect object or word naming (McConkie & Zola, 1979; Rayner, McConkie, & Zola, 1980).

In contrast to these findings, the study of Deubel, Schneider, and Bridgeman (1996) suggested that very precise position information from the previous fixation is available after the saccade, but that it can be used only under certain conditions. In this study, subjects had to saccade to a small target that appeared in the periphery. In some trials, the target was displaced during the saccade, but remained on the display. In other trials, the target was blanked for a time interval that ranged between 0 and 250 ms, and reappeared at a displaced position. It turned out that in the blanking condition, that is when no visual information about the target location was available in a critical temporal period immediately after the end of the saccade, participants were considerably more sensitive to transsaccadic target displacements. This phenomenon was termed the "blanking effect".

According to Deubel et al. (1996), this finding indicates that the absence of position information at the end of the saccade allows the visual system to use extraretinal signals and information about the egocentric target location stored in transsaccadic memory in order to compute a veridical prediction of the postsaccadic target location. If, however, visual information is present at the end of the saccade, the stored position information is not considered for perceptual localisation. Instead, the visual system uses the visual information that it finds after the eye movement, by relying on the "built-in assumption" that the visual world does not change during the saccade. Given the short durations of saccadic eye movements, this *null-hypothesis* of visual stability is certainly a very reasonable assumption which is correct in most cases.

It also turned out that the presence of a stimulus after the saccade largely determines whether *other* objects in the field are seen as stable or as displaced across the saccade. This was demonstrated in experiments with two stimuli, a target and a distractor (Deubel, Bridgeman, & Schneider, 1998). One of the manipulations included a short intra- and postsaccadic blanking of one of the stimuli, while the other stimulus was displaced during the saccade. Even when the blank was very short (e.g., 50 ms), the blanked object was perceived as moving across the saccade, while the moved (but continuously present) object was perceived as stable. This was true whether the object had originally been defined as the saccade goal or as the distractor. The fact that this striking illusion even occurred for object displacements of up to half of the size of the saccade illustrates that under these conditions perceptual stability is determined not by extraretinal signals, but by the object that is found when the eyes land – this object serves as a spatial reference. The blanked object is then seen as displaced because its position is judged relative to the landmark object, whose position is assumed to be stable.

The role of landmarks was elaborated further in the study by Deubel, Schneider, and Bridgeman (2002) showing that what is stored in transsaccadic memory is the relational spatial information of the presaccadic stimulus configuration. This allocentric representation is then used, after the saccade, to generate predictions of where the stimuli should appear in the visual field. Deubel (2004) analysed the spatial range within which distractors become effective as transsaccadic references. For this purpose, landmark objects (pairs of small rectangles) were placed at various locations in the vicinity of the target. The landmark objects were present when the saccade landed, while the target was blanked for 200 ms. It turned out that the effect of landmarks is spatially selective, being highest when the distractors appear close to the target. Under these conditions, the landmarks indeed largely determined transsaccadic localisation: More than 70 % of a distractor displacement was reflected in the induced target mislocalisation. This implies that when visual structure is available close to the target, the efference copy signal of eye position after the saccade plays only a minor role in transsaccadic localisation; rather, transsaccadic displacement judgements and perceived visual stability are based on the evaluation of postsaccadic landmark objects. The effect of the landmark falls off within 2–3° of visual angle away from the target.

Thus, it seems that landmark objects found when the eyes land after a saccade are of fundamental importance for the transsaccadic localisation of targets. The present investigation extends the previous findings by studying in more detail the relative contribution of allocentric and egocentric information stored across the saccade, and the way how the postsaccadic information is integrated with the stored representations

1. Experiment 1

As discussed before, previous findings suggested that both allocentric and egocentric representations of target position are stored across the saccade and interact in the process of postsaccadic localisation. In Experiment 1, we investigated the spatial range in which the presentation of a postsaccadic distractor would dominate the localisation of the target. For this purpose, participants were asked to saccade to a spatial arrangement of a saccade target and a distractor. The distractor was displaced during the saccade, and we determined the effect of the size of this displacement on the localisation of the (temporarily blanked) target.

1.1. Method

The stimuli were presented on a 21″ video-monitor at a frame rate of 100 Hz. Screen background luminance was 2.2 cd/m^2; the luminance of all presented stimuli was 25 cd/m^2. The participants viewed the display binocularly from a distance of 80 cm. Head movements were restricted by a chin and forehead rest. Eye movements were measured with an SRI Generation 5.5 Purkinje-image eye-tracker (Crane & Steele, 1985) and sampled at a rate of 400 Hz. Further details of computer control, calibration, and triggering of the saccade-contingent display change are given in Deubel et al. (1996).

As shown in Figure 1a, each block started with the presentation of a fixation cross (size: 0.2°). After a random delay of 500–1000 ms the fixation cross-disappeared, and a peripheral target appeared together with a distractor. The target was similar in size and shape to the fixation cross and was always presented on the horizontal meridian at a distance of 6° to the left or the right of the current fixation. The distractor was a vertically elongated rectangle (size: 0.11° × 0.22°). It appeared directly above the target, at a vertical distance of 0.5° from the horizontal meridian.

Triggered by the onset of the saccade, the distractor was displaced horizontally, either in the same or in the opposite direction of the initial target step. Displacement size D varied randomly between $-1.6, -0.8, 0, +0.8$, and $+1.6°$, where positive values indicate an onward displacement, that is a shift in the direction of the saccade, while negative values indicate a backward shift. Moreover, starting with the time of saccade onset, the target was blanked for 100 ms and reappeared with a displacement T from the presaccadic target position.

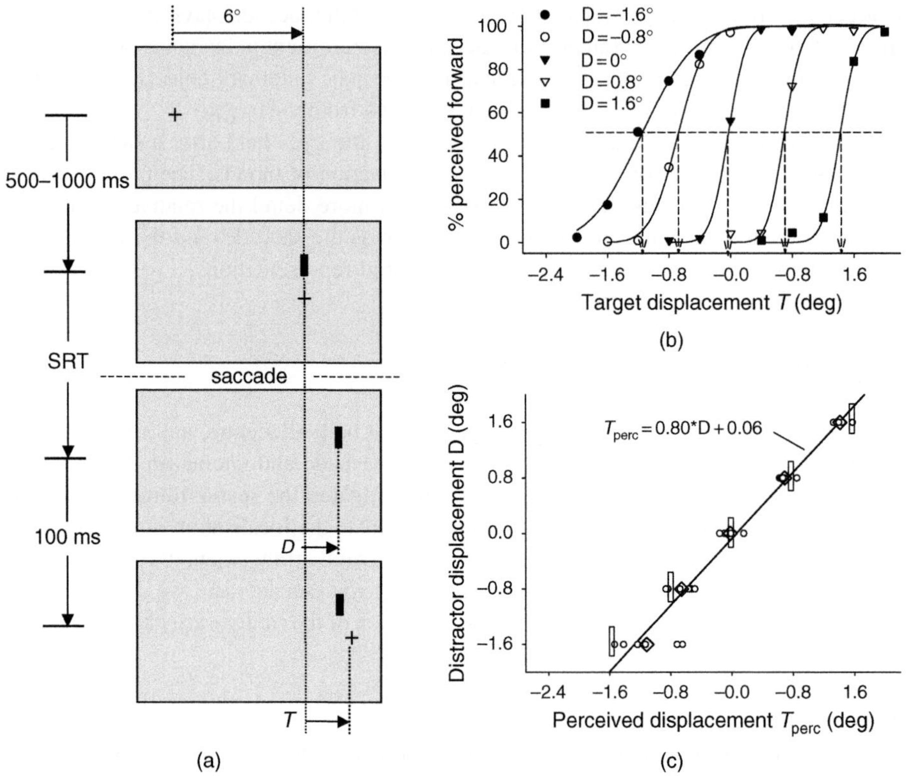

Figure 1. (a) Stimulus sequence of a typical trial in Experiment 1. Initially, the participant fixated a small cross. After a random delay, the target jumped left or right by 6° to elicit a saccade. Simultaneously with the target jump, a distractor appeared directly above the target. Triggered by the onset of the saccade, the target disappeared. The distractor remained continuously visible, but was shifted by the displacement *D*. The target reappeared 100 ms after the saccade onset with a displacement *T*, relative to the presaccadic target position. (b) Percent perceived "forward" judgements of the target as a function of the size of the target displacement *T*. Parameter is the magnitude and the direction of the intrasaccadic distractor displacement *D*. Each set of data points is fitted with a cumulative Gaussian function. The values indicated by the vertical arrows are the 50% neutral positions, that is the target displacements where the target is perceived as stable. (c) Small circles indicate induced subjective target displacements T_{perc} of each of the six participants, for the different distractor displacements. The larger diamonds indicate means across the subjects. Note that the dependent variable is represented on the x-axis, for graphical purposes. The solid line is the linear regression through the data points. Postsaccadic distractor positions are indicated by the vertical rectangles.

From the findings of our former experiments, we assumed that the distractor would act as a landmark, and that the participant would therefore expect that a stable target would reappear at or close to the (displaced) distractor location. Therefore, we selected the positions of the reappearing target at and around the postsaccadic distractor position, with the target reappearing with a displacement of -1.6, -0.8, 0, $+0.8$, or $+1.6°$ relative to the postsaccadic distractor position. This allowed us to compute psychometric functions and to estimate the perceived transsaccadic target displacements T_{perc} (see Figure b). Both the size of T_{perc} and the displacement size T of the target are measured with respect to the position where the target initially appeared. If, as we anticipated from the former findings, participants would expect the target to reappear after the saccade in alignment with the distractor, then the value of T_{perc} should be found to be equal to the distractor displacement D.

At the end of each trial the participant's task was to report, by pressing one of two buttons, whether the intrasaccadic displacement of the saccade target had occurred in the *same* or the *opposite* direction to the initial target step (forward or backward displacement). The final target position served as the starting position for the next trial. Depending on whether the final target position was rather at the centre or close to the right or left border of the display, the initial target step of the next trial could occur into the same or into the opposite direction of the previous saccade.

Each of the 25 different postsaccadic target–distractor arrangements was presented six times per experimental block, in a random order. We did not distinguish between trials with leftward and rightward initial steps; the data from these symmetrical conditions were averaged in the data analysis. Each participant performed six blocks (including 900 trials), distributed over two days. Six paid participants (5 female, 1 male) participated in the experiment. All participants had normal vision or vision that was corrected to normal by contact lenses. Their age ranged from 23 to 25 years. The participants were naive with respect to the aim of the study, but were experienced with the laboratory equipment from other eye-movement-related tasks.

1.2. Results

We excluded from the analysis all trials in which the eye-tracker registered an eye blink or lost the participant's pupil. Also, the trials in which saccadic latency was below 140 ms or above 400 ms were discarded from further analysis. This amounted to 2.7% of all trials (146 trials).

Figure 1b displays what participants perceive in this task, dependent on the size and the direction of the intrasaccadic distractor shift. The graph shows the percentage at which participants indicate a forward displacement of the target as a function of the size of the actual intrasaccadic target shift T, and for the five different distractor displacements D. In order to obtain a measure for the perceived target displacement that was induced by the distractor manipulation, we fitted, for each of the five distractor displacements, a cumulative Gaussian function to each set of data points and then computed the thresholds (for a criterion level of 50%).[1] These values are indicated by the vertical arrows in the

[1] We used a bootstrap procedure to estimate the induced target mislocalisation. This "C" weighted linear regression, a cumulative Gaussian psychometric function to a set of binary data. It then computes the threshold (for a given criterion level of performance) and the gradient. The bootstrap procedure is similar to that given by Foster & Bischof (1991), but a more robust procedure has been adopted in that standard deviations are co-program is freely available from http://www.cs.ualberta.ca/~wfb/software.html. It fits, by computed from centiles, assuming a normal distribution.

figure and signify the perceived target displacement induced by the distractor shift; in other words, they represent the spatial target positions at which the participants would have seen a perfectly stable target.

The relative horizontal displacements of the five psychometric functions in Figure 1b clearly demonstrate that the intrasaccadic distractor shift has a dominant effect on perceived target displacement. So, when the distractor is shifted forward by 1.6°, the psychometric function is displaced by about 1.2° in the forward direction. This indicates that the participant experiences a quite dramatic illusion: In order to be perceived as stable, the target has now to move by about 1.2°, in the same direction as the distractor shift! A one-way ANOVA confirmed a significant effect of the postsaccadic distractor position on perceived target displacement, $F(4, 20) = 99.984$; $p < 0.001$.

The diamonds in Figure 1c show the means of the induced subjective target displacements T_{perc}, as calculated from the psychometric functions for the different intrasaccadic distractor shifts. The small dots indicate the individual values for each of the six participants. The postsaccadic distractor positions are indicated by the vertical rectangles. Note that here and in the following data graphs, the dependent variable is represented on the x-axis for illustrative purposes. The graph also displays the regression line through the data points, yielding a slope of 0.80 and a value of the correlation coefficient r of 0.98. The slope of this regression line is a direct measure of the effectiveness of the distractor in this task, describing the extent to which a certain distractor displacement results in a perceived target displacement. For this experiment, the result indicates that a distractor displacement of 1° will lead to an induced target mislocalisation of 0.8°.

The high correlation coefficient suggests that the relation between the size of the intrasaccadic distractor displacement and the induced illusion is largely linear. In order to test for linearity, we computed a one-way ANOVA with an unknown population parameter. The ANOVA yielded a non-significant deviation of the data from the linear regression, $F(3, 25) = 1.56$; $p = 0.223$. This implies that, independent of the size of the intrasaccadic distractor displacement, the size of the induced target shift as perceived by the participant is always around 80% of the distractor displacement. Thus, at least for the range of distractor displacements tested in the experiment ($\pm 1.6°$ which is $\pm 26.6\%$ of the saccade amplitude), the efficiency of the postsaccadic distractor to act as a spatial landmark was constant.

The results from this experiment confirm our previous findings that a distractor which appears close to the saccade target, though irrelevant for the participant's task, acts as a highly efficient landmark for the transsaccadic localisation of the target (Deubel, 2004). When the target is blanked after the saccade, the target location is evaluated with reference to the continuously present distractor. Quantitatively, the results also agree well with the efficiency of landmarks as analysed by Deubel (2004) for smaller distractor displacements. In this former study, two vertical lines appearing collinearly above and below the target were used as landmarks. The data analysis was similar to the one applied here and yielded a distractor efficiency of 0.73.

This experiment demonstrates that, at least within the range of $\pm 1.6°$ around the presaccadic target positions, continuously presented distractors are highly efficient as landmarks.

This indicates that allocentric information about the presaccadic target–distractor configuration is important in postsaccadic localisation: A blanked target is expected to reappear at the same relative location with respect to distractor as in the presaccadic configuration. Also, the results show that the *null-hypothesis* of stability of the visual world holds also in the present task; the continuously present object (the distractor) is, by default, taken as stable; therefore, it acts as a landmark. However, note that the assumed process of landmark-based localisation requires the visual system to re-identify the landmark in the postsaccadic display in order to allow for an unambiguous mapping of pre- and postsaccadic information. In the next experiment, we study a situation in which an unequivocal mapping of pre- and postsaccadic information is no longer possible. For this purpose, the presaccadic distractor was replaced by two spatially separate stimuli, visually identical to the presaccadic distractor.

2. Experiment 2

In Experiment 2, a discrepancy between the pre- and postsaccadic stimulus layout was introduced by presenting, after the saccade, two spatially separate distractors instead of one, with similar visual features as the presaccadic distractor. As in Experiment 1, we asked how the distractor configuration would affect the transsaccadic localisation of the saccade target. We had three possible hypotheses about the effect of the mapping ambiguity: (1) As a first possible scenario, the introduced inconsistency between pre- and postsaccadic information due to the appearance of the additional object in the postsaccadic scene could potentially lead to the rejection of the *null-hypothesis* (of a stable world), just like in simpler tasks in which no target is found due to blanking (Deubel et al., 1996). Similar to the blanking paradigm, the visual system would then possibly use a stored egocentric representation of the target position and an efference copy of the eye position for postsaccadic target localisation; as in the *blanking* experiments, this would result in a higher sensitivity to detect the veridical intrasaccadic target displacement. (2) Alternatively, the visual system could possibly use the one of both distractors as a landmark that is found closer to the original target position, or closer to the landing position of the eyes. (3) Finally, it is possible that both distractors may be considered with equal weights in the task; in this case the induced target motion should result from a geometrical centre of both postsaccadic stimuli.

2.1. Method

The experimental procedure is displayed in Figure 2a. The presaccadic display and the timing of the presentation were identical to Experiment 1. Now, however, triggered with the onset of the saccade and simultaneous with the target blanking, the presaccadic distractor line was replaced by a pair of lines, both visually identical to the presaccadic distractor. The two lines had a horizontal spatial separation of 1.6°. The postsaccadic distractor pair was presented at various spatial positions which, for convenience, are depicted by the

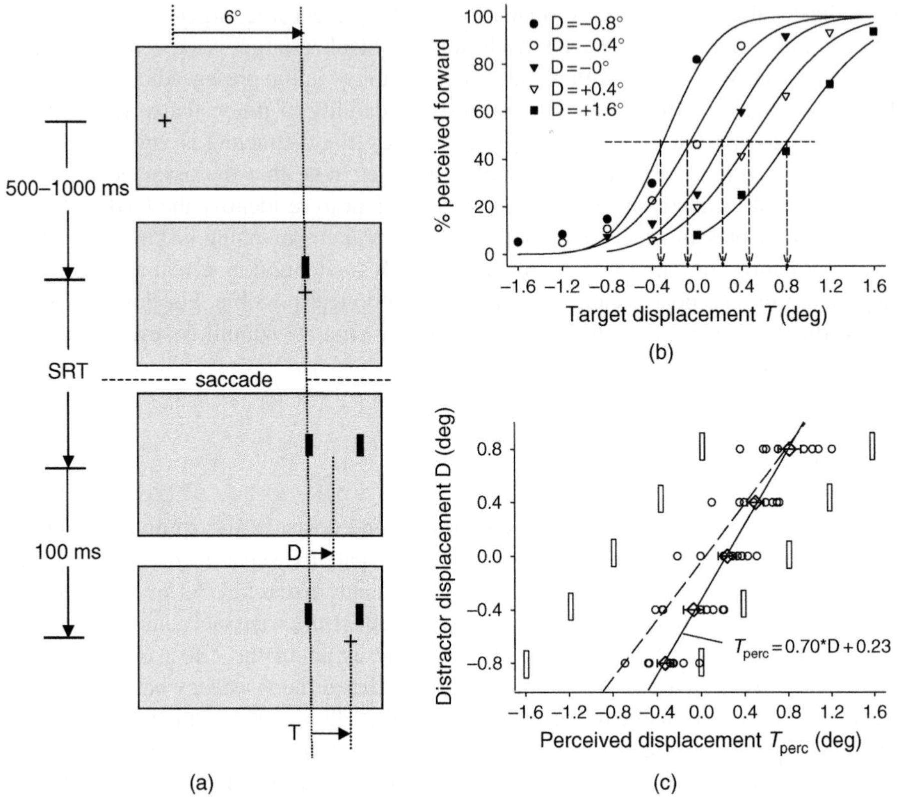

Figure 2. (a) Stimulus sequence in Experiment 2. At saccade onset, the presaccadic distractor line was replaced by a pair of distractors with a horizontal separation of 1.6°. The position of the distractor pair is described by the distance D between the initial target position and the spatial centre of the pair. The saccade target reappeared after 100 ms with a displacement T, relative to the presaccadic target position. (b) Percentage of perceived "forward" displacement as a function of the target displacement and distractor position. (c) Induced subjective target displacements T_{perc} as a function of distractor displacement. The horizontal bars show standard errors. The positions of the postsaccadic distractor pairs are indicated by the pairs of vertical rectangles. The dashed line shows a line with a slope of 1. The solid line represents the empirically found regression.

distance D of the spatial centre of the distractor pair from the presaccadic target location. D was selected at random from −0.8, −0.4, 0, +0.4, and +0.8°. After a blanking interval of 100 ms the target reappeared at one of five distances (−0.8, −0.4, 0, +0.4, or +0.8°), relative to the spatial centre of the postsaccadic distractors.

Altogether, this resulted in 25 different spatial arrangements of the distractor pair and the target. The different conditions were each presented six times per experimental block, in a random order. The participants performed six blocks in two separate sessions (totalling 900 trials per participant). Eight paid participants (7 female, 1 male) participated in this experiment. Their age ranged from 22 to 25 years.

2.2. Results

Of all trials (353 trials), 4.9% were excluded from the analysis because the eye-tracker registered an eye blink or lost the participant's pupil, or because saccadic latency was below 140 ms or above 400 ms.

As in Experiment 1, we first analysed the perceived direction of target displacement as a function of the position of the postsaccadic distractor (given as the spatial centre of the pair of lines). Figure 2b displays the participants' percentage of "forward" judgements for each of the distractor displacements, and the psychometric functions fitted through the data points. The relative displacements of the curves with varying distractor displacement *D* clearly demonstrate that the position of the postsaccadic distractor pair had a strong effect on the location where the target was expected to reappear. Obviously, the first of our hypotheses that, due to the ambiguity of the mapping of pre- and postsaccadic information, the visual system would ignore the distractor and judge the target position independently was not confirmed.

In order to analyse whether the perceived target shift would be predominantly determined by one of the two distractor lines (as assumed in hypothesis 2), we quantitatively determined, as in the previous experiment, the amount of induced target mislocalisation for each value of *D* by means of a bootstrap procedure. The result is plotted in Figure 2c. In this graph, the locations of the two distractor lines (relative to the presaccadic target/distractor position) are displayed by the small vertical bars.

A repeated-measures one-way analysis of variance confirmed a significant effect of the factor distractor displacement D on the perceived target displacement T_{perc}, $F(4, 28) = 72.416$; $p < 0.001$. If, as assumed in hypothesis (2), postsaccadic localisation were largely determined by the distractor line that appeared closest to the presaccadic target position, then the postsaccadic target should be localised at the position of one of the two distractor lines. For example, for the case where the pair of lines appeared at 0 and +1.6°, perceived target displacement should be 0. Obviously, this is not the case. Rather, the data reveal that both distractors contribute about equally to the induced effect: Participants now perceive an induced target displacement which is in between both line locations. Distractor effectiveness can again be determined by computing the linear regression through the data. If both distractor lines would contribute with equal weights and would completely determine postsaccadic localisation, we would expect the induced target shifts to be equal to *D*, that is, the target should be relocalised at the geometrical centre of the distractors. This prediction is indicated by the dashed line, having a slope of 1. As in the first experiment, the data, however, reveal that the effectiveness of the distractor is incomplete, yielding a slope smaller than 1. The solid line in the graph displays this regression line; analysis yields a slope of 0.71, an intercept of 0.23, and a correlation coefficient *r* of 0.87. This indicates that, for the present condition with two postsaccadic distractor lines, about 71% of the size of the intrasaccadic distractor displacement is reflected in the induced target shift. This value is in perfect agreement with previous reports for a single distractor (Deubel, 2004), and close to the value of 80% as found in the first experiment. Additionally, the intercept of 0.23 indicates that both distractor lines are not equally weighted, suggesting that the spatial centre of gravity of the postsaccadic arrangement is not identical to the geometrical centre of the distractor pair.

To conclude, the results of Experiment 2 suggest that in the (ambiguous) case of two spatially separate postsaccadic distractors, the visual system uses, as a first approximation, the centre of gravity of both stimuli as the landmark for postsaccadic localisation – the induced mislocalisation is about the same as if a single distractor had appeared at this location.

3. Experiment 3

The previous experiment suggested that when no unequivocal mapping between pre- and postsaccadic distractor configuration is possible, it is the global postsaccadic stimulus configuration that determines the effective landmark position – under this condition, a centre of gravity of the postsaccadic arrangement is used. The question arises what would happen if an unambiguous mapping of the presaccadic distractor to the postsaccadic configuration would be made possible – would then only the postsaccadic stimulus that is congruent to the presaccadic distractor be taken as a landmark, while the other distractor would be discarded? Therefore, in the next experiment, again a combination of two distractors was presented after the saccade. Now, however, one appeared above and the other below the horizontal meridian. The two vertical positions of the distractors were easy to discriminate for the participant.

3.1. Method

The experimental procedure is illustrated in Figure 3a and b. In this experiment, the presaccadic distractor was presented with equal probability either above or below the target, at a vertical distance of 0.5° from the horizontal meridian. Like in Experiment 2, the single line was replaced at the onset of the saccade by two postsaccadic distractor lines that appeared simultaneously. The distractor lines again had a horizontal separation of 1.6°. Now, however, one of the two lines appeared above and the other below the horizontal meridian. The horizontal distance *D* of the geometrical centre of the distractor pair and the presaccadic target position was varied between −0.8, 0, and +0.8°. After a blanking interval of 100 ms the target reappeared at one of five distances (−0.8, −0.4, 0, +0.4, or +0.8°), relative to the spatial centre of the postsaccadic distractor lines.

These different arrangements were presented with equal probability in each block, yielding 60 different conditions (2 presaccadic distractor arrangements × 2 postsaccadic vertical distractor arrangements × 3 distractor displacements × 5 relative target displacements). In each block, each condition was presented twice. Each participant performed 7 of these experimental blocks, distributed over three days (totalling 840 trials per participant). Six participants (all female, aged 22–25 years) took part in the experiment.

In this experiment, we expected that the distractor line that is congruent with the presaccadic vertical distractor position should be mainly considered for the postsaccadic

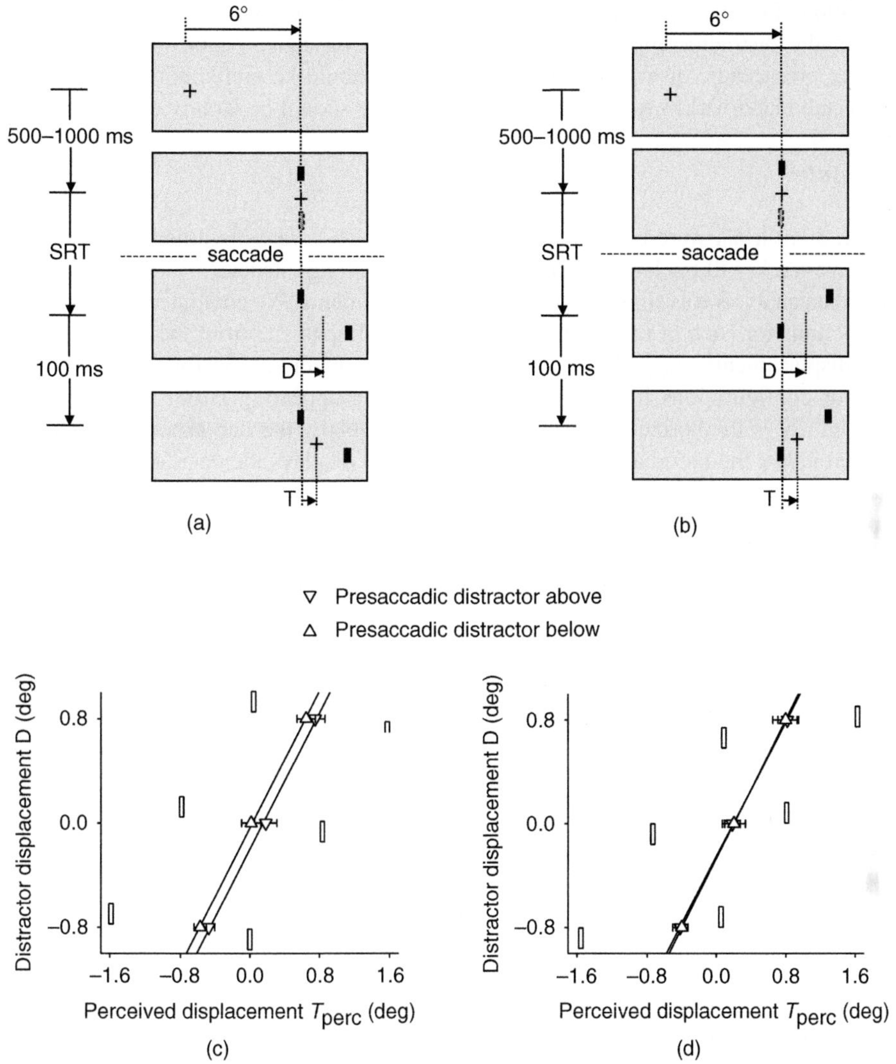

Figure 3. (a, b) Two types of stimulus sequences of Experiment 3. With the jump of the saccadic target, the distractor was presented either directly above the target (filled rectangle) or below the target (dashed rectangle). With saccade onset, the single distractor is replaced by two distractors, one above and the other below the horizontal meridian. The distractor above the horizontal meridian was either closer to the initial fixation cross (Figure 3a) or farther away from the fixation cross (Figure 3b). (c, d) Induced subjective target displacements T_{perc} as a function of the distractor displacement. The positions of the postsaccadic distractor pairs are indicated by the pairs of vertical rectangles. Figure 3c shows the results for the configuration where the line *above* the horizontal meridian was closer to initial fixation cross. Conversely, Figure 3d displays the data for the condition where the line *below* the horizontal meridian was closer to initial fixation cross. In both graphs, the data are plotted separately for the trials where the presaccadic distractor was shown above the target (triangles up) and below the target (triangles down), respectively.

localisation. For example, in the example presented in Figure 3a, the less eccentric postsaccadic distractor appeared above the target. Consequently, in combination with an upper presaccadic distractor, this distractor line should be preferentially taken as the postsaccadic landmark, while the other distractor line should be largely neglected.

3.2. Results

In total, 5040 trials were run; of these, 187 trials (3.7%) were discarded due to failures of the eye-tracker to track the eye or too short or too long latencies.

The data analysis was similar to the previous experiments. We computed the psychometric functions for each of the different distractor conditions and from these the perceived target displacements, T_{perc}. The results are displayed in Figure 3c for the postsaccadic distractor configurations in which the distractor line appearing *closer* to fixation was presented *above* the horizontal meridian (and, accordingly, the line *farther* from fixation appeared *below* the meridian). Conversely, Figure 3d displays the data for the conditions where the closer distractor line appeared *below* the meridian. In both graphs, the data are plotted separately for the trials where the presaccadic distractor was shown above the target (upright triangles) and below the target (inverted triangles), respectively.

It becomes obvious that the target is again localised at a position in between both distractor lines, indicating that, in contrast to our expectation, *both* distractor elements are considered in postsaccadic localisation. As in the previous experiment, the visual system uses the centre of gravity of both stimuli as landmark for postsaccadic localisation. A two-factor ANOVA (vertical arrangement × distractor displacement *D*) revealed a significant main effect of D, $F(1, 5) = 170.31$; $p < 0.001$, but a non-significant effect of the vertical arrangement, $F(1, 5) = 0.029$; $p = 0.1005$, and a non-significant interaction, $F(2, 10) = 0.0004$; $p = 0.793$.

4. Experiment 4

The previous experiment demonstrated that the (vertical) spatial position of the presaccadic distractor – above or below the target – has no effect on the postsaccadic landmark computation. This is even the case when only one of the two postsaccadic distractor lines appears at a (vertically) congruent position – even then both distractor lines are considered about equally in the postsaccadic localisation process. The question therefore arises whether the presaccadic spatial arrangement of target and distractor is considered at all in postsaccadic localisation. Alternatively, the visual system may simply tend to take the position of any localised object found after the saccade as the assumed target location. Some evidence for the latter assumptions came from a recent experiment studying the effect of an irrelevant object presented only after the saccade on target localisation (Deubel et al., 2002, second experiment). As in the present study, the target reappeared after a blanking period. It turned out that the location of the distractor, present when the eyes landed after the primary saccade, was indeed taken by the visual system as the position of the (presaccadic) target. In order to test whether the presaccadic allocentric information is indeed stored across the saccade and whether it would affect the localisation performance, we varied, in the final experiment, the horizontal distance between target and distractor in the *presaccadic* display.

4.1. Method

The sequence of stimulus presentation was similar to that of Experiment 2, except that the presaccadic distractor now appeared either with a forward or a backward displacement pD of 0.4° with respect to the target position (Figure 4a). The postsaccadic distractor pair was again presented at various positions depicted by the distance D of the spatial centre of the distractor pair from the presaccadic target location. D was selected at random from −0.8, 0, and +0.8°. After a blanking interval of 100 ms the target reappeared at one of five distances (−0.8, −0.4, 0, +0.4, or +0.8°), relative to the spatial centre of the postsaccadic distractors.

Altogether, this resulted in 30 different spatial arrangements of the distractor pair and the target (2 presaccadic distractor positions $pD \times 3$ distractor displacements $D \times 5$ relative target displacements). The different conditions were each presented six times per experimental block, in a random order. The participants performed four blocks in two separate sessions (totalling 720 trials per participant). Six paid participants (5 female, 1 male) participated in this experiment. Their age ranged from 20 to 25 years.

4.2. Results

Altogether, 4320 trials were run; of these, 112 trials (2.6%) were discarded due to failures of the eye-tracker to track the eye or too short or too long latencies.

Figure 4b displays the size of the perceived target displacement T_{perc} induced by the different distractor displacements D. The data are plotted separately for both presaccadic target/distractor arrangements (for $pD = -0.4$ and $+0.4°$, respectively) and fitted with first-order regression lines. The dashed line in the graph represents the regression line through the data points of Experiment 2, that is for $pD = 0°$. It is obvious from the graph that the presaccadic distractor position relative to the target affects postsaccadic target localisation. So, when the presaccadic target appeared behind the distractor (i.e., the distractor was presented closer to the fixation, $pD = -0.4°$), there was also a strong tendency to expect the reappearing target behind the centre of gravity of the postsaccadic arrangement. Vice versa, for the positive value of pD (here the presaccadic saccade target appeared in front of the single distractor line), the perceived target displacements are shifted to closer locations. Quantitatively, the spatial distance between both presaccadic distractor positions of 0.8° yields a postsaccadic difference of target localisation of about 0.5°, indicating that about 60% of the presaccadic target–distractor distance is reflected in postsaccadic localisation.

A two-way ANOVA with pD and D revealed a significant main effect of the presaccadic target–distractor displacement pD, $F(1,5) = 73.758$; $p < 0.001$, and a significant

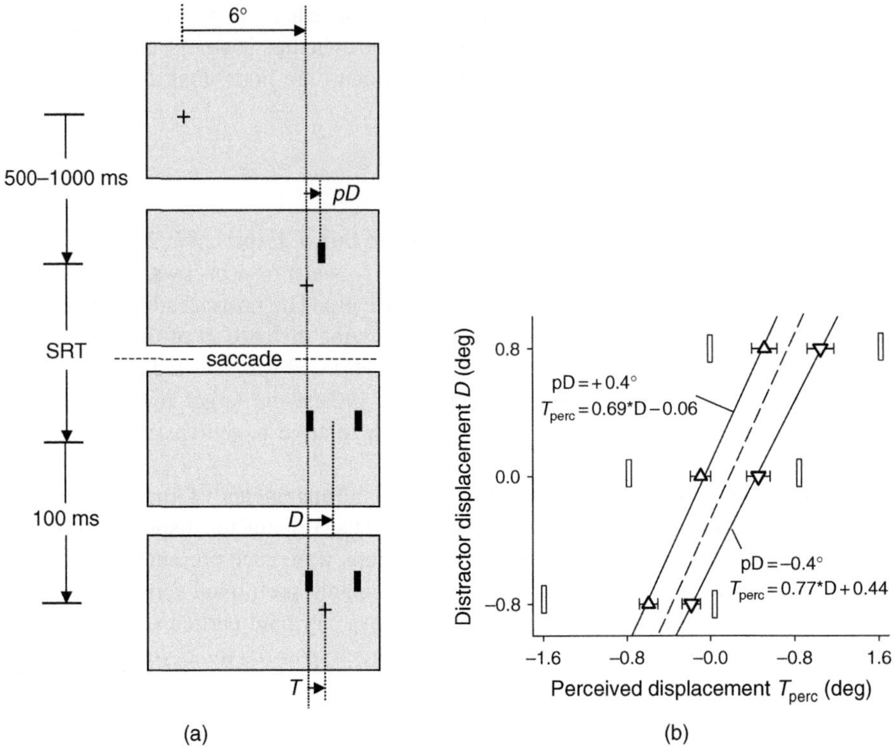

Figure 4. (a) Stimulus sequence of a typical trial in Experiment 4 in which the presaccadic horizontal arrangement of target and distractor was varied. Simultaneously with the presaccadic target jump, a distractor appeared at a presaccadic distractor position *pD* of either $+0.4°$ (forward displacement with respect to the target position) or $-0.4°$ (backward displacement). Triggered by the onset of the saccade, the target disappeared and a distractor pair was presented at one of the three positions $D(-0.8,\ 0, \text{or}\ +0.8°)$. The target reappeared 100 ms after the saccade onset with a displacement *T*, relative to the presaccadic target position. (b) Induced subjective target displacements T_{perc} as a function of the distractor displacement *D*, plotted separately for the two values of *pD*. The solid lines represent the regressions for $pD = +0.4°$ and $pD = -0.4°$, respectively. The dashed line shows the result from Experiment 2 where the presaccadic distractor was presented directly above the saccade target ($pD = 0°$).

main effect of distractor displacement *D*, $F(2, 10) = 172.86$; $p < 0.001$. Moreover, the interaction of both factors was significant, $F(2, 10) = 6.558$; $p < 0.05$. The interaction reflects the fact that the slope of the regression lines, indicating the efficiency of the distractor pair, is slightly different for both presaccadic distractor positions. The slope of the regression line is 0.77 at $pD = -0.4°$, and 0.69 for $pD = +0.4°$. This indicates that the distractor is somewhat more effective if it appears between the fixation cross and the saccade target than if it is presented beyond the target. The results from this experiment show that transsaccadic localisation is dependent on relational information from before the saccade. If available, relational information in the presaccadic scene is a major determinant of postsaccadic localisation.

5. General discussion

In general, the results of the experiments presented here are well in accordance with our previous findings (Deubel et al., 1998, 2002; Deubel, 2004). They clearly demonstrate that the spatial layout found by the visual system after the saccade forms a major source of information used to establish space constancy. Found objects serve as anchor points to determine the expectations where objects such as the blanked target should appear after a saccade. The results, however, also reveal a few quite unexpected findings which will hopefully deepen our understanding of the perceptual processes that provide visual stability across saccadic eye movements.

The results from our first experiment demonstrate that stimuli found at saccade end serve as landmarks, even if they appear at a considerable distance from their presaccadic position. More specifically, we found that the efficiency of the landmark was about constant as long as it appeared with 25% (1.6°) of the saccade size. Unfortunately, our experiments did not extend this range farther to test the limits of the system. However, some of the data reported in Deubel et al. (1998, Experiment 2) suggest that the spatial range within which postsaccadic landmarks are effective may extend even to half of the size of the saccade. This implies that landmarks probably dominate whenever available, overruling the information based on efference copies that would signal a displacement.

In the second experiment, the visual system was confronted with two distractor lines instead of one in the postsaccadic display. Since the visual features of the postsaccadic distractor pair were identical to the presaccadic distractor, it was not possible to decide which of both corresponded to the presaccadic distractor. It turns out that under this condition the visual system neither uses extraretinal signals in combination with the stored position information for the postsaccadic localisation nor does it take one of the distractor lines as the preferred landmark. Rather, the effective landmark position results from a combination of both distractors, suggesting that the spatial centre of gravity of the postsaccadic arrangement is used for the target localisation. This is the case even in Experiment 3 where, due to the vertical spatial arrangement of the postsaccadic distractor pair, pre- and postsaccadic distractors can be easily related. Obviously, the mechanism which detects the landmark position after the end of a saccade is not particularly selective about the geometric characteristics of the stimuli. The spatial layout of the reference object is rather unimportant when searching for the postsaccadic reference.

The finding that postsaccadic localisation results from a centre of gravity of both distractors is reminiscent of the so-called "global effect" in saccadic eye-movement programming. The *global effect* relates to the observation that, when a fast saccade is performed to a configuration of a target and a distractor, the eyes often land at a position intermediate to both objects, considering relative target properties such as size and brightness (Deubel, Wolf, & Hauske, 1984; Findlay, 1982). This suggests that the saccadic amplitude computation is based on stimulation integrated over a rather wide area of visual space and involves integration of the visual signals in a relatively "raw" form. It has been suggested that the physiological substrate of saccade averaging may be the superior colliculus (SC) in the midbrain which uses a distributed population code to represent visual and oculomotor direction. So, Robinson (1972) and others have shown that simultaneous stimulation of two separate locations in the SC will produce a saccade which is a vector average, much in line with the *global effect*. The SC is now also known to be part of an attentional network, to draw visual attention efficiently to visual onsets in the periphery (e.g., Kustov & Robinson, 1996; Krauzlis, Liston, & Carello, 2004). So, it may be assumed that the locus of visual attention finally provides the spatial pointer that is also responsible for transsaccadic localisation.

An important question addressed in Experiment 4 was whether the visual localisation system would at all use the relational information of target and distractors from before the saccade, stored across the eye movement, to generate an expectation of the postsaccadic target location. Alternatively, the visual system may simply tend to take the position of any localised object or of a configuration found after the saccade as the assumed target location. The results show that the latter is clearly not the case. In Experiment 4, presaccadic (horizontal) distractor–target distance was varied. The data clearly demonstrate that perceptual localisation was indeed strongly affected by the relative arrangement of target and distractor stored across the saccade. It can be concluded that presaccadic, relational information in the direction of the saccadic eye movement is a major determinant of postsaccadic localisation. Only if this information is missing, as in the case of a landmark appearing only after the saccade (Deubel et al., 2002), the postsaccadic landmark itself is taken as the anchor for the localisation process. These findings imply that quite accurate relational information about the relative positions of a few objects in the visual field is stored in a transsaccadic memory and used after a saccade. Independent evidence for a precise transsaccadic memory of relative spatial positions have been provided in studies on the effect of contextual cues on the transsaccadic coding of objects. So, Verfaillie and De Graef (2000) showed that displacements of a target that brought it towards another object were easier to detect than changes that moved the target away from another object. Currie, McConkie, Carlson-Radvansky and Irwin (2000) found that detection of a location change of a saccade target across an eye movement can be made on the basis of both the target's change of absolute position, and a change in the spatial relations formed by the target and its neighbours. Carlson-Radvansky (1999) demonstrated that relational information in scenes composed of geometrical figures are encoded before the saccade and retained in a transsaccadic memory.

Our findings have some implications for theories of visual stability around saccades. A few theories of visual stability have emphasised the special role of the postsaccadic visual layout for perceptual stability across saccades (e.g., Currie et al., 2000; Deubel et al., 1996). Deubel et al. (1984) were the first to propose that a transsaccadic memory representation of the saccade target may serve to relocate visual objects across saccades. In more recent work, we (Deubel & Schneider, 1994; Deubel et al., 1996, 1998) developed a

"Reference Object Theory" which assumed that pre- and postsaccadic visual "snapshots" are linked by means of the visual target structure which is assumed by the visual system as being stable. The theory states that with each new fixation the visual system runs through a sequence of processing steps that starts with the selection of one object as the target for the next saccade. Particular features of the saccade target and a few surrounding objects are selected and stored in a transsaccadic memory to facilitate its re-identification at the start of the next fixation. When the eye has landed after the saccade, the visual system searches for the critical target features within a limited region around the landing site. If the target object is found, the relationship between its retinal location and its mental representation is compared in order to coordinate these two types of information. If the postsaccadic target localisation fails, however (e.g., because the intrasaccadic target shift was too large or the target was absent), the assumption of visual stability is abandoned. As a consequence, a target displacement is perceived.

The present findings allow the specification of the theory in two important aspects. The first specification concerns the finding of this investigation and also of our previous studies on the landmark effect that not just the saccade target, but also other (distractor) objects can serve as spatial references (Deubel et al., 2002; Deubel, 2004). Whether an object is defined as the target or as a distractor before the saccade seems to play little role in the postsaccadic determination of the reference object. More critical for the selection of a postsaccadic object as a reference is a temporal constraint, namely its presence right at the time when the eyes land. This demonstrates that temporal continuity of an object is more important even than selection as a saccade target in establishing a reference object. The second important specification is related to the here-established fact that, when more than just a single stimulus is present at saccade end, the spatial centre of gravity is taken as the landmark position. Obviously, the mechanism postulated in the theory that searches for the critical features is far from being selective. This may imply that the mapping of pre- and postsaccadic information is based on spatially crude, low spatial frequency information.

Acknowledgements

This work was supported by the Deutsche Forschungsgemeinschaft (Grant De336/2). We wish to thank Birgitt Assfalg for her valuable help in running the experiments, and Ben Tatler for very insightful comments on an earlier version of the paper.

References

Bridgeman, B., Hendry, D., & Stark, L. (1975). Failure to detect target displacement of the visual world during saccadic eye movements. *Vision Research, 15*, 719–722.

Bridgeman, B., van der Heijden, A. H. C., & Velichkovsky, B. M. (1994). A theory of visual stability across saccadic eye movements. *Behavioural and Brain Research, 17*, 247–292.

Burr, D. C., Morrone, M. C., & Ross, J. (1994). Selective suppression of the magnocellular visual pathway during saccadic eye movements. *Nature, 371*, 511–513.

Carlson-Radvansky, L. A. (1999). Memory for relational information across saccadic eye movements. *Perception & Psychophysics, 61*, 919–934.

Crane, D., & Steele, C. (1985). Generation-V Dual-Purkinje-Image eyetracker. *Applied Optics, 24*, 527–537.

Currie, C. B., McConkie, G. W., Carlson-Radvansky, L. A., & Irwin, D. E. (2000). The role of the saccade target object in the perception of a visually stable world. *Perception & Psychophysics, 62*, 673–683.

Deubel, H., & Schneider, W. X. (1994). Perceptual stability and postsaccadic visual information: Can man bridge a gap? *Behavioral and Brain Sciences, 17*, 259–260.

Deubel, H., Schneider, W. X., & Bridgeman, B. (1996). Postsaccadic target blanking prevents saccadic suppression of image displacement. *Vision Research, 36*(7), 985–996.

Deubel, H., Bridgeman, B., & Schneider, W. X. (1998). Immediate post-saccadic information mediates space constancy. *Vision Research, 38*, 3147–3159.

Deubel, H., Schneider, W. X., & Bridgeman, B. (2002). Transsaccadic memory of position and form. In J. Hyönä, D. Munoz, W. Heide & R. Radach (Eds.), *The brain's eye: Neurobiological and clinical aspects of oculomotor research* (pp. 165–180). Amsterdam: Elsevier Science.

Deubel, H., Wolf, W., & Hauske, G. (1984). The evaluation of the oculomotor error signal. In A. G. Gale & F. Johnson (Eds.), *Theoretical and applied aspects of eye movement research* (pp. 55–62). Amsterdam: Elsevier Science.

Deubel, H. (2004). Localisation of targets across saccades: Role of landmark objects. *Visual Cognition, 11*, 173–202.

Findlay, J. M. (1982). Global processing for saccadic eye movements. *Vision Research, 22*, 1033–1045.

Foster, D. H., & Bischof, W. F. (1991). Thresholds from psychometric functions: superiority of bootstrap to incremental and probit variance estimators. *Psychological Bulletin, 109*, 152–159.

Kleiser, R., Seitz, R. J., & Krekelberg, B. (2004). Neural correlates of saccadic suppression in humans. *Current Biology, 14*, 386–390.

Krauzlis, R. J., Liston, D., & Carello, C. D. (2004). Target selection and the superior colliculus: goals, choices and hypotheses. *Vision Research, 44*, 1445–1451.

Kustov, A. A., & Robinson, D. L. (1996). Shared neural control of attentional shifts and eye movements. *Nature, 384*, 74–77.

Macknik, S. L., Fisher, B. D., & Bridgeman, B. (1991). Flicker distorts visual space constancy. *Vision Research, 31*(12), 2057–2064.

McConkie, G. W., & Zola, D. (1979). Is visual information integrated across successive fixations in reading? *Perception & Psychophysics, 25*(3), 221–224.

Rayner, K., McConkie, G. W., & Zola, D. (1980). Integrating information across eye movements. *Cognitive Psychology, 12*(2), 206–226.

Robinson, D. A. (1972). Eye movement evoked by collicular stimulation in the alert monkey. *Vision Research,* 12, 1795–1808.

Ross, J., Morrone, M. C., Goldberg, M. E., & Burr, D. C. (2001). Changes in visual perception at the time of saccades. *Trends in Cognitive Sciences, 24*(2), 113–121.

Sperry, R. W. (1950). Neural basis of the spontaneous optokinetic response produced by visual inversion. *Journal of Comparative Physiological Psychology, 43*, 482–489.

Verfaillie, K., & De Graef, P. (2000). Transsaccadic memory for position and orientation of saccade source and target. *Journal of Experimental Psychology: Human Perception and Performance, 26*, 1243–1259.

von Holst, E., & Mittelstaedt, H. (1950). Das Reafferenzprinzip. Wechselwirkungen zwischen Zentralnervensystem und Peripherie. *Naturwissenschaften, 20*, 464–467.

Chapter 10

TRANSSACCADIC MEMORY: BUILDING A STABLE WORLD FROM GLANCE TO GLANCE

DAVID MELCHER

Oxford Brookes University, UK, and University of Trento, Italy

CONCETTA MORRONE

San Raffaele University and Institute of Neuroscience of the National Research Council, Italy

Eye Movements: A Window on Mind and Brain
Edited by R. P. G. van Gompel, M. H. Fischer, W. S. Murray and R. L. Hill

Abstract

During natural viewing, the eye samples the visual environment using a series of jerking, saccadic eye movements, separated by periods of fixation. This raises the fundamental question of how information from separate fixations is integrated into a single, coherent percept. We discuss two mechanisms that may be involved in generating our stable and continuous perception of the world. First, information about attended objects may be integrated across separate glances. To evaluate this possibility, we present and discuss data showing the transsaccadic temporal integration of motion and form. We also discuss the potential role of the re-mapping of receptive fields around the time of saccades in transsaccadic integration and in the phenomenon of saccadic mislocalization. Second, information about multiple objects in a natural scene is built up across separate glances into a coherent representation of the environment. Experiments with naturalistic stimuli show that scene memory builds up across separate glances in working memory. The combination of saccadic re-mapping, occurring on a timescale of milliseconds, and a medium-term scene memory, operating over a span of several minutes, may underlie the subjective impression of a stable visual world.

1. Introduction

While the input to the visual system is a series of short fixations interleaved with rapid and jerky shifts of the eye, the conscious percept is smooth and continuous. The eyes move but the world is normally perceived as stable (unless, for example, one passively moves the eyeball with one's fingers). This sharp contrast between the input from the eye and the naïve visual experience has long puzzled philosophers and, more recently, psychologists and neuroscientists. One of the first to notice this problem was the Persian scholar Alhazen (965–1039 AD): "For if the eye moves in front of visible objects while they are being contemplated, the form of every one of the objects facing the eye . . . will move on the eyes as the latter moves. But sight has become accustomed to the motion of the objects' forms on its surface when the objects are stationary, and therefore does not judge the objects to be in motion" (Alhazen, 1989).

There have been several attempts to solve the problem of visual stability. One suggested mechanism is a transsaccadic memory buffer to support the fusion of separate images (Irwin, Yantis, & Jonides, 1983; Jonides, Irwin, & Yantis, 1982). The basic assumption, which followed the computational tradition of Marr, was that the goal of the visual system was to build a metric internal representation of the world. Perception was defined as the creation of three-dimensional volumes that could be matched to similar representations in memory. It follows from this assumption that our subjective perception of a continuous world, which is perceived to extend beyond the limits of the fovea, might depend on the storage of a virtual three-dimensional space in the brain that is eye-invariant. This proposal has been attacked on various fronts, including the philosophical argument that visual perception does not involve viewing the world in an internal "Cartesian theater" (Dennett, & Kinsbourne, 1992). A critical challenge for an internal, metric representation that integrates shape information across saccades is the fact that objects move and rotate with respect to the viewer. It is impossible, for example, to represent in metric detail multiple viewpoints of a face as it turns towards or away from the viewer, unless one is viewing a painting by Picasso. Objects in an internal theater would have to be continually updated in terms of position and viewpoint, which would be both computationally expensive and practically impossible given the limited acuity of peripheral vision.

The extreme opposite view to the Cartesian theater is the proposal that perceptual stability depends, paradoxically, on the lack of internal representation of the world (O'Regan & Noë, 2001). Proponents of "active vision" have suggested that detailed visual information can be gleaned by making an eye movement "just in time", when it is needed, rather than storing large amounts of unnecessary information (Ballard, Hayhoe, & Pelz, 1995; Findlay & Gilchrist, 2003). This can be seen as an extension of the "ecological approach" to vision, which focuses on the visual information readily available from the world rather than on internal representations. In practice, however, it is still necessary for the brain to know where to look for the information that it needs, since eye movements are not random and are rarely wasted in natural tasks (Land & Hayhoe, 2001; Land, Mennie, & Rusted, 1999). We do not need to see the whole world in clear detail in any particular glance, but we need to be able (have the potential) to see any particular area of the world in clear detail in the next glance. That "potential" may be a critical part of naïve perception, but it requires knowledge. Thus, details about the layout of the scene and the position of important objects must be represented, raising the spectre of the internal theater yet again.

The basic hypothesis underlying the experiments reported in this chapter was that in order to yield a subjective impression of a stable and detailed scene, without actually building a detailed representation in the mind, there must be at least two different mechanisms of scene memory operating at different timescales. The first timescale is that of a single saccade, in which local, photographic detail is replaced by new visual input (Irwin et al., 1983; McConkie & Zola, 1979; Tatler, 2001), while more invariant information survives and combines with relevant input from the new fixation (Loftus, 1972; Melcher, 2001; Tatler, Gilchrist, & Rusted, 2003). The second time frame is over a period of minutes or seconds, in which the observer interacts with the immediate visual environment. In order to perceive a stable, immediately available world, we have suggested that it is important to be able to learn about the identity and location of previously fixated objects without resorting to a time-consuming search. A number of studies that have shown that information about the identity and location of objects in a scene persists (Germys, De Graef, & Verfaillie, 2002; Henderson & Hollingworth, 2003; Pollatsek, Rayner, & Collins, 1984) and accumulates (Loftus, 1972; Melcher, 2001, 2006; Melcher & Kowler, 2001; Tatler et al., 2003) across multiple glances. We have suggested that this information is available in a "medium-term" memory store that involves the long-term memory system but does not necessarily require consolidation into permanent memory if the object and/or environment is not sufficiently salient or fails to be repeatedly viewed (Melcher, 2001, 2006).

2. Combining basic visual information across saccades

We have investigated the mechanisms involved in learning about objects in a scene across a saccade. It is clear that integration of two separate views cannot be anything like overlaying two photographs. We thought that a likely place to start looking for memory that survives saccades is at the next step of visual processing, in which consistent and predictive information about an object is extracted over time.

2.1. *Transsaccadic integration of motion*

To look for evidence of transsaccadic integration, we chose to study visual motion since its long integration times can exceed typical fixation durations (Burr & Santoro, 2001; Neri, Morrone, & Burr, 1998). The output of individual motion detectors is combined over both space and time, allowing even a weak signal to be seen given a sufficiently long view. The visual system is extremely sensitive to regularities and coincidences hidden in noise, making even a few dots moving coherently in a noise "snowstorm" detectable after several hundred milliseconds of viewing. In the case of complex motion patterns and biological motion, the time period of integration can exceed one second (Burr & Santoro, 2001; Neri et al., 1998). It is important to note that the improvement in motion sensitivity as a function of stimulus duration cannot be explained by information summation, but involves true integration of motion information over time (Burr & Santoro, 2001).

At the same time, it has been argued that the visual "buffer" is "refreshed" with each new fixation (McConkie & Zola, 1979; Tatler, 2001), implying that long motion integration times must either continue across saccades or must be a laboratory phenomenon limited to artificially stable viewing conditions. We tested temporal integration of simple translational motion to see if it would continue even when a saccadic eye movement was made during the stimulus presentation (Melcher & Morrone, 2003). The stimulus was a cluster of black and white random noise dots (6° diameter) presented on a gray background. During each trial there were two periods of coherent horizontal motion (150 ms each) embedded in 10 s of random noise (dots re-plotted randomly on every frame), allowing for the presence of a saccadic eye movement in between the two motion signals (Figure 1, top). Motion coherence sensitivity was measured using a staircase procedure (for details, see Melcher & Morrone, 2003).

The delay between the two motion signals was varied across blocks of trials. At brief delays between the two motion signals, coherence sensitivity was not affected, showing full temporal integration. For larger delays, sensitivity decreased as a function of delay duration, up to the point at which the delay reached the temporal limit of integration. Beyond a certain delay duration, performance was equivalent to that found with a single period of motion signal, as expected from previous studies showing temporal limits on motion integration (Burr & Santoro, 2001).

On some blocks of trials, an eye movement (12°) was made from above to below the stimulus during the interval between the motion, while in the other blocks of trials the observer maintained fixation either above or below the stimulus. Temporal integration of coherent motion signals was not reduced by saccades. Rather, the two motion signals were combined across the saccade (Figure 1, bottom). It is important to note that the motion signals that were integrated across saccades were often, by themselves, below the level of conscious discrimination threshold. Thus, the methodology used excluded cognitive strategies or verbal recoding since the observer could not detect each motion signal – only by combining the two subthreshold signals could motion be correctly discriminated.

Interestingly, integration durations were actually higher for trials with saccades compared to maintained fixation. A comparable effect has been shown for the McCollough aftereffect, in which trials with eye movements produced longer aftereffect durations than those with fixation (Ross & Ma-Wyatt, 2004). These findings suggest that transsaccadic integration may involve a specific memory mechanism that is not active (or less active) in the case of maintained fixation. One possible explanation for this phenomenon is described in a later section (5.3).

Why might information be integrated across separate glances? Ironically, the ability to perceive relevant change (in location) may depend on the capacity to ignore irrelevant change (noise) while keeping track of what is consistent and unchanging (coherent motion

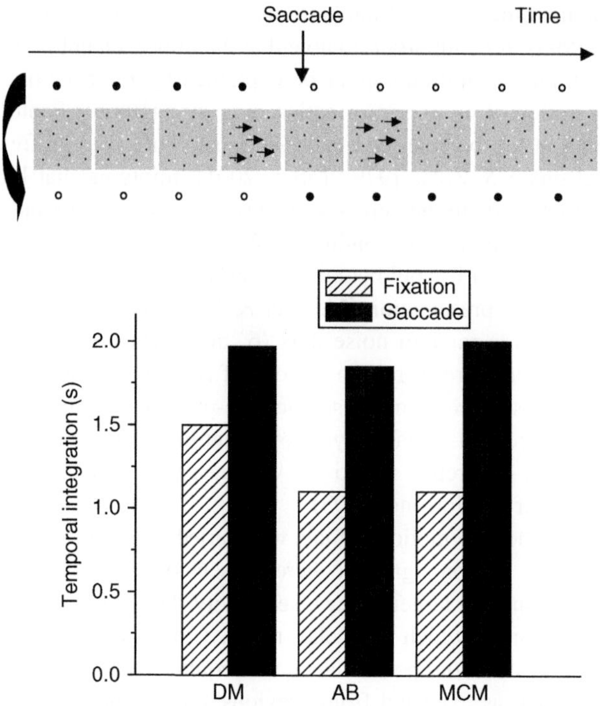

Figure 1. Transsaccadic integration of motion. *Top*: Schematic illustration of the stimuli comprising two short intervals of coherent translation motion embedded in 2 seconds of random noise. The subjects performed motion discrimination task while fixating the top circles or making a saccade from top to bottom circles in the interval between the two coherent motion. *Bottom*: Estimates of temporal integration for motion direction discrimination obtained by measuring coherence thresholds as function of the temporal separation between the two coherent motion signals (see Melcher & Morrone, 2003) for three subjects (DM, AB and MCM).

in one direction). In a complex and ever-changing environment, it is critical that the brain filters out the noise and identifies constancies and patterns in the perceptual input.

2.2. *Transsaccadic integration of form*

The temporal integration of motion found across saccades might be unique to motion processing or, conversely, might reflect a more widespread mechanism in visual perception. Since different features are processed in separate areas of cerebral cortex, the motion stimulus used in the previous study may involve a different processing stream (the dorsal "action/where" system) from perception of form (the "what" system) (Goodale & Milner, 1992; Mishkin, Ungerleider & Macko, 1983). Thus, it would premature to make any generalization about object processing based on our findings with motion, which is usually associated with the "where/action" stream. It is also important to note that we can usually recognize an object in a single glance, raising the question of whether temporal integration (as found with motion) would be necessary. There are, however, exceptions in which form identification changes over time including slit-viewing (Anstis & Atkinson, 1967; Moore, Findlay & Watt, 1982; Nishida, 2004; Yin, Shimojo, Moore & Engel, 2002), form-from-motion (Domini, Vuong, & Caudek, 2002; Ullman, 1983), object priming (Germys, De Graef, & Verfaillie, 2002; Pollatsek, Rayner, & Collins, 1984), the interpretation of ambiguous figures (for an interesting review see Piccolino & Wade, 2006) and form adaptation (Blakemore & Campbell, 1969; Kohler & Wallach, 1944; Melcher, 2005; Purkinje, 1823).

We modified the stimulus used in the previous experiment in order to test form coherence (Braddick, O'Brien, Wattam-Bell, Atkinson, & Turner, 2000). The basic logic was to take a complex form (a radial or circular pattern) and to show a small portion of that pattern (embedded in random dots) in each single frame. Over a series of frames, the observer begins to perceive the global shape of the pattern. The observer's task was to indicate whether the pattern was radial or circular (Figure 2, top). Individual dots did not move coherently across separate frames. There were two short periods ("coherent form stimulus") containing the form pattern embedded in a longer period of dynamic random noise with no pattern. As in the first experiment, the stimulus was 6° in diameter and observers were instructed to either make a 12° saccade from the fixation point above the stimulus to the one below the stimulus (or vice-versa) or to maintain fixation during the entire trial (in separate blocks). The trial lasted for 4 s in total, with 100 dark or white dots plotted against a background of mean gray. A particular trial contained either one or two periods of coherent form. As in the first experiment, coherence sensitivity was measured using a staircase procedure for each condition (single or double signal) and delay duration (0–800 ms). The duration of the coherent form signal was 100 ms, compared to 150 ms for the first experiment (motion). The use of a shorter stimulus period was motivated by the need to keep performance well below ceiling.

As expected, form coherence sensitivity decreased as the two motion signals were separated in time beyond the critical limit (Figure 2, bottom). The pattern of form integration was similar for trials with saccades (triangles) and trials with maintained fixation (squares), demonstrating transsaccadic integration of form. Overall, the temporal integration time (around 700 ms) was about half of that found for motion for the same observer (more than 1500 ms). Yet the finding of transsaccadic integration remained the same for the two tasks, implying that transsaccadic integration also occurs for shape information.

Of course, the radial/circular stimulus did include a motion component since the dots were dynamically replotted on each frame, even if the pattern did not move in a consistent direction. Thus, it was important to also test transsaccadic perception without any motion cues whatsoever. Further evidence that form processing is not strictly retinotopic comes from studies of shape adaptation aftereffects (Melcher, 2005). In these experiments, four different adaptation aftereffects were tested: contrast, tilt, complex form and face adaptation. We found that the magnitude of the aftereffect was modulated by whether or not the adapter and test were in the same spatial location. While contrast adaptation

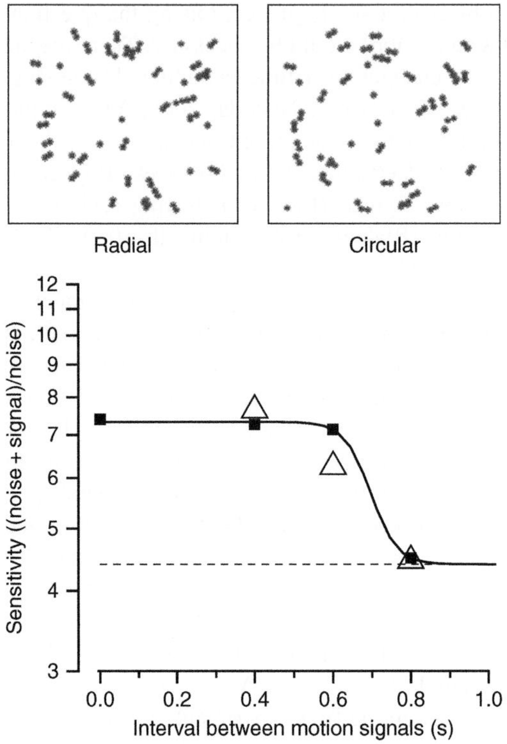

Figure 2. Temporal integration of form. *Top*: Example of the stimuli used for form discrimination. The stimuli sequence was the same as shown in Figure 1 (top), except that the coherent motion intervals were substituted with coherent form intervals. *Bottom*: Sensitivity, the inverse of the proportion of dots displayed along the radial or circular form to the total number of dots at coherence threshold, as function of the separation between the two coherent form interval measured during fixation (filled square) and during saccade (open triangles). The dotted line shows the sensitivity when only one coherent form interval is presented.

showed no transsaccadic aftereffects, the other conditions showed increasing aftereffect size as the stimulus increased in complexity. This suggests that shape processing becomes increasingly eye-independent along the visual processing pathway (see Melcher, 2005 for details).

3. Transsaccadic accumulation of memory for natural scenes

One simple way to understand what people learn across separate glances would be to show them a natural scene and allow them to freely scan the picture. Then, the observer would be stopped at a certain point and asked about what they had just viewed. This logic had worked well in an initial set of studies of scene memory (Melcher, 2001; Melcher and

Kowler, 2001). In those experiments, the task was to freely recall the names of objects in computer-generated rooms. Viewing time ranged from 250 ms to 4 s. The number of items recalled increased linearly as a function of viewing time.

One novel aspect of the recall experiments was that some displays were displayed more than once to test for memory savings across separate views. Surprisingly, the number of items recalled in these retest displays also increased at the same rate as a function of total viewing time. In other words, a display shown twice for 2 s each would lead to the same performance as a display shown once for 4 s. This memory accumulation occurred despite the fact that the retest displays were separated by an average of 4 other scenes. Previously, such temporal integration had only been found for brief views (400 ms) of scenes tested by a recognition test (Loftus, 1981). The finding of memory accumulation across repeated presentations of the same display suggested that visual working memory was capable of keeping in mind several scenes (each full of many objects), not just several objects.

A new set of experiments tested transsaccadic memory under more natural viewing conditions with photographs and pictures (Melcher, 2006). Instead of free recall, observers (23) were asked questions about the color, location or identity of specific objects in the display (Examples include the following: (1) What color is the tablecloth? peach, white or blue (2) Where is the teacup? bottom right, bottom left or center (3) What food is in the middle plate of the three-tiered plate stand? cake, sandwiches or sausage rolls?). Stimuli were shown for up to 20 s, in order to look for ceiling effects in memory performance. Critically, some images were not followed by a memory test after the first presentation but only after being shown a second time. This manipulation served to measure whether the memory accumulation across repeats of the same stimuli found previously for object recall extended also to questions about specific object attributes. Re-test trials were shown 4–6 trials after the initial display of that stimulus. Images included drawings of realistic scenes, reproductions of paintings and photographs of both indoor and outdoor natural scenes. The memory test contained a series of written questions on the screen about specific objects in the scene and a list of three choices (see Melcher, 2006 for more detail).

On each trial, the stimulus display was presented for a time period of 5, 10 or 20 s. The 5 s and 10 s trials were run in the same blocks (since they also contained re-test trials), while the 20 s trials (with no re-tests) were run in separate blocks. After each trial, participants were either given a memory test or instructed to continue on to the next trial. Participant responses were given by keyboard press and recorded for later analysis. The order of conditions was randomized across observers (see Melcher, 2006 for more detail).

Figure 3 shows the average percent correct performance for all types of questions as a function of total viewing time. The solid circles show performance after 5, 10 or 20 s on the non-repeat trials. Open squares show re-test trials, in which the memory test was given only after the second time that the picture had been shown. The leftmost square shows performance after seeing a display twice for 5 s each time on repeat trials, for a total viewing time of 10 s. The rightmost square shows percent correct response after seeing a 10-s and a 5-s display for a total viewing time of 15 s (on half of those trials,

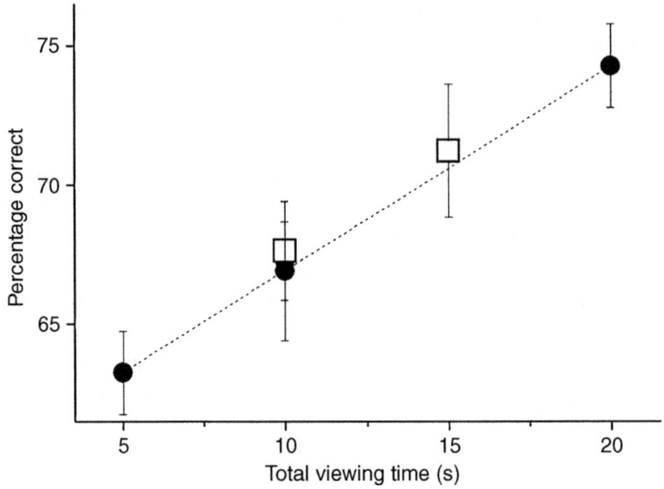

Figure 3. Performance in answering questions about the content of natural scenes as a function of the display duration. Filled circles show performance on continuous presentation trials, while open squares show percentage correct when the stimulus was shown twice before the test (10 s total viewing time = viewed twice for 5 s each time.

the 10-se duration trial preceded the 5-s trial, while the other half of trials contained the opposite pattern).

Overall, performance improved for longer views. Performance improved linearly as a function of total viewing time, with no difference between re-test trials and normal trials of the same total viewing time, consistent with our previous study with free recall. Similar results were found for each type of question (color, location or identity).

This study extends the results with object perception across single saccades and provides further evidence that information about the visual properties of objects persists across saccades (Germys et al., 2002; Henderson & Hollingworth, 2003; Loftus, 1972; Polatsek et al., 1984; Tatler et al., 2003). Moreover, these results show that many scenes, each containing multiple objects, are maintained in memory over a period of minutes. In other words, it is not only objects that persist across saccades, but the scene context as well.

4. The cost of transsaccadic integration

In real life, we are typically surrounded by many objects, many of which are not being directly attended at a given point in time. The results of the first three experiments suggest the existence of an active, perhaps default, integration mechanism that combines information across saccades. Given the fact that the world does change over time, this suggests that we also need a dedicated mechanism to detect these changes in order to avoid incorrectly integrating information from different (or no longer visible) objects.

The visual system appears to depend on the detection of visual transients to notice changes in scenes (for review, see Kanai & Verstraten, 2004). Under laboratory conditions in which changes occur during saccades, or simulated saccades, observers perform poorly at detecting changes that occur without accompanying visual transients (Blackmore, Brelstaff, Nelson, & Troscianko, 1995; Grimes, 1996; Rensink, O'Regan, & Clark, 1995). Perhaps this failure reflects a natural tendency to integrate, rather than compare, images across saccades.

Consider the case of transsaccadic localization. When the saccadic target changes position during the saccade, observers often fail to report shifts in target position. This can be considered a form of change blindness. When the target is blanked during the saccade and reappears only after the saccade has landed, then observers regain their ability to notice the change in location (Deubel, Bridgeman, & Schneider, 1998). Surprisingly, it requires a blank delay of around 200 ms for observers to regain their best performance.

We hypothesized that for change detection a similar temporal delay might be required after the interruption in the display (saccade, blank screen or blink) in order to accurately detect the change on all trials. We tested the ability of observers to detect the number of changes (either one or two) that occurred to a display of 1–10 items. Items could change color (red, green, blue) or shape (circle, square, rectangle, ellipse) and all changes were easy to detect when they occurred without a blank intervening interval. There were five separate experiments, with different instructions. The first experiment was a simple change detection task with 1–10 items. The second experiment, with 3–6 items, required observers to discriminate the number of changes (one or two) that had occurred during the blank delay. If two changes occurred, they always happened to different items, so that the same item did not change twice. The remaining experiments all used four item displays. In the third experiment, a temporal delay was introduced between the first and second change, such that the second change occurred without a blank screen preceding it. Thus, the second change was visually obvious, even if it was not always detected. Observers in the fourth and fifth experiment were instructed to attend either to the first change only (immediately after the blank delay) or to wait and report the presence or absence of a change to an item that occurred after the stimulus display had reappeared.

Normally, change detection is fairly good for small displays of less than about four items (Figure 4a). For discriminating number of changes, however, observers were surprisingly poor even with a display of only three items (Figure 4b). It is often assumed that change detection involves placing the pre-change display items in visual short-term memory (VSTM) for comparison with the post-change display. The VSTM limit of about 4 items is inferred from change detection performance, such that VSTM as a concept (and debates about its properties) has become largely defined by how people perform in change detection tasks (for example, Alvarez & Cavanagh, 2004; Luck & Vogel, 1997). There are two possible ways to reconcile the failure to detect number of changes in a three-item display with previous studies of VSTM. The first option is to posit an additional step or mechanism involved in comparing post-stimulus to pre-stimulus displays when more than one change needs to be counted. The second, more problematic, possibility is that performance on change detection task is not based on comparing post-change items to

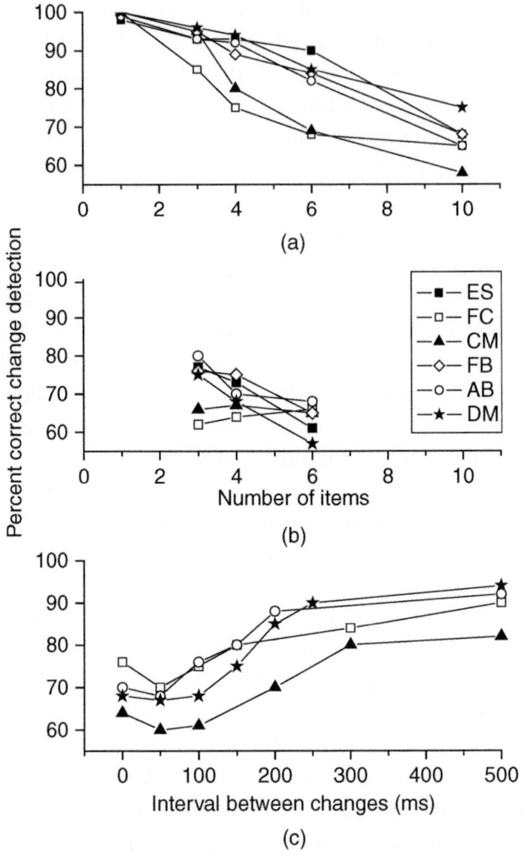

Figure 4. Performance in detecting a single change (a) or double change (b) to the objects presented in a complex display as function of the number of the item present in the display. Different symbols show performance for different subjects. (c): Performance in detecting number of changes as a function of the temporal separation between the two changes.

VSTM on an object-by-object basis at all. Change detection might involve comparing the entire pre-change display to the whole post-change display as a pattern rather than as a collection of independent objects. In the latter case, change detection would not be a valid methodology to determine the number of items stored in VSTM but rather the complexity of information stored in VSTM.

It might have been expected, based on the principle of chunking, that change detection would be easier when the two changing items were near to each other. On the contrary, the average distance between the two changing items was similar for correct and incorrect trials (AB: 5.3° vs 5.3°; CM: 5.7° vs 5.8°; DM: 6.3° vs 6.2°; FB: 5.3° vs 5.8°). Thus, the ability to discriminate number of changes did not depend only on the spatial distance between the two changing targets.

As expected, performance improved as a function of the duration of the temporal delay added between the first and the second change. Figure 4C shows that the time to reach optimal performance, equal to that found for detection of one vs no changes, ranged from 150 to 300 ms for a four-item display. When the two changes were separated in time, subjects were better able to detect the second change. Again, it is important to note that the visual transient created by the second change was not masked by a blank screen, blink or saccade, making the inability to correctly detect both changes surprising.

The remaining experiments tested the influence of instruction on the ability to detect one of the two changes. On four-item trials in which subjects were told to attend only to the second change, detection for that second change was nearly perfect, even when the second change was delayed by only 50 ms (ranging from 90 to 100% correct), demonstrating that low-level visual masking was not responsible for these effects. Thus, it was possible to ignore the change that occurred across the blank delay and focus only on whether or not a change occurred in the post-blank display. It was more difficult, however, to attend only to the first change and ignore the second. When subjects were instructed to ignore a second change following 50 ms later, detection for the first change was poorer than when there was no second change at all (AB: 80% vs 90%; DM: 76% vs 94%; FB: 76% vs 89%). This suggests that the second change, even when ignored, influenced the ability to consciously report the existence of the first change.

The current findings complement other studies showing failures of awareness when attention must be quickly shifted between two or more different items, time periods or locations, such as the attentional blink (Raymond, Shapiro, & Arnell, 1992). Our results are in agreement with the idea that change detection requires focused attention to only one object or cluster of items at a time (Rensink, 2000). When the results of this experiment are viewed in the context of transsaccadic integration, it suggests that the change detection task may be difficult because it interferes with the default strategy of memory accumulation. Change detection involves an attempt to compare two representations rather than combine them (Becker, Pashler, & Anstis, 2000). In the real world, such as walking through a crowded city sidewalk, the brain is faced with a myriad of changes across fixations. Rather than concentrating on the continual flux, the visual system searches for constancies and patterns. Thus, "blindness" to changes in parts of the scene that are not currently relevant, along with the build-up of information about aspects of the scene that are consistent or important to current goals, may be the natural strategy for the visual system in everyday life.

5. Discussion

5.1. Principles of transsaccadic memory

Human memory is often described in terms of limits: a limited capacity in terms of items or a limit in duration. In the case of transsaccadic memory, it is certainly true that not everything is remembered (such as in the case of change blindness), and it is equally true that not everything is forgotten (visual details are retained in memory). The question, then, is why some information is forgotten while other information is remembered.

The experiments reported here, for both a single saccade and across minutes, emphasize the balanced cooperation of learning and forgetting to separate the wheat from the chaff. An efficient system should retain information that is predictive and invariant across views but not remember transient details (Melcher, 2005). Trans-saccadic memory is not just about putting more and more information into storage; it is equally critical to forget information from previous fixations when that information is no longer relevant. We believe that this principle holds for both of the time periods studied here, a single saccade or multiple saccades within the same scene. In the case of motion and form integration, the information that is integrated over time and combined across the saccade is not a particular visual detail or set of dots, but a trend in the overall pattern of stimulation. The results described in this chapter emphasize the way in which what is seen in each new fixation depends on what has been seen in previous fixations – perception does not start over from scratch in each glance. The studies of motion integration (Melcher & Morrone, 2003), form aftereffects (Melcher, 2005) and scene memory (Melcher, 2001, 2006) all suggest that transsaccadic memory is not merely a static storage of abstract information, but rather an active factor that modifies processing of subsequent visual information.

For complete scenes with multiple objects, we have argued for a central role of forgetting as part of an efficient memory (Melcher, 2001; Melcher, 2006; Melcher & Kowler, 2001). The traditional "cognitive psychology" model is to view memory acquisition as a one-way process of moving information through bottlenecks from sensory to short-term to long-term memory. We have argued, on the contrary, that it would be grossly inefficient to have a single short-term memory/long-term memory (STM/LTM) distinction, since it would mean that every attended item, after a few additional fixations, would have to either be forgotten entirely or transferred permanently into LTM. Instead, we have argued, there is a "medium-term" period of proto-LTM that involves keeping in mind information about a particular environment for a particular task over a period of minutes, without all of that information necessarily being consolidated into LTM (Melcher, 2001; Melcher, 2006; Melcher & Kowler, 2001). While the exact nature of visual memory remains a matter of debate (Germys et al., 2002; Hollingworth, 2004; Tatler, Baddeley, & Land, 2005), the medium-term memory theory can account for a number of aspects of memory that are difficult to explain with a dichotomous STM/LTM theory (Melcher, 2001, 2006; Pierrot-Deseilligny, Muri, Rivaud-Pechoux, Gaymard, & Ploner, 2001).

A second theoretical consideration is that the environment, and our interaction with it, is structured in certain regular ways (Gibson, 1979; O'Regan & Noë, 2001). Thus, we might expect our transsaccadic memory to take advantage of the inherent structure of the environment in the efficient encoding and retrieval of scene information. There is now considerable evidence that the "object" is an organizing principle for perception and cognition (Blaser, Pylyshyn, & Holcombe, 2000; Duncan, 1984; Melcher, Papathomas, & Vidnyanszky, 2005). One possibility is that the "scene", like the object, serves as a structure to organize information in memory (Biederman, 1972; Gould, 1976; Melcher, 2001; Tatler et al., 2003). Perhaps one reason that the gist and layout of a scene are gleaned in a single glance (Potter, 1976) is that such information is used to organize the data from subsequent fixations. One piece of evidence for this hypothesis comes from the finding that it is more difficult to learn objects in a scene when the background context has previously been associated with a different set of objects (Melcher, 2001; Melcher & Kowler, 2001). Thus, an object and its background context may naturally and automatically linked, as has been suggested since the time of ancient Greece (Cicero, 1942/55 B.C.).

5.2. Remapping of receptive fields across a single saccade

Can the transsaccadic integration effects reported here be simulated using simple and basic operations? If so, are these operations biologically plausible? Based on the known receptive field properties of the neurons in the visual areas involved in motion and form processing, we hypothesize that integration across a single saccade may involve two mechanisms working together: (1) peri-saccadic changes in retinotopic visual receptive fields, and (2) neurons with eye-independent receptive fields.

Based on electrophysiological recordings, it is now well established that only in a few visual areas (probably only V1 and V2) do receptive fields (RF) fail to show peri-saccadic changes (Duhamel, Colby, & Goldberg, 1992). Many neurons start to respond to stimuli positioned at locations different from their direct retinal afference before the eye movement begins. These effects are particularly evident in lateral intraparietal area (LIP) (Duhamel et al., 1992), but also very common in V3 and V3A and V4 (Nakamura & Colby, 2002). There are several ways that receptive fields can change around the time of the saccade. For some of these neurons the size of the RF can shrink, such as cells in V4 (Tolias et al., 2001), or enlarge by extending in the opposite direction of the saccade (Kubischik, 2002). Retinotopic RF can undergo re-mapping before the saccade, becoming de facto craniotopic from the moment of the initial remapping up to the next intention to move the eye compensating for the retinal shift induced by the eye movement.

During the re-mapping of the RF, which can be intuitively visualized as a moving RF in cortical space, it is plausible to suggest that the interested neuronal network is still able to carry out the visual analysis that it normally performs during fixation (Figure 5). The analysis could be form, contrast, motion or location, depending on the specificity of the area. For example a V3A neuron that participates in the analysis of motion would continue to participate to the motion analysis, but would now respond to a signal at a different eccentricity of the retinal input that correspond to the same external spatial location. This means that for a limited time period, this neuron is able to integrate visual information about form, colour and motion independently of eye position. This integration would, presumably, be performed just as during normal fixation, without particular loss of acuity or changes in RF size.

The existence of changes in RF around the time of saccades cannot, by itself, account for our data because the remapping is fast and transient: a complete predictive shift of the RF take place in less than 100 ms for a range of saccadic size from 10 to 20°. (Kusunoki & Goldberg, 2003; Nakamura & Colby, 2002). Our results for form and

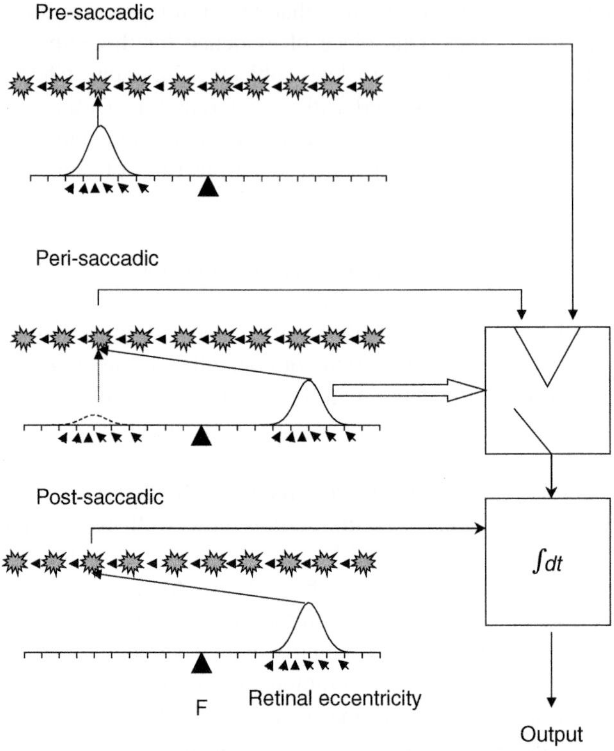

Figure 5. Schematic diagram of the proposed model of transsaccadic integration. Three separate instants are represented: top, time during fixation, middle the time around saccadic onset, bottom, the time period after the saccade when fixation is again stabilized. In each panel, the bottom graph represent the retinotopic activation map of an area of motion-sensitive cortex, with the big arrow indicating the foveal representation and the little arrows the motion selectivity of the neurons. The top layers represent an assembly of neurons that construct a receptive field anchored to the head position or to external space. Each individual neuron will necessarily select different retinotopic input from the lower layer depending on the gaze position (the input is shown by the arrows from the activity function to the selected cell). During fixation, the output of the same craniotopic neuron is integrated. Across saccades, the integration of the pre- and post-saccadic fixation is contingent (white arrow to the switch) on appropriate post-saccadic activity by the motion-selective neuron.

motion show integration between events that are separated from saccadic onset as far as 800 ms, a separation too large for the re-mapping process. To explain the transsaccadic integration we need a second stage of craniotopic RF (citations) that remap their retinal input in preparation for the saccade (see Figure 5).

To explain how this two-stage model might function in transsaccadic perception, it is worth examining motion integration as a specific test case. It has been proposed that LIP neurons support the long integration times observed psychophysically by temporally integrating the decision biases of ensembles of MT cells which are directionally selective

(Mazurek, Roitman, Ditterich, & Shadlen, 2003). In this view, the craniotopic/spatiotopic neuron acts like a non-specific memory buffer that accumulates information from the earlier retinotopic visual area. But which area is selecting the appropriate input to be integrated? How would it be possible that form is integrated with specific form information and not, for example, with irrelevant motion information? An important role for the selection could be performed by the highly selective, manly retinotopic areas that undergo transient remapping. Those neurons know in advance of the eye movement what will be in the new receptive field, and thus would be in the position to know whether the information before and after the saccade should be integrated (and sent to the craniotopic RF).

Figure 5 illustrates the proposed mechanism, with three time periods and an integrator with an open/closed switch. In the case of motion, a specific direction-selective neuron in MT or V3A feeds its activity to a set of craniotopic neurons that would integrate this activity over time. If its activation during the remapping is constant, then the information will continue to be passed to the same craniotopic RF that will integrate the information between the two fixations. However, if the low-level neuron terminates its response, the switch will open and the craniotopic RF will no longer integrate the information. If the switch is open, then the craniotopic RF will accumulate new information starting afresh after the saccade, with the pre-saccadic information being lost. This would happen in situations in which the two images cannot be integrated, as has been suggested in studies of transsaccadic perception of ambiguous figures (Ross & Ma-Wyatt, 2004) and change detection across saccades (Becker, Pashler, & Anstis, 2000; Grimes, 1996).

Interestingly, the re-mapping of receptive fields has also been implicated in the mis-localization of object position during saccades (Ross, Morrone, Goldberg & Burr, 2001; Burr & Morrone, 2004). Thus, integration across saccades and mislocalization during the saccade might involve the same mechanisms. This simple idea could explain several how information is combined across separate glances without fusion of local, pixel-like visual detail.

5.3. Physiological underpinnings of scene perception and memory

In real-world scenes, memory involves information about the location and the identity of specific objects, in addition to gist and layout of the scene (Biederman, 1972; Loftus, 1972; Potter, 1976). What brain areas might underlie the type of scene memory studied here? A distinction is often made between object and location working memory (WM), with the former involving temporal regions and the latter parietal regions (Smith & Jonides, 1997). Natural scenes, however, provide an interesting case in which both object and location information are equally important. This suggests that both systems might work simultaneously to represent "what" and "where", or, perhaps, they might work under different conditions or in differing time frames.

There is a body of evidence suggesting that the type of scene memory studied here may depend in particular on temporal regions. For example, remembering scene-based location, in addition to recognizing object identity, critically involves temporal regions (Goh et al., 2004; Lee et al., 2005; Rolls, Xiang, & Franco, 2005). Thus, memory for object location within the scene has been correlated with both temporal and parietal regions, which may work together in scene representation. Another possibility, however, is that the parietal system may be limited to the earliest stages of attentional orienting to scene location rather than a more lasting memory of object location. Studies of visual-spatial WM have suggested that a transition occurs from parietal to temporal regions over time (for review, see Pierrot-Deseilligny et al., 2002). It is possible that the scene memory described in this chapter depends on this maintenance of information in temporal cortex that allows for further consolidation into LTM. Given that objects can change their retinotopic coordinates as a result of a saccade, it is important in future studies to examine whether peri-saccadic changes also occur in object- and scene-processing areas in temporal cortex and to what extent craniotopic and spatiotopic representations are used for complex scenes.

It is interesting to note that the frontal-parietal attentional network, rather than brain areas involved in learning about complex scenes, has been implicated in change detection and "change blindness" (Todd & Marois, 2004; Vogel & Machizawa, 2004). The dissociation between brain areas involved in change detection and in scene and object memory may help to explain the dissociations between change detection and memory performance (Henderson & Hollingworth, 2003). If this theorized dissociation is correct, then it suggests that caution should be taken before using VSTM tasks to make generalizations about the operation of memory in everyday life.

6. Conclusions

The experiments described in this chapter provide further evidence that visually detailed, yet not metrically exact, information is retained across saccades and used to influence perception and action. The model described here implies that the brain searches for consistent information across saccades and uses this to allow predictive and invariant information to build up over time. The combination of saccadic re-mapping, occurring on a timescale of milliseconds, and a more durable scene memory, operating over a span of several minutes, may underlie the subjective impression of a stable visual world.

References

Alhazen, I. (1083). Book of optics. In A. I. Sabra (Ed.) *The Optics of Ibn al-Haytham.* London: Warburg Institute, 1989.

Alvarez, G. A. & Cavanagh, P. (2004). The capacity of visual short-term memory is set both by visual information load and by number of objects. *Psychological Science*, 15, 106–111.

Anstis, S., & Atkinson, J. (1967). Distortions in moving figures viewed through a stationary slit. *American Journal of Psychology*, 80, 572–785.

Ballard, D. H., Hayhoe, M. M., & Pelz, J. B. (1995). Memory representations in natural tasks. *Journal of Cognitive Neuroscience*, 7, 66–80.

Becker, M. W., Pashler, H., & Anstis, S. M. (2000). The role of iconic memory in change detection tasks. *Perception*, 29, 273–286.

Biederman, I. (1972). Perceiving real-world scenes. *Science*, 177, 77–80.

Blackmore, S. J., Brelstaff, G., Nelson, K., & Troscianko, T. (1995). Is the richness of our visual world an illusion – Transsaccadic memory for complex scenes. *Perception*, 24, 1075–1081.

Blaser, E., Pylyshyn, Z. W., & Holcombe, A. O. (2000). Tracking an object through feature space. *Nature*, 408, 196–199.

Braddick, O. J., O'Brien, J. M., Wattam-Bell, J., Atkinson, J., & Turner, R. (2000). Form and motion coherence activate independent, but not dorsal/ventral segregated, networks in the human brain. *Current Biology*, 10, 731–734.

Burr, D., & Morrone, M. C. (2004). Visual perception during saccades In J. Werner (Ed.) *The visual neurosciences* (pp. 1391–1401). Boston: MIT Press.

Burr, D. C., & Santoro, L. (2001). Temporal integration of optic flow, measured by contrast thresholds and by coherence thresholds. *Vision Research*, 41, 1891–1899.

Cicero, M. T. (1942/55 B.C.). *De Oratore*. Loeb Classical Library.

Dennett, D. C. & Kinsbourne, M. (1992). Time and the observer. *Behavioural and Brain Science*, 15, 183–247.

Deubel, H., B. Bridgeman, B., & Schneider, W. X. (1998). Immediate post-saccadic information mediates space constancy. *Vision Research*, 38, 3147–3159.

Findlay, J. M., & Gilchrist, I. D. (2003). *Active vision: The psychology of looking and seeing*. Oxford: Oxford University Press.

Domini, F., Vuong, Q. C., & Caudek, C. (2002). Temporal integration in structure from motion. *Journal of Experimental Psychology: Human Perception & Performance*, 28, 816–38.

Duhamel, J. R., Colby, C. L., & Goldberg, M. E. (1992). The updating of the representation of visual space in parietal cortex by intended eye movements. *Science*, 255, 90–92.

Duncan, J. (1984). Selective attention and the organization of visual information. *Journal of Experimental Psychology: General*, 113, 501–517.

Germys, F. De Graef, P., & Verfaillie, K. (2002). Transsaccadic perception of saccade target and flanker objects. *Journal of Experimental Psychology: Human Perception & Performance*, 28, 868–883.

Gibson, J. J. (1979). *The Ecological approach to visual perception*. Boston: Houghton-Mifflin.

Goh, J. O. S., Siong, S. C., Park, D., Gutchess, A., Hebrank, A., & Chee, M. W. L. (2004). Cortical areas involved in object, background and object-background processing revealed with functional magnetic resonance adaptation. *Journal of Neuroscience*, 24, 10223–10228.

Goodale, M. A., & Milner, A. D. (1992). Separate pathways for perception and action. *Trends in Neuroscience*, 15, 20–25.

Gould, J. D. (1976). Looking at pictures. In R. A. Monty, & J. W. Senders (Eds.) *Eye movements and psychological processes.* Hillsdale, NJ: Lawrence Erlbaum, pp. 323–346.

Grimes, J. (1996). On the failure to detect changes in scenes across saccades. In K. Atkins (Ed.), *Perception: Vancouver studies in cognitive science* (Vol. 2, pp. 89–110). New York: Oxford University Press.

Henderson, J. M., & Hollingworth, A. (2003). Eye movements and visual memory: Detecting changes to saccade targets in scenes. *Perception & Psychophysics*, 65, 58–71.

Hollingworth, A. (2004). Constructing visual representations of natural scenes: The roles of short- and long-term visual memory. *Journal of Experimental Psychology: Human Perception and Performance*, 30, 519–537.

Irwin, D. E., Yantis, S., & Jonides, J. (1983). Evidence against visual integration across saccadic eye movements. *Perception & Psychophysics*, 34, 49–57.

Jonides, J., Irwin, D. E., & Yantis, S. (1982). Integrating visual information from successive fixations. *Science*, 215, 192–194.

Kanai, R. & Verstraten, F. A. J. (2004). Visual transients without feature changes are sufficient for the percept of a change. *Vision Research 44*, 2233–2240.

Kubischik, M. (2002). Dynamic spatial representations during saccades in the macaque parietal cortex. Bochum: Ruhr-Universitaet Bochum.

Kusunoki, M., & Goldberg, M. E. (2003). The time course of perisaccadic receptive field shifts in the lateral intraparietal area of the monkey. *Journal of Neurophysiology*, 89, 1519–1527.

Land, M. F., Mennie, N., & Rusted, J. (1999). The roles of vision and eye movements in the control of activities of daily living. *Perception*, 28, 1311–1328.

Land, M. F., & Hayhoe, M. M. (2001). In what ways do eye movements contribute to everyday activities? *Vision Research*, 41, 3559–3565.

Lee, A. C. H., Buckley, M. J., Pegman, S. J., Spiers, H., Scahill, V. L., Gaffan, D. et al. (2005). Specialization of the medial temporal lobe for processing objects and scenes. *Hippocampus*, 15, 782–797.

Loftus, G. R. (1972). Eye fixations and recognition memory for pictures. *Cognitive Psychology*, 3, 525–551.

Loftus, G. R. (1981). Tachistoscopic simulations of eye fixations on pictures. *Journal of Experimental Psychology: Human Learning and Memory*, 7, 369–376.

Luck, S. J., & Vogel, E. K. (1997). The capacity of visual working memory for features and conjunctions. *Nature*, 390, 279–281.

Mazurek, M. E., Roitman, J. D., Ditterich, J., & Shadlen, M. N. (2003). A role for neural integrators in perceptual decision making. *Cerebral Cortex*, 13, 1257–1269.

McConkie, G. W. and D. Zola (1979). Is visual information integrated across succesive fixations in reading? *Perception and Psychophysics*, 25, 221–224.

Melcher, D. (2001). Persistence of visual memory for scenes. *Nature*, 412, 401.

Melcher, D., & Kowler, E. (2001). Scene memory and the guidance of saccadic eye movements. *Vision Research*, 41, 3597–3611.

Melcher, D., & Morrone, M. C. (2003). Spatiotopic integration of visual motion across saccadic eye movements. *Nature Neuroscience*, 6, 877–881.

Melcher, D. (2005). Spatiotopic transfer of visual form adaptation across saccadic eye movements. *Current Biology*, 15, 1745–1748.

Melcher, D. (2006). Accumulation and persistence of memory for natural scenes, *Journal of Vision*, *6*, 8–17.

Melcher, D., Papathomas, T. V., & Vidnyanszky, Z. (2005). Implicit attentional selection of bound visual features. *Neuron*, 46, 723–729.

Mishkin, M., Ungerleider, L. G., & Macko, K. A. (1983). Object vision and spatial vision: two cortical pathways. *Trends in Neurosciences*, 6, 414–417.

Morgan, M. J., Findlay, J. M., & Watt, R. J. (1982). Aperture viewing: A review and a synthesis. *Quarterly Journal of Experimental Psychology* A, *34*, 211–233.

Nakamura, K., & Colby, C. L. (2002). Updating of the visual representation in monkey striate and extrastriate cortex during saccades. *Procedures of the National Academy of Sciences USA*, 99, 4026–4031.

Neri, P., Morrone, M. C., & Burr, D. C. (1998). Seeing biological motion. *Nature*, 394, 894–896.

Nishida, S. (2004). Motion-based analysis of spatial patterns by the human visual system. *Current Biology, 14*, 830–839.

O'Regan, J. K. , & A. Noë (2001). A sensorimotor account of vision and visual consciousness. *Behavioral Brain Science, 24*, 939–973.

Piccolino, M., & Wade, N. J. (2006). Flagging early examples of ambiguity. *Perception, 35*, 861–864.

Pierrot-Deseilligny, C., R. Muri, M., Rivaud-Pechoux, S., Gaymard, B., & Ploner, C. J. (2002). Cortical control of spatial memory in humans: the visuooculomotor model. *Annals of Neurology, 52*, 10–19.

Pollatsek, A., Rayner, K., & Collins, W. E. (1984). Integrating pictorial information across eye movements. *Journal of Experimental Psychology: General, 113*, 426–442.

Potter, M. C. (1976). Short-term conceptual memory for pictures. *Journal of Experimental Psychology: Human Learning and Memory, 2*, 509–522.

Purkinje, J. (1823). Beobachtungen und Versuche zur Physiologie der Sinne. Beiträge zur Kenntniss des Sehens in subjectiver Hinsichts. Prague: Calve.

Raymond, J. E., Shapiro, K. L., & Arnell, K. M. (1992). Temporary suppression of visual processing in an RSVP task: An attentional blink? *Journal of Experimental Psychology: Human Perception & Performance, 18*, 849–860.

Rensink, R. A. (2000). Seeing, sensing, and scrutinizing. *Vision Research*, 40, 1469–1487.

Rensink, R. A., O'Regan, J. K., & Clark, J. J. (1995). Image Flicker is as good as saccades in making large scene changes invisible. *Perception, 24*(suppl.), 26–27.

Ross, J. & A. Ma-Wyatt (2004). Saccades actively maintain perceptual continuity. *Nature Neuroscience, 7*, 65–69.

Rolls, E. T., Xiang, J., & Franco, L. (2005). Object, space and object-space representations in primate hippocampus. *Journal of Neurophysiology, 94*, 833–844.

Ross, J., Morrone, M. C., Goldberg, M. E., & Burr, D. (2001). Changes in visual perception at the time of saccades. *Trends in Neuroscience, 24*, 111–121.

Smith, E. E., & Jonides, J. (1997). Working memory: A view from neuroimaging. *Cognitive Psychology, 33*, 5–42.

Tatler, B. W. (2001). Characterising the visual buffer: Real-world evidence for overwriting early in each fixation. *Perception, 30*, 993–1006.

Tatler, B. W., Gilchrist, I. D., & Land, M. F. (2005). Visual memory for objects in natural scenes: From fixations to object files. *Quarterly Journal of Experimental Psychology Section A-Human Experimental Psychology, 58*, 931–960.

Tatler, B. W. , Gilchrist, I. D., & Rusted, J. (2003). The time course of abstract visual representation. *Perception, 32*, 579–592.

Todd, J. J. & Marois, R. (2004). The capacity limit of visual short-term memory in human posterior parietal cortex. *Nature, 428*, 751–753.

Tolias, A. S., Moore, T., Smirnakis, S. M., Tehovnik, E. J., Siapas, A. G., & Schiller, P. H. (2001). Eye movements modulate visual receptive fields of V4 neurons. *Neuron, 29*, 757–767.

Ullman, S. (1983). Computational studies in the interpretation of structure and motion: summary and extension. In J. Beck, B. Hope, and A. Rosenfeld (Eds.) *Human and Machine Vision*. New York: Academic Press.

Vogel, E. K. & Machizawa, M. G. (2004). Neural activity predicts individual differences in visual working memory capacity. *Nature, 428*, 748-751.

Yin, C., Shimojo S., Moore, C., & Engel, S. A. (2002). Dynamic shape integration in extrastriate cortex. *Current Biology, 12*, 1379–1385.

PART 4

MODELLING OF EYE MOVEMENTS

Edited by

WAYNE S. MURRAY

Chapter 11

MODELS OF OCULOMOTOR CONTROL IN READING: TOWARD A THEORETICAL FOUNDATION OF CURRENT DEBATES

RALPH RADACH

Florida State University, USA

RONAN REILLY

National University of Ireland, Maynooth, Ireland

ALBRECHT INHOFF

State University of New York, Binghamton, USA

Eye Movements: A Window on Mind and Brain
Edited by R. P. G. van Gompel, M. H. Fischer, W. S. Murray and R. L. Hill

Abstract

This chapter begins with a review and classification of the range of current approaches to the modeling of eye movements during reading, discussing some of the controversies and important issues arising from the variety of approaches. It then focuses on the role and conceptualization of visual attention inherent in these models and how this relates to spatial selection, arguing that it is important to distinguish visual selection for the purpose of letter and word recognition from visual selection for the purpose of movement preparation. The chapter concludes with some proposals for model testing and evaluation and some challenges for future model development.

1. Introduction: Models of oculomotor control in reading

The study of complex cognitive processes via the measurement and analysis of eye movements is a flourishing field of scientific endeavor. We believe that three reasons can account for this continuing attractiveness (Radach & Kennedy, 2004). First, reading provides a domain in which basic mechanisms of visual processing (e.g. "selective attention" or "intersaccadic integration") can be studied in a highly controlled visual environment during a meaningful and ecologically valid task. Second, reading can be studied as an example of complex human information processing in general. This involves the explicit description of the relevant levels and modules of processing and the specification of their dynamic interactions. Third, understanding links between cognition and eye-movement control in reading is one of the backbones of modern psycholinguistics, where oculomotor methodology has become a standard tool of studying the processing of written language at the level of words, sentences and integrated discourse. Excellent discussions of theoretical and methodological issues in this important subfield of reading research can be found in Murray (2000) and Clifton, Straub, and Rayner (Chapter 15, this volume).

In the context of the description given above, the development of computational models is relevant for all three levels of ongoing research on reading. An important theoretical starting point for any model of this kind is the idea that continuous reading involves two concurrent streams of processing. Quite obviously, the primary task is the processing of written language, where the acquisition of orthographically coded linguistic information feeds into the construction of a cognitive text representation. At the same time, the targeting and timing of saccadic eye movements serves to provide adequate spatio-temporal conditions for the extraction of text information. To understand the coordination and integration of these two processing streams is the main motivating force for the development of computational models of oculomotor control in reading.

This view on modeling reading differs slightly from the position taken in the commentary chapter by Grainger (2003) in the book edited by Hyönä, Radach, & Deubel (2003). In his thoughtful discussion of relations between research on single-word recognition and continuous reading, he emphasizes a distinction between a "model of reading" and a "model of a task" used to study reading. In this context, measuring eye movements is one particular task, with a role similar to a lexical decision or perceptual identification task. All possible tasks emphasize different aspects of the process and the "functional overlap" between them will be instrumental in understanding the true nature of reading. Although this view provides a useful strategy for research on specific hypotheses about word processing, it neglects one fundamental point: Eye movements are not just an indicator of cognition, they are part and parcel of visual processing in reading, just as "active vision" (Findlay & Gilchrist, 2003) is part and parcel of perception and information processing in general. The eyes are virtually never static and words always compete with other visual objects for processing on multiple levels and for becoming the target of the ensuing saccade. Thus, coordinating eye and mind is a key part of the natural process of reading rather than a level of complexity added by just another task.[1]

In the mid-1990s, the arena of (pre-computational) models was dominated by the debate about the degree to which "eye" and "mind" are linked during reading. In this discussion, the prototypical adversary on the cognitive side was the attention-based control model originally proposed by Morrison (1984) and reformulated by Rayner & Pollatsek (1989). On the other visuomotor end of the spectrum was the "strategy and tactics" theory by O'Regan (1990), claiming that the eyes were driven by a global scanning strategy in combination with local (re)fixation tactics. However, it soon became clear that these extremes could not account for the full range of phenomena, so that both sides acknowledged the need to include ideas from both points of view (O'Regan, Vitu, Radach, & Kerr, 1994; Rayner & Raney, 1996). As evident in the description of the current E-Z reader model in Chapter 12, the model implements both the refixation tactics suggested by O'Regan and the metrical principles of saccade-landing positions proposed by McConkie, Kerr, Reddix, & Zola, (1988). On the other hand, Glenmore, a model that has been developed out of the tradition of low-level theories (Reilly & O'Regan, 1998; Radach & McConkie, 1998) incorporates a word-processing module that, together with visual processing and oculomotor constraints, determines the dynamics of eye movements during reading.

Today the most comprehensive computational models of visuomotor control in reading are E-Z reader and the SWIFT model, which, on a comparable level of model complexity, account for an impressively wide range of empirical phenomena. These include not just basics such as effects of word frequency and word length on spatial and temporal parameters, but also such intricate phenomena as the modulation of parafoveal preprocessing by foveal processing, the generation of regressions and the so-called "inverted optimal viewing position effect". Interestingly, there are a number of similarities between both families of models, such as the idea of a labile and a non-labile phase of saccade programming and the implementation of saccade amplitude generation based on McConkie et al. (1988). Major differences concern two central questions about the nature of the eye–mind link. While in E-Z reader every interword saccade is assumed to be triggered by a specific word-processing event, saccades are triggered in SWIFT by an autonomous generator which in turn is modulated (delayed) by the mental load of foveal linguistic processing. The second important difference is the degree of spatial and temporal overlap in the processing of words within the perceptual span. In all sequential attention shift (SAS) models, a one-word processing beam, referred to as "attention", moves in strictly sequential fashion. In contrast, linguistic processing

[1] Measuring eye movements in continuous reading situations is certainly not the only way to understand the processing of written language. Single-word recognition paradigms make extremely valuable contributions to the understanding of word processing under rigorously controlled conditions. These paradigms have, in many respects, laid the foundation for experimental reading research as a whole (see Jacobs & Grainger, 1994, for an informative review).

encompasses several words within a gradient of attention or "field of activation" in SWIFT.

The idea that there is a certain degree of parallel word processing is shared in one or another way by a number of recent models, which are often referred to as PG (processing gradient) models (Reilly & Radach, 2006; Inhoff, Eiter, & Radach, 2005) or GAG (guidance by attentional gradient) (e.g. Engbert, Nuthmann, Richter, & Kliegl, 2006) models. We prefer the term "processing gradient" because it avoids reference to the notion of "attention", which, as we will discuss below in some detail, may be applied to a variety of diverse phenomena and hence might lead to misunderstandings.

Looking at the recent models listed above, it is clear that all can be readily classified along the axes just mentioned, autonomous saccade generation vs cognitive control on the one hand and sequential vs parallel word processing on the other hand. It is also interesting to note that both dimensions are indeed necessary for a meaningful classification. This can be illustrated using the example of Mr. Chips, an ideal observer model developed by Legge, Klitz, & Tjan (1997; see also Legge, Hooven, Klitz, Mansfield, & Tjan, 2002). Here, letter and word processing is parallel across words boundaries within a fixed visual span but saccade control is exclusively determined by cognitive processing.

Within the family of PG models, there is substantial variation in the degree to which (more or less) parallel processing is associated with cognitive modulation of saccade triggering. Today, nearly everyone in the field acknowledges the overwhelming evidence that linguistic processing is reflected in temporal and spatial aspects of saccade control. However, it is nonetheless fascinating to see how far one can go with models that rely almost exclusively on low-level visual processing and oculomotor constraint. One such model is the SERIF model (McDonald, Carpenter, & Shillcock, 2005), another is the Competition-Interaction model (Yang, 2006; Yang & McConkie, 2001). In both models, there is little (and rather indirect) cognitive influence on saccade control, while, in contrast, ongoing linguistic processing has a substantial impact on saccade triggering in SWIFT and Glenmore.

As mentioned above, at the other end of the spectrum there is the assumption by the authors of the E-Z reader model that each saccadic eye movement is triggered by a single cognitive processing event. Given the number and diversity of opposing approaches, it may seem that the sequential attention assumption is a "minority position" or an extreme viewpoint. However, such an impression is unjustified for two reasons. First, the E-Z reader model is actually not at the real end of the spectrum, as it incorporates a lot of low-level machinery. Much more extreme versions of cognitive control have been proposed in the past (e.g. Just & Carpenter, 1988) but did not receive much empirical support. Second, and more importantly, when conceiving a spatially distributed mode of processing, there are many ways by which one can design and implement such a system. However, in the case of a sequential processing system, there are much tighter constraints such as the need for a precisely defined trigger event for saccades, driving the design of a model in a certain direction. It is therefore no surprise that there is only one family of E-Z reader models and a relatively large number of competitors. Engbert, Nuthmann,

Richter, & Kliegl (2005) have recently expressed this relation of opposing approaches by referring to a sequential control mechanism as one special case within a space of possible solutions along a dimension from massive to very limited parallel processing.

By occupying one of the extreme positions on this continuum, the E-Z reader model gains a unique quality: It becomes so specified that falsification of core mechanisms within the model becomes feasible (see Jacobs, 2000, for a detailed discussion). We believe that it is primarily for this reason that this type of model has provoked an enormous amount of empirical work. Therefore, even if some of the central assumptions of the sequential architecture turn out to be incorrect, the model will have contributed more to the field than many of its less traceable competitors.

Our goal for this commentary chapter is not to discuss the state of the art in the entire field of computational modeling of continuous reading. An excellent overview has been provided in the review by Reichle, Rayner, & Pollatsek (2003), summarizing key features of virtually all existing types of models. Publications on the most recent versions of the SWIFT model (Engbert, Nuthmann, Richter, & Kliegl, 2005) and the E-Z reader model (Pollatsek, Reichle, & Rayner, 2006) include detailed discussions of their background and also point to some controversial points in the ongoing theoretical debate. A special issue of Cognitive Systems Research edited by Erik Reichle includes new or updated versions of no less than six different models: the SWIFT model (Richter, Engbert, & Kliegl, 2006), the E-Z Reader model (Reichle, Pollatsek, & Rayner, 2006), the Glenmore model (Reilly & Radach, 2006), the Competition/Interaction model (Yang, 2006) and the SHARE model (Feng, 2006).

Rather than giving another descriptive overview concerning design principles, mechanisms and implementations of the existing models, this chapter will attempt to provide a relatively detailed discussion of some rather fundamental theoretical ideas that are implicit in many modeling approaches but only rarely made explicit. We will focus this discussion on aspects of visual processing often subsumed under the concept of "attention". This discussion will include quite a few references to research on oculomotor control outside the domain of reading, which appears particularly appropriate in a volume that reflects the state of the art in oculomotor research as a whole. The idea is to explore the question to what extent current modeling in the field of reading is grounded in basic visuomotor research and how it corresponds to evidence from neighboring domains. Since there is only one integrated human information processing system, any model of reading should be seen as special case of a more general theory of visual processing and oculomotor control. Processing mechanisms proposed for reading should be in harmony with the mainstream of visuomotor research, and models should, in principle, be able to generalize to other domains such as scene perception and visual search. After looking in some detail at answers to a number of key questions about "visual attention", we will consider consequences for models of reading. In the final part of the chapter we will discuss some issues for future model developments and point to important problems for model comparison and evaluation.

2. The role of visual attention for theories and models of continuous reading

2.1. Definitions of visual selective attention

Until recently, in most cases the notion of "attention" was used in the literature on eye movements in reading without providing or referring to any explicit definition. The reason for this lack of precision may be that some authors have shared the popular view that "everyone knows what attention is" (James, 1890), so that a definition was considered unnecessary.[2] Alternatively, authors may have avoided definitions because they were aware of the problems that still exist in the present literature on attention with providing a precise and unambiguous specification of the concept. In recent publications of the E-Z reader model, Reichle et al. have specified their understanding of attention by referring to the classic work by Posner (1980, see below). Other authors have linked their theoretical ideas to a gradient conception of attention (e.g. Engbert, Nuthmann, Richter, & Kliegl, 2005) or deliberately avoided the term "attention" altogether (Reilly & Radach, 2006). In this section we intended to deepen the ongoing discussion by discussing some fundamental questions of research about "attention" that should from the base of more detailed consideration within the framework of certain models on eye-movement control in reading.

There is an enormous body of literature on attention and for non-specialists it is quite difficult to keep track with the dynamic and complex development in the field (see, e.g., Chun & Wolfe, 2001; Egeth & Yantis, 1997 and Pashler, 1998, for reviews). Following Groner (1988), we can differentiate different facets of "attention" as follows: On a very abstract level, one can first distinguish general attention in terms of alertness, arousal or general activation from specific attention. Within the domain of specific attention, the notion is used to describe issues related to "divided" attention, the processing of information on competing channels, further, the "orienting of attention", and finally, problems related to the "allocation of attentional resources". Zooming in on attentional orienting, the distinction of overt vs covert orienting refers to whether "orienting" is observable in terms of a behavioral act (e.g. an eye movement) or is hypothetical, for example as an "attentional movement" inferred indirectly from behavioral data. A further distinction can be drawn between orienting that is controlled externally, for example by a sudden visual onset, "capturing" attention, or internally, for example by inducing expectations about the location of a to-be-displayed stimulus. In this chapter we will concentrate primarily on attentional orienting, and, more specifically, visual selection,

[2] The explanation given by James (1890) reads as follows: "It is the taking possession by the mind, in clear and vivid form, of one out of what seem several simultaneously possible objects or trains of thought. Focalization, concentration, of consciousness are of its essence. It implies withdrawal from some things in order to deal effectively with others, and is a condition which has a real opposite in the confused, dazed, scatterbrained state which in French is called distraction, and Zerstreutheit in German" (http://psychclassics.yorku.ca/James/Principles/prin11.htm). Interestingly, although this definition appears straightforward, it captures only one aspect of attentional processing, which is of rather peripheral importance to our discussion.

although here and there problems of divided processing and limited resources may also play a role.

Principal viewpoints on the role of attention in human information processing appear to oscillate between two extremes: Posner & Petersen (1990) postulate the existence of an integrated attention system that is anatomically separate from the various data processing systems performing operations on specific inputs. This attention system is composed of a network of anatomical areas that carry out different functions; attention is thus a property neither of a single center nor of the brain as whole. Posner (1990) organizes the attention system into three subsystems responsible for: (a) orienting to sensory events; (b) detecting signals for focal (conscious) processing; and (c) maintaining a vigilant or alert state.

An alternative view is advocated, for example, by Allport (1992), who rejects the idea of one integrated functional system. Instead, he emphasizes the existence of a multitude of qualitatively different mechanisms that are involved in implementing visual-attentional selectivity. Allport discusses two questions that were at the core of dispute for more than 25 years: the question of early vs late selection and the question of which processes require attention (e.g. automatic vs controlled). He arrives at the conclusion that the quest for unified answers to these questions is hopeless, as "... there is a rich diversity of neuropsychological control mechanisms of many different kinds (and no doubt many yet to be discovered) from whose cooperative and competitive interactions emerge the behavioral manifestations of attention" (p. 203). This general statement also applies to the domain of reading, where, as we will see below, different mechanisms of selection and preferred processing co-exist and serve different specific purposes.

In our discussion we will focus on processes and mechanisms tapped by (a) Posner's functions and (b) visual selection of information and processing of selected information at the expense of other potential visual input. Even within this limited domain, it is interesting to observe subtle but theoretically important differences between theorists' positions on what the term "attention" shall refer to. We can illustrate this using the example of attention as a directable resource vs attention as a controller, a mechanism or an assembly of mechanisms for visual selection (Johnston & Dark, 1986).

According to a popular, almost canonical, definition, selective attention is a limited resource, something that can be "directed" to or "focused" on a certain location or object. Posner & Peterson (1990) not only specify operations of disengagement, movement and re-engagement of attention but also make suggestions about the neural circuitry underlying visual attention shifts. "The parietal lobe first *disengages* attention from its present focus, then the midbrain area acts to *move* the index of attention to the area of the target and the pulvinar is involved in *reading out data* from the indexed locations" (pp. 28–29, our italics). A second important component of this view is that attentional processing is confined to a relatively narrow area, often described as having a focus and a sharp margin (James, 1890). Posner (1980, p. 172) introduced the metaphor of "a spotlight that enhances the efficiency of the detection of events within its beam".

From a slightly different angle, attention can be conceptualized as a controller that directs or allocates a limited resource. For example, Yantis (1998) begins his review on the control of attention by emphasizing that the visual system accomplishes object recognition by visually selecting a relevant or salient part of the image and then concentrates processing there before moving on to another part. He continues by stating that "... The mechanism that accomplishes selection is called visual attention ... " (p. 223). This view is also quite popular in the attention literature and it is not unlikely that one is confronted with multiple references to the concept of visual attention within the same publication: First, as a *mechanism that does selection* and then as a *entity that is moved* as a result of this selection, or, similarly, as a beam of enhanced processing due to resource allocation.

Given these (and other) different ways to specify the term "attention", we will adopt a minimalist definition, using "attention" as a synonym for visual spatial selection and preferred processing of visual input, leaving details unspecified until specifications are required and become useful for our discussion. At some points in this chapter it will become difficult to draw a clear line between visual processing terminology (e.g. letter identification, lexical processing) and attention terminology (attention allocation, shift of attention). This is a consequence of the custom of using the concept of attention simply to describe perceptual and cognitive processing. Also, in many current discussions, the necessary distinction between attention as a description of what has been observed vs an agent that causes what can be observed is hard to establish, a difficulty that our chapter shares with much of the literature in the field (Johnston & Dark, 1986; Neumann, 1987).

Below we will raise a number of questions around the issues of attention and spatial selection in reading. We hope that it will become clear that, although these questions may appear straightforward, attempts to give clear answers can turn out to be quite complicated. With these issues in mind we review some basic attention research as well as relevant work within the domain of reading itself.

2.2. *What is the purpose of spatial selection in reading?*

It is basic knowledge in the field that the efficiency of processing during reading across the functional visual field is not simply a function of visual factors like acuity and lateral masking (see Radach, Kennedy & Rayner, 2004, for a recent compilation of empirical work with a focus on parafoveal processing in reading). As the classic studies of Rayner, McConkie and their colleagues have shown, the "span of letter identification" during a given fixation extents about 8–9 letters to the right and about 4 letters to the left. The fact that this asymmetry reverses when Hebrew readers read from right to left is seen as a consequence of the dynamic allocation of visual processing resources or "attention" (see Rayner & Sereno, 1994 and Rayner, 1998, for comprehensive reviews). An intriguing question is whether the types of spatial selection relevant for linguistic processing and oculomotor control are the same or different. If they turn out to be the same, this would raise the further question of how attention is divided between the two tasks (Hoffman, 1998). If they are different, then we need to ask further questions about the nature of this difference and about what specific mechanisms accomplish selection to serve both linguistic and visuomotor processing.

A look into the relevant literature suggests that there is solid empirical evidence in support of different control processes for the two types of selections. In fact the fundamental architecture of the human visual system suggests the existence of this division. Ungerleider & Mishkin (1982) proposed that the processing of an object's qualities and its spatial location rest on different neurobiological grounds. They distinguished a "ventral stream" of projections from the striate cortex to the inferior temporal cortex from a "dorsal stream" of projections terminating in the posterior parietal cortex. This distinction is widely known as the "what" vs "where" – systems of visual processing. An alternative perspective on modularity in the cortical visual system was put forward subsequently by Goodale & Milner (1992), placing less emphasis on informational input and taking more account of output requirements, the *purposes* of visual processing. They refer to the ventral and dorsal processing streams as two visuomotor systems, named the "what" vs "how" system. In this dichotomy, the dorsal system is concerned with processing for the purpose of motor control, including, of course, the preparation and execution of eye movements. Goodale & Milner (1992) note that "attention needs to be switched to particular locations and objects whenever they are the targets either for intended action or for identification . . . " (p. 23) and emphasize that ". . . spatial attention is physiologically non-unitary and may be as much associated with the ventral system as with the dorsal" (p. 24).

On an information processing level, the neurobiological segregation of a ventral from a dorsal system has its analogy in the distinction between "selection for object recognition" (e.g. La Berge & Brown, 1989) or "selection for perception" (Deubel, Schneider, & Paprotta, 1998; Kowler, Anderson, Dosher, & Blaser, 1995) vs "selection for action" (Allport, 1987; Neumann, 1987). The latter mechanism can be seen as solving the problem of parameter specification: Space-based motor actions like saccadic eye movements require a mechanism that supplies the motor system with the spatial parameters of the intended target object. Therefore, selection-for-space-based-motor-action is considered a central function of visual attention (Schneider, 1995). Following this view, we can assume as a working hypothesis that, in normal reading too, two qualitatively different forms of selection are operational. Whether the term "attention" is used to characterize either form or whether it is being reserved for describing selection for the purpose of recognition (i.e. linguistic processing) is merely a question of terminological preference.

2.3. What are the units spatial selection is operating on?

McConkie & Zola (1987) characterize the process of word-selection as being based on a two-dimensional representation of the stimulus array provided by early visual processing. This stimulus structure can be conceived as a four-level object hierarchy (letters, words, lines, pages) consisting of objects that are relatively homogenous in size and shape. Within this rapid process of parsing of the available visual information, low spatial frequency word-objects, bounded by empty spaces, are assigned the role of potential saccade targets. This proposal can be seen as the first specification of "preattentive" processing in the domain of research on continuous reading.

Initially, eye movements were assumed to be aimed at bringing the eyes to a location close to the border of the current perceptual span (e.g. McConkie, 1979). However, when looking closer at local fixation patterns, it became obvious that word boundaries play an important role in guiding the eyes. One example is that distributions of fixation locations over adjacent words are often multimodal. This can occur when two relatively short words follow to the right of the current fixation, leading to one peak at the next word (word $N+1$) and one peak at word $N+2$ including saccades that have skipped $N+1$ (McConkie & Zola, 1984). Another example is the situation where progressive saccades depart from one of the first letters of a relatively long word, resulting in one peak of landing positions within the second half of that word (refixations) and one peak on locations after the word boundary as a consequence of interword saccades (Radach & McConkie, 1998).

Looking more closely at the spatial distribution of saccade landing positions within words, there is a strong tendency for landing positions of initial (first), progressive (left to right) saccades into words to cluster about halfway between word beginning and word center. Rayner (1979) named this phenomenon the preferred viewing position and it has become one of the most intensely researched effects in oculomotor reading research. McConkie, Kerr, Reddix, & Zola (1988) have first shown that landing position distributions can be decomposed into component distributions depending on the launch distance of the saccade relative to the target word and the length of that word (see Figure 2, left panel). The preferred viewing position is primarily determined by the spatial layout of potential target words and visuomotor constraints.[3] Comparing saccade landing site distributions for two legal spellings of German compound words ("Gehirnchirurg" vs "Gehirn-Chirurg"), Inhoff, Radach, & Eiter (2003) found that the introduction of a hyphen shifts the initial landing site distribution to the left by an amount that corresponds precisely to the distribution for a six-letter word. Looking at the distributions for within word refixations, the unified spelling leads to a distribution with a peak at the word center, while the hyphenated spelling causes two distinct sub-distributions with peaks in the centers of either substring.

The preferred viewing position needs to be kept separate from the "convenient" or "optimal" position first described by O'Regan (1984; see O'Regan, 1990, for a detailed discussion) located near the center of words that is most effective in terms of word recognition and/or oculomotor effort. Taken together, the available evidence indicates that the spatial component of eye movement control in reading consists of two processes, one that selects a specific word as a saccade target and the other that supplies saccade parameters aiming at the selected word. The actual saccade landing positions are best described as the result of an attempt to get the eyes to the center of the selected words. However, due to specific oculomotor and perceptual constraints, there are systematic deviations leading to the dissociation between the "optimal" and the actual, "preferred" landing positions.

[3] There is a relatively small effect on landing positions of orthographic properties of parafoveal words (e.g. Hyönä, 1995, White & Liversedge, 2004), which appears to be due to a graded modulation of saccade amplitudes rather than a pop put of unusual letter clusters (Radach, Heller, & Inhoff, 2004).

The important point for our current discussion is that the selection of saccade target locations appears to be strictly word-based. That is the oculomotor system always aims for a specific word-object (McConkie et al., 1988, Radach & McConkie, 1998).

If visuomotor selection is strictly word-based, how about selection for the purpose of linguistic processing? This question has been addressed in an elegant experiment by McConkie & Zola (1987). They selected pairs of target words that differed in only one critical letter such that each word would fit appropriately into a short passage of text. Thirty-two participants were asked to read 312 passages, half in an experimental and half in a control condition. In the experimental condition the critical letter was changed during each saccade as soon as the eyes had entered a region near the target word. As a result, the target word was different during each successive fixation, alternating between the two members of a pair. The following sentence is an example of one item (p. 389, our italics).

John does not store his tools in the shed any more because it *leaks* (vs leans) so much.

After reading a passage, four words were presented in succession and the subject indicated whether or not the respective word had been in the passage. In addition to the critical words, this presentation included two additional words that also differed from each other by a single letter. A letter perception index (LPI) was computed on the basis of all observations where a participant had identified one, and only one, of the critical words. For each fixation position within a region from 10 letters left to 5 letters right of the critical letters, the relative frequency of critical letter identifications was computed. Thus, the data indicate where the eyes were centered when a given letter was processed.

The data presented in Figure 1 indicate that within the letter identification span there is a sharp drop-off in letter processing performance as a function of distance between fixation

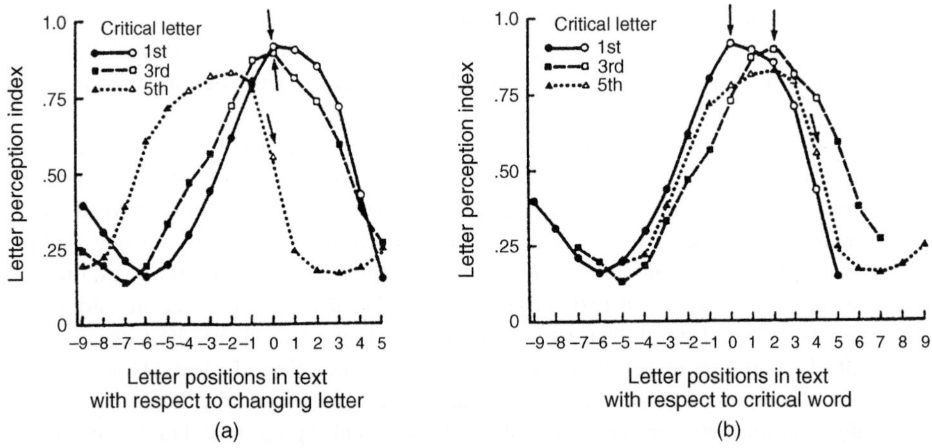

Figure 1. Letter perception index curves for first, central and final letters within 5-letter words. Left panel: LPI with respect to actual fixation positions. Right panel: LPI relative to critical letter positions within target words – beginning with the first letter of the 5-letter target. Arrows indicate fixated letter positions. Reprinted with permission from McConkie & Zola (1987).

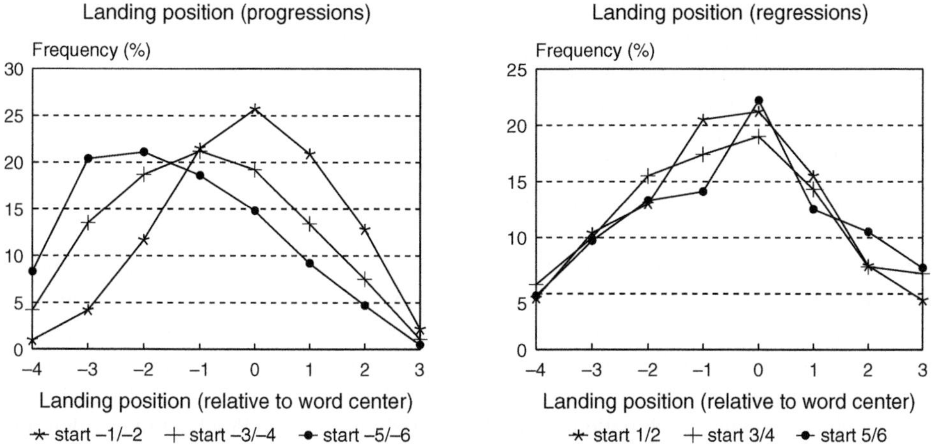

Figure 2. Distribution of saccade-landing positions into words while reading a classic novel in German. Left panel: Landing positions of initial progressive (left to right) saccades into 7-letter words coming from different launch sites relative to the word beginning. Right panel: Landing positions of regressive (right to left) saccades coming from different launch sites relative to the word ending. Data from four participants reading a classic novel in German (see Radach & McConkie, 1998, for a detailed discussion).

position and critical letter. As could be expected, the well-known asymmetry in the extent of the letter identification span that we have mentioned above is also clearly expressed in these data. More importantly, as shown in the left panel of Figure 2, the LPI curves are displaced horizontally as a function of the position of the critical letter within the target word. This is particularly striking for letters at the end of the word, which are processed with maximum accuracy when the eyes fixate several letter positions away. This indicates that the position of a letter within a word strongly modulates processing. Accordingly, the curves move much closer together when plotted relative to the boundaries of the target word (right panel).

McConkie & Zola (1987) concluded from their data that "text is attended in word units during skilled reading, both for identification and eye-guidance purposes" (p. 393). At the same time, although their curves show that word boundaries constrain letter processing, they also indicate that some processing takes place across these demarcations. This finding is in harmony with a large body of evidence indicating that information from neighboring words is often acquired during parafoveal preprocessing within the perceptual span (Rayner, 1998). As indicated by the basic fact that many words are not fixated at all during reading, this parafoveal processing often includes whole words, especially when these words are short and close to the current fixation position. From here, two related questions originate.

The first is whether and under what circumstances, more than one word may be processed during a given fixation. A second related question concerns the time course of spatially selective linguistic processing. Can linguistic information from two words be processed concurrently. That is starting from the beginning of a fixation, or does "perceptual attention" need to be shifted away from its initial focus and re-allocated at a new location? This controversial point, which, as pointed out in the initial part of this chapter, is central to current models of eye-movement control in reading, will be considered in the following section.

2.4. What is the time course of spatial selection?

Estimations of how long it takes attention to "move" from one stimulus location to another are often based on the slopes of search functions in serial search tasks, used to compute the amount of time spent per item in the visual display.[4] The time per unit suggested by this procedure is usually in the order of about 25–75 ms (Egeth & Yantis, 1997), with 50 ms as the value initially reported in the classic study by Treisman & Gelade (1980). Other techniques, such as the attentional blink paradigm, have yielded estimates that are in the order of 150 to well over 300 ms and are thus similar to typical fixation durations (Horowitz, Holcombe, Wolfe, Arsenio, & Dimase, 2004; Theeuwes, Godijn & Pratt, 2004; Ward, 2000). In a recent review, Wolfe (2003) discussed the apparent contradiction between these vastly different estimates using the metaphor of a car wash: Cars enter the car wash in a sequence, but at any point in time several cars are being washed in parallel. Using this metaphor, the 50-ms-per-item slope corresponds to the rate at which "cars" pass though the system. However, it may still take 300 ms for every car to get washed, leading to a 300 ms "attentional dwell time". Logan (2005) recently used a clever experimental technique to separate cue-encoding time from the time it takes to switch attention. He estimated a time interval of 67–74 ms for cue encoding and 76–101 ms for attention switching.

Eriksen (1990) concluded from a large body of research using attentional cueing paradigms that an enhancement of processing at a cued location (taken to indicate a shift of attention) starts to develop after 50 ms and continues to grow up to an asymptote at about 200 ms after the cue. As Kinchla (1992) put it: "In terms of the spotlight metaphor, it is as if the spotlight went off at one point and then gradually came on again at the target location" (p. 726). Most interesting in this context is the conclusion that the shift of attention itself takes time in addition to the time for the processing event that triggers that shift. Taken together, there appears to be consensus in the field that, if an attention shift is assumed to take place, there must be a specific time interval assigned to its preparation and execution.

A related question that is central to our line of discussion concerns the point in time during an eye fixation in reading when letters and words are being "attended" within and outside the focus of foveal vision. For one type of visual selection in reading, selection

[4] Of course, this approach rests on the assumption that visual search is indeed strictly serial. Alternatives involving mechanisms of parallel processing have been proposed, however, for example, by Duncan & Humphries (1989), Wolfe (1994) and others. Findlay & Gilchrist (1998) have provided empirical evidence that makes it unlikely that search displays are scanned item-by-item by an attentional beam.

for the purpose of visuomotor processing (a special case of "selection for action"), the answer to this question appears straightforward. Pollatsek & Rayner (1982) reported an experiment where spaces between words were filled at various times contingent upon the onset of a reading fixation. Filling the first space to the right of the current fixation position had no effect when it occurred later than 50 ms after the beginning of the fixation, suggesting that basal information on the visual configuration is acquired very early. This is in harmony with the idea advocated above that early vision provides an array of potential saccade targets (initially perhaps in the form of low spatial frequency word-objects) among which the processing system chooses on the basis of visuomotor constraints and/or linguistic processing.

More interesting is the demonstration by Rayner, Inhoff, Morrison, Sloviaczek, & Bertera, (1981) that reading can proceed quite smoothly when the entire text is masked after 50 ms. Recently, Rayner, Liversedge, White, & Vergilino-Perez (2003) showed that there are robust word frequency effects when fixated words disappear after 60 ms. These findings can be taken to suggest that "readers acquire the visual information necessary for reading during the first 50–70 ms of a fixation . . . " (Rayner, 1998, p. 378). This appears to be in contradiction with results by Blanchard, McConkie, Zola, & Wolverton (1984). They changed the word displayed during a fixation in sentence reading at various intervals (50, 80 and 120 ms from fixation onset) and received responses by readers who reported to have seen only the first word, both or only the last word. These results are compatible with a model that distinguishes the registration of information by the visual system from its utilization for linguistic processing (McConkie, Underwood, Zola, & Wolverton, 1985). The idea is that, whenever language processing requires new input during a fixation, the visual information that is currently present will be used. Within this model one could accommodate Rayner et al.'s findings by assuming that visual information is registered throughout the fixation (including the first 50–70 ms) and can then be stored for cognitive utilization in terms of cognitive processing. The data presented by Blanchard et al. pose a problem for models assuming a sequential word-to-word movement of attention, because late during a fixation there should be no report of the replacement word, as attention is supposed to have moved away already.

2.5. What is the nature of the relation between attention and eye movements?

It has often been claimed that attentional orienting precedes eye movements to a specific location. However, observations that allow such a conclusion to be drawn are not necessarily informative about the question whether this relation is one of causality or rather one of coincidence, as covert and overt orienting will usually happen to have the same goal (Hoffmann, 1998). The question whether attention and eye movements can operate concurrently on different regions or objects has been studied in dual task paradigms. Participants are asked to move their eye to a specific location and, at the same time, to detect or identify a visual target located at varying distances relative to the saccade target. Performance in the identification task is taken as an indirect measure of attention allocation. If it is possible to spatially dissociate attention and saccades, subjects should be able to attend to one location (i.e. show a high identification performance) while making a saccade to another.

Studies using this methodology found ample evidence in favor of close links between attention and saccades. For example, Hoffman & Subramaniam (1995) instructed subjects to make a saccade to one of four fixation boxes located left, right, above or below a central fixation point as soon as they heard a tone. The direction of the saccade target was held constant throughout the experiment. Before each trial, an arrow was presented that pointed to one of the four possible locations at which a target letter could be presented for a very short time interval. The arrow was used to direct attention to a specific location, and, as indicated by a detection-only control condition, this manipulation did indeed enhance performance in cue-target match trials. The key result of this study is that targets were detected better in the dual task situation when they occurred in a location to which a saccade was being prepared even in comparison to locations the central arrow was pointing to. This can be taken as evidence that visuospatial attention is required to execute a saccade, rendering the endogenous "attentional" cue ineffective (Hoffman, 1998; Hoffman & Subramaniam, 1995).

Deubel & Schneider (1996) reported experiments using stimulus materials that were similar to a reading situation in that they presented a horizontal string of letters separated by blanks. A central cue designated a specific member of the string as the saccade target. Before the onset of a saccade toward this goal, a discrimination stimulus was presented briefly within the string. Their results, again, indicated a high degree of selectivity: discrimination performance was nearly perfect when the saccade was directed to the critical item, but close to chance level when the target was located only one item to the left or right of the discrimination stimulus. Importantly, conditions designed to provoke a decoupling of "recognition" attention from saccade programming failed to alter the principal results.

Taken together, these studies indicate that object-specific coupling of saccade programming and object discrimination is strong and mandatory, at least in paradigms using dual tasks with explicit oculomotor and recognition instructions.[5] This has led to the currently dominant view that a pre-saccadic shift of attention is functionally equivalent to the selection of the peripheral item as the target of a subsequent eye movement, thus constituting the initial step in the programming of a saccade (see Deubel, O'Regan, & Radach, 2000, for a detailed discussion). However, it remains to be seen whether this conclusion extends to other experimental conditions, and, most importantly, whether it is an adequate description of what happens in normal reading.

[5] One study claiming that voluntary saccades can be programmed independently of attention shifts is by Stelmach et al. (1997) who used a temporal order judgment as the indicator for attention allocation. However, in agreement with most other studies, they found that under conditions of exogenous cuing, attention clearly "moves" much faster than the eyes. Their finding that under endogenous cueing more than 300 ms is needed to redirect attention to the parafovea is very interesting but perhaps irrelevant to a possible role of attention for reading saccades, as these need to be programmed within a much shorter time interval.

Elaborating on the research described above, Godijn & Theeuwes (2003) recently studied the association of attention with eye movements in a paradigm involving the execution of eye movements to a sequence of targets. They used a dual task paradigm to independently manipulate saccadic target selection and the allocation of attention to a particular location in space. In one experiment, the primary task required the execution of a sequence of saccades to two simultaneously cued object locations of a circular display with eight locations, and the secondary task required the identification of a letter that was presented for one of four time intervals from 47 to 129 ms following the saccade cue. Letters at the locations of both the first and the second saccade target were recognized more accurately than letters presented at a non-cued location, and this occurred even at the shortest dual-saccade cue-letter presentation interval. Five experiments using different task variations and controls provide compelling evidence for the view that attention was allocated (in parallel) to more than one distinct object location, and that parallel attention allocation preceded saccade targeting.

3. Consequences for visual processing and oculomotor control in reading

A number of important conclusions can be drawn from our discussion above for theories and models of oculomotor control in reading. First, it has become obvious that we need to distinguish visual selection for the purpose of letter and word recognition from visual selection for the purpose of movement preparation. As illustrated in Chapter 12, current SAS models assume that "attention" (for the purpose of lexical processing) moves sequentially in a word-by-word fashion. This attention shift is triggered by the completion of lexical processing on the current word. However, a preliminary stage of processing, referred to as L1, is assumed to trigger the programming of an eye movement to the neighboring word. As the authors emphasize, in their theory, "attention is decoupled from eye movement control". As noted above, the idea that saccades are programmed toward locations that are not (yet) "attended" for the purpose of recognition puts SAS models at odds with the mainstream of research on relations between attention and eye-movement control (Deubel, O'Regan, & Radach, 2000).

But what exactly does "decoupling" mean? It may mean that in fact there is a dissociation between attention for letter recognition and attention for oculomotor control, in which case one specific form of attention shift would be associated with either stream of processing. It may also mean that the *selection for movement* is left to pre-attentive processing, an idea that brings the SAS theory a step closer to many proposals made within the family of PG models.

The problem of how "covert" (perceptual) and "overt" (visuomotor) attentional processing is being coordinated in reading has received little emphasis in prior theoretical discussions. However, the example of interword regressions shows that this question is far from trivial. Here, SAS theory suggests that many regressions occur when attention is still at word N while the eyes have already advanced to word $N+1$ (Reichle, Rayner, & Pollatsek, 2003). If word N turns out be more difficult to process than expected on the basis of L1 processing, the eyes go back to the *attended* region. Does this involve a different, non-decoupled (or better: coupled) mode of control? Our data presented in Figure 2 can be taken as evidence for such a view. The left panel of the figure shows the well-known effect of saccade launch distance on initial progressive saccade-landing position, commonly referred to as saccade distance effect (McConkie, Kerr, Reddix, & Zola, 1988). For launch distances further away from the target word the resulting distribution of landing positions is shifted to the right. The right panel depicts landing positions for regressive saccade coming from position to the right of the target word. Quite strikingly, the saccadic range error is no longer present and saccades from all launch sites attain the word center with remarkable precision (see Radach & McConkie, 1998, for a detailed discussion). This is in line with the high spatial accuracy observed in attention-saccade coupling experiments such as Deubel & Schneider (1996), suggesting that it may be the aid of perceptual attention that eliminates the range error. This, in turn, points to the possibility that "normal" progressive saccades are in fact not coupled with perceptual attention.

This argument reveals an interesting similarity between SAS theory and many PG models. As an example, Glenmore, in line with the theoretical framework proposed by Findlay & Walker (1999), assumes that word-objects are represented as potential saccade targets on a vector of saliency. When a fixate center triggers the execution of a saccade, it will go to the target with the highest saliency, co-determined by visual and linguistic processing. This process of target specification within the saliency map is already equivalent to the initial phase of saccade programming. In this model, eye movements are not "coupled" to a separate mechanism of attentional selection because such a mechanism does not play a role in the automated routine model of oculomotor control. This may be different in the preparation of interword regressions which are assumed to rely on spatial memory for prior fixation positions (Inhoff, Weger, & Radach, 2005). Hence, for quite different reasons, both E-Z reader and Glenmore appear compatible with the effect pattern shown in Figure 2.

Looking at models emphasizing low-level aspects of processing, the problem of spatial selection is approached in an essentially non-cognitive way. Here, spatial selection follows rather simple heuristic rules (like "attain the largest word within a 20-letter window", Reilly & O'Regan, 1998) or may be based on a more sophisticated form of "educated guessing" in the spirit of Brysbaert & Vitu (1998; see also Brysbaert, Drieghe, & Vitu, 2005). The latter route was taken with impressive success in the SERIF model by McDonald, Carpenter, & Shillcock (2005), who nonetheless acknowledge that reading involves much more than moving eyes across a pager and that such a low-level "shortcut-mechanism" will, in later versions of their model, need to be supplemented with or replaced by a word-processing module.

In line with most other models, the educated guessing mechanism in the SERIF model uses low-level word unit information obtained via parafoveal visual processing for saccade target selection. In our discussion we have repeatedly pointed to the overwhelming empirical evidence in favor of word-based visuomotor control. A theoretical conception that casts doubt on the importance of word units in eye guidance has been suggested by Yang & McConkie (2004). These authors report experiments in which they used gaze contingent display changes to mask lines of text for the duration of some critical fixations so that (among other changes) word boundary information was not available. As it turned out, only relatively long fixations were affected by manipulations of this kind and distributions of saccade landing sites were quite similar in conditions with and without the presence of word boundaries. In their contribution to this book, Yang & Vitu claim that the planning of some saccades are word based while others seems not to be influenced by low spatial frequency word boundary information. We suggest caution when applying data from masking studies to the question of word-based control. Yang & McConkie (2004) themselves discuss a number of alternative word-based hypothesis that might explain part of their results. A factor that may play a critical role is that word length information was only removed during a single critical fixation and then immediately restored. If, as suggested by McConkie, Kerr, Reddix, & Zola (1988), low spatial frequency information serves as the basis for saccade targeting, it is reasonable to assume that this information is accumulated over consecutive fixations at least within the total perceptual span. Thus information acquired during earlier fixations may have supported eye guidance when word spaces were temporarily filled during some fixations.

Rayner, Fischer, & Pollatsek (1998) have shown that the permanent removal of word space information interferes with both word-processing and eye-movement control, again providing solid support for word-based eye guidance. We recently examined eye movements while reading Thai, an alphabetic writing system with no spatial segmentation at word boundaries (Reilly, Radach, Corbic, & Luksaneeyanawin, 2005). Analyses of local eye-movement patterns revealed the existence of an attenuated preferred viewing position phenomenon position in Thai reading. Interestingly, the steepness of the Gaussian distribution of initial saccade-landing positions (see Figure 1, left panel) was a function of the frequency with which specific letters occur at the beginning and end of words. We concluded that, in the absence of visual cues for word segmentation, orthographic information can serve as a base for the parafoveal specification of saccade target units. This mode of oculomotor control requires a substantial degree of distributed (and perhaps interactive) processing where word segmentation may be a result of rather than a precondition for lexical access. From a more fundamental point of view, these observations can be taken as an intriguing example of the principle that, in the interests of optimal resource allocation for linguistic processing, oculomotor control is as low level as possible and as cognitive as necessary (Underwood & Radach, 1998).

With respect to the question of sequential vs parallel processing of visual target objects within the functional field of view, the traditional battlefield for this issue within the domain of reading is the highly debated evidence for the so-called "parafovea-on-fovea effects", where properties of neighboring words appear to affect viewing duration on the currently fixated word. A recent study by Kliegl, Nuthmann, & Engbert (2006) can be taken as an example. Using one of the largest available data bases, where 222 participants each read 144 sentences, they showed that linguistic properties of both the prior and the next word in the sentence influenced the viewing time of a fixated word. Such effects of a parafoveal word on a fixated (foveal) word should not occur if the attention-controlled linguistic processing of words is strictly serial and if the saccade from the fixated word to the next word is programmed before a corresponding shift of attention takes place (for similar findings, see, e.g., Inhoff, Radach, Starr, & Greenberg, 2000; Inhoff, Starr, & Schindler, 2000; Kennedy & Pynte, 2005; Schroyens, Vitu, Brysbaert, & d'Ydewalle, 1999; Starr & Inhoff, 2004; Underwood, Binns, & Walker, 2000). However, it should be noted that parafovea-on-fovea effects have not always been found and that methodological objections can be raised against some studies (see Rayner & Juhasz, 2004, for a recent discussion).

In fact, Pollatsek, Reichle, & Rayner (2003) mustered a very clever defense against this type of evidence, by pointing out that during reading many saccades do not land on the intended target word. Therefore, a fixation falling on one of the last letters of word n might have been intended to land at word $n+1$, producing a spurious effect of word $n+1$ on word n. The issue of mislocated fixations in reading has assumed prominence also as a result of efforts to account for the inverted optimal viewing position (IOVP) effect observed for fixation durations in fluent reading (Vitu, McConkie, Kerr, & O'Regan, 2001; Nuthmann, Engbert, & Kliegl, 2005).

Chapter 14 of this book provides an elegant analytic solution to the problem of estimating the number of misplaced fixations on a word given that the landing site distribution comprises both intended landings and landings aimed at neighboring words. The surprising result is that the average rate of misplaced fixations is over 20 % and significantly greater in cases where the intended targets are shorter words. The authors' estimate is considerably larger than some researchers have thus far assumed (e.g., Reichle, Rayner, & Pollatskek, 2003, p. 510). This chapter provides an excellent starting point for taking a much closer look at the potential impact of misplaced fixations on the reading process. The primary motivation for Engbert et al.'s analysis was to account for the IOVP phenomenon, which they do here and elsewhere (Nuthmann, Engbert, & Kliegl, 2005) with remarkable success. The broader and no less interesting issue is what impact the estimated high-incidence of misplaced fixations might have on the various classes of reading model. In the context of the discussion about spatially distributed and temporally overlapping word processing, the techniques developed by Engbert et al. may provide a way to estimate whether, as suggested by Pollatsek, Reichle, & Rayner (2003), parafovea-on-fovea effects are indeed compromised by misplaced fixations.[6]

[6] As noted before, a distinguishing feature among current models of reading is the relative importance each ascribes to linguistic and oculomotor factors in eye guidance. In models where linguistic or lexical processes play a significant role in driving eye-movement control, the prospect of over a fifth of landings on a word being misdirected clearly requires to be taken account of. One would expect that E-Z reader, for example, might have problems in dealing with saccades that undershoot the intended target word resulting in either a refixation of the current word or a fixation on a word that was intended to be skipped. Given the importance to E-Z reader of early lexical processing, landing on an already-processed word is likely to have some disruptive effects. This problem should be less serious in models that permit the simultaneous processing of several words. In any case these models also need to account for misplaced fixations and the inverted OVP effect (see, e.g., Engbert, Nuthmann, Richter, & Kliegl, 2005; McDonald, Carpenter & Shillcock, 2005).

In addition to the somewhat indirect argumentation based on parafovea-on-fovea effects, there have also been attempts to examine the issue of sequential vs parallel word processing more directly. Inhoff, Eiter, & Radach, (2005) examined several preview conditions in a sentence reading experiment involving a display change occurring 150 ms after *fixation onset* on a pretarget word. When this technique was used to allow a preview of the subsequent target word exclusively during the initial part of the pretarget fixation, a 24 ms preview benefit emerged relative to a control condition involving a target fully masked by a pseudoword. This effect was not very large relative to the 90 ms full preview benefit, and it was also smaller than the benefit from an end of fixation preview. However, it supported the view that there is some temporal overlap between the processing of subsequent words and hence some degree of parallel processing.

Experimental evidence in support of the sequential processing position has been presented by Rayner, Juhasz, & Brown (in press), who used a saccade contingent display change technique where an invisible boundary was set either at the end of a pretarget word (word $n-1$) or at the end of word $n-2$. Replicating a large number of studies on this issue (see Rayner, 1998, for a review), they obtained a substantial parafoveal preview benefit when the boundary was located at word $n-1$. Importantly, no such benefit was obtained when the boundary was set at word $n-2$. This result is in line with the assumption common to all SAS models that word processing is restricted to exactly one word at a time and contradicts the prediction of any processing gradient model that some preview benefit should occur when letter information from a word two positions to the right of the current fixation is available during prior fixations.

Eventually, the position one takes in debates of this kind will also depend on the more general issue of whether information processing and oculomotor control in reading is seen as task specific or whether reading is considered to be a special case of universal processing mechanisms involved in "active vision" (Findlay & Gilchrist, 2003). If the latter view is adopted, the multiple lines of evidence reviewed in the previous section will play their role in assessing the viability of theoretical conceptions and conceptual models in reading.

One area where generalization from the mainstream of basic research on visual processing is particularly important is the time course of attentional orienting. As our discussion has shown, any approach that takes the notion of an attention shift seriously will need to allocate a certain amount of time for the triggering and execution of the attentional "movement". Pollatsek, Reichle, & Rayner, (2006) have noted that the assumption of an "instantaneous attention shift" in their model is not very plausible. They suggest that in later versions of the E-Z reader the time it takes to shift attention may be counted toward the duration of the L2 phase of lexical processing. However, if lexical processing and the shift are considered to be sequential, this would mean that the time allowed for L2 would need to be reduced by at least as much as 50 ms. To avoid this erosion of the sequential time line, it could be stated that part of L2 is equivalent to a "latency" for the shift, which is equivalent to positing that both occur in parallel (see also Reichle, Pollatsek, & Rayner, for a discussion of possible interpretations of L1 and L2). However, this eliminates the idea that the *completion of lexical access* is the trigger for moving attention, which in our view would constitute a major change in the philosophy and architecture of the model.

4. Problems of comparing and evaluating models

As evident from our discussions above, there is now a rich diversity of approaches to modeling information processing and eye-movement control during reading. In the concluding section of this chapter we would like to point to some problems related to the comparison between and evaluation of these competing models. We will try to avoid repeating the points made in prior discussions of these problems by Reichle, Rayner, & Pollastsek (1998) and Jacobs (2000). Our intention is to supplement their views by contributing a few remarks that may help raising awareness for what we believe to be a major deficit in the current state of the field. The question is how much has changed since Jacob's refusal to compare three computational models presented in the volume edited by Kennedy, Radach, Heller, & Pynte (2000), based on his impression that these models differed in so many different respects that a fair comparison was impossible.

4.1. Levels of description for computational models

As we pointed out in the introduction, from a theoretical point of view these models can be classified along the two axis of oculomotor vs cognitive control and sequential vs parallel word processing. Looking at the existing models from a more technical point of view, there are a number of additional aspects that can provide useful classifications. On a *conceptual level*, computational models are necessarily abstractions from a larger phenomenon, described by a theory, of which the researcher is seeking to gain a deeper understanding. The conceptual aspects of the model comprise the building blocks or conceptual units that the modeler considers essential to the theoretical account of the target phenomenon. For competing models to be comparable, there must be some degree of agreement between theorists regarding the core conceptual units of a model. Of course, the choice and scope of these units and of testable data is an issue for debate among theorists. Fortunately, as noted by Grainger (2003), the development of models in the field of continuous reading is characterized by a broad consensus regarding the critical phenomena that are to be explained. A useful compilation of mostly well-replicated and unquestionable facts about eye movements during reading that serve as accepted benchmarks for modeling can be found in Reichle, Rayner, & Pollatsek (2003).

Those aspects of a model that are *formally* described mathematically represent its core. In a complex, cognitive science domain such as reading, such formalizations will necessarily be partial. A key feature of the formal components of a model, or more correctly, their computational realization, is their success or otherwise in fitting empirical data. Since one can fit any data given enough free parameters, a measure of a model's power is the extent to which it can formally account for the data with the fewest free parameters. Any complex model will, inevitably, be unable to propose a complete, integrated formal account of all its conceptual components. In particular, it may not be possible to adequately account for how the various components interact. We refer to the non- or semi-formal framework that is used to integrate the various formal components of the model as the model's *architecture*. Usually, this architecture is provided by the model's computer implementation. So, for example, a computational model of eye-movement control in reading will comprise several distinct equations each describing the behavior of a conceptual unit of the model (word recognition, saccade triggering, etc.). These equations will be integrated within an algorithmic structure that can be used to generate data, which in turn can be compared to empirical observations.

The *dynamical aspects* of a model refer to the temporally extended behavior of the model. This is a function of the interaction between the model's formally realized components and its architecture. Ultimately, it is the dynamics of the model that generate testable data. Computational models that can generate moment-to-moment data on the time course of, for example, saccade generation in reading will provide the most testable and convincing accounts of the process, as they (should) generate principally refutable predictions.

4.2. Implicit and explicit model assumptions

While the above-mentioned dimensions are largely uncontroversial and most computational models can be readily characterized in this way, it is more interesting to explore the boundary between what aspects of a particular dimension are explicitly highlighted by the modeler as theoretically crucial and what aspects are left implicit. This can often be the main arena for comparison between models, since what is made explicit is usually what the model designer considers testable and potentially refutable, whereas the implicit aspects are assumed to be uncontroversial and not critical to the explanatory status of the model. However, what is uncontroversial for one researcher may be a key battleground for another.

For all of the dimensions of the model described above, we can identify implicit assumptions that may or may not be well founded, but which are necessary in order to get the model to work. Beneath the exterior of any computational model there is a considerable amount of superstructure built upon a foundation of varying solidity. For example, in the case of reading, there is a common implicit assumption that the visual segmentation of a word should occur before it can be identified. However, an examination of reading data from non-spaced texts such as Thai suggests that this assumption may not be well founded, and that when reading in this type of script there may be an interactive process of segmentation and lexical identification involving multiple word candidates and multiple segmentation hypotheses (Reilly, Radach, et al., 2005). This raises an interesting question about which of the existing models could, in principle, survive exposure to alternative writing systems including Chinese and the different Japanese systems.

This is an example of a conceptual level assumption. However, implicit assumptions can be made at the formal, architectural and dynamical levels as well. For example, in the case of the well-known Interactive Activation (IA) model of word recognition

(Rumelhart & McClelland, 1982), the designers made an architectural decision to represent letters in separate banks of letter-position channels. Certainly, there is no evidence from Rumelhart and McClelland's description of their model that this decision was anything other than a computational convenience in order to get the model to function. Nonetheless, this did not prevent some researchers choosing to test the model on the basis of this particular architectural feature (e.g. Mewhort & Johns, 1988).[7]

As computational models become more complex, it will become increasingly difficult to delineate those aspects that one wishes to stand for empirically from those that are less central to theory testing. Ultimately one wishes to test the theory, not the model. If spurious tests of the model are to be avoided, one needs to find a principled rather than an *ad hoc* way of indicating those aspects of the model that are of central theoretical significance and those that are not. The proposals in the next section aim to go some way toward this.

4.3. Some methodological proposals

To go some way to avoid problems relating to the testing of implicit and explicit aspects of a model, we propose a set of methodological approaches to the modeling exercise. Our methodological proposals fall under three headings: (1) the facilitation of the comparison of the structural and functional assumptions of competing models; (2) the grounding of models in the neuroscience of vision and language; and (3) the establishment of data sets for model comparison and benchmarking.

(1) With regard to the comparison of the structure and function of models, this could be facilitated by using a common implementation framework comprising a set of reusable software components (Schmidt & Fayad, 1997). In software engineering terms, a framework is a reusable, "semicomplete" application that can be specialized to produce particular applications or, in this case, particular models. The components would need to be fine-grained enough to accommodate the range of model types and model instances that are to be considered. If one could develop an acceptable and widely adopted modeling framework, it would be possible to establish a common basis on which to implement a variety of models. This would make the models more directly comparable in terms of not only their ability to account for data, but also their underlying theoretical assumptions. The modeling environment could provide a semi-formal language within which a model's structures and process functions could both be unambiguously articulated. This would aid both the task of designing the models and communicating the design to other researchers.

[7] It is somewhat ironic that a key conceptual component of the IA model, namely the concept of interaction between letter and word units and its supposed central role in mediating the word-superiority effect proved not to be as crucial as the model's designers first thought. As Norris (1992) demonstrated, it is possible to produce the word-superiority effect in a feed-forward network, without feedback connections, and without explicit interaction. Nonetheless, despite this lack of specificity the IA model still stands as a tour de force of cognitive modeling with an impressive set of empirical findings to its credit.

(2) Functionalist computational models, of which E-Z reader is an excellent example, are inherently underdetermined in terms of their relationship to the brain mechanisms that underlie them. For example, one could envisage a family of E-Z reader–like models with quite different combinations of parameters and/or parameter values that would be capable of providing an equally good fit to the empirical data (e.g., Engbert & Kliegl, 2001). One way to reduce this lack of determinism is to invoke a criterion of biological plausibility when comparing models. There is an increasingly rich set of data emerging from the field of cognitive neuroscience which could be used to augment the traditional behavioral sources of constraint on computational models. An excellent example for this approach is the use of ERP analyses to delineate the time course of lexical access (e.g. Sereno, Rayner & Posner, 1998; Hauk & Pulvermüller, 2004). Another, not unrelated, factor in assessing competing models is to take account of the evolutionary context in which our visual system evolved. Because it evolved for purposes quite different from reading, we need to beware of too easy recourse to arguments of parsimony, particularly when they are couched solely in terms of the reading process itself. A model with the minimum of modifiable parameters may be parsimonious on its own terms but fail the test of biological realism when compared with, say, a model that comprises an artificial neural network with many hundreds of adjustable parameters. While evolution is parsimonious in the large, when we look at brain subsystems in isolation, such as those involved in reading, we need to be careful how we wield Occam's razor.

(3) Finally, the issue of appropriate data sets with which to test and compare computational models of eye-movement control needs closer attention than it has been given to date. For example, the Schilling et al. (1998) data set used to parameterize and test E-Z reader and several other models is not particularly extensive. A good case can be made for establishing a range of publicly accessible data sets against which any proposed model can be tested. This would be similar to what has been done, for example, in machine learning, in data mining and most notably in the field of language acquisition (MacWhinney, 1995). Furthermore, the corpus of benchmark data should be extended to include corpora with common specifications in a variety of languages, alphabets and scripts. An excellent first step in this direction is the development of the Dundee-Corpus in English and French (Kennedy, 2003) that has been used to develop and test the SERIF model (McDonald, Carpenter, & Shillcock, 2005). In the long term, the more successful models will be those that can readily generalize beyond just one language and one writing system.

5. Challenges for future model developments

As the present literature, including the following three chapters of this book, shows, progress in the area has been impressive since the mid-1990s. There is now a rich spectrum of competing theories and models. Old debates about oculomotor vs cognitive control models have been replaced with much more complex approaches covering a theoretical middle ground in which both ends of the spectrum have their place. The scope of these models is still limited, partly for purposes of tractability, so that they cover a range of core phenomena, centered on the coordination of "eye" and "mind" during reading.

However, a close look at the state of the art in modeling eye-movement control in reading shows a striking deficit: So far no model has been published that is capable of accommodating inter-individual differences and intra-individual variations in reading. An obvious candidate for the latter would be variations along the axis of superficial vs careful reading (O'Regan, 1992) as can be induced by changing the reading task to induce a rather shallow vs deep linguistic processing during reading. Examples of inter-individual variation are the development of eye-movement control from childhood to skilled reading at adult age (Feng, 2006) and changes that occur in ageing readers (e.g. Kliegl, Grabner, Rolfs, & Engbert, 2004).

In addition, there are several aspects of the reading process itself that appear under-specified in current models. We would like to point to four areas that in our view deserve consideration. This list includes processes and mechanisms that manifest themselves relatively clearly via measurement of eye movements. As an example, a key component of reading that is likely to modulate oculomotor control but is not readily traceable in standard data sets is the processing of phonological information, both at the level of word processing (e.g. Lee, Binder, Kim, Pollatsek, & Rayner, 1999) and phonological working memory (Inhoff, Connine, Eiter, Radach, & Heller, 2004).

5.1. Binocular coordination

One of the most solid (and rarely challenged) implicit assumptions in research about eye movements is that both eyes behave essentially in the same way. Eye movements measured from one eye are routinely generalized to both and the few existing studies on binocular coordination have thus far received relatively little attention. As demonstrated by Heller & Radach (1999) there are systematic differences in the amplitude of saccades made by both eyes, resulting in mean disparities in the order of 1–1.5 letters, which in turn are partly offset by low convergence movements (Hendricks, 1996). A comprehensive metrical description of binocular coordination has been provided by Liversedge, White, Findlay & Rayner (2006). Critical for our discussion is Juhasz, Liversedge, White & Rayner (2006), who have shown that variation in word frequency does not affect fixation disparity or any other aspect of binocular coordination. Recent data from our laboratory confirm these observations for a sample of elementary school students who showed larger disparities than adults but again no effects of word frequency. The fact that binocular coordination is essentially a physiological phenomenon (see Colleweijn, Erkelens, & Steinman, 1988, for a seminal discussion) with no sensitivity to local variation of cognitive workload is good news for the modeling community. At this point we do not see the inclusion of this aspect of oculomotor behavior in future computational models as a priority. An important exception is the family of "split fovea models" (e.g. McDonald, Carpenter, & Shillcock, 2005). These make strong claims about both linguistic processing and eye movement control during reading based on the fact that the information entering the left vs right visual hemifields is projected to opposite brain hemispheres. In this context the disparity between both eyes needs to be accounted for, as a retinal split will feed different information to the hemispheres when the eyes exhibit uncrossed or crossed disparity.

5.2. Letter processing within the perceptual span

Current models do a relatively good job in approximating the visual and informational (e.g. orthographic) constraints for word processing within the perceptual span. In SWIFT, the rate with which letter level input is processed is approximated using a Gaussian distribution, and in the current version of Glenmore a gamma distribution scales the input to the saliency map. In the SERIF model a stochastic selection mechanism is based on the (Gaussian) extended optimal viewing position described by Brysbaert & Vitu (1998). However, as shown by McConkie & Zola (1987, see Figure 1), letter discrimination performance around the current fixation position is more complex, as word boundaries play a modulating role. In recent years there has been a lively debate on the role of letter position coding in word recognition (e.g. Peressotti & Grainger, 1999; Stevens & Grainger, 2003) and dynamic reading (e.g. Inhoff, Radach, Eiter, & Skelly, 2003). We anticipate that results from these lines of research will eventually lead to major refinements in the way letter recognition within the perceptual span is understood and implemented in computational models of reading.

5.3. Orthographic and lexical processing

In most current models, such as E-Z reader or SWIFT, there are no explicit mechanisms to simulate the microlevel of word processing. Instead, the time course of word processing is approximated on the bases of parameters like word length, frequency and contextual predictability. One exception is the Glenmore model, which includes a relatively realistic connectionist processing module, where the dynamics of activation and inhibition on the level of letter and word nodes determines the flow of linguistic processing. Although this is a step in the right direction, a more comprehensive approach would be to combine a model of continuous reading with one of the existing computational models of single-word recognition. These models presently ignore the dynamic aspects of continuous reading but provide detailed and plausible accounts of sub-lexical and lexical aspects of word processing. Candidates that could be considered for such a combination include the revised activation-verification model (Paap, Johansen, Chun, & Vonnahme, 2000), the multiple read-out model (Jacobs, Graf, & Kinder, 2003) and the cascaded dual route model of visual word recognition (Coltheart, Rastle, Perry, Langdon, & Ziegler, 2001).

5.4. Sentence-level processing

So far, computational models of eye movements during reading have eschewed consideration of specific sentence-level factors. Instead, in a number of models, supra-lexical knowledge is captured by a generic factor, word "predictability", empirically established in cloze tasks. This approach is taken to a new level in Chapter 12 of this book, in an attempt to study how sentence context might affect the lexical processing of ambiguous words and the decision about when to move the eyes from one word to the next. They used the existing set of equations specifying the time course of L1 and L2 processing in the E-Z reader model, adding meaning dominance as a lexical factor and disambiguating context as a sentence level factor affecting word predictability. Their simulations lead to interesting conclusions about how well different hypotheses can account for the data pattern observed by Duffy, Morris, & Rayner (1988) in the processing framework of their model.

In general, examination of specific sentence-level effects often involved analysis of multiword units with a wide range of oculomotor measures. Oculomotor effects have thus often been diffuse and relatively intricate (see Murray, 2000, for a review). For instance, in a study by Rayner, Warren, Juhasz, & Liversedge (2004), gaze durations were increased on impossible subject complement, for example on "large carrots" in the phrase "He used a pump to inflate large carrots", but the same complement did not receive longer gazes when it is implausible, for example "He used an axe to chop the large carrots". Readers of implausible constructions tended to move the eyes to an earlier text section which increased the complement's go-past time. As Clifton, Staub, and Rayner indicate in their chapter in this book, a principled account of such heterogeneous sentence-level effects on eye movements is desirable, and this may require more explicit syntactic and semantic/pragmatic theories of sentence processing than are currently available. Given these complexities on both the methodological and theoretical level, we suspect that an integration of complex sentence-level models with computational accounts of visual processing and oculomotor control is not part of the immediate agenda for future developments in our field.

References

Allport, D. A. (1992). Attention and control: have we been asking the wrong questions? A critical review of twenty-five years. In D. E. Meyer, & S. M. Kornblum (Eds.), *Attention and performance XIV: Synergies in experimental psychology, artificial intelligence, and cognitive neuroscience*. Cambridge, MA: MIT Press.

Blanchard, H. E., McConkie, G. W., Zola, D., & Wolverton, G. S. (1984). The time course of visual information utilization during fixations in reading. *Journal of Experimental Psychology: Human Perception and Performance, 10*, 75–89.

Brysbaert, M., & Vitu, F. (1998). Word skipping: implications for theories of eye movement control in reading. In G. Underwood (Ed.), *Eye guidance in reading and scene perception*.(pp. 125–148), Amsterdam: Elsevier.

Brysbaert, M., Drieghe, D., & Vitu, F. (2005). Word skipping: Implications for theories of eye movement control in reading. In G. Underwood (Ed.), *Cognitive processes in eye guidance*. Oxford: Oxford University Press.

Chun, M. M., & Wolfe, J. M. (2001). Visual attention. In E. B. Goldstein (Ed.), *Blackwell's handbook of perception*.(pp. 272–310), Oxford, UK: Blackwell.

Coltheart M., Rastle K., Perry C., Langdon R., & Ziegler J. (2001). DRC: A dual route cascaded model of visual word recognition and reading aloud. *Psychological review, 108*, 204–256.

Deubel, H., & Schneider, W. X. (1996). Saccade target selection and object recognition: Evidence for a common attentional mechanism. *Vision research, 36*, 1827–1837.

Deubel, H., Mokler, A., Fischer, B., & Schneider, W. X. (1999). *Reflexive Saccades are not Preceded by Shifts of Attention: Evidence From an Antisaccade task.* Paper presented at the 10th European Conference on Eye Movements, Utrecht, September 23–25.

Deubel, H., O'Regan, K., & Radach, R. (2000). Attention, information processing and eye movement control. In A. Kennedy, R. Radach, D. Heller, & J. Pynte (Eds.), *Reading as a perceptual process.* (pp. 355–376) Oxford: Elsevier.

Deubel, H., Schneider, W. X., & Paprotta, I. (1998). Selective dorsal and ventral processing: Evidence for a common attentional mechanism in reaching and perception. *Visual Cognition, 5*, 81–107.

Duffy, S. A., Morris, R. K., & Rayner, K. (1988). Lexical ambiguity and fixation times in reading. *Journal of Memory and Language, 27*, 429–446.

Duncan, J., & Humphreys, G. (1989). Visual search and stimulus similarity. *Psychological Review, 96*, 433–458.

Egeth, H. E., & Yantis, S. (1997). Visual attention: control, representation, and time course. *Annual Review of Psychology, 48*, 269–297.

Engbert, R., & Kliegl, R. (2001). Mathematical models of eye movements in reading: A possible role for autonomous saccades. *Biological Cybernetics, 85*, 77–87.

Engbert, R., Nuthmann, A., Richter, E., & Kliegl, R. (2005). SWIFT: A dynamical model of saccade generation during reading. *Psychological Review, 112*, 777–813.

Eriksen, C. H. (1990). *Attentional search of the visual field.* In D. Brogan (ed.) *Visual Search,* (pp. 3–19). Hove: Taylor and Francis.

Feng, G. (2006). Eye movements as time-series random variables: A stochastic model of eye movement control in reading. *Cognitive Systems Research, 7*, 70–95.

Findlay, J. M., & Gilchrist, I. D. (1998). Eye guidance and visual search. In Underwood, G. (Eds.), *Eye guidance in reading and scene perception.* Oxford: Elsevier.

Findlay, J. M., & Gilchrist, I. D. (2003). *Active vision: The psychology of looking and seeing.* Oxford: Oxford University Press.

Findlay, J. M., & Walker, R. (1999). A model of saccade generation based on parallel processing and competitive inhibition. *Behavioral & Brain Sciences, 22*(4), 661–721.

Fischer, B., & Weber, H. (1992). Characteristics of "anti" saccades in man. *Experimental Brain Research, 89*, 415–424.

Godijn, R., & Theeuwes, J. (2003). Parallel allocation of attention prior to the execution of saccade sequences. *Journal of Experimental Psychology: Human Perception and Performance, 29*, 882–896.

Goodale, M. A., & Milner, A. D. (1992). Seperate visual pathways for perception and action. *Trends in Neurosciences, 15*, 20–25.

Grainger, J. (2003). Moving eyes and reading words: How can a computational model combine the two? In Hyönä, J., Radach, R., & Deubel, H. (Eds.). *The mind's eye: Cognitive and applied aspects of eye movement research.* (pp. 457–470) Oxford: Elsevier.

Groner, R. (1988). Eye movements, attention and visual information processing: some experimental results and methodological considerations. In G. Lüer, U. Lass, & J. Shallo-Hoffmann (Eds.), *Eye movement research:. Physiological and psychological aspects.* Göttingen: Hogrefe.

Hauk, O., & Pulvermüller, F. (2004). Effects of word length and frequency on the human event-related potential. *Clinical Neurophysiology, 115*, 1090–1103.

Hoffman, J. E. (1998). Visual attention and eye movements. In H. Pashler (Ed.), *Attention.* (119–149). Hove: Taylor and Francis.

Hoffman, J. E., & Subramaniam, B. (1995). The role of visual attention in saccadic eye movements. *Perception & Psychophysics, 57(6)*, 787–795.

Horowitz, T. S., Holcombe, A. O., Wolfe, J. M., Arsenio, H. C., & DiMase, J. S. (2004). Attentional pursuit is faster than attentional saccade. *Journal of Vision, 20*, 585–603.

Hyönä, Y. (1995). Do irregular combinations attract readers' attention? Evidence from fixation locations in words. *Journal of Experimental Psychology: Human Perception and Performance, 21*, 142–152.

Hyönä, J., Radach, R., & Deubel, H. (Eds.), (2003). *The mind's eye: Cognitive and applied aspects of eye movements.* Oxford: Elsevier Science.

Inhoff, A. W., Connine, C., Eiter, B., Radach, R., & Heller, D. (2004). Phonological representation of words in working memory during sentence reading. *Psychonomic Bulletin and Report, 11*, 320–325.

Inhoff, A. W., Eiter, B. M., & Radach, R. (2005). The time course of linguistic information extraction from consecutive words during eye fixations in reading. *Journal of Experimental Psychology: Human Perception and Performance, 31*, 979–995.

Inhoff, A. W., Radach, R., Eiter, B., & Juhasz, B. (2003). Distinct subsystems for the parafoveal processing of spatial and linguistic information during eye fixations in reading. *Quarterly Journal of Experimental Psychology, 56A*, 803–827.

Inhoff, A., Radach, R., Eiter, B., Skelly, M. (2003). Exterior letters are not privileged in the early stage of visual word recognition. *Journal of Experimental Psychology. Learning, Memory & Cognition, 29*, 894–899.

Inhoff, A. W., Radach, R., Starr, M., & Greenberg, S. (2000). Allocation of visuo-spatial attention and saccade programming in reading. In A. Kennedy, R. Radach, D. Heller and J. Pynte (Eds.), *Reading as a perceptual process*. Oxford: Elsevier, pp. 221–246.

Inhoff, A. W., Starr, M., & Shindler, K. (2000). Is the Processing of Words during Eye Fixations in Reading Strictly Serial? *Perception and Psychophysics, 62(7)*, 1474–1484.

Inhoff, A. W., Weger, U. W., Radach, R. (2005). Sources of information for the programming of short- and long-range regressions during reading. In Underwood, G. (Ed.), *Cognitive processes in eye guidance*. Oxford University Press.

Jacobs, A. M., & Grainger, J. (1994). Models of visual word recognition – Sampling the state of the art. *Journal of Experimental Psychology: Human Perception and Performance, 20*, 1311–1334.

Jacobs, A. M. (2000). Five questions about cognitive models and some answers from three models of reading. In A. Kennedy, R. Radach, D. Heller and J. Pynte (Eds.), *Reading as a perceptual process*. Oxford: Elsevier, pp. 721–732.

Jacobs, A. M., Graf R., Kinder, A. (2003). Receiver operating characteristics in the lexical decision task: evidence for a simple signal-detection process simulated by the multiple read-out model. *Journal of Experimental Psychology: LMC, 29*, 481–488.

James, W. (1890). *The principles of psychology*. New York: Holt.

Johnston, W. A., & Dark, V. J. (1986). Selective attention. *Annual Review of Psychology, 37*, 43–75.

Juhaz, B. J., Liversedge, S. P., White, S. J., & Rayner, K. (2006). Binocular coordination of the eyes during reading: Word frequency and case alternation affect fixation duration but not fixation disparity. *Quarterly Journal of Experimental Psychology, 59*, 1614–1625.

Just, M. A., & Carpenter, P. (1988). Reading and spatial cognition: Reflections from eye fixations. In G. Lüer, U. Lass, & J. Shallo–Hoffmann (Eds.), *Eye movement research: Physiological and psychological aspects*. Göttingen: Hogrefe, 193–213.

Kennedy, A. (2003). *The Dundee corpus*. [CD-ROM]. Dundee, Scotland. University of Dundee. Department of Psychology.

Kennedy, A., Pynte, J. (2005). Parafoveal-on-foveal effects in normal reading. *Vision Research, 45*, 153–68.

Kennedy, A., Radach, R., Heller, D., and Pynte, J., (Eds.), *Reading as a perceptual process*, Oxford: Elsevier.

Kinchla, R. A. (1992). Attention. *Annual Review of Psychology, 43*, 711–743.

Kliegl, R., Nuthmann, A., & Engbert, R. (2006). Tracking the mind during reading: The influence of past, present, and future words on fixation durations. *Journal of Experimental Psychology: General, 135*, 12–35.

Kowler, E., Anderson, E., Dosher, B., & Blaser, E. (1995). The role of attention in the programming of saccades. *Vision Research, 35*, 1897–1916.

LaBerge, D., & Brown, V. (1989). Theory of attentional operations in shape identification. *Psychological Review, 96*, 101–124.

Legge, G. E., Klitz, T. S., & Tjan, B. S. (1997). Mr. Chips: An ideal observer model of reading. *Psychological Review, 104*, 524–553.

Legge, G. E., Hooven, T. A., Klitz, T. S., Mansfield, J. S., & Tjan, B. S. (2002). Mr. Chips 2002: New insights from an ideal-observer model of reading. *Vision Research, 42*, 2219–2234.

Lee, Y., Binder, K., Kim, J., Pollatsek, A., Rayner, K. (1999). Activation of phonological codes during eye fixations in reading. *Journal of Experimental Psychology: Human Perception and Performance, 25*, 948–964.

Liversedge, S. P., White, S. J., Findlay, J. M., & Rayner, K. (2006). Binocular coordination of eye movements during reading. *Vision Research, 46*, 2363–2374.

MacWhinney, B. (1995). The CHILDES Project: Tools for Analyzing Talk, 2nd edition. Hillsdale, NJ: Erlbaum.

McConkie, G. W. (1979). On the role and control of eye movements in reading. In P. A. Kolers, M. E. Wrolstad, and H. Bouma (Eds.), *Processing of visible language. Vol. I.* (pp. 37–48) New York: Plenum Press.

McConkie, G. W., & Zola, D. (1987). Visual attention during eye fixations while reading. In M. Coltheart (Ed.), *Attention and performance (Vol. 12)*, London, NJ: Erlbaum, pp. 385–401.

McConkie, G. W., Kerr, P. W., Reddix, M. D., and Zola, D. (1988). Eye movement control during reading: I. The location of initial eye fixations on words. *Vision Research, 28*, 1107–1118.

McConkie, G. W., Underwood, N. R., Zola, D., & Wolverton, G. S. (1985). Some temporal characteristics of processing during reading. *Journal of Experimental Psychology: Human Perception and Performance, 11*, 168–186.

McConkie, G. W., & Yang, S. (2003). How cognition affects eye movements during reading. In Hyönä, J., Radach, R., & Deubel, H. (Eds.), *The mind's eye: Cognitive and applied aspects of eye movement research.* (pp. 413–428), Oxford: Elsevier.

McConkie, G. W., & Zola, D. (1984). Eye movement control during reading: The effects of word units. In W. Prinz, and A. T. Sanders (Eds.), *Cognition and Motor Processes.* (pp. 63–74) Berlin: Springer.

McConkie, G. W., & Zola, D. (1987). Visual attention during eye fixations while reading. In M. Coltheart (Ed.), *Attention and performance XII, The psychology of reading.* (pp. 385–401), Hove: Erlbaum.

McDonald, S. A., Carpenter, R. H., Shillcock, R. C. (2005). An anatomically constrained, stochastic model of eye movement control in reading. *Psychological Review, 112*, 814–840.

Mewhort, D., & Johns, E. E. (1988). Some tests of the interactive activation model for word recognition. *Psychological Research, 50*, 135–147.

Morrison, R. E. (1984). Manipulation of stimulus onset delay in reading: Evidence for parallel programming of saccades. *Journal of Experimental Psychology: Learning, Memory, and Cognition, 21*, 68–81.

Murray, W. S. (2000). Sentence Processing: Issues and Measures. In A. Kennedy, R. Radach, D. Heller, & J. Pynte (Eds.), *Reading as a perceptual process* (pp. 649–664). Oxford: Elsevier.

Neumann, O. (1987). Beyond capacity: A functional view of attention. In H. Heuer, and A. F. Sanders (Eds.), *Perspectives on selection and action.* Hillsdale, NJ: Erlbaum.

Nuthmann, A., Engbert, R., & Kliegl, R. (2005). Mislocated fixations during reading and the inverted optimal viewing position effect. *Vision Research, 45*, 2201–2217.

O'Regan, J. K. (1984). How the eye scans isolated words. In Gale, A. G., and Johnson, F. (Eds.), *Theoretical and applied aspects of eye movement research.* (pp. 159–168) Amsterdam: North – Holland.

O'Regan, J. K. (1990). Eye movements in reading. In E. Kowler (Ed.), Eye movements and their role in visual and cognitive processes. Elsevier: Amsterdam, pp. 395–453.

O'Regan, J. K., Vitu, F., Radach, R., & Kerr, P. (1994). Effects of local processing and oculomotor factors in eye movement guidance in reading In Ygge, J., & G. Lennerstrand (Ed.), Eye movements in reading. Pergamon, New York, pp. 323–348.

Paap, K. R., Johansen, L. S., Chun, E., Vonnahme, P. (2000). Neighborhood frequency does affect performance in the Reicher task: encoding or decision? *Journal of Experimental Psychology: HPP, 26*, 1691–720.

Pashler, J. (1998). *Attention.* Hove: Taylor and Francis.

Peressotti, F., & Grainger, J. (1999). The role of letter identity and letter position in orthographic priming. *Perception & Psychophysics, 61*, 691–706.

Pollatsek, A., & Rayner, K. (1982). Eye movement control in reading: The role of word boundaries. *Journal of Experimental Psychology: Human Perception and Performance, 8*, 817–833.

Pollatsek, A., & Rayner, K. (1989). Reading. In Posner, M. I. (Eds.), *Foundations of cognitive science.* Cambridge, MA: MIT Press.

Pollatsek, A., Reichle, E. D., & Rayner, K. (2003). Modeling eye movements in reading: Extentions of the E-Z reader model. In Hyönä, J., Radach, R., & Deubel, H. (Eds.)., *The mind's eye: Cognitive and applied aspects of eye movement research.* (pp. 361–390) Oxford: Elsevier.

Pollatsek, A., Reichle, E. D., & Rayner, K. (2006). Tests of the E-Z Reader model: Exploring the interface between cognition and eye-movement control. *Cognitive Psychology, 52*, 1–56.

Posner, M. I. (1980). Orienting of attention. *Quarterly Journal of Experimental Psychology, 32*, 3–25.
Posner, M. I., & Petersen, S. E. (1990). The attention system of the human brain. *Annual Review of Neuroscience, 13*, 25–42.
Radach, R., Heller, D., & Inhoff, A. W. (2004). Orthographic regularity gradually modulates saccade amplitudes in reading. *European Journal of Cognitive Psychology, 16*, 27–51.
Radach, R., & Kennedy, A. (2004). Theoretical perspectives on eye movements in reading: Past controversies, current issues, and an agenda for the future. *European Journal of Cognitive Psychology, 16(1/2)*, 3–26.
Radach, R., Kennedy, A., & Rayner, K. (2004). *Eye movements and information processing during reading.* Hove: Psychology Press.
Radach, R., & McConkie, G. W. (1998). Determinants of fixation positions in reading. In G. Underwood (Ed.), *Eye guidance in reading and scence perception.* (pp. 77–100) Oxford: Elsevier.
Rayner, K. (1979). Eye guidance in reading: Fixation locations within words. *Perception, 8,* 21–30.
Rayner, K. (1998). Eye movements in reading and information processing: 20 years of research. *Psychological Bulletin, 124*, 372–422.
Rayner, K., Fischer, M. H., Pollatsek, A. (1998). Unspaced text interferes with both word identification and eye movement control. *Vision Research, 38*, 1129–1144.
Rayner, K., & Juhasz, B. J. (2004). Eye movements in reading: Old questions and new directions. *European Journal of Cognitive Psychology, 16*, 340–352.
Rayner, K., Juhasz, B. J., & Brown, S. J. (in press). Do readers obtain preview benefit from word n+2? A test of serial attention shift versus distributed lexical processing models of eye movement control in reading. *Journal of Experimental Psychology: Human Perception and Performance.*
Rayner, K., Liversedge, S. P., White, S. J., & Vergilino-Perez, D. (2003). Reading disappearing text: Cognitive control of eye movements. *Psychological Science, 14*, 385–388.
Rayner, K., & Pollatsek, A. (1989). *The psychology of reading.* Boston: Prentice–Hall.
Rayner, K. & Raney, G. E. (1996). Eye movement control in reading and visual search: Effects of word frequency. *Psychonomic Bulletin and Review*, 3, 238–44.
Rayner, K., & Sereno, S. C. (1994). Eye movements in reading: Psycholinguistic studies. In M. Gernsbacher (Ed.), *Handbook of psycholinguistics.* (pp. 57–82) New York: Academic Press.
Rayner, K, Inhoff, A. W., Morrison, R. E., Slowiaczek, M. L., & Bertera, J. H. (1981). Masking of foveal and parafoveal vision during eye fixations in reading. *Journal of Experimental Psychology: Human Perception and Performance, 7*, 167–179.
Rayner, K., Sereno, S. C., & Raney, G. E. (1996). Eye movement control in reading: A comparison of two types of models. *Journal of Experimental Psychology: Human Perception and Performance, 22*, 1188–1200.
Reichle, E. D., Pollatsek, A., & Rayner, K. (2006). E–Z Reader: A cognitive-control, serial-attention model of eye-movement behavior during reading. *Cognitive Systems Research, 7*, 4–22.
Rayner, K., Warren, T., Juhasz, B. J., Liversedge, S. P. (2004). The effect of plausibility on eye movements in reading. *Journal of Experimental Psychology. Learning, Memory & Cognition, 30*, 1290–1301.
Reichle, E. D., Rayner, K., & Pollatsek, A. (2003). The E-Z Reader model of eye movement control in reading: comparisons to other models. *Behavioral and Brain Sciences, 26*, 445–476.
Reilly, R. G., & O'Regan, J. K. (1998). Eye movement control during reading: A simulation of some word-targeting strategies. *Vision Research, 38*, 303–317.
Reilly, R., & Radach, R. (2006). Some empirical tests of an interactive activation model of eye movement control in reading. *Cognitive Systems Research, 7*, 34–55.
Reilly, R., Radach, R., Corbic, D., & Luksaneeyanawin, S. (2005): *Comparing reading in English and Thai: The role of spatial word unit segmentation for distributed processing and eye movement control.* Talk presented a the 13th European Conference on Eye Movements, Berne, Switzerland.
Richter, E. M., Engbert, R. and Kliegl, R. (2006). Current advances in SWIFT. *Cognitive Systems Research, 7*, 23–33.
Rumelhart, D. E., & McClelland, J. L. (1982). An Interactive Activation Model of Context Effects in Letter Perception: Part 2. The Contextual Enhancement Effect and Some Tests and Extensions of the Model, *Psychological Review, 89*, 60–94.

Schilling, H. E. H., Rayner, K., & Chumbley, J. I. (1998). Comparing naming, lexical decision, and eye fixation times: Word frequency effects and individual differences. *Memory & Cognition, 26*, 1270–1281.

Schmidt, D. C., & Fayad, M. (1997). Object-oriented application frameworks. *Communications of the Association for Computing Machinery, 40*, 32–38.

Schneider, W. X. (1995). VAM: A neuro-cognitive model for visual attention control of segmentation, object recognition and space-based motor actions. *Visual Cognition, 2*, 331–375.

Schroyens, W., Vitu, F., Brysbaert, M., & d'Ydewalle, G. (1999). Eye movement control during reading: foveal load and parafoveal processing. *The Quarterly Journal of Experimental Psychology, 52A(4)*, 1021–1046.

Sereno, S. C., Rayner, K., & Posner, M. I. (1998). Establishing a timeline of processing during reading: Evidence from eye movements and event-related potentials. *NeuroReport, 9*, 2195–2200.

Starr, M. S., & Inhoff, A. W. (2004). Attention allocation to the right and left of a fixated word: Use of orthographic information form multiple words during reading. *European Journal of Cognitive Psychology, 16*, 203–225.

Stelmach, L. B., Campsall, J. M., & Herdman, C. M. (1997). Attentional and ocular eye movements. *Journal of Experimental Psychology: Human Perception and Performance, 23*, 823–844.

Stevens, M., & Grainger, J. (2003). Letter visibility and the viewing position effect in visual word recognition. *Perception & Psychophysics, 65*, 133–151.

Theeuwes, J., Godijn, R., & Pratt, J. (2004). A new estimation of the duration of attentional dwell time. *Psychonomic Bulletin & Review, 11*, 60–64.

Treisman, A. M., & Gelade, G. (1980). A feature-integration theory of attention. *Cognitive Psychology, 12*, 97–136.

Underwood, G., Binns, A., & Walker, S. (2000). Attentional demands on the processing of neighbouring words. In A. Kennedy, R. Radach, D. Heller and J. Pynte (Eds.), *Reading as a perceptual process* (pp. 247–268). Oxford: Elsevier.

Ungerleider, L. G., & Mishkin, M. (1982). In D. J. Ingle, M. A. Goodale, & R. J. W. Mansfield (Eds.), *Analysis of visual behavior*, Cambridge, MA: MIT Press.

Van der Stigchel S., Theeuwes J. (2005). The influence of attending to multiple locations on eye movements. *Vision Research, 45*, 1921–7.

Vitu, F., McConkie, G. W., Kerr, P., & O'Regan, J. K. (2001). Fixation location effects on fixation durations during reading: an inverted optimal viewing position effect. *Vision Research, 41*, 3513–3533.

Ward, R. (2001). Visual attention moves no faster than the eyes. In K. Shapiro (Ed.), *The limits of attention: Temporal constrains in human information processing.* Oxford: University Press.

White, S. J., & Liversedge, S. P. (2004). Orthographic familiarity influences initial fixation positions in reading. *European Journal of Cognitive Psychology, 16*, 52–78.

Wolfe, J. M. (1994). Guided search 2.0: A revised model of visual search. *Psychonomic Bulletin & Review, 1*, 202–238.

Wolfe, J. M. (2003). Moving towards solutions to some enduring controversies in visual search. *TRENDS in Cognitive Sciences, 7*, 70–76.

Yang, S. (2006). An oculomotor-based model of eye movements in reading: The competition/interaction model. *Cognitive Systems Research, 7*, 56–69.

Yang, S.-N., & McConkie, G. W. (2001). Eye movements during reading: A theory of saccade initiation times. *Vision Research. 41*, 3567–3585.

Yantis, S. (1998). Control of visual attention. In H. Pashler (Ed.), *Attention.* East Sussex: Psychology Press.

Chapter 12

MODELING THE EFFECTS OF LEXICAL AMBIGUITY ON EYE MOVEMENTS DURING READING

ERIK D. REICHLE

University of Pittsburgh, USA

ALEXANDER POLLATSEK and KEITH RAYNER

University of Massachusetts, Amherst, USA

Eye Movements: A Window on Mind and Brain
Edited by R. P. G. van Gompel, M. H. Fischer, W. S. Murray and R. L. Hill

Abstract

The eye movements of readers are affected by both the meaning dominance of an ambiguous word and whether prior sentence context disambiguates the meaning (Duffy, Morris, & Rayner, 1988). These findings have been used to support claims about the role of linguistic constraints on word identification. Unfortunately, the adequacy of these claims has not been evaluated; that is, it is not known if the assumptions that have been made to account for the effects of lexical ambiguity resolution on eye movements are sufficient to explain these effects. In this chapter, the *E-Z Reader* model of eye-movement control during reading (Pollatsek, Reichle, & Rayner, 2006; Rayner, Ashby, Pollatsek, & Reichle, 2004; Reichle, Pollatsek, Fisher, & Rayner, 1998; Reichle, Rayner, & Pollatsek, 2003) is used as a framework to evaluate the theoretical assumptions that have been developed to explain lexical ambiguity resolution. This modeling exercise is informative because it demonstrates that not all accounts of lexical ambiguity are equally viable, and because it suggests further constraints on the types of cognitive processes that drive eye movements during reading.

It is a truism that both spoken and written language is full of ambiguities. For example, many words have more than one distinct meaning (Duffy, Kambe, & Rayner, 2001), and most words have several (sometimes subtly) different shades of meaning (Klein & Murphy, 2001). The prevalence of such ambiguities, along with the fact that readers often seem to have little difficulty understanding text containing them, suggests that the overall meaning of whatever it is that is being read can ameliorate the problems that readers might otherwise experience because of lexical ambiguities. By examining how readers do this, one might be able to gain a better understanding of how the processes that are necessary to identify printed words interact with the various higher-level processes that are necessary to understand language. During the last 20 years, researchers have started addressing this question by measuring the eye movements of readers as they encounter ambiguous words in text (Rayner, 1998). These eye-tracking experiments have proved useful in elucidating the variables that affect the resolution of lexical ambiguity (Duffy et al., 1988) and in developing theories about the mechanisms involved in ambiguity resolution (Pacht & Rayner, 1993). Both of these endeavors have also contributed to our more general understanding of language processing during reading.

Despite all of these advances, however, the theories that are used to explain the results of experiments that have informed our understanding of lexical ambiguity resolution have not themselves been directly evaluated. In other words, the theories about how readers resolve lexical ambiguity have not been embedded within a more general framework of a theory of eye-movement control to determine if – and if so, how well – various theoretical assumptions can explain the patterns of eye movements that have been observed when readers encounter ambiguous words. The goal of this chapter is to do this. More specifically, we will embed the key assumptions of the theories that have been proposed to explain lexical ambiguity within the framework of the *E-Z Reader* model of eye-movement control during reading (Pollatsek et al., 2006; Rayner et al., 2004; Reichle et al., 1998, 2003) to evaluate the explanatory adequacy of the theories. Before we do this, however, a few caveats are necessary.

First, because the literature on lexical ambiguity is a large one and because many variables are known to influence ambiguity resolution (for a review, see Pacht & Rayner, 1993), the present discussion will only focus on two variables: (1) the *meaning dominance* or relative frequencies of the two meanings of a lexically ambiguous word, and (2) whether the prior sentence context biases the interpretation of the ambiguous words (Dopkins, Morris, & Rayner, 1992; Duffy et al., 1988; Pacht & Rayner, 1993; Rayner & Duffy, 1986; Rayner & Frazier, 1989; Sereno, O'Donnell, & Rayner, 2006; Sereno, Pacht, & Rayner, 1992). In line with this first caveat, we will not provide a definitive statement about how the E-Z Reader model accounts for the patterns of eye movements that are observed when readers encounter ambiguous words in and out of context. Our goal is instead much more modest – to simply specify the factors that any model of eye-movement control must consider to provide such an account. We believe that this exercise will be useful because it will provide a means to evaluate both models of eye-movement control during reading and the various theories that have been offered to explain the resolution of lexical ambiguity during reading. Before we review the basic findings related to lexical ambiguity and how its resolution affects eye movements during reading, it is first important to provide a brief overview of the E-Z Reader model.

1. E-Z Reader

Figure 1 is a schematic diagram of the E-Z Reader model. As the figure shows, word identification begins with a pre-attentive stage of visual processing (labeled "V" in Figure 1) that allows the visual features from the printed page to be propagated from the retina to the brain. This early stage of processing is "pre-attentive" in that information is extracted in parallel from across the entire visual field; however, the quality of this information decreases the further it is from the fovea. The low-spatial frequency information (e.g., word boundaries) that is obtained from this stage is used to select the targets for upcoming saccades. The high-spatial frequency information (e.g., letter shapes) that is obtained from this stage can serve as fodder for lexical processing on whatever word is being attended. In contrast, it is assumed that (a) subsequent lexical processing requires attention and (b) that attention is allocated serially to one word at a time. Finally, although this early stage of pre-attentive visual processing takes some time to complete (i.e., 50 ms, or the duration of the "eye-to-brain lag"; Clarke, Fan, & Hillyard, 1995; Foxe & Simpson, 2002; Mouchetant-Rostaing, Gaird, Bentin, Aguera, & Pernier, 2000; Van Rullen & Thorpe, 2001), the effects of this delay are predicted by the model to be negligible in most normal reading situations because lexical processing continues during each saccade using

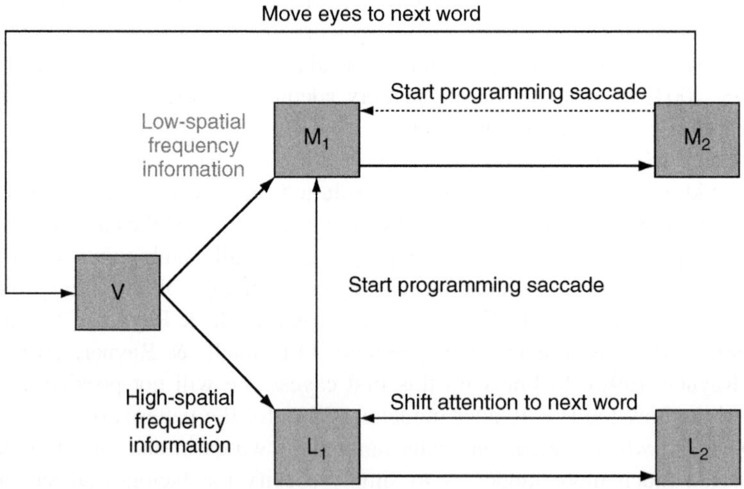

Figure 1. Schematic diagram of the *E-Z Reader* model of eye-movement control during reading. "V" corresponds to the pre-attentive stage of visual processing; "L_1" and "L_2" correspond to the first and second stages of lexical processing, respectively; and "M_1" and "M_2" correspond to the labile and non-labile stages of saccadic programming.

whatever information was obtained from the preceding viewing location. (For a complete discussion of how this pre-attentive stage of visual processing is related to attention and lexical processing, see Pollatsek et al., 2006.)

The second stage of word identification in the E-Z Reader model (i.e., the first stage after the visual stage), called the "familiarity check" or L_1, ends when processing of word n has almost been completed, and provides a signal to the oculomotor system to program a saccade to move the eyes to word $n+1$. (Note that here and elsewhere, we will refer to the word attended at the beginning of a fixation as word n; this is usually also the fixated word except if there has been a saccadic error in targeting the word.) Processing of word n then continues until the word has been identified to the point where attention can be disengaged from it and reallocated to the next word. When this final stage of word identification, "completion of lexical access" or L_2, is finished, a signal to shift attention from word n to word $n+1$ is sent to the attentional system; thus, the shifting of attention is decoupled from the programming of saccades. In the current version of the model, the mean times required to complete L_1 and L_2 on a word, $t(L_1)$ and $t(L_2)$, are an additive function of that word's frequency of occurrence in printed text (freq) and its predictability within its local sentence context (pred), as specified by Equations 1 and 2. For L_1, there is some probability ($p =$pred) that the word will be "guessed" from it's context, so that no time is needed to complete L_1 on that word; however, for most words and in most cases ($p = 1 - \text{pred}$), the time needed to complete L_1 is given by Equation 1.

$$t(L_1) = \alpha_1 - \alpha_2 \ln(\text{freq}) - \alpha_3 \,\text{pred} \tag{1}$$

$$t(L_2) = \Delta[\alpha_1 - \alpha_2 \ln(\text{freq}) - \alpha_3 \,\text{pred}] \tag{2}$$

In Equation 1, the free parameters α_1 (= 122 ms), α_2 (= 4 ms), and α_3 (= 10 ms) are the best-fitting values that have been used with all of the simulations involving E-Z Reader 9 (Pollatsek et al., 2006). The *frequency* with which a word occurs in printed text is determined through corpora norms (e.g., Francis & Kucera, 1982) and its *predictability* is set equal to the mean probability of guessing the word from its prior sentence context, as determined using cloze-task norms. Both equations define the *mean* times to complete L_1 and L_2, respectively. The actual times for a given Monte Carlo run of the model are found by sampling random deviates from gamma distributions having means equal to the values specified by Equations 1 and 2, and standard deviations equal to 0.22 of their means.

Note that the time to complete L_1, $t(L_1)$, is also modulated by visual acuity, resulting in $t(L_1)'$ (see Equation 3, below). The free parameter ε(= 1.15) modulates the effect of visual acuity, which in our model is defined in terms of the mean absolute *distance* (i.e., number of character spaces) between the current fixation location and each of the letters in the word being processed (N is the number of letters in the word). The value of ε was selected so that the rate of L_1 completion would decrease by factors of 1.15, 1.32, and 1.52 when the first letter of 3-, 5-, and 7-letter words (respectively) is fixated (relative to when a 1-letter word is directly fixated). It thus takes more time to identify long words and words that are farther from the fovea (i.e., the current fixation location). Both of these predicted outcomes are consistent with empirical results; word identification is slower and less accurate in peripheral vision (Lee, Legge, & Ortiz, 2003; Rayner & Morrison, 1981) and longer words do take longer to identify than shorter words (Just & Carpenter, 1980; Rayner & McConkie, 1976; Rayner, Sereno, & Raney, 1996). The assumption that L_2 is *not* affected by visual acuity is consistent with L_2 being a later stage of processing that is operating on information provided by L_1 rather than on lower-level perceptual information.

$$t(L_1)' = t(L_1)\varepsilon^{\Sigma|\text{distance}|/N} \tag{3}$$

In the most recent version of E-Z Reader (Pollatsek et al., 2006 Reichle et al., 2006), the mean total minimum and maximum times to complete L_1 on words (i.e., the sum of the visual pre-processing stage and L_1, ignoring the affects of visual acuity) that are not completely predictable from their sentence contexts are 117 and 172 ms, respectively. Similarly, the mean total minimum and maximum times to complete L_2 on a word (i.e., the sum of the previously mentioned times and L_2) are 151 and 233 ms, respectively. (These values are based on values of several free parameters that were selected to maximize the model's overall goodness-of-fit to a corpus of sentences used by Schilling, Rayner, and Chumbley, 1998.) Note that the added time to complete L_2 is a function of word frequency and predictability (see Equation 2), which has two important consequences, both of which are best explained by reference to Figure 2, which shows the mean times to complete L_1 and L_2 on a word, along with the time that is needed to program a saccade to move the eyes off of that word. Because the time needed to complete L_1 on word n is a function of that word's processing difficulty, and because the time that is needed to complete L_2 is (on average) some fixed proportion of the time that is needed to complete L_1, the amount of time that can be spent processing word $n+1$ when fixating on word n is a function of the difficulty of processing word n. This is because the time that is

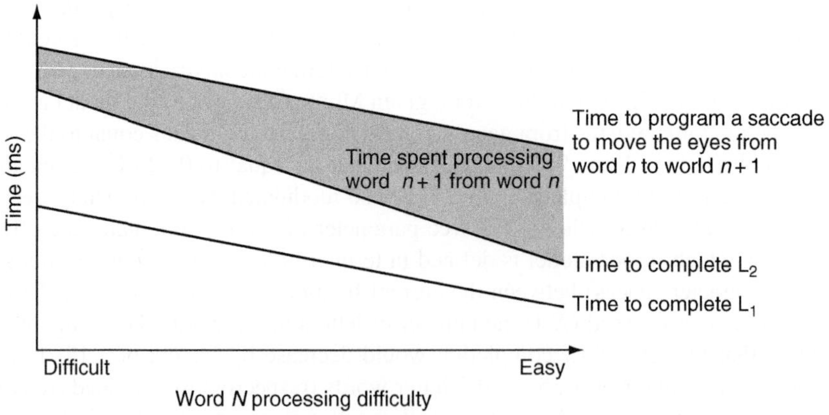

Figure 2. Means times to complete L_1 and L_2 on word n as a function of its processing difficulty, and the mean time to program a saccade to move the eyes from word n to word $n+1$.

needed to program a saccade is (on average) constant. Thus, the amount of time that can be allocated to the parafoveal processing of word $n+1$ (i.e., the processing of word $n+1$ from word n) is limited to the interval of time between when processing (i.e., L_2) of word n has completed and when the saccadic program to move the eye from word n to word $n+1$ has completed. This interval is indicated in Figure 2 by the gray area.

Two important consequences emerge from the assumption that the amount of parafoveal processing that can be completed on word $n+1$ diminishes as word n becomes more difficult to process. The first is simply that the amount of preview benefit, or the degree to which preventing normal parafoveal processing of word $n+1$ slows its identification, will be affected by the processing difficulty of word n. The second is that the processing difficulty of word n can "spill over" onto word $n+1$ and inflate the fixation duration on that word. Both of these outcomes are consistent with empirical results; several experiments have shown that parafoveal preview benefit is attenuated by foveal processing load (Henderson & Ferreira, 1990; Kennison & Clifton, 1995; Schroyens, Vitu, Brysbaert, & d'Ydewalle, 1998; White, Rayner, & Liversedge, 2005), and "spillover" effects have been well documented. The capacity to simulate these two related findings is – at least in part (see Pollatsek et al., 2006) – what motivated the distinction between L_1 and L_2 in the E-Z Reader model[1].

This finishes our description of the "front end" of the model, or its assumptions about how cognition affects the "decisions" about when and where to move the eyes. All of the remaining assumptions are related to saccade programming and execution and are based on the work of others. Briefly, these assumptions are as follows. First, using the findings of Becker and Jürgens (1979), we assume that saccadic programming is completed in two stages: an earlier labile stage that can be canceled if another saccadic program is subsequently started ("M_1" in Figure. 1), followed by a second, non-labile stage that cannot be canceled ("M_2" in Figure 1). By allowing later saccade programs to cancel earlier ones, the model can explain skipping: From word n, the completion of the L_1 for word $n+1$ will cause a saccade program to move the eyes to word $n+2$ to be initiated; this program will cancel the program (if it is still in its labile stage) that would otherwise move the eyes to word $n+1$, thereby causing that word to be skipped.

The remaining assumptions about saccades concern their execution. Building upon prior work (McConkie, Kerr, Reddix, & Zola, 1988; McConkie, Zola, Grimes, Kerr, Bryant, & Wolff, 1991; O'Regan, 1990, 1992; O'Regan & Lévy-Schoen, 1987; Rayner, 1979; Rayner et al., 1996), we adopted the assumptions that saccades are always directed toward the centers of their intended word targets, but that saccades often miss their intended targets because of both random and systematic motor error. These assumptions are sufficient for the model to generate fixation landing-site distributions that are approximately normal in shape, and that become more variable as the fixation duration on the launch-site word decreases, and as the length of the saccade between the launch site and

[1] It is noteworthy that at least two of the alternative models of eye-movement control during reading also include two stages of lexical processing (*SWIFT*: Engbert, Longtin, & Kliegl, 2002; Engbert, Nuthmann, Richter, & Kliegl, 2005; *EMMA*: Salvucci, 2001).

landing site increases. Finally, the model includes an assumption that, upon fixating a word, the oculomotor system makes a "decision" about whether or not to initiate a saccade to move the eyes to a second viewing position on the word. This decision is based on the distance between the initial landing position and the center of the word, under the assumption that initial fixations near the beginnings and endings of words afford a poor view of the word (due to limited visual acuity) and hence are more likely to result in a corrective refixation to move the eyes to a better viewing location. Because this decision is based on the quality of the visual information from the initial viewing location, the decision can only be made after enough time has elapsed to provide feedback about the distance between the initial landing position and the center of the intended word target. These assumptions about corrective refixation saccades allow the model to account for some of the findings related to the inverted optimal viewing position effect (Vitu, McConkie, Kerr, & O'Regan, 2001), although additional assumptions will be necessary to completely explain these phenomena.

With this overview of the E-Z Reader model, it is now possible to demonstrate how it can be used as a framework to examine the different theories that have been developed to explain the patterns of eye movements that are observed when readers encounter lexically ambiguous words in text. Before we do this, however, we will first briefly review what has been learned about how lexical ambiguity affects readers' eye-movement behavior, and then briefly review the various accounts of this behavior.

2. Lexical Ambiguity

Most words in English are polysemous and have two or more (sometimes subtly) different meanings (Klein & Murphy, 2001). Many words are even more ambiguous in that they have two or more distinct meanings (Duffy et al., 2001). Often, the meanings of these words differ dramatically in terms of their *meaning dominance*, or the frequency with which the different meanings are encountered in written or spoken language. For example, with a *balanced* ambiguous word such as *case*, the two meanings of the word (one related to legal proceedings, the other related to containers) are approximately equally prevalent in the language. In contrast, with *biased* ambiguous words, like *port*, the *dominant* meaning (the one that is a synonym of *harbor*) is much more prevalent or common in the language than its second, *subordinate* meaning (the one that is a type of wine).

Given that two or more quite distinct meanings can be associated with a single word form, the question then becomes one of trying to understand which meaning or meanings of an ambiguous word are encoded when it is encountered in text. For example, when an ambiguous word is encountered when the prior text does not bias or support a particular interpretation of the word (e.g., "Actually the port . . . "), do readers access both meanings of the word, or is only one of the meanings (likely the most dominant one) accessed? Similarly, when the ambiguous word is preceded by context that supports one meaning of the word (e.g., "Even though it had a strange flavor, the port . . . "), how does this context influence or bias the meaning of the word that is accessed? These questions are important because their answers may provide a better understanding of how words are processed during reading and thus may have more general implications for theories of word identification.

A number of experiments (many of them involving eye tracking) have been conducted to address these questions (for a review, see Duffy et al., 2001). One of the first of these studies (Duffy et al., 1988) orthogonally varied the meaning dominance of ambiguous words and whether or not specific meanings of the words (in this case, the subordinate meaning) were supported by prior context (see also Rayner & Duffy, 1986; Rayner & Frazier, 1989). Each type of ambiguous word had an unambiguous control word that had the same frequency and length as the orthographic form of the ambiguous word and fit into the sentence frame equally well. Thus, participants in the Duffy et al. experiment read a sentence that contained either an ambiguous word or its control word, and the sentence either contained a prior disambiguating context (that supported the subordinate meaning of the biased ambiguous word) or had no such prior supporting context. Example sentences from each of the experimental conditions are shown in Table 1. The mean gaze durations that were observed in each condition are shown in Table 2. (The mean first-fixation durations followed the same qualitative pattern as the gaze durations, but were not reported by Duffy et al. in the interest of brevity.)

The key findings from the Duffy et al. (1988) experiment can best be described by first considering the two "no conflict" conditions. In the condition where the balanced ambiguous words were preceded by supporting context (i.e., *balanced-prior-context*), the gaze durations were the same as those that were observed on the unambiguous control words (mean difference = 0 ms; see Table 2). One interpretation of this finding is that, although the meaning that was supported by the context is as well represented or available as the non-supported meaning, the former can be identified more rapidly because the context influences the order in which alternative word meanings are activated. Thus, although the

Table 1
Example sentences used by Duffy, Morris, and Rayner (1988) for each condition

	Prior context		No prior context	
	Ambiguous	Control	Ambiguous	Control
Balanced	Although it was wrinkled and worn, his *case* attracted much attention.	Although it was wrinkled and worn, his *face* attracted much attention.	Of course his *case* attracted attention although it was wrinkled and worn.	Of course his *face* attracted attention although it was wrinkled and worn.
Biased	Even though it had a strange flavor, the *port* was popular.	Even though it had a strange flavor, the *soup* was popular.	Actually the *port* was popular even though it had a strange flavor.	Actually the *soup* was popular even though it had a strange flavor.

Note: Ambiguous and unambiguous (control) target words are italicized.

Table 2
Mean observed (Duffy et al., 1988) and simulated target-word gaze durations as a function of meaning dominance (balanced vs biased) and disambiguating context (present vs absent)

	prior context			No prior context		
	Ambiguous	Control	Difference	Ambiguous	Control	Difference
Observed						
Balanced	264	264	0	279	261	18
Biased	276	255	21	261	259	2
Simulation 1						
Balanced	246	246	0	246	246	0
Biased	248	248	0	248	248	0
Simulation 2						
Balanced	247	246	1	247	246	1
Biased	260	248	12	248	248	0
Simulation 3						
Balanced	248	248	0	267	248	19
Biased	270	249	21	250	249	1

Note: Times are in milliseconds.

words in the balanced-prior-context condition have two meanings, there is no "conflict" between these meanings because the disambiguating context somehow lends an advantage to one of those meanings.

In the second "no conflict" condition (*biased-no-prior-context*), the biased ambiguous words were not preceded by disambiguating context, and the gaze durations were again about the same as those on the unambiguous control words (mean difference = 2 ms). One interpretation of this result is that, because the dominant meaning of the ambiguous word is rapidly available, the time that is needed to assign a meaning to the word is about the same as with the frequency-matched control word. However, in this case, when the subsequent context instantiates the subordinate meaning, there is quite a large cost in processing when this context is reached, indicating that the "wrong" meaning had been initially accessed and further processing was necessary to repair the error.

The two remaining conditions of the Duffy et al. (1988) experiment can be considered "conflict" conditions. That is, in the condition where the biased ambiguous words were preceded by context that supported the subordinate meaning of the words (i.e., *biased-prior-context*), the gaze durations were longer than those on the control words (mean difference = 21 ms). Similarly, in the condition where the balanced ambiguous words were not preceded by disambiguating context (*balanced-no-prior-context*), the gaze durations were also longer than on the control words (mean difference = 18 ms). One interpretation of both of these findings is that they result from a "conflict" that arises because the two word meanings are somehow competing or interfering with each other. In the biased-prior-context condition, the supporting context favors the subordinate meaning of the word, which – because it is less common and hence less well represented – takes longer to be retrieved from memory. The increased difficulty associated with retrieving the subordinate meanings of these words thus results in longer gaze durations. In the balanced-no-prior-context condition, both meanings of the ambiguous words are equally common and hence equally well represented in memory. Because both meanings are equally available, and because there is context to lend an advantage to one of these meanings, the two meanings somehow interfere with each other. The conflict that results from having two competing meanings available results in longer gaze durations. We will examine the assumptions of this model in greater detail below. Before doing so, however, we will first briefly review the theories that have been proposed to explain lexical ambiguity resolution (for a more comprehensive review, see Pacht & Rayner, 1993).

Theories of lexical ambiguity resolution fall along a continuum with respect to the role that higher-level linguistic processing is posited to play in lexical access. On one end of this continuum, *autonomous access models* posit that all meanings of words are automatically and exhaustively accessed at a rate that is proportional to their frequency of occurrence, irrespective of the context in which they occur. At the other end of the continuum, *selective access models* posit that sentence context plays a very pronounced role in lexical access, and that only the contextually appropriate meanings of ambiguous words are accessed, again at a rate that is proportional to their frequency of occurrence. Two other models are situated between these two extreme views. The first is the *re-ordered* access model, in which higher-level linguistic processing can influence the order in which the meanings of ambiguous words are accessed (Duffy et al., 2001). The second is the *integration model*, which can be viewed as being an extension of the autonomous access models in that higher-level linguistic processing does not guide lexical processing (which is assumed to be exhaustive), but instead only affects the speed of post-lexical processing (e.g., how rapidly a word's meaning can be integrated into the overall meaning of a sentence; Rayner & Frazier, 1989).

To further examine how well each of these theories can explain the pattern of results reported by Duffy et al. (1988), we attempted to insert what we took to be the key assumptions of the theories into the word-processing component of the E-Z Reader model. One goal in doing this was to determine whether such a model is sufficient to explain the pattern of results that were observed by Duffy et al. (1988) experiment. A second goal is to provide a conceptual "scaffolding" for thinking about the role that ongoing sentence processing plays in influencing the time course of lexical processing and the decision about when to move the eyes from one word to the next.

3. Simulations

With the above goals in mind, we completed three simulations in which various assumptions of the re-ordered model were added into the basic framework of the E-Z Reader model. Each simulation was completed using 1000 statistical subjects and the 48 sentences from the Schilling et al. (1998) corpus. These sentences were used as "frames" to examine how differences in meaning dominance and the presence vs absence of prior disambiguating context would influence the viewing times on the ambiguous words embedded within these sentences. In the simulation, the ambiguous target words were assigned to the ordinal word positions of the high- and low-frequency word targets that were used by Schilling et al. We attempted to make the simulation as true to the Duffy et al. (1988) experiment as possible[2]. For example, the frequency of the target words was set equal to the mean values of the targets used by Duffy et al. (1988). For the balanced condition and its unambiguous control, the frequency was set equal to 94 per million (based on the norms of Francis & Kucera, 1982); in the biased condition and its unambiguous control, the frequency was set equal to 61 per million. The length of the target words was set equal to five letters for all of the conditions, and the predictability was set to zero in the all of the conditions.

Finally, it is important to note that all of the simulations are predicated on the assumption that the variables that have been shown to affect how long readers look at ambiguous words (i.e., meaning dominance and whether or not the words are preceded by disambiguating sentence context) influence the duration of L_1 in the E-Z Reader model. This assumption was necessary because, in the model, the duration of L_1 largely determines how long a given word is fixated[3]. One implication of this assumption is that the completion of L_1 corresponds to some aspect of the processing of a word's *meaning* because both meaning dominance and its interaction with overall sentence meaning (such as is posited to happen in the integration model) are by definition related to the processing of meaning. Although we have preferred to remain agnostic about the precise interpretation of L_1 (see Rayner, Pollatsek, & Reichle, 2003 for three possible interpretations of the distinction between L_1 and L_2), our new assumption that the variables that influence fixation durations on ambiguous words do so by influencing the duration of L_1 forces us to refine our conceptualization of L_1 and to acknowledge that – at a minimum – this stage of lexical processing has something to do with the processing of word meaning. We will return to this issue and the larger question of what the two stages of lexical processing (L_1 vs L_2) in E-Z Reader correspond to in the last section of this chapter.

[2] We did not attempt to find optimal model parameter values for the simulation because doing so is prohibitively labor intensive, requiring one to calculate a number of different dependent measures for each word, and well as each word's frequency, length, and mean cloze-task predictability. We therefore opted to use the same parameter values (unless otherwise noted) that have been used in our previous simulations (Pollatsek et al., 2006; Reichle et al., 2006) and to limit our efforts to making predictions about the target words in the various conditions of the Duffy et al. (1988) experiment.

[3] Other factors can also affect fixation durations in the model. For example, fixations on words preceding or following skips tend to be longer than those preceding or following other fixations (Pollatsek, Rayner, & Balota, 1986; Rayner et al., 2004; Reichle et al., 1998; cf. Kliegl & Engbert, 2005). The fixation duration on a word can also be affected by properties of the preceding or following word in cases involving saccadic error (i.e., cases where the eyes either undershot or overshot their intended targets; Rayner, White, Kambe, Miller, & Liversedge, 2003). We will ignore these factors in our discussion because their effects on fixation durations are negligible in comparison to properties of the words themselves.

3.1. Simulation 1

The first simulation was completed using the "standard" version of the E-Z Reader model that was described in the preceding section and has no additional assumptions. As Table 2 shows, the model failed to capture the effect of ambiguity or its interaction with disambiguating context; the only effect that was predicted was a small effect of word frequency that was due to the small difference in the mean frequency of the target words in the balanced vs biased conditions. Although all of the predicted gaze durations were shorter than those that were actually observed, this discrepancy is due to the fact that the model's parameter values were not adjusted to provide the best fit to the Duffy et al. (1988) data (see Footnote 2). If one ignores this minor discrepancy, then the results of Simulation 1 can be used as a "benchmark" to evaluate the consequences of including additional theoretical assumptions (Simulations 2 and 3) that have been proposed to explain the Duffy et al. results.

3.2. Simulation 2

The next simulation was done to evaluate one potential explanation for the ambiguity effects – that the effects reflect a simple retrieval mechanism that is based on the meaning dominance or frequency of a particular word meaning. For example, for a biased ambiguous word, the frequency of the dominant meaning is close to that of the unambiguous control word, whereas the frequency of the subordinate meaning is a lot less frequent than that of the control. Thus, from this kind of frequency analysis, it qualitatively makes sense that gaze durations on biased ambiguous words will be similar to those of the unambiguous controls when given no prior context (assuming the dominant meaning is accessed) but longer than the controls given prior context instantiating the subordinate meaning. (This raises the question of how the dominant meaning would be suppressed so that it isn't retrieved; we will return to that later.) By the same logic, for balanced ambiguous words, each meaning has a frequency somewhat less than the meaning of the unambiguous control word, so that one would expect gaze durations on the ambiguous words to be somewhat longer than those on the control words. Finally, although it is not clear that anything in such a meaning-frequency mechanism can explain why prior context reduces the size of the ambiguity effect with balanced ambiguous words, it is still of interest to see how well a model that relies solely on an access mechanism where retrieval time is based solely on the frequency of the particular word meaning can explain the lexical ambiguity data that were reported by Duffy et al. (1988).

To evaluate a simple meaning-frequency-based mechanism, Simulation 2 was completed using Equation 4 (which is a modification of Equation 1, above). In Equation 4, the amount of time that is needed to identify a given word is modulated by two variables. The first variable is the word's frequency of occurrence in printed text (freq); that is, its (orthographic) token frequency as tabulated in the Francis and Kucera (1982) norms. The second variable is the meaning dominance of the contextually appropriate or "supported" meaning of the word, $p(meaning_{\text{supported}})$, where the supported meaning is the one that is consistent with prior disambiguating context or – in the absence of such context – whatever meaning happens to be the most common one. The specific meaning-dominance values of the supported meanings were estimated using the conditional probabilities of giving one or the other meaning of an ambiguous word as the dominant meaning of the word in the absence of biasing context. These conditional probabilities were taken directly from the norms collected by Duffy et al. (1988) for each of the different conditions of that experiment: $p(meaning_{\text{supported}}) = 0.57$ for balanced ambiguous words in and out of context; $p(meaning_{\text{supported}}) = 0.07$ and 0.93 for biased ambiguous words in and out of context, respectively; and $p(meaning_{\text{supported}}) = 1$ for the unambiguous control words. The second term of Equation 4 can thus be interpreting as representing the frequency with which the contextually supported (or in the absence of prior disambiguating context, the dominant) meaning of an ambiguous word occurs in printed text.

$$\mathrm{t}(\mathrm{L}_1) = \alpha_1 - \alpha_2 \ln[\mathrm{freq} * p(meaning_{\text{supported}})] - \alpha_3 \,\mathrm{pred} \tag{4}$$

Table 2 shows the results of Simulation 2. If one compares the results of this simulation to the results reported by Duffy et al. (1988; the "observed" values that are also shown in Table 2), then it is clear that the frequency-based mechanism can predict at least some of the cost that is observed with gaze durations when biased ambiguous words follow disambiguating context: 12 ms vs 21 ms for the simulated vs observed costs, respectively. The model simulates this subordinate bias effect because the sentence context in this condition is consistent with the less frequent, subordinate meanings of the ambiguous words, so that these words take longer to identify than their frequency-matched control words. Of course, the model is completely silent about *how* context selects the appropriate meaning of the ambiguous word. For example, what happens to the dominant meaning of the ambiguous word in this condition? Is it accessed but then "ignored", or does the context instead somehow inhibit the dominant meaning? A complete (process model) account of the Duffy et al. results seemingly requires some type of explanation for what happens to the meanings of ambiguous words that are not congruent with prior disambiguating context.

Table 2 also shows the second simulation also produced the correct pattern of gaze durations for the two no-conflict conditions. In the balanced-prior-context condition, this is due to the fact that the ambiguous words have meaning frequencies that are comparable to those of the unambiguous control words. Similarly, in the biased-no-prior-context condition, the dominant meanings (which is supported by the context) of ambiguous words are also nearly as frequent as those of the control words. Finally, it is not surprising that the model failed to simulate the cost (1 ms) that was observed (18 ms) in the balanced-no-prior-context condition. This failure was not surprising because the model effectively handles balanced ambiguous words the same way both in and out of context: In both cases, the dominant meanings are assumed to be accessed, with no cost associated with having two alternative, "competing" word meanings. This failure thus indicates that the simple frequency-based retrieval mechanism is not sufficient to account for the full pattern of results that were reported by Duffy et al. (1988).

3.3. Simulation 3

A second possible mechanism to explain ambiguity effects is that they reflect a slowing down in lexical processing that results from an active competition between the alternative meanings of ambiguous words. The basic idea behind such a mechanism is that retrieval time of a word meaning is not solely (or possibly not at all) based on the frequency of the individual meanings of an ambiguous word, but is instead based on the relative frequencies of two competing meanings. That is, in such a model, the strength of the competition is greater, the more similar the strengths of the competing meanings are. We will discuss the specific form of how we model this conflict below; however, for the moment, it is sufficient to say that there will be a term in the equation for the L_1 time that will be larger, the smaller the difference in meaning frequency between the two meanings of an ambiguous word. Thus, without prior context, there would be greater conflict (and slower retrieval time) for balanced ambiguous words than for biased ambiguous words. This suggests that the elevated gaze durations for balanced ambiguous words in the absence of prior context (relative to unambiguous controls) may be solely due to this kind of interference and may not be dependent on the fact that both meanings of such ambiguous words are less frequent than the meaning of the control word.

To evaluate this competition-based mechanism, we completed a third simulation using Equation 5 (see below). Notice that, in this equation, the time that is needed to identify an ambiguous word is no longer a direct function of the word's meaning dominance (cf. Equations 4 and 5); instead, the final term represents the amount by which the active competition between alternative meanings of ambiguous words increases the time that is necessary to complete L_1 on those words. In the equation, $\alpha_4 (= 7)$ is a scaling parameter that is used to modulate the absolute amount of competition that can result from meaning ambiguity. The variables labeled "$meaning_{dom}$" and "$meaning_{sub}$" are the conditional probabilities of human subjects giving the dominant and subordinate meanings (respectively) of the ambiguous words in the absence of any biasing context. The values of these variables were taken from the Duffy et al. (1988) norms (as in Simulation 2). The parameter $\phi (\phi = .36$ with prior context; $\phi = 0$ with no prior context) modulates the degree to which the prior sentence context alters the strengths of the alternative word meanings. As a result, qualitatively, this should explain the decrease in the ambiguity effect for balanced ambiguous words with biasing prior context because context should push the frequencies of the two meanings further apart. Likewise, this mechanism should predict an increase in the ambiguity effect with prior biasing context for the biased ambiguous words because it should make the strength of the two meanings more equal. Finally, a z-transform is used to convert the probability difference scores in the denominator of the last term of Equation 5 into z-values so that values close to zero and one become very large[4].

$$t(L_1) = \alpha_1 - \alpha_2 \ln(\text{freq}) - \alpha_3 \text{ pred} + \alpha_4 / \{z[p(meaning_{dom} + \phi)] - z[p(meaning_{sub} - \phi)]\} \quad (5)$$

[4] The terms $(meaning_{dom} + \phi)$ and $(meaning_{sub} - \phi)$ were restricted to the range of .001 to .999

The results of Simulation 3 are also shown in Table 2. As can be seen there, the model nicely captured the pattern of results that were reported by Duffy et al. (1988). That is, it predicted a 21 ms cost for the biased ambiguous words when they followed disambiguating context, a 19 ms cost for the balanced ambiguous words without context, and no costs for the ambiguous words in the two "no conflict" conditions. This final simulation thus demonstrates that the E-Z Reader model, when augmented with a few fairly simple assumptions about how meaning dominance and disambiguating context influence the ease with which the alternative meanings of ambiguous words are accessed, can explain the pattern of fixation durations that were observed by Duffy et al. (1988). (Our simulation would also qualitatively predict that there would be later disruption in the biased-no-prior-context condition because the dominant meaning would be accessed and thus would be inconsistent with the subsequent context.)

Perhaps not surprisingly, the assumptions that are encapsulated in Equation 5 (above) are similar to the assumptions of the re-ordered access model, and are similar to – but not identical with – the assumptions that were used in the simulations by Duffy et al. (2001). Their simulation used an interactive-constraint model adapted from Spivey and Tanenhaus (1998) in which the weights for meaning dominance and context are adjusted after each cycle, and the predictions are in terms of the number of cycles it takes for a meaning to exceed a predetermined threshold. Their simulation did not predict the pattern of results all that closely. It did predict virtually no difference in the number of cycles between the ambiguous words and the control words in the two no-conflict situations and a substantial difference in the two conflict situations. However, it predicted almost three times as many cycles in the biased-prior-context condition as in the balanced-no-prior-context condition, even though the observed differences in the cost in the gaze duration data were not very different in the two conflict conditions. In the final section of this chapter, we will use these simulations as frameworks for evaluating various assumptions about how context is affecting the word identification process, and we will return to the question of what L_1 and L_2 correspond to.

4. Discussion

In our final simulation, the disambiguating sentence context was allowed to influence the times that were needed to identify ambiguous words. We did this by allowing context either to increase or to attenuate the relative disparity between the "strengths" of alternative word meanings and to thereby affect the amount of competition that resulted from the two interpretations (see Equation 5). To be a bit more precise, the basic mechanism tacitly assumed in Equation 5 for the effect of context is that it enhances the strength of the representation of meanings that are consistent with the prior text relative to the strength of the representation of meanings that are inconsistent with the prior text. This assumption, by itself, is a satisfactory explanation of the Duffy et al. (1988) results. In the balanced-prior-context condition, the constraint provided by the prior disambiguating context largely eliminates the competition that would otherwise slow lexical processing, making the time for completing L_1 about as rapid as for the control word. Conversely, in the biased-prior-context condition, the disambiguating context increases the competition between alternative word meanings, thereby slowing lexical processing and making the time that is needed for completing L_1 longer than the control word. In the biased-no-prior-context condition, the dominant meaning of the word is readily available and there is minimal competition between meanings, which results in the rapid completion of L_1 and a rapid eye movement off of the word. Finally, in the balanced-no-prior-context condition, the two word meanings are equally available, which results in competition between meanings, the slower completion of L_1, and a less rapid saccade off of the word.

As already noted, the assumptions of Simulation 3 (Equation 5) are similar to those of the re-ordered access theory; both assume that the overall meaning of the sentence can change the order in which alternative meanings of a word become available. Other interesting points of contrast can be gained by comparing the simulations in this chapter to the core assumptions of the theories that have been proposed to explain lexical ambiguity resolution. For example, although both the selective access theory and the frequency-based retrieval mechanism of Simulation 2 (Equation 5) can account for the subordinate bias effect by positing that the frequency of subordinate meaning of an ambiguous word in context is effectively equal to that of low-frequency control word (Sereno et al., 1992), neither account can say why cost is observed in the balanced-no-prior-context condition. The cost for ambiguous words in this condition seemingly requires some type of competition between more-or-less equally matched word meanings, such as that provided by the competition mechanism in Simulation 3.

Similarly, one might be inclined to implement a variant of the integration model (Rayner & Frazier, 1989) by first adding some type of post-lexical integration process to the frequency-based retrieval mechanism (Equation 4) and then assuming that this integration process can only begin when the word meaning that is consistent with the prior sentence has been accessed[5]. Although such a model would undoubtedly predict more cost in the biased-prior-context condition and thereby cause the simulated subordinate bias effect to be more in line with the effect size reported by Duffy et al. (1988), this model would still fail to predict the cost that is observed in the balanced-no-prior context condition. The reason for this is that any frequency-based retrieval mechanism will still make the dominant meaning of a balanced ambiguous word available for integration about as quickly as the meaning of the unambiguous control word. This again suggests the need for some type of active competition between alternative interpretations of ambiguous words.

Finally, although the competition-between-meanings mechanism of Simulation 3 (Equation 5) is admittedly a very coarse way of handling the effect of disambiguating

[5] Of course, this model would also require very rapid word identification and integration times, such as the duration of L_1 in E-Z Reader (e.g., 117–172 ms). An alternative assumption that post-lexical processing corresponds to some later stage of processing (e.g., L_2 in E-Z Reader) would not work because fixation durations on ambiguous words would not be affected by the integration process.

context, it is conceivable that some combination of syntactic, semantic, and/or pragmatic constraints are sufficient to influence the processing of ambiguous words, making whatever meaning of the word that is being supported more or less accessible than it otherwise would be by decreasing or increasing the amount of competition between alternative word meanings. Within the framework of the E-Z Reader model, this implies that these contextual constraints are rapid enough to influence the earliest stage of lexical processing – L_1. How reasonable is this assumption? A recent study by Rayner, Cook, Juhasz, and Frazier (2006) demonstrated that these constraints can also be produced rapidly: an adjective immediately preceding a biased ambiguous word that disambiguated it resulted in longer fixations on the target word when the subordinate meaning was instantiated.

A recent *event-related potentials* (ERP) experiment also provides some evidence that context can influence the early stages of word identification (Sereno, Brewer, & O'Donnell, 2003). In this experiment, ambiguous words were embedded in sentences contexts that were either neutral or supported the words' subordinate meanings. This manipulation affected an early ERP component (the N1, which occurs 132–192 ms post-stimulus onset) that has been interpreted as an index of early lexical processing (Sereno, Rayner, & Posner, 1998). In the condition involving neutral sentence context, the N1 component that was associated with processing ambiguous words resembled the N1 component that was associated with processing high-frequency unambiguous control words. However, in the condition involving subordinate-biasing context, the N1 associated with ambiguous words resembled the N1 for low-frequency unambiguous control words. The presence of prior disambiguating context thus affected an early electrophysiological signature of lexical processing, suggesting that context can influence the stage of word identification that, in the E-Z Reader model, corresponds to L_1. Moreover, it is worth noting that this early effect of context on lexical processing occurred even in a paradigm in which there was no parafoveal preview of the word because the words were displayed one at a time.

The competition mechanism may also provide a useful way of thinking about how other aspects of higher-level linguistic processing affect word identification. In addition to explaining the lexical ambiguity phenomenon, the competition assumption also seems like an obvious mechanism to explain the increased fixation durations when a word is anomalous given the prior context (Rayner et al., 2004). However, what needs to be explicated in a more serious model is how this mechanism would work. That is, such a model should specify what set of words would be excited by or inhibited by prior context and offer a plausible mechanism for it. For example, such a model should explain why gaze durations on the anomalous word are lengthened but that the effect of context for words that are only implausible given the prior context occurs only on "spillover" fixations. Another open question is whether one has to posit that alternative meanings compete with each other such that one meaning needs to be "significantly stronger" than its competitors in order for meaning access to take place. This would be similar to the competition assumed in many *interactive-activation* models of word identification

(McClelland & Rumelhart, 1981)[6]. The above discussion suggests that such an assumption may not be necessary if one allows prior context to inhibit certain meanings sufficiently.

This challenge brings us to the question of what L_1 and L_2 – the core theoretical constructs in the E-Z Reader model – actually correspond to. Answering this question will not only allow us to further refine how we conceptualize our model, it may provide a basis for evaluating all of the various models that have been developed to explain eye-movement control in reading (for a review, see Reichle et al., 2003). Given all of this, what can be said about L_1 and L_2 based on this modeling exercise?

At this time, it seems most likely that L_1 corresponds to some type of rapidly available "sense" of word familiarity that is sensitive to both orthographic and semantic information. L_1 has to be rapid because it has to be completed early enough to affect decisions about when to move the eyes to a new viewing location. L_1 has to be sensitive to orthographic information if the E-Z Reader model is to explain how variation in the orthographic form of a word can influence the initial fixation on the word but not the amount of spillover that is observed (Reingold & Rayner, 2006). Similarly, it has to be sensitive to the meaning of the word and how this meaning is related to higher-level linguistic information if the model is going to handle the host of effects related to such information, including the effects of word predictability (Balota, Pollatsek, & Rayner, 1985; Ehrlich & Rayner, 1981; Kliegl, Grabner, Rolfs, & Engbert, 2004; Rayner et al., 2004; Rayner & Well, 1996), thematic role assignment (Rayner, Warren et al., 2004), and the resolution of lexical ambiguity (Dopkins et al., 1992; Duffy et al., 1988; Pacht & Rayner, 1993; Rayner & Duffy, 1986; Rayner & Frazier, 1989; Sereno et al., 1992, 2006), to name just a few (for a recent review, see Rayner, 1998). However, this leaves open the question of what information L_1 is *directly* responding to. Given the available data, there are two possibilities. The first is that it is only sensitive to access the meaning of a word but that the access of the meaning is dependent on completion of prior stages of orthographic and phonological access (i.e., a serial stage model). The second is an interactive model, in which familiarity is being computed from an evaluation of the relative completion of various types of processing (e.g., orthographic, phonological, semantic; for an example of how this might work, see Reichle & Laurent, 2006). One way to test this would be to factorially vary various factors, such as lexical ambiguity and visibility. If these factors have roughly additive effects on gaze durations, then it would be evident that L_1

[6] Mutual inhibition among lexical representations has also been incorporated as a central part of the *Glenmore model* of eye-movement control during reading (Reilly & Radach, 2003, 2006). In this connectionist model, word units that are within a fixed window of attention provide mutual inhibition to each other, and this word-level activation feeds back to letter units, which then influence the activation of a "saliency map" that determines where the eyes move. Because the activation of the letter units can also inhibit impending saccades, it is conceivable that the Glenmore model would provide a natural account of the Duffy et al. (1988) if the model were modified so that top-down effects of higher-level sentence processing are allowed to modulate the activation of the word units. If the model were successful explaining the effects of ambiguity resolution on eye movements, then this success would be paradoxical because it would violate the basic spirit of the model by allowing higher-level cognitive processes to exert a very rapid influence on the decisions about when to move the eyes during reading.

is only sensitive to later, meaning, stages of word processing, but that these stages are in turn dependent on the completion of earlier stages. On the other hand, if the effects are strongly interactive, then it would suggest that L_1 is directly influenced by both earlier and later stages. However, we should point out that such additive factors logic (Sternberg, 1969) can not literally be applied to gaze duration as gaze duration is not simply equal to L_1 duration. Nonetheless, such a pattern would be strongly suggestive of an interactive mechanism, and one could further evaluate this hypothesis through simulations using the E-Z Reader model as a framework.

Acknowledgments

This research was supported by a grant R305H030235 from the Department of Education's Institute for Education Sciences and grant HD26765 from the National Institute of Health. We would like to thank Wayne Murray and two anonymous reviewers for their helpful comments on an earlier version of this chapter.

References

Balota, D. A., Pollatsek, A., & Rayner, K. (1985). The interaction of contextual constraints and parafoveal visual information in reading. *Cognitive Psychology, 17*, 364–390.

Becker, W. & Jürgens, R. (1979). An analysis of the saccadic system by means of double step stimuli. *Vision Research, 19*, 967–983.

Clark, V. P., Fan, S., & Hillard, S. A. (1995). Identification of early visual evoked potential generators by retinotopic and topographic analyses. *Human Brain Mapping, 2*, 170–187.

Clifton, C., Staub, A., & Rayner, K. (2005). Eye movements in reading words and sentences. This volume.

Dopkins, S., Morris, R. K., & Rayner, K. (1992). Lexical ambiguity and eye fixations in reading: A test of competing models of lexical ambiguity resolution. *Journal of Memory and Language, 31*, 461–476.

Duffy, S. A., Kambe, G., & Rayner, K. (2001). The effect of prior disambiguating context on the comprehension of ambiguous words: Evidence from eye movements. In D. S. Gorfein (Ed.), *On the consequences of meaning selection* (pp. 27–43). Washington, DC: APA Books.

Duffy, S. A., Morris, R. K., & Rayner, K. (1988). Lexical ambiguity and fixation times in reading. *Journal of Memory and Language, 27*, 429–446.

Ehrlich, S. F., & Rayner, K. (1981). Contextual effects on word perception and eye movements during reading. *Journal of Verbal Learning and Verbal Behavior, 20*, 641–655.

Engbert, R., Longtin, A., & Kliegl, R. (2002). A dynamical model of saccade generation in reading based on spatially distributed lexical processing. *Vision Research*, 42, 621–636.

Engbert, R., Nuthmann, A., Richter, E., & Kliegl, R. (2005). SWIFT: A dynamical model of saccade generation during reading. *Psychological Review, 112*, 777–813.

Foxe, J. J. & Simpson, G. V. (2002). Flow of activation from V1 to frontal cortex in humans: A framework for defining "early" visual processing. *Experimental Brain Research, 142*, 139–150.

Francis, W. N. & Kucera, H. (1982). *Frequency analysis of English usage: Lexicon and grammar.* Boston: Houghton Mifflin.

Henderson, J. M., & Ferreira, F. (1990). Effects of foveal processing difficulty on the perceptual span in reading: Implications for attention and eye movement control. *Journal of Experimental Psychology: Learning, Memory, and Cognition, 16*, 417–429.

Just, M. A. & Carpenter, P. A. (1980). A theory of reading: From eye fixations to comprehension. *Psychological Review, 87*, 329–354.

Kennison, S. M., & Clifton, C. (1995). Determinants of parafoveal preview benefit in high and low working memory capacity readers: Implications for eye movement control. *Journal of Experimental Psychology: Learning, Memory, and Cognition, 21*, 68–81.

Kliegl, R. & Engbert, R. (2005). Fixation durations before word skipping in reading. *Psychonomic Bulletin & Review, 12*, 132–138.

Kliegl, R., Grabner, E., Rolfs, M., & Engbert, R. (2004). Length, frequency, and predictability effects of words on eye movements in reading. *European Journal of Cognitive Psychology, 16*, 262–284.

Klien, D. E. & Murphy, G. L. (2001). The representation of polysemous words. *Journal of Memory and Language, 45*, 259–282.

McConkie, G. W., Kerr, P. W., Reddix, M. D., & Zola, D. (1988). Eye movement control during reading: I. The location of initial eye fixations in words. *Vision Research, 28*, 1107–1118.

McConkie, G. W., Zola, Grimes, J., D., Kerr, P. W., Bryant, N. R., & Wolff, P. M. (1991). Children's eye movements during reading. In. J. F. Stein (Ed.) *Vision and Visual Dyslexia 13*. London: MacMillan.

Mouchetant-Rostaing, Y., Giard, M.-H., Bentin, S., Aguera, P.-E., & Pernier, J. (2000). Neurophysiological correlates of face gender processing in humans. *European Journal of Neuroscience, 12*, 303–310.

O'Regan, J. K. (1990). Eye movements and reading. In E. Kowler (Ed.), *Eye movements and their role in visual and cognitive processes* (pp. 395–453). Amsterdam: Elsevier.

O'Regan, J. K. (1992). Optimal viewing position in words and the strategy-tactics theory of eye movements in reading. In K. Rayner (Ed.), *Eye movements and visual cognition: Scene perception and reading* (pp 333–354). Springer-Verlag.

O'Regan, J. K., & Lévy-Schoen, A. (1987). Eye movement strategy and tactics in word recognition and reading. In M.Coltheart (Ed.), *Attention and performance XII: The psychology of reading* (pp 363–383). Erlbaum.

Pacht, J. M. & Rayner, K. (1993). The processing of homophonic homographs during reading: Evidence from eye movement studies. *Journal of Psycholinguistic Research, 22*, 251–271.

Pollatsek, A., Rayner, K., & Balota, D. A. (1986). Inferences about eye movement control from the perceptual span in reading. *Perception & Psychophysics, 40*, 123–130.

Pollatsek, A., Reichle, E. D., & Rayner, K. (2005). Tests of the E-Z Reader model: Exploring the interface between cognition and eye-movement control. *Cognitive Psychology, 52*, 1–56.

Rayner, K. (1979). Eye guidance in reading: Fixation locations in words. *Perception, 8*, 21–30.

Rayner, K. (1998). Eye movements in reading and information processing: 20 years of research. *Psychological Bulletin, 124*, 372–422.

Rayner, K., Ashby, J., Pollatsek, A., & Reichle, E. (2004). The effects of frequency and predictability on eye fixations in reading: Implications for the E-Z Reader model. *Journal of Experimental Psychology: Human Perception and Performance, 30*, 720–732.

Rayner, K., Cook, A. E., Juhasz, B. J., & Frazier, L. (2006). Immediate disambiguation of lexically ambiguous words during reading: Evidence from eye movements. *British Journal of Psychology*, 97, 467–482.

Rayner, K. & Duffy, S. M. (1986). Lexical complexity and fixation times in reading: Effects of word frequency, verb complexity, and lexical ambiguity. *Memory & Cognition, 14*, 191–201.

Rayner, K. & Frazier, L. (1989). Selection mechanisms in reading lexically ambiguous words. *Journal of Experimental Psychology: Learning, Memory, and Cognition, 15*, 779–790.

Rayner, K. & McConkie, G. W. (1976). What guides a reader's eye movements. *Vision Research, 16*, 829–837.

Rayner, K., Pollatsek, A., & Reichle, E. D. (2003). Eye movements in reading: Models and data. *Behavioral and Brain Sciences, 26*, 507–526.

Rayner, K., Sereno, S. C., & Raney, G. E. (1996). Eye movement control in reading: A comparison of two types of models. *Journal of Experimental Psychology: Human Perception and Performance, 22*, 1188–1200.

Rayner, K. Warren, T., Juhasz, B. J., & Liversedge, S. P. (2004). The effect of plausibility on eye movements in reading. *Journal of Experimental Psychology: Learning, Memory, and Cognition, 30*, 1290–1301.

Rayner, K. & Well, A. D. (1996). Effects of contextual constraint of eye movements in reading: A further examination. *Psychonomic Bulletin & Review, 3*, 504–509.

Rayner, K., White, S. J., Kambe, G., Miller, B., & Liversedge, S. p. (2003). On the processing of meaning from parafoveal vision during eye fixations in reading. In J. Hyönä, R. Radach, & H. Deubel (Eds.), *The mind's eye: Cognitive and applied aspects of eye movement research* (pp. 213–224). Oxford, England: Elsevier.

Reichle, E. D. & Laurent, P. (2006). Using reinforcement learning to understand the emergence of "intelligent" eye-movement behavior during reading. *Psychological Review, 113*, 390–408.

Reichle, E. D., Pollatsek, A., Fisher, D. L., & Rayner, K. (1998). Toward a model of eye movement control in reading. *Psychological Review, 105*, 125–157.

Reichle, E. D., Rayner, K., & Pollatsek, A. (1999). Eye movement control in reading: Accounting for initial fixation locations and refixations within the E-Z reader model. *Vision Research, 39*, 4403–4411.

Reichle, E. D., Rayner, K., and Pollatsek, A. (2003). The E-Z Reader model of eye movement control in reading: Comparison to other models. *Behavioral and Brain Sciences, 26*, 445–476.

Reilly, R., & Radach, R. (2003). Foundations of an interactive activation model of eye movement control in reading. In J. Hyona, R. Radach, & H. Deubel (Eds.), *The mind's eye: Cognitive and applied aspects of eye movement research* (pp. 429–455). Oxford: Elsevier.

Reilly, R. G. & Radach, R. (2006). Some empirical tests of an interactive activation model of eye movement control in reading. *Cognitive Systems Research, 7*, 34–55.

Reingold, E. M. (2003). Eye-movement control in reading: Models and predictions. *Behavioral and Brain Sciences, 26*, 445–526.

Reingold, E. M., & Rayner, K. (2006). Examining the word identification stages hypothesized by the E-Z Reader model. *Psychological Science, 17*, 742–746.

Salvucci, D. D. (2001). An integrated model of eye movements and visual encoding. *Cognitive Systems Research, 1*, 201–220.

Schilling, H. E. H., Rayner, K., & Chumbley, J. I. (1998). Comparing naming, lexical decision, and eye fixation times: Word frequency effects and individual differences. *Memory & Cognition, 26*, 1270–1281.

Schroyens, W., Vitu, F., Brysbaert, M., & d'Ydewalle, G. (1999). Eye movement control during reading: Foveal load and parafoveal processing. *Quarterly Journal of Experimental Psychology, 52A*, 1021–1046.

Sereno, S. C., Brewer, C. C., & O'Donnell, P. J. (2003). Context effects in word recognition: Evidence for early interactive processing. *Psychological Science, 14*, 328–333.

Sereno, S. C., O'Donnell, P. J., & Rayner, K. (2006). Eye movements and lexical ambiguity resolution: Investigating the subordinate bias effect. *Journal of Experimental Psychology: Human Perception and Performance, 32*, 335–350.

Sereno, S. C., Pacht, J. M., & Rayner, K. (1992). The effect of meaning frequency on processing lexically ambiguous words: Evidence from eye fixations. *Psychological Science, 3*, 296–300.

Sereno, S. C., Rayner, K., & Posner, M. I. (1998). Establishing a time-line of word recognition: Evidence from eye movements and event-related potentials. *NeuroReport, 9*, 2195–2200.

Sternberg, S. (1969). Memory scanning: Mental processes revealed by reaction-time experiments. *American Scientist, 57*, 421–457.

Van Rullen, R., & Thorpe, S. (2001). The time course of visual processing: From early perception to decision-making. *Journal of Cognitive Neuroscience, 13*, 454–461.

Vitu, F., McConkie, G. W., Kerr, P., & O'Regan, J. K. (2001). Fixation location effects on fixation durations during reading: an inverted optimal viewing position effect. *Vision Research, 41*, 3513–3533.

White, S. J., Rayner, K., & Liversedge, S. P. (2005). Eye movements and the modulation of parafoveal processing by foveal processing difficulty: A re-examination. *Psychonomic Bulletin & Review, 12*, 891–896.

Chapter 13

DYNAMIC CODING OF SACCADE LENGTH IN READING

SHUN-NAN YANG

Smith-Kettlewell Eye Research Institute, USA

FRANÇOISE VITU

CNRS, Université de Provence, France

Eye Movements: A Window on Mind and Brain
Edited by R. P. G. van Gompel, M. H. Fischer, W. S. Murray and R. L. Hill

Abstract

It is commonly assumed that eye movements in reading are determined with respect to word entities, and that the distribution of initial landing sites in words derives from a strategy of aiming at specific parts of peripherally located target words. The current study investigated the oculomotor processes responsible for the determination of saccade length and initial landing sites. Distributions of saccade length during pseudo-reading and normal reading were analyzed. Strategy-based, visually guided and corrective saccades were qualitatively identified after examining the influence of launch site, word length and saccade latency. Gaussian mixture models were implemented to approximate the frequency of the three groups of saccades. The distributions of saccade length in pseudo-reading and normal reading were simulated; the percentages of saccade frequency for the three groups varied in relation to saccade latency, word length and launch distance. Both simulated and empirical data showed that strategy-based saccades of a relatively constant length were favored at early time intervals whereas visually guided saccades became more likely at later times during a fixation; the former were more frequent in reading compared with pseudo-reading. It was concluded that eye guidance in reading is the result of dynamic coding of saccade length instead of cognitively based aiming strategies.

Saccadic eye movements have long been considered ballistic and goal-oriented (Jurgens, Becker, & Kornhuber, 1990; Robinson, 1975). In reading research, many models of eye-movement control assume such properties (e.g., Engbert, Nuthmann, Richter, & Kliegl, 2006; Just & Carpenter, 1980; Morrison, 1984; Reichle, Pollatsek, Fisher, & Rayner, 1998; Reichle, Rayner, & Pollatsek, 2003). These models postulate that the gaze is continuously shifted from one processed word to the next, aiming specifically at certain parts of peripheral target words for various reasons (McConkie, Kerr, Reddix, & Zola, 1988; O'Regan & Lévy-Schoen, 1987; Reichle, Rayner, & Pollatsek, 1999). Variations in the targeting strategy may occur when there is a need to refixate a word, to skip a word, or to make a regressive saccade to a previously fixated word. We will refer to this as "word-based eye guidance".

The most cited finding for supporting word-based eye guidance is the observation of the "preferred viewing position" effect. The effect can be seen when one plots the distribution of initial landing sites in a word. The typical Landing Site (LS) curve shows that, despite a great variability of initial landing sites, the eyes most often land at a position slightly to the left of the center of words, with the amount of deviation being a function of word length (McConkie et al., 1988; O'Regan, 1979; Rayner, 1979; Vitu, O'Regan, & Mittau, 1990). This phenomenon has been repeatedly demonstrated in numerous studies and few researchers question the influence of word length on initial landing sites.

Other phenomena further argue for word-based eye guidance (see Inhoff, Radach, Eiter, & Juhasz, 2003). These include the influence of launch site (or distance of the eyes to the beginning of a word) on initial landing site (McConkie, Kerr, Reddix, Zola, & Jacobs, 1988; Radach & McConkie, 1998). The effect shows that the eyes tend to overshoot the center of words when they are launched from a close distance and undershoot the center with distant launch sites. In addition, when mean initial landing sites are expressed with respect to the center of words and are plotted against center-based launch site (the distance of the launch site to the center of the word), they exhibit a linear relationship (mean center-based landing site increases with center-based launch distance) and the effect of word length is strongly reduced.

Commonly held explanations for the LS curve and the launch site effect rely on the assumption that readers aim for the center (or the optimal viewing position) of peripherally located target words in order to expedite word processing (O'Regan, 1990, 1992; O'Regan & Lévy-Schoen, 1987; McConkie et al., 1988). A word is more easily recognized when the eyes initially fixate near the center of the word than when they fixate at the beginning or end of the word (Brysbaert, Vitu, & Schroyens, 1996; O'Regan, 1990; O'Regan & Jacobs, 1992; O'Regan, Lévy-Schoen, Pynte, & Brugaillère, 1984; Nazir, O'Regan, & Jacobs, 1991). The great variability of initial landing sites in words and the fact that many saccades do not land near the center of words would result from oculomotor aiming errors. In particular, the influence of launch site would result from systematic oculomotor range error (Kapoula, 1985; McConkie et al., 1988; Radach & McConkie, 1998; but see Vitu, 1991a–b), that the eyes would be biased towards making saccades of a constant length (i.e., the length that corresponds to the center of the range of saccade lengths associated with the task).

Two different views are held with respect to the mechanism responsible for determining the saccade-target word in peripheral vision. The first view envisages saccade-target selection as the result of low-level visuomotor processes, assuming that it is not related to ongoing word identification processes. This will be referred to as the "stimulus-based" hypothesis. In this, words are nothing more than visual blobs and the selected target word is either the next word (McConkie et al., 1988; O'Regan, 1990; Radach & McConkie, 1998) or the next long word (Reilly & O'Regan, 1998).

Alternatively, as proposed in recent models of eye-movement control, the saccade-target word may be defined with respect to ongoing word identification processes; this view will be referred to as the "processing-based" hypothesis. For instance, the E-Z reader model proposes that in reading, words are processed serially due to sequential attention shifts (Reichle et al., 2003). A saccade within the fixated word (refixation) is first programmed as a default and the refixation probability depends on word length. If an intermediate level of processing associated with the fixated word (i.e., word familiarity check) is reached before the refixation saccade is ready to be triggered, the refixation is cancelled and the programming of a saccade to the next word begins. However, again, if processing of the next word reaches the intermediate processing level before saccade programming enters a stage of no return, the saccade is cancelled and a new saccade is planned toward the following word, hence skipping the initial target word. Thus, in this model, a processing-based targeting mechanism is used to guide the eyes. The SWIFT model is another example of processing-based models, although it relies on a slightly different saccade-target selection mechanism. In this model, words are processed in parallel, and each word/letter receives a certain amount of attractiveness depending on the level of processing associated with the word at a given point in time (Engbert et al., 2006; Kliegl et al., 2003). When a word is completely processed, the weight of each letter in this word is set to zero, while the attractiveness of each letter in a word that is not yet identified increases as the processing of that word progresses. A saccade is sent to the letter with maximal attractiveness when a random waiting time for saccade initiation is reached. Although these two versions of the processing-based hypothesis differ in the unit of targeting (word vs letter), they are similar in that the level of linguistic processing for a word determines where the eyes are sent. Both assume that information from the word-processing mechanism can directly determine the location of the saccade target. The oculomotor system merely executes the decision of language processes, although subject to oculomotor errors.

In recent studies, alternatives of the above-mentioned word-based hypotheses were proposed. These proposals reject the assumption that eye guidance in reading relies predominantly on a saccade-targeting mechanism. For instance, Vitu (2003) in a commentary of the E-Z reader model pointed out the great similarity of the effects of launch site on the likelihood of word skipping and the distributions of initial landing sites in words. This particular finding is inconsistent with a processing-based hypothesis that attributes variations in both word skipping and initial landing sites to different mechanisms, namely ongoing word identification processes and oculomotor aiming errors. She proposed that the eyes move toward the center of gravity of the peripheral text configuration, thus without aiming for a specific target location. The eyes would be pulled by the visual stimuli in the forward peripheral region and the stimuli closer to the fovea would have greater weight (see also, Vitu, 1991a,b).

Yang and McConkie (2001, 2004) proposed a different view on the determination of saccade length, although sharing the assumption that word-based eye guidance is not the general rule. Their rationale was the following: If the detection of word boundaries, and in the extreme case the identification of peripheral words, were necessary for eye guidance in reading, as assumed in processing- and stimulus-based hypotheses, reading should be impaired in the absence of clear visual boundaries between individual words. Without proper word segmentation, word identification would become extremely laborious and its processing time very lengthy. In addition, the process of isolating visual word units for the purpose of aiming at any particular part of words would also be particularly difficult and time consuming. Reading under those conditions should break down somehow. Yang and McConkie (2004) showed that this is not the case. In their study, during randomly selected saccades, the normal text was replaced with un-spaced nonwords or unbroken homogenous letter strings for a single (critical) fixation and the normal text was returned during the immediate following saccade. Thus, during the critical fixation, no word boundaries were available. If the localization of the to-be-fixated word was essential for guiding the eyes, saccade initiation should be greatly postponed or at least altered in this condition. Results showed instead that lacking clearly defined word boundary for a single fixation did not significantly alter the length of the immediate following saccade. Only a slight shift of the distribution of saccade length toward smaller lengths was observed; the length of many late saccades (initiated 225 ms or later during the critical fixation) was shortened by one or two letters, but the length of earlier saccades remained unaffected. Thus, it seemed that the lack of spatial segmentation did not systematically disrupt or postpone saccade initiation.

To account for the above observations, Yang and McConkie proposed the competition/interaction (or C/I) model of eye-movement control in reading. In this model, the oculomotor system encodes the impending saccade using a population-coding scheme. Many neurons optimally representing different saccadic metrics (direction and length) are activated to collectively signal the precise landing location (Lee, Rohrer, & Sparks, 1988). The variability of saccade length results from the varying distribution of movement-coding activity from saccade to saccade. Two main factors affect the coding of saccade length: At earlier time intervals, saccades occur independently of the visual input; eye guidance is propelled by strategy-based activation encoded in the oculomotor system. The current visual input has its influence at later time intervals; it changes the pattern of activation in population coding. This influences in turn saccade metrics and allows the eyes to be sent to a specific location (such as the center of words, or a position slightly left of it) without employing any specific aiming strategy. The term "strategy-based activation" here refers to the tendency for the oculomotor system to generate experience-dependent activation in the neural region that actually computes the metrics of saccadic eye movements. In reading English, readers move their eyes with an average forward saccade length of 7–8 letters. Yang and McConkie hypothesized that, for skilled readers, at the beginning of each fixation there is an anticipatory activity that is built up gradually in the neural regions to signify the preferred saccade length. Recent oculomotor research has estimated that the onset time of the anticipatory activity is around 60–70 ms after fixation onset (Dorris et al., 1999). This strategy-based activation persists throughout the fixation, and the coding of the movement itself is continuously updated. As the new information about current visual stimuli becomes available, it can be integrated into the ongoing movement computation. Any time when the threshold for movement is reached, a saccade based on the currently coded movement activity is initiated. Effectively, the C/I model is a revision of stimulus-based models. It adds the strategy-based activation to the oculomotor system to allow saccades to occur despite the lack of any useful visual segmentation. The C/I model accepts that saccade length is affected by the visual configuration or even by the saliency of individual words because of their frequency in the language; however, it rejects the notion that eye guidance is uniquely controlled by a saccade-targeting mechanism which systematically sends the eyes to a specific location in a specific word depending on visual or ongoing language processing.

The current study was an attempt to account for the frequency distribution of saccade length in both pseudo-reading and normal reading based on the C/I model. It aimed at testing the idea that strategy-based activation is responsible for eye guidance during the initial part of a fixation and that it continues to influence eye guidance even when visual input becomes available. Three predictions on the "where" decision in reading arise from the C/I model. First, saccades initiated at earlier times during fixation (before visual input is available for computing saccade length) are mostly strategy-based and have a constant length. The landing site of the resulting saccades should deviate from the visually preferred location (e.g., the center of a word) and it should be greatly influenced by launch site. Second, saccades with a later onset time should be sent closer to the preferred location in a word; these should be relatively independent of the launch site, as the location of word/letter units becomes available for the computation of saccade amplitude. However, due to the continuing existence of strategy-based activation, the landing sites may still be biased toward a preferred saccade length. Third, when late-onset saccades are not made accurately enough, they are more likely corrected with a following saccade. The correction is based on the comparison between the current landing site in a word and the visually preferred landing site in relation to that word. Early-onset saccades with the same amount of deviation would be less likely to result in correction, as they are triggered by strategy-based activation and do not initially aim at a specific location (therefore requiring no correction).

By adopting the oculomotor range error assumption, processing- and stimulus-based hypotheses also predict that the length of some saccades (most likely the early ones) is independent of the location of the target word (Engbert et al., 2006; Reichle et al., 2003). Saccade programming would favor the execution of saccades of a preferred length (Kapoula, 1985; McConkie et al., 1988; Radach & McConkie, 1998); however, the bias should not be toward the mean length of saccades usually observed in reading as proposed by the C/I model but toward the center of the range of saccade lengths executed in the experiment.

1. Study one: Pseudo-reading

In reading, the length of words varies greatly and often one or even two words are skipped. For any particular fixation, it is difficult to assert whether the eyes landed on the intended word and whether they landed at the aimed-for location in that word. When refixations on the same word occur, it is also difficult to examine whether they are oculomotor- or linguistically based. To discern the oculomotor mechanism underlying the "where" decision from other confounding factors, we conducted a pseudo-reading study designed to examine the oculomotor processes hypothesized in the C/I model. In the study, subjects were asked to read pages of multiple-line pseudo-text (homogenous X-letter strings). Participants were asked to shift their gaze horizontally along the lines (from left to right) as in normal reading and to fixate the center of each letter string (hereafter pseudo-word). Each pseudo-word consisted of 11 X letters, with either the central letter or the whole letter string being capitalized. Examples of both center-uppercased and all-uppercased text are shown in Figure 1a. The length of X-letter strings was chosen so as to create noticeable difference between the average saccade length usually observed in normal reading (7–8 letters) and the expected length of visually guided saccades in the current study (12-letter distance between the centers of adjacent pseudo-words), and hence to favor the occurrence of landing position errors and the subsequent oculomotor responses to such error. The C/I model predicts that early saccades would be biased toward an average length of 7–8 letters, while both processing- and stimulus-based hypotheses predict a bias toward the average length of 12 letters in the task, with the latter being due to systematic oculomotor range effect.

Thirty-one adults who were native English speakers with normal or corrected-to-normal vision participated in the experiment. They were seated in a quiet room with controlled fluorescent light and their eyes were about 80 cm from the screen. At this distance, there were 2.8 letters per degree of visual angle. An SR Eyelink eye-tracker was mounted onto the head of the participant and a chin-rest was used to help stabilize the head. Participants used a button box to control the progress of the experiment.

1.1. Results

All forward saccades, excluding the first and last ones on a line, were analyzed. Figure 1b shows the frequency distributions of saccade length regardless of launch distance or fixated location in a pseudo-word. It reveals that, when reading both center- and all-uppercased text, the frequency distribution of saccade length was bimodal, with the frequency of smaller saccades peaking at 3 letters and larger saccades peaking at 11 letters. This is quite different from the usual unimodal distribution observed in reading, which usually peaks at around 7- to 8-letter saccade lengths. The similarity of the distributions of saccade length in center- and all-uppercase conditions suggests that the oculomotor system does not need a well-defined visual cue to fixate at the center of pseudo-words. This could also reflect the fact that localization of the target position was not crucial in

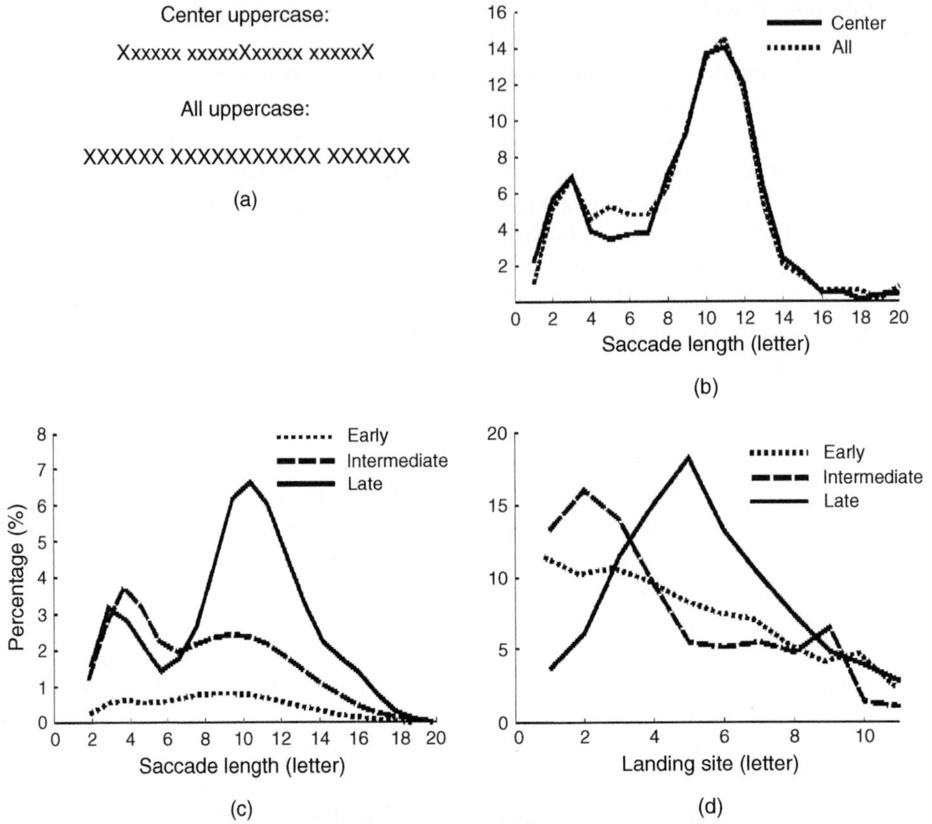

Figure 1. Frequency distributions of saccade length and landing positions in pseudo-reading. (a) Examples of pseudo-words with the central letter or all letters capitalized. (b) The overall distributions for reading text with the center letter capitalized or with all letter capitalized. (c) Distributions of saccade length for three different latency groups (early: 0–150 ms; intermediate: 175–225 ms; late: 300–400 ms). (d) Distributions of landing sites in pseudo-words for three groups of saccade latency; saccades were launched from the center of the previous pseudo-word.

the process of driving the eyes from one string to the next. In the following analysis, data from these two text conditions were pooled.

1.2. Effect of saccade latency on saccade length

The C/I model assumes that earlier saccades are driven by strategy-based activation, hence predicting that early saccades do not land at the center of pseudo-words but rather have the preferred saccade length that is usually observed in normal reading. To test this prediction, we plotted the frequency distribution of saccade length for three saccade latency groups separately (early: 0–150 ms; intermediate: 175–225 ms; late: 300–400 ms). The time intervals were selected based on previous observations suggesting that saccades made at these three time intervals are determined by different mechanisms (Yang & McConkie, 2001). The duration of the intervals was defined to make the saccade frequency of the latter two groups roughly equal (there were far fewer cases in the early saccade group). These groups were also separated from each other in time so as to enhance the distinction between the groups. Figure 1c shows the resulting distributions of saccade length for the three saccade-latency groups. In the figure, the frequency of saccade length in each time interval was not divided by the total frequency of each group but by the combined frequencies of the three groups. This allowed direct comparisons of saccade frequency for different groups at each time interval. Similar calculation was applied to other pseudo-reading figures. Figure 1c reveals that when saccades were initiated within the first 150 ms of fixation their frequency of length had a broad unimodal distribution that peaked at 7- to 8-letter saccade lengths. The distribution of late saccades showed one main peak at a length of 11 letters, and a minor peak at 3- to 4-letter lengths. The frequency of intermediate-latency saccades, meanwhile, formed a bimodal distribution, having a peak at 8- to 9-letter saccade lengths and the other at 3- to 4-letter saccade lengths respectively.

The above results show that, in reading pseudo-words, early-triggered saccades were more variable in length than later saccades, and their length did not reflect the distance between the centers of adjacent pseudo-words but was more similar to that in normal reading. Later saccades were more likely to have the predicted saccade length of 12 letters, consistent with earlier findings that late saccades are more accurate (Coëffé & O'Regan, 1987; McConkie et al., 1988; Radach & Heller, 2000). In addition, there was an overall rightward shift of saccade lengths (for the major peak of the distributions) as saccade latency increased.

To ensure that this was not due to launch site being not controlled for, the distributions of initial landing sites in pseudo-words were plotted for the three latency intervals, but only in instances where saccades were launched from the central region of the previous pseudo-word (letter positions 9–12 in the strings). This also served to exclude most occurrences of refixations within pseudo-words. As shown in Figure 1d, early saccades landed at the very beginning of the pseudo-word, or on the space in front of it, instead of its center. For saccades of an intermediate latency, the LS curve was shifted further to the right, peaking at a distance of 3–4 letters from the center of the pseudo-word. For late saccades, the LS curve peaked very close to the center of pseudo-words. Thus, the landing position distribution shifted toward the center of pseudo-words as saccade latency increased, even when launch site was controlled for and refixations were minimized. The later a saccade was made, the more accurately it landed around the center of the pseudo-word.

1.3. Corrective saccades and saccade latency

As shown in Figure 1b, there were many small-length saccades in the groups of intermediate- and late-latency saccades. Since there was no benefit from refixating pseudo-words for linguistic reasons and since there were no pseudo-words shorter than 11 letters, these small-length saccades most likely were made in order to correct "inaccurate" saccades sent to an off-center position. Previous studies have suggested that corrective saccades having a very short latency are based on corollary discharge computed in the cerebellum (Optican & Robinson, 1980; May, Hartwich-Young, Nelson, Sparks, & Porter, 1990); visually guided corrective saccades require a longer latency, usually longer than 150 ms (Vitu, McConkie, & Yang, 2003). Most likely, the small-length saccades observed in the present study were visually driven as they occurred mainly after 175 ms.

In the C/I model, visually guided corrective saccades occur following a saccade that aims at a specific location but actually fails to land at that location; thus, they are presumably more likely following long-latency saccades since these allow the encoding of the target location. In contrast, early saccades do not rely on visual input; they should not be corrected. In other words, the likelihood of corrective saccades *N* should be a function of how frequently the eyes landed away from the target location following a late saccade $N-1$. To test this prediction, we selected cases . . . where saccade *N* was preceded by either a short- (shorter than 150 ms) or long-latency (300–400 ms) saccade $N-1$ and plotted the frequency of length for the three latency groups of saccade *N*. The results revealed that the percentage of small-length saccades was greatly reduced when preceded by short-latency saccades $N-1$ (a 32% reduction for all three groups combined compared to those reported in Figure 1b). In contrast, when saccade $N-1$ had a longer latency (300–400 ms), the frequencies of small-length saccades *N* for the three latency groups of saccade *N* increased by about 24%. Thus, consistent with the C/I model, small-length saccades were more likely when the latency of the preceding saccade was longer.

These results are at odd with previous reports suggesting an increase of saccade accuracy with saccade latency in general (Coëffé & O'Regan, 1987; Viviani & Swensson, 1982). This would predict less corrective saccades *N* following lately triggered saccades $N-1$. To examine the respective contributions of latency and accuracy of saccade $N-1$ to the likelihood that saccade *N* was a corrective saccade, the percentage of small-length saccades *N* (2–5 letters long) for individual subjects was plotted against the percentage of saccades $N-1$ that landed in the central region of pseudo-words (letter positions between 9 and 12 in the strings) and were initiated at earlier (200–300 ms) or later (longer than 300 ms) time interval, as shown in Figure 2. These two time intervals were chosen to include comparable frequencies of center-fixating saccades; according to the C/I model, the earlier time interval should contain much less visually guided saccades than the later time interval. If saccades in these two latencies were both visually guided, the relation between the frequency of corrective saccades *N* and the frequency of center-fixating saccade $N-1$ should be the same regardless the latency of saccade $N-1$. Figure 2 reveals that when saccade $N-1$ had a latency longer than 300 ms, the percentage of center-fixating saccades $N-1$ was negatively correlated with the percentage of small-length saccades *N* (2- to 5-letter length), $r = .711$, $p < .0005$. In contrast, there was no significant correlation when the latency of center-fixating saccade $N-1$ was between 200 and 300 ms, $r = .152$, $p = .94$. This shows that the likelihood of corrective saccades following shorter-latency saccades $N-1$ (200–300 ms) was unrelated to the landing site of saccade $N-1$ in relation to the center of pseudo-words, whereas the likelihood of corrective saccades following long-latency saccades depended on it. Therefore, the occurrence of small-length saccades

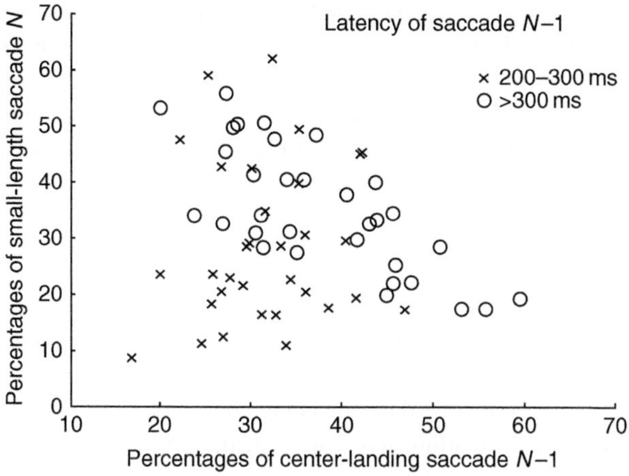

Figure 2. Correlation Diagram for the percentage of large-length saccades $N-1$ (9–12 letters) and the percentage of small-length (2–5 letters) saccades N; the latency of saccades $N-1$ was either intermediate (200–300 ms) or late (>300 ms).

in the pseudo-reading task was related to both the landing position of the preceding saccade and its latency, consistent with the prediction of the C/I model.

To sum up, the results of the pseudo-reading study indicated that, even when participants were instructed to fixate at a clearly marked location within a pseudo-word, there were many off-center saccades and visually based corrective saccades. Saccade length increased as the latency became longer, with the landing site moving from the beginning of the pseudo-word to its center, even when the saccade's launch position was controlled for (i.e., launched from the center of the previous pseudo-word). Corrective saccades were most likely when the preceding saccade was characterized with a long latency or quite likely when the preceding saccade was aimed at a given target location but actually failed to reach it.

2. Study 2: Normal reading

One would likely argue that the mechanisms that determine saccade length in pseudo-reading are quite different from those in normal reading. For instance, it is evident that the bimodal frequency distribution of saccade length in the current pseudo-reading task is different from the unimodal distribution typically observed in normal reading. However, the discrepancy could simply result from the occurrence of more corrective saccades, due to the specific instruction to fixate at the center of pseudo-words and the unusually long pseudo-words.

To examine whether the mechanisms inferred from the pseudo-reading study also determine saccade length in normal reading, a posteriori analyses of eye-movement data from a normal-reading study were conducted. These were based on a previously reported dataset (Kerr, 1992; Vitu, McConkie, Kerr, & O'Regan, 2001). In the study, four adults read long sections of normal text (Gulliver's Travels) presented one line at a time on a computer screen while their eye movements were being monitored using the Generation IV Dual Purkinje Image Eye-tracker (for further details on Materials, Procedure and Apparatus, see Vitu et al., 2001).

2.1. Results

To examine the relationship between saccade latency and saccade length, the frequency distributions of forward saccade length were plotted separately for various time intervals of the preceding fixation, but regardless of launch site and word length (see Figure 3a). In contrast with the pseudo-reading data, all distributions displayed a single peak. For saccades initiated between 100 and 250 ms, there was a gradual shift of the distributions of saccade length toward larger lengths. Beyond 250 ms, the distributions were very similar to those obtained in the 200–250 ms interval.

In Figures 3b and 3c, the distributions of the landing sites of forward saccades were plotted separately for different launch sites, and in instances the prior fixation duration was short (between 150 and 200 ms) or long (between 250 and 300 ms). This classification gave reasonable *n*'s in the different time intervals. Launch sites and landing sites were expressed relative to the beginning of Word $N+1$ that is next to Word N from which the saccade was launched. Positive values on the X-axis indicate landing sites on Word $N+1$ or on a following word while a value of zero indicates landing sites on the space between Words N and $N+1$. Negative values correspond to landing sites on Word N (hence refixations). For instance, -3 would indicate that the gaze landed at the third letter from the end of Word N. Word length was not controlled for in the analysis; however, if there was a preference for landing at the center of words, saccades launched from various distances should form distributions fairly close to each other and the main peak should be to the right of the space between Word N and Word $N+1$. As shown in Figure 3b, the landing site of early-triggered saccades was greatly affected by launch site. The distributions shifted further leftward as launch distance increased, thus suggesting little influence of word locations on saccade length. In contrast, Figure 3c shows that the distributions of late-triggered saccades overlapped to a greater extent, with the peak concentrating at about the same region of text despite different launch sites. However, for distant launch sites (greater than 6 letters), the percentage of small-length saccades increased, forming an increasingly "bimodal" distribution. Thus, the present normal-reading data are consistent with the pseudo-reading data: Up to a certain point in time, the eyes moved with saccades of relatively constant length, irrespectively of their initial distance to a given word. At later time intervals, saccades were more precisely aimed at a given location in words, although still with great variability. Furthermore, when fixation duration was long enough (250–300 ms in this case) and the launch site was more distant, small-length saccades formed a distinct distribution.

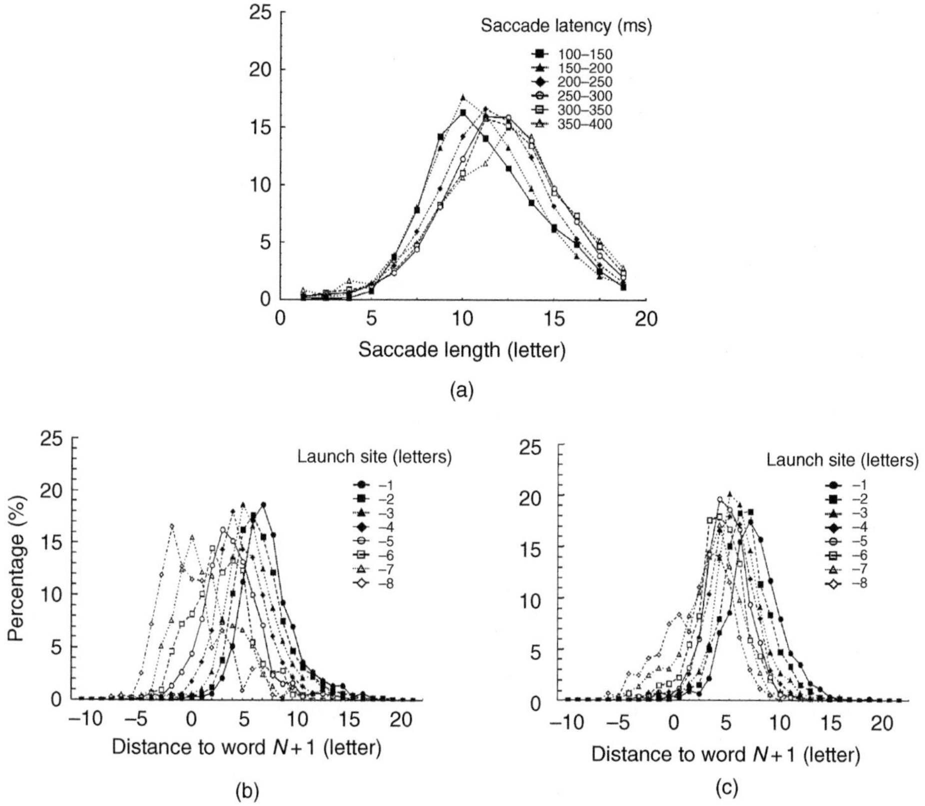

Figure 3. Distributions of saccade length and landing site for various saccade latencies or launch distances. (a) Distributions of saccade length for different saccade latency intervals. (b) Distributions of the landing sites of forward saccades characterized with a latency between 150 and 200 ms and launched from various distances. (c) Distributions of the landing sites of forward saccades characterized with a latency between 250 and 300 ms and launched from various distances. Landing sites and launch sites were expressed relative to the beginning of Word $N+1$, while the saccades were launched from Word N. Word length was not controlled.

To examine the influence of variables related to word processing on forward saccade length in normal reading, we also plotted the distribution of saccade length as a function of the length of the parafoveal word (using 3- to 8-letter words), with the frequency of the parafoveal word being controlled for. As shown in Figure 4a, for lower-frequency parafoveal words (their frequency was within the first quartile of word frequencies in the corpus), the distributions overlapped greatly across word lengths; there was only a slight shift toward larger saccade lengths for longer parafoveal words. For parafoveal words of higher frequency (within the fourth quartile of word frequencies in the corpus), the distributions for the different word lengths again overlapped, as shown in Figure 4b. Thus, saccade length was only slightly modulated by word length and word frequency.

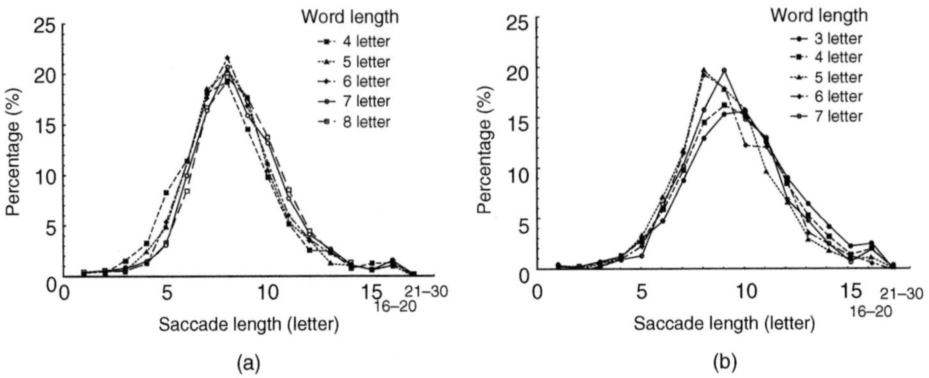

Figure 4. Distributions of saccade length in normal reading. (a) Saccades landed on low-frequency words between 4 and 8 letters. (b) Saccades landed on high-frequency words between 3 and 7 letters.

The present results showed that the landing sites of the saccades initiated between 150 and 200 ms in reading were greatly affected by factors such as launch site: Off-center landing sites were more frequent when the launch site was more distant. Later saccades displayed a smaller effect of launch site, hence suggesting that they aimed for the center of words. These results are consistent with the finding in pseudo-reading. The main difference was that small-length saccades were not as frequent. This could result from words in normal reading being on average shorter than pseudo-words (11 letters) and strategy-based preferred saccade length being more appropriate to guide the eyes to the center of words. It could also result from the fact that in normal reading fewer saccades aim at specific locations. Overall, these results are in line with the conclusions we drew from the pseudo-reading data.

3. Simulations

The following simulations attempted to account quantitatively for the frequency distributions of saccade length observed in pseudo- and normal-reading studies. In line with the C/I model, we assumed that three groups of saccades are included in the usually observed distribution of saccade length in reading: strategy-based, visually guided, and corrective saccades. To be consistent with the above text, we refer to these groups as small-length (corrective), medium-length (strategy-based) and large-length (visually guided) saccades. Frequency distributions of these groups were simulated using a Gaussian mixture model; this gave an estimation of mean and variance of each distribution as well as the percentage of the three groups of saccades. In addition, the same model was used to estimate the percentages of the three groups of saccades when initiated at different time intervals. The Gaussian distribution is expressed as in the equation below:

$$g(x|\mu, \sigma^2) = 1/(2\pi\sigma^2)^{1/2} * \exp\{-(x-\mu)^2/2\sigma^2\} \quad (1)$$

Here $g(x|\mu, \sigma^2)$ is the Gaussian distribution, with μ and σ^2 corresponding respectively to estimated mean and variance of the Gaussian distribution for the corresponding groups of saccades. The Gaussian mixture models of saccade length include nine parameters, as shown below:

$$[\theta] = \{\mu_k, \omega_k, \sigma_k \mid k = 1, \ldots, K\}, \quad K = 3, \Sigma\omega_K = 1 \tag{2}$$

Here θ represents the matrix of estimated parameters; ω indicates the weight (percentage) of Gaussian distributions for respective groups. The nine parameters were approximated using the least square method. Resulting values for estimated parameters and goodness of fits (GIFs) are shown in Table 1.

3.1. Pseudo-reading

We first approximated the frequency of each group of saccades in pseudo-reading. Parameters were estimated to fit the overall distribution of saccade length. The resulting simulated and observed frequency distributions were plotted in Figure 5a, together with the estimated distribution of each group of saccades, weighted by their percentage of frequencies. The model estimated the percentages of saccade frequencies for the three groups as 11% (small-length), 39% (medium-length) and 50% (large-length). The effect of saccade latency on the distribution of saccade length was estimated by fitting the same model to the distributions from Figure 1b; the percentages of frequency for the saccade groups in the three time intervals were estimated separately (all GIFs < 0.95). Figure 5b shows the observed and simulated frequency distributions for the three time intervals. These were obtained without changing the mean and variance estimated for the three saccade-length groups for the overall frequency distributions. The embedded figure on the right shows the percentages of saccade groups for the three time intervals. It indicates that small-length saccades were most frequent in the intermediate time interval

Table 1
Estimated parameters and the resultant goodness of fit for the Gaussian mixture models.

	σ	μ	ω	R^2	GIF
Models					
Pseudo Reading					
Small	1.02	3	.11		
Intermediate	4.10	7	.39		
Large	1.61	11	.50		
				.99	.9951
Normal Reading					
Small	1.00	3	.00		
Intermediate	2.1	8	.67		
Large	2.41	11	.33		
				.99	.9985

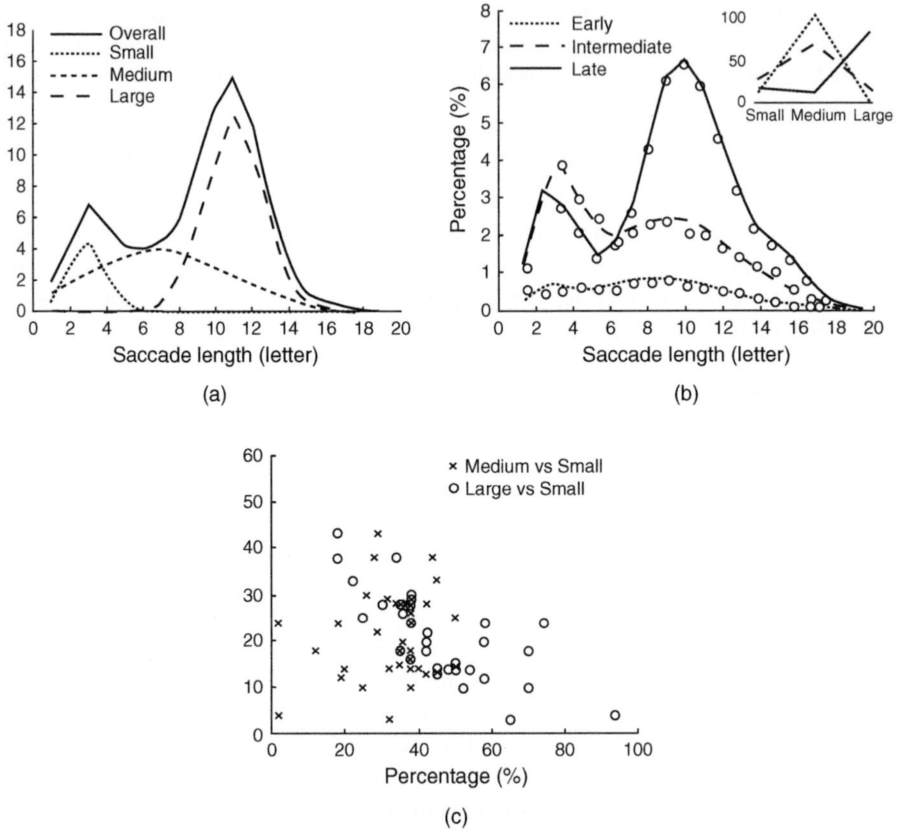

Figure 5. Observed and simulated saccade length in pseudo-reading. (a) Observed and simulated distributions of saccade length in pseudo-reading; the latter is based on the combined frequency of the three saccade-length groups (small, medium and large) and regardless of word length and launch distance. (b) Observed and simulated distributions of saccade length in pseudo-reading based on the three saccade-length groups for saccades launched at different time intervals (early, intermediate and late); the embedded graph shows the percentages of the three saccade-length groups in early, intermediate and late time intervals (marked with the same line style as in the main figure). (c) A diagram of correlation between the percentage of large-length saccades $N-1$ and small-length saccades N (circles), and between the percentage of medium-length saccades $N-1$ and small-length saccades N (Xs).

(25%); large-length saccades had the largest percentage in the late time interval (75%) and medium-length saccades had the largest percentage in the early time interval (90%). The percentage of medium-length saccades decreased as saccade latency became longer.

In the analysis of pseudo-reading data, it was shown that the percentage of small-length saccades N was negatively correlated with the percentage of center-fixating saccades $N-1$, when saccades $N-1$ were of a long latency (see Figure 2). To test whether this relationship could be simulated with the model, we estimated the percentages of the three groups of saccades for each subject. The Gaussian mixture model was used to fit the distributions of saccade length for individual subjects (all GIFs > 0.90). Figure 5c shows the resulting correlation diagrams between the percentage of large- and small-length saccades, $r_{33} = -.757$, $p < .0005$, and between the percentage of medium- and small-length saccades, $r_{33} = .253$, $p = .44$. The results are consistent with the observed data.

3.2. Normal reading

The Gaussian mixture model was also implemented to simulate the overall distribution of saccade length for the normal-reading data reported above. Resulting estimates are presented in Table 1. Note that the percentage of small-length saccades was negligible; medium-length saccades accounted for 67% of all saccades and large-length saccades accounted for only 33%. The percentage of large-length saccades was much smaller than that observed in pseudo-reading (49% vs 33%). It suggests that a greater portion of saccades in normal reading remained uninfluenced by word boundary information. This was expected as the mean latency of saccades was shorter in normal compared with pseudo-reading (219 vs 284 ms). Shorter-latency saccades were influenced to a lesser extent by the visual input.

To estimate the effect of saccade initiation times on the distribution of saccade length, we fitted the model to the observed distributions for saccades initiated at different time intervals. Figure 6a shows the estimated and observed distributions of saccade length for different time intervals. Estimated percentages of medium- and large-length saccades for different time intervals were plotted in Figure 6c. It clearly shows that, as saccade latency increased, the percentage of medium-length saccades decreased while that of large-length saccades increased.

To simulate the contribution of launch site, we approximated the percentage of the three groups of saccades for different launch sites and different time intervals. Figure 6b shows the results of observed (line symbols) and simulated (circle symbols) frequencies for different launch distances during the earlier (150–200 ms, thick lines) and later (250–300 ms, thin lines) time intervals. The model approximated the frequency distributions of length for saccades initiated at different times by alternating the percentage of the three groups of saccades. As launch site increased, the distribution of landing sites associated with short-latency saccades shifted to the left; there was no shift for saccades of longer latency. Figure 6d reveals the percentages of small-length (triangle symbols), medium-length (X symbols) and large-length (circle symbols) saccades derived from the simulated distributions shown in Figure 6b; the curves for earlier saccades (150–200 ms) were represented by the solid lines and those for later saccades (250–300 ms) were represented by the dashed lines. It reveals that for all launch sites, medium-length saccades were more numerous when the latency was shorter. The percentage of small saccades slightly increased when the launch distance increased (from 0 to 9%). When the launch site was 6-letter to the left of a word, the percentage of small-length saccades was slightly greater for saccades of longer latency (6% for early saccades and 9% for late saccades). However, when the launch site was 7 letters away from the beginning of the word (not

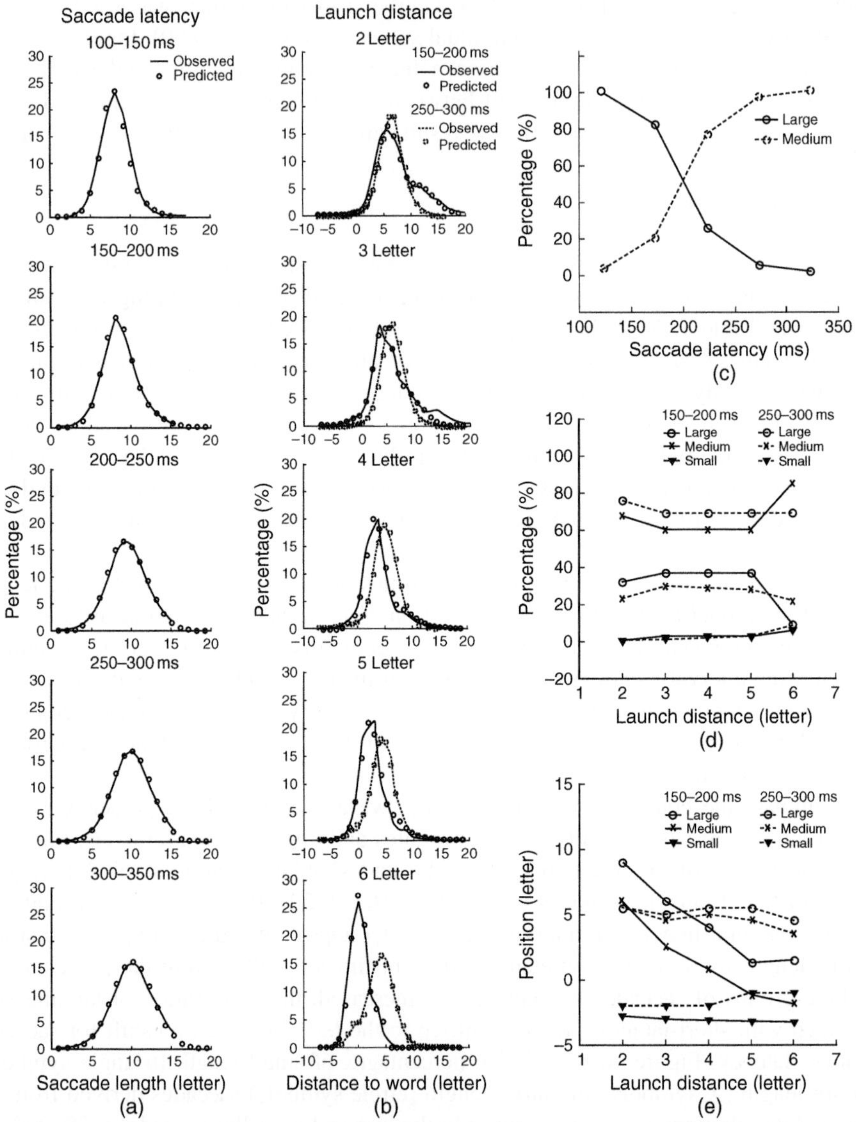

Figure 6. Observed and simulated distributions of saccade length in relation to different saccade latencies and launch sites. (a) Observed and simulated distributions of saccade length for different saccade latency intervals. (b) Observed and simulated distributions of landing sites for different launch sites, and separately for saccades initiated at earlier (150–200 ms) and later (250–300 ms) time intervals. (c) Predicted percentages of medium- and large-length saccades that were launched at different time intervals. (d) Predicted percentages of small-, medium- and large-length saccades that were launched from different distances from the beginning of Word *N* and at different time intervals. (e) Predicted landing sites (in word *N*) of small-, medium- and large-length saccades that were launched from different distances from Word *N* and at different time intervals.

shown in the figure), the estimated percentage of small saccades jumped to 21% for later saccades but only 7% for earlier saccades. The mean length of small-length saccades (3 letters) and its variance (1.00) suggests that more than half of them landed on the previously fixated word when launched from 4 or more letters away from the beginning of the next word.

Finally, Figure 6e shows mean landing sites relative to the beginning of Word $N+1$ (the word following the word the saccade was launched from) as a function of launch site (also expressed relative to the beginning of Word $N+1$). It shows that medium- and large-length saccades of short latencies landed closer to the beginning of Word $N+1$ as launch site increased. This is consistent with the prediction by the C/I model that the length of visually guided (large-length) saccades is affected by the launch site at earlier time intervals. When medium- and large-length saccades were initiated later they landed at about the same location regardless of their launch distance. Correspondingly, the landing site of small saccades that were initiated later shifted slightly to the right.

To sum up, the simulations of the distributions of saccade length in reading and pseudo-reading provided quantitative evidence that (a) the distribution of saccade length in both tasks is composed of three groups of saccades characterized with different amplitudes; (b) the three groups of saccades are initiated at different "preferred" time intervals, with large-length saccades being often characterized with longer latency and greater accuracy than medium-length saccades; (c) the effect of launch site on initial landing sites results from early-triggered saccades, and hence mainly from medium-length saccades, these being initiated at earlier times than large-length saccades. (d) the frequency of small-length saccades increases as launch distance increase; these saccades would reflect visually based correction. In normal reading, these saccades should be very rare as saccade latency usually is not long enough to induce such correction.

4. Discussion

The mechanisms underlying eye guidance in reading have been extensively studied and many theoretical and computational models have been proposed (Engbert et al., 2006; Reichle et al., 1999, 2003; Reilly & Radach, 2003, 2006). These models share the assumption that the eyes are sent to specific locations in peripherally located target words. In addition, some of them assume that selection of the saccade-target word is determined based on ongoing word identification processes (e.g., E-Z reader and SWIFT), hence suggesting that the oculomotor system is merely a subordinate to the strict control of language processes.

This chapter examined whether eye movements in reading can be accounted for by strategy- and visually based movement activity without the strict control of language processes. The analyses were based on the C/I model of eye-movement control (Yang & McConkie, 2004). The results showed that the distribution of saccade length and the landing site curve in reading can be empirically accounted for and computationally simulated based on three hypothesized groups of saccades. We have temporarily called them small-, medium- and large-length saccades. Medium-length saccades mainly occur at early time intervals and their length is independent of launch site. Large-length saccades are more greatly influenced by the visual configuration formed by words/letters and their length depends on the launch site; the percentage of these saccades increases as saccade latency becomes longer. Small-length saccades, which are initiated 150 ms or more after fixation onset, are probably of a corrective nature; their frequency increases as the launch distance becomes greater. They are more likely to be preceded by an off-center, long-latency saccade. Thus, the latency of saccade initiation affects the percentages of these three groups of saccades. Finally, word frequency and word length themselves have little influence on the distribution of saccade length.

It is important to note that visually guided corrective saccades are different from corrective saccades that have a very short latency (less than 150 ms). These short-latency corrective saccades are probably based on corollary discharge within the oculomotor system. A great percentage of them are regressive saccades (see Vitu et al., 2003). In the current study, only the frequency of forward saccade length was modeled. This explains why the current simulation can account for the frequency of saccade length adequately without accounting for the occurrence of these corrective saccades.

In the following sections, we discuss the nature of these groups of saccades, how they are related to previous models, and what they reveal about the nature of eye-movement control in reading.

4.1. Categories of saccades in reading

Several earlier computational studies suggested that saccades in reading be divided into different groups. Suppes (1989) proposed a model that describes the fixation duration distribution as a mixture of two saccade-triggering states. The system would be in one state at the beginning of each fixation and the other state would begin at some later time; both states would exist simultaneously and a saccade would be triggered in response to one of both states. A mixture parameter determines the number of saccades that are triggered from either state. McConkie and colleagues (Dyre & McConkie, 2000) proposed an alternative, sequential two-state model based on an analysis of fixation durations represented as hazard functions. This model assumes that when a fixation begins, the system is in one state, which can be conceptualized as a race between two processes, one that would trigger a saccade, and the other that would trigger a transition to the other state. The first state would represent the period prior to processing of currently available visual information in the fixation; saccades in this period could result from the processing of previous information or from random activity in the oculomotor system. The second state would represent the period after the system has begun to make use of visual information. Both computational models are able to account for the frequency distribution of saccade latency extremely well (Dyre & McConkie, 2000); however, they do not specify the nature of the mechanisms that generate these groups of saccades and they do not make specific predictions about resulting saccade lengths.

In this chapter, we were able to identify three groups of saccades and to infer the source of these saccades based on the C/I model. First, the hypothesized strategy-based activation in the C/I model can form the basis for both medium-length saccades in the current simulation and the saccades resulting from the first state in McConkie et al.'s (2000) or Suppes' (1989) model. Two properties of the medium-length saccades are consistent with the hypothesized strategy-based activation. First, the length distribution for medium-length saccades is relatively independent of location and length of the to-be-fixated word, suggesting that these saccades may not be the result of a saccade-target mechanism. Actually, medium-length saccades often failed to reach the saccade-target location in pseudo-reading. Second, the latency of these saccades tends to be short; they started occurring after fixation onset and were reduced to quite a lower frequency after the first 200 ms, which is in agreement with Dyre and McConkie's (2000) simulation of the latency for the transition from the first to the second state (202 ms). Therefore, medium-length saccades are probably initiated before detailed visual information is integrated in the coding of saccade length. Indeed, visual information about word location in reading does not seem to be available before at least 150–175 ms from fixation onset (Vitu et al., 2003; Figure 6c, current study).

Saccades influenced by visual information related to word boundaries would correspond to large-length saccades in the simulation and probably second-stage saccades in McConkie et al.'s model (1994). These are triggered at later time intervals, when visual information is available for computing the location of word/letter units (see also Coëffé & O'Regan, 1987). Large-length saccades are more likely directed at specific locations in specific words, but they are much less frequent than medium-length saccades in normal reading (only 33%) and they present substantial variability when preceded by fixation durations shorter than 300 ms (see Figures 3b–c)

Finally, corrective saccades, or the small-length saccades in the simulation, are more likely when they are preceded by an inaccurate saccade (or off-center fixation) of prolonged latency. Earlier computational analyses did not have the third group of saccades, probably because the frequency of corrective saccades in reading is so low (estimated between 0 and 9% in normal reading, depending on launch distance and latency, Figure 6d). In the pseudo-reading study, however, the lack of these small-length saccades would make the simulation inadequate (GIF = 0.74). The frequent occurrence of corrective saccades in the pseudo-word study was likely due to the fact that participants were asked to fixate specifically at the center of pseudo-words, and that pseudo-words were longer than words in reading, hence increasing the launch distance and the likelihood of landing at off-center locations.

The above categorization does not rule out the influence of language processes on eye guidance. However, as linguistic information would take as long as 250 ms to influence the triggering of forward eye movements (McConkie & Yang, 2003), the frequency of cognitively triggered saccades may be too low to require an additional state or group of saccades in computational terms. In addition, many of these saccades may be regressive ones that are not accounted for in the current model. In certain situations, such as reading incongruent sentences, the three-group model may not fit the data as well because of the need of an additional group of saccades to account for cognitive influences.

Models of eye-movement control in reading that posit word-based eye guidance may be able to account for the relationship between launch site and landing site for medium-length saccades (Engbert et al., 2006; Reichle et al., 2003; Reilly & Radach, 2003, 2006). The assumption of oculomotor range error, that oculomotor aiming errors result from a bias toward the center of the range of saccade lengths in the task, would predict a preference for the same saccade length at different launch sites (Kapoula, 1985; McConkie et al., 1988; Radach & McConkie, 1998). However, these models would have difficulty accounting for three findings in the present experiments. First, both pseudo- and normal reading favored an early bias toward saccades of about 7–8 letters (estimated mean length of medium-length saccades), whereas the length of pseudo-words was actually greater than the length of words in normal reading. This result is in contradiction with the prediction of oculomotor range error. Second, the distributions of initial landing sites in words and pseudo-words presented similar shifts over the period of a fixation, suggesting that effects of saccade latency were unrelated to ongoing word-identification processes. Third, the rather small percentage of late-latency (or large-length) saccades in reading and the small percentage of corrective saccades severely question the importance of a saccade-target mechanism in reading. Finally, it should be noted that in this chapter, we favored the C/I model and the distinction between strategy-based and visually based activation, as it accounts for the dissociation between medium- and large-length saccades or between early and late saccades. However, the present findings are also consistent with the assumption that in reading, the eyes move forward in a center-of-gravity manner without aiming for specific target locations (Vitu, 2003). As the launch site moves further away from the beginning of a word, the center of gravity of the peripheral configuration would move accordingly, shifting the distribution of initial landing sites, but producing only negligible changes on the distribution of saccade length (see also Vitu, 1991a–b). It would be only at later times, when visual information becomes more detailed and center-of-gravity influence are reduced (Coëffé & O'Regan, 1987; Findlay, 1982) that the eyes would attempt to reach predefined target locations. However, as suggested by the present data, this would occur only occasionally.

4.2. Word-based gaze guidance

As we have seen above, the term "word-based guidance" can be interpreted from a linguistic or a visual perspective and implemented based on word- or letter-units. Some may consider this chapter as arguing radically against the proposal that eye movements in reading are word-based; this would be a misunderstanding of the C/I model. What the C/I model proposes is that eye guidance is not mainly controlled by the processing of the content of a word, such as its identification; cognitive/linguistic influences may exist but they would come in late and they would be relatively infrequent. The C/I model suggests that the visual configuration of words/letters serves as the basis for visual guidance. The proposal of strategy-based activation does not exclude the use of visual information; it suggests that both processes can influence the decision of saccade length via population coding. The strategy-based activation provides a way for saccades to be triggered ahead of the processing of visual or linguistic information and when lacking useful visual segmentation.

Some recently proposed models of eye-movement control in reading adapted a more compromised view on how the content of word processing can influence the computation of saccade length. For instance, the SWIFT model added letter-based influence to its originally word-based targeting, allowing the activation level of each letter to influence the landing site. Word frequency was used to compute the activation of the word itself. In this model, the completion of word identification deactivates the word's representation, allowing more distant words to become the next saccade target. Letter-based activation determines where in a word the eyes will land. The Glenmore model proposes a similar mechanism for exerting time-related change in saccade length. These models allow the selected saccade-target location to shift gradually from left to right, as observed in this chapter. However, they cannot account for the findings reported by Yang and McConkie (2004) or Vitu, O'Regan, Inhoff and Topolski (1995) that saccade length was not greatly affected even when the visual and/or linguistic information was lacking. They may also have difficulty explaining the variations of saccade length observed in the current pseudo-reading study, as it involved no lexical processing at all.

In summary, we have shown that phenomena such as the preferred viewing position effect and the launch site effect can be accounted for by low-level visuomotor processes. These phenomena cannot be taken as direct evidence for specific strategies that aim for specific locations in words; neither are these evidence of direct linguistic control. Rather, by simply assuming the strategy-based movement activity and the influence of visual input based on the configuration of word/letter units in the peripheral area, the same patterns of landing site can be predicted. The properties of words may affect the coding of saccade length, but their effect likely is relatively small and subtle. This conclusion is not meant to exclude any influence of linguistic processing on eye guidance; cognitive eye guidance could also exist but might not be as critical as previous models have postulated in normal reading.

Acknowledgements

We would like to thank G. W. McConkie and P. Kerr for letting us use the eye movement data corpus they collected. We would also like to thank Ralf Engbert and an anonymous reviewer for their very helpful comments on an earlier version of the manuscript.

References

Brysbaert, M., Vitu, F., & Schroyens, W. (1996). The right visual field advantage and the Optimal Viewing Position Effect: On the relation between foveal and parafoveal word recognition, *Neuropsychologia, 10*, 385–395.

Coëffé, C., & O'Regan, J. K. (1987). Reducing the influence of non-target stimuli on saccade accuracy: Predictability and latency effects. *Vision Research, 27*, 227–240.

Dorris, M. C., Taylor, T. L., Klein, R. M., & Munoz, D. P. (1999). Influence of previous visual stimulus or saccade on saccadic reaction times in monkey. *Journal of Neurophysiology, 81*(5), 2429–2436.

Engbert, R., Nuthmann, A., Richter, E., & Kliegl, R. (2006). SWIFT: A dynamical model of saccade generation during reading. *Psychological Review, 112*(4), 777–813.

Findlay, J. M. (1982). Global visual processing for saccadic eye movements. *Vision Research, 22*, 1033–1045.

Inhoff, A. W., Radach, R., Eiter, B. M. & Juhasz, B. (2003). Distinct subsystems for the parafoveal processing of spatial and linguistic information during eye fixations in reading. *Quarterly Journal of Experimental Psychology, 56A(5)*, 803–827.

Jurgens, R., Becker, W., & Kornhuber, H. (1990). Natural and drug-induced variation of velocity and duration of human saccadic eye movements: Evidence for control of the neural pulse generator by local feedback. *Biological Cybernetics, 39*, 87–96.

Just, M. A., & Carpenter, P. A. (1980). A theory of reading: From eye fixations to comprehension. *Psychological Review, 87*(4), 329–354.

Kapoula, Z. (1985). Evidence for a range effect in the saccadic system. *Vision Research, 25*(8), 1155–1157.

Kerr, P. W. (1992). *Eye movement control during reading: The selection of where to send the eyes*. Unpublished doctoral dissertation (Doctoral dissertation, University of Illinois at Urbana-Champaign, 1992). Dissertation Abstracts International, #9305577.

Kliegl, R. & Engbert, R. (2003). SWIFT explorations. In J. Hyönä, R. Radach & H. Deubel (Eds.), *The mind's eyes: Cognitive and applied aspects of eye movements* (pp. 391–411). Elsevier: Oxford.

Lee, C., Rohrer, W. H., & Sparks, D. L. (1988). Population coding of saccadic eye movements by neurons in the superior colliculus. *Nature, 332*, 357–360.

May, P. J., Hartwich-Young, R., Nelson, J., Sparks, D. L. & Porter, J. D. (1990). Cerebellotectal pathways in the macaque: implications for collicular generation of saccades. *Neuroscience, 36*, 305–324.

McConkie, G. W., & Dyre, B. P. (2000). Eye fixation durations in reading: Models of frequency distributions. In A. Kennedy, D. Heller, and J. Pynte (Ed.), *Reading as a perceptual process* (pp. 683–700). Oxford: Elsevier.

McConkie, G. W., & Yang, S.-N. (2003). How cognition affects eye movements during reading. In J. Hyönä, R. Radach, and H. Deubel (Eds.), *The mind's eye: Cognitive and applied aspects of eye movement research* (pp. 413–427). Oxford, UK: Elsevier.

McConkie, G. W., Kerr, P. W., & Dyre, B. P. (1994). What are 'normal' eye movements during reading: Toward a mathematical description. In J. Ygge & Lennerstrand (Eds.), *Eye Movements in Reading* (pp. 331–343), Oxford, England: Elsevier.

McConkie, G. W., Kerr, P. W., Reddix, M. D., & Zola, D. (1988). Eye movement control during reading: I. The locations of initial eye fixations in words. *Vision Research, 28*(10), 1107–1118.

Morrison, R. E. (1984). Manipulation of stimulus onset delay in reading: Evidence for parallel programming of saccades. *Journal of Experimental Psychology: Human Perception & Performance, 10*(5), 667–682.

Nazir, T., O'Regan, J. K., & Jacobs, A. M. (1991). On words and their letters. *Bulletin of the Psychonomic Society, 29*, 171–174.

O'Regan, J. K. & Levy-Shoen, A. (1987). Eye-movement strategy and tactics in word recognition and reading. In M. Coltheart (Ed.), *Attention and performance 12: The psychology of reading* (pp. 363–383). Hove, England UK: Erlbaum.

Optican, L. M. & Robinson, D. A. (1980). Cerebellar-dependent adaptive control of primate saccadic system. *Journal of Neurophysiology, 44*, 1058–1076.

O'Regan, J. K. (1990). Eye movements and reading. In: E. Kowler (Ed.), *Eye movements and their role in visual and cognitive processes* (pp. 395–453). Oxford, UK: Elsevier.

O'Regan, J. K. (1992). Optimal viewing position in words and the strategy-tactics theory of eye movements in reading. In K. Rayner (Ed.), *Eye movements and visual cognition: Scene Perception and Reading* (pp. 333–354). New York: Springer-Verlag.

O'Regan, J. K., & Jacobs, A. (1992). Optimal viewing position effect in word recognition: A challenge to current theories. *Journal of Experimental Psychology: Human Perception and Performance, 18*(1), 185–197.

O'Regan, J. K., Lévy-Schoen, A., Pynte, J., & Brugaillère, B. (1984). Convenient fixation location within isolated words of different length and structure. *Journal of Experimental Psychology: Human Perception and Performance, 10*, 250–257.

O'Regan, J. K. (1979). Eye guidance in reading. Evidence for the linguistic control hypothesis. *Perception and Psychophysics, 25*, 501–509.

O'Regan, J. K., & Jacobs, A. (1992). Optimal viewing position effect in word recognition: A challenge to current theories. *Journal of Experimental Psychology: Human Perception and Performance, 18*(1), 185–197.

Radach, R., & Heller, D. (2000). Relations between spatial and temporal aspects of eye movement control. In A. Kennedy, R. Radach, D. Heller, & J. Pynte (Eds.), *Reading as a perceptual process* (pp. 65–192). Oxford, UK: Elsevier.

Radach, R., & McConkie, G. W. (1998). Determinants of fixation positions in words during reading. In G. Underwood (Ed.), *Eye Guidance in Reading and Scene Perception* (pp. 77–100). Oxford, UK: Elsevier.

Rayner, K. (1979). Eye guidance in reading: fixation location within words. *Perception, 8*, 21–30.

Reichle, E. D., Pollatsek, A., Fisher, D. L., & Rayner, K. (1998). Toward a model of eye movement control in reading. *Psychological Review, 105*(1), 125–157.

Reichle, E. D., Rayner, K., & Pollatsek, A. (1999). Eye movement control in reading: Accounting for initial fixation locations and refixations within the E-Z Reader model. *Vision Research, 39*(26), 4403–4411.

Reichle, E. D., Rayner, K., & Pollatsek, A. (2003). The E-Z Reader model of eye movement control in reading: comparisons to other models. *Behavioral and Brain Sciences, 26*, 445–526.

Reilly, R. & Radach, R. (2006). Some empirical tests of an interactive activation model of eye movement control in reading. *Cognitive systems research, 7*, 34–55.

Reilly, R., & O'Regan, J. K. (1998). Eye movement control in reading: A simulation of some word-targeting strategies. *Vision Research, 38*, 303–317.

Reilly, R., & Radach, R. (2003). Foundations of an interactive activation model of eye movement control in reading. In J. Hyönä, R. Radach, & H. Deubel (Eds.), *The mind's eye: Cognitive and applied aspects of eye movements*. Oxford, UK: Elsevier.

Robinson, D. A. (1975). Oculomotor control signals. In G. Lennerstrand, & P. Bach-y-Rita (Eds.), *Basic mechanisms of ocular motility and their clinical implications* (pp. 337–374). Oxford: Pergamon Press.

Suppes, P. (1989). Eye-movement models for arithmetic and reading performance. In E. Kowler (Ed.), *Eye Movements and their Role in Visual and Cognitive Processes* (pp. 455–477), Amsterdam: Elsevier.

Vitu, F. (1991a). The existence of a center of gravity effect during reading, *Vision Research, 31*(7/8), 1289–1313.

Vitu, F. (1991b). Research note: Against the existence of a range effect during reading, *Vision Research, 31*(11), 2009–2015.

Vitu, F. (2003). The basic assumptions of E-Z reader are not well-founded. *Behavioral and Brain Sciences*, 26(4), 506–507.

Vitu, F., McConkie, G. W., & Yang, S.-N. (2003). Readers' oculomotor responses to intra-saccadic text shifts. *Perception, 32*, 24.

Vitu, F., McConkie, G. W., Kerr, P., & O'Regan, J. K. (2001). Fixation location effects on fixation durations during reading: An inverted optimal viewing position effect. *Vision Research, 41*(25–26), 3511–3531.

Vitu, F., O'Regan J. K., & Mittau, M. (1990). Optimal landing position in reading isolated words and continuous text. *Perception & Psychophysics, 47*(6), 583–600.

Vitu, F., O'Regan, J. K., Inhoff, A., & Topolski, R. (1995). Mindless reading: Eye movement characteristics are similar in scanning strings and reading texts. *Perception and Psychophysics, 57*, 352–364.

Viviani, P., & Swensson, R. G. (1982). Saccadic eye movements to peripherally discriminated visual targets. *Journal of Experimental Psychology: Human Perception and Performance, 8*, 113–126.

Yang, S.-N. & McConkie, G. W. (2004). Saccade generation during reading: Are words necessary? *European Journal of Cognitive Psychology, 16*(1/2), 226–261.

Yang, S.-N., & McConkie, G. W. (2001). Eye movements during reading: a theory of saccade initiation times. *Vision Research, 41*, 3567–3585.

Chapter 14

AN ITERATIVE ALGORITHM FOR THE ESTIMATION OF THE DISTRIBUTION OF MISLOCATED FIXATIONS DURING READING

RALF ENGBERT, ANTJE NUTHMANN, and REINHOLD KLIEGL
University of Potsdam, Germany

Eye Movements: A Window on Mind and Brain
Edited by R. P. G. van Gompel, M. H. Fischer, W. S. Murray and R. L. Hill

Abstract

During reading, oculomotor errors not only produce considerable variance of within-word landing positions, but can even lead to mislocated fixations, that is fixations on unintended words. Recently, we proposed a new quantitative approach for the estimation of the proportion of mislocated fixations from experimental data (Nuthmann, Engbert, & Kliegl, 2005). Here, we present an advanced algorithm, which iteratively decomposes observed landing position distributions into mislocated and well-located contributions. The algorithm is checked with numerical simulations of the SWIFT model (Engbert, Nuthmann, Richter, & Kliegl, 2005). Finally, we outline the link between mislocated fixations and the Inverted-Optimal Viewing Position (IOVP) effect.

1. Introduction

Most research on eye movements in reading is based on the implicit assumption that every saccade lands on an intended target word. The validity of this assumption requires that oculomotor errors are small compared to the spatial extension of words. It was noticed early, however, that within-word landing distributions are rather broad and even that preferred viewing locations (Rayner, 1979) show a systematic leftward bias away from word centers. Consequently, McConkie, Kerr, Reddix, and Zola (1988), who showed in an influential paper that saccadic error can be decomposed into systematic and random contributions, suggested more than 15 years ago that "... the combination of systematic and random error in landing site distributions on a word can lead to eye fixations that were destined for one word but are actually centered on another" (p. 1117). Thus, it is surprising that no quantitative approach to the study of mislocated fixations has been proposed so far.[1] If there is a substantial contribution of mislocated fixations, this might be a challenge for theories of eye-movement control during reading, which assume that a specific target word is selected for each saccade.

Recently, we proposed a method for the quantitative estimation of the proportion of mislocated fixations during reading (Nuthmann et al., 2005). We became interested in mislocated fixations as a possible explanation of the Inverted-Optimal Viewing Position (IOVP) effect, discovered by Vitu, McConkie, Kerr, and O'Regan (2001): Fixation durations are lowest, rather than highest, near word boundaries – contrary to predictions derived from visual acuity limitations. As first noticed by McConkie et al. (1988; p. 1117), "... mislocated fixations would occur most frequently at the beginnings and ends of words." Thus, we suggested that – because of the greater frequency of mislocated fixations near word boundaries – an error-correction strategy in response to mislocated fixations which produces shorter than average saccade latencies should selectively reduce average fixation durations at the beginning and end of words. Thus, error-correction of mislocated fixations might be an important source for the IOVP effect.

In this chapter we develop an advanced algorithm for the estimation of the proportion of mislocated fixations based on an iterative approach (Section 2). Since there is currently no means to identify mislocated fixations experimentally, numerical simulations are necessary to check the algorithm. Using the latest version of the SWIFT model (Engbert et al., 2005) we can directly compare the estimated distributions of mislocated fixations with the exact distributions generated in computer simulations (Section 3). Additionally, we perform various analyses on different classes of mislocated fixations and investigate in depth how the SWIFT model responds to these types of oculomotor errors. Finally, we study the relation between mislocated fixations and the IOVP effect for fixation durations (Section 4).

[1] However, there were some speculations on the role of mislocated fixations in reading (Rayner, Warren, Juhasz, & Liversedge, 2004).

2. Estimation of mislocated fixations from data: An iterative approach

Distributions of within-word landing positions are relatively broad with a leftward shift away from word center (Rayner, 1979). The observation of these broad landing position distributions produced by saccadic errors with systematic and random components (McConkie et al., 1988) implies that a saccade can even be strongly misguided and land on an unintended word. Such an event, which might be potentially more dramatic for word processing than within-word variability in landing positions, was termed a mislocated fixation by McConkie et al. (1988). Since we do not have experimental access to the intended target word for a specific saccade, at least in continuous reading, it is generally assumed that mislocated fixations can be neglected. Recently, however, we developed a numerical algorithm for the estimation of the proportion of mislocated fixations (Nuthmann et al., 2005), which will be refined in this section.

The experimentally observed distribution of relative frequencies of landing positions represents an estimate of the probability $P_L^{\exp}(x)$ that a saccade lands on letter x of a word of length L. This probability can be decomposed into the well- and mislocated contributions,

$$P_L^{\exp}(x) = W_L(x) \cdot (1 - M_L(x)), \tag{1}$$

where $W_L(x)$ is the probability that a saccade lands on letter position x of a word of length L, *given* that the word was the intended target (i.e., well-located fixation) and $M_L(x)$ is the probability (or proportion) that a fixation on position x within a word of length L is mislocated (i.e., a different word was the intended saccade target). A common assumption underlying almost all studies on eye-movement control during reading is that mislocated fixations are rare, that is $P_L^{\exp}(x) \approx W_L(x)$ and $M_L(x) \approx 0$. We show that this assumption is invalid and that results on landing positions change strongly, when the distributions are corrected for mislocated fixations.

2.1. *Extrapolation of landing position distributions*

As a first step in deriving a constructive procedure for the distribution of mislocated fixations, we introduce the conditional probability $p(x, n|m)$ that a saccade lands on letter x of word n *given* that word m was the intended saccade target.[2] For simplicity, we assume that each word is a target once during the reading process, so that the sequence of *intended* target words is $\ldots, n-2, n-1, n, n+1, n+2, \ldots$ The two most important contributions to mislocated fixations on word n arise from *overshoot* of word $n-1$ and *undershoot* of word $n+1$. The first case will lead to an unintended skipping of word

[2] Here words n and m can be the same ($n = m$) or different words ($n \neq m$).

$n-1$ or a failed refixation of word $n-1$, while the latter case represents an unintended refixation of word n or a failed skipping of word n. Formally we can write

$$\text{overshoot}: p_{n-1}^{+}(x) = p(x, n|n-1), \tag{2a}$$

$$\text{undershoot}: p_{n+1}^{-}(x) = p(x, n|n+1). \tag{2b}$$

Using this notation, the probability for a mislocated fixation on letter position x of word n can be approximated by

$$p_n^{mis}(x) = \frac{p_{n-1}^{+}(x) + p_{n+1}^{-}(x)}{p(x, n|n) + p_{n-1}^{+}(x) + p_{n+1}^{-}(x)}, \tag{3}$$

where $p(x, n|n)$ is the well-located landing probability on word n (i.e., word n was intended and realized saccade target). From this quantity, we can compute averages over all words of length L to obtain the desired probability $M_L(x)$ for mislocated fixations on letter x of words of length L,

$$M_L(x) = \langle p_n^{mis}(x) \rangle_L, \tag{4}$$

where $\langle . \rangle_L$ denotes the average over all words of length L.

For numerical calculations, we assume that (i) physical word length L is the only property of a word which modulates its landing probability distribution and that (ii) all landing distributions can be approximated by normal distributions $N(\mu,\sigma; x)$ with mean $\mu(L)$ and standard deviation $\sigma(L)$ as a function of word length L. Using these simplifying assumptions, the probability for a well-located fixation at letter position x on word n can be written as

$$p(x, n|n) = N(\mu(L_n), \sigma(L_n); x), \tag{5}$$

where L_n is the length (i.e., number of letters) of word n. In general, we will observe a mislocated contribution to the landing distribution on word n from a target word $m \neq n$, that is

$$p(x, n|m) = N(\mu(L_m), \sigma(L_m); \Delta_{n,m}(x)), \tag{6}$$

where $\Delta_{n,m}(x)$ is a linear transformation of x to move the metric of within-word positions of word m to within-word position of word n. The overlapping contributions $p_{n-1}^{+}(x)$ and $p_{n+1}^{-}(x)$ can be estimated by extrapolation of the corresponding landing position distributions from words $n-1$ and $n+1$,

$$p_{n-1}^{+}(x) = N(\mu(L_{n-1}), \sigma(L_{n-1}); \Delta_{n,n-1}(x)) \quad \text{with } \Delta_{n,n-1}(x) = x + L_{n-1}, \tag{7a}$$

$$p_{n+1}^{-}(x) = N(\mu(L_{n+1}), \sigma(L_{n+1}); \Delta_{n,n+1}(x)) \quad \text{with } \Delta_{n,n+1}(x) = -x, \tag{7b}$$

Note that the distribution parameters μ and σ are word-length dependent. Therefore, the probability for mislocated fixations on word n can be written as

$$p_n^{mis}(x) = \frac{N(\mu(L_{n-1}), \sigma(L_{n-1}); x + L_{n-1}) + N(\mu(L_{n+1}), \sigma(L_{n+1}); -x)}{N(\mu(L_n), \sigma(L_n); x) + N(\mu(L_{n-1}), \sigma(L_{n-1}); x + L_{n-1}) + N(\mu(L_{n+1}), \sigma(L_{n+1}); -x)}, \tag{8}$$

and the average over all words of length L gives an estimate for $M_L(x)$, Equation (4).

2.2. The iterative algorithm

Using the estimation of the probability for mislocated fixations $M_L(x)$ and Equation (1), we can iteratively decompose the experimentally observed probability $P_L^{\exp}(x)$ into the probabilities for mislocated and well-located fixations (Table 1). As a starting point (see

Table 1
Iterative algorithm for the estimation of mislocated fixations.

0. Initialization ($i = 0$): Obtain estimations for $\mu^{(0)}(L)$ and $\sigma^{(0)}(L)$ based on a Gaussian fit of the distribution $W_L^{(0)}(x)$ and on the assumption that all experimentally observed fixations are well-located, i.e. $W_L^{(0)}(x) \approx P_L^{\exp}(x)$ and $M_L^{(0)}(x) \approx 0$.

1. Compute the probability distributions for mislocated fixations $p_L^{mis}(x)$ at letter position x for all words n, Eq. (8),

$$p_n^{mis}(x) = \frac{N(\mu^{(i)}(L_{n-1}), \sigma^{(i)}(L_{n-1}); x + L_{n-1}) + N(\mu^{(i)}(L_{n+1}), \sigma^{(i)}(L_{n+1}); -x)}{N(\mu^{(i)}(L_n), \sigma^{(i)}(L_n); x) + N(\mu^{(i)}(L_{n-1}), \sigma^{(i)}(L_{n-1}); x + L_{n-1}) + N(\mu^{(i)}(L_{n+1}), \sigma^{(i)}(L_{n+1}); -x)}$$

2. Compute the average distribution of mislocated fixations $M_L^{(i+1)}(x)$ for words of length L, Eq. (4),

$$M_L^{(i+1)}(x) = \langle p_n^{mis}(x) \rangle_L .$$

3. Update the distribution of well-located fixations,

$$W_L^{(i+1)}(x) = P_L^{\exp}(x) \cdot \left(1 - M_L^{(i+1)}(x)\right).$$

4. Goto to step 1 until the distribution $M_L^{(i+1)}(x)$ has converged, i.e.,

$$\sum_L \sum_x \left(M_L^{(i+1)}(x) - M_L^{(i)}(x)\right)^2 < \varepsilon \text{ with a pre-defined } \varepsilon > 0.$$

initialization in step 0 of the algorithm), we assume that all fixations are well-located (cf., Nuthmann et al., 2005). Based on this zero-order approximation, we obtain our zero-order estimate of the distribution parameters $\mu^{(0)}(L)$ and $\sigma^{(0)}(L)$ for the probability distribution of well-located fixations. Using the procedure discussed above, we compute the first estimate of the probability distribution of mislocated fixations, $M_L^{(1)}(x)$ (steps 1 and 2 of the algorithm). The fact that well- and mislocated fixations sum up to the experimentally observed distribution $P_L^{\text{exp}}(x)$ can then be exploited to calculate the first estimate of the probability distribution of well-located fixations, $W_L^{(1)}(x) = P_L^{\text{exp}}(x) \cdot (1 - M_L^{(1)}(x))$ (step 3). Obviously, the parameters $\mu^{(1)}(L)$ and $\sigma^{(1)}(L)$ obtained from fitting Gaussian distribution to $W_L^{(1)}(x)$ will deviate from the zeroth-order estimates. Applying this idea many times, we end up with an iterative estimation scheme. A self-consistency check of this approach is the asymptotic convergence of the probability distributions for well- and mislocated fixations to values $W_L^{(\infty)}(x)$ and $M_L^{(\infty)}(x)$, respectively (step 4).

2.3. *Numerical results*

After the theoretical derivation of the iterative algorithm, we carried out numerical simulations based on experimentally observed distributions of landing positions (see Nuthmann et al., 2005, for materials and methods). We computed all distributions for word lengths ranging from 3 to 8 letters. First, it is important to note that for iteration step 1 (word-based probabilities) we excluded the first and last fixations of each sentence. Second, the calculation of averages over word lengths were limited to a maximum value of a word length of 8 letters for reasons of statistical power (i.e., number of cases). We chose a value of $\varepsilon = 10^{-2}$ as a criterion of convergence (step 4), which resulted in a termination of the algorithm after only 4 iterations.[3]

A glance at the numerical results illustrated in Figure 1 indicates that our algorithm reduces the standard deviation of the landing position distributions considerably. Because mislocated fixations occur most likely near word boundaries, the removal of the contributions from mislocated fixations mainly affects the tails of the normal distributions. This effect can be derived from McConkie et al. (1988)'s argument. Overall, the probabilities of a word receiving a mislocated fixation depends strongly on word length. Probabilities for word lengths from 3 to 8 ranged from 25.3 to 5.9%. In the next section, we check the algorithm using numerical simulations of the SWIFT model.

3. Mislocated fixations: Model simulations

3.1. *The SWIFT model*

The SWIFT model (Engbert et al., 2005; see also Engbert, Longtin, & Kliegl, 2002; Engbert, Kliegl, & Longtin, 2004; Laubrock, Kliegl, & Engbert, 2006; Richter, Engbert, &

[3] A second check with $\varepsilon = 10^{-6}$ took 8 iterations.

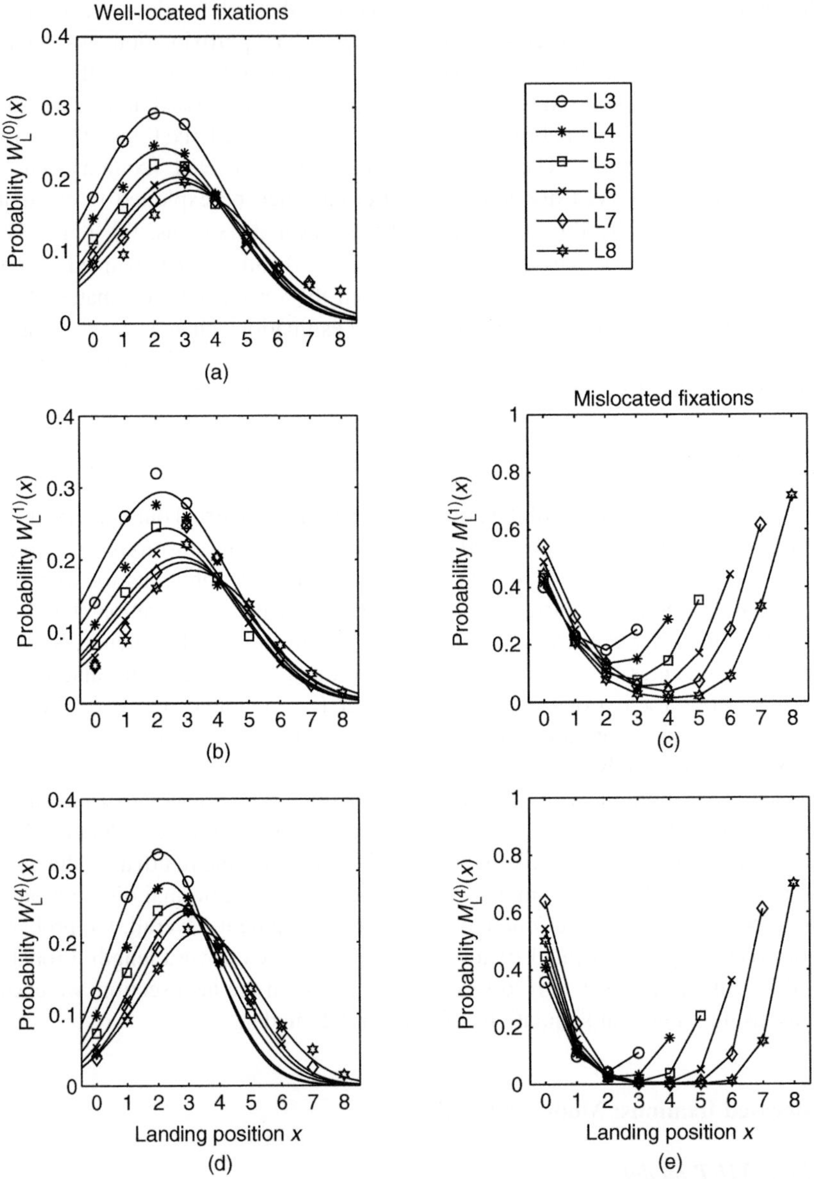

Figure 1. Iterative algorithm to estimate proportions of mislocated fixations from empirical data. Left panels show landing position distributions for well-located fixations as a function of word length. Letter 0 corresponds to the space to the left of the word. Also presented is the best-fitting normal curve for each distribution. Right panels display the proportion of mislocated fixations as a function of word length and landing position. First row of panels: zeroth iteration step; second row: first iteration step; third row: fourth iteration step where the algorithm converged (for $\varepsilon = 0.01$).

Kliegl, 2006) is currently among the most advanced models of eye-movement control, because it reproduces and explains the largest number of experimental phenomena in reading, in particular, the IOVP effect (Section 4; see also Nuthmann et al., 2005). Here, we mainly exploit the fact that the SWIFT model accurately mimics the experimentally observed landing distributions. This has been achieved by implementing both systematic (saccade range error) and random error components in saccade generation, as suggested by McConkie et al. (1988). First, we would like to underline that SWIFT operates on the concept of spatially distributed processing of words. As a consequence, mislocated fixations mainly affect the model's processing rates for particular words, but they do not have a major impact on the model's dynamics. This might be different for models based on the concept of sequential attention shifts (SAS; e.g., Engbert & Kliegl, 2001, 2003; Reichle et al., 2003; Pollatsek et al., 2006), because these models rely on processing of words in serial order. Mislocated fixations, however, are a potential source of violation of serial order. However, basic work on the coupling of attention and saccade programming (e.g., Deubel & Schneider, 1996) suggests that the locus of attention might be more stable than the saccade landing position. Thus, even in a serial model, a mislocated fixation might simply induce a situation where locus of attention and realized saccade target must be distinguished. However, we would expect processing costs in such a situation, because the intended target word must be processed from a non-optimal fixation position. Second, in SWIFT, temporal and spatial aspects of processing, that is *when* vs *where* pathways in saccade generation, are largely independent of each other. In addition to the neurophysiological plausibility of this separation (Findlay & Walker, 1999), this concept equips the model with considerable stability against oculomotor noise.

3.2. Testing the algorithm using SWIFT simulations

Different from experiments, we are always in a perfect state of knowledge during computer simulations of numerical models: For every fixation, we can decide whether the incoming saccade hit the intended or an unintended word. Therefore, model simulations are an ideal tool to investigate the problem of mislocated fixations (see also Engbert et al., 2005, for a discussion of mislocated fixations). Here, we check the accuracy of our algorithm for the estimation of the distributions of mislocated fixations. First, the model output generates distributions of within-word landing positions (Figure 2a), which can be used as input for our algorithm to construct the corresponding distributions of mislocated fixations (in exactly the same way as for experimentally observed landing position distributions). Second, from the knowledge of intended target words and mislocated fixations (for every single saccade), we can directly compute the exact distributions of mislocated fixations (Figure 2b, dotted lines).

Numerical results from SWIFT simulations demonstrate that our algorithm provides estimates for the probabilities of mislocated fixations (Figure 2b, solid lines), which nicely match the exact results (Figure 2b, dotted lines). The implications of this result are twofold. First, the iterative algorithm is a powerful tool to estimate the probabilities for

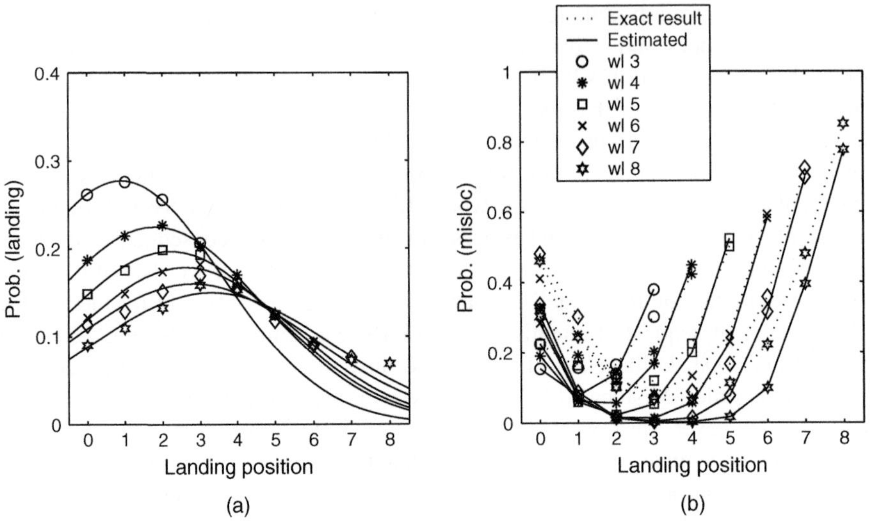

Figure 2. Testing the algorithm with SWIFT simulations. Panel (a) shows landing position distributions. Also presented is the best-fitting normal curve for each distribution. Panel (b) displays the proportion of mislocated fixations. The estimated curves are calculated from extrapolations of the distributions in (a). Exact results are directly computed from model simulations.

mislocated fixations (and well-located fixations). Second, since SWIFT also mimics the experimentally observed landing position distributions (compare Figure 1a with Figure 2a or see the discussion of within-word landing positions in Engbert et al., 2005), our results lend support to the conclusion that mislocated fixations occur frequently (up to more than 20% for short words) and play a major role in normal reading.

3.3. Exploring the magnitude of different cases of mislocated fixations

Using numerical simulations of the SWIFT model, we now investigate the prevalence of different types of mislocated fixations (Figure 3). After removing the first and last fixation in a sentence it turned out that 23.2% of all fixations generated by the model (200 realizations) were mislocated. Table 2 provides a complete synopsis of all mislocated fixation cases.

We distinguish between mislocated fixations due to an undershoot vs overshoot of the intended target word. The undershoot cases comprise failed skipping, unintended forward refixation, and undershot regression (Figure 3, cases I to III). For example, a forward refixation is unintended (case II) if the eyes actually planned to leave the launch word (n) and move to the next word ($n+1$) but instead remained on the launch word. Undershoot cases cover 69% of all mislocated fixations. Note that failed skipping (case I, 41.8%) is by far the most frequent case of mislocated fixations. Of all fixations, 3.3% are unintended forward refixations (case II). Somewhat surprisingly, the SWIFT model also produces a

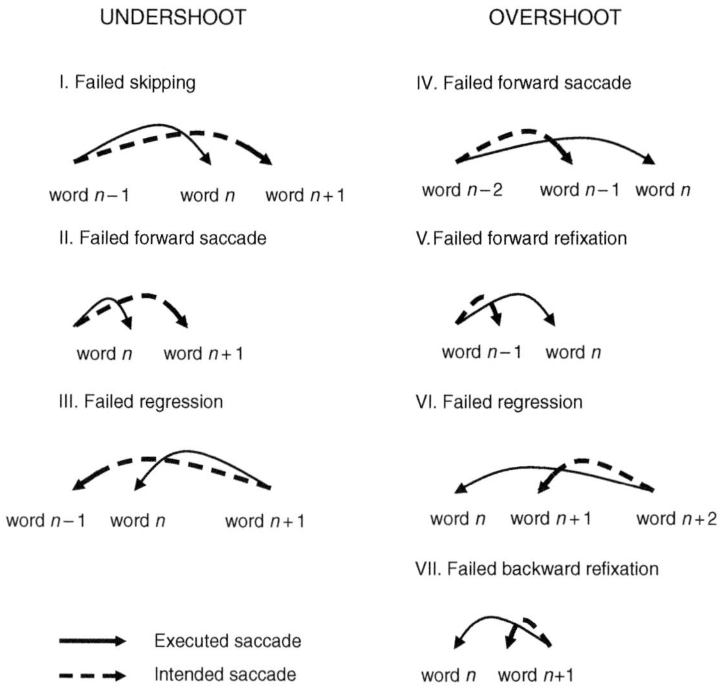

Figure 3. Most important cases of mislocated fixations, if the realized saccade target shows a maximum deviation of one word from the intended saccade target. According to the nomenclature used in the figure, word *n* is the realized target word. The dashed lines indicate the intended saccades, which are misguided due to saccadic error (solid lines represent executed saccades).

certain amount of undershot regressions (case III, 3.0% of all fixations). This effect is related to the fact that a saccade range error applies also to regressions in SWIFT. We expect, however, that this behavior of the model can easily be optimized by changing the oculomotor parameters for regressions.

In summary, the SWIFT model predicts a rather specific pattern of mislocated fixations. The overshoot cases consist of unintended skipping, failed forward refixation, overshot regression, and failed backward refixation (Figure 3, cases IV to VII). Overshot inter-word regressions (case VI) as well as failed backward refixations (case VII) hardly ever occur. However, 4.3% of all fixations are failed forward refixations (case V) while 2.3% of all fixations are unintended skippings (case IV).

3.4. Exploring responses to mislocated fixations

The exploration of mislocated fixations raises the interesting question how the eye-movement control system *responds* to a mislocated fixation. This is obviously another problem that can be studied with the help of model simulations. In the SWIFT model,

Table 2

Mislocated fixation cases.

	x/misloc	*x*/all
Undershoot		
failed one-word skipping (case Ia)	22.7%	5.3%
other types of failed skipping (case Ib)	19.1%	4.4%
failed forward saccade = unintended forward refixation (case II)	14.4%	3.3%
undershot regression (case III)	12.8%	3.0%
undershoot	68.9%	16.0%
Overshoot		
failed forward saccade = unintended skipping (case IV)	9.9%	2.3%
failed forward refixation (case V)	18.4%	4.3%
overshot regression (case VI)	2.1%	0.5%
failed backward refixation (case VII)	0.7%	0.2%
overshoot	31.1%	7.2%
all cases	100%	23.2%

Note: Percentages are provided relative to all mislocated fixations (column 2) and/or relative to all fixations (column 3); see Figure 3 for an illustration of cases.

we mainly addressed the problem of mislocated fixations because of its relation to the IOVP effect (Engbert et al., 2005; Nuthmann et al., 2005; see Section 4 in this chapter).

To establish a reference for what follows, we first consider the model's response to well-located initial fixations (Figure 4). Every word-length category is represented with two panels. The upper panel displays the corresponding *landing position distribution* while the lower panel shows the proportions of subsequent types of eye movements in *response* to well-located initial fixations as a function of landing position.

The landing position distribution panels show that the SWIFT model nicely replicates the preferred viewing position phenomenon. As for the lower panels it is important to note that, for a given landing position, the displayed data points sum up to 1. For any word length and any landing position, an inter-word forward saccade is the most frequent response after initially landing within a word. For longer words, the corresponding position-dependent curve develops an inverted *u*-shape. Thus, toward either end of the word, the frequency of responding with an inter-word forward saccade decreases, as compared to the center of the word. This behavior is clearly compensated by the *u*-shaped refixation curve which originates when forward refixations (line with circles) and backward refixations (line with squares) are jointly considered as is emphasized with the gray-shaded area under the curve. In addition, the proportion of skipping saccades strongly increases with increasing initial landing position (*-line). Interestingly, the inter-word regression response (*x*-line) appears not to be modulated by initial landing position.

How does this picture change when mislocated fixations are considered? Failed skipping is the most frequently occurring type of mislocated fixations (Table 2). In this case, the eyes undershot their intended target word. We predict that these undershoot saccades

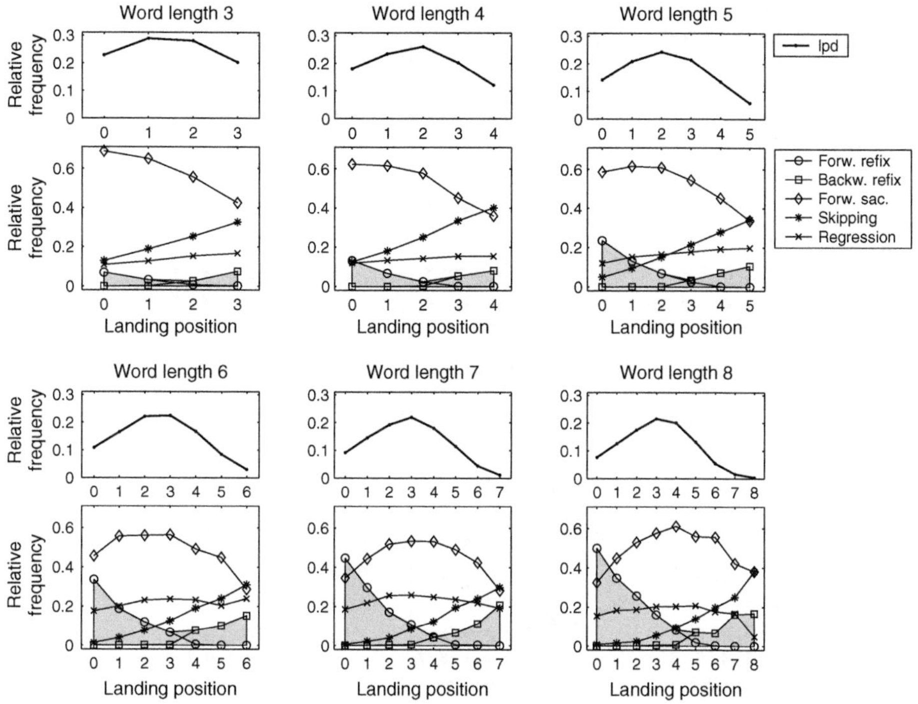

Figure 4. Simulations with the SWIFT model. Every word-length category is represented with two panels. Upper panels display landing position distributions; lower panels show the proportion of different saccade types after well-located initial fixations on a word. Shaded areas indicate *u*-shaped refixation curves.

predominantly land at the *end* of the current word (in case of an intended *one*-word skip) or on any word to the left of the intended target word (in case of an intended *multiple*-word skip).[4] Indeed, landing position distributions for failed skipping saccades are clearly shifted to the right, having their maximum at the last letter of the word (Figure 5, upper panels). Thus, these saccades predominantly land at the end of words.

Two types of saccades apparently play a major role when failed skipping occurs (Figure 5, lower panels): The SWIFT model primarily responds with a skipping of the next word(s) or simply a forward saccade to the next word. Refixations and regressions seem to play a minor role.[5] For illustration, let us consider the case of a failed one-word skipping. If the misguided saccade is followed by a saccade to the next word (one-word forward saccade), the error of initially not hitting this word is corrected.

[4] Please note that this prediction does not imply that most fixations landing at the end of a word are mislocated in the sense that they undershot the intended target word. Rather, fixations at the end of words also comprise saccades that did land on the selected target word but overshot the center of this word.

[5] Note that results for responses to landing positions at word *beginnings* are not very stable since these landing positions are infrequent.

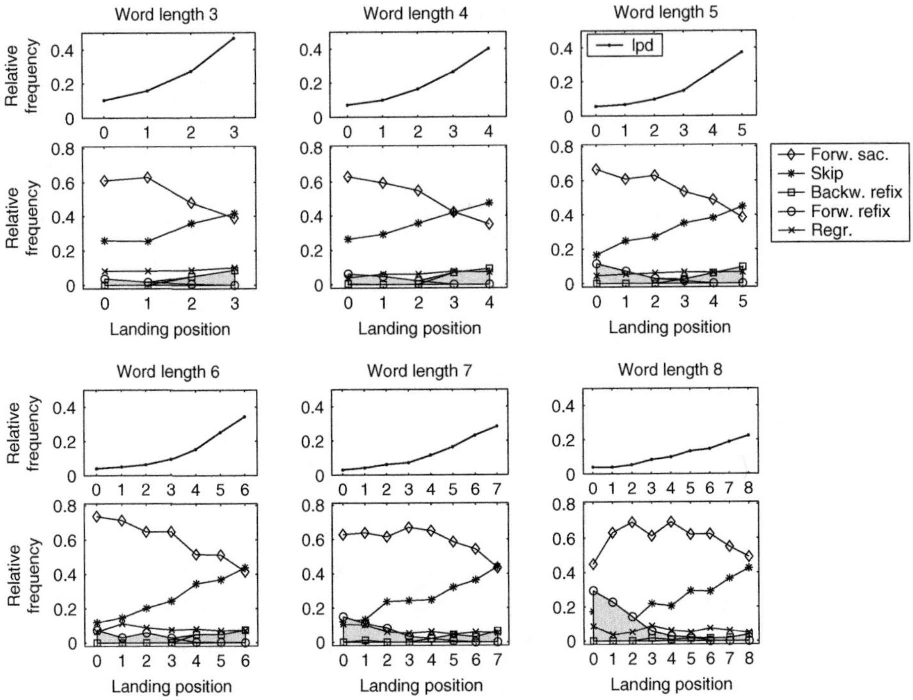

Figure 5. Simulations with the SWIFT model. Landing positions (upper panels) and proportions of subsequent saccade types after failed skipping (i.e., the most frequently occurring case of mislocated fixations; lower panels).

However, it is possible that this particular word was fully processed from the parafovea during the mislocated fixation. In this case, a corrective saccade to this word is no longer necessary. Instead, the eyes proceed with skipping the word.

4. Exploring mislocated fixations and the IOVP effect in the SWIFT model

The observation that mislocated fixations most often happen near word boundaries suggests a theoretical link to the IOVP effect: If we assume that an error-correction saccade-program is immediately started as a response to a mislocated fixation, this might reduce the average fixation durations near word boundaries (Nuthmann et al., 2005). This conjecture is valid as long as initiations of new saccade programs occur on average with a finite interval after a fixation begins. Next, we investigate this issue in the SWIFT model.

4.1. Error correction of mislocated fixations in SWIFT

In SWIFT (Engbert et al., 2005), the assumption of an immediate start of an error-correction saccade-program is implemented as Principle VI: If there is currently no labile saccade program active, a new saccade program is immediately started. The target of this saccade will be determined at the end of the labile saccade stage according to the general rule (Principle IV).

In eye-movement research in reading, fixation durations are generally interpreted in terms of latencies for subsequent saccades[6] (cf., Radach & Heller, 2000). Thus, it is assumed that the next saccade program is started close to the beginning of the current fixation. In SWIFT, however, saccade programs are generated autonomously and in parallel (see Fig. 19.2 in Kliegl & Engbert, 2003, for an illustration). This has several implications. First, different from the E-Z Reader model (Reichle et al., 2003), saccades are not triggered by a cognitive event. Second, fixation durations are basically realizations of a random variable; saccade latency is randomly sampled from a gamma distribution. However, lexical activation can delay the initiation of a saccade program via the inhibition by foveal lexical activity (Principle III). Third, saccade latency is not equal to fixation duration. In general, there are two possibilities for the initiation of a saccade program terminating the current fixation. Either the program is initiated within the current fixation or it was initiated even before the start of the current fixation. In addition, we now introduced a corrective saccade program in response to a mislocated fixation (see Engbert et al., 2005, for a discussion of the neural plausibility of an immediate triggering process for error-correcting saccades). If there is currently no labile saccade program active, a new saccade program is initiated at the beginning of the (mislocated) fixation. Thus, in case of a mislocated fixation, SWIFT's autonomous saccade timer is overruled. As a consequence, the program for the saccade terminating a mislocated fixations is initiated earlier (on average) than in the case of a well-located fixation, which leads to the reduced fixation duration for mislocated fixations. This reduction, however, is only achieved if there is a substantial proportion of saccade programs initiated after the start of the current fixation.

In Figure 6, we consider saccade programs that were executed and thus not canceled later in time. Computed is the time difference (in ms) between the start of such a saccade program and the start of the current fixation. When considering well-located fixations, we obtain a relatively broad distribution with most saccade programs initiated during the current fixation. However, the picture drastically changes for mislocated fixations. First, there are saccade programs that were initiated *before* the start of these mislocated fixations (ca. 15% of all cases, Figure 6b), reflecting the situation that there was a labile saccade program active in the moment the eyes landed on an unintended word. However, most of the time this was not the case leading to the immediate start of a new saccade program (Principle VI). Therefore, the curve presented in Figure 6a has a very pronounced peak at the start of the current fixation. As simulations suggest that mislocated fixations comprise about 23% of all fixations, this peak, though attenuated, reappears when all fixations are considered.

It is important to note that the correction mechanism cannot increase fixation duration since the corrective saccade program is only initiated if there is currently no labile saccade program active. In addition, as long as the probability for an active saccade program at

[6] Saccade latency is defined as the time needed to plan and execute a saccade.

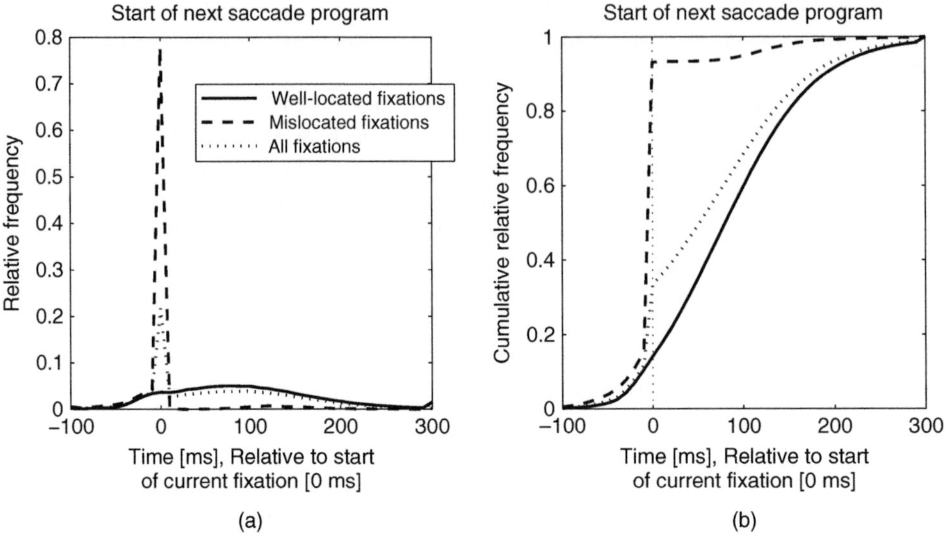

Figure 6. Start of next saccade program, relative to the start of the current fixation for well-located vs mislocated fixations. The 0 value on the x-axis marks the start of the current fixation. Thus, negative values represent saccade programs that were started before the start of the current fixation. (a) Relative frequencies. (b) Cumulative relative frequencies.

the beginning of a fixation is smaller than one (which is the case in SWIFT simulations), the mechanism can only *decrease* but not increase the mean duration of mislocated fixations.

4.2. Fixation duration IOVP effects in SWIFT

Theoretically, all types of fixations can be mislocated. The explanation of the IOVP effect suggested in Nuthmann et al. (2005) is based on the assumption that mislocated fixations (often) trigger a new saccade program immediately. Thus, we predict and observe an IOVP effect for durations of single fixations, first of multiple fixations, and second of multiple fixations (cf., Vitu et al., 2001). Our empirical estimates, however, were based on *all* fixations. Consequently, the suggested IOVP mechanism is not able to differentiate between different types of fixations (e.g., single, first, second).

Again, we employed simulations with the SWIFT model to investigate and reproduce quantitative differences between various IOVP functions. The correction mechanism for mislocated fixations, implemented as Principle VI, was able to reproduce the IOVP effect for single fixation durations (Engbert et al., 2005, Fig. 15). However, to reproduce the IOVP effect for the first of multiple fixations as well as the fixation duration trade-off effect for two-fixation cases, the model had to be furnished with Principle (VII): It is assumed that saccade latency is modulated by the amplitude of the intended saccade

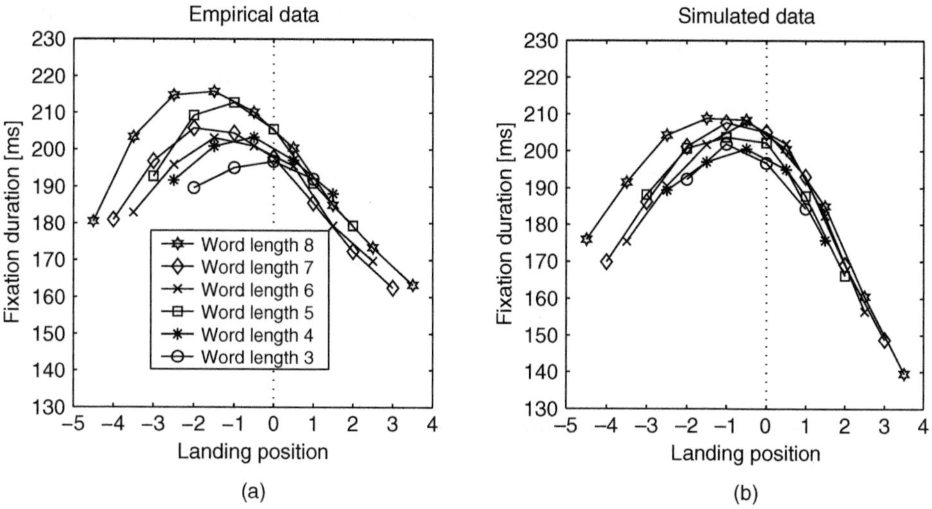

Figure 7. Mean fixation duration as a function of word length and landing position within a word, based on all fixations except first and last fixations in a sentence. Empirical data (a) vs simulated (b) data.

(see Engbert et al., 2005, Appendix B with an incremental model comparison). The joint implementation of both principles is able to reproduce the IOVP effect for all fixations (Figure 7).

5. Discussion

We investigated the prevalence of mislocated fixations during normal reading. First, using an advanced iterative algorithm, we estimated the distributions of mislocated fixations from experimental data. This procedure is a refinement of a previous approach (Nuthmann et al., 2005), which corresponds to the first-order estimation (first iteration step) of the framework presented here. We showed that our algorithm converges after only a few iteration steps and generates self-consistent estimates of the distributions of mislocated and well-located fixations – as a decomposition of the experimentally observed landing distributions. The results indicate that mislocated fixations occur frequently. In our model simulations, about 20% of all fixations turned out to be mislocated. In experimental data, estimates for mislocated fixations as a function of word length range from a few percent for long words up to a third of all fixations for short words (see also Nuthmann et al., 2005).

Second, using numerical simulations of the SWIFT model, we checked the iterative algorithm. From model simulations, we computed the exact distributions of mislocated fixations. Furthermore, the distributions of mislocated fixations were reconstructed from simulated landing distributions by our algorithm in the same way as for the experimental data. Because the exact and reconstructed distributions of mislocated fixations were in good agreement, we conclude that our iterative algorithm precisely estimates mislocated fixations from landing position data.

Third, we investigated different cases of mislocated fixations occurring in SWIFT simulations. These analyses showed that patterns of mislocated fixations are very specific, with failed skipping being a major source of errors. Finally, we outlined the relation between mislocated fixations and the IOVP effect. We conclude that mislocated fixations play an important role in eye-movement control during reading.

Acknowledgements

We would like to thank Wayne Murray and Keith Rayner for valuable comments, which helped to improve the quality of the manuscript. This research was supported by grants from Deutsche Forschungsgemeinschaft (KL 955/3 and 955/6).

References

Deubel, H., & Schneider, W. X. (1996). Saccade target selection and object recognition: Evidence for a common attentional mechanism. *Vision Research, 36*, 1827–1837.

Engbert, R., & Kliegl, R. (2001). Mathematical models of eye movements in reading: a possible role for autonomous saccades. *Biological Cybernetics, 85*, 77–87.

Engbert, R., & Kliegl, R. (2003). Noise-enhanced performance in reading. *Neurocomputing, 50*, 473–478.

Engbert, R., Kliegl, R., & Longtin, A. (2004). Complexity of eye movements in reading. *International Journal of Bifurcation and Chaos, 14*, 493–503.

Engbert, R., Longtin, A., & Kliegl, R. (2002). A dynamical model of saccade generation in reading based on spatially distributed lexical processing. *Vision Research, 42*, 621–636.

Engbert, R., Nuthmann, A., Richter, E., & Kliegl, R. (2005). SWIFT: A dynamical model of saccade generation during reading. *Psychological Review, 112*, 777–813.

Findlay, J. M., & Walker, R. (1999). A model of saccade generation based on parallel processing and competitive inhibition. *Behavioral and Brain Sciences, 22*, 661–721.

Kliegl, R., & Engbert, R. (2003). SWIFT explorations. In J. Hyönä, R. Radach, & H. Deubel (Eds.), *The mind's eyes: Cognitive and applied aspects of oculomotor research* (pp. 103–117), Oxford: Elsevier.

Laubrock, J., Kliegl, R., & Engbert, R. (2006). SWIFT explorations of age differences in reading eye movements. *Neuroscience and Biobehavioral Reviews, 30*, 872–884.

McConkie, G. W., Kerr, P. W., Reddix, M. D., & Zola, D. (1988). Eye movement control during reading: I. The location of initial eye fixations on words. *Vision Research, 28*, 245–253.

Nuthmann, A., Engbert, R., & Kliegl, R. (2005). Mislocated fixations during reading and the inverted optimal viewing position effect. *Vision Research, 45*, 2201–2217.

Pollatsek, A., Reichle, E. D., & Rayner, K. (2006). Tests of the E-Z Reader model: Exploring the interface between cognition and eye-movement control. *Cognitive Psychology, 52*, 1–56.

Radach, R., & Heller, D. (2000). Spatial and temporal aspects of eye movement control. In A. Kennedy, R. Radach, D. Heller, & J. Pynte (Eds.), *Reading as a perceptual process* (pp. 165–191). Oxford: Elsevier

Rayner, K. (1979). Eye guidance in reading: Fixation locations within words. *Perception, 8*, 21–30.

Rayner, K., Warren, T., Juhasz, B. J., & Liversedge, S. P. (2004). The effect of plausibility on eye movements in reading. *Journal of Experimental Psychology: Learning, Memory and Cognition, 30*, 1290–1301.

Reichle, E. D., Rayner, K., & Pollatsek, A. (2003). The E-Z Reader model of eye movement control in reading: Comparisons to other models. *Behavioral and Brain Sciences, 26*, 445–526.

Richter, E. M., Engbert, R., & Kliegl, R. (2006). Current advances in SWIFT. *Cognitive Systems Research, 7*, 23–33.

Vitu, F., McConkie, G. W., Kerr, P., & O'Regan, J. K. (2001). Fixation location effects on fixation durations during reading: an inverted optimal viewing position effect. *Vision Research, 41*, 3513–3533.

PART 5

EYE MOVEMENTS AND READING

Edited by

ROBIN L. HILL

Chapter 15

EYE MOVEMENTS IN READING WORDS AND SENTENCES

CHARLES CLIFTON, Jr., ADRIAN STAUB, and KEITH RAYNER
University of Massachusetts, Amherst, USA

Eye Movements: A Window on Mind and Brain
Edited by R. P. G. van Gompel, M. H. Fischer, W. S. Murray and R. L. Hill

Abstract

Word recognition processes seem to be reflected quite straightforwardly in the eye movement record. In contrast, eye movements seem to reflect sentence comprehension processes in a more varied fashion. We briefly review the major word identification factors that affect eye movements and describe the role these eye movement phenomena have played in developing theories of eye movements in reading. We tabulate and summarize 100 reports of how syntactic, semantic, pragmatic, and world-knowledge factors affect eye movements during reading in an initial attempt to identify order in how different types of challenges to comprehension are reflected in eye movements.

Readers move their eyes through a text in order to acquire information about its content. Measurements of the duration and location of the fixations they make have taught researchers a great deal about how people acquire information from the printed text, how they represent it, and how they integrate it in the course of understanding a text (see Rayner, 1978, 1998, for extensive overviews). Much of the systematic variance in fixation duration and location can be attributed to processes of recognizing the individual words in the text. Understanding of the relation between word recognition and eye movements has progressed to the point where several formal and implemented models of eye movements exist. Many of these models are described in detail, as well as compared and evaluated, by Reichle, Rayner, and Pollatsek (2003; more recent descriptions of new or updated models can be found in Engbert, Nuthmann, Richter, & Kliegl, 2005; Feng, 2006; McDonald, Carpenter, & Shillcock, 2005; Pollatsek, Reichle, & Rayner, 2006; Reichle, Pollatsek, & Rayner, 2006; Reilly & Radach, 2006; Yang, 2006). In our opinion, the most successful models are those that link the word recognition process to the time when an eye moves from one fixation to the next and the target of the saccade that accomplishes this movement. Our favored model, the E-Z Reader model (Pollatsek et al., 2006; Rayner, Ashby, Pollatsek, & Reichle, 2004; Reichle et al., 1998; Reichle et al., 1999), predicts a large proportion of the variance in eye movement measures on the basis of variables whose effect on word recognition has been independently established.

Despite their success, word recognition-based models of eye movement control do not yet provide fully satisfactory answers about all aspects of eye movements during reading. In the E-Z Reader model, a distinction made between two phases of recognizing a word (which are assumed to control different aspects of programming a saccade and the shifting of attention) has been criticized as being not fully compelling (see the replies to Reichle et al., 2003). No model fully specifies the nature of the mental representations of words (e.g., their orthographic or phonological or morphological content) nor does any model fully specify how information that specifies these different representations is acquired foveally vs parafoveally. No model fully specifies how the sequence in which orthographic symbols that appear in a printed word is mentally represented. And, even though it has been clear at least since Frazier and Rayner (1982; Rayner et al., 1983) that higher-level factors such as syntactic parsing and semantic integration can influence fixation durations and eye movements, no existing model adequately accounts for their effects.

In the first section of this chapter, we will briefly review some of the well-understood effects of word recognition on eye movements and comment on the extensions of these effects that are discussed in the chapters that appear in the Eye Movements and Reading part of this volume. In the next part, we go on to analyze the effects of syntactic, semantic, and pragmatic factors on eye movements, and discuss one basis of the difficulty of modeling, namely the apparently variable way that these factors find expression in eye movements. We begin this section with a discussion of one case study of how different measurements of eye movements can provide very different pictures of how some high-level factors influence reading and language comprehension (Clifton, 2003). We continue with an extensive survey of published articles that investigate the effects of high-level factors on eye movements, attempting to find some order in what kinds of effects appear in which measures of eye movements.

1. Word recognition and eye movements

Perhaps the two most robust findings in studies of eye movements and reading are that (1) fixation time on a word is shorter if the reader has a valid preview of the word prior to fixating it, and (2) fixation time is shorter when the word is easy to identify and understand. The chapters in this section largely deal with these two issues. Johnson's Chapter 19 provides further information regarding the specifics of preview information and demonstrates (see also Johnson, Perea, & Rayner, 2006) that transposed letters are more efficient previews than substituted letters. This result indicates that specific letter identities (probably converted into abstract letter codes) are important in preview benefit. Bertram and Hyönä's Chapter 17 deals with the extent to which morphological information from Finnish words can be processed parafoveally. Consistent with research on English (Inhoff, 1989; Kambe, 2004; Lima, 1987), they find little evidence for morphological preview benefit. Interestingly, research on Hebrew has demonstrated morphological preview benefits (Deutsch, Frost, Pollatsek, & Rayner, 2000, 2005; Deutsch, Frost, Peleg, Pollatsek, & Rayner, 2003). White's Chapter 19 deals with the effect of foveal load on skipping words. Prior research (Henderson & Ferreira, 1990; Kennison & Clifton, 1995; Schroyens, Vitu, Brysbaert, & d'Ydewalle, 1999; White, Rayner, & Liversedge, 2005) has demonstrated that preview benefit is reduced with greater foveal load. In her chapter, White shows that foveal load does not influence word skipping. The other two chapters in this section largely provide further evidence for the conclusion that difficulty of processing or accessing the meaning of a word strongly influences how long readers look at it. Juhasz (Chapter 16) deals with the effect of transparency of compound words on eye movements.

In the remainder of this section, we will briefly review findings which have demonstrated effects due to (1) word frequency, (2) word familiarity, (3) age-of-acquisition, (4) number of meanings, (5) morphology, (6) contextual constraint, and (7) plausibility. Our interest in this section is in how a word is identified as distinct from how it is integrated into the sentence that carries it. However, we recognize that this distinction between recognition and integration needs a great deal of theoretical refinement. It may prove best to recognize that in addition to factors inherent to an individual word, factors involving the word's relation to other words may affect how it is read. It may prove best to draw theoretical distinctions at points other than recognition vs integration (cf., the E-Z Reader's distinction between two stages of accessing a word; Reichle et al., 1998). At some points, we hedge our bets on whether the effect of some factor, e.g., plausibility, is best discussed in connection with word recognition or sentence integration, and discuss the data about the effect of the factor in both sections of this chapter.

We will focus on the measures most commonly used to investigate the process of identifying a word: first fixation duration (the duration of the first fixation on a word, provided that the word was not skipped), single fixation duration (the duration of fixation on a word when only one fixation is made on the word), and gaze duration (the sum of all fixations on a word prior to moving to another word). In the following section, we will concentrate on how integrating an identified word into syntactic and semantic structures affects eye movements. Since some of the factors to be discussed in the first section may affect both word identification and integration, we will revisit their effects in the second section.

Word Frequency. How long readers look at a word is clearly influenced by how frequent the word is in the language (as determined from corpus data). Rayner (1977) first anecdotally noticed that readers look longer at infrequent words than frequent words and Just and Carpenter (1980) reported a similar frequency effect via a regression analysis. However, frequency and word length are invariably confounded in natural language. Rayner and Duffy (1986) and Inhoff and Rayner (1986) therefore controlled for word length and demonstrated that there was still a strong effect of frequency on fixation times on a word. These researchers reported first fixation and gaze duration measures. The size of the frequency effect in Rayner and Duffy was 37 ms in first fixation duration and 87 ms in gaze duration; in Inhoff and Rayner it was 18 ms in first fixation duration and 34 ms in gaze duration (when the target word processing had not been restricted in any way). Since these initial reports, numerous studies have demonstrated frequency effects on the different fixation measures (see Rayner, 1998; Reichle et al., 2003 for summaries). One interesting finding is that the frequency effect is attenuated as words are repeated in a short passage (Rayner, Raney, & Pollatsek, 1995) so that by the third encounter of a high or low frequency word, there is no difference between the two. The durations of fixations on low frequency words decreases with repetition; the durations of fixations on high frequency words also decreases, but not as dramatically as for low frequency words.

Word familiarity. Although two words may have the same frequency value, they may differ in familiarity (particularly for words that are infrequent). Whereas word frequency is usually determined via corpus counts, word familiarity is determined from rating norms in which participants have to rate how familiar they are with a given word. Effects of word familiarity on fixation time (even when frequency and age-of-acquisition are statistically controlled) have been demonstrated in a number of recent studies (Chaffin, Morris, & Seely, 2001; Juhasz & Rayner, 2003; Williams & Morris, 2004).

Age-of-acquisition. Words differ not only in frequency and familiarity but also in how early in life they were acquired, and this variable influences how long it takes to process a word (Juhasz, 2005). Age-of-acquisition is determined both by corpus counts and by subjective ratings. Juhasz and Rayner (2003, 2006) demonstrated that there was an effect of age-of-acquisition above and beyond that of frequency on fixation times in reading. Indeed, in the Juhasz and Rayner studies, the effect of age-of-acquisition tended to be stronger than that of word frequency.

Number of meanings. A very interesting result is that there are clear effects of lexical ambiguity on fixation times. Rayner and Duffy (1986), Duffy, Morris, and Rayner (1988), and Rayner and Frazier (1989) first demonstrated these effects, which have subsequently been replicated a number of times (most recently by Sereno, O'Donnell, &

Rayner, 2006). The basic finding is that when a balanced ambiguous word (a word with two equally likely but unrelated meanings) is encountered in a neutral context, readers look longer at it than an unambiguous control word matched on length and frequency, whereas they do not look any longer at a biased ambiguous word (a word with one dominant meaning) in a neutral context than an unambiguous control word. In the latter case, apparently the subordinate meaning is not registered; however, if the later-encountered disambiguating information makes clear that the subordinate meaning should be instantiated, then there is considerable disruption to reading (long fixations and regressions). When the disambiguating information precedes the ambiguous word, readers do not look any longer at the balanced ambiguous word than the control word. Apparently, the context provides sufficient information for the reader to choose the contextually appropriate meaning. However, in the case of biased ambiguous words when the subordinate meaning is instantiated by the context, readers look longer at the ambiguous word than the control word. This latter effect has been termed "the subordinate bias effect." Rayner, Cook, Juhasz, and Frazier (2006) recently demonstrated that a biasing adjective preceding the target noun is sufficient to produce the effect.

An interesting study by Folk and Morris (2003) suggests, however, that effects of lexical ambiguity interact with syntactic context. Folk and Morris found that the subordinate bias effect disappears when a biased ambiguous word has one noun meaning and one verb meaning (e.g., *duck*) and only the subordinate meaning provides a syntactically legal continuation of the sentence. In a second experiment, Folk and Morris preceded balanced ambiguous words with a context that allowed a noun continuation, but not a verb continuation. They found increased reading times on target words with two noun meanings, but not on target words that were ambiguous between noun and verb meanings. A possible moral of the two experiments, taken together, is that assignment of a word's syntactic category precedes access to meaning. As a result, when a word's two meanings are associated with different syntactic categories and only one of these categories can legally continue the sentence, competition between the two meanings does not occur. It is an open question how the results obtained by Folk and Morris should be reconciled with cross-modal priming results obtained by Tanenhaus and colleagues (Seidenberg, Tanenhaus, Leiman, & Bienkowski, 1982; Tanenhaus & Donenwerth-Nolan, 1984; Tanenhaus, Leiman & Seidenberg, 1979), who reach the conclusion that syntactic context does not prevent access to inappropriate meanings. It is worth noting that the eye-tracking paradigm, due to its naturalness, may be less likely to introduce strategic effects or task demands.

The nature of the mechanisms underlying these effects is still under debate. However, the experiments described here demonstrate that, in general, the number of meanings a word has influences how long readers will look at it. Likewise, words that are phonologically ambiguous (like *tear* and *wind*) also yield differential looking times (Carpenter & Daneman, 1981), and words with two different spellings but the same pronunciation (and two different meanings, such as *beech–beach, soul–sole,* and *shoot–chute*) also produce differing fixation times (Folk, 1999; Jared, Levy, & Rayner, 1999; Rayner, Pollatsek, & Binder, 1998).

Finally, it is interesting to note that Frazier and Rayner (1987) reported that words with syntactic category ambiguity (*desert trains* can be a noun–noun compound or a noun and a verb) resulted in delayed effects in contrast to lexical ambiguity, which results in immediate effects. Pickering and Frisson (2001) likewise reported delayed effects with verbs that are ambiguous in meaning. Also, Frazier and Rayner (1990) found that in contrast to nouns with two meanings (which are typically used in lexical ambiguity studies) reading time is not slowed for words with two senses (such as the two senses of *newspaper*).

Morphological effects. Traditionally, most recent research on word recognition has dealt with rather simple mono-morphemic words. This tradition has also been largely true of research on eye movements and word recognition. More recently, however, a fair number of studies have examined processing of morphemically complex words. This newer tradition (Hyönä & Pollatsek, 1998; Pollatsek, Hyönä, & Bertram, 2000) started with the processing of Finnish words (which by their very nature tend to be long and morphologically complex). Hyönä and Pollatsek (1998) found that the frequency of the first morpheme (and, to a lesser extent, the second morpheme) in two-morpheme words influenced how long readers fixated on the word, even when the overall word frequency was controlled, implying that recognition of the word decomposed it into its component morphemes. Morphological decomposition of compound words has recently been demonstrated with English words (Andrews, Miller, & Rayner, 2004; Juhasz, Starr, Inhoff, & Placke, 2003). Pollatsek and Hyönä (2005) recently demonstrated that transparency had no effect on fixation times on morphologically complex words. In her chapter in the present volume, Juhasz did find a main effect of transparency in gaze durations. However, both semantically transparent and opaque compound words also exhibited morphological decomposition supporting Pollatsek and Hyönä's main conclusion that both types of compounds are decomposed during word recognition.

Contextual constraint. Like word familiarity, word predictability is determined via norming studies (after experimenters have prepared sentence contexts such that certain target words are either predictable or unpredictable from the context). Cloze scores are then used to confirm the experimenter's intuitions as to how constrained a word is by the context. Considerable research has demonstrated that words that are predictable from the preceding context are looked at for less time than words that are not predictable. This result was first demonstrated by Ehrlich and Rayner (1981) and was confirmed a number of times, most notably by Balota, Pollatsek, and Rayner (1985) and Rayner and Well (1996), and most recently by Rayner, Ashby et al. (2004) and Ashby, Rayner, and Clifton (2005). Not only are fixation time measures shorter on high predictable words than low predictable words, readers also skip over high predictable words more frequently than low predictable words (Ehrlich & Rayner, 1981; Rayner & Well, 1996).

Plausibility effects. Although plausibility clearly affects sentence interpretation and integration, we discuss it here because it may also affect word recognition. Several studies have examined whether manipulations of plausibility or anomaly have effects on eye movements that are immediate enough to suggest that the manipulations may affect word recognition (Murray & Rowan, 1998; Ni, Crain, & Shankweiler, 1996; Ni, Fodor, Crain, &

Shankweiler, 1998; Rayner, Warren, Juhasz, & Liversedge, 2004). We discuss Rayner et al. in some detail in the second section of this chapter. Briefly, they showed that an outright anomaly (e.g., *John used a pump to inflate the large carrots . . .*) affected time to read the critical word (*carrots*). However, the effect did not appear on the first fixation measure, which is ordinarily sensitive to word recognition difficulty, but only on gaze duration. A simple implausibility (*. . . used an axe to chop the large carrots . . .*) only affected the go-past measure (described in the next section of this chapter as an arguably-late measure) and gaze duration on the word following the critical word, suggesting that its effects are limited to processes of integrating the implausible word into the sentence context.

Interim summary. Up to this point, we have reviewed some basic findings of how certain variables arguably related to word recognition mechanisms manifest themselves in the eye movement record. In general, the primary assumption is that lexical factors play a large role in influencing when the eyes move, and these effects appear in first fixation and first pass measures. And, as we noted earlier, the most successful models of eye movement control are based on the premise that how long readers look at a word is influenced by the ease or difficulty associated with accessing the meaning of the word. Factors that presumably affect word recognition are currently utilized in the models, including our favored E-Z Reader model (Reichle et al., 1998) to predict fixation times. These factors include word frequency, morphological complexity (Pollatsek, Reichle, & Rayner, 2003), and number of meanings (Reichle, Pollatsek, & Rayner, this volume). We ended this section by suggesting that two higher-order "relational" factors (contextual constraint and plausibility) may affect word recognition under some conditions, e.g., when their operation can be triggered before the target word is fixated (as in a predictable word) or when their manipulation is strong enough (as in anomaly). We turn now to the more difficult issues of the effect of high-order factors on eye movements.

2. Effects of syntactic, semantic, and pragmatic factors

While single fixation, first fixation, and gaze duration are the measures of choice for studying the time course of word identification, a wider variety of measures is commonly used in measuring how factors that guide integration of text affect eye movements. For the most part, authors of the experiments that we will discuss in this section identify critical regions of text, sometimes consisting of as many as three or four words (occasionally even more), and then examine how long it takes readers to read the regions of interest. The standard measures are: first pass reading time (the sum of all fixations in a region from first entering the region until leaving the region, given that the region was fixated at least once), go-past or regression path duration (the sum of all fixations in a region from first entering the region until moving to the right of the region; fixations made during any regressions to earlier parts of the sentence before moving past the right boundary of the region are thus included in this measure, again given that the region was fixated), regressions-out (the probability of regressing out a region, generally limited to the first pass reading of that region), second pass reading time (the sum of all fixations in a region following the initial first pass time, including zero times when a region is not refixated), and total reading time (the sum of all fixations in a region, both forward and regressive movements, again given that the region was fixated). First fixation durations are also sometimes reported, especially when the disambiguating region is short or when the researcher is interested in spillover effects from the previous region, but when regions are long and the disambiguating material is not likely to be included in the initial fixation, the first fixation measure is inappropriate. Measures such as first pass time (and first fixation time) are often referred to as "early" measures; measures such as second pass time (and total time, to the extent that it reflects second pass time rather than first pass time) are referred to as "late" measures (Rayner, Sereno, Morris, Schmauder, & Clifton, 1989). The go-past and regressions out measures have been described as both early and late measures. The occurrence of a regression reflects some difficulty in integrating a word when it is fixated, arguably an early effect. The go-past measure reflects this effect, and also the cost of overcoming this difficulty, which may well occur late in processing. The terms "early" and "late" may be misleading, if they are taken to line up directly with first-stage vs second-stage processes that are assumed in some models of sentence comprehension (Frazier, 1987; Rayner et al., 1983). Nonetheless, careful examination of when effects appear may be able to shed some light on the underlying processes. Effects that appear only in the "late" measures are in fact unlikely to directly reflect first-stage processes; effects that appear in the "early" measures may reflect processes that occur in the initial stages of sentence processing, at least if the measures have enough temporal resolving power to discriminate among distinct, fast-acting, processes.

As argued in the first section of this chapter, a clear, if incomplete, picture seems to be developing about how lexical factors control eye movements. The same is not true about high-level factors. The earliest eye-movement research on such factors (Frazier & Rayner, 1982) held promise that syntactic factors would have sharp and understandable influences on eye movements. Frazier and Rayner examined the reading of sentences like (1) and (2) given below, and found that the very first fixations on the disambiguating region (presented in bold face in the examples) were slowed, compared to earlier fixations, when they resolved a temporary ambiguity in favor of the theoretically unpreferred reading (in 4, when *this* was absent). This disruption persisted through the next several fixations, and also appeared as an increased frequency of regressions. Eye movements appeared to provide a clear window onto syntactic "garden-pathing" (Bever, 1970).

(1) Since Jay always jogs a mile and a half (this) **seems like** a very short distance to him.

(2) (The lawyers think his/His) second wife will claim the entire family inheritance (./**belongs to** her.)

This early research was open to some criticisms. The disruption in (1) appeared in a region that followed the absence of an arguably obligatory comma (or prosodic break); the disruption in (2) appeared in a sentence-continuation that had no counterpart in the non-disruptive control condition. But the force of the missing-comma criticism is compromised by the fact that an equally obligatory comma is missing in the control condition, with no effect on reading times, and the lack of a closely matched control in (2) has been corrected in later research (Rayner & Frazier, 1987).

On balance, it appeared that syntactic processing difficulty could be identified by quickly appearing disruptions in the eyetracking record. Rayner et al. (1983) provided evidence for a similar conclusion when an initial syntactic misanalysis is signaled by a semantic anomaly. They found increased first pass reading times for sentences like (3b), where the first noun is semantically anomalous under the presumably preferred initial analysis, compared to sentences like (3a). The effect appeared in the initial fixations in the disambiguating region, where it was significant when averaged over the first three fixations, and apparent on the first fixation (and significantly longer than the previous fixation).

(3a) The kid hit the girl with a **whip before he got off the subway.**

(3b) The kid hit the girl with a **wart before he got off the subway**.

Later research, unfortunately, has not always demonstrated such clear, immediate, and regular effects of syntactic and semantic factors on eye movements. We will briefly describe one example of how a manipulation of syntactic and semantic factors can have apparently very different results, depending on what eye movement measures one looks at (this analysis was presented by Clifton, 2003).

An early demonstration of syntactic effects on eye movements was presented by Ferreira and Clifton (1986), who showed disruption in the disambiguating region of sentences like (4) when they were temporarily ambiguous (when the *who/that was* phrase was absent) compared to when they were not ambiguous (when the *who/that was* phrase was present). The effect appeared both when the initial noun was animate (4a) and when it was inanimate (4b) and implausible as the subject of the following verb.

(4a) The defendant (who was) examined **by the lawyer** proved to be unreliable.

(4b) The evidence (that was) examined **by the lawyer** proved to be unreliable.

The disruption appeared in first pass reading time measures, and was taken to show that the semantic implausibility of the presumably preferred main clause analysis in (4b) did not override initial syntactic parsing preferences. This conclusion was challenged by Trueswell, Tanenhaus, and Garnsey (1994), who argued that some of the Ferreira and Clifton items that were claimed to semantically block the preferred main clause reading did not do so. Trueswell et al. prepared two more adequate sets of materials, carefully normed, and showed that any effect of ambiguity on first pass reading time was nonsignificant (nearly zero, in one experiment) in materials like (4b), where semantic preferences weighed against the main clause analysis. They concluded that their experiment did demonstrate that semantic factors could overturn syntactic preferences, favoring an interactive, constraint-satisfaction model over the modular serial model favored by Ferreira and Clifton.

Clifton et al. (2003) revisited the question, using materials taken from Trueswell et al. (1994). In two experiments they varied parafoveal preview of the disambiguating information and examined the effects of participants' reading span. Abstracting from these factors (which for the most part did not affect the magnitude of the disruption triggered by a temporary ambiguity), the first pass time measures were similar to those reported by Trueswell et al. (1994). Semantic biases reduced the first pass reading time measure of the temporary ambiguity effect to nonsignificance in sentences like (4b) (although, similar to the findings of Trueswell et al., the interaction of semantic bias and temporary ambiguity was not fully significant, and, unlike the findings of Trueswell et al. the ambiguity effect did not go fully to zero). However, a very different pattern of results was observed for the go-past (or regression path duration) and proportion of first pass regressions-out measures (Figure 1). These measures showed disruptive effects of temporary ambiguity that were as large in semantically biased, inanimate-subject sentences like (4b) as in animate-subject sentences like (4a) where no semantic bias worked against the presumed preference for a main clause analysis.

Clifton et al. (2003) concluded that a full examination of the eye movement record indicated that initial syntactic parsing preferences were not overcome by semantic biases, although such biases clearly affected overall comprehension difficulty for both temporarily

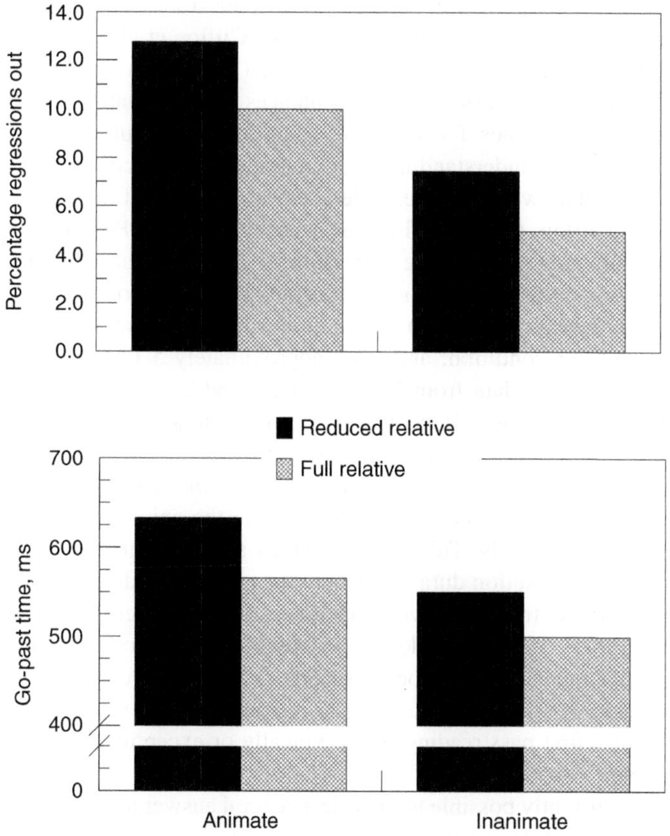

Figure 1. Regression-out probability and Go-past time, Disambiguating region; Data from Clifton et al., 2003, Pooled over Experiments 1 and 2.

ambiguous and unambiguous sentences. However, this conclusion does leave several salient questions unanswered. The first is, why did Ferreira and Clifton (1986) find first pass time garden-path effects for both animate and inanimate-subject sentences while later research found nonsignificant first pass time effects for inanimate-subject sentences? Perhaps their sentences were inadequately controlled, as suggested by Trueswell et al. (1994). Examination of the effects for individual items in the Ferreira and Clifton data, however, does not support this claim: First pass effects were observed both for items that Trueswell et al. later found to be acceptable and for inadequately biased items. A more likely cause is that Ferreira and Clifton used a display that presented only 40 characters on a row, frequently necessitating a line break before the beginning of the disambiguating by-phrase region. This would have prevented parafoveal preview of the by-phrase (although we note that absence of parafoveal preview in the boundary-change conditions of Clifton et al. (2003) did not affect the size of the ambiguity effect), it could have encouraged a commitment to the apparent subject-verb structure of the material on the first line, and it could have discouraged regressions from the disambiguating region (which would have had to cross lines of text, unlike Clifton et al. (2003)).

A second question is, why in the Clifton et al. (2003) data did significant garden-path effects appear in first pass times for sentences with animate subjects but only in regressions and go-past times for sentences with inanimate subjects? Answering this question requires a better understanding of the relation between comprehension difficulty and eye movements than we now have. A detailed examination of the Clifton et al. (2003) data (reported by Clifton, (2003)) did not answer the question. Perhaps the most salient result of this examination is that while regression frequency increased in the syntactically ambiguous conditions, regressions from the disambiguating region were quite infrequent and the increase in regression frequency was quite small (from approximately 10 to 13% for the animate-subject condition, and from approximately 5 to 8% for the inanimate-subject condition, pooling data from Experiments 1 and 2; see Figure 1). The increase in the size of the garden-path effect in the inanimate-subject condition from first pass to go-past times thus has to be attributed to eye movement events that take place on a very small minority of the trials. It is even possible that first pass fixation durations may have been increased by temporary ambiguity, even in the animate-subject condition, on only a small minority of trials. This would contrast sharply with what is true of effects of lexical frequency on fixation duration, where the entire time distribution appears to be shifted upwards for low frequency words (Rayner, 1995; Rayner, Liversedge, White, & Vergilino-Perez, 2003). To our knowledge, no existing research on syntactic garden-paths provides data on a large enough number of sentences to permit a convincing distributional analysis to be made. It remains a challenge to researchers to devise a way of asking the question of whether first pass reading times typically or exceptionally increase upon the resolution of a garden-path.

Even if it is not currently possible to provide a general answer to the question of whether syntactic (and other high-level) factors affect eye movements on many or on few trials, it may be possible to make some progress toward understanding how high-level factors affect eye movements by examining the existing literature (see Boland, 2004, for related discussion). As suggested above, some of the early research indicated that syntactic or semantic anomaly slowed eye movements essentially immediately. Other, more recent, research suggests that under some conditions, such anomalies may trigger regressive eye movements rather than affecting fixation durations. Still other research suggests that effects of anomaly may in some instances appear only later in the eye movement record. Given the frequently stated desire to use eye movements to make inferences about the immediacy of various levels of processing in language comprehension (Rayner, Sereno, Morris, Schmauder, & Clifton, 1989; Rayner & Sereno, 1994), we believe it may be useful to take stock of just when and how a wide variety of high-level factors impact the eye movement record.

Survey of eyetracking articles. We identified 100 articles that used eye movements to explore the effects of syntactic, semantic, and pragmatic factors on sentence comprehension (listed in Table 1). We attempted to include all such articles that had been

Table 1
100 Articles on the effect of higher-order processes on eye movements

1. Adams, B. C., Clifton, C., Jr., & Mitchell, D. C. (1998). Lexical guidance in sentence processing? *Psychonomic Bulletin & Review, 5*, 265–270.
2. Altmann, G. T. M., Garnham, A., & Dennis, Y. (1992). Avoiding the garden path: Eye movements in context. *Journal of Memory and Language, 31*, 685–712.
3. Altmann, G. T. M., Garnham, A., & Henstra, J. A. (1994). Effects of syntax in human sentence parsing: Evidence against a structure-based proposal mechanism. *Journal of Experimental Psychology: Learning, Memory, and Cognition, 20*, 209–216.
4. Altmann, G. T. M., van Nice, K. Y., Garnham, A., & Henstra, J.-A. (1998). Late closure in context. *Journal of Memory and Language, 38*, 459–484.
5. Ashby, J., Rayner, K., & Clifton, C. J. (2005). Eye movements of highly skilled and average readers: Differential effects of frequency and predictability. *Quarterly Journal of Experimental Psychology, 58A*, 1065–1086.
6. Binder, K., Duffy, S., & Rayner, K. (2001). The effects of thematic fit and discourse context on syntactic ambiguity resolution. *Journal of Memory and Language, 44*, 297–324.
7. Birch, S., & Rayner, K. (1997). Linguistic focus affects eye movements during reading. *Memory & Cognition,* 25, 653-660.
8. Boland, J. E., & Blodgett, A. (2001). Understanding the constraints on syntactic generation: Lexical bias and discourse congruency effects on eye movements. *Journal of Memory and Language, 45*, 391–411.
9. Braze, D., Shankweiler, D., Ni, W., & Palumbo, L. C. (2002). Readers' eye movements distinguish anomalies of form and content. *Journal of Psycholinguistic Research, 31*, 25–44.
10. Britt, M. A., Perfetti, C. A., Garrod, S., & Rayner, K. (1992). Parsing in discourse: Context effects and their limits. *Journal of Memory and Language, 31*, 293–314.
11. Brysbaert, M., & Mitchell, D. C. (1996). Modifier attachment in sentence parsing: Evidence from Dutch. *Quarterly Journal of Experimental Psychology, 49A*, 664–695.
12. Carreiras, M., & Clifton, C. Jr. (1999). Another word on parsing relative clauses: Eyetracking evidence from Spanish and English. *Memory and Cognition, 27*, 826–833.
13. Clifton, C. (1993). Thematic roles in sentence parsing. *Canadian Journal of Experimental Psychology, 47*, 222–246.
14. Clifton, C., Jr., Speer, S., & Abney, S. (1991). Parsing arguments: Phrase structure and argument structure as determinants of initial parsing decisions. *Journal of Memory and Language, 30*, 251–271.

(*continued on next page*)

Table 1

(Continued)

15. Clifton, C., Jr., Traxler, M., Mohamed, M. T., Williams, R. S., Morris, R. K., & Rayner, K. (2003). The use of thematic role information in parsing: Syntactic processing autonomy revisited. *Journal of Memory and Language, 49*, 317–334.
16. Desmet, T., & Gibson, E. (2003). Disambiguation preferences and corpus frequencies in noun phrase conjunction. *Journal of Memory and Language, 49*, 353–374.
17. Deutsch, A., & Bentin, S. (2001). Syntactic and semantic factors in processing gender agreement in Hebrew: Evidence from ERPs and eye movements. *Journal of Memory and Language, 45*, 200–224.
18. Ehrlich, S. F., & Rayner, K. (1981). Contextual effects on word perception and eye movements during reading. *Journal of Verbal Learning and verbal Behavior, 20*, 641–655.
19. Ferreira, F., & Clifton, C. (1986). The independence of syntactic processing. *Journal of Memory and Language, 25*, 348–368.
20. Ferreira, F., & Henderson, J. M. (1990). Use of verb information in syntactic parsing: Evidence from eye movements and word-by-word self-paced reading. *Journal of Experimental Psychology: Learning, Memory, and Cognition, 16*, 555–568.
21. Ferreira, F., & Henderson, J. M. (1993). Reading processes during syntactic analysis and reanalysis. *Canadian Journal of Experimental Psychology, 47*, 247–275.
22. Ferreira, F., & McClure, K. K. (1997). Parsing of garden-path sentences with reciprocal verbs. *Language & Cognitive Processes, 12*, 273–306.
23. Filik, R., Paterson, K. B., & Liversedge, S. P. (2004). Processing doubly quanitified sentences: Evidence from eye movements. *Psychonomic Bulletin & Review, 11*, 953–959.
24. Frazier, L., & Clifton, C., Jr. (1998). Comprehension of sluiced sentences. *Language and Cognitive Processes, 13*, 499–520.
25. Frazier, L., Clifton, C. J., Rayner, K., Deevy, P., Koh, S., & Bader, M. (2005). Interface problems: Structural constraints on interpretation. *Journal of Psycholinguistic Research, 34*, 201–231.
26. Frazier, L., Munn, A., & Clifton, C., Jr. (2000). Processing coordinate structures. *Journal of Psycholinguistic Research, 29*, 343–370.
27. Frazier, L., Pacht, J. M., & Rayner, K. (1999). Taking on semantic commitments, II: Collective vs distributive readings. *Cognition, 70*, 87–104.
28. Frazier, L., & Rayner, K. (1982). Making and correcting errors during sentence comprehension: Eye movements in the analysis of structurally ambiguous sentences. *Cognitive Psychology, 14*, 178–210.
29. Frazier, L., & Rayner, K. (1987). Resolution of syntactic category ambiguities: Eye movments in parsing lexically ambiguous sentences. *Journal of Memory and Language, 26*, 505–526.
30. Frazier, L., Carminati, M. N., Cook, A. E., Majewski, H., & Rayner, K. (2006). Semantic evaluation of syntactic structure: Evidence from eye movements. *Cognition*, 99, B53–B62.
31. Frenck-Mestre, C., & Pynte, J. (1997). Syntactic ambiguity resolution while reading in second and native languages. *Quarterly Journal of Experimental Psychology, 50A*, 119–148.
32. Frenck-Mestre, C., & Pynte, J. (2000). 'Romancing' syntactic ambiguity: Why the French and the Italians don't see eye to eye. In: Kennedy, A., Radach, R., & Heller, D. (Eds.), *Reading as a perceptual process* (pp. 549–564). Amsterdam: Elsevier.
33. Frisson, S., & Frazier, L. (2005). Carving up word meaning: Portioning and grinding. *Journal of Memory and Language, 53*, 277–291.
34. Frisson, S., & Pickering, M. J. (1999). The processing of metonomy: Evidence from eye movements. *Journal of Experimental Psychology: Learning, Memory, and Cognition, 25*, 1366–1383.
35. Frisson, S., Rayner, K., & Pickering, M. J. (2005). Effects of Contextual Predictability and Transitional Probability on Eye Movements During Reading. *Journal of Experimental Psychology: Learning, Memory, and Cognition, 31*, 862–877.
36. Garnsey, S. M., Perlmutter, N. J. Myers, E., & Lotocky, M. A. (1997). The contributions of verb bias and plausibility to the comprehension of temporarily ambiguous sentences. *Journal of Memory and Language, 37*, 58–93

Table 1

(Continued)

37. Hoeks, J. C. J., Vonk, W., & Schriefers, H. (2002). Processing coordinated structures in context: The effect of topic-structure on ambiguity resolution. *Journal of Memory and Language, 46*, 99–119.
38. Holmes, V. M., & O'Regan, J. K. (1981). Eye fixation patterns during the reading of relative-clause sentences. *Journal of Verbal Learning and Verbal Behavior, 20*, 417–430.
39. Hyönä, J., & Hujanen, H. (1997). Effects of case marking and word order on sentence parsing in Finnish: An eye fixation analysis. *Quarterly Journal of Experimental Psychology, 50A*, 841–858.
40. Hyönä, J., & Vainio, S. (2001). Reading morphologically complex clause structures in Finnish. *European Journal of Cognitive Psychology, 13*, 451–474.
41. Kemper, S., Crow, A., & Kemtes, K. (2004). Eye-fixation patterns of high- and low-span young and older adults: Down the garden path and back. *Psychology and Aging, 19*, 157–170.
42. Kennedy, A., Murray, W. S., Jennings, F. & Reid, C. (1989). Parsing complements: Comments on the generality of the principal of minimal attachment. *Language and Cognitive Processes, 4*, 51–76.
43. Kennison, S. M. (2001). Limitations on the use of verb information during sentence comprehension. *Psychonomic Bulletin and Review, 8*, 132–138.
44. Kennison, S. M. (2002). Comprehending noun phrase arguments and adjuncts. *Journal of Psycholinguistic Research, 31*, 65–81.
45. Konieczny, L., & Hemforth, B. (2000). Modifier attachment in German: Relative clauses and prepositional phrases. In: Kennedy, A., Radach, R., & Heller, D. (Eds.), *Reading as a perceptual process* (pp. 517–527). Amsterdam: Elsevier.
46. Konieczny, L., Hemforth, B., Scheepers, C., & Strube, G. (1997). The role of lexical heads in parsing: Evidence from German. *Language and Cognitive Processes, 12*, 307–348.
47. Lipka, S. (2002). Reading sentences with a late closure ambiguity: Does semantic information help? *Language and Cognitive Processes, 17*, 271–298.
48. Liversedge, S. P., Paterson, K. B., & Clayes, E. L. (2002). The influence of *only* on syntactic processing of "long" relative clause sentences. *Quarterly Journal of Experimental Psychology: Human Experimental Psychology, 55A*, 225–240.
49. Liversedge, S. P., Pickering, M. J., Branigan, H. P., & Van Gompel, R. P. G. (1998). Processing arguments and adjuncts in isolation and context: The case of by-phrase ambiguities in passives. *Journal of Experimental Psychology: Learning, Memory, and Cognition, 24*, 461–475.
50. Liversedge, S. P., Pickering, M., Clayes, E. L., & Brannigan, H. P. (2003). Thematic processing of adjuncts: Evidence from an eye-tracking experiment. *Psychonomic Bulletin & Review, 10*, 667–675.
51. Mak, W. M., Vonk, W., & Schriefers, H. (2002). The influence of animacy on relative clause processing. *Journal of Memory and Language, 47*, 50–68.
52. Mauner, G., Melinger, A., Koenig, J-P., & Bienvenue, B. (2002). When is schematic participant information encoded? Evidence from eye-monitoring. *Journal of Memory and Language, 47*, 386–406.
53. McDonald, S. A., & Shillcock, R. (2003). Eye movements reveal the on-line computation of lexical probabilities during reading. *Psychological Science, 14*, 648–652.
54. Meseguer, E., Carreiras, M., & Clifton, C. (2002). Overt reanalysis strategies and eye movements during the reading of mild garden path sentences. *Memory & Cognition, 30*, 551–561.
55. Murray, W. S., & Rowan, M. (1998). Early, mandatory, pragmatic processing. *Journal of Psycholinguistic Research, 27*, 1–22.
56. Ni, W., Crain, S., & Shankweiler, D. (1996). Sidestepping garden paths: Assessing the contributions of syntax, semantics, and plausibility in resolving ambiguities. *Language and Cognitive Processes, 11*, 283–334.
57. Ni, W., Fodor, J. D., Crain, S., & Shankweiler, D. (1998). Anomaly detection: Eye movement patterns. *Journal of Psycholinguistic Research, 27*, 515–539.

(continued on next page)

Table 1

(*Continued*)

58. Paterson, K. B., Liversedge, S. P., & Underwood, G. (1999). The influence of focus operators on syntactic processing of short relative clause sentences. *Quarterly Journal of Experimental Psychology: Human Experimental Psychology, 52A*, 717–737.
59. Pearlmutter, N. J., Garnsey, S. M., & Bock, K. (1999). Agreement processes in sentence comprehension. *Journal of Memory and Language, 41*, 427–456.
60. Pickering, M. J., & Traxler, M. J. (1998). Plausibility and recovery from garden paths: An eye-tracking study. *Journal of Experimental Psychology: Learning, Memory, and Cognition, 24*, 940–961.
61. Pickering, M. J., & Traxler, M. J. (2001). Strategies for processing unbounded dependencies: Lexical information and verb-argument assignment. *Journal of Experimental Psychology: Learning, Memory, and Cognition, 27*, 1401–1410.
62. Pickering, M. J., & Traxler, M. J. (2003). Evidence against the use of subcategorisation frequencies in the processing of unbounded dependencies. *Language and Cognitive Processes, 18*, 469–503.
63. Pickering, M. J., Traxler, M. J., & Crocker, M. W. (2000). Ambiguity resolution in sentence processing: Evidence against frequency-based accounts. *Journal of Memory and Language, 43*, 447–475.
64. Pynte, J., & Colonna, S. (2001). Competition between primary and non-primary relations during sentence comprehension. *Journal of Psycholinguistic Research, 30*, 569–599.
65. Rayner, K., Carlson, M., & Frazier, L. (1983). The interaction of syntax and semantics during sentence processing: Eye movements in the analysis of semantically biased sentences. *Journal of Verbal Learning and Verbal Behavior, 22*, 358–374.
66. Rayner, K., & Frazier, L. (1987). Parsing temporarily ambiguous complements. *The Quarterly Journal of Experimental Psychology, 39A*, 657–673.
67. Rayner, K., Garrod, S., & Perfetti, C. A. (1992). Discourse influences during parsing are delayed. *Cognition, 45*, 109–139.
68. Rayner, K., Kambe, G., & Duffy, S. (2000). The effect of clause wrap-up on eye movements during reading. *Quarterly Journal of Experimental Psychology, 53A*, 1061–1080.
69. Rayner, K., & Sereno, S. C. (1994). Regressive eye movements and sentence parsing: On the use of regression-contingent analyses. *Memory & Cognition, 22*, 281–285.
70. Rayner, K., Sereno, S., Morris, R., Schmauder, R., & Clifton, C. J. (1989). Eye movements and on-line language comprehension processes. *Language and Cognitive Processes, 4*, SI 21–50.
71. Rayner, K., Warren, T., Juhasz, B. J., & Liversedge, S. P. (2004). The effect of plausibility on eye movements in reading. *Journal of Experimental Psychology: Learning, Memory and Cognition, 30*, 1290–1301.
72. Rayner, K., & Well, A. D. (1996). Effects of contextual constraint on eye movements in reading: A further examination. *Psychonomic Bulletin & Review, 3*, 504–509.
73. Rinck, M., Gamez, E., Diaz, J. M., & de Vega, M. (2003). The processing of temporal information: Evidence from eye movements. *Memory & Cognition, 31*, 77–86.
74. Schmauder, A. R., & Egan, M. C. (1998). The influence of semantic fit on on-line sentence processing. *Memory & Cognition, 26*, 1304–1312.
75. Schmauder, A. R., Morris, R. K., & Poynor, D. V. (2000). Lexical processing and text integration of function and content words: Evidence from priming and eye fixations. *Memory & Cognition, 28*, 1098–1108.
76. Speer, S. R., & Clifton, C., Jr. (1998). Plausibility and argument structure in sentence comprehension. *Memory & Cognition, 26*, 965–978.
77. Spivey, M. J., & Tanenhaus, M. K. (1998). Syntactic ambiguity resolution in discourse: Modeling the effects of referential context and lexical frequency. *Journal of Experimental Psychology: Learning, Memory, and Cognition, 24*, 1521–1543.
78. Staub, A. (2005). Effects of syntactic category predictability on lexical decision and eye movements. Unpublished Masters Thesis, University of Massachusetts, Amherst, Massachusetts, USA.

Table 1

(*Continued*)

79. Staub, A., & Clifton, C., Jr. (2006). Syntactic prediction in language comprehension: Evidence from *either... or*. *Journal of Experimental Psychology: Learning, Memory, and Cognition, 32*, 425–436.
80. Staub, A., Clifton, C., Jr., & Frazier (2006). Heavy NP shift is the parser's last resort: Evidence from eye movements. *Journal of Memory and Language, 54*, 389–406.
81. Sturt, P. (2003). The time course of the application of binding constraints in reference resolution. *Journal of Memory and Language, 48*, 542–562.
82. Sturt, P., & Lombardo, V. (2005). Processing coordinated structures: Incrementality and connectedness. *Cognitive Science, 29*, 291–305.
83. Sturt, P., Scheepers, C., & Pickering, M. J. (2002). Syntactic resolution after initial misanalysis: The role of recency. *Journal of Memory and Language, 46*, 371–390.
84. Traxler, M. J., Bybee, M., & Pickering, M. J. (1997). Influence of connectives on language comprehension: Eye-tracking evidence for incremental interpretation. *Quarterly Journal of Experimental Psychology, 50A*, 481–497.
85. Traxler, M., Foss, D. J., Seely, R. E., Kaup, B., & Morris, R. K. (2000). Priming in sentence processing: Intralexical spreading activation, schemas, and situation models. *Journal of Psycholinguistic Research, 29*, 581–596.
86. Traxler, M., McElree, B., Williams, R. S., & Pickering, M. (2005). Context effects in coercion: Evidence from eye movements. *Journal of Memory and Language, 53*, 1–26.
87. Traxler, M., Morris, R. K., & Seely, R. E. (2002). Processing subject and object relative clauses: Evidence from eye movements. *Journal of Memory and Language, 47*, 69–90.
88. Traxler, M. J., & Pickering, M. J. (1996). Case-marking in the parsing of complement sentences: Evidence from eye movements . *Quarterly Journal of Experimental Psychology, 49A*, 991–1004.
89. Traxler, M. J., & Pickering, M. J. (1996). Plausibility and the processing of unbounded dependencies: An eye-tracking study . *Journal of Memory and Language, 35*, 454–475.
90. Traxler, M. J., Pickering, M. J., & Clifton, C. (1998). Adjunct attachment is not a form of lexical ambiguity resolution. *Journal of Memory and Language, 39*, 558–592.
91. Traxler, M., Pickering, M. J., & McElree, B. (2002). Coercion in sentence processing: Evidence from eye movements and self-paced reading. *Journal of Memory and Language, 47*, 530–548.
92. Traxler, M., Williams, R. S., Blozis, S. A., & Morris, R. K. (2005). Working memory, animacy, and verb class in the processing of relative clauses. *Journal of Memory and Language, 53*, 204–224.
93. Trueswell, J., Tanenhaus, M. K., & Kello, C. (1993). Verb specific constraints in sentence processing: Separating effects of lexical preference from garden-paths. *Journal of Experimental Psychology: Learning, Memory, and Cognition, 19*, 528–533.
94. Trueswell, J. C., Tanenhaus, M. K., & Garnsey, S. M. (1994). Semantic influences on parsing: Use of thematic role information in syntactic disambiguation. *Journal of Memory and Language, 33*, 285–318.
95. Vainio, S., Hyönä, J., & Pajunen, A. (2003). Facilitatory and inhibitory effects of grammatical agreement: Evidence from readers' eye fixation patterns. *Brain and Language, 85*, 197–202.
96. Van Gompel, R. P. G., Pickering, M. J., Pearson, J., & Liversedge, S. P. (2005). Evidence against competition during syntactic ambiguity resolution. *Journal of Memory and Language, 52*, 284–307.
97. Van Gompel, R. P. G., & Pickering, M. J. (2001). Lexical guidance in sentence processing: A note on Adams, Clifton, and Mitchell (1998). *Psychonomic Bulletin & Review, 8*, 851–857.
98. Van Gompel, R. P. G., Pickering, M. J., & Traxler, M. J. (2001). Reanalysis in sentence processing: Evidence against current constraint-based and two-stage models. *Journal of Memory and Language, 45*, 225–258.
99. Wiley, J., & Rayner, K. (2000). Effects of titles on the processing of text and lexically ambiguous words: Evidence from eye movements. *Memory & Cognition, 28*, 1011–1021.
100. Zagar, D., Pynte, J., & Rativeau, S. (1997). Evidence for early-closure attachment on first-pass reading times in French. *Quarterly Journal of Experimental Psychology, 50A*, 421–438.

published in peer-reviewed journals at the time of writing.[1] We did not include articles where the main factor of interest involved discourse structure, text properties, inferences, or anaphora (although we did include articles where the effects of discourse structure, etc., on the effect of some syntactic or semantic property of a sentence were studied). We generally did not include papers published as chapters in edited books, but we did include a very few that struck us as making a unique contribution. We did not include any unpublished papers, apart from a few of our own. The 100 articles under consideration are those indicated by a number in Table 1. Our following discussion refers to these articles by this number.

We examined each of these articles, categorizing the experiments they contained in several ways. The results of this categorization appear in Tables 2 and 3. The final "ALL" column of these tables lists the numbers (see Table 1) of all the articles that fall in a given category. These tables indicate a variety of properties of the experiments, including a specification of the first region in which an effect of the primary manipulated factor appears in each reported eyetracking measure. The measures are FF (first fixation), FP (first pass), GP (go-past), SP/TT (either second pass or total time, whichever was reported), and RO (regressions out).

If an experiment involved the resolution of a temporary syntactic ambiguity, it is listed in Table 2. In this table, Region D indicates the region in which the disambiguation first appeared (and D+1 the next region). SP/TT effects are reported if they occurred in any region. If an experiment did not involve temporary ambiguity, but instead involved factors hypothesized to affect predictability, plausibility, complexity, or similar properties of sentences, it appears in Table 3. In this table, Region C indicates the critical region, the region involving the predictable/plausible/etc. word or words. In both tables, brief descriptions of the temporary ambiguity or the primary manipulated factor appear in the first column. In Table 2, the second column indicates the nature of the disambiguating material. "Category" means that disambiguating information was conveyed by the syntactic category of the disambiguating phrase (e.g., in the SCO/MCS ambiguity, an ambiguity between subordinate clause object and main clause subject – see Notes to Table 2 – "category" means that the disambiguation was conveyed by the fact that the main verb of the sentence followed the NP that was temporarily ambiguous between an object and a subject). The number of an article in Table 1 appears in the earliest column of each measure for which a statistically significant effect was reported. Note that experiments differ substantially in the length of the critical or disambiguating region, and that experiments where this region is very short may tend to yield effects that emerge only on the following region. Note further that few experiments included reports of all measures, so the absence of an article-number in a column does not mean that there was a null effect; it may simply mean that the effect was not reported for the measure in question.

Some multi-experiment articles appear in multiple categories of a table, or in both tables. In a great many cases, we have abstracted away from the factors of most interest

[1] If we have missed any, we apologize to the authors, and ask them to accept our oversight as an error, not as a snub.

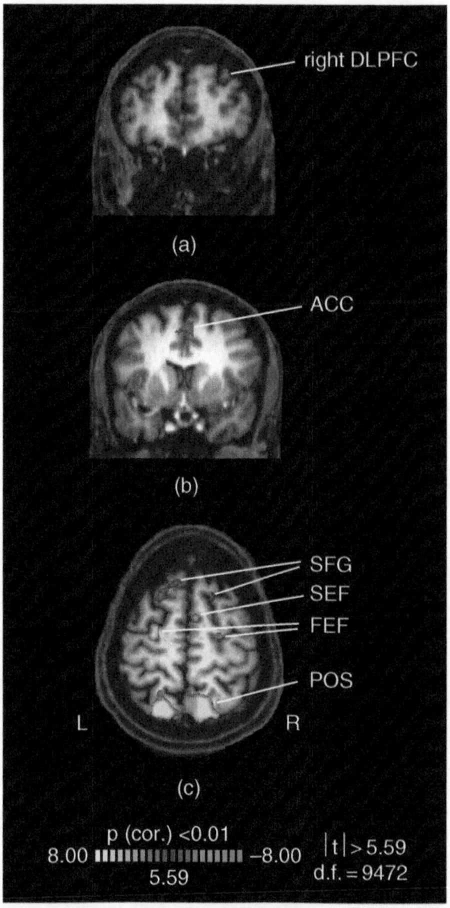

Color Plate 1. Group statistical activation maps generated from the general linear model (GLM) contrast comparing block of anti-saccades to blocks of pro-saccades from 10 subjects. *Red* and *yellow* regions exhibited significantly more BOLD activation for anti-saccades than for pro-saccades. *Blue* and *green* regions exhibited significantly more BOLD activation for pro-saccades than for anti-saccades. Bonferroni-corrected $p < 0.01$. L and R denote left and right. Maps obey neurological conventions. (a) *DLPFC*, dorsolateral prefrontal cortex; (b) *ACC*, anterior cingulate cortex; (c) *SFG*, superior frontal gyrus; *SEF*, supplementary eye fields; *FEF*, frontal eye fields; *POS*, parieto-occipital sulcus. (*See Figure 2, Chapter 6, p. 132.*)

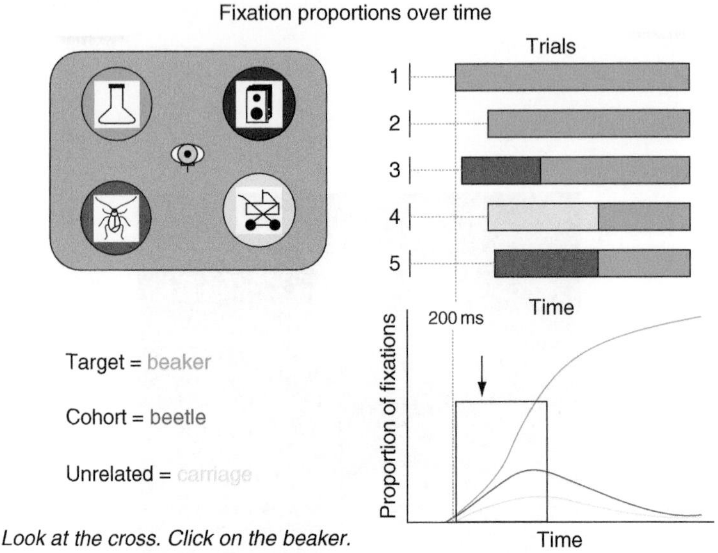

Color Plate 2. Schematic illustrating proportion of fixation curves. (*See Figure 1, Chapter 20, p. 449.*)

Color Plate 3. Top left: Original scene. Top middle: Model-determined salient regions in the scene. Top right: Fixation locations from all participants. Bottom: Scene with salient regions and participant fixations overlaid. Red dots show participant fixations within a salient region. Red tails mark saccade paths that originated in a non-salient region. Green dots denote participant fixations outside of the salient regions. (*See Figure 1, Chapter 25, p. 543.*)

Color Plate 4. An example of a congruent indoor picture (top panel) and a congruent outdoor picture (bottom panel). (*See Figure 1a, Chapter 26, p. 568.*)

Color Plate 5. An example of an incongruent indoor picture (top panel) and an incongruent outdoor picture (bottom panel). (*See Figure 1b, Chapter 26, p. 569.*)

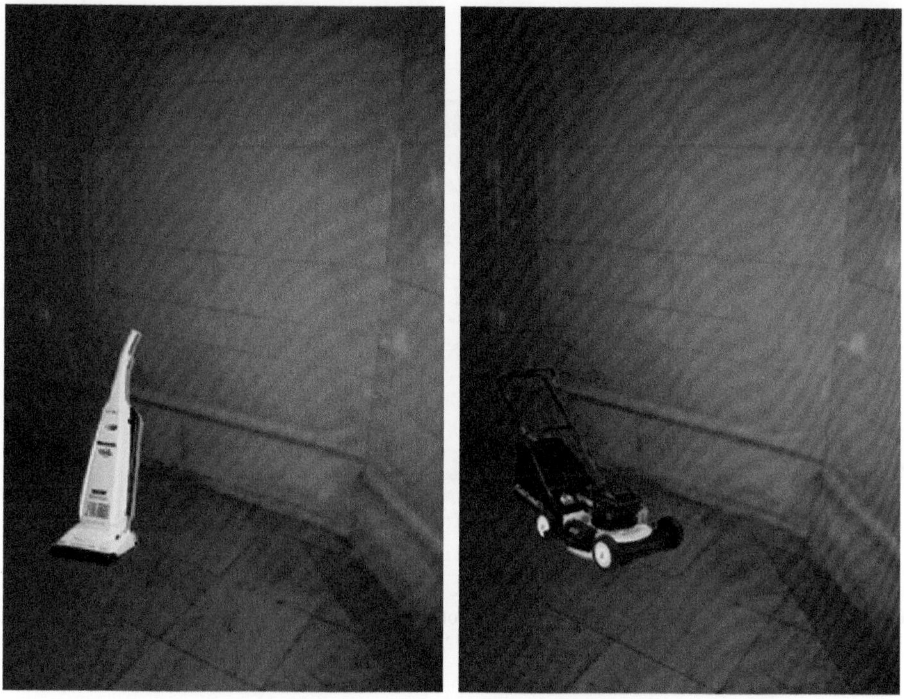

Color Plate 6. An example of a neutral picture containing an indoor object (left panel) and a neutral picture containing an outdoor object (right panel). (*See Figure 1c, Chapter 26, p. 570.*)

Color Plate 7. A sample picture as used in the plausible condition with the skier wearing the pink jacket as the target object. The fixations of one participant are superimposed on the stimulus in this example, with lines indicating the movements between fixations that were themselves shown as circles. Duration of fixation is indicated by the size of the circles with larger circles indicating longer fixation durations than smaller ones. (*See Figure 2a, Chapter 26, p. 575.*)

Color Plate 8. A sample picture with fixation patterns of one viewer as seen in the implausible condition with a snowman appearing as the target object. (*See Figure 2b, Chapter 26, p. 575.*)

Color Plate 9. A sample picture with fixation patterns of one viewer as seen in the bizarre condition with the cow as the target object. (*See Figure 2c, Chapter 26, p. 576.*)

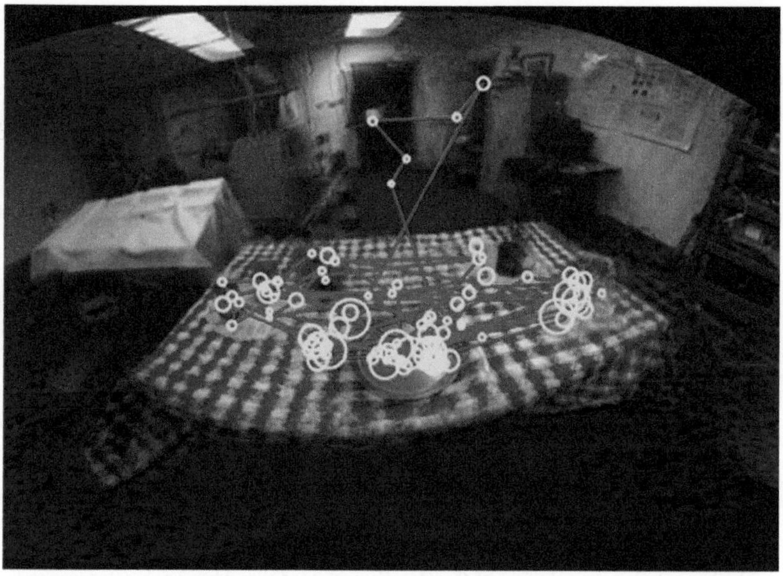

Color Plate 10. Fixations made by an observer while making a peanut butter and jelly sandwich, indicated by yellow circles. Images were taken from a camera mounted on the head, and a composite image mosaic was formed by integrating over different head positions using a method described in Rothkopf and Pelz (2004) et al. (The reconstructed panorama shows artifacts because the translational motion of the subject was not taken into account.) Fixations are shown as yellow circles, with a diameter proportional to fixation duration. The red lines indicate the saccades. Note that almost all fixations fall on task relevant objects. (*See Figure 1, Chapter 30, p. 644.*)

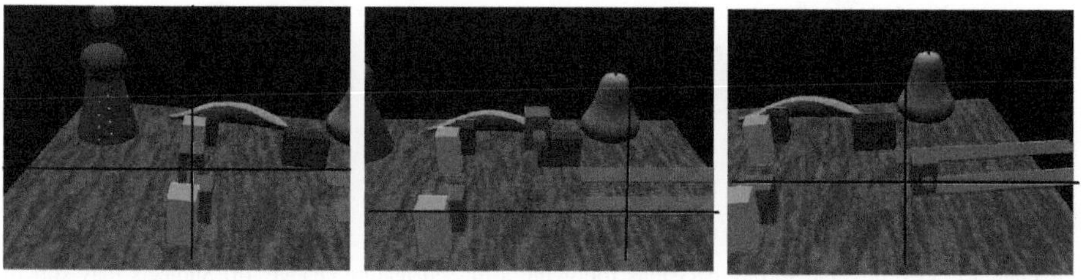

(a) (b) (c)

Color Plate 11. View of the virtual work-space as a subject (a) picks up, (b) carries, and (c) places a brick on a conveyor belt. The dot visible in front of the lifted brick is one of the fingers. The cross-hair shows fixation. Adapted from Triesch et al. (2003). (*See Figure 5, Chapter 30, p. 652.*)

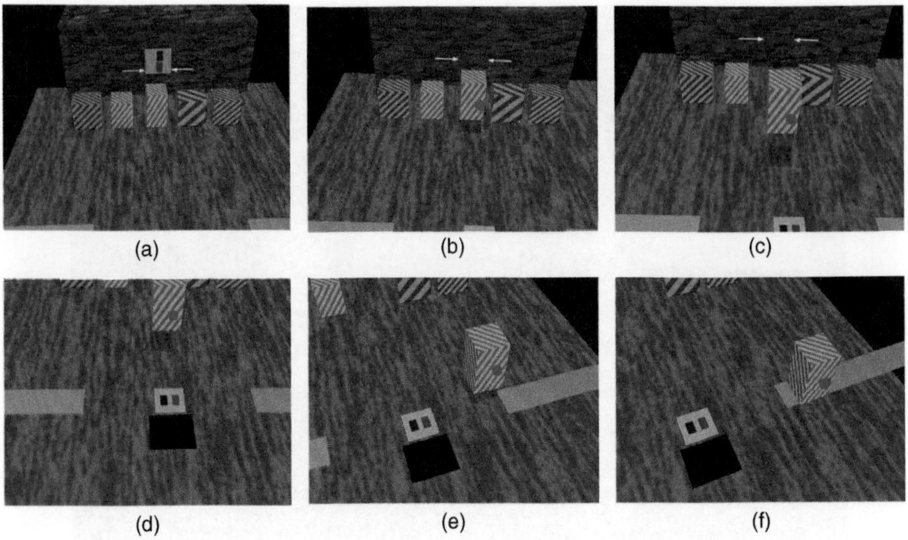

Color Plate 12. Scene during a single trial of the *One Feature* condition when brick color was task relevant. Fingertips are represented as small red spheres. In a single trial, a subject (a) selects a brick based on the pick-up cue, (b) lifts the brick, (c) brings it towards themselves, (d) decides on which conveyor belt the brick belongs based on a put-down cue, (e) guides the brick to the conveyor belt, (f) sets the brick on the belt where the brick is carried off. In other trials, subjects may have used width, height, or stripes for the pick-up or put-down decision. Adapted from Droll et al (2005). (*See Figure 6, Chapter 30, p. 654.*)

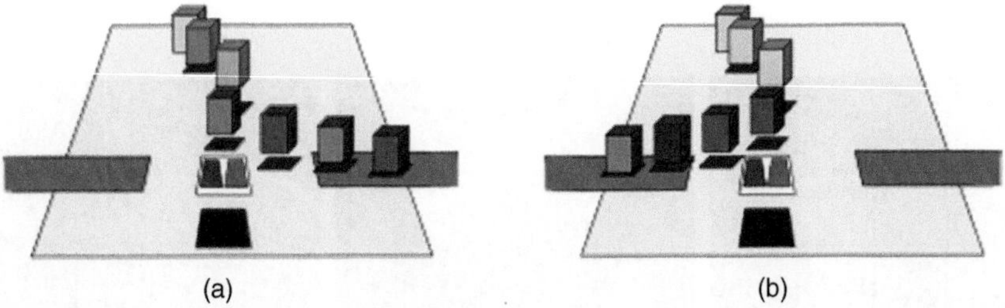

Color Plate 13. Two possible sorting decisions following a missed feature change. (a) Subjects may sort the brick by the old, pre-change, feature, in which case changes are missed due to a failure to update the new visual information. (b) Subjects may sort also the brick by the new, post-change, feature, in which case changes are missed due to a failure to maintain visual information. When subjects performed blocks of trials using the same feature for the put-down decision, missed changes were most often sorted by the old feature (85%). (*See Figure 7, Chapter 30, p. 654.*)

Table 2

Classification of articles examining effects of temporary ambiguity

Structure	Disambiguation	FF – D	FF D+1	FP D	FP D+1	GP D	GP D+1	SP/TT (any)	RO – D	RO – D+1	ALL
SCO/MCS	category	1, 13, 28, 97		1, 13, 21, 22, 28, 60, 97		29, 63, 64, 97		29, 47, 63, 64	13, 21, 29, 31, 47, 60, 97	63, 64	1, 13, 21, 22, 28, 29, 31, 47, 60, 63, 64, 97
SCO/MCS	Transitivity					97		1		97	1, 97
MC/RC	Category	58, 65		6, 15, 19, 41, 56, 58, 65, 67, 69[a], 77, 94		15		6, 19, 41, 48, 58, 77, 94	15, 41, 48, 56, 65, 67		6, 15, 19, 41, 48, 56, 58, 65, 67, 69, 77, 94
NP/S comp	Category	20, 28, 65		28, 36, 42[b], 43, 65, 83, 93	63	83		63, 74, 83, 93	42[b], 65		20, 28, 36, 42, 43, 63, 65, 74, 83, 93
NP/S comp	casemarking	88		88							88
PP Attach	category			19				19			19
PP Attach	plausibility	31[c], 65[d]		10, 14, 31, 45, 46, 65, 76	98	46	98	10, 46,	14, 31, 76	98	10, 14, 31, 45, 46, 56, 65, 76, 98
adverb attachment	morphology						54	54		54	54
adverb attachment	time adverb (plausibility)			4, 96		96			4		4, 96
RC attachment	plausibility			12[e]		45				90, 96[f]	12, 45, 90, 96
RC attachment	morphology (gender)			11, 100		11, 100		100			11, 32, 100
argument/ adjunct	plausibility			49			50	44, 50			44, 49, 50
LDD	category + plausibility	89		89		62		61, 62		62[g]	61, 62, 89

(continued on next page)

Table 2

(*Continued*)

Structure	Disambiguation	FF − D	FF D+1	FP D	FP D+1	GP D	GP D+1	SP/TT (any)	RO − D	RO − D+1	ALL
NP/S coordination	category		80[h]	37, 80		80				80	37, 80
S/O RC	category			87		92		92	87, 92		87, 92
S/O RC	morphology			51			51[i]		38		38, 51
SC/RC	category			2, 3							2, 3

Note 1: SCO/MCS = Initial subordinate clause object vs main clause subject; MC/RC = Main clause vs reduced relative clause; NP/S Comp = Direct object NP vs sentence complement; PP attach = Attach PP to verb or noun; adverb attach = attach adverb high or low; RC attach = Attach relative clause to N1 or N2 in N1 of N2 construction; argument adjunct = Analyze phrase as argument vs adjunct e.g. agentive-by vs locative-by; LDD = Long distance dependency (filler/gap); NP/S coordination = Coordinate phrases as NP or as S; S/O RC = Subject vs object extracted relative clause.

Note 2: The following syntactic disambiguation experiments are unclassified:
8 lexical/syntactic category bias, FP effect, later modulated by context
16 NP conjunction, biased by pronoun or parallelism, late effects
24 sluicing, marginally faster with two than one possible antecedent
25 semantic (quantifier presupposition), FP and TT effects on D+1
26 conjoined NP, facilitated by parallelism, TT and marginal FP effects on D
29 Noun–noun compound vs noun–verb predication, delayed effect
39 Finnish, normal FP SVO preference overcome by casemarking
64 SCO/MCS French, really anaphora
82 apparent immediate interpretation of anaphor in coordinated VP, GP effect before end of VP.

Note 3: Footnotes from Table 2 follow:
[a] Reanalysis of 67, regression-contingent
[b] S-Comp slow regardless of ambiguity
[c] Second language only
[d] Significant when pooled over first three fixations
[e] English only
[f] Ambiguous easier than disambiguated
[g] Effect appeared at region D+2
[h] D is the phrase "or NP" and D+1 is the following, truly disambiguating, region
[i] Effect appeared at region D+2.

Table 3
Classification of Articles Involving High-Order Effects other than Temporary Ambiguity (C = Critical Region)

Category	FF – C	FF C+1	FP C	FP C+1	GP C	GP C+1	SP/TT (any)	RO – C	RO – C+1	ALL
Lexical predictability	35, 53, 72		5, 18[a], 35, 53, 72		5		72	5		5, 18, 35, 53, 72
semantic/pragmatic anomaly	55	71	9, 71, 85	57	71		71, 55, 85	57	71	9, 55, 57, 71, 85
syntactic anomaly			9, 17	59			59	9, 57, 59		9, 17, 57, 59
lexical semantics	33		33, 34	33, 34, 91	86		34, 85, 91	33, 34	86	33, 34, 86, 91
complexity	70, 80		40[b], 70, 80, 95		80		95	80		40, 70, 80, 95
syntactic category			78	78	78		75			75, 78
semantic interpretation – phrase and clause			7, 84, 99	27	84	52	23, 52, 73	7, 52	27	7, 23, 27, 52, 73, 84, 99

Note 1: Footnotes from table follow.
[a] Also increased skipping of predictable word.
[b] Also more fixations of complex region.

to the authors of an experiment. For instance, the authors may have been interested in the effect of plausibility or context on how a syntactic garden-path is resolved. In these tables, since we are interested primarily in what aspects the eyetracking record reflect what types of processing difficulty, we simply categorize the experiment on the basis of the type of garden-path and how it was eventually resolved, and report the earliest appearances of the resolution in any condition of the experiment.

Tables 2 and 3 are presented largely to stimulate a deeper examination of how eye movements reflect sentence processing. These tables, by themselves, cannot present all relevant information about an experiment. For instance, authors of different experiments on the same topic commonly differ in how they divide their sentences into regions, which clearly can affect where an eyetracking effect can appear. However, even a superficial examination of these tables supports some generalizations. Consider first Table 2, the "garden-path" table. It is clear that few first fixation effects appear. This is largely because few authors report such effects. This is justified when the disambiguating region contains multiple words and the first word of the region does not contain disambiguating material. No first fixation effect should be expected if the first fixation does not land on critical material. However, in cases where the first (or only) word of a disambiguating region was of a syntactic category that reversed a strong initial preference for one interpretation (e.g., the SCO/MCS, or the MC/RC, main clause/relative clause, ambiguity), first fixation effects have been reported. The only instance of a first fixation effect on the following region appears in article 80, where it is probably a spillover effect. First pass effects are very common, certainly where disambiguation is carried by syntactic category, and also in some cases (e.g., ambiguous attachment of a prepositional phrase, PP, to a verb or a noun, as in (6), cited earlier), sheer implausibility of the initially preferred interpretation can result in first pass effects. One can conclude that semantic interpretation (at least, of the initially preferred alternative) is done very quickly indeed, and can appear quickly in the eyetracking record. Note, however, as discussed above, currently available data does not allow us to decide whether such an effect occurs on most or all trials, or only on a possibly small subset of trials.

There are rather few cases where an effect shows up in go-past but not in first pass (as was the case for the inanimate-subject items in Clifton et al., 2003), but there are some. The majority of these are in the object/subject "late closure" (SCO/MCS) ambiguity, but some appear in PP attachment, long distance dependencies (LDDs; filler-gap constructions), and subject vs object extracted relative clauses. The appearance of effects in percentage of regressions out (RO) of the disambiguating region that did not appear in first pass time probably reflects a similar dynamic. These can be seen in some of the cases just discussed. A very late effect sometimes appears as regressions out of the region following the disambiguating region (e.g., in some cases of PP, adverb, or relative clause attachment). Second pass/total time effects in the absence of effects already discussed (first pass, go-past, regressions out) are almost non-existent, appearing only in articles 1, 44, 54, and 61. We note that the generally low power of experiments on sentence processing leaves open the possibility that these experiments simply failed to detect a real effect in the earlier measures. We further note that late effects in the absence of early effects are reported for some conditions of the experiments reviewed, e.g., the unambiguous inanimate-subject conditions of Clifton et al. (2003) (15). In these cases, they may reflect some general comprehension difficulty, not associated with the resolution of a syntactic ambiguity.

One final point is worth bringing up, even though it is not reflected in Table 2. We examined all the relevant articles for evidence about whether reading is slowed in the ambiguous region, compared to an unambiguous control condition. Models of sentence processing that posit a time-consuming process of competition among alternative analyses apparently predict such slowing. These models include MacDonald, Pearlmutter, and Seidenberg (1994), and McRae, Spivey-Knowlton, and Tanenhaus (1998), and Spivey and Tanenhaus (1998); see Gibson and Pearlmutter, 2000; Lewis, 2000; and Traxler, Pickering, and Clifton, 1998, for some discussion.

It turns out that there are very few instances of such a slowing in the ambiguous region. In most cases, if the data are presented, no slowing appears, and, in fact, there are several reports of a speedup in an ambiguous region as compared to an unambiguous region (see Articles 28, 90, 96, and 98). There are a very few specious cases of apparent slowing (e.g., Articles 13, 19, 36, 47, 60), but they all appear to be due to semantic implausibility of the preferred interpretation rather than a slowdown due to ambiguity per se. However, there are a few cases of apparent slowing due to ambiguity. Most of these involve the main clause/reduced relative ambiguity, and include Articles 15, 56, and 58. The apparent slowing in an ambiguous phrase also appears in Articles 43 and 74, which examined the direct object/sentence complement ambiguity. Several of these (15, 43, 58, 74) could be dismissed as simply reflecting fast reading after the highly frequent series of words in the disambiguating condition *that was* and the slowdown in Article 43 could actually reflect the semantic implausibility of attaching a prepositional phrase as a modifier of the preceding verb, a normally preferred analysis. However, Article 56 (Ni et al., 1996) cannot be dismissed so easily since disambiguation in that case was carried by the morphology of the otherwise-ambiguous verb (e.g. *The horse raced* vs *ridden*...), but note that different sentences with different lexical items and different content were used in the ambiguous and unambiguous conditions, making direct comparison uncertain. None of the remaining studies of the main clause/reduced relative clause ambiguity reported slower reading in the ambiguous region. It is possible that the experiments that did not detect slowing in the ambiguous region simply did not have enough power to detect the effect, but it is also possible that some of the participants in the experiments that did report the effect became aware of the occasional ambiguity and deliberately read it cautiously.

Table 3 encompasses articles examining effects on eye movements generated by a range of factors other than syntactic ambiguity. Many of these articles examine effects on word processing, but we discuss them here, as well (in some cases) as in the first section of this chapter, because these articles focus on the syntactic, semantic, or pragmatic relationship between a word and its context.

Relatively few eyetracking studies have examined the effect on eye movements of encountering a syntactically or semantically anomalous word in printed text. It is somewhat surprising that of the four studies (9, 17, 57, 59) that have explicitly examined responses to syntactic anomaly (e.g., agreement errors), only two (9, 17) found effects appearing on the anomalous word. On the other hand, four (9, 55, 71, 85) of the five studies of semantic or pragmatic anomaly have found increased first fixation duration or gaze duration on the offending word (57 reported only a late effect). Of course, it is possible that which measure an effect first appears in reflects the magnitude of the processing disruption occasioned by the effect, and not simply the timing of the processes that the effect reflects.

It is interesting to contrast the paucity of eyetracking studies of anomaly with the profusion of event-related potentials (ERP) studies that have focused on brain responses to anomalous words. The earliest reported electrophysiological response to a syntactic word category violation (the early left anterior negativity, or ELAN; Hahne & Friederici, 1999; Neville, Nicol, Barss, Forster, & Garrett, 1991) occurs 150–200 ms after the onset of the critical word, while the typical response to a semantic violation (the N400, first identified by Kutas & Hillyard, 1980) peaks about 400 ms after word onset. However, whether agreement violations trigger an early effect is not certain, with some studies reporting such an effect (Coulson, King, & Kutas, 1998; Deutsch & Bentin, 2001; Osterhout & Mobley, 1995) and others reporting only a much later effect, the P600 (Hagoort, Brown, & Groothusen, 1993; Munte, Heinze, & Mangun, 1993; Osterhout, McKinnon, Bersick, & Corey, 1996). In sum, the overall picture from both ERP and eye movement research suggests that the question of exactly when syntactic and semantic anomalies each affect language comprehension is still to be settled.

A study by Rayner et al. (2004; 71), which was mentioned in the earlier discussion of plausibility effects on word recognition, suggests that semantic anomaly is probably not a unitary phenomenon with respect to its effect on eye movements. Rayner et al. had participants who read sentences such as:

(5) John used a knife to chop the large carrots for dinner last night.

(6) John used an axe to chop the large carrots for dinner last night.

(7) John used a pump to inflate the large carrots for dinner last night.

In all sentences, the target word is *carrots*. Sentence (5) is a normal control condition; in (6), the target word is an implausible theme given the combination of verb and instrument; and in (7), the target word is an anomalous theme of the verb. Rayner et al. found that while (6) only caused mild disruption to reading, appearing in the go-past measure on the target word and in gaze duration on the following word, (7) caused an earlier disruption, appearing as an increase in gaze duration on the target word. Given that this relatively subtle difference in the type of implausibility produces a clear difference in the eye movement record, it is not surprising that when semantic anomaly or implausibility has been used as a means of disambiguation in studies of syntactic ambiguity processing, the time course of its effect has varied considerably.

Even within the range of words that are not semantically anomalous given the preceding context, there are, as was discussed above, early effects on eye movements of the word's semantic fit. Five studies (5, 18, 35, 53, 72) have examined the effect of a word's predictability or "contextual constraint"; in general, this is defined in terms of the word's cloze probability (i.e., the probability that informants will produce the target word as the likely next word in the sentence, given the sentence up to that word). The basic finding is that when a word is highly predictable, the first fixation duration or gaze duration on the word is decreased. One article (53) has reported that the transitional probability between two words in corpora has an independent facilitatory effect on processing of the second word, though a second article (72) has reported results suggesting that when overall predictability is well controlled, transitional probability does not have an independent effect.

In a recent study (78) that is related to the issue of predictability, we demonstrated that when a word's syntactic category is predictable, though the word itself is not, gaze duration is reduced either on the word itself or on the next word, depending on the syntactic construction. For example, after a determiner (e.g., *the*), gaze duration is shorter on a noun (which must occur in the phrase beginning with the determiner) than on an adjective (which is legal but optional), even when factors such as length, frequency, and lexical predictability held constant. Another study that specifically examined syntactic category effects (75) found that readers tend to refixate function words more frequently than content words, though somewhat surprisingly this study did not find any significant earlier effects of this distinction between word classes.

In Table 3 we have identified four studies (33, 34, 86, 91) that manipulated the nature of the semantic processing that is required on a word, under the heading of *lexical semantics*. Since these studies focused on a range of types of semantic processing, it is not surprising that there is considerable variation in the time course with which the manipulations affected eye movements. At one extreme, Frisson and Frazier (2005; 33) found that when a mass noun appears in the plural (e.g., *some beers*), or a count noun appears in the singular with a plural determiner (e.g., *some banana*), the first fixation duration on the critical word is lengthened. On the other hand, in a study by Traxler, Pickering, and McElree (2002; Article 91; see also Pickering, McElree, & Traxler, 2005) that examined so-called coercion, where a noun with no intrinsic temporal component must be interpreted as an event (as in the phrase *finish the book*), the earliest significant effects were on the word after the critical word.

We identified four articles (40, 70, 80, 95) that investigated the effect of syntactic complexity of a phrase in the absence of syntactic ambiguity. All of these reported an effect of increased complexity on first fixation duration or first pass time in the critical region. To cite just one example (80), we have recently conducted a study of the processing of so-called Heavy NP Shift (Ross, 1967), in which a verb's direct object appears at the end of the sentence rather than adjacent to the verb. The experiments varied the point in the sentence at which the reader had to construct this complex syntactic analysis. At the point at which the Heavy NP Shift analysis had to be constructed, readers consistently slowed down, made regressive eye movements, or both.

Finally, we also included a number of articles examining semantic processing effects on linguistic structures larger than a single word (7, 23, 25, 27, 52, 73, 84, 99). Again, a rather diverse collection of phenomena are investigated in these articles. Several of the studies report early effects of their manipulations, but two report only second pass or total time effects. We suspect, in addition, that this may be one area in which researchers have obtained various null effects that have remained unpublished.

3. Conclusions

Measuring eye movements during reading has greatly enhanced the understanding of how people identify words and comprehend sentences. The early impact of linguistic variables such as lexical frequency and age-of-acquisition on eye movements has shown that eye movements quite directly reflect linguistic processing. In turn, the speed of eye movements, and their tight linkage to at least some parts of the reading process, has provided convincing support for the thesis that language processing is often essentially immediate, at least in the sense that a word is typically interpreted and integrated into the communicated message while the eyes are still fixated on it (see Marslen-Wilson, 1973, for an early statement of this thesis in the domain of listening). Eye movements have allowed researchers to probe the early stages of reading in a clear and direct fashion that is exceeded by no other technique we know of.

In the domain of word identification, eye movement data have been extremely clear and orderly. Intrinsic lexical factors generally have their effect on very early measures of eye movements, including first fixation and gaze duration, and some relational factors that may affect word identification do as well. The basic phenomena seem to be sufficiently consistent and replicable to support the development of theories in a "bottom-up" fashion. To be sure, there is still plenty of room for theorists to argue about the best way to interpret data (see the discussion following Reichle et al., 2003). But the strategy of first identifying solid empirical phenomena and then building formal theories that account for them has paid off very well in this domain.

The domain of sentence comprehension is similar in some ways, but very different in others. Eye tracking measures have shown that much, if not quite all, of sentence comprehension is nearly immediate. Reflections of syntactic or semantic anomaly or complexity sometimes can appear very quickly in the eye movement record, as do effects of recovering from garden-paths. Eyetracking measures have also shown that syntactic knowledge and at least some kinds of semantic, pragmatic, and real-world knowledge have effects even during fixations on the phrase that provides access to this knowledge. But our survey of the literature shows that the effects of sentence comprehension factors are more variable than the effects that word identification factors, such as lexical frequency and lexical ambiguity, have on eyetracking measures.

Some of this variability may reflect experimental limitations more than deep-seated differences between lexical processing and sentence integration. For instance, the greater variability in the length of critical regions in studies of sentence integration than in the length of words that constitute critical regions in studies of lexical processing certainly gives rise to more variability in where an effect will appear in the eyetracking record. Further, we suspect sentence integration and comprehension processes are more sensitive than word recognition processes to the task and goals given to the reader, leading to greater variability across studies.

On the other hand, the variability in effects of sentence comprehension factors may be more fundamental. A reader has more options about how to deal with processing difficulty when the difficulty is occasioned by plausibility or complexity or syntactic misanalysis than when it is occasioned by difficulty recognizing a word. In the latter case, the only option the reader has is to continue looking at the word (or giving up, or guessing). In the former case, the reader may go back into earlier text to try to identify problems, or continue thinking about the phrase that made the difficulty apparent, or plunge ahead, hoping that later information will resolve the issue. Furthermore, in contrast to normal word recognition, a wide range of factors contributes to sentence comprehension. We are far from understanding how these factors are coordinated (a topic of raging disagreement which we have largely avoided in our review) and whether their coordination is modulated by differences in a reader's abilities and strategies. Suffices it to say that the greater flexibility in dealing with sentence comprehension difficulty and the wide range of factors that affect it could mean that high-level processing shows up in the eye movement record in a variety of different ways, with any one effect appearing only occasionally.

In our view, the "high-level" variables that affect sentence interpretation are much more complex, both in their definition and in their effect, than the variables that govern much of the variation in word identification. We suspect that understanding how these high-level variables operate is not something that can be induced from observations of eyetracking phenomena (as we claim has been true in large part in the domain of word identification). Rather, we suspect that understanding must be guided by the development of more explicit theories than existing now of how syntactic, semantic, pragmatic, and real-world knowledge guide language comprehension. We hold hope that development of such theories will help make sense of the empirical variability that we have illustrated in Tables 2 and 3.

Acknowledgments

The preparation of this chapter, and some of the research reported in it, was supported in part by Grants HD18708, HD26765, and HD17246 to the University of Massachusetts. We would like to thank Alexander Pollatsek, Roger Van Gompel, and the anonymous reviewer for their comments on earlier versions of this chapter.

References

Andrews, S., Miller, B., & Rayner, K. (2004). Eye movements and morphological segmentation of compound words: There is a mouse in mousetrap. *European Journal of Cognitive Psychology, 16*, 285–311.

Ashby, J., Rayner, K., & Clifton, C. J. (2005). Eye movements of highly skilled and average readers: Differential effects of frequency and predictability. *Quarterly Journal of Experimental Psychology,* 58A, 1065–1086.

Balota, D. A., Pollatsek, A., & Rayner, K. (1985). The interaction of contextual constraints and parafoveal visual information in reading. *Cognitive Psychology, 17*, 364–388.

Bever, T. G. (1970). The cognitive basis for linguistic structures. In J. R. Hayes (Ed.), *Cognition and the development of language* (pp. 279–352). New York: Wiley.

Boland, J. (2004). Linking eye movements to sentence comprehension in reading and listening. In M. Carreiras & C. Clifton, Jr. (Eds.), *The on-line study of sentence comprehension* (pp. 51–76). New York: Psychology Press.

Carpenter, P. A., & Daneman, M. (1981). Lexical retrieval and error recovery in reading: A model based on eye fixations. *Journal of Verbal Learning and Verbal Behavior, 28*, 138–160.
Chaffin, R., Morris, R. K., & Seely, R. E. (2001). Learning new word meanings from context: A study of eye movements. *Journal of Experimental Psychology: Learning, Memory and Cognition, 27*, 225–235.
Clifton, C., Jr. (2003). On using eyetracking data to evaluate theories of on-line sentence processing: The case of reduced relative clauses. Invited talk, 12th European Conference on Eye Movements, Dundee, Scotland, August, 2003.
Clifton, C. J., Traxler, M., Mohamed, M. T., Williams, R. S., Morris, R. K., & Rayner, K. (2003). The use of thematic role information in parsing: Syntactic processing autonomy revisited. *Journal of Memory and Language, 49*, 317–334.
Coulson, S., King, J. W., & Kutas, M. (1998). Expect the unexpected: Event-related brain response to morphosyntactic violations. *Language and Cognitive Processes, 13*, 21–58.
Deutsch, A., & Bentin., S. (2001). Syntactic and semantic factors in processing gender agreement in Hebrew: Evidence from ERPs and eye movement. *Journal of Memory and Language, 45*, 200–224.
Deutsch, A., Frost, R., Pollatsek, A., & Rayner, K. (2000). Early morphological effects in word recognition: Evidence from parafoveal preview benefit. *Language and Cognitive Processes, 15*, 487–506.
Deutsch, A., Frost, R., Pollatsek, A., & Rayner, K. (2005). Morphological parafoveal preview benefit effects in reading: Evidence from Hebrew. *Language and Cognitive Processes, 20*, 341–371.
Deutsch, A., Frost, R., Peleg, S., Pollatsek, A., & Rayner, K. (2003). Early morphological effects in reading: Evidence from parafoveal preview benefit in Hebrew. *Psychonomic Bulletin & Review, 10*, 415–422.
Duffy, S. A., Morris, R. K., & Rayner, K. (1988). Lexical ambiguity and fixation times in reading. *Journal of Memory and Language, 27*, 429–446.
Ehrlich, S. F., & Rayner, K. (1981). Contextual effects on word perception and eye movements during reading. *Journal of Verbal Learning and Verbal Behavior, 20*, 641–655.
Engbert, R., Nuthmann, A., Richter, E., & Kliegl, R. (2005). SWIFT: A dynamical model of saccade generation during reading. *Psychological Review, 112*, 777–813.
Feng, G. (2006). Eye movements as time series random variables: A stochastic model of eye movement control in reading. *Cognitive Systems Research, 7*, 70–95.
Ferreira, F., & Clifton, C., Jr. (1986). The independence of syntactic processing. *Journal of Memory and Language, 25*, 348–368.
Folk, J. R. (1999). Phonological codes are used to access the lexicon during silent reading. *Journal of Experimental Psychology: Learning, Memory, and Cognition, 25*, 892–906.
Folk, J. R., & Morris, R. K. (2003). Effects of syntactic category assignment on lexical ambiguity resolution in reading: An eye movement analysis. *Memory & Cognition, 31*, 87–99.
Frazier, L. (1987). Sentence processing: A tutorial review. In M. Coltheart (Ed.), *Attention and performance* (pp. 559–586). Hillsdale, NJ: Lawrence Erlbaum Associates.
Frazier, L., & Rayner, K. (1982). Making and correcting errors during sentence comprehension: Eye movements in the analysis of structurally ambiguous sentences. *Cognitive Psychology, 14*, 178–210.
Frazier, L., & Rayner, K. (1987). Resolution of syntactic category ambiguities: Eye movements in parsing lexically ambiguous sentences. *Journal of Memory and Language, 26*, 505–526.
Frazier, L., & Rayner, K. (1990). Taking on semantic commitments: Processing multiple meanings vs multiple senses. *Journal of Memory and Language, 29*, 181–200.
Frisson, S., & Frazier, L. (2005). Carving up word meaning: Portioning and grinding. *Journal of Memory and Language, 53*, 277–291.
Gibson, E., & Pearlmutter, N. J. (2000). Distinguishing serial and parallel processing. *Journal of Psycholinguistic Research, 29*, 231–240.
Hagoort, P., Brown, C., & Groothusen, J. (1993). The syntactic positive shift (SPS) as an ERP measure of syntactic processing. *Language and Cognitive Processes, 8*, 439–484.
Hahne, A., & Friederici, A. D. (1999). Electrophysiological evidence for two steps in syntactic analysis: Early automatic and late controlled processes. *Journal of Cognitive Neuroscience, 11*, 194–205.
Henderson, J. M., & Ferreira, F. (1990). Effects of foveal processing difficulty on the perceptual span in reading: Implications for attention and eye movement control. *Journal of Experimental Psychology: Learning, Memory, and Cognition, 16*, 417–429.

Hyönä, J., & Pollatsek, A. (1998). Reading Finnish compound words: Eye fixations are affected by component morphemes. *Journal of Experimental Psychology: Human Perception and Performance, 24*, 1612–1627.

Inhoff, A. W. (1989). Lexical access during eye fixations in reading: Are word access codes used to integrate lexical information across interword fixations? *Journal of Memory and Language, 28*, 444–361.

Inhoff, A. W., & Rayner, K. (1986). Parafoveal word processing during eye fixations in reading: Effects of word frequency. *Perception & Psychophysics, 40*, 431–439.

Jared, D., Levy, B. A., & Rayner, K. (1999). The role of phonology in the activation of word meanings during reading: Evidence from proofreading and eye movements. *Journal of Experimental Psychology: General, 128*, 219–264.

Johnson, R., Perea, M., & Rayner, K. (2007, in press). Transposed letter effects in reading: Evidence from eye movements and parafoveal preview. Journal of Experimental Psychology: Human Perception and performance.

Juhasz, B. J. (2005). Age-of-acquisition effects in word and picture processing. *Psychological Bulletin, 131*, 684–712.

Juhasz, B. J., & Rayner, K. (2003). Investigating the effects of a set of intercorrelated variables on eye fixation durations in reading. *Journal of Experimental Psychology: Learning, Memory, and Cognition, 29*, 1312–1318.

Juhasz, B. J., & Rayner, K. (2006). The role of age-of-acquisition and word frequency in reading: Evidence from eye fixation durations. *Visual Cognition*, 13, 846–863.

Juhasz, B. J., Starr, M. S., Inhoff, A. W., & Placke, L. (2003). The effects of morphology on the processing of compound words: Evidence from naming, lexical decisions and eye fixations. *British Journal of Psychology, 94*, 223–244.

Just, M. A., & Carpenter, P. (1980). A theory of reading: From eye fixations to comprehension. *Psychological Review, 85*, 109-130.

Kambe, G. (2004). Parafoveal processing of prefixed words during eye fixations in reading: Evidence against morphological influences on parafoveal preprocessing. *Perception & Psychophysics, 66*, 279–292.

Kennison, S. M., & Clifton, C. (1995). Determinants of parafoveal preview benefit in high and low working memory capacity readers: Implications for eye movement control. *Journal of Experimental Psychology: Learning, Memory, and Cognition, 21*, 68–81.

Kutas, M., & Hillyard, S. A. (1980). Reading between the lines: Event-related brain potentials during natural sentence processing. *Brain and Language, 11*, 354–373.

Lewis, R. L. (2000). Falsifying serial and parallel parsing models: Empirical conundrums and an overlooked paradigm. *Journal of Psycholinguistic Research, 29*, 241–248.

Lima, S. D. (1987). Morphological analysis in reading. *Journal of Memory and Language, 26*, 84–99.

MacDonald, M. C., Pearlmutter, N. J., & Seidenberg, M. S. (1994). The lexical nature of syntactic ambiguity resolution. *Pyschological Review, 101*, 676–703.

Marslen-Wilson, W. D. (1973). Linguistic structure and speech shadowing at very short latencies. *Nature, 244*, 522–523.

McDonald, S. A., Carpenter, R. H. S., & Shillcock, R. C. (2005). An anatomically-constrained, stochastic model of eye movement control in reading. *Psychological Review, 112*, 814–840.

McRae, K., Spivey-Knowlton, M. J., & Tanenhaus, M. K. (1998). Modeling the influence of thematic fit (and other constraints) in on-line sentence comprehension. *Journal of Memory and Language, 38*, 283–312.

Munte, T. F., Heinze, H. J., & Mangun, G. R. (1993). Dissociation of brain activity related to syntactic and semantic aspects of language. *Journal of Cognitive Neuroscience, 5*, 335–344.

Murray, W. S., & Rowan, M. (1998). Early, mandatory, pragmatic processing. *Journal of Psycholinguistic Research, 27*, 1–22.

Neville, H., Nicol, J. L., Barss, A., Forster, K. I., & Garrett, M. F. (1991). Syntactically based sentence processing classes: Evidence from event-related brain potentials. *Journal of Cognitive Neuroscience, 3*, 152–165.

Ni, W., Crain, S., & Shankweiler, D. (1996). Sidestepping garden paths: Assessing the contributions of syntax, semantics and plausibility in resolving ambiguities. *Language and Cognitive Processes, 11*, 283–334.

Ni, W., Fodor, J. D., Crain, S., & Shankweiler, D. (1998). Anomaly detection: Eye movement patterns. *Journal of Psycholinguistic Research, 27*, 515–540.

Osterhout, L., McKinnon, R., Bersick, M., & Corey, V. (1996). On the language specificity of the brain response to syntactic anomalies: Is the syntactic positive shift a member of the P300 family? *Journal of Cognitive Neuroscience, 8*, 507–526.

Osterhout, L., & Mobley, L. A. (1995). Event-related brain potentials elicited by failure to agree. *Journal of Memory and Language, 34*, 739–773.

Pickering, M. J., & Frisson, S. (2001). Processing ambiguous verbs: Evidence from eye movements. *Journal of Experimental Psychology: Learning, Memory, and Cognition, 27*, 556–573.

Pickering, M., McElree, B., & Traxler, M. (2005). The difficulty of coercion: A response to de Almeida. *Brain and Language, 93*, 1–9.

Pollatsek, A., & Hyönä, J. (2005). The role of semantic transparency in the processing of Finnish compound words. *Language and Cognitive Processes, 20*, 261–290.

Pollatsek, A., Hyönä, J., & Bertram, R. (2000). The role of morphological constituents in reading Finnish compound words. *Journal of Experimental Psychology: Human Perception and Performance, 26*, 820–833.

Pollatsek, A., Reichle, E. D., & Rayner, K. (2003). Modeling eye movements in reading: Extensions of the E-Z Reader model. In J. Hyönä, R. Radach, & H. Deubel (Eds.), *The mind's eye: Cognitive and applied aspects of eye movement research* (pp. 361–390). Amsterdam: North Holland.

Pollatsek, A., Reichle, E., & Rayner, K. (2006). Tests of the E-Z Reader model: Exploring the interface between cognition and eye-movement control. *Cognitive Psychology, 52*, 1–56.

Rayner, K. (1977). Visual attention in reading: Eye movements reflect cognitive processes. *Memory & Cognition, 4*, 443–448.

Rayner, K. (1978). Eye movements in reading and information processing. *Psychological Bulletin, 85*, 618–660.

Rayner, K. (1995). Eye movements and cognitive processes in reading, visual search, and scene perception. In J. M. Findlay, R. Walker, & R. W. Kentridge (Eds.), *Eye movement research: Mechanisms, processes and applications* (pp. 3–22). Amsterdam: North-Holland Press.

Rayner, K. (1998). Eye movements in reading and information processing: 20 years of research. *Psychological Bulletin, 124*, 372–422.

Rayner, K., Ashby, J., Pollatsek, A., & Reichle, E. (2004). The effects of frequency and predictability on eye fixations in reading: Implications for the E-Z Reader model. *Journal of Experimental Psychology: Human Perception and Performance, 30*, 720–732.

Rayner, K., Carlson, M., & Frazier, L. (1983). The interaction of syntax and semantics during sentence processing: Eye movements in the analysis of semantically biased sentences. *Journal of Verbal Learning and Verbal Behavior, 22*, 358–374.

Rayner, K., Cook, A. E., Juhasz, B. J., & Frazier, L. (2006). Immediate disambiguation of lexically ambiguous words during reading: Evidence from eye movements. *British Journal of Psychology, 97*, 467–482.

Rayner, K., & Duffy, S. (1986). Lexical complexity and fixation times in reading: Effects of word frequency, verb complexity, and lexical ambiguity. *Memory & Cognition, 14*, 191–201.

Rayner, K., & Frazier, L. (1989). Selection mechanisms in reading lexically ambiguous words. *Journal of Experimental Psychology: Learning, Memory, and Cognition, 15*, 779–790.

Rayner, K., Liversedge, S. P., White, S. J., & Vergilino-Perez, D. (2003). Reading disappearing text: Cognitive control of eye movements. *Psychological Science, 14*, 385–389.

Rayner, K., Pollatsek, A., & Binder, K. (1998). Phonological codes and eye movements in reading. *Journal of Experimental Psychology: Learning, Memory, and Cognition, 24*, 476–497.

Rayner, K., Raney, G., & Pollatsek, A. (1995). Eye movements and discourse processing. In R. F. Lorch & E. J. O'Brien (Eds.), *Sources of coherence in reading* (pp. 9–36). Hillsdale, NJ: Erlbaum.

Rayner, K., & Sereno, S. C. (1994). Eye movements in reading: Psycholinguistic studies. In M. A. Gernsbacher (Ed.), *Handbook of psycholinguistics* (pp. 57–81). San Diego: Academic Press.

Rayner, K., Sereno, S., Morris, R., Schmauder, R., & Clifton, C. J. (1989). Eye movements and on-line language comprehension processes. *Language and Cognitive Processes, 4*, SI 21–50.

Rayner, K., Warren, T., Juhasz, B. J., & Liversedge, S. P. (2004). The effect of plausibility on eye movements in reading. *Journal of Experimental Psychology: Learning, Memory and Cognition, 30*, 1290–1301.

Rayner, K., & Well, A. D. (1996). Effects of contextual constraint on eye movements in reading: A further examination. *Psychonomic Bulletin & Review, 3*, 504–509.

Reichle, E. D., Pollatsek, A., Fisher, D. F., & Rayner, K. (1998). Toward a model of eye movement control in reading. *Psychological Review, 105*, 125–156.

Reichle, E. D., Pollatsek, A., & Rayner, K. (2006). E-Z Reader: A cognitive-control, serial-attention model of eye-movement behavior during reading. *Cognitive Systems Research, 7*, 4–22.

Reichle, E. D., Rayner, K., & Pollatsek, A. (1999). Eye movement control in reading: Accounting for initial fixation locations and refixations within the E-Z Reader model. *Vision Research, 39*, 4403–4411.

Reichle, E. D., Rayner, K., & Pollatsek, A. (2003). The E-Z Reader model of eye-movement control in reading: Comparisons to other models. *Behavioral and Brain Sciences, 26*, 445–476.

Reilly, R., & Radach, R. (2006). Some empirical tests of an interactive activation model of eye movement control in reading. *Cognitive Systems Research, 7*, 34–55.

Ross, J. R. (1967). *Constraints on Variables in Syntax*, Unpublished Dissertation, MIT, Cambridge, Massachusetts, USA.

Schroyens, W., Vitu, F., Brysbaert, M., & d'Ydewalle, G. (1999). Eye movement control during reading: Foveal load and parafoveal processing. *Quarterly Journal of Experimental Psychology, 52A*, 1021–1046.

Seidenberg, M. S., Tanenhaus, M., Leiman, J., & Bienkowski, M. (1982). Automatic access of the meanings of ambiguous words in context: Some limitations of knowledge-based processing. *Cognitive Psychology, 4*, 489–537.

Sereno, S. C., O'Donnell, P., & Rayner, K. (2006). Eye movements and lexical ambiguity resolution: Investigating the subordinate bias effect. *Journal of Experimental Psychology: Human Perception and Performance, 32*, 335–350.

Spivey, M. J., & Tanenhaus, M. K. (1998). Syntactic ambiguity resolution in discourse: Modeling the effects of referential context and lexical frequency. *Journal of Experimental Psychology: Learning, Memory, and Cognition, 24*, 1521–1543.

Tanenhaus, M. K., & Donnenwerth-Nolan, S. (1984). Syntactic context and lexical access. *Quarterly Journal of Experimental Psychology, 36A*, 649–661.

Tanenhaus, M. K., Leiman, J. M., & Seidenberg, M. S. (1979). Evidence for multiple stages in the processing of ambiguous words in syntactic contexts. *Journal of Verbal Learning and Verbal Behavior, 18*, 427–440.

Traxler, M., Pickering, M., & Clifton, C., Jr. (1998). Adjunct attachment is not a form of lexical ambiguity resolution. *Journal of Memory and Language, 39*, 558–592.

Traxler, M., Pickering, M. J., & McElree, B. (2002). Coercion in sentence processing: Evidence from eye movements and self-paced reading. *Journal of Memory and Language, 47*, 530–548.

Trueswell, J. C., Tanenhaus, M. K., & Garnsey, S. M. (1994). Semantic influences on parsing: Use of thematic role information in syntactic disambiguation. *Journal of Memory and Language, 33*, 285–318.

White, S. J., Rayner, K., & Liversedge, S. P. (2005). Eye movements and the modulation of parafoveal processing by foveal processing difficulty: A re-examination. *Psychonomic Bulletin & Review, 12*, 891–896.

Williams, R. S., & Morris, R. K. (2004). Eye movements, word familiarity, and vocabulary acquisition. *European Journal of Cognitive Psychology, 16*, 312–339.

Yang, S. (2006). An oculomotor-based model of eye movements in reading: The competition/activation model. *Cognitive Systems Research, 7*, 56–69.

Chapter 16

THE INFLUENCE OF SEMANTIC TRANSPARENCY ON EYE MOVEMENTS DURING ENGLISH COMPOUND WORD RECOGNITION

BARBARA J. JUHASZ

Wesleyan University, USA

Eye Movements: A Window on Mind and Brain
Edited by R. P. G. van Gompel, M. H. Fischer, W. S. Murray and R. L. Hill

Abstract

Compound words present an opportunity to study the organization of the mental lexicon. Decomposition of compound words occurs in multiple languages. Most studies of compound words have used semantically transparent compounds. It is therefore difficult to localize the level at which decomposition occurs. The present study explored the role of semantic transparency for English compound words. Transparent and opaque compounds were embedded into sentences and the frequencies of their lexemes were manipulated. Analysis of gaze durations revealed main effects of lexeme frequency and transparency. Transparency did not interact with lexeme frequency, suggesting decomposition occurs for both transparent and opaque compounds.

1. Introduction

One important question for reading researchers regards the nature of representations for words in the mental lexicon and how those representations are accessed. Recording readers' eye movements has been a particularly useful way to study the word recognition process. Numerous studies have demonstrated that characteristics of the word currently being fixated affect its reading time, such as the word's frequency or familiarity (e.g., Juhasz & Rayner, 2003; 2006; Rayner & Duffy, 1986; Rayner, Sereno, & Raney, 1996; Schilling, Rayner, & Chumbley, 1998; Williams & Morris, 2004), age-of-acquisition (e.g., Juhasz & Rayner, 2003; 2006), concreteness (e.g. Juhasz & Rayner, 2003), and length (e.g., Juhasz & Rayner, 2003; Rayner et al., 1996). One can assume, therefore, that these variables affect the accessibility of the word's stored representation in the mental lexicon. Any complete model of word recognition or eye movements during reading should therefore strive to account for the effects of these variables. However, the fact that these variables influence word recognition speed does not necessarily provide direct information regarding the organization of the mental lexicon.

The study of morphologically complex words provides a good opportunity to investigate the organization of the mental lexicon. One central question in this research is whether complex words are represented by only a single lexical entry, whether they are only recognized by being decomposed into their parts, or whether they are represented at multiple levels. There are several types of morphologically complex words including prefixed words, suffixed words, and compound words. A compound is a word composed of (at least) two free lexemes that, when combined, refer to a new concept (e.g., *blackbird*, *farmhouse*).[1] Research that has recorded readers' eye movements as they read sentences containing compound words has been informative regarding the nature of the representations of compound words in the mental lexicon.

As mentioned above, multiple eye movement studies have demonstrated that high-frequency words are fixated for a shorter time than low-frequency words. Several studies have used this fact to examine whether compound words are decomposed into their lexemes during recognition. In a compound word it is possible to manipulate the frequencies of the constituents independent of the overall word frequency. If compound words are decomposed during recognition, then compounds with a high-frequency lexeme should produce shorter fixation durations compared with compounds containing a low-frequency lexeme. If compound words are not decomposed during recognition, then it is the frequency of the overall compound word that should matter, and the frequency of the compound's constituents should not affect fixation durations.

In one eye-movement study, Hyönä and Pollatsek (1998) observed decomposition for bilexemic Finnish compound words. The frequency of the beginning lexeme influenced

[1] In this chapter, the term "compound word" refers to compounds that are written without an interword space. For a discussion of the role of spacing for compound words and how this affects eye fixation behavior, see Juhasz, Inhoff, and Rayner (2005).

first fixation durations as well as gaze durations (a measure that takes refixations on a word into account) on the compound. Pollatsek, Hyönä, and Bertram (2000) also observed effects of second lexeme frequency and whole compound frequency in gaze durations. Based on these results, Pollatsek et al. proposed a dual-route theory of compound word recognition, according to which there is both a direct route to compound word meaning as well as a route through the decomposed lexemes.

Recent eye-movement experiments with English compound words have also found evidence of morphological decomposition. Juhasz, Starr, Inhoff, and Placke (2003) orthogonally manipulated the frequencies of beginning and ending lexemes in familiar compound words. A small, non-significant effect of beginning lexeme frequency was observed on first fixation durations as well as a larger, more robust effect of ending lexeme frequency in gaze duration. Based on these findings, Juhasz et al. concluded that the ending lexeme may have a privileged role in the recognition of English compound words.

Andrews, Miller, and Rayner (2004) also examined the processing of English compound words in an eye-movement study where lexeme frequencies were manipulated. Unlike the Juhasz et al. study, they observed large effects of beginning lexeme frequency in gaze duration and a smaller effect of ending lexeme frequency. Similar to the Finnish data, Andrews et al. also observed a whole-word frequency effect for compound words in a regression analysis. Thus, these results support the hypothesis that compounds are decomposed during recognition, but also have a whole-word representation. It appears that some of the difference between the Juhasz et al. findings and Andrews et al. findings in terms of importance of lexeme location may be due to the length of the compound words used in the two studies. In the Juhasz et al. study, all compound words were 9 letters in length. In the Andrews et al. study, compound word lengths varied between 6 and 11 characters. Recent results by Bertram and Hyönä (2003) have suggested less of an influence for the beginning lexeme frequency for shorter compound words (7–9 letters) in Finnish. Thus, the addition of longer compounds in the Andrews et al. study may be one of the reasons why larger beginning lexeme effects were observed compared to the Juhasz et al. study.

The clear result from the above studies is that compound words are decomposed at some point during their recognition. In addition, compound words appear to be represented as a whole word in the lexicon as well. What the studies cannot speak to is how the decomposed lexemes are being used to access the compound word representation. In the studies mentioned above, most of the compound words were semantically transparent. A semantically transparent compound word is one in which both lexemes contribute to the meaning of the compound word (e.g., *farmhouse*). This can be contrasted with a semantically opaque compound word where the lexemes do not contribute to the meaning of the compound (e.g., *snapdragon*). For semantically transparent compound words, the lexemes in the compound word are related to the compound word on both a morphological level as well as a semantic level.

In one influential view of morphology (Marslen-Wilson, Tyler, Waksler, & Older, 1994), the role of morphemes is to provide a relationship between word forms and word meanings. According to this view, only semantically transparent complex words are connected to the representations of their morphological constituents. However, recent work using the masked priming technique has called this view into question. In this technique, a quickly presented prime word is masked. After the prime word appears a target word is presented for some type of response. If the duration of the prime word is short enough, participants are not consciously aware of it. Recently, Rastle, Davis, and New (2004) demonstrated significant priming using this technique for words with semantically transparent morphological relationships (*cleaner/clean*), and words with an apparent morphological relationship but no semantic relationship (*corner/corn*), but no priming for words consisting of an embedded word and an illegal suffix (*brothel/broth*). These results suggest a very early morphological decomposition process that operates at a level in the lexicon prior to semantics (see also Christianson, Johnson, & Rayner, 2005).

There is also evidence using compound word stimuli for an early non-semantically mediated decomposition process. The majority of work investigating the role of semantic transparency for compound words has used priming in a lexical decision task. In this task, participants are provided with a string of letters and they must decide whether these letters make up a word. Monsell (1985) presented participants with English transparent compound words (e.g., *tightrope*), opaque compound words (e.g., *butterfly*), or pseudo-compounds (e.g., *furlong*). In the prime phase, constituents of the three types of stimuli were presented for lexical decision. Following this, the actual compound words were presented for lexical decision. There was significant constituent repetition priming for transparent compounds, opaque compounds, and pseudocompounds in this experiment, suggesting that all compound words were decomposed at some level.

In contrast to Monsell (1985), Sandra (1990) did not find equivalent priming for transparent and opaque compound words. In a series of lexical decision experiments using Dutch stimuli, Sandra failed to obtain priming for the lexemes in opaque compounds and pseudocompounds but did obtain significant priming for transparent compound words. However, Sandra (1990) used a semantic prime for the constituents in the compounds (e.g., *bread* priming *butterfly*), as opposed to a repetition prime. The importance of this difference is highlighted by Zwitserlood (1994), who investigated the role of semantic transparency for German compound words. In one experiment, Zwitserlood used compound words that were transparent, partially opaque (where one constituent was related in meaning to the compound and the other one was not), or fully opaque (where neither constituent of the compound was related in meaning to the compound expression). These compounds served as primes for their first or second constituents in a lexical decision task. Significant priming was observed for all types of compounds in this experiment. In a second experiment, transparent, partially opaque, fully opaque, and pseudocompounds served as primes for a word semantically associated with one of their constituents. There was no priming of semantic associates for fully opaque compounds or pseudocompounds. However, significant semantic priming occurred for the transparent compounds and the partially opaque compounds. Based on these findings, Zwitserlood argued that all compound words (and pseudocompound words) are represented as morphologically complex at some level in the lexicon, and that this information is represented at a stage prior to semantics.

Jarema, Busson, Nikolova, Tsapkini, and Libben (1999) also found significant constituent repetition priming in a lexical decision task for fully transparent, partially opaque, and fully opaque French compound words. For Bulgarian, significant constituent repetition priming was observed for transparent and partially opaque compounds, but not for fully opaque compounds. Using English compounds, Libben, Gibson, Yoon, and Sandra (2003) found significant constituent priming for transparent, partially opaque, and fully opaque compound words.

These studies support a fast decomposition process that is not tied to semantics. However, all of the studies have used the lexical decision task. This task is sensitive to other factors, such as the makeup of the stimulus list (see Andrews, 1986). The recording of eye movements while reading compound words provides a more natural way to study the role of semantic transparency in morphological decomposition. To date, there are only a small set of eye-movement studies examining the semantic transparency of compound words. Underwood, Petley, and Clews (1990) embedded semantically transparent and opaque compound words in sentences. They observed a significant effect of transparency on gaze durations, with longer gazes in the case of opaque compounds. These results can be contrasted with those of Pollatsek and Hyönä (2005), who manipulated the frequency of the first lexeme in Finnish bilexemic transparent and opaque compound words. They observed a significant effect of beginning lexeme frequency, but no significant effects of semantic transparency on gaze duration. Semantic transparency of the compound also did not interact with the frequency of the beginning lexeme. Thus, both semantically transparent and semantically opaque compound words were decomposed, supporting an early decomposition process.

In the present experiment, semantically transparent and semantically opaque English compound words were embedded in neutral sentence contexts. As mentioned above, two published eye-movement studies examining semantic transparency have yielded contradictory results as to whether transparency influences gaze durations. Therefore, one purpose of the present experiment was to test the hypothesis that semantic transparency influences gaze durations. In addition, the frequencies of the first and second lexemes were orthogonally manipulated to index morphological decomposition. This extends the work of Pollatsek and Hyönä (2005), since in that study only the frequency of the beginning lexeme was manipulated. Finally, this study also explored an additional measure compared to previous studies, go-past durations. This measure is particularly suited to examine how words are integrated with the beginning of the sentence. The hypothesis currently under examination is that both transparent and opaque compound words are decomposed initially due to a fast initial decomposition process that is not tied to the meaning of the compound word. When words are integrated into the sentence, the meaning of the word becomes particularly important. Thus, it is possible that different effects will be observed in a measure indexing sentence integration compared to initial word recognition.

2. Method

2.1. Participants

Twenty-four University of Massachusetts community members participated in exchange for extra course credit or eight dollars. All participants were native speakers of American English and had normal vision or wore contacts.

2.2. Apparatus

Eye movements were recorded via a Fourward Technologies Dual-Purkinje eye-tracker (Generation V) interfaced with an IBM compatible computer. Viewing was binocular; however, eye movements were only recorded from the right eye. The eye tracker has a resolution of less than 10 min. of arc. The sentences were displayed on a 15-inch NEC MultiSync 4FG monitor. The monitor was set 61 cm from the participant. At this distance, 3.8 characters equal one degree of visual angle.

2.3. Procedure

Participants were given instructions detailing the procedure upon arrival to the experiment. A bite bar was prepared and head rests were also used to stabilize the head. Prior to starting the experiment the eye-tracker was calibrated. This calibration was checked after each sentence throughout the experiment, and redone as necessary. Comprehension questions were asked on 10–15% of the trials by providing participants with a question that they could verbally answer "yes" or "no".

2.4. Materials

Forty transparent and 40 opaque English bilexemic compound words were selected as the stimuli. Opaque and transparent compounds were initially chosen based on experimenter intuitions, with transparent compounds classified as those where both lexemes in the compound contribute to the overall meaning of the compound (e.g., *dollhouse*) and opaque compounds classified as compounds where the meaning of the compound word was not easily computable from the meaning of the two lexemes (e.g., *pineapple)*. These intuitions were checked in an experiment where eight University of Massachusetts community members who did not take part in the eye-tracking experiment were asked to rate the compound words on a 1–7 scale in terms of how transparent the meaning of the compound words were, with higher numbers signaling greater transparency of meaning. There was a clear dissociation between the transparent and opaque compounds. On average, opaque compounds received a rating of 2.86 while transparent compounds received an average rating of 5.79. This difference between opaque and transparent compounds was significant ($t(78) = -21.74$, $p < .001$). Average transparency ratings did not differ as a function of lexeme frequency (t's <1).

In addition to semantic transparency, the frequencies of the first and second lexeme were also orthogonally manipulated, resulting in four conditions for each transparency class (eight conditions in total). These conditions consisted of words where both lexemes were high in frequency (HH), where the first lexeme was high in frequency and the second lexeme was low in frequency (HL), where the first lexeme was low in frequency and the second was high in frequency (LH), and where both lexemes were low in frequency (LL).[2] Lexeme frequency was calculated as each lexeme's frequency of occurrence as an individual word. Frequencies were calculated from the CELEX English database written frequencies and scaled to be out of 1 million (Baayen, Piepenbrock, & Gulikers, 1995). High-frequency lexemes had a frequency of greater than 53 per million (range 53.18–1009.55) while low-frequency lexemes had a frequency of less than 39 per million (range 0–38.21 per million). Beginning lexeme frequency did not vary significantly as a function of ending lexeme frequency or transparency (t's < 1). Ending lexeme frequency did not vary significantly as a function of beginning lexeme frequency or transparency (t's < 1). Compound words were also controlled on overall word frequency, which did not vary significantly as a function of transparency or lexeme frequency (p's $> .1$). Compound words ranged from 8 to 11 characters. Length of the compound words, and the length of their lexemes did not vary significantly as a function of transparency or lexeme frequencies (t's < 1). In addition, the number of unspaced compounds occurring in the CELEX database (Baayen et al., 1995) with the same beginning lexeme was also calculated for each compound word to get a measure of morphological family size for the beginning lexeme.[3] The average number of compounds containing the same beginning lexeme was 8.29 (range 1–106). As would be expected, this number was significantly greater for compounds containing a high-frequency lexeme ($t(78) = 2.72$, $p < .01$). Morphological productivity did not significantly vary as a function of ending lexeme frequency or transparency (t's < 1). Table 1 provides information about the conditions and Table 2 provides examples of the materials.

Compound words were embedded in neutral sentence frames, with the condition that the compound could not occupy the first two or last two positions in the sentence. Each compound word was fit into its own neutral sentence frame. Each target compound was preceded and followed by a word of 5–8 characters. The pre-target word was carefully selected to be of a mid-range frequency (ranging from 21 to 72 words per million in the CELEX database, Baayen et al., 1995), while the post-target word frequency was allowed to vary (ranging from 11 to 180 words per million). Importantly, the frequency of the pre-target or post-target word did not significantly differ with respect to the frequency of lexemes, or semantic transparency of the compounds (all p's $> .30$). Ten University of

[2] Initially there were 10 items per condition. However, two items were incorrectly classified as LH transparent compound words. The items (*battlefield* and *minefield*) should actually have been classified as HH transparent compound words. Prior to analyzing the data, these items were reclassified, resulting in 12 items in the HH transparent condition and 8 items in the HL transparent condition. Importantly, the pattern of significant results did not change when these items were reclassified.

[3] Plurals and derivations of the compounds that occurred in the database were not counted as separate compounds.

Table 1
Stimulus characteristics as a function of compound type (opaque or transparent) and the frequency of the beginning and ending lexeme

Type	Lex1	Lex2	Compound Length	Compound Freq.	Lex1 Length	Lex1 Freq.	Lex2 Length	Lex2 Freq.
Opaque	High	High	9.2	1.1	4.7	219.6	4.5	214.1
Opaque	High	Low	9.1	1.6	4.5	235.2	4.6	13.1
Opaque	Low	High	8.8	1.9	4.3	21.0	4.5	302.7
Opaque	Low	Low	9.4	0.7	4.4	6.1	5.0	6.2
Transparent	High	High	9.0	2.6	4.1	203.5	4.9	235.5
Transparent	High	Low	9.4	1.2	4.7	298.0	4.7	7.7
Transparent	Low	High	9.1	0.6	4.8	12.9	4.4	265.8
Transparent	Low	Low	8.8	0.7	4.6	9.3	4.2	6.8

Note: Lex1 = beginning lexeme, Lex2 = Ending lexeme, Freq = average Celex written frequency per million (Baayen et al., 1995). Lengths are expressed in average number of letters per condition.

Table 2
Examples of the compound words and sentences used in the experiment

Type	Lex1	Lex2	Example
Opaque	High	High	I know that the massive **pocketbook** belongs to my mom.
Opaque	High	Low	The very modest **wallflower** smiled when asked to dance.
Opaque	Low	High	The honest **deckhand** returned the lost money.
Opaque	Low	Low	Her valuable **heirloom** broke yesterday morning.
Transparent	High	High	We had a new wooden **headrest** added to the chair.
Transparent	High	Low	The police found a bloody **doorknob** prior to entering the house.
Transparent	Low	High	A vile and angry **swearword** killed the good mood.
Transparent	Low	Low	Our nation's highest **flagpole** broke in the sudden storm.

Massachusetts community members performed a cloze task where they were presented with the sentence up to the pre-target word and were asked to provide the next word in the sentence. Target predictability was very low, amounting to 2.5% or less on average for each condition. In addition, cloze task performance did not significantly differ as a function of transparency or lexeme frequency (all p's > .12). All sentences were less than 80 characters in length and occupied only a single line on the computer monitor. Each participant viewed 80 experimental sentences along with 72 filler sentences.

3. Results

In order to obtain a complete picture of the time-course of compound word processing, a number of dependent measures were examined. The primary measure was gaze duration, which is a sum of all fixation durations on the compound word prior to the eyes leaving the word to the right or left during the first time it was read. In addition, to investigate early processing of the compound, the duration of the first fixation was examined. Further, to investigate integration of the compound word with the sentence, go-past duration was also examined. Go-past duration is the sum of all fixations on the compound plus the duration of any regressions back to the beginning of the sentence before the reader moves their eyes to the right of the compound. This measure gives an indication of the ease with which the compound meaning can be integrated in the sentence. Table 3 displays the participant means for the eight conditions.

Data were analyzed using 2 (semantic transparency) × 2 (beginning lexeme frequency) × 2 (ending lexeme frequency) Analyses of Variance (ANOVAs). Error variance was computed over participants and items. All variables were considered within participants and between items. In English, there are a limited number of compound words. The compounds in this experiment were matched on many variables. According to Raaijmakers, Schrijnemakers, and Gremmen (1999), when items are selected so as to be controlled on several variables correlated with the dependent measures, the traditional F1 statistics (where participants are treated as the only random variable) are the correct ones to use, as F2 analyses may underestimate the true effects. Also, in this experiment, each word was fit into its own sentence frame, which adds variability into the items analysis. Therefore, the current items analyses did not produce many results at the standard $p < .05$ level. The items analyses are reported for completeness, and to be consistent with other published work on compound words. However, the results from the participants' analyses will be stressed in the following.

Fixations shorter than 80 ms that were within one character of another fixation were combined. Trials were removed from analysis due to track losses or blinks on the target, pre-target, or post-target regions. This led to the removal of approximately 5% of trials. In addition, fixations shorter than 100 ms and longer than 1000 ms were eliminated from the data by the data analysis software.

Table 3
First Fixation Duration (FFD), Gaze Duration (GD) and Go-Past Duration (GP) as a function of compound type (opaque or transparent) and the frequency of the beginning and ending lexeme.

Type	Lex1	Lex2	FFD	GD	GP
Opaque	High	High	322	427	496
Opaque	High	Low	321	433	520
Opaque	Low	High	322	430	475
Opaque	Low	Low	339	472	521
Transparent	High	High	323	385	444
Transparent	High	Low	316	400	438
Transparent	Low	High	312	444	480
Transparent	Low	Low	340	441	496

Note: All durations are in milliseconds.

3.1. Gaze duration

There was a main effect of transparency on gaze durations. Transparent compounds received 24 ms shorter gaze durations on average compared to opaque compounds ($F1(1,23) = 10.36$, $MSe = 2747$, $p = .004$; $F2(1,72) = 2.14$, $MSe = 4824$, $p = .148$). There was an effect of beginning lexeme frequency, with compounds containing a high-frequency beginning lexeme receiving 36 ms shorter gaze durations than compounds with a low-frequency beginning lexeme ($F1(1,23) = 13.67$, $MSe = 4388$, $p = .001$; $F2(1,72) = 5.34$, $MSe = 4824$, $p = .024$). There was also a main effect of ending lexeme frequency in the analysis by participants. Compounds containing a high-frequency ending lexeme received 16 ms shorter gaze durations on average compared to compounds containing a low-frequency ending lexeme ($F1(1,23) = 7.83$, $MSe = 1448$, $p = .01$; $F2 < 1$). None of the interactions reached significance in this measure (all p's $> .15$).

3.2. First fixations

First fixations landed 3.79 characters on average into the compound words. There were no significant main effects or interactions on first fixation position (all p's $> .1$).There was an effect of beginning lexeme frequency on first fixation duration, with compounds containing a high-frequency beginning lexeme receiving 8 ms shorter first fixation durations ($F1(1,23) = 4.89$, $MSe = 720$, $p = .037$; $F2(1,72) = 1.77$, $MSe = 702$, $p = .188$). There was also a marginal effect of ending lexeme frequency ($F1(1,23) = 4.14$, $MSe = 1294$, $p = .054$; $F2(1,72) = 2.39$, $MSe = 701$, $p = .126$) and a significant interaction between beginning and ending lexeme frequencies ($F1(1,23) = 5.01$, $MSe = 1653$, $p = .035$; $F2(1,72) = 4.52$, $MSe = 702$, $p = .037$). Follow-up comparisons demonstrated that the nature of the interaction was that compounds with two low-frequency lexemes received significantly longer first fixations (340 ms) compared with compounds in the other three conditions (321, 318, 316; all p's$<.02$ except comparing HH to LL for items where $t2(40) = -1.97$, $p = 0.056$), which did not differ significantly from each other (all t's < 1).

3.3. Go-past duration

For go-past duration there was again a significant main effect of transparency, with opaque compounds taking 40 ms longer on average ($F1(1,23) = 8.20$, $MSe = 9461$, $p = .009$; $F2(1,72) = 3.21$, $MSe = 9079$, $p = .078$). In addition, there was a significant effect of ending lexeme frequency in the participants analysis, with compounds with high-frequency second lexemes producing 20 ms shorter go-past durations ($F1(1,23) = 5.56$, $MSe = 3612$, $p = .027$; $F2 < 1$). Transparency also significantly interacted with beginning lexeme frequency by participants ($F1(1,23) = 6.69$, $MSe = 5066$, $p = .016$; $F2(1,72) = 1.81$, $MSe = 9079$, $p = .182$). Specifically, while there was a significant 46 ms effect of beginning lexeme frequency for transparent compounds ($t1(23) = -2.67$, $p = 0.014$; $t2(38) = -1.58$, $p = .122$), there was no significant effect for opaque compounds (t's < 1). These effects in go-past durations were not due to a difference in the percentage of regressions out of the compounds during their first pass, as percentage of regressions did not significantly vary as a function of transparency or lexeme frequency (all p's > 0.25).

4. Discussion

Similar to past eye-movement studies examining compound word recognition, there was a small effect of beginning lexeme frequency on first fixation durations in the present experiment (e.g. Andrews et al., 2004; Hyönä & Pollatsek, 1998; Juhasz et al., 2003). Somewhat differing from these studies, this main effect was qualified by an interaction due to the fact that only compounds composed of two low-frequency lexemes differed significantly in terms of their first fixation durations. In gaze durations in the present study there was a large effect of beginning lexeme frequency (36 ms) and a smaller effect of ending lexeme frequency (16 ms). In terms of previous published experiments with English compound words, these results appear consistent with Andrews et al. and call into question the possibility suggested by Juhasz et al. that the ending lexeme has a privileged role in English compound word recognition. Instead, as Bertram and Hyönä (2003) suggested, it may be that the length of the compounds is what determines the pattern of lexeme frequency effects. The inclusion of longer compound words in the present experiment may have resulted in more robust beginning lexeme effects.

There was no effect of transparency on first fixation durations in the present experiment. This is not surprising since a reader presumably cannot tell whether a compound is opaque or transparent until the second lexeme and/or whole compound is identified. On gaze durations there were main effects of transparency that did not interact with lexeme frequency. Finally, on go-past durations there were again main effects of transparency and second lexeme frequency. However, in this measure transparency did interact with beginning lexeme frequency.

The finding of a main effect of transparency on gaze durations supports the results of Underwood et al. (1990) using English compound words but differs from the results of Pollatsek and Hyönä (2005) using Finnish compound words. One may be tempted to explain this difference as a function of the two languages. However, Frisson, Niswander-Klement, and Pollatsek (2006) recently also examined the role of semantic transparency for English compound words and failed to find any effects. They paired words where both lexemes were transparently related in meaning to the compound word with either partially or fully opaque words. Frisson et al. did not find any effects of transparency on fixation durations or percentage of regressions out of the compound words. Given these discrepancy in results, one may wish to view the present main effect of transparency somewhat cautiously. As mentioned previously, opaque and transparent words were embedded in different sentence frames in the present experiment, although every attempt was made to keep the sentences as uniform as possible.

However, the main effect of transparency is not the most important finding from the gaze duration analyses. Instead, the finding that the frequency of both lexemes influences gaze durations and that these effects do not interact with the transparency of the compound suggests that both transparent and opaque compound words are decomposed in a similar fashion during the early stages of recognition. These findings support the main conclusions from Pollatsek and Hyönä (2005) and Frisson et al. (2006) that semantic transparency of compound words does not mediate the decomposition process. In contrast to the view of morphology discussed in the introduction, the results from these studies do not support the notion that only semantically transparent morphologically complex items are linked to the representations of their morphemes.

The present results from the go-past measure add an important caveat to this conclusion. The go-past duration measure can be conceptualized as a measure of how easily the compound word meaning is integrated with the meaning of the sentence. In this measure there were main effects of transparency and ending lexeme frequency. However, beginning lexeme frequency only had a significant effect for transparent compounds. The difference between lexeme frequency effects as a function of semantic transparency in gaze duration compared to go-past duration is somewhat analogous to the difference between results with lexeme repetition and semantic priming in the lexical decision task highlighted in the Introduction. These results can be explained if transparent compound words are tied to the representations of their lexemes at two different levels in the lexicon. Libben (1998) described a model of compound word recognition that incorporates this hypothesis. His model is composed of three different levels: stimulus, lexical, and conceptual. The compound word is encountered at the stimulus level and is decomposed. If the compound word is not novel, it is connected to both of its decomposed lexemes as well as its whole-word representation at the lexical level. Facilitory links exist at the lexical level between the lexemes and the whole-word form. Finally, the lexical whole-word form is also linked to the conceptual whole-word representation of the compound as well as the conceptual representation for any lexemes it shares meaning with. For example, the transparent compound *blueberry* will be connected to its lexeme representations on both the lexical and the conceptual level. In contrast, the partially opaque compound word *strawberry* will be connected to both of its lexemes on a lexical level, but will only be connected to its ending lexeme *berry* on the conceptual level.

Libben's (1998) theory can incorporate the dual-route model of Pollatsek et al. (2000), which has been very useful for understanding compound word processing. Specifically, the whole-word lexical representation of the compound word can be accessed through either the decomposed lexemes (which have facilitory links to the whole-word lexical representation) or through the entire compound. Which route dominates could be a function of the length of the compound, as suggested by Bertram and Hyönä (2003). The idea that a dual-route to the lexical (non-semantic) representation of the compound word exists was also suggested by Pollatsek and Hyönä (2005).

The Libben (1998) model can also provide an explanation for the interaction between transparency and beginning lexeme frequency in go-past durations in the present study if one assumes that high-frequency concepts are easier to integrate into sentences. Transparent compounds are related in meaning to their lexemes, and are therefore linked to those lexemes on the conceptual level. If these concepts are high in frequency, this may aid the integration process. Opaque compounds are not related in meaning to their lexemes, so they are not linked to them on a conceptual level and therefore the frequency of their lexemes makes no difference at the conceptual level. Of course, this conceptualization of the model would predict interactions of transparency with both lexeme frequencies. In the present case only an interaction was observed for beginning lexeme frequency. This is most likely due to the fact that the opaque compounds used in the present study were not always completely opaque, as defined by Libben. It is possible that on average in the present stimuli the opaque compounds were more related in meaning to their ending lexeme.

The present results coupled with the results from other previous studies provide a convincing case for an early decomposition process that operates at a level prior to semantics. The next question to address is how this quick "pre-semantic" decomposition operates. One possibility would be that information about the morphological status of the upcoming word is processed in the parafovea. However, in English no evidence of morphological processing in the parafovea has been obtained (e.g., Kambe, 2004; Lima, 1987), although evidence has been obtained in Hebrew (Deutsch, Frost, Pelleg, Pollatsek, & Rayner, 2003). Therefore, at least for English this pre-semantic decomposition process must begin once the word is fixated. Seidenberg (1987) suggested that the frequencies associated with letter clusters around morpheme boundaries could provide a possible low-level segmentation cue. However, the only sentence-reading experiment to explicitly examine this with compound words did not find any evidence for an effect of bigram frequency (Inhoff, Radach, & Heller, 2000). There is some evidence from Finnish (Bertram, Pollatsek, & Hyönä, 2004) that other types of information such as vowel harmony can provide useful segmentation cues for decomposition. This suggests that readers can make use of some low-level cues. However, these cues are not language universal, so they cannot be the only way the pre-semantic decomposition operates.

Libben (1994; Libben, Derwig, & de Almeida, 1999) has suggested that all morphologically legal decompositions are fed to the lexical level for analysis. This conclusion is reached through work with ambiguous novel compound words such as clamprod, where both parses (*clam-prod* vs *clamp-rod*) appear to be activated. Work with Finnish compound words, however, may call this hypothesis into question. Hyönä, Bertram, and Pollatsek (2004) presented readers with non-spaced bilexemic Finnish compound words where the ending lexeme was masked prior to fixation. Masking the ending lexeme had no effect on fixation times on the beginning lexeme. This suggests that the lexemes are accessed sequentially in compound words and the morphological legality of the ending lexeme is not checked prior to the access of the beginning lexeme's lexical representation.

In conclusion, the present results provide evidence for an early pre-semantic decomposition mechanism in reading. However, it is still an open question as to how this mechanism operates. There are now quite sophisticated computational models of eye movements during reading. The majority of these models do not take into account the morphological complexity of the word currently being fixated. One exception to this is Pollatsek, Reichle, and Rayner (2003), who adjusted the E-Z Reader model (Reichle, Pollatsek, Fisher, & Rayner, 1998) to account for the pattern of compound frequency effects observed in Finnish. Basically, in this version of the model, the lexemes in a compound are accessed serially and then the meaning of the compound is "glued" together with a formula that is sensitive to the whole-word frequency. However, as Bertram and Hyönä (2003) point out, this model could most likely not account for their finding that the length of the compound modified the pattern of decomposition. This model also does not attempt to address how the decomposition process occurs. While modeling morphological processes may present somewhat of a challenge for modelers, hopefully future eye-movement models will also attempt to model how morphological processes affect eye movements and will be able to take these aspects into account.

Acknowledgements

This research was supported by Grant 16745 from the National Institute of Mental Health. I would like to thank Matt Starr for all of his help with various aspects of this experiment, Sarah Brown for her help in collecting ratings, as well as Keith Rayner, Rebecca Johnson, Robin Hill, Jukka Hyönä, and Sarah White for their helpful comments on the manuscript. Requests for reprints or the materials used in this study should be sent to Barbara J. Juhasz, Department of Psychology, 207 High Street, Wesleyan University, Middletown, CT, 06459, USA.

References

Andrews, S. (1986). Morphological influences on lexical access: Lexical or nonlexical effects? *Journal of Memory and Language, 25*, 726–740.

Andrews, S., Miller, B., & Rayner, K. (2004). Eye movements and morphological segmentation of compound words: There is a mouse in mousetrap. *European Journal of Cognitive Psychology, 16*, 285–311.

Baayen, R. H., Piepenbrock, R., & Gulikers, L. (1995). The CELEX Lexical Database. [CD-ROM]. Philadelphia: Linguistic Data Consortium, University of Pennsylvania.

Bertram, R., & Hyönä, J. (2003). The length of a complex word modifies the role of morphological structure: Evidence from eye movements when reading short and long Finnish compounds. *Journal of Memory and Language, 48*, 615–634.

Bertram, R., Pollatsek, A., & Hyönä, J. (2004). Morphological parsing and the use of segmentation cues in reading Finnish compounds. *Journal of Memory and Language, 51*, 325–345.

Christianson, K., Johnson, R. L., & Rayner, K. (2005). Letter transpositions within and across morphemes. *Journal of Experimental Psychology: Learning, Memory & Cognition, 31*, 1327–1339.

Deutsch, A., Frost, R., Pelleg, S., Pollatsek, A., & Rayner, K. (2003). Early morphological effects in reading: Evidence from parafoveal preview benefit in Hebrew. *Psychonomic Bulletin & Review, 10*, 415–422.

Frisson, S., Niswander-Klement, E., & Pollatsek, A. (2006). The role of semantic transparency in the processing of English compound words. *Manuscript submitted for publication.*

Hyönä, J., Bertram, R., & Pollatsek, A. (2004). Are long compound words identified serially via their constituents? Evidence from an eye-movement-contingent display change study. *Memory & Cognition, 32*, 523–532.

Hyönä, J. & Pollatsek, A. (1998). Reading Finnish compound words: Eye fixations are affected by component morphemes. *Journal of Experimental Psychology: Human Perception and Performance, 24*, 1612–1627.

Inhoff, A. W., Radach, R., & Heller, D. (2000). Complex compounds in German: Interword spaces facilitate segmentation but hinder assignment of meaning. *Journal of Memory and Language, 42*, 23–50.

Jarema, G., Busson, C., Niklova, R., Tsapkini, K., & Libben, G. (1999). Processing compounds: A cross-linguistics study. *Brain and Language, 68*, 362–369.

Juhasz, B. J., Inhoff, A. W., & Rayner, K. (2005). The role of interword spaces in the processing of English compound words. *Language & Cognitive Processes, 20*, 291–316.

Juhasz, B. J., & Rayner, K. (2006). The role of age-of-acquisition and word frequency in reading: Evidence from eye fixation durations. *Visual Cognition, 13*, 846–863.

Juhasz, B. J., & Rayner, K. (2003). Investigating the effects of a set of inter-correlated variables on eye fixation durations in reading. *Journal of Experimental Psychology: Learning, Memory & Cognition, 29*, 1312–1318.

Juhasz, B. J., Starr, M., & Inhoff, A. W., & Placke, L. (2003). The effects of morphology on the processing of compound words: Evidence from naming, lexical decisions, and eye fixations. *British Journal of Psychology, 94*, 223–244.

Kambe, G. (2004). Parafoveal processing of prefixed words during eye fixations in reading: Evidence against morphological influences on parafoveal processing. *Perception & Psychophsyics, 66*, 279–292.

Libben, G. (1994). How is morphological decomposition achieved? *Language and Cognitive Processes, 9*, 369–391.

Libben, G. (1998). Semantic transparency in the processing of compounds: Consequences for representation, processing, and impairment. *Brain and Language, 61*, 30–44.

Libben, G., Derwing, B. L., & de Almeida, R. G. (1999). Ambiguous novel compounds and models of morphological parsing. *Brain and Language, 68*, 378–386.

Libben, G., Gibson, M., Yoom, Y. B., & Sandra, D. (2003). Compound fracture: The role of semantic transparency and morphological headedness. *Brain and Language, 84*, 50–64.

Lima, S. D. (1987). Morphological analysis in sentence reading. *Journal of Memory and Language, 26*, 84–99.

Marslen-Wilson, W. D., Tyler, L. K., Waksler, R., & Older, L. (1994). Morphology and meaning in the English mental lexicon. *Psychological Review, 101*, 3–33.

Monsell, S. (1985). Repetition and the lexicon. In A. W. Ellis (Ed.), *Progress in the psychology of language* (Vol. 1). Hove, UK: Erlbaum.

Pollatsek, A. & Hyönä, J. (2005). The role of semantic transparency in the processing of Finnish compound words. *Language and Cognitive Processes, 20*, 261–290.

Pollatsek, A., Hyönä, J., & Bertram, R. (2000). The role of morphological constituents in reading Finnish compound words. *Journal of Experimental Psychology: Human Perception and Performance, 26*, 820–833.

Pollatsek, A., Reichle, E., & Rayner, K. (2003). Modeling eye movements in reading. In J. Hyönä, R. Radach, and H. Deubel (Eds.). *The mind's eyes: Cognitive and applied aspects of eye movement research* (pp. 361–390). Amsterdam: Elsevier.

Raaijmakers, J. G. W., Schrijnemakers, J. M. C., & Gremmen, F. (1999). How to deal with "the language-as-fixed effect fallacy": Common misconceptions and alternative solutions. *Journal of Memory and Language, 41*, 416–426.

Rastle, K., Davis, M. H., & New, B. (2004). The broth in my brother's brothel: Morpho-orthographic segmentation in visual word recognition. *Psychonomic Bulletin & Review, 11*, 1090–1098.

Rayner, K., & Duffy, S. A. (1986). Lexical complexity and fixation times in reading: Effects of word frequency, verb complexity, and lexical ambiguity. *Memory & Cognition, 14*, 191–201.

Rayner, K., Sereno, S. C., & Raney, G. E. (1996). Eye movement control in reading: A comparison of two types of models. *Journal of Experimental Psychology: Human Perception & Performance, 22*, 1188–1200.

Reichle, E. D., Pollatsek, A., Fisher, D. L., Rayner, K. (1998). Toward a model of eye movement control in reading. *Psychological Review, 105*, 125–157.

Sandra, D. (1990). On the representation and processing of compound words. Automatic access to constituent morphemes does not occur. *The Quarterly Journal of Experimental Psychology, 42a*, 529–567.

Schilling, H. E., Rayner, K., & Chumbley, J. I. (1998). Comparing naming, lexical decision, and eye fixation times: Word frequency effects and individual differences. *Memory & Cognition, 26*, 1270–1281.

Seidenberg, M. S. (1987). Sublexical structures in visual word recognition: Access units or orthographics redundancy? In M. Coltheart (Ed.), *Attention and performance 12: The psychology of reading* (pp. 245–263). Hillsdale, NJ: Erlbaum.

Underwood, G., Petley, K., & Clews, S. (1990). Searching for information during sentence comprehension. In Groner, R., d'Ydewalle, G. et al. (Eds.), *From eye to mind: Information acquisition in perception, search, and reading*. (pp. 191–203). Oxford, England: North-Holland.

Williams, R. S., & Morris, R. K. (2004). Eye movements, word familiarity, and vocabulary acquisition. *European Journal of Cognitive Psychology, 16*, 312–339.

Zwitserlood, P. (1994). The role of semantic transparency in the processing and representation of Dutch compounds. *Language and Cognitive Processes, 9*, 341–368.

Chapter 17

THE INTERPLAY BETWEEN PARAFOVEAL PREVIEW AND MORPHOLOGICAL PROCESSING IN READING

RAYMOND BERTRAM and JUKKA HYÖNÄ

University of Turku, Finland

Eye Movements: A Window on Mind and Brain

Edited by R. P. G. van Gompel, M. H. Fischer, W. S. Murray and R. L. Hill

Abstract

This study investigated whether a morphological preview benefit can be obtained in Finnish with the eye contingent display change paradigm. For that purpose, we embedded in sentences compound words with long or short 1st constituents, while manipulating the amount of information available of the 1st constituent before the compound word was fixated. In the change conditions, the first 3–4 letters were made available parafoveally, which constituted the whole 1st constituent in the case of the short, but only part of the 1st constituent in the case of the long 1st constituents. The results showed that the change manipulation was equally effective for long and short 1st constituent compounds, indicating that in reading Finnish there is no morphological preview benefit. On the other hand, 1st constituent length affected several eye movement measures, indicating that the role of morphology in lexical processing is constrained by visual acuity principles.

When readers fixate on a word, they spend a certain amount of time to access and identify that very word. At the same time, in languages like Finnish or English, they often retrieve information from the word to the right of the fixated word. There is an ongoing debate about the nature of the information that is picked up from the word next to the fixated word. It is widely agreed that low-level features like word length and letter identity of the first few letters are parafoveally processed. However, whereas some scholars claim that parafoveal processing is predominantly visual-orthographic in nature (e.g., Rayner, Balota, & Pollatsek, 1986), others posit that effects of parafoveal processing extend to higher linguistic levels (e.g., Kennedy, 2000; Murray, 1998). In the face of the current empirical evidence, it can be argued that evidence for semantic preprocessing is marginal at best and often restricted to tasks that only mimic natural reading to some extent (Rayner, White, Kambe, Miller, & Liversedge, 2003).

On one higher linguistic level, namely the morphological level, the current evidence is contradictory. It seems that depending on the language, morphological preprocessing may or may not take place. More specifically, Lima (1987), Inhoff (1989), and Kambe (2004) did not observe any preview benefits for morphological units in English. In contrast, Deutsch and her colleagues have observed in several studies that word processing in Hebrew significantly benefits from having a parafoveal preview of the word's morphological root (Deutsch, Frost, Pelleg, Pollatsek, & Rayner, 2003; Deutsch, Frost, Pollatsek, & Rayner, 2000, 2005). The differences between Hebrew and English are notable in many respects (script, reading direction, non-concatenative vs concatenative morphology), which have led Kambe (2004) and Deutsch et al. (2003) to argue that the morphological richness of Hebrew in comparison to English may well be the main reason for the earlier and more prominent impact of morphological structure on word processing in Hebrew. Kambe explicitly states also that in a highly inflected language like Finnish (with concatenative morphology, and the same script and reading direction as in English), morphological preview benefits may be found. In the current study we explored by means of the eye contingent change paradigm (Rayner, 1975) whether this is indeed the case. For long compound words with either a short 1st constituent (HÄÄ/SEREMONIA 'wedding ceremony') or a long 1st constituent (KONSERTTI/SALI 'concert hall'), either a full preview (i.e., the whole word, also coined the 'no change condition') or a partial preview of 3–4 letters (also coined 'the change condition') was provided. In the partial preview condition, the first 3–4 letters amounted to the complete 1st constituent when the constituent was short, but to only part of the constituent when it was long. We reasoned that the partial preview would be more beneficial for short 1st constituent compounds, if Finnish readers indeed accessed the morphological code in the parafovea. If, on the other hand, the preview benefit is purely orthographic in nature, a change effect of similar size should be found for long and short 1st constituent compounds.

The use of compound words is a good way to study morphological preview effects in Finnish, since the processing of Finnish compound words has been quite intensively studied. In fact, there are two compound word studies (Hyönä & Bertram, 2004; Hyönä & Pollatsek, 1998) that have touched upon the issue of parafoveal processing in

Finnish. In the next section we briefly discuss these studies. In the subsequent section we summarize those Finnish compound studies in which the focus is on foveal processing, but in which the investigated variables are at the intersection of morphology and visual acuity.

1. Parafoveal processing of constituents in compound words

Hyönä and Pollatsek (1998) compared the processing of long 1st constituent compounds (e.g., MAAILMAN/SOTA 'world war') with that of short 1st constituent compounds (e.g., YDIN/REAKTORI 'nuclear reactor') by matching them on overall length and frequency. They argued that if the constituents are processed similarly to separate words, the location of initial fixation location would be located further in the word when the 1st constituent is long, since the preferred landing position on words is shown to be somewhat left of the word centre (O'Regan, 1992; Rayner, 1979). Because a saccade into a word is programmed when the eyes are fixated on a preceding word, a constituent length effect on first fixation location would imply morphological parafoveal preprocessing. Hyönä and Pollatsek found that first fixation location was not dependent on the length of the 1st constituent, but second fixation location was further in the word when the 1st constituent was long. These findings imply that morphological structure is recognized once the word is fixated, but not before that (i.e. parafoveally). However, it is possible that a morphological preview effect does not show itself in the initial fixation location (i.e., that morphological processing does not affect where the eyes are sent).

In principle, it is possible that a morphologically complex word appearing in the parafovea affects the processing of the preceding word, Word *N*–1 (a so-called parafoveal-on-foveal effect). Hyönä and Bertram (2004) reanalysed the data of five compound word experiments in order to find out whether such parafoveal-on-foveal effects exist. Even though they found some impact of the frequency of the 1st constituent of the parafoveally presented compound words, their results were inconsistent in that sometimes a positive, sometimes a negative, and sometimes no frequency effect was found; in addition, the generalizability was questionable in that often the item analyses remained non-significant. However, post-hoc regression analyses showed that the 1st constituent of parafoveally presented short compound words might affect the processing of Word *N*–1, with a low frequency constituent attracting a fixation towards it more readily than a high frequency 1st constituent, leading to a shorter gaze duration on Word *N*–1 in the former compared to the latter case.

In sum, both studies assessed the issue of morphological preprocessing to some extent. One of them did not reveal a morphological preview effect, the other hinted at such an effect for short compound words, but the paradigm and the dependent measures used may not have been the most optimal to reveal morphological preview effects. The eye contingent change paradigm as employed in the present study more directly tested morphological preprocessing either by presenting a morphological preview or not.

2. Foveal processing of compound words and the visual acuity principle

Two-constituent compounds can be (a) long (e.g., TULEVAISUUDEN/SUUNNITELMA 'future plan'); (b) short (e.g., HÄÄ/YÖ 'wedding night'); (c) relatively long but with a short 1st constituent (e.g., YDIN/REAKTORI 'nuclear reactor'); or (d) relatively long with a long 1st constituent (e.g., MAAILMAN/SOTA 'world war'). It is quite clear that the four types of compound words described here differ greatly as to which morpheme would be submitted to detailed visual analysis during the initial fixation on the word. For word type (a) it may be unlikely to have all characters of the initial constituent in foveal vision; word type (b) will have all characters of both constituents in foveal vision; word type (c) will have all characters of the 1st constituent in foveal vision in the great majority of initial fixations; and word type (d) will have the 1st constituent in foveal vision as long as the first fixation is around the centre of the 1st constituent. Even though it may seem obvious that the length of a complex word or a morphemic unit within a word may significantly modulate the role morphology plays in lexical processing, it is only recently taken into consideration when assessing the role of morphology in word recognition.

Hyönä and Pollatsek (1998), Bertram and Hyönä (2003), and Hyönä, Bertram, and Pollatsek (2004) have demonstrated solid effects of 1st constituent frequency on the identification of relatively long two-constituent compounds (12 or more characters). These 1st constituent frequency effects appeared from the first fixation onwards, implying that the 1st constituent is already involved in the early stages of the word recognition process. For short compounds (7–8 characters), the identification process is much more holistic with early and late effects of whole-word frequency, but only a later and statistically marginal effect for 1st constituent frequency (Bertram & Hyönä, 2003). On the basis of these results, Bertram and Hyönä argued that compound-word length indeed modulates the role of morphology due to visual acuity constraints of the eye. Thus, all letters of short words fall into foveal vision and can be identified in a single fixation. Since whole-word access is visually possible and in principle faster than decomposed access (for one thing, one does not need to parse out the individual constituents in case of whole-word access), the holistic access procedure dominates the identification process of short compounds. In contrast, when reading long compound words, a second fixation is needed (typically 80–90% of the time) to make the letters of the 2nd constituent available for foveal inspection. This, in combination with solid 1st constituent frequency effects from the first fixation onwards, suggests that recognition of long compound words is achieved by first accessing the initial constituent followed by the access of the 2nd constituent and that of the whole-word form.

Bertram, Pollatsek, and Hyönä (2004) showed that the identification of a relatively long 1st constituent (7–9 characters) is problematic, when there are no cues to indicate the constituent boundary, but identification can proceed smoothly when a boundary cue is present – for instance, when two vowels of different quality appear at the constituent boundary. In Finnish, vowels of the same type (e.g., all vowels are front vowels) always appear in monomorphemic, inflected and derived words, but vowel quality can differ in compound words across the constituents (e.g., there can be a front vowel /y/ and a back vowel /o/ at the constituent boundary). Bertram et al. (2004) found that reading long 1st constituent compounds with vowels of the same quality around the constituent boundary yielded 114 ms longer gaze durations than reading compounds with vowels of different quality around the constituent boundary. However, short 1st constituent compounds with vowels of the same vowel quality elicited similar gaze durations as those with different vowel quality. Bertram et al. reasoned that in case of short 1st constituents the first letters are practically always in foveal vision allowing an easy access to the 1st constituent so that a cue signalling constituent boundary is of no help. In the case of long 1st constituents, on the other hand, the access of the 1st and 2nd constituent benefits from a segmentation cue.[1]

One question we have not yet clearly answered with our previous studies is what factors govern long compound recognition when 1st constituents are short and thus presumably easily accessible. Are they processed in a similar vein as long compounds with relatively long 1st constituents? In the current study we wished to obtain a clearer insight of the role of 1st constituent length. Specifically, we wanted to investigate whether constituent frequency and whole word frequency play a similar role in processing long compounds with a relatively short versus long first constituent.

3. The current study

What is clear from the aforementioned studies is that constituents are functional processing units in compound-word processing, but that the role they play is modulated by visual acuity constraints. With respect to parafoveal morphological preprocessing, if the 1st constituent is accessed parafoveally, it is quite clear that it has to be short. Due to visual acuity constraints and also in line with our previous study (Hyönä & Bertram, 2004), it is unlikely that a long 1st constituent (or the 2nd constituent) would be accessed parafoveally. In the present study we set out to examine this question. We investigated this question by embedding compounds with long and short 1st constituents in sentences, while manipulating the amount of parafoveal information prior to foveal inspection of the compounds. The second issue we wanted to explore in more detail was the role of 1st constituent length in compound-word processing once the compound is fixated. To that end, we conducted post-hoc regression analyses with 1st constituent frequency, 2nd constituent frequency, and whole word frequency as predictor variables of gaze duration separately for short and long first constituent compounds (with the overall word length being approximately the same). If processing would go along the same lines, one would expect a similar pattern of results for the two types of compounds.

[1] Due to space limitations, our review on compound processing is somewhat limited. A more comprehensive review on compound word identification and morphological processing can be found in Hyönä, Bertram, & Pollatsek (2005).

4. Method

4.1. Participants

Twenty-four university students took part in the experiment as part of a course requirement. All were native speakers of Finnish.

4.2. Apparatus

Data were collected by the EyeLink eyetracker manufactured by SR Research Ltd. The eyetracker is an infrared video-based tracking system combined with hyperacuity image processing. There are two cameras mounted on a headband (one for each eye) including two infrared LEDs for illuminating each eye. The headband weighs 450 g in total. The cameras sample pupil location and pupil size at the rate of 250 Hz. Registration is monocular and is performed for the selected eye by placing the camera and the two infrared light sources 4–6 cm away from the eye. The spatial accuracy is better than 0.5°. Head position with respect to the computer screen is tracked with the help of a head-tracking camera mounted on the centre of the headband at the level of the forehead. Four LEDs are attached to the corners of the computer screen, which are viewed by the head-tracking camera, once the subject sits directly facing the screen. Possible head motion is detected as movements of the four LEDs and is compensated for on-line from the eye position records.

4.3. Materials

The experiment was conducted in Finnish. The target words were embedded in single sentences that extended a maximum of three lines (the target word always appeared sentence-medially in the first text line). An eye movement contingent display change paradigm (Rayner, 1975) was employed. An invisible boundary was set three character spaces to the left from the target word (see also Balota, Pollatsek, & Rayner, 1985; Pollatsek, Rayner, & Balota, 1986), as readers typically do not fixate on the last two letters of a word. When the eyes crossed this boundary, the target word was changed into its intended form. It took an average of about 13 ms to implement the change once the eyes crossed the invisible boundary. Thus, when the target was foveally inspected, it always appeared in the correct form. Two parafoveal preview conditions were created for each short and long 1st constituent compound: (a) a full preview (when crossing the boundary, the word was replaced by itself), and (b) a partial preview (the initial 3–4 letters were preserved, but the rest was replaced with random letters). In the partial preview condition, the short 1st constituent was visible in its entirety, whereas only about half of the letters of the long 1st constituents were parafoveally available. In Table 1 an example of all four conditions is presented.

Fifty compounds with a short (3–4 letters) initial constituent were paired with fifty compounds with a long (8–11 letters) initial constituent. A sentence frame was prepared

Table 1
An example of a target sentence pair

1st constituent length	Sentence beginning	Full preview	Partial preview	Sentence end
Short (työ)	Johnin tämänhetkinen	työympäristö	työzjkfffgtx	on parempi kuin...
	'John's current working environment is better than...'			
Long (toiminta)	Johnin tämän-hetkinen	toimintakyky	toizxddidqyh	on alhaisempi kuin...
	'John's current physical capacity is lower than...'			

for each pair so that the sentence was identical up to the target word; the length of Word *N*+1 was matched. Two stimulus lists were prepared so that in one list a given compound word appeared in the full preview condition and in the other list in the partial preview condition. Thus both lists contained 25 short and 25 long initial constituent compounds in each of the two preview conditions. The presentation of the stimulus lists was counterbalanced across participants. Using the newspaper corpus of *Turun Sanomat* comprising 22.7 million word forms (Laine & Virtanen, 1999), the two conditions were matched for whole word frequency, word length, average bigram frequency, and initial trigram frequency, but it was not possible to perfectly match for the frequency of the 1st and 2nd constituent (in Finnish, and probably in any language, short lexemes tend to be much more frequent than long ones). The lexical-statistical properties of the two conditions are found in Table 2. The sentences were presented in 12 point, Courier font.

Table 2
Lexical-statistical properties of the two compound conditions

Lexical-statistical property	Long1c[a]	Short1c[a]
Whole word freq. per $1 * 10^6$	6.1	6.4
1st constituent freq. per $1 * 10^6$	166.8	598.1
2nd constituent freq. per $1 * 10^6$	521.0	138.8
Word length in characters	12.6	12.6
1st constituent length in characters	8.7	3.8
Average bigram freq. per 1000	8.0	7.3
Initial trigram freq. per 1000	1.1	0.9
Final trigram freq. per 1000	1.0	1.2

[a] Long1c = compounds with a long 1st constituent; short1c = compounds with a short 1st constituent.

With a viewing distance of about 60 cm, one character space subtended approximately 0.3° of visual angle.

4.4. Procedure

Prior to the experiment, the eye-tracker was calibrated using a 9-point calibration grid that extended over the entire computer screen. Prior to each sentence, the calibration was checked by presenting a fixation point in an upper-left position of the screen; if needed, calibration was automatically corrected, after which a sentence was presented to the right of the fixation point. Participants were instructed to read the sentences for comprehension at their own pace. They were further told that periodically they would be asked to paraphrase the last sentence they had read to ensure that they attended to what they read. It was emphasized that the task was to comprehend, not to memorize the sentences. A short practice session preceded the actual experiment.

5. Results

A total of 11.4% of the data was discarded from the statistical analyses. Almost all the discarded trials were excluded because the display change occurred too late (i.e., a fixation was initiated on the target word region before the display change was completed). Analyses of variance (ANOVAs) for participants and items were computed on several eye fixation measures, with display change (change vs no change) as a within-subject and within-item factor and 1st constituent length (short vs long) as a within-subject factor but a between-item factor. The condition means for the analyzed eye movement parameters are shown in Table 3.

5.1. Gaze duration

Gaze duration is the summed duration of fixations made on a word before fixating away from it (either to the left or right). An effect of 1st constituent length was significant in both analyses, but the display change effect was only significant in the participant analysis: 1st constituent length, $F_1(1, 23) = 39.96$, MSE = 668, $p < 0.001$, $F_2(1, 98) = 4.02$, MSE = 10109, $p < 0.05$, and display change, $F_1(1, 23) = 4.42$, MSE = 890, $p < 0.05$; $F_2(1, 98) = 1.86$, MSE = 4385, $p = 0.18$. Gaze duration was 33 ms longer for compounds with a long 1st constituent. The display change effect amounted to 13 ms. Most importantly, the 1st Constituent Length×Display Change interaction was not significant, $F_1\ (1, 23) = 1.30$, MSE = 871, $p > 0.2$, $F_2. < 1$. If anything, gaze duration showed a non-significant trend in the opposite direction than was expected. That is, the display change effect was 20 ms for the short 1st constituent compounds and 6 ms for the long 1st constituent compounds, whereas we expected to find a larger benefit (in the form of a reduced preview effect) of the first 3–4 letters when these letters would constitute a real constituent as was the case for the short 1st constituent compounds. However, it should

Table 3
Eye fixation measures as a function of 1st constituent length and parafoveal preview

Eye fixation measure	Short 1st constituent			Long 1st constituent		
	Full preview	Partial preview	Difference	Full preview	Partial preview	Difference
Gaze duration[a]	415	435	20	455	461	5
First fixation duration[a]	212	218	6	200	202	2
Second fixation duration[a]	191	192	1	205	197	−8
Third fixation duration[a]	181	181	0	193	200	7
Average number of fixations	2.04	2.13	0.09	2.26	2.32	0.06
Initial fixation location[b]	4.05	3.99	−0.06	4.00	3.97	−0.04
Location of second fixation[b]	8.47	8.26	−0.21	9.37	8.96	−0.41
Duration of previous fixation[a]	190	186	−4	193	191	−2

[a] in milliseconds; [b] in character spaces.

be noted that separate t tests to assess the change effect were non-significant for both long and short 1st constituent compounds.[2]

5.2. *First fixation duration*

The duration of first fixation made on the word was reliably affected by 1st constituent length, $F_1(1, 23) = 32.45$, MSE $= 145$, $p < 0.001$, $F_2(1, 98) = 17.51$, MSE $= 690$, $p < 0.001$. First fixation was 14 ms longer when the initial constituent was short. The display change effect did not reach significance, $F_1(1, 23) = 2.69$, $p = 0.12$, MSE $= 130$, F_2. < 1.1. The interaction was clearly non-significant, $F_{1,2} < 1$.

[2] We made a peculiar observation that the difference in effect size may be related to some extent to long 1st constituents having on average slightly higher frequency initial trigrams. If ten items of each length group were excluded in order to perfectly match on initial trigram frequency, while preserving the matching on word frequency and length, the change effect on gaze duration was identical (15 ms) for long and short 1st constituent compounds. However, for both type of compounds the initial trigrams were identical in the change and no change condition, so that a straightforward explanation for this observation cannot be given. One possibility is that low initial trigrams attract attention somewhat earlier, making the display change slightly more salient.

5.3. *Second fixation duration*

The duration of second fixation on the target was to some extent influenced by the length of the 1st constituent, although the item analysis remained non-significant, $F_1\,(1, 23) = 8.93$, MSE = 249, $p < 0.01$, $F_2(1, 98) = 1.30$, MSE = 1076, $p > 0.2$. The second fixation was 10 ms longer for the compounds with a long 1st constituent. The display change effect did not reach significance, $F_1(1, 23) = 1.18$, $p > 0.2$, MSE = 281, $F_2. = 2.23$, MSE = 756, $p = 0.14$. The interaction was clearly non-significant, $F_{1,2} < 1$.

5.4. *Third fixation duration*

Items and participants with less than two observations per cell were excluded. Only 16 participants and 61 items contributed to the analyses. The duration of third fixation on the target was to some extent influenced by the length of the 1st constituent, although the item analysis remained non-significant, $F_1(1, 15) = 10.50$, MSE = 353, $p < 0.01$, $F_2(1, 98) = 1.68$, MSE = 1976, $p > 0.2$. The third fixation was 15 ms longer for the compounds with a long 1st constituent. The display change effect and the interaction were clearly non-significant, $F_{1,2} < 1$.

5.5. *Number of fixations*

In the number of first-pass fixations, both main effects proved significant in the participant analysis, but the display change effect was not significant in the item analysis: 1st constituent length – $F_1(1, 23) = 43.03$, MSE = 0.02, $p < 0.001$, $F_2(1, 98) = 7.86$, MSE = .24, $p < 0.01$; and display change – $F_1(1, 23) = 5.60$, MSE = 0.02, $p = 0.03$, $F_2(1, 98) = 2.23$, MSE = 0.13, $p = 0.13$. Readers made on average 0.21 fixations more on long than on short 1st constituent compounds. The display change effect amounted to 0.08 fixations. The 1st Constituent Length × Display Change interaction did not reach significance, $F_{1,2} < 1$. Analyses of the probability of a second and third fixation yielded exactly the same results.

5.6. *Initial fixation location*

There were no statistically significant effects, all Fs < 1.

5.7. *Location of second fixation*

The second fixation was located 0.80 character spaces further into the word when the initial constituent was long, $F_1(1, 23) = 26.14$, MSE = .59, $p < 0.001$, $F_2(1, 98) = 18.88$, MSE = 2.20, $p < 0.001$. The main effect of display change was marginal in the participant analysis, but significant in the item analysis, $F_1(1, 23) = 3.13$, MSE = .75, $p = 0.09$, $F_2(1, 98) = 6.20$, MSE = 1.36, $p = 0.01$. The second fixation was positioned 0.31 characters further into the target word when there was no display change. The interaction remained non-significant.

5.8. *Duration of fixation prior to fixating the target*

The duration of the final fixation prior to fixating the target word showed no reliable effects. There was a non-significant 4 ms difference between the short and long 1st constituent compounds, F_1 (1, 23) = 2.99, MSE = 169, $p = 0.10$, $F_2 < 1$. The previous fixation was slightly longer before fixating a long 1st constituent compound. Other effects were non-significant, all Fs < 1.2.

6. Additional analyses

6.1. *Launch site analyses*

We tested the possibility of finding evidence for morphological preprocessing when the target word is processed parafoveally from a close distance prior to inspecting it foveally. We did this by conducting a 3-way ANOVA including launch site as a within factor. Trials were categorized into near (within six letters of the target word, 42% of data) and far (more than six letters from the target word, 58% of data) launch site. Apart from main effects of launch site in several measures (e.g., 18 ms longer gaze durations and initial fixation 2.8 characters closer to the word beginning for trials launched from far away), no significant interactions between launch site and constituent length were found (all ps >0.13). This was also true for other categorizations (e.g., defining near as within three or four letters to the left from the invisible boundary). In other words, similarly to Kambe (2004), there was no sign of a morphological preview benefit even when the eyes were fixated close to the target compound prior to its foveal inspection. The only interaction approaching significance in the item analysis was a Preview × Launch Site interaction, indicating that the preview effect was larger for near launch site (24 ms) than for far launch site (8 ms), $F_1(1, 22) = 1.38$, MSE = 5440, $p = 0.25$, $F_2(1, 55) = 2.83$, MSE = 7431, $p < 0.10$.

6.2. *Factors that predict gaze duration on the target compound*

Using the regression analyses technique proposed by Lorch and Myers (1990), we set out to determine what factors predict gaze durations on short and long 1st constituent compounds. Separate analyses were conducted for the two types of compounds since 1st and 2nd constituent frequency distributions were different for short and long 1st constituent compounds. The regression analysis was performed separately for each participant, after which one-sample t tests on the unstandardized regression coefficients were conducted for four predictor variables. The four predictor variables were initial trigram frequency of 1st constituent, 1st constituent frequency, 2nd constituent frequency, and whole word frequency (log-transformed values were used for lexical frequencies). For long 1st constituent compounds, both the 1st constituent frequency and initial trigram frequency turned out to be significant predictors, whereas whole word frequency was a marginally significant predictor (1st constituent frequency, $t(23) = -5.35$, $p < 0.001$; initial trigram frequency, $t(23) = 2.35$, $p < 0.05$; whole word frequency, $t(23) = -2.03$, $p < 0.06$). However, 2nd constituent frequency did not predict gaze duration for long 1st constituent compounds, $t < 1$. In contrast, for short 1st constituents compounds, 2nd constituent frequency and whole word frequency, but not initial trigram frequency or 1st constituent frequency, turned out to be significant predictors of gaze duration (1st constituent frequency, $t < 1$; initial trigram frequency, $t < 1$; 2nd constituent frequency, $t(23) = -4.10$, $p < 0.001$; whole word frequency, $t(23) = -2.89$, $p < 0.01$). The mean coefficients of all variables and their standard errors in the one-sample t tests can be found in Table 4. It appears from Table 4 that, for instance, for long 1st constituent compounds an increase of 1 log unit in 1st constituent frequency results in a decrease of 54.84 ms in gaze duration.

7. Summary of results

To sum up, preserving the first 3–4 letters of relatively long compound words while replacing the remaining letters with random letters brought about a 13 ms preview effect in gaze duration, which is in line with other studies in which the first few letters were preserved and the remaining letters were replaced with visually dissimilar letters (e.g., Inhoff, 1989; Kambe, 2004). However, the above analyses revealed no evidence supporting parafoveal morphological preprocessing. As morphological preprocessing seemed possible only for the short initial constituent compounds, a partial preview was expected to be more beneficial for short than for long 1st constituent compounds (e.g., the difference between full and partial preview was predicted to be smaller for short 1st constituent compounds). Our data did not provide support for this prediction. In all measures the change effect was moderate and not modulated by the length of the 1st constituent. Even

Table 4
Estimates of mean unstandardized regression coefficients and their standard errors for gaze duration of long 1st constituent compounds (e.g., KONSERTTI/SALI 'concert hall') and short 1st constituent compounds (HÄÄ/SEREMONIA 'wedding ceremony')

	Short 1st constituent		Long 1st constituent	
Variable	*B*	SE	*B*	SE
Initial trigram frequency	0.89	5.86	22.05*	9.40
1st constituent frequency	−4.97	10.13	−54.84*	10.26
2nd constituent frequency	−30.55*	7.45	7.65	8.46
Whole word frequency	−25.29*	8.75	−19.74+	9.70

*$p < 0.05$; +$p < 0.06$

when only fixations that are launched from a near position were considered, the 1st constituent length did not modulate the display change effect. Other evidence against the morphological preprocessing hypothesis comes from the duration of fixation prior to fixating the target and from the first fixation location; neither of them was affected by the length of the 1st constituent or by the type of parafoveal preview. That 1st constituent length did not influence first fixation location compares favourably with the results of Experiment 1 of Hyönä and Pollatsek (1998).

Effects of 1st constituent length on foveal processing replicated the results of Hyönä and Pollatsek (1998) also in other respects. First, the second fixation location was further into the word when the 1st constituent was long. This lengthening of the first within-word saccade is needed when attempting to optimize the fixation location on the 2nd constituent. The constituent length effect in gaze duration was significant in the present study (an effect of 33 ms), but only marginal in Hyönä and Pollatsek (an effect of 10 ms). The trade-off in the duration of first versus second fixation (a longer first fixation but a shorter second fixation for the short 1st constituent compounds) was also observed in Hyönä and Pollatsek. In addition, there was a hint for the third fixation duration to be shorter for short 1st constituent compounds. Another indication that long 1st constituent compounds are processed differently from short 1st constituent compounds came from the regression analyses. These analyses showed that 1st constituent frequency but not 2nd constituent frequency was a reliable predictor of processing long 1st constituent compounds, whereas 2nd constituent frequency, but not 1st constituent frequency, predicted gaze duration on short 1st constituent compounds. For both types of compounds the whole word frequency turned out to be a reliable secondary predictor.

Finally, the finding that display change increased the number of fixations (the effect was not quite significant in the item analysis) may be taken to suggest that more preprocessing was done in the full preview condition, thus diminishing slightly the need for making additional refixations on the target word.

8. Discussion

In contrast to the prediction of Kambe (2004), a morphological preview benefit in Finnish seems hard to obtain. Thus, even though Finnish is a morphologically rich and highly productive language, readers do not seem to make use of parafoveally available morphological codes. Extracting a morphological unit out of a multimorphemic word is not an easy task, not even during foveal processing. As mentioned in the Introduction, Bertram et al. (2004) found evidence that morphological parsing of long 1st constituent compound words without a clear morpheme boundary cue is a time-consuming process. This is most probably so because foveal vision is needed to locate a morpheme boundary that is not clearly marked. When the to-be-located morpheme boundary is in the parafoveally presented word, the task may become practically impossible. Apart from that, assuming that morphological codes become activated after orthographic codes, there simply may not be enough time for a morphological unit to accumulate enough activation when it is available in the parafovea. Evidence in support of this hypothesis comes from Kambe's (2004) second experiment, in which she failed to find a morphological preview benefit, even though she demarcated morpheme boundaries in a highly salient way by using capital *X*s (e.g., reXXXX). Even when she considered only fixations launched from a near position, the morphological preview benefit failed to show up. This implies that, at least in English, a morphological preview benefit may be impossible to obtain. The current study indicates also that, in Finnish, parafoveal processing is restricted to the visual-orthographic level. The fact that the preview effect, which appeared mainly in gaze duration and in the number of first-pass fixations, was small (though not smaller in size than in other studies using a similar manipulation) implies that the first 3–4 letters are of greatest importance in the word identification process (as also argued by Briihl & Inhoff, 1995). However, the finding that a change effect is nevertheless observable when the first 3–4 letters were preserved while the remaining letters were replaced by random letters entails that also in Finnish non-initial letters are coded in the parafovea to some extent.

In line with previous studies, the present study showed that once a compound is fixated, morphological structure is recognized and affects processing. In addition, constituent length and the length of the whole compound seem to play a crucial role in how the compound word is processed. We observed that compounds of equal length, but with different 1st and thus 2nd constituent lengths, elicited differences in gaze duration, as well as in first, second and third fixation duration, and second fixation location. A short 1st constituent compound elicited longer first fixation durations but shorter second, third, and gaze durations than a long 1st constituent compound. In addition, for short 1st constituent compounds the second fixation was located around 0.8 characters closer to the word beginning and gaze duration was mainly determined by 2nd constituent frequency, whereas for long 1st constituent compounds gaze duration varied primarily as a function of 1st constituent frequency. This pattern of results implies that a reader can deal more easily with a long compound, when it has a short 1st constituent. Perhaps orthographic preprocessing of the initial trigram in addition to the whole 1st constituent being in clear foveal vision once the compound word is fixated makes the 1st constituent readily accessible. This may even lead to a situation where a 1st constituent frequency effect, normally observed for long Finnish compounds, is undermined. The results of Juhasz, Starr, Inhoff, & Placke, (2003) are compatible with this line of reasoning. Juhasz et al. found a much more prominent role for 2nd constituent frequency than for 1st constituent frequency in English compounds with a 1st constituent of 3–5 characters (average 4.6). Their claim is that a compound may be accessed via the 2nd constituent, since that is the main carrier of compound meaning. However, the question is then how to morphologically decompose a compound so that lexical-semantic information of the 1st constituent is not accessed or is even totally neglected. It seems more plausible to assume that due to visual acuity reasons a short 1st constituent is more readily available than a long 1st constituent, and that this allows fast lexical access to the first constituent, even when 1st constituent frequency is not so high. Thus it seems that visual acuity constraints and a word's morphological structure interact in interesting ways during the foveal processing of long Finnish compound words. However, to obtain a more detailed insight into the interaction of 1st constituent length and compound processing, more systematic experimentation is needed.

The most important conclusion of this chapter is nevertheless that despite the morphological productivity of the Finnish language, readers of Finnish do not seem to profit from a parafoveal morphological preview. Hence, readers of Hebrew remain unique in being able to crack the morphological code parafoveally.

Ackowledgements

This study was financially supported by the Academy of Finland (grant to the second author).

References

Balota, D. A., Pollatsek, A., & Rayner, K. (1985). The interaction of contextual constraints and parafoveal visual information in reading. *Cognitive Psychology, 17*, 364–390.

Bertram, R., & Hyönä, J. (2003). The length of a complex word modifies the role of morphological structure: Evidence from eye movements when reading short and long Finnish compounds. *Journal of Memory and Language, 48*, 615–634.

Bertram, R., Pollatsek, A., & Hyönä, J. (2004). Morphological parsing and the use of segmentation cues in reading Finnish compounds. *Journal of Memory and Language, 51*, 325–345.

Briihl, D., & Inhoff, A. W. (1995). Integrating information across fixations during reading: The use of orthographic bodies and exterior letters. *Journal of Experimental Psychology: Learning, Memory & Cognition, 21*, 55–67.

Deutsch, A., Frost, R., Pelleg, S., Pollatsek, A., & Rayner, K. (2003). Early morphological effects in reading: Evidence from parafoveal preview benefit in Hebrew. *Psychonomic Bulletin & Review, 10*, 415–422.

Deutsch, A., Frost, R., Pollatsek, A., & Rayner, K. (2000). Early morphological effects in word recognition in Hebrew: Evidence from parafoveal preview benefit. *Language and Cognitive Processes, 15*, 487–506.

Deutsch, A., Frost, R., Pollatsek, A., & Rayner, K. (2005). Morphological parafoveal preview benefit effects in reading: Evidence in Hebrew: Evidence from parafoveal. *Language and Cognitive Processes, 20*, 341–371.

Hyönä, J., & Bertram, R. (2004). Do frequency characteristics of nonfixated words influence the processing of fixated words during reading? *European Journal of Cognitive Psychology, 16*, 104–127.

Hyönä. J., Bertram, R., & Pollatsek, A. (2004). Are long compounds identified serially via their constituents? Evidence from an eye-movement contingent display change study. *Memory and Cognition, 32*, 523–532.

Hyönä. J., Bertram, R., & Pollatsek, A. (2005). Identifying compound words in reading: An overview and a model. In G. Underwood (Ed.), *Cognitive processes in eye guidance* (pp. 79–104). Oxford: University Press.

Hyönä, J., & Pollatsek, A. (1998). Reading Finnish compound words: Eye fixations are affected by component morphemes. *Journal of Experimental Psychology: Human Perception & Performance, 24*, 1612–1627.

Inhoff, A. W. (1989). Parafoveal processing of words and saccade computation during eye fixations in reading. *Journal of Experimental Psychology:Human Perception and Performance, 15*, 544–555.

Juhasz, B. J., Starr, M. S., Inhoff, A. W., & Placke, L. (2003). The effects of morphology on the processing of compound words: Evidence from naming, lexical decisions and eye fixations. *British Journal of Psychology, 94*, 223–244.

Kambe, G. (2004). Parafoveal processing of prefixed words during eye fixations in reading: Evidence against morphological influences on parafoveal processing. *Perception & Psychophysics, 66*, 279–292.

Kennedy, A. (2000). Parafoveal processing in word recognition. *Quarterly Journal of Experimental Psychology, 53A*, 429–455.

Laine, M., & Virtanen, P. (1999). *WordMill lexical search program.* Center for Cognitive Neuroscience, University of Turku, Finland.

Lima, S. D., (1987). Morphological analysis in sentence reading. *Journal of Memory and Language, 26*, 84–99.

Lorch, R. F., & Myers, J. L. (1990). Regression analyses of repeated measures data in cognitive research. *Journal of Experimental Psychology; Learning, Memory, and Cognition,* 16, 149–157.

Murray, W. S. (1998). Parafoveal pragmatics. In G. Underwood (Ed.), *Eye guidance in reading and scene perception* (pp. 181–199). Oxford: Elsevier Science.

O'Regan, J. K. (1992). Optimal viewing position in words and the strategy-tactics theory of eye movements in reading. In K. Rayner (Ed.), *Eye movements and visual cognition: scene perception and reading* (pp. 334–354). New York: Springer-Verlag.

Pollatsek, A., Rayner, K., & Balota, D. A. (1986). Inferences about eye movement control from the perceptual span in reading. *Perception & Psychophysics, 40*, 123–130.

Rayner, K. (1975). The perceptual span and peripheral cues in reading. *Cognitive Psychology, 7*, 65–81.

Rayner, K. (1979). Eye-guidance in reading: Fixation location within words. *Perception, 8*, 21–30.

Rayner, K., Balota, D. A., & Pollarsek, A. (1986). Against parafoveal semantic preprocessing during eye fixations in reading. *Canadian Journal of Psychology, 40*, 473–483.

Rayner, K., White, S. J., Kambe, G., Miller, B., & Liversedge, S. P. (2003). On the processing of meaning from parafoveal vision during eye fixations in reading. In J. Hyönä, R. Radach, & H. Deubel (Eds.), *The mind's eye: cognitive and applied aspects of eye movement research* (pp. 213–234). Oxford: Elsevier Science.

Chapter 18

FOVEAL LOAD AND PARAFOVEAL PROCESSING: THE CASE OF WORD SKIPPING

SARAH J. WHITE

University of Leicester, UK

Eye Movements: A Window on Mind and Brain

Edited by R. P. G. van Gompel, M. H. Fischer, W. S. Murray and R. L. Hill

Abstract

Three experiments showed that localised foveal load does not modulate the probability of skipping the following 4–6 letter parafoveal word (word $n+1$). In Experiments 1 and 2 the preview of the word $n+1$ was always correct. In Experiment 3 the preview of the word $n+1$ was either correct or incorrect. Localised foveal difficulty did not significantly modulate the effect of preview on the probability of skipping the word $n+1$. The results suggest that the processes that produce modulations of parafoveal preprocessing by foveal load on reading time measures may not apply to the control of word skipping.

As we read, we preprocess text that has not yet been fixated. Such preprocessing results in a greater probability of skipping words that are short, frequent or predictable compared to words that are long, infrequent or unpredictable (for reviews see Brysbaert & Vitu, 1998; Rayner, 1998). There are two different approaches to explaining the mechanisms that control which words are fixated or skipped during reading. The first is to suggest that the processes that determine word skipping are the same or similar to those that influence reading time. The second is to suggest that the processes that determine reading times and word skipping are qualitatively different. The present study investigates this issue by examining whether foveal load modulates parafoveal preprocessing in the same way for both reading times and word skipping.

Studies have suggested that the amount of parafoveal preprocessing, as shown by reading times, is limited by foveal processing difficulty[1] (Henderson & Ferreira, 1990; Kennison & Clifton, 1995; Schroyens, Vitu, Brysbaert, & d'Ydewalle, 1999; White, Rayner, & Liversedge, 2005). These studies used the boundary saccade contingent change technique, which involves altering the parafoveal preview (which may be correct or incorrect) such that the word is correct when it is subsequently fixated (Rayner, 1975).

For example, Henderson and Ferreira (1990) compared reading times on critical words (e.g. *despite*) when the preview of that word was correct (e.g. *despite*) or incorrect (e.g. *zqdioyv*) and when the word prior to the critical word was either frequent (e.g. *chest*) or infrequent (e.g. *trunk*). The difference in reading times when the preview is correct or incorrect gives a measure of the extent to which preprocessing of the correct preview facilitates processing once the word is fixated, known as preview benefit (Rayner & Pollatsek, 1989). Henderson and Ferreira showed that preview benefits for the critical word were larger when the previous word was frequent compared to when it was infrequent. That is, parafoveal preprocessing was reduced (preview benefits were smaller) when foveal processing was difficult compared to when it was easy.

Two different accounts have been proposed to explain the finding that parafoveal preprocessing is limited by foveal load. One is based on serial processing of words and a second is based on parallel processing of multiple words. Critically, both these accounts suggest that foveal processing difficulty influences parafoveal preprocessing as shown by both reading times and word skipping.

Serial attention shift models, such as the E-Z reader model (Reichle, Pollatsek, Fisher, & Rayner, 1998; Reichle, Rayner, & Pollatsek, 1999, 2003), have adopted an architecture in which shifts of attention and programming of eye movements are de-coupled. After saccade programming to the following word (word $n+1$) has begun, linguistic processing of the fixated word (word n) continues. The time to process word n is influenced by foveal load. When processing of word n has finally been completed, and usually before saccade programming is complete, attention shifts to word $n+1$ so that it can be preprocessed. Due to the de-coupling of saccade programming and attention, the time to preprocess word

[1] These studies used incorrect preview conditions in which multiple letters were incorrect. Other studies which have used incorrect previews containing a single internal incorrect letter have not shown any modulation of preprocessing by foveal load (Drieghe, Rayner, & Pollatsek, 2005; White & Liversedge, 2006b).

$n+1$ (whilst fixating word n) is restricted by the time required to complete processing of word n. Critically, the time to attend to word $n+1$ influences the extent to which word $n+1$ is preprocessed and therefore the amount of preview benefit for word $n+1$ (as shown by reading times on word $n+1$). Similarly, once attention has moved to word $n+1$, if there is sufficient time before the saccade is executed, and if word $n+1$ is identified quickly enough, then the saccade programme may be re-programmed to skip word $n+1$. Therefore both reading times and word skipping are determined by the same mechanism which is influenced by foveal load (time to process word n). Consequently Reichle et al. predict that foveal load should reduce the probability of skipping the following word.

In contrast, in their Glenmore model, Reilly and Radach (2003) suggest that multiple words can be processed in parallel and that there is competition between words for activation related to linguistic processing. Consequently greater processing of the fixated word reduces processing of other words. As a result, foveal load reduces preview benefit for the following word. Reilly and Radach also suggest that each word has a salience value, such that saccades are directed to the word with greatest salience. These salience values are influenced by the same linguistic activation system that influences reading times. Therefore, although not explicitly stated, Reilly and Radach's account also appears to suggest that word skipping is modulated by foveal load.

To summarise, empirical evidence suggests that foveal load modulates parafoveal preprocessing as shown by reading time preview benefit. Accounts based on both serial processing and parallel processing have been proposed that explain this phenomenon. Both these models also predict that foveal load modulates word skipping in a similar way as for preview benefits. However, other studies have suggested that the processes that determine when and where the eyes move can be different (Radach & Heller, 2000; Rayner & McConkie, 1976; Rayner & Pollatsek, 1981). Indeed, a number of models of eye movement control in reading have been developed in which different mechanisms determine when and where the eyes move. These models either do not predict that foveal load modulates preprocessing as in the case of SWIFT (Engbert, Longtin, & Kliegl, 2002; Engbert, Nuthmann, Richter & Kliegl, 2005; Kliegl & Engbert, 2003) and the Competition/Interaction model (Yang & McConkie, 2001), or they predict that both the when and the where systems are modulated by foveal load, as in Glenmore (Reilly & Radach, 2003). Nevertheless, accounts which differentiate between mechanisms that determine when and where the eyes move highlight the possibility that although foveal load modulates preprocessing as shown by reading times, foveal load may not necessarily modulate where the eyes move (as shown by the probability of word skipping).

The issue of whether foveal load modulates the probability of word skipping is therefore critical not only for evaluating the architecture of current models but also for assessing the fundamental question of whether the mechanisms that determine when and where the eyes move are the same or different. Drieghe et al. (2005) investigated whether foveal load modulated the probability of skipping parafoveal three-letter words (see also Kennison & Clifton, 1995). For cases in which there was a correct preview of the parafoveal word, Drieghe et al. showed no significant effect of foveal load on the probability of skipping the following word. Despite the non-significant result, skipping rates were numerically higher when there was low, compared to high, foveal load which is suggestive of the possibility that foveal load may modulate word skipping. Therefore it is important to examine whether foveal load does reliably influence word skipping. Also, as Drieghe et al. only tested the probability of skipping three letter words, it is important to test whether foveal load modulates the probability of skipping slightly longer words.

The present study includes three experiments that test whether localised foveal load influences the probability of skipping four- to six-letter parafoveal words. The manipulations of foveal load include orthographic regularity, spelling and word frequency. These manipulations are intended to influence the ease with which a specific word can be processed. Importantly, this study does not test whether general processing load modulates word skipping. General processing load may be modified by text difficulty (e.g. contextual factors) or reading strategy. For example, general processing load could modulate global parameters for eye movement control such that increased load might increase fixation durations or shorten saccade lengths (see Yang & McConkie, 2001). Within each of the experiments presented here, such general factors are controlled by using the same sentence beginnings up until the critical words across each of the experimental conditions.

For all of the experiments presented here, the analyses include only cases in which a single fixation was made on the foveal word and no regressions were made out of the foveal word. Refixations on the foveal word could modify factors which might influence word skipping, such as launch site and the quality of the parafoveal preview. Therefore, restricting the analyses to cases in which single fixations were made on the foveal word ensures that any differences in skipping probabilities could not be accounted for by differences in refixation probabilities. Overall, if foveal load influences word skipping then the probability of skipping the parafoveal word (word $n+1$) should be greater when the foveal word (word n) is easy, compared to difficult, to process.

1. Experiment 1

In Experiment 1, foveal load was manipulated by orthographic regularity. The foveal word (word n) was either orthographically regular (low foveal load e.g. *miniature*) or orthographically irregular (high foveal load e.g. *ergonomic*) and was followed by the parafoveal word (word $n+1$) (e.g. *chairs*).

Note that the foveal processing load manipulation in Experiment 1 has been shown to have a small (less than 0.5 character) but reliable influence on initial fixation positions on these words (White & Liversedge, 2006a). Therefore in the present study, initial fixations land nearer to the beginning of the foveal words that are difficult to process (orthographically irregular) than the foveal words that are easy to process (orthographically regular). Consequently, the launch site prior to skipping or fixating word $n-1$ may have been slightly further away for the high foveal load words compared to the low foveal load words. Launch site may influence skipping probabilities such that saccades launched from further away may be less likely to skip word $n+1$. Importantly, note that the direction of these effects would have facilitated an effect of foveal load on word skipping such that when foveal load was high, word $n+1$ would be less likely to be skipped. This additional factor would therefore have to be taken into account if orthographic modulations of foveal load were to modulate word skipping.[2]

1.1. Method

1.1.1. Participants

One hundred and four native English speakers at the University of Durham were paid to participate in the experiment. The participants all had normal or corrected to normal vision and were naïve in relation to the purpose of the experiment.

1.1.2. Materials and design

The foveal word n had orthographically regular (low foveal load) or irregular (high foveal load) word beginnings and these two conditions were manipulated within participants and items. The parafoveal word $n+1$ was identical for each of the conditions within each item. The foveal words were nine or ten letters long and the parafoveal words were five or six letters long. Sixty of the participants read half of the sentences entirely in upper case, this variable was not included in the analysis.[3] There were 24 critical words in each condition. Word n and $n+1$ were embedded roughly in the middle of the same sentential frame up to and including word $n+1$. Each of the sentences was no longer than one line of text (80 characters). See Table 1 for examples of experimental sentences and critical words. Full details regarding the nature of the orthographic regularity manipulation and the construction of the stimuli lists can be found in White and Liversedge (2006a).

Table 1
Examples of experimental sentences and critical words for each condition in Experiment 1. Word n is shown in italics. For each sentence frame, version "a" is the low foveal load (regular beginning word) condition and version "b" is the high foveal load (irregular beginning word) condition

1a. Last friday the modern *miniature* chairs were placed in the dolls house.
1b. Last friday the modern *ergonomic* chairs were transported to the shops.
2a. He hated the heavy *primitive* tools that the farmer gave him to use.
2b. He hated the heavy *pneumatic* tools that were used to dig up the road.
3a. He knew that the clever *candidates* would produce impressive answers.
3b. He knew that the clever *auctioneer* would ask him about the valuable lots.
4a. She knew that the modern *extension* would add value to the house.
4b. She knew that the modern *ointments* would work if she could get them in time.

[2] Similar reasoning also applies in Experiment 2. The misspellings used in Experiment 2 were also found to modulate saccade targeting (White & Liversedge, 2006b).
[3] In Experiment 1, type case did not significantly influence reading times on the foveal word n.

1.1.3. Procedure

Eye movements were monitored using a Dual Purkinje Image eye tracker. Viewing was binocular but only the movements of the right eye were monitored. The letters were presented in light cyan on a black background. The viewing distance was 70 cm and 3.5 characters subtended 1° of the visual angle. The resolution of the eye tracker is less than 10 min of arc and the sampling rate was every millisecond.

Participants were instructed to understand the sentences to the best of their ability. A bite bar and a head restraint were used to minimise head movements. The participant completed a calibration procedure and the calibration accuracy was checked after every few trials during the experiment. After reading each sentence the participants pressed a button to continue and used a button box to respond "yes" or "no" to comprehension questions.

1.1.4. Analyses

Fixations shorter than 80 ms that were within one character of the next or previous fixation were incorporated into that fixation. Any remaining fixations shorter than 80 ms and longer than 1200 ms were discarded. Five percent of trials were excluded due to either no first pass fixations on the sentence prior to word $n-1$ or tracker loss or blinks on first pass reading of word $n-1$ or n.

1.2. Results and discussion

Paired-samples t tests were undertaken with participants (t_1) and items (t_2) as random variables. Eleven percent of trials were excluded due to first pass regressions made out of the foveal word and 30 percent of trials were excluded due to skipping or multiple first pass fixations on the foveal word.

1.2.1. Single fixation duration word n

Single fixation durations on word n were significantly longer in the high foveal load orthographically irregular condition ($M = 366$, SD = 135) than in the low foveal load orthographically regular condition ($M = 314$, SD = 100), $t_1(103) = 9.63$, $p < 0.001$; $t_2(23) = 6.4$, $p < 0.001$. The manipulation of the orthographic regularity of word n was clearly effective.[4]

[4] For all three of the experiments the foveal load manipulation for word n also significantly influenced first fixation durations and gaze durations on word n.

1.2.2. *Skipping probability word* n + 1

There was no difference in the probability of skipping word $n+1$ on first pass between the high foveal load orthographically irregular condition (0.14) and the low foveal load orthographically regular condition (0.15), $t_1(103) = 1.62$, $p = 0.107$; $t_2(23) = 1.02$, $p = 0.319$. Regardless of whether the foveal word had caused reduced (e.g. *miniature*) or increased (e.g. *ergonomic*) foveal processing difficulty, the probability of skipping the following parafoveal word (e.g. *chairs*) was the same. The findings from Experiment 1 show that although orthographic regularity clearly increased processing time on word n, this had no effect on the probability of skipping the following word.

2. Experiment 2

Orthographic regularity significantly influenced single fixation durations on the foveal word in Experiment 1. Nevertheless, perhaps this manipulation of foveal difficulty was not sufficiently strong to influence subsequent word skipping. Experiment 2 used a stronger manipulation of foveal difficulty. Previous studies have shown that misspellings cause disruption to reading due to the difficulty associated with understanding non-words as words (e.g. Zola, 1984). Therefore in Experiment 2 foveal difficulty was manipulated by spelling. The foveal words were either spelled correctly (low foveal load, e.g. *performer*) or incorrectly (high foveal load, e.g. *pwrformer*) and these were followed by the parafoveal words (e.g. *stood*). The method was the same as for Experiment 1 except where noted below.

2.1. *Method*

2.1.1. *Participants*

Forty-four native English speakers at the University of Durham participated in the experiment.

2.1.2. *Materials and design*

The foveal word n was either spelled correctly (low foveal load) or misspelled (high foveal load) and these two conditions were manipulated within participants and items. The foveal words were all nine or ten letters long and the parafoveal words were five or six letters long. Half the foveal words were preceded by a frequent word and half by an infrequent word, this variable was not included in the analysis. Except for word $n-1$ (which had high or low word frequency), word n and word $n+1$ were embedded in sentence frames which were otherwise identical for each condition. See Table 2 for examples of experimental sentences and critical words. Full details regarding the nature of the misspelling manipulation, the word frequency manipulation, and construction of the stimuli lists can be found in White and Liversedge (2006b, Experiment 1).

Table 2
Examples of experimental sentences and critical words for each condition in Experiment 2. Word *n* is shown in italics. For each sentence frame, version "a" is the low foveal load (correctly spelled) condition and version "b" is the high foveal load (misspelled) condition

1a. After the circus act the famous *performer* stood to receive the applause.
1b. After the circus act the famous *pwrformer* stood to receive the applause.
2a. At the meeting the whole *committee* voted against the planning application.
2b. At the meeting the whole *ctmmittee* voted against the planning application.
3a. The tourists enjoyed talking to the young *traveller* about his many experiences.
3b. The tourists enjoyed talking to the young *tlaveller* about his many experiences.
4a. The brave explorers knew that the great *endeavour* would need a lot of effort.
4b. The brave explorers knew that the great *ezdeavour* would need a lot of effort.

2.1.3. *Procedure*

Participants were instructed that some sentences would contain misspellings but that they should read and understand the sentences to the best of their ability.

2.1.4. *Analyses*

Seven percent of trials were excluded.

2.2. Results and discussion

The results were analysed in the same manner as for Experiment 1. Sixteen percent of trials were excluded due to first pass regressions made out of the foveal word and 30 percent of trials were excluded due to skipping or multiple first pass fixations on the foveal word.

2.2.1. *Single fixation duration word* n

Single fixation durations on word *n* were significantly longer in the high foveal load misspelled condition ($M = 376$, $\mathrm{SD} = 162$) than in the low foveal load correctly spelled condition ($M = 307$, $\mathrm{SD} = 92$), $t_1(43) = 7.59$, $p < 0.001$; $t_2(47) = 6.98$, $p < 0.001$. The manipulation of spelling accuracy on word *n* was clearly effective.

2.2.2. *Skipping probability word $n + 1$*

There was no difference in the probability of skipping word $n + 1$ on first pass between the high foveal load misspelled condition (0.3) and the low foveal load correctly spelled condition (0.3), ($ts < 1$). Regardless of whether the foveal word had caused reduced (e.g. *performer*) or increased (e.g. *pwrformer*) foveal processing difficulty, the probability of skipping the following parafoveal word (e.g. *stood*) was the same. Therefore, similar to Experiment 1, Experiment 2 showed no effect of foveal load on the probability of skipping the following word, even when a very strong manipulation of foveal processing difficulty was used.

3. Experiment 3

Foveal difficulty was manipulated by orthographic regularity in Experiment 1 and by spelling in Experiment 2. The findings of both these experiments suggest that localised foveal load does not modulate the probability of skipping the following word. In order to ensure that this finding is robust across a range of different types of localised foveal load, Experiment 3 used word frequency to modulate foveal processing difficulty. In addition, both Experiments 1 and 2 used long foveal words. Experiment 3 therefore tested whether the findings held for short foveal words. Furthermore, Experiment 3 manipulated the nature of the preview of the parafoveal word.

In Experiment 3 the foveal words (word n) had high word frequency (low foveal load, e.g. *happy*) or low word frequency (high foveal load, e.g. *agile*) and these were followed by the parafoveal word (e.g. *girl*). The boundary saccade contingent change technique (Rayner, 1975) was used such that the preview of the parafoveal word was either correct (e.g. *girl*) or incorrect (e.g. *bstc*). The reading time data for the parafoveal word in Experiment 3 are reported in White et al. (2005) (Group 1 data). When foveal load was low (e.g. *happy*), there was 47 ms gaze duration preview benefit for the subsequent word (e.g. *girl*) whereas when foveal load was high (e.g. *agile*) there was only 1 ms gaze duration preview benefit. Similar to Henderson and Ferreira (1990) these results show that the difficulty of the foveal word modulates preprocessing (preview benefit) for the following word. Therefore, in Experiment 3, foveal load is clearly modulating parafoveal preprocessing, at least as shown by when the eyes move.

If foveal load modulates the probability of skipping the following word then the correct preview of word $n+1$ will be more likely to be skipped when the foveal word is easy (high frequency) compared to difficult (low frequency) to process. If this is the case then the influence of foveal load on the probability of skipping the visually dissimilar incorrect preview should provide further insight into the nature of such an effect. First, foveal load may influence the probability of skipping word $n+1$ regardless of the characteristics of word $n+1$. That is, foveal load should have the same effect on the probability of skipping word $n+1$ both when the preview is correct (e.g. *girl*) and incorrect (e.g. *bstc*). Second, it might be argued that words should usually only be skipped if they are familiar (Reichle et al., 1998, 1999, 2003). Consequently the incorrect preview of word $n+1$ (e.g. *bstc*) should be skipped only very rarely because it is an unfamiliar non-word. That is, there should be an interaction such that foveal load modulates the probability of skipping the correct, but not the incorrect, previews of word $n+1$. The method for Experiment 3 is the same as for Experiment 1 except where noted below.

3.1. Method

3.1.1. Participants

Thirty-two students at the University of Massachusetts were paid or received course credit to participate in the experiment.

3.1.2. Materials and design

Two variables, foveal processing difficulty (word n) and parafoveal preview (word $n+1$), were manipulated within participants and items. The foveal word n was easy to process (high frequency, *happy*) or difficult to process (low frequency, *agile*). The preview of word $n+1$ before it was first fixated was correct (*girl*) or incorrect (*bstc*). The foveal word n was either five or six letters long and the parafoveal word was always four letters long. Full details regarding the nature of the materials and construction of the stimuli lists can be found in White et al. (2005). See Table 3 for examples of experimental sentences and critical words.

3.1.3. Procedure

The eye contingent boundary technique was used (Rayner, 1975); the display changes occurred within 5 ms of detection of the boundary having been crossed. Sentences were displayed at a viewing distance of 61 cm and 3.8 characters subtended 1° of visual angle.

3.1.4. Analyses

Trials were excluded due to: (a) display changes happening too early, (b) tracker loss or blinks on first pass reading of words n or $n+1$ and (c) zero reading times on the first part of the sentence. Seventeen percent of trials were excluded.[5]

Table 3

Examples of experimental sentences and critical words for each condition in Experiment 3. Word n is shown in italics. For each sentence frame, version "a" is the low foveal load (high frequency) condition and version "b" is the high foveal load (low frequency) condition. The incorrect preview of word $n+1$ is shown in parentheses

1a. Outside the school the *happy* girl (bstc) skipped around the other children.
1b. Outside the school the *agile* girl (bstc) skipped around the other children.
2a. The supporters cheered when the *local* team (wtdr) finally won the match.
2b. The supporters cheered when the *inept* team (wtdr) finally won the match.
3a. The cook ordered the *daily* food (gkhn) from the local market.
3b. The cook ordered the *bland* food (gkhn) from the local market.
4a. The child pestered the *green* fish (jbws) that was hiding behind the pondweed.
4b. The child pestered the *timid* fish (jbws) that was hiding behind the pondweed.

[5] Note that a larger proportion of data was excluded in Experiment 3 compared to Experiments 1 and 2 because the display contingent change technique requires that additional data must be excluded due to display changes happening too early.

3.2. Results and discussion

Fourteen percent of trials were excluded due to first pass regressions made out of the foveal word and 22 percent of trials were excluded due to skipping or multiple first pass fixations on the foveal word. A series of 2 (word n foveal load: frequent, infrequent) by 2 (word $n+1$ preview: correct, incorrect) repeated measures Analyses of Variance (ANOVAs) were undertaken with participants (F_1) and items (F_2) as random variables. Table 4 shows the mean single fixation durations on word n and the probability of skipping word $n+1$ for Experiment 3.

3.2.1. Single fixation duration word n

Single fixation durations on word n were significantly longer in the high foveal load infrequent condition than in the low foveal load frequent condition, $F_1(1,31) = 11.99$, $p < 0.01$; $F_2(1,40) = 18.37$, $p < 0.001$. There was no effect of the preview of word $n+1$, $F_1(1,31) = 1.16$, $p = 0.29$; $F_2 < 1$, and no interaction between the frequency of word n and the preview of word $n+1$, $F_1(1,31) = 1.81$, $p = 0.188$; $F_2(1,40) = 2.99$, $p = 0.092$, for single fixation durations on word n. Therefore the manipulation of foveal load on word n was clearly effective and the preview of word $n+1$ did not significantly influence reading times on word n.

3.2.2. Skipping probability word n+1

The parafoveal words were more likely to be skipped when the preview was correct (.185) compared to when it was incorrect (.135), this effect was significant across participants, $F_1(1,31) = 4.91$, $p = 0.03$, but not items, $F_2(1,40) = 2.63$, $p = 0.113$. Importantly, the effect of preview on the probability of word skipping indicates that the linguistic characteristics of parafoveal words influences whether they are subsequently fixated. The parafoveal word $n+1$ was skipped on 17 percent of trials when the foveal word n was high frequency and 15 percent of trials when the foveal word was low frequency. Foveal load had no significant effect on the probability of skipping the parafoveal word (F's < 1)

Table 4
Experiment 3. Single fixation durations on word n. Standard deviations in parentheses. Probability of skipping word $n+1$

		Word n	Word $n+1$
Word n	Preview of word $n+1$	Single fixation duration	Skipping probability
Frequent	Correct	277 (83)	0.18
	Incorrect	295 (103)	0.16
Infrequent	Correct	321 (112)	0.19
	Incorrect	311 (110)	0.11

and there was no interaction between foveal load and the parafoveal preview, $F_1 < 1$; $F_2(1, 40) = 1.51$, $p = 0.226$.

As in Experiments 1 and 2, there was no difference in the probability of skipping a correct preview of word $n+1$ when there was low (0.18) compared to high (0.19) foveal load. Therefore, regardless of whether the foveal word had caused reduced (e.g. *happy*) or increased (e.g. *agile*) foveal processing difficulty, the probability of skipping the following parafoveal word (e.g. *girl*) was the same. However, note that the word $n+1$ was numerically less likely to be skipped when there was high foveal load and an incorrect parafoveal preview, compared to the other conditions. The nature of this interactive pattern is similar to that shown by Drieghe et al. (2005) (see Drieghe et al. for an extended discussion of possible explanations). Critically, the absence of any effect of foveal load on the probability of skipping the correctly spelled word $n+1$ suggests that whatever might have caused, the numerical effect for incorrect previews does not hold during normal reading of correctly spelled text.

4. General discussion

All three of the experiments presented here show no effect of localised foveal load on the probability of skipping four to six letter words. The results are consistent with Drieghe et al.'s (2005) finding that for correctly spelled words, there was no significant difference between the probability of skipping three-letter words when there was high, compared to low, foveal load. Although Drieghe et al. showed a numerical difference in skipping probabilities, the fact that none of the experiments here showed more than a 0.01 difference in skipping probabilities for correctly spelled words suggests that there is no reliable effect of foveal load on the probability of word skipping when reading normal text. These findings contrast with studies which demonstrate that foveal load modulates parafoveal preprocessing as shown by preview benefit (Henderson & Ferreira, 1990; Kennison & Clifton, 1995; Schroyens et al., 1999; White et al., 2005). Furthermore, the fact that the same experiment demonstrated such effects for reading times (as reported in White et al.) but shows no such effects for word skipping (Experiment 3) is particularly poignant. Together, these findings provide strong evidence for the notion that localised foveal load modulates preview benefits but not the probability of skipping the following word.

Although the experiments presented here suggest that localised foveal load does not influence the probability of word skipping, as noted in the Introduction this does not preclude the possibility that general processing load may have a global influence on skipping rates. Indeed, note that the skipping probabilities are higher in Experiment 2 than in either Experiment 1 or 3. This could be because the sentence beginnings included context relevant to the remainder of the sentence in Experiment 2, and because word $n+1$ tended to occur later in the sentence in Experiment 2 compared to the other experiments. Such differences could have influenced general processing load, which may have modulated skipping probabilities.

The current study suggests that localised foveal load modulates parafoveal preprocessing as shown by reading times, but not as shown by word skipping. It is possible that foveal load may influence word skipping, but this effect may be so very small that it is undetectable in standard reading experiments. Alternatively, reading times and word skipping may be influenced by qualitatively different processes. The latter suggestion would be inconsistent with accounts which suggest that foveal load influences reading times and word skipping by the same mechanism (Reichle et al., 1998, 1999, 2003) or due to a common input (Reilly & Radach, 2003). However, if necessary, such models might be adapted such that the word skipping mechanism operated independent of foveal load.

The possibility that reading times and word skipping are controlled by qualitatively different processes supports the notion that different processes might determine when and where the eyes move (Rayner & McConkie, 1976). Indeed, White and Liversedge (2006b) also showed that foveal difficulty does not modulate parafoveal orthographic influences on where words are first fixated. This finding suggests that saccade targeting to a word is also independent of foveal processing load. However, note that the processes that determine word skipping and saccade targeting may be different to other types of "where" decisions such as refixations and regressions.

The findings presented here indicate that words may be preprocessed qualitatively differently for the mechanisms that determine reading times and word skipping. Parafoveal preprocessing that is limited by foveal load influences the mechanisms that determine reading times. In contrast, word skipping mechanisms may be influenced by parafoveal preprocessing that occurs regardless of foveal load. For example, the processes that determine reading times may be sensitive to the progress of word recognition and sentence comprehension processes. Such language comprehension processes might be limited by processing load such that a parafoveal word is preprocessed to a lesser extent when there is high, compared to low, foveal load. Therefore, reading time preview benefits for parafoveal words reflect reduced preprocessing of words when there is high, compared to low, foveal load. In contrast, the processes that determine word skipping may acquire information from parafoveal text in an automatic manner, independent of comprehension difficulty, such that words can be skipped even if they have not been recognised.

The processes that determine word skipping may be different from those which determine reading times because word skipping may only be influenced by parafoveal information. Note that as a result, the eyes may sometimes become "out of sync" with the location of attention (the progress of sentence comprehension). For example, when reading the phrase "*agile girl*", the high frequency word *girl* may be skipped before the low frequency word *agile* has been fully processed. Consequently processing of skipped words may continue on subsequent fixations. This suggestion is consistent with the finding that there are more regressions following skips, compared to first pass fixation, of words (Vitu, McConkie, & Zola, 1998). Such an automatic word targeting mechanism may sometimes move the eyes away from what needs to be processed. However, a system based on simple parafoveal linguistic processing may be most optimal for selecting which words to fixate given the very limited time periods available for saccade programming.

To summarise, the results suggest that qualitatively different mechanisms might determine parafoveal preprocessing as shown by reading times (preview benefit) and word skipping. Future accounts of eye movement control in reading may need to adopt an architecture in which there is separate processing of, and possibly inputs to, the mechanisms that determine preview benefits and word skipping.

Acknowledgements

The findings reported here were presented at the Sixth European Workshop on Language Comprehension, St. Pierre d'Olèron, France (2004). The author acknowledges the support of a University of Durham Studentship; Experiments 1 and 2 form part of the author's PhD thesis. Experiment 3 was undertaken whilst the author was on a research visit at the University of Massachusetts, Amherst, supported by a Study Visit Grant from the Experimental Psychology Society. This research was also supported by a Biotechnology and Biological Sciences Research Council Grant 12/S19168. Thanks to Simon P. Liversedge, Hazel I. Blythe, Alan Kennedy and Keith Rayner for their comments on an earlier version of this chapter. Other data from the Experiments are reported elsewhere. See White and Liversedge (2006a, 2006b) and White et al. (2005) for further details of Experiments 1, 2 and 3 respectively.

References

Brysbaert, M., & Vitu, F. (1998). Word skipping: Implications for theories of eye movement control in reading. In G. Underwood (Ed.), *Eye guidance in reading and scene perception* (pp. 125–148). Oxford, UK: Elsevier.

Drieghe, D., Rayner, K., & Pollatsek, A. (2005). Eye movements and word skipping during reading revisited. *Journal of Experimental Psychology: Human Perception and Performance, 31*, 954–969.

Engbert, R., Longtin, A., & Kliegl, R. (2002). A dynamical model of saccade generation in reading based on spatially distributed lexical processing. *Vision Research, 42*, 621–636.

Engbert, R., Nuthmann, A., Richter, E. M., & Kliege, R. (2005). SWIFT: A dynamical model of saccade generation during reading. *Psychological Review, 112*, 777–813.

Henderson, J. M., & Ferreira, F. (1990). Effects of foveal processing difficulty on the perceptual span in reading: Implications for attention and eye movement control. *Journal of Experimental Psychology: Learning, Memory and Cognition, 16*, 417–429.

Kennison, S. M., & Clifton, C. (1995). Determinants of parafoveal preview benefit in high and low working memory capacity readers: Implications for eye movement control. *Journal of Experimental Psychology: Learning Memory and Cognition, 21*, 68–81.

Kliegl, R., & Engbert, R. (2003). SWIFT explorations. In J. Hyönä, R. Radach, & H. Deubel (Eds.), *The mind's eye: cognitive and applied aspects of eye movement research* (pp. 391–411). Amsterdam: Elsevier.

Radach, R., & Heller, D. (2000). Relations between spatial and temporal aspects of eye movement control. In A. Kennedy, R. Radach, D. Heller, & J. Pynte (Eds.), *Reading as a perceptual process* (pp. 165–191). Netherlands: Elsevier North Holland.

Rayner, K. (1975). The perceptual span and peripheral cues in reading. *Cognitive Psychology, 7*, 65–81.

Rayner, K. (1998). Eye movements in reading and information processing: 20 years of research. *Psychological Bulletin, 124*, 372–422.

Rayner, K., & McConkie, G. W. (1976). What guides a reader's eye movements? *Vision Research, 16*, 829–837.

Rayner, K., & Pollatsek, A. (1981). Eye movement control during reading: Evidence for direct control. *Quarterly Journal of Experimental Psychology, 33A*, 351–373.

Rayner, K., & Pollatsek, A. (1989). *The psychology of reading*. Englewood Cliffs, NJ: Prentice Hall.

Reichle, E. D., Pollatsek, A., Fisher, D. L., & Rayner, K. (1998). Toward a model of eye movement control in reading. *Psychological Review, 105*, 125–157.

Reichle, E. D., Rayner, K., & Pollatsek, A. (1999). Eye movement control in reading: accounting for initial fixation locations and refixations within the E-Z reader model. *Vision Research, 39*, 4403–4411.

Reichle, E. D., Rayner, K., & Pollatsek, A. (2003). The E-Z Reader model of eye movement control in reading: Comparisons to other models. *Behavioral and Brain Sciences, 26*, 445–526.

Reilly, R. G., & Radach, R. (2003). Foundations of an interactive activation model of eye movement control in reading. In J. Hyönä, R. Radach, & H. Deubel (Eds.), *The mind's eye: cognitive and applied aspects of eye movement research* (pp. 429–455). Amsterdam: Elsevier.

Schroyens, W., Vitu, F., Brysbaert, M., & d'Ydewalle, G. (1999). Eye movement control during reading: foveal load and parafoveal processing. *Quarterly Journal of Experimental Psychology, 52A*, 1021–1046.

Vitu, F., McConkie, G. W., & Zola, D. (1998). About regressive saccades in reading and their relation to word identification. In G. Underwood (Ed.), *Eye guidance in reading and scene perception* (pp. 101–124). Oxford, UK: Elsevier.

White, S. J., & Liversedge, S. P. (2006a). Linguistic and non-linguistic influences on the eyes' landing positions during reading. *Quarterly Journal of Experimental Psychology, 59*, 760–782.

White, S. J., & Liversedge, S. P. (2006b). Foveal processing difficulty does not modulate non-foveal orthographic influences on fixation positions. *Vision Research, 46*, 426–437.

White, S. J., Rayner, K., & Liversedge, S. P. (2005). Eye movements and the modulation of parafoveal processing by foveal processing difficulty: A re-examination. *Psychonomic Bulletin & Review, 12*, 891–896.

Yang, S.-N., & McConkie, G. W. (2001). Eye movements during reading: A theory of saccade initiation time. *Vision Research, 41*, 3567–85.

Zola, D. (1984). Redundancy and word perception during reading. *Perception & Psychophysics, 36*, 277–284.

Chapter 19

THE FLEXIBILITY OF LETTER CODING: NONADJACENT LETTER TRANSPOSITION EFFECTS IN THE PARAFOVEA

REBECCA L. JOHNSON

University of Massachusetts, Amherst, USA

Eye Movements: A Window on Mind and Brain

Edited by R. P. G. van Gompel, M. H. Fischer, W. S. Murray and R. L. Hill

Abstract

Previous experiments have shown that transposed-letter (TL) nonwords (e.g., *jugde* for *judge*) produce significant priming relative to orthographic controls (e.g., *jupte*). In fact, masked priming experiments indicate that TL effects exist even when the letter manipulations are nonadjacent, as long as the transposed letters are both consonants (Perea & Lupker, 2004a). This chapter presents data from a new study in which nonadjacent TL effects and the differential effects of vowels and consonants are explored during sentence reading using an eye-contingent display change paradigm. Results indicate that TL effects exist when nonadjacent letter positions are manipulated, suggesting that the coding of letter identities within a word is not specific to the absolute letter position, but is, instead, much more flexible. However, unlike the results of Perea and Lupker, those from the present study indicate that vowels and consonants pattern similarly.

Many current models of visual word recognition assume that letter positions are encoded very early in visual word recognition, even before the encoding of letter identities. Such models include the Multiple Read-Out Model (Grainger & Jacobs, 1996), the Dual Route Cascaded model (Coltheart, Rastle, Perry, Ziegler, & Langdon, 2001), the Interactive Activation Model (McClelland & Rumelhart, 1981), and the Activation Verification Model (Paap, Newsome, McDonald, & Schvaneveldt, 1982). These models all assume a "channel-specific" coding scheme for the processing of letter identities. That is, letter positions are encoded first, followed by the encoding of letter identities within each specific letter position.

One sharp criticism that has been made against these models is that they fail to account for the fact that transposed-letter (TL) nonwords (e.g., *jugde*) have been found to be more similar to their base words (e.g., *judge*) than nonwords in which two letters are substituted with other letters (e.g., *jupte*). This transposed-letter effect is well documented across a number of tasks including naming (Andrews, 1996; Christianson, Johnson, & Rayner, 2005), lexical decision (Andrews, 1996; Chambers, 1979; Forster, Davis, Schoknecht, & Carter, 1987; Holmes & Ng, 1993; O'Connor & Forster, 1981; Perea & Lupker, 2003a, 2003b, 2004a, 2004b; Perea, Rosa, & Gómez, 2005; Schoonbaert & Grainger, 2004), semantic categorization (Taft & van Graan, 1998), and normal silent reading (Johnson, Perea, & Rayner, 2007). Models employing a channel-specific coding scheme incorrectly predict that these two nonwords are equally similar to one another, because in both cases, three of the five letters are in their correct letter position. Findings from such experiments have helped to argue against models of word recognition that suggest a "channel-specific" encoding of letters. It appears that the encoding of letter identities within a word is not dependent upon absolute letter position, but is much more flexible.

While the majority of these studies have found TL effects at the foveal level (i.e., where all stimuli fell within 2° of visual angle around the point of fixation), these effects have recently been found to exist in normal silent reading (Johnson et al., 2007) in which transpositions occurred in the parafovea (i.e., the area extending 4° to the left and 4° to the right beyond the foveal area). Johnson et al. used an eye-contingent display change technique (the boundary paradigm, Rayner, 1975) to manipulate the parafoveal preview readers received prior to fixating on a given target word (Figure 1). The stimuli from Perea and Lupker (2003a) were embedded into sentences and the prime conditions served as parafoveal previews of the target word. Parafoveal previews fell into one of five conditions: (1) identical to the target word (*clerk* as the preview of *clerk*), (2) a transposition of two internal letters (*celrk*), (3) a substitution of two internal letters (*cohrk*), (4) a transposition of the two final letters (*clekr*), or (5) a substitution of the two final letters (*clefn*). Johnson et al. found that the TL effects obtained in masked priming also exist during normal silent reading, where the potential priming information is located to the right of fixation in the parafovea. That is, parafoveal previews involving a transposition of two adjacent letters led to shorter fixation durations than previews involving a substitution of two adjacent letters. For short (five-letter) words, this pattern was true for both internal and final-letter manipulations, but for longer (seven-letter) words, there was no difference

```
Greg put the wild flewor in a vase at his grandmother's house.
  *
Greg put the wild flewor in a vase at his grandmother's house.
      *
Greg put the wild flewor in a vase at his grandmother's house.
               *
Greg put the wild flower in a vase at his grandmother's house.
                     *
Greg put the wild flower in a vase at his grandmother's house.
                               *
Greg put the wild flower in a vase at his grandmother's house.
                                                *
Greg put the wild flower in a vase at his grandmother's house.
                                                          *
```

Note: The asterisk located below each sentence indicates the reader's fixation location. At the onset of the sentence, the target word (here shown in bold) is replaced with one of the three parafoveal previews (in this example, the transposed-letter preview, *flewor*). When the reader's eyes cross the invisible boundary (located just to the left of the space immediately preceding the target word), the parafoveal preview of the target word changes to the target word (here, *flower*) and remains as such until the participant indicates that they have finished reading the sentence.

Figure 1. Example sentence employing the boundary paradigm.

between the transposed and substituted letter (SL) conditions at the word-final position, likely due to acuity constraints.

Thus, it appears that letter identity information can be extracted from the fovea (and the parafovea from the first five letters of the word to the right of fixation) independent of absolute letter position. These experiments also suggest that the encoding of specific letter positions follows some time after the encoding of letter identities. What is unclear, however, is the extent to which letter position does not matter. In the experiments presented so far, all of the TL conditions involved a single transposition of two adjacent letters.[1]

In a series of experiments using masked-priming techniques, Perea and Lupker (2004a) explored the nature of TL effects with nonadjacent letter manipulations in Spanish. In addition, manipulations involved either vowels (e.g., *anamil* as the prime for *animal*) or consonants (e.g., *caniso* as the prime for *casino*). The results indicated that when the letter manipulations involved consonants, TL primes led to faster lexical

[1] Interestingly, research indicates that transposing the first two letters in a word causes disruption during normal silent reading (Johnson, Perea, & Rayner, 2007) and in masked-priming (Chambers, 1979). Chambers found that word-initial TL nonwords (e.g., *omtor* for *motor*) were less similar to their base words than word-internal TL nonwords (*liimt* for *limit*).

decision times than SL primes. However, when the letter manipulations involved vowels, there was no significant TL effect. Thus, TL-nonwords involving nonadjacent transpositions can activate the lexical representation of their base words, but only under certain circumstances.

The differential patterning of TL effects among consonants and vowels was further explored by Perea and Lupker (2004b) using English stimuli. Form priming effects were obtained for adjacent consonant transpositions (e.g., *hosre–horse* vs *honce–horse*) and for adjacent consonant–vowel transpositions (*brcik–brick* vs *brsok–brick*), but there were no priming effects for adjacent vowel transpositions (*draem–dream* vs *droim–dream*). These results, then, also support the differences in TL effects across vowels and consonants.

However, in English, the spelling-to-sound correspondences for consonants are much more regular than those for vowels. It would follow that consonants should be coded and processed more rapidly than vowels. This has led many researchers to hypothesize that these two types of letters are processed differently in reading (Berent & Perfetti, 1995). In fact, there has been much data from response time tasks (Perea & Lupker, 2004a, 2004b), silent reading tasks (Lee, Rayner, & Pollatsek, 2001, 2002), and brain-damaged patients (Caramazza, Chialant, Capasso, & Miceli, 2000) that suggest that vowels and consonants do play different roles in visual word recognition. For example, Berent and Perfetti (1995) and Lee et al. (2001, 2002) have data suggesting that at the foveal level, consonants play a greater role than vowels in the early stage of visual word recognition. The contribution of vowel information is just as strong as that of consonants, but plays a role much later in lexical identification.

In light of this previous research, the goal of the current experiment was twofold. First, I sought to investigate whether TL effects exist during normal silent reading when letter manipulations involve nonadjacent letter positions. Although letter identity can be encoded independent of absolute letter position while reading sentences, it could be the case that letter identities can only be encoded outside of their correct letter position when they are displaced one letter position to the left or right (to positions $N-1$ and $N+1$). If, however, readers are able to extract useful identity information from the parafovea that falls outside of this region (i.e., in this case, two character positions from the correct location, to positions $N-2$ and $N+2$), we would expect to find shorter fixation durations on target words (e.g., *flower*) preceded by a TL parafoveal preview (e.g., *flewor*) rather than a SL parafoveal preview (e.g., *flawur*). Such findings would provide even more support against models suggesting channel-specific encoding strategies.

Secondly, I sought to explore the differential patterning of TL effects that has been found in masked priming lexical decision tasks among vowels and consonants using a parafoveal preview experiment. All previous research on the differential roles of vowels and consonants in visual word recognition has addressed the patterning of these two letter groups when the stimuli (and experimental manipulations) were presented in foveal vision. If vowels and consonants are also processed differently in the parafovea, we might expect to see different patterns of TL-effects for these two types of letters. In the present experiment, TL nonwords (e.g., *flewor* and *fosert*) and SL nonwords (e.g., *flawur* and *fonewt*) were presented as parafoveal previews of their base words (e.g., *flower* and *forest*, respectively) to explore the role of vowels and consonants in parafoveal processing.

1. Method

1.1. Participants

Thirty-three members of the University of Massachusetts Amherst community who were native speakers of American English participated in the experiment. All participants had normal vision or wore soft contact lenses and were naïve to the purpose of the experiment. At the completion of the experiment, they received course credit or monetary compensation for their time.

1.2. Apparatus

Single-line sentences appeared one at a time on a 15-inch NEC MultiSync 4FGe monitor. Participants were seated 61 cm from the monitor, and at this distance, 3.8 letters equaled 1° of visual angle. The display was refreshed every 5 ms. Eye movements were recorded using a Generation V Fourward Technologies Dual Purkinje Eyetracker interfaced with a Pentium computer. Although reading took place binocularly, eye movements were sampled every millisecond from only the reader's right eye.

1.3. Stimuli

Thirty-six six-letter target words were embedded into single-line sentences no longer than 76 characters. Target words never occupied the sentence-initial or sentence-final word position and represented a variety of word classes and word frequencies. Three parafoveal preview conditions were created for each target word. In the identity condition, the parafoveal preview was identical to the target word (e.g., *flower* as the preview of *flower*). In the TL condition, the preview involved the transposition of the third and fifth letters (e.g., *flewor*). Finally, in the SL condition, the preview involved the substitution of the third and fifth letters (e.g., *flawur*). The replacement letters in the SL condition were visually similar to the two transposed letters. That is, vowels were substituted with vowels, consonants were substituted with consonants, ascending letters were substituted with ascending letters, and descending letters were substituted with descending letters. The TL condition and SL condition always maintained the overall word shape as presented in Courier font. For example, *fosert* was used as a TL nonword for *forest* (both the letters *s* and *r* are neither ascending nor descending) but *furute* was not used as a TL nonword for *future* (the letter *r* is neither ascending nor descending but the letter *t* is ascending).

In addition to the three parafoveal preview conditions, two types of target words were used. Target words either included (1) vowels at letter positions 3 and 5 (e.g., *flower*), or (2) consonants at letter positions 3 and 5 (e.g., *forest*). The two target word groups were matched for word frequency using both the Francis and Kučera (1982) frequency count and the Celex Lexical Database (Baayen, Piepenbrock, & Gulikers, 1995). Francis and Kučera frequencies for the 18 vowel words ranged from 1 to 340 per million (mean = 76). For the 18 consonant words, frequencies ranged from 1 to 301 per million (mean = 76). The frequencies of these two groups did not differ significantly from each other using either of the two frequency counts (t's < 1).[2]

Previous research has found that when words are highly predictable from their previous context, they are often skipped (Rayner, 1998). Thus, in order to maximize the likelihood that target words would be fixated, the context leading up to each target word was neutral. In a predictability norming procedure, ten participants were presented with the beginning part of each sentence (up to the target word) and asked to predict the next word in the sentence. All target words were found to be unpredictable from their previous context (mean predictability score = 4.7%). There was also no significant difference in the predictability scores across the two word types ($t < 1$).

In order to ensure that all of the target words fit well within their sentence context, the sentences were also normed for understandability. Ten participants were asked to rate from one (not understandable) to seven (very understandable) how well each target word fit within its sentence frame. All target words were judged to be highly understandable (mean = 6.5). In addition, there were no significant differences in understandability across the two word types ($t < 1$). The experimental sentences (including the three parafoveal preview conditions) for each of the two word types are presented in the Appendix.

1.4. Design and procedure

In order to reduce head movements during the experiment, a bite bar and a forehead rest were used. The initial calibration then took place (which lasted roughly 5 min), followed by a practice session involving eight sentences. The experimental session then followed. Each experimental sentence appeared one at a time (in random order) along the center row of the monitor. Readers were told to read each sentence silently at a comfortable pace and to press a response key when finished. In order to investigate the amount of parafoveal information the readers are gaining about a target word before fixating it, the boundary paradigm (Rayner, 1975) was used (see Figure 1). Prior to the presentation of the sentence, a fixation box appeared at the leftmost part of the screen. The experimenter then initiated the onset of the trial in which the sentence appeared on the screen with the first letter of the sentence at the location of the fixation box.

[2] Another possible difference between the two word type conditions or the three parafoveal preview conditions includes the mean bigram textual frequency and the mean trigram textual frequency. Although the mean bigram and mean trigram type frequencies of the identity previews (56.93 and 10.24, respectively) were significantly greater than those of the TL (40.38 and 3.84) and SL previews (32.72 and 3.03), there were no significant differences in the mean bigram or trigram type frequencies across the two word type conditions (p's > 0.24).

The target word appeared in one of the three preview conditions. When the readers moved their eyes to fixate on the target word (crossing the invisible boundary located just to the left of the space immediately preceding the target word), the display changed so that the preview changed to the target word. The display change occurred during the saccade, and the target word then remained throughout the remainder of the trial. Between each trial, the accuracy of the initial calibration was checked before the experimenter initiated the next trial.

Parafoveal preview was a within-subject and within-item variable; word type was a within-subject and between-item variable. Each participant read all 36 experimental sentences (18 of which included a consonant target word and 18 of which included a vowel target word). Items were counterbalanced so that there were 12 sentences in each of the three preview conditions. Thus, there were three counterbalancing conditions. Experimental sentences were presented in random order along with 78 filler sentences. Comprehension questions followed 16% of the trials to ensure that participants were carefully reading the sentences. All of the readers scored above 89% accuracy on the questions (mean = 97%). The entire experimental procedure took less than 30 min.

2. Results

The amount of time spent fixating a word is thought to reflect the time it takes to process that word (Rayner, 1998; Rayner & Pollatsek, 1989). Given that readers can extract useful information from the parafovea prior to fixating a word, parafoveal previews that provide more useful information will lessen the subsequent time the reader spends directly fixating the target word. If the different parafoveal preview conditions provide more or less useful information, we would expect to see differences in fixation times on the target words themselves. Three common measures of the amount of time spent on the target word are first fixation duration, single fixation duration, and gaze duration. First fixation duration is the amount of time spent on the initial fixation of the target word, regardless of whether there is more than one fixation on it. In contrast, single fixation duration is the amount of time spent on the initial fixation of the target word given that there was only one fixation on the first pass reading of the word. Gaze duration is the sum of all fixation durations on the target word before the reader leaves the word.

Trials were eliminated from data analysis if (1) the display change was triggered too early, (2) tracker loss occurred during a trial, or (3) the participant blinked while fixating the pre-target word, target word, or post-target word. In cases in which adjacent fixations fell within one character of one another, and one of the fixations was short (less than 80 ms), the two fixations were pooled (see Rayner, 1998). In addition, extremely short (less than 80 ms) isolated fixations and extremely long (greater than 800 ms) fixations were eliminated from the data. Altogether, 15.8% of the data were eliminated. The mean first fixation durations, single fixation durations, and gaze durations for each of the three parafoveal preview conditions in each of the two word type conditions are shown in Table 1.

Table 1
Means as a function of word type and parafoveal preview (standard errors in parentheses)

	First fixation		Single fixation		Gaze duration	
	V	C	V	C	V	C
Identity	281 (8.9)	278 (8.1)	283 (8.7)	278 (8.3)	302(10.9)	294 (7.9)
TL	283 (7.8)	288 (8.8)	290 (7.9)	296 (9.3)	323 (9.8)	319 (9.5)
SL	301 (8.2)	310 (8.6)	312 (7.5)	321 (9.6)	332 (8.4)	336 (11.3)

Note: All durations for first fixation duration, single fixation duration, and gaze duration are in ms. Word type involved the manipulation of either vowels (V) or consonants (C).

For each of the three dependent fixation duration measures, a 2 (word type: vowels or consonants) by 3 (parafoveal preview: identity control, transposed letters, or substituted letters) Analysis of Variance (ANOVA) was conducted on the data. Error variance was calculated over participants ($F1$) and over items ($F2$). In addition, planned comparisons were run to compare fixation duration in the TL condition to the respective identity condition and SL condition across the two word types.

The main effect of parafoveal preview was highly significant both by participants and by items across all three viewing duration measures (first fixation: $F1(2, 64) = 6.62, p < 0.01$; $F2(2, 68) = 7.46, p < 0.01$; single fixation: $F1(2, 64) = 13.87, p < 0.001$; $F2(2, 68) = 12.90, p < 0.001$; gaze duration: $F1(2, 64) = 10.33, p < 0.001$; $F2(2, 68) = 7.50, p < 0.01$). For first fixation duration and single fixation duration, this main effect was due to significantly longer viewing durations on target words preceded by SL previews when compared to both identity previews (first fixation: $t1(32) = 3.20, p < 0.01$; $t2(35) = 3.43, p < 0.01$; single fixation: $t1(32) = 4.75, p < 0.001$; $t2(35) = 4.58, p < 0.001$) and TL previews (first fixation: $t1(32) = 3.32, p < 0.01$; $t2(35) = 2.87, p < 0.01$; single fixation: $t1(32) = 4.45, p < 0.001$; $t2(35) = 3.14, p < 0.01$). For first fixation duration, there was no significant difference between the identity condition and the TL condition (both t's < 1), and for single fixation duration, the difference between these two conditions was significant only by items ($t1(32) = 1.64, p = 0.11$; $t2(35) = 2.15, p < 0.05$).

In contrast, for gaze duration, this main effect was the result of significantly shorter viewing durations for identity previews when compared to both TL previews ($t1(32) = 3.29, p < 0.01$; $t2(35) = 2.73, p < 0.01$) and SL previews ($t1(32) = 4.00, p < 0.001$; $t2(35) = 3.77, p < 0.001$). The difference between the TL condition and the SL condition was not significant either by participants or by items (both p's > 0.12).

The main effect of word type was not significant across any of the viewing duration measures (all F's < 1, all p's > 0.5). Critically, the interaction between parafoveal preview and word type was also not significant across any of the dependent measures (all F's < 1, all p's > 0.55). That is, the same pattern of parafoveal preview facilitation was seen in words in which vowels were transposed as in words in which consonants were transposed.

For vowels, the 18 ms TL effect for first fixation duration was significant by participants ($t1(32) = 2.22$, $p < 0.05$) and marginally significant by items ($t2(17) = 1.6$, $p = 0.064$). The 22 ms TL effect for single fixation duration was significant both by participants and by items ($t1(32) = 2.90$, $p < 0.01$; $t2(17) = 1.76$, $p < 0.05$). However, the 9 ms TL effect for gaze duration was not significant in either analysis (both p's > 0.19). For consonants, a similar pattern arose. The 22 ms TL effect for first fixation was significant by participants and by items ($t1(32) = 2.46$, $p < 0.01$; $t2(17) = 2.42$, $p < 0.05$). The 25 ms TL effect for single fixation was significant by participants and by items ($t1(32) = 2.79$, $p < 0.01$; $t2(17) = 2.62$, $p < 0.01$). The 17 ms TL effect for gaze duration was marginally significant by participants ($t1(32) = 1.33$, $p = 0.096$) and not significant by items ($p = 0.1$).

3. Discussion

The results from the current experiment are straightforward. When looking at early measures of visual word recognition (i.e., first fixation duration and single fixation duration), viewing duration measures were shortest for the identity condition and the TL condition and significantly longer for the SL condition. These findings indicate that letter identity is encoded early in visual word recognition. Previews that contained accurate letter identity information (i.e., the identity condition and the TL condition) led to significantly shorter early viewing duration times than previews containing inaccurate letter identity information (i.e., the SL condition). Furthermore, the fact that TL previews led to shorter viewing durations than SL previews indicates that letter identities can be encoded outside of their specific letter position. Encoding of letter identities is thus much more flexible than channel-specific models assume. These results are consistent with previous research showing TL effects during visual word recognition tasks and extend the work of Johnson et al. (2007) to show that TL effects also exist with nonadjacent transpositions.

In the analysis of a later measure of visual word recognition (namely, gaze duration), it becomes apparent that the role of accurate letter position is important in the encoding process. While the TL effect seen in earlier measures was no longer significant, the identity condition led to significantly shorter viewing durations than either of the other two conditions. These findings, thus, also challenge the aforementioned models of visual word recognition, which suggest that letter positions are encoded prior to letter identities. It appears the importance of letter identity occurs earlier in visual word recognition than the role of letter position. Others, too, have postulated that letter positions take longer to encode than letter identities (Adams, 1979). As these results indicate, it is quite likely that when a reader encounters a TL nonword such as *caniso*, he or she begins to encode and activate the lexical representations of each of the component letters (including the *n* and *s*, which are in their incorrect letter location). At some later point, however, the processing system must reorder the letters (i.e., those that are in incorrect letter positions) in order to accurately match the lexical representation. This reordering comes with a cost, and so processing suffers a latency penalty.[3]

Models assuming a channel-specific coding scheme for letter identities cannot account for the current findings. Furthermore, as Perea and Lupker (2004a, 2004b) pointed out, the presence of TL effects in nonadjacent manipulations causes especially great problems for interactive-activation models, which rest on the assumption that the bottom-up encoding of letter identities and letter positions is non-noisy. Perea and Lupker argued that another form of coding scheme, rather than a channel-specific coding scheme, is needed to be able to fully account for TL effects. Some more recent models of visual word recognition can account for the TL effects found in previous studies as well as those in the current experiment because they allow for "noise" in the system, although in differing ways. Three such models are the SOLAR model (Davis, 1999), the SERIOL model (Whitney, 2001), and the Overlap model (Gómez, Perea, & Ratcliff, 2003).

The SOLAR model (Davis, 1999) employs a spatial coding scheme to assign letter positions different activation levels according to their location within the letter string. According to the model, the first letter position receives the highest level of activation, followed by the second letter position and so forth. Letter strings with different letter identities in different letter positions, then, receive different patterns of activation. This leads to the successful distinction between anagrams like *stop*, *pots*, *opts*, *post*, *tops*, and *spot*. The model can account for TL effects because it also includes a separate parameter to measure the amount of similarity in the set of letter nodes. Since the nonword *caniso* shares the same set of six letters as *casino*, it is more similar to its base word than the nonword *caviro*.

The SERIOL model (Whitney, 2001) also assigns varying activation levels to successive letter positions. In addition, it relies on the activation of bigram nodes to encode words. The word *casino*, for example, can be broken down into 15 bigrams (*ca*, *cs*, *ci*, *cn*, *co*, *as*, *ai*, etc.). The nonword *caniso* shares 12 of these bigrams, while the nonword *caviro* shares only 6 of these. Thus, nonadjacent TL nonwords are successfully predicted to be more similar to their base words than SL nonwords.[4]

[3] As one reviewer noted, the fact that the TL condition differed from the identity condition in gaze duration (a pattern not seen in first fixation or single fixation) could indicate that in order for absolute (or even relative) letter position encoding to occur, more than one fixation on the word is necessary. While this is a possible explanation for the pattern of data in this experiment, the current results are inconclusive to address this hypothesis because gaze durations also include trials in which only a single fixation occurred. In the current experiment, only 6% of the target words received more than one fixation on the first pass reading of the word, making subanalyses of these items impossible. Furthermore, although the first fixation durations and single fixation durations between the TL condition and the identity condition were not always significant, the pattern of means suggests a trend in the same direction as that seen in gaze duration. Therefore, the differences seen in early versus late measures are likely due to the increased amount of time spent fixating on the target word rather than on the number of fixations the target word receives.

[4] As reported by Perea and Lupker (2004a), the SOLAR model would predict that the identity condition (e.g., *casino*) would yield the highest similarity score to its base word (1.00), followed by the TL condition (*caniso*, 0.83), and lastly the SL condition (*caviro*, 0.54). Likewise, the SERIOL model predicts the same pattern of results with the following similarity scores: identity, 1.00; TL, 0.88; SL, 0.49.

Finally, the Overlap Model (Gómez, Perea, & Ratcliff, 2003) can account for TL effects because it assumes that letter representations extend into neighboring letter positions. The encoding activation of a given letter is greatest at its correct letter position, but this activation also encompasses other letter positions as well. Specifically, the encoding activation is represented as a normal distribution with activation decreasing as a letter appears further from its correct letter position. The Overlap Model would also successfully predict the pattern of results from the current experiment since activation of a given letter not only is greatest at its correct letter position (thus predicting greatest facilitation from the identity condition), but also extends out to neighboring letter positions (thus predicting more facilitation from TL nonwords than SL nonwords).

As indicated by the lack of a significant interaction, this pattern of results was nearly identical when consonants were manipulated as when vowels were. This suggests that, at least at the parafoveal level, the processing of consonants and vowels is similar in terms of the encoding of letter identities and letter positions. This is in contrast to the previous literature that has found differences between the role of consonants and vowels when stimuli are presented in foveal vision (Berent & Perfetti, 1995; Lee et al., 2001, 2002; Perea & Lupker, 2003b, 2004a, 2004b). Based on several foveal TL experiments manipulating vowels and consonants in both English and Spanish, Perea and Lupker (2004a) concluded that the absolute letter position of vowels may be more important than the absolute letter position of consonants. While this may be the case for foveally presented stimuli, the current results indicate that the flexibility of letter coding in the parafovea extends to include both vowels and consonants. It is thus likely that the parafoveal transposed letter effects found in the current study and also by Johnson et al. (2007) are likely to occur at a very low level of visual word recognition, before the encoding of a vowel/consonant label and the phonological attachment of letters to sounds. Clearly, more research should be conducted to assess the role of consonants and vowels at the parafoveal level to evaluate whether these two letter types have differing roles prior to fixation.

While the current study indicates that TL effects exist when nonadjacent letter positions are manipulated, there are still many questions that need to be addressed to explore the limitations of letter identity and letter position coding. Whether these parafoveal priming effects would exist (and how strong they would be) if manipulations were made further than two letter positions away (e.g., *cnsiao* for *casino*) or if multiple manipulations were made (e.g., *csanio* for *casino*) is currently unknown. Knowing exactly how far these TL effects extend and which factors mediate these effects can be helpful for testing models and informing certain parameters. For example, in the Overlap Model, knowing exactly how far TL effects extend would help to set the parameters of the encoding activation at neighboring letter positions to better define the shape and spread of the bell-shaped curves.

In summary, although nonadjacent TL effects have previously been found only to exist at the foveal level when consonants are manipulated, the current findings indicate that these effects hold for both consonants and vowels presented in the parafovea. Although consonants and vowels may play different roles in foveal visual word recognition, the evidence presented here indicates that these two letter groups pattern similarly at the parafoveal level. The extraction of letter identity from the parafovea is not strictly dependent upon specific letter position, but rather, letter identity is encoded prior to letter position. These findings challenge models of visual word recognition that assume a channel-specific letter coding scheme and support models that allow more flexibility in the coding of letter identities. Specifically, the current findings support the SOLAR model, the SERIOL model, and the Overlap model. Future studies, then, should be designed to directly test opposing predictions that these three models make.

Acknowledgements

This research was supported by Grant HD26765 from the National Institute of Health to Keith Rayner. Thanks to Keith Rayner, Barbara Juhasz, Simon Liversedge, and one anonymous reviewer for their helpful comments on an earlier draft.

Appendix

A.1. Vowels

Greg put the wild (flower, flewor, flawur) **flower** in a vase at his grandmother's house.

Some athletes shy away from a tough (league, leugae, leogie) **league** with harsh competition.

Consider the age and intelligence level of your typical (reader, reedar, reodur) **reader** when writing.

Octavian attained the supreme (status, stutas, stotis) **status** as the first Roman Emperor in 27 BC.

The magazine sought a careful (editor, edotir, edater) **editor** and great headline writer for the job.

Samantha put the sharp (weapon, weopan, weepun) **weapon** outside of the young child's reach.

Our high school soccer team finished their amazing (season, seosan, seisun) **season** last weekend.

In Roman circuses, the most popular (animal, anamil, anomul) **animal** for arena use was the lion.

Jesus appointed Simon Peter as the first notable (leader, leedar, leodur) **leader** of the church.

Lowering the risk of disease is a worthy (reason, reosan, reusin) **reason** to breastfeed your baby.

Last night, a brown (weasel, weesal, weosil) **weasel** ran across the highway in front of our car.

When the rich (deacon, deocan, deucen) **deacon** began abusing his authority, he was asked to step down.

The candles and lovely (violet, vielot, vialit) **violet** blossoms added a nice touch to the reception.
The Johnsons have a tall (spiral, sparil, sporel) **spiral** staircase in the foyer of their home.
It is possible to have a droopy upper (eyelid, eyiled, eyalud) **eyelid** following cataract surgery.
The most useful (shovel, shevol, shuval) **shovel** for digging up fossils has a semi-pointed front edge.
After driving on the rough (gravel, greval, grivul) **gravel** road for an hour, our car was filthy.
Kimberly cut out the latest (coupon, coopun, coapen) **coupon** from the paper to use at the grocery.

A.2. Consonants

The animal hid in the deep (forest, fosert, fonewt) **forest** until it was safe for it to come out.
The inappropriate racial (remark, reramk, recaxk) **remark** led to much anger and violent reactions.
The child's sudden (desire, derise, dewice) **desire** for seclusion and inactivity concerned his mom.
I bought a fresh (bottle, boltte, boktbe) **bottle** of cough medicine when the expiration date arrived.
Among many things, the Census collects data on family (income, inmoce, inroxe) **income** and housing.
While the boy was in surgery, the anxious (parent, panert, pamect) **parent** paced in the waiting room.
Especially in older adults, extreme (stress, stsers, stcens) **stress** can cause many health problems.
We brought some pure (spring, spnirg, spmicg) **spring** water on our hiking trip to the Rockies.
After much searching, Jen bought the rare (record, rerocd, rewosd) **record** from an online auction.
Ross plotted the path of the rocket from the initial (moment, monemt, morect) **moment** it took off.
Cassie pulled the warm (waffle, walffe, watfke) **waffle** out of the toaster and added syrup and jam.
The popular hangout for young adults is the busy (tavern, tarevn, tanemn) **tavern** on the corner.
Is it better to be a silent (coward, corawd, cosamd) **coward** or an outspoken radical?
Lyla said she had a quick (errand, ernard, ersacd) **errand** to run before going to dinner with us.

Doctors recommend that you maintain a healthy (intake, inkate, inlabe) **intake** of dietary fiber.

For their honeymoon, Matt and Barb stayed at a nearby (resort, rerost, recomt) **resort** on the beach.

Theodore poured the thick (cement, cenemt cesert) **cement** to secure the basketball goal in place.

After I saw the empty (pantry, partny, pawtvy) **pantry** and bare refrigerator, I went to the grocery.

References

Adams, M. J. (1979). Models of word recognition. *Cognitive Psychology, 11*(2), 133–176.

Andrews, S. (1996). Lexical retrieval and selection processes: Effects of transposed-letter confusability. *Journal of Memory and Language, 35*, 775–800.

Baayen, R. H., Piepenbrock, R., & Gulikers, L. (1995). *The CELEX lexical database (CD-ROM)*. Philadelphia: Linguistic Data Consortium, University of Pennsylvania.

Berent, I., & Perfetti, C. A. (1995). A rose is a REEZ: The two cycles model of phonology assembly in reading English. *Psychological Review, 102*, 146–184.

Caramazza, A., Chialant, D., Capasso, D., & Miceli, G. (2000). Separable processing of consonants and vowels. *Nature, 403*, 428–430.

Chambers, S. M. (1979). Letter and order information in lexical access. *Journal of Verbal Learning and Verbal Behavior, 18*, 225–241.

Christianson, K., Johnson, R. L., & Rayner, K. (2005). Letter transpositions within and across morphemes. *Journal of Experimental Psychology: Learning, Memory, and Cognition, 31*, 1327–1339.

Coltheart, M., Rastle, K., Perry, C., Ziegler, J., & Langdon, R. (2001). DRC: A Dual Route Cascaded model of visual word recognition and reading aloud. *Psychological Review, 108*, 204–256.

Davis, C. J. (1999). *The Self-Organising Lexical Acquisition and Recognition (SOLAR) model of visual word recognition*. Unpublished doctoral dissertation, University of New South Wales, Australia.

Forster, K. I., Davis, C., Schoknecht, C., & Carter, R. (1987). Masked priming with graphemically related forms: Repetition or partial activation? *The Quarterly Journal of Experimental Psychology: Human Experimental Psychology, 39A*(2), 211–251.

Francis, W., & Kučera, H. (1982). *Frequency analysis of English usage: Lexicon and grammar*. Boston: Houghton Mifflin.

Gómez, P., Perea, M., & Ratcliff, R. (2003). *A model of the coding of letter positions: The overlap model*. Poster presented at the 45th Annual Meeting of the Psychonomic Society, Vancouver, BC, Canada.

Grainger, J., & Jacobs, A. M. (1996). Orthographic processing in visual word recognition: A multiple read-out model. *Psychological Review, 103*, 518–565.

Holmes, V. M., & Ng, E. (1993). Word-specific knowledge, word-recognition strategies, and spelling ability. *Journal of Memory and Language, 32*, 230–257.

Johnson, R. L., Perea, M., & Rayner, K. (2007). Transposed-Letter Effects in Reading: Evidence from Eye Movements and Parafoveal Preview. *Journal of Experimental Psychology: Human Perception and Performance*.

Lee, H-W., Rayner, K., & Pollatsek, A. (2001). The relative contribution of consonants and vowels to word identification during reading. *Journal of Memory and Language, 44*, 189–205.

Lee, H-W., Rayner, K., & Pollatsek, A. (2002). The processing of consonants and vowels in reading: Evidence from the fast priming paradigm. *Psychonomic Bulletin & Review, 9*, 766–772.

McClelland, J. L., & Rumelhart, D. E. (1981). An interactive activation model of context effects in letter perception: Part I. An account of basic findings. *Psychological Review, 88*, 375–407.

O'Connor, R. E., & Forster, K. I. (1981). Criterion bias and search sequence bias in word recognition. *Memory & Cognition, 9*(1), 78–92.

Paap, K. R., Newsome, S. L., McDonald, J. E., & Schvanereldt, R. W. (1982). An activation-verification model for letter and word recognition: The word superiority effect. *Psychological Review, 89*, 573–594.

Perea, M., & Lupker, S. J. (2003a). Does jugde activate COURT? Transposed-letter similarity effects in masked associative priming. *Memory and Cognition, 31*, 829–841.

Perea, M., & Lupker, S. J. (2003b). Transposed-letter confusability effects in masked form priming. In S. Kinoshita & S. J. Lupker (Eds.), *Masked priming: The state of the art* (pp. 97–120). New York, NY: Psychology Press.

Perea, M., & Lupker, S. J. (2004a). Can CANISO activate CASINO? Transposed-letter similarity effects with nonadjacent letter positions. *Journal of Memory and Language, 51*, 231–246.

Perea, M., & Lupker, S. J. (2004b, November). *The effect of transposed letter stimuli in visual word recognition.* Poster presented at the 45th Annual Meeting of the Psychonomic Society, Minneapolis, USA.

Perea, M., Rosa, E., & Gómez, C. (2005). The frequency effect for pseudowords in the lexical decision task. *Perception and Psychophysics, 67*, 301–314.

Rayner, K. (1975). The perceptual span and peripheral cues in reading. *Cognitive Psychology, 7*, 65–81.

Rayner, K. (1998). Eye movements in reading and information processing: 20 years of research. *Psychological Bulletin, 124*(3), 372–422.

Rayner, K., McConkie, G. W., & Zola, D. (1980). Integrating information across eye movements. *Cognitive Psychology, 12*(2), 206–226.

Rayner, K., & Pollatsek, A. (1989). *The psychology of reading.* Englewood Cliffs, NJ: Prentice Hall.

Schoonbaert, S., & Grainger, J. (2004). Letter position coding in printed word perception: Effects of repeated and transposed letters. *Language and Cognitive Processes, 19*, 333–367.

Taft, M., & van Graan, F. (1998). Lack of phonological mediation in a semantic categorization task. *Journal of Memory and Language, 38*, 203–224.

Whitney, C. (2001). How the brain encodes the order of letters in a printed word: The SERIOL model and selective literature review. *Psychonomic Bulletin & Review, 8*(2), 221–243.

PART 6

EYE MOVEMENTS AS A METHOD FOR INVESTIGATING SPOKEN LANGUAGE PROCESSING

Edited by

ROGER P. G. VAN GOMPEL

Chapter 20

EYE MOVEMENTS AND SPOKEN LANGUAGE PROCESSING

MICHAEL K. TANENHAUS
University of Rochester, USA

Eye Movements: A Window on Mind and Brain
Edited by R. P. G. van Gompel, M. H. Fischer, W. S. Murray and R. L. Hill

Abstract

This chapter provides an overview of recent research that uses eye movements to investigate both spoken language comprehension and language production. Issues of data analysis and linking hypotheses are addressed in what is now commonly referred to as the "visual world" paradigm, including issues that arise in comparing results to those from eye-tracking reading studies. It is argued that eye-tracking reading studies primarily use fixation duration as a processing load measure, whereas visual world studies use the location and timing of fixations as a representational measure. Considerations are raised about how the presence of a visual scene and use of actions might influence results from visual world studies.

Eye movements have been one of the most widely used response measures in psycholinguistic studies of reading for more than a century (see Ferreira & Henderson, 2004; Rayner, 1998). In contrast, it is only within the last decade that eye movements have become a commonly used response measure in studies of spoken language processing. In these studies, participants' eye movements to real world objects or to pictures in a display or scene are monitored, typically using a head-mounted eye-tracker, as the participants follow instructions, listen to sentences, or generate utterances about the "visual world."

Psycholinguists now use the visual-world eye-movement method to study both language production and language comprehension, in studies that run the gamut of current topics in language processing. Eye movements are a response measure of choice for studies addressing many classical questions about spoken language processing in psycholinguistics: e.g., is the processing of stop consonants categorical? (McMurray, Tanenhaus, & Aslin, 2002); does context influence the earliest moments of temporary lexical and syntactic ambiguity resolution? (Dahan & Tanenhaus, 2004; Spivey, Tanenhaus, Eberhard, & Sedivy, 2002); what is the locus of frequency effects in spoken word recognition? (Dahan, Magnuson, & Tanenhaus, 2001); what factors influence the time course with which anaphoric expressions, such as pronouns, are resolved? (Arnold, Eisenband, Brown-Schmidt, & Trueswell, 2000) and, for bilingual speakers, does a word spoken in one language activate the lexical representations of similar sounding words in the other language (Spivey & Marian, 1999; Ju & Luce, 2004).

The use of eye movements has also opened up relatively uncharted territory in language comprehension and language production. In comprehension these include real-time sentence processing in children (Trueswell, Sekerina, Hill, & Logrip, 1999); the role of common ground in online processing (Hanna, Tanenhaus, & Trueswell, 2003; Keysar, Barr, Balin, & Brauner, 2000); how listeners make use of disfluencies in real-time language processing (Arnold, Tanenhaus, Altmann, & Fagnano, 2004; Baily & Ferreira, 2003, this volume); and how participants in a conversation coordinate their referential domains (Brown-Schmidt, Campana, & Tanenhaus, 2005; Tanenhaus, & Brown-Schmidt, to appear). In production these include, the locus of disfluency effects (Griffin, 2004b), and the interface between message formulation and utterance planning (Bock, Irwin, Davidson, & Levelt, 2003; Brown-Schmidt & Tanenhaus, in press; Griffin & Bock, 2000). Finally, the visual-world approach has spawned a new family of studies investigating the interface between action and language and between vision and language (Altmann & Kamide 2004; Chambers, Tanenhaus & Magnuson, 2004; Knoeferle, Crocker, Scheepers, & Pickering, 2005; Spivey & Geng, 2001; Spivey et al., 2002).

Why is the visual world paradigm becoming so widely used? First, in contrast to reading, time-locked, relatively natural measures of spoken language processing have been hard to come by. Many of the most widely used tasks for studying spoken language comprehension present only a snapshot of processing at a single point in time, require meta-linguistic judgments, and interrupt the flow of the speech input. In contrast, eye movements provide a sensitive, implicit measure of spoken language processing in which the response is closely time-locked to the input without interrupting the speech stream. Second, the eye movement paradigm can be used with simple natural tasks such as picking up and moving objects, making it well suited for studies with young children (Trueswell et al., 1999) and with special populations (Novick, Trueswell, & Thompson-Schill, 2005; Yee, Blumstein, & Sedivy, 2000). It also makes the paradigm well suited to investigations of language within an embodiment framework (e.g., Spivey, Richardson, & Fitneva, 2004). Third, the coupling of a visual world with language makes it possible to ask questions about real-time interpretation, especially questions about reference that would be difficult to address, and perhaps intractable, if one were limited to measures of processing complexity for written sentences or spoken utterances (cf. Sedivy, Tanenhaus, Chambers, & Carlson, 1999). It also makes it possible to examine questions at the interface between language, perception, and action (see chapters in Henderson & Ferreira, 2004 and Trueswell & Tanenhaus, 2005). Fourth, eye movements can be used to study issues about the relationship between real-time message planning and utterance planning (Bock, Irwin, & Davidson, 2004; Brown-Schmidt & Tanenhaus, in press; Griffin, 2004a).

Finally, the paradigm allows one to study real-time language production and comprehension in natural tasks involving conversational interaction. This makes it possible to bridge the two dominant traditions in language processing research: the "language-as-action" tradition, which has focused on natural interactive conversation while generally ignoring questions about the time course of real-time language processing, and the "language-as-product" tradition, which has focused on the time course of processing while being primarily limited to de-contextualized language (Clark, 1992; Tanenhaus & Trueswell, 2005).

As with any new paradigm, excitement about novel findings and new arenas of investigation must be tempered with concerns about the nature of the paradigm itself, including task-specific strategies, and the assumptions that link the behavioral measure to the hypothesized underlying mechanisms. This chapter is divided into four sections. The first section provides an introduction to how eye movements are used to study spoken language processing, beginning with a brief review of some of the foundational studies and concluding with an overview of how eye movement data are analyzed. The second section makes some observations about similarities and differences between how eye movements are used to study similar psycholinguistic issues in reading and spoken language comprehension. The third section raises issues about when using a circumscribed visual world might introduce distortions that limit the degree to which results from visual world studies can be generalized to language processing in less constrained settings. The chapter then concludes with a brief overview of the other contributions to this part of the volume.

1. Some foundational studies

1.1. Comprehension

The use of eye movements as a tool for studying spoken language comprehension was pioneered by Roger Cooper (1974) in a remarkable article, presciently titled *The control of eye fixation by the meaning of spoken language: a new methodology for the real-time investigation of speech perception, memory and language processing.* Cooper tracked participants' eye movements as they listened to stories while looking at a display of pictures. He found that participants initiated saccades to pictures that were named in the stories, as well as pictures associated to words in the story. Moreover, fixations were often generated before the end of the word.

Tanenhaus, Spivey-Knowlton, Eberhard, and Sedivy (1995) initiated the recent surge of interest in what they dubbed the "visual world paradigm", now sometimes referred to as the *action-based* version of the visual world paradigm. Taking advantage of the advent of accurate lightweight head-mounted eye-trackers, they examined eye movements as participants followed instructions to perform simple tasks with objects in a workspace. They found that varying the number of potential referents for a temporarily ambiguous prepositional phrase (e.g., *Put the apple on the towel in the box* in a display containing four objects, a towel, a box, an apple on another towel, and either a second apple, or another object such as a pencil) determined whether the ambiguous prepositional phrase (*on the towel*) was initially parsed as a goal argument (where to put the apple) or as a modifier (the location of the apple to be moved), as predicted by Altmann and Steedman (1988). (A more complete report of the results from Tanenhaus et al. (1995) is presented in Spivey, Tanenhaus, Eberhard, & Sedivy, 2002.)

Trueswell et al. (1999) replicated the Tanenhaus et al. (1995) study with adults and with five- and eight-year-old children, finding important developmental differences. The developmental differences, and the fact that the paradigm could be adapted for use with young children, have laid the foundation for studies of online sentence processing in preliterate children.

Eberhard, Spivey-Knowlton, Sedivy, & Tanenhaus (1995) demonstrated that fixations to entities referred to in an instruction are remarkably time-locked to the unfolding utterance, with fixations to a target referent in a display of competitors occurring as soon as continuous integration of constraints provided by the unfolding speech and the visual display could, in principle, distinguish the referent from its competitors. These results were obtained both for simple instructions (*Touch the starred red square*) and for complex instructions (*Put the five of hearts that's below the eight of clubs above the three of diamonds*). This "point-of disambiguation" logic is widely used in studies of reference resolution (e.g., Sedivy et al., 1999; Chambers, Tanenhaus, Eberhard, Filip, & Carlson, 2002). As in studies of eye movements in natural tasks in vision (Hayhoe & Ballard, 2005), for the complex instructions, a high proportion of the saccades are generated to entities at the point in the instruction where they become task relevant.

Building on initial results by Spivey-Knowlton (1996), Allopenna, Magnuson, & Tanenhaus (1998) demonstrated that the timing of fixations to a pictured referent, and competitors with different types of phonological overlap, was sufficiently time-locked to the input to trace the time course of lexical access. Allopenna et al. also showed that a simple linking hypothesis could be used to map fixations onto computational models of lexical activation, thus laying the foundation for the growing body of work that uses the visual world paradigm to study spoken word recognition.

Altmann and Kamide (1999) made an important addition to the visual world paradigm by demonstrating linguistically mediated anticipatory eye movements using a task like Cooper's in which the participants' primary task was to listen to a description of an upcoming event involving entities depicted in a display. As participants heard sentences such as *the boy will eat the cake*, they made anticipatory eye movements to a picture of cake before the offset of *eat*, when the other depicted objects were not eatable. Anticipatory eye movements are now widely used as a dependent measure, typically with this so-called, *passive listening*, variant of the visual-world paradigm.

1.2. Production

The foundation for eye movement studies of language production comes from two studies: one by Meyer, Sleiderink, and Levelt (1998), the other by Griffin and Bock (2000). Meyer et al. had participants name sequences of objects. Eye gaze was tightly coordinated with the speech. Participants fixated a to-be-named object about 1 s prior to the onset of naming. This eye–voice lag is similar to the time it takes to initiate naming an object in isolation (Rossion & Pourtois, 2004; Snodgrass & Yuditsky, 1996), suggesting that the eye–voice delay reflects word preparation. About 150 ms before the onset of speech, participants launched a saccade to the next object.

Griffin and Bock (2000) presented participants with a simple event rendered as a line drawing that could be described with either an active or passive sentence, such as a woman shooting a man, or lightening striking a church. The sequence of eye movements reflected the order of constituents in the utterance. Speakers looked at pictured objects about 800 ms to 1 s before naming them. Once speaking began, the sequence and timing of fixations was controlled by the utterance, rather than perceptual properties of the input, suggesting that the speaker had completed message planning prior to beginning to speak (also see Bock, Irwin, Davidson, & Levelt, 2003).

The contribution by Wheeldon, Meyer, and van der Meulen (this volume) builds on the line of research initiated by Meyer et al. to ask whether fixation patterns differ for correct pronunciations and pronunciations with anticipation errors. They concluded that the timing of fixations does not differ for correctly and incorrectly produced names.

2. Data analysis and linking assumptions

In order to briefly describe how eye movement data are analyzed in comprehension, I will use Experiment 1 from Allopenna et al. (1998). This experiment will also prove useful for subsequent discussion of some of the methodological concerns that arise in visual world studies. Allopenna et al. (1998) evaluated the time course of activation for lexical competitors that shared initial phonemes with the target word (e.g., beaker and beetle) or that rhymed with the target word (e.g., beaker and speaker). Participants were instructed to fixate a central cross and then followed a spoken instruction to move one of four objects displayed on a computer screen with the computer mouse (e.g., *Look at the cross. Pick up the beaker. Now put it above the square*).

2.1. Data analysis

A schematic of a sample display of pictures is presented in Figure 1, upper left. The pictures include the target (the beaker), the cohort (the beetle), a picture with a name that rhymes with the target (speaker), and the unrelated picture (the carriage). For purposes of illustrating how eye movement data are analyzed, we will restrict our attention to the target, cohort, and unrelated pictures. The particular pictures displayed are used to exemplify types of conditions and are not repeated across trials.

Five hypothetical trials are shown in the upper right portion of the figure. The 0 ms point indicates the onset of the spoken word *beaker*. The grey line begins at about 200 ms – the earliest point where one would expect to see signal-driven fixations. On trial one, the hypothetical participant initiated a fixation to the target about 200 ms after the onset of the word, and continued to fixate on it (typically until the hand brings the mouse onto the target). On trial two, the fixation to the target begins a bit later. On trial three, the first fixation is to the cohort, followed by a fixation to the target. On trial four, the first fixation is to the unrelated picture. Trial five shows another trial where the initial fixation is to the cohort. The graph in the lower right hand portion of the figure illustrates the proportion of fixations over time for the target, cohort, and unrelated pictures, averaged across trials and participants. These fixation proportions are obtained by determining the

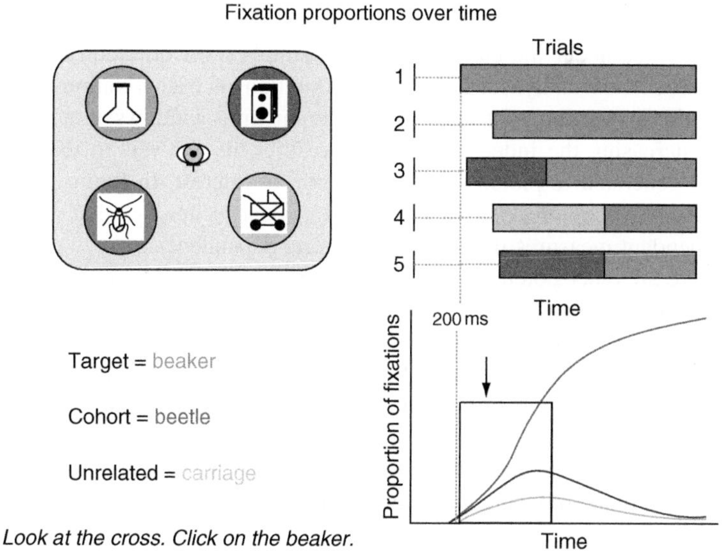

Figure 1. Schematic illustrating proportion of fixation curves. (*See Color Plate 2.*)

proportion of looks to the alternative pictures at a given time slice and they show how the pattern of fixations change as the utterance unfolds. The fixations do not sum to 1.0 as the word is initially unfolding because participants are often still looking at the fixation cross.

Proportion of fixation curves would seem to imply that eye movements provide a continuous measure – a misconception that my colleagues and I may have sometimes contributed to in some of our papers. It is more accurate to say that eye movements provide an approximation to a continuous measure. For example, the assumption behind linking fixations to continuous word recognition processes is that as the instruction unfolds, the probability that the listener's attention will shift to a potential referent of a referring expression increases with the activation of (evidence for) its lexical representation, with a saccadic eye movement typically following a shift in visual attention to the region in space where attention has moved. Because saccades are rapid, low cost, low-threshold responses, a small proportion of saccades will be generated by even small increases in activation, with the likelihood of a saccade increasing as activation increases. Thus, while each saccade is a discrete event, the probabilistic nature of saccades ensures that with sufficient numbers of observations, the results will begin to approximate a continuous measure. For an insightful discussion, including the strengths and weaknesses of eye movements compared to a truly continuous measure, tracking the trajectories of hand movements, see Spivey, Grosjean, and Knoblich (2005) and Magnuson (2005).

Researchers often define a window of interest, as illustrated by the rectangle in the proportion graph. For example, one might want to focus on the fixations to the target and cohort in the region from 200 ms after the onset of the spoken word to the point in the speech stream where disambiguating phonetic information arrives. The proportion of fixations to pictures or objects and the time spent fixating on the alternative pictures (essentially the area under the curve, which is a simple transformation of proportion of fixations), can then be analyzed. Because the duration of each fixation is likely to be 150–250 ms, the proportion of fixations in different time windows is not independent. One way of increasing the independence is to restrict the analysis to the proportion of new saccades generated to pictures within a region of interest. In future research it will also be important to explore using additional statistical techniques that are designed to deal with dependent measures that contain temporal dependencies.

Finally, there are other potential measures that can provide additional information. For example, the duration of fixations to a picture or entity in a scene has proved particularly useful in production studies. Comprehension researchers have occasionally restricted analyses to the initial saccade, especially in tasks where the participant is likely to be looking at a fixation point when the critical input arrives. In addition, as discussed later, one can examine a variety of *contingent* fixations.

To date, there has been relatively little debate about the relative merits of different measures, especially in comparison to eye movement reading studies (but, cf. Altmann & Kamide, 2004). One likely reason is that the visual world paradigm is new enough so that there have been few, if any, studies that I am aware of where investigators have examined similar issues, but come to different conclusions, using different measures.

A second reason is that examining timing may be more straightforward in visual world studies than in reading studies for reasons that we will discuss later.

In Figure 2, the graph in the upper left quadrant shows the data from the Allopenna et al. (1998) experiment. The figure plots the proportion of fixations to the target, cohort, rhyme, and unrelated picture. Until 200 ms, nearly all of the fixations are on the fixation cross. These fixations are not shown. The first fixations to pictures begin at about 200 ms after the onset of the target word. These fixations are equally distributed between the target and the cohort. These fixations are remarkably time-locked to the utterance: input-driven fixations occurring 200–250 ms after the onset of the word are most likely programmed in response to information from the first 50–75 ms of the speech signal. At about 400 ms after the onset of the spoken word, the proportion of fixations to the target began to diverge from the proportion of fixations to the cohort. Subsequent research has established that cohorts and targets diverge approximately 200 ms after the first phonetic input that provides probabilistic evidence favoring the target, including coarticulatory information in vowels (Dahan, Magnuson, Tanenhaus, & Hogan, 2001; Dahan & Tanenhaus, 2004).

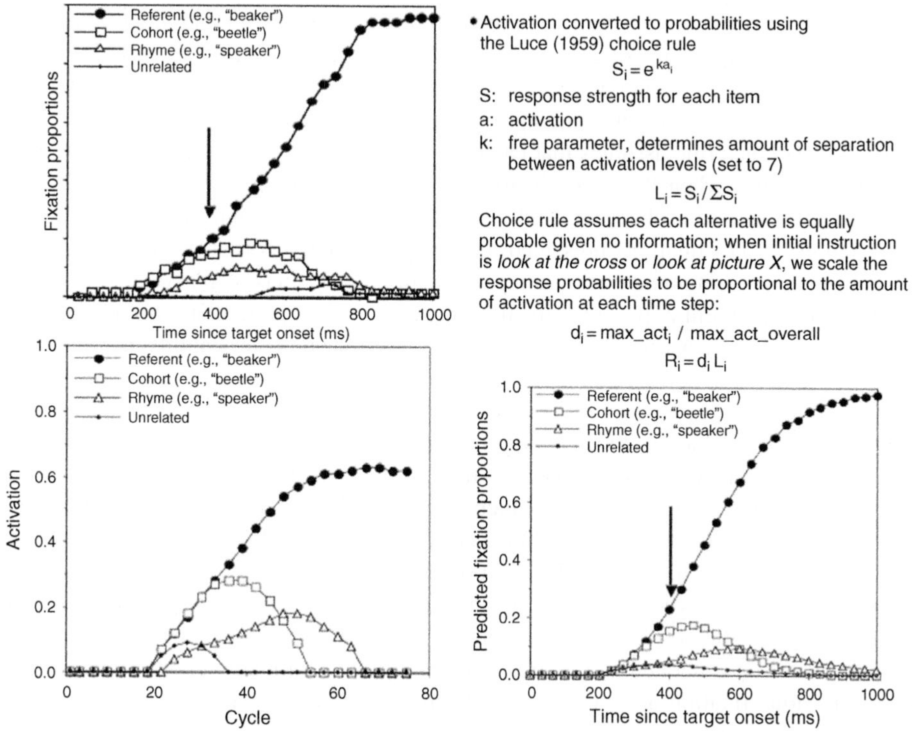

Figure 2. Top left: data from Allopenna, et al., (1998); Bottom left: Trace simulations; Top right: linking hypothesis; Bottom right predicted proportion of fixations.

Shortly after fixations to the target and cohort begin to rise, fixations to rhymes start to increase relative to the proportion of fixations to the unrelated picture. This result discriminates between predictions made by the cohort model of spoken word recognition and its descendents (e.g., Marslen-Wilson, 1987, 1990, 1993), which assume that any feature mismatch at the onset of a word is sufficient to strongly inhibit a lexical candidate, and continuous mapping models, such as TRACE (McClelland & Elman, 1986), which predict competition from similar words that mismatch at onset (e.g., rhymes). The results strongly confirmed the predictions of continuous mapping models.

2.2. Formalizing a linking hypothesis

We can now illustrate a simple linking hypothesis between an underlying theoretical model and fixations. The assumption providing the link between word recognition and eye movements is that the activation of the name of a picture determines the probability that a subject will shift attention to that picture and thus make a saccadic eye movement to fixate it. Allopenna et al. formalized this linking hypothesis by converting activations into response strength, following the procedures outlined in Luce (1959). The Luce choice rule is then used to convert the response strengths to response probabilities. The graph in the lower left quadrant of Figure 2 shows the activation values for beaker, beetle, carriage, and speaker, generated by a TRACE simulation. The equations used in the linking hypothesis are shown in the upper right hand quadrant of the figure.

The Luce choice rule assumes that each response is equally probable when there is no information. Thus when the initial instruction is *look at the cross* or *look at picture X*, we scale the response probabilities to be proportional to the amount of activation at each time step using the equations presented in the top right hand corner of the figure, where max_act is the maximum activation at a particular time step, max_act_overall is a constant equal to the maximum expected activation (e.g., 1.0), i is a particular item and d_t is the scaling factor for each time step. Thus the predicted fixation probability is determined both by the amount of evidence for an alternative and the amount of evidence for that alternative compared to the other possible alternatives.

Finally, we introduce a 200 ms delay because programming an eye movement takes approximately 200 ms (Matin, Shao, & Boff, 1993). In experiments without explicit instructions to fixate on a particular picture, initial fixations are randomly distributed among the pictures. Under these conditions, the simple form of the choice rule can be used (see Dahan, Magnuson, & Tanenhaus, 2001; Dahan, Magnuson, Tanenhaus, & Hogan 2001). Note that the Allopenna et al. formalization is only an approximation to a more accurate formalization of the linking hypothesis which would predict the probability that a saccade would be generated at a particular point in time, contingent upon (a) the location of the previous fixation and perhaps the several preceding fixations; (b) time from the onset of the last fixation and (c) the current goal state of the listener's task – which can be ignored in a simple "click" task like the Allopenna et al. paradigm. In a more complex task, such as assembling a piece of furniture or preparing a recipe, goal states might have a more complex structure, with several sub-goals competing for the capture of local attention at any point in time.

When the linking hypothesis is applied to TRACE simulations of activations for the stimuli used by Allopenna et al., it generates the predicted fixations over time shown in Figure 2, bottom right hand corner. Note that the linking hypothesis transforms the shape of the functions because it introduces a non-linear transformation. This highlights the importance of developing and using explicit linking hypotheses (see Tanenhaus, 2004), when evaluating the goodness of fit with behavioral data, as opposed to merely assuming a monotonic relationship. The fixations over time to the target, the cohort competitor, and a rhyme competitor closely match the predictions generated by the hypothesis linking activation levels in TRACE to fixation proportions over time.

2.3. Action-contingent analyses

One useful feature of the action-based approach is that the behavioral responses reveal the participants' interpretation. This allows for *interpretation-contingent* analyses in which fixations are analyzed separately for trials on which participants choose a particular interpretation. Two recent applications illustrate how interpretation-contingent analyses can be used to distinguish between competing hypotheses.

McMurray et al. (2002; McMurray, Aslin, Tanenhaus, Spivey, & Subik, 2005) used a variation on the Allopenna et al. task to investigate the hypothesis that lexical processing is sensitive to small-within category differences in Voice-Onset Time (VOT). The stimuli were synthesized minimal pairs that differed only in voicing, such as *bomb/palm* and *peach/beach*. VOT varied in steps sizing from 0 to 40 ms (voiced sounds, such as /*b*/ have shorter VOTs than unvoiced sounds such as /*p*/). McMurray et al. found gradient increases in looks to the cross-category competitor as the VOT moved closer to the category boundary, with clear linear trends even when the trials with VOTs abutting the category boundary were excluded. While these results are consistent with the hypothesis that lexical processing is sensitive to within-category variation, the results could also be accounted for without abandoning the traditional assumption that within-category variation is quickly discarded.

The categorical interpretation would go as follows. If we make the plausible assumption that there is noise in the system, then as VOT approaches the category boundary, listeners are more likely to incorrectly categorize the input. Assume a category boundary of approximately 18 ms, which is what McMurray et al. found with their synthesized stimuli. For trials with a VOT of 20 ms, perhaps 20% of the stimuli might be perceived as having a VOT of less than 18 ms. With a VOT of 25 ms, the percentage might drop to 12%, compared to 8% for trials with a VOT of 30 ms and 4% for a VOT of 35 ms etc. Thus, the proportion of looks to the cross-category competitor might increase as VOT approaches the category boundary because the data will include increasingly more trials where the target word was misheard as the cross-category competitor and not because the underlying system responds in a gradient manner.

McMurray et al. were able to rule out this alternative explanation by filtering any trials where the participant clicked on the cross-category picture. For example, if the VOT was 25 ms, and the participant clicked on the picture of the bomb, rather than the palm, then the eye movement data from that trial would be excluded from the analyses. McMurray et al. found that looks to the cross-category competitor increased as VOT approached the category boundary, even when all "incorrect" responses were excluded from the analyses, thus providing strong evidence that the system is indeed gradient.

A second illustration comes from recent studies by Runner and his colleagues (e.g., Runner, Sussman, & Tanenhaus, 2003; in press) investigating the interpretation of reflexives and pronouns in so-called picture noun phrases with possessors, e.g., *Harry admired Ken's picture of him/himself.* Participants were seated in front of a display containing three male dolls, Ken, Joe, and Harry, each with distinct facial features. Digitized pictures of the doll's faces were mounted in a column on a board directly above each individual doll. The participant was told that each doll "owned" the set of pictures directly above him; that is, the three pictures in the column above Joe were Joe's pictures, the pictures in the column above Ken were Ken's pictures, etc.

Binding theory predicts that the reflexive, *himself,* will be interpreted as referring to Ken's picture of Ken in instructions such as *Pick up Harry. Now have Harry touch Ken's picture of himself.* Runner et al. found that looks to both the binding-appropriate and inappropriate referents began to increase compared to an unrelated picture in the same row, beginning about 200 ms after the onset of the reflexive. This result suggests that both binding-appropriate and inappropriate referents are initially considered as potential referents for a reflexive. Moreover, participant's choices showed frequent violations of classic binding for reflexives: on approximately 20% of trials with reflexives, participants had Harry touch Ken's picture of Harry. However, one might argue that the early looks to binding-inappropriate referents came from just those trials on which the participant mistakenly arrived at the "incorrect" interpretation. Runner et al. were able to rule out this interpretation by analyzing just those trials where the participant made the binding-appropriate response, finding that there was still an increase in looks to the inappropriate referent, compared to controls. Thus, both the binding-appropriate and inappropriate referents compete, with the binding- appropriate referent (probabilistically) preferred.

One clear limitation of the action-based paradigm is that the linguistic stimuli must be embedded in instructions, which can limit the experimenter's degrees of freedom. Beginning with Altmann and Kamide (1999), a number of researchers using eye movements to study spoken language comprehension have begun to use variations on the original Cooper procedure in which participants simply listen to the input (e.g., Arnold et al., 2000; Boland, 2005; Knoeferle et al., 2005). This procedure places fewer constraints on both the experimenter and the participant. Without an explicit action, the participant's attention is less constrained, thus increasing the likelihood of anticipatory eye movements, which are extremely useful for inferring expectations generated by the listener. Some important applications of listening without an explicit action include Kaiser and Trueswell (2004), who examined expectations driven by different word orders, Boland

(2005), who compared verb-based expectations for adjuncts and arguments, the line of research initiated by Altmann and colleagues (e.g., Altmann & Kamide, 1999; Kamide, Altmann, & Haywood, 2003) and the effects of visually based information on expectation about thematic role assignment initiated by Knoeferle and her colleagues (Knoeferle et al., 2005; Knoeferle, this volume).

Few studies have directly compared the action-based and non-action-based versions of the paradigm. However, to a first approximation, it appears that when anticipatory eye movements are excluded, the timing of fixations to potential referents may be slightly delayed in passive listening tasks compared to action-based tasks. The data from action-based tasks is somewhat cleaner than the data from listening tasks, perhaps because a high proportion of the fixations are task-relevant.

2.4. Other contingent analyses

Investigators have also used a variety of other contingent analyses. For example, one can compare the time to launch a saccade to a referent contingent upon the participant having fixated that referent or a potential competitor during preview (Dahan & Tanenhaus, 2005; Dahan et al., this volume). Likewise, one can compare the time to launch a fixation from a competitor, once disambiguating information arrives.

In production, one can also examine the timing of the onset of an utterance, or the form of an utterance, contingent upon whether or not the participant has looked at a critical part of the display. For example, Brown-Schmidt and Tanenhaus (in press) examined the timing and form of referential descriptions with contrast (e.g., *the little peach*) contingent upon whether or not, and, if so, when the participant (first) fixated the contrast referent (e.g., another bigger peach). Griffin (2004b) has compared the duration of fixations to pictures as a function of whether or not the picture was named correctly, a strategy also adopted by Wheeldon et al., this volume to compare fluent and disfluent productions.

3. Comparing visual world and eye movement reading studies

Many of the issues that have been investigated for decades using eye movements in reading, in particular issues in lexical processing and sentence processing are now being investigated using eye movements with spoken language. Although some aspects of these processes will differ in reading and spoken language because of intrinsic differences between the two modalities, scientists investigating issues such as syntactic ambiguity resolution and reference resolution using eye movements in reading and eye movements in spoken language believe they are testing theoretical claims about these processes that transcend the modality of the input. Thus, the psycholinguistic community will increasingly be faced with questions about how to integrate results from visual world studies with results from studies of eye movements in reading and sometimes how to reconcile conflicting results.

This section addresses some of these issues of interpretation by working through two examples of how similar questions would be asked in eye-tracking reading and eye-tracking listening studies. We then turn to some cases where the results support different conclusions. This will lead us into a discussion about ways of addressing closed set issues within the visual world paradigm.

3.1. Processing load vs representational measures

In considering these examples, it will be useful to make a distinction between behavioral measures of language processing which measure *processing difficulty* and measures that *probe representations*. The distinction is more of a heuristic than a categorical distinction because many response measures combine aspects of both. Processing load measures assess transient changes in processing complexity, and then use these changes to make inferences about the underlying processes and representations. Representational measures examine when during processing a particular type of representation emerges and then use that information to draw inferences about the underlying processes and representations. Neither class of measure or its accompanying experimental logic is intrinsically preferable to the other; the nature of the question under investigation determines which type of response measure is more appropriate.

The majority of studies that use eye movements to examine reading make use of eye movements as a processing load measure. The primary dependent measure is fixation duration. The linking hypothesis between fixation duration and underlying processes is that reading times increase when processing becomes more difficult. In contrast, the majority of studies that use eye movements with spoken language processing use eye movements as a representational measure. The primary dependent measure is when and where people fixate as the utterance unfolds.

3.2. Lexical ambiguity

In a well-known series of studies, Rayner and colleagues (e.g., Rayner & Duffy, 1986) have examined whether multiple senses of homographs, such as *bank, ball* and *port* are accessed during reading, and if so, what are the effects of prior context and the frequency with each sense is used. Processing difficulty compared to an appropriate control is used to infer how ambiguous words are accessed and processed. For "balanced" homographs with two more or less equally frequent senses, fixation duration is longer compared to frequency-matched controls, resulting in the inference that the multiple senses are competing with one another. This ambiguity "penalty" is reduced or eliminated for biased homographs when a "dominant" sense is far more frequent than a "subordinate" sense and when the context strongly favors either one of two equally frequent senses or the more frequent sense. Note that while these results do not provide clear evidence about time course per se, the overall data pattern allows one to infer that multiple senses are accessed, with the dominant sense accessed more rapidly. One can get crude time course information by separately analyzing the duration of the initial fixation and using that as a measure of relatively early processes. More detailed information about time course can be obtained by using fixation duration as a measure, but using variations on the fast priming methods, introduced by Sereno and Rayner (1992).

Studies using the visual world paradigm adopt a similar approach to that used by Allopenna et al. Potential referents associated with the alternative senses are displayed and the time course of looks to these referents is used to infer degree of activation and how it changes over time. For balanced homophones, looks are found to the referents of both senses. For biased homophones, looks to the more frequent sense begin earlier than looks to the less frequent sense (Huettig & Altmann, 2004). This pattern is similar to those obtained in classic studies using cross-modal priming from the 1970s and early 1980s (Swinney, 1979; Tanenhaus, Leiman, & Seidenberg, 1979; for review see Simpson, 1984 and Lucas, 1999). These results do not provide direct information about processing difficulty, though one might infer that competing senses would result in an increase in complexity.

Thus, while the eye movements reading studies do not provide direct information about time course and visual world studies do not provide direct information about processing difficulty, the results from reading studies that use a processing load strategy and visual world studies that probe emerging representations converge on the same conclusions.

3.3. Syntactic ambiguity

Beginning with the classic article by Frazier and Rayner (1982), eye tracking in reading has been the response measure of choice for psycholinguists interested in syntactic processing. Frazier and Rayner's approach was to examine the processing of temporarily ambiguous sentences, using reading times within pre-defined regions to infer if and when the reader had initially pursued the incorrect interpretation. For syntactic ambiguities, which involved disambiguating the phrase that could be "attached" to a verb phrase in a subordinate clause (the initially preferred interpretation) as the subject of the main clause, Frazier and Rayner found an increase in fixation duration and an increase in regressive eye movements from the disambiguating region. For current purposes we will focus on fixation duration because it is most clearly a processing load measure. The question of how to interpret regressions is more complex and beyond the scope of this chapter. The increase in fixation duration was interpreted as evidence that processing had been disrupted, thereby leading to the inference that readers had initially chosen the argument interpretation. Frazier and Rayner also introduced several different measures that divided fixations within a region in different ways. For example, "first pass" reading times include all fixations beginning with the first fixation within a region until a fixation that leaves a region, and are often used as a measure of early processing.

Studies examining syntactic ambiguity resolution with the visual world paradigm use the timing of looks to potential referents to infer, if and, if so, when, a particular analysis is under consideration. For example, in one-referent contexts (an apple on a towel, a towel, a box and a pencil) and instructions such as *Put the apple on the towel in the box*, Spivey et al. (2002) found that looks to the false goal (the towel without the apple) began to increase several hundred milliseconds after the onset of *towel*. In contrast, in two-referent contexts (two apples, one on a towel and one on a napkin), fixations to the apple on the towel begin to increase several hundred milliseconds after the onset of *towel*. This pattern of results suggests that the prepositional phrase *on the towel* is initially considered a goal argument in the one-referent context and a noun-phrase modifier in the two-referent context. Information about time course is relatively straightforward with the visual world logic because fixations can be aligned with the input, allowing strong inferences about what information in the input was likely to have triggered the fixation. The reason one can align fixations and the input is, of course, that the input unfolds sequentially. Nonetheless, as will be discussed shortly, timing is complicated by issues of when and how the participant has encoded information from the display or scene.

Timing is less straightforward in eye-tracking reading when the measure is fixation duration and the fixations are divided into multiple word regions. Most of the complexities in inferring time course in reading studies arise because the sequence of fixations need not correspond to the linear order of the words, including when they have first been encountered. This is especially the case when one considers that arguments about timing often depend on defining regions of text and then partitioning fixations into categories in ways that separate the measure from when the input was first encountered.

One way to appreciate this is to compare results from a recent study by Clifton et al. (2003) with Trueswell, Tanenhaus, and Garnsey (1994). Both studies examined the effects of thematic bias on reading times to temporarily ambiguous, reduced-relatives clauses compared to unambiguous full-relative baselines. Trueswell et al. concluded that thematic constraints had immediate effects, whereas Clifton et al. argued that initial syntactic processing was unaffected by thematic context. When the same measures are compared, e.g., first pass reading times, Clifton et al. replicate the results found by Trueswell et al. However, when fixations are examined using a different measure (regression path), the evidence is somewhat more consistent with delayed effects of context. Which conclusion one endorses depends on the measure one emphasizes. Note, however, that it is difficult to make strong time course claims unless one includes only those fixations that preserve the order in which the information is first encountered (i.e. fixations on a word after a subsequent word has been fixated would be excluded and the regions would correspond to single words). Nonetheless, if we exclude the effects of context, which we will return to shortly, the conclusions that emerge for reading and visual world studies are similar for the prepositional phrase attachment structures which have been examined using both methods. The preferred analysis that is inferred from looks to potential referents is similar to the preferred analysis that is inferred by fixation durations.

4. Effects of display

On the basis of the two ambiguity cases we have compared thus far, it might seem that I am arguing that visual world studies have intrinsic advantages over reading studies when it comes to asking questions about the time course with which interpretations arise because of two factors: (1) fixations and the input can be easily aligned and (2) the logic of representational measures is more direct. However, we have not yet considered the single factor that most complicates the interpretation of visual world studies of language processing – the need to use a display. The use of a visual display is what allows one to use fixations to probe emerging representations. However, it introduces two kinds of complexities. First, the encoding of the display can introduce contingencies. For example, the timing of looks to a potential referent at point *t* could be affected by whether or not that referent has been fixated on during time *t-x*, either during preview or as the sentence unfolds. Thus the likelihood of a fixation may be contingent on both the input and the pattern of prior fixations. This, of course, has the potential to complicate inferences about time course, in much the same way that re-reading after a regression can complicate the interpretation of fixation duration data in eye movement reading studies. The contribution to this volume by Dahan, Tanenhaus, and Salverda begins to address this issue by examining how having fixated a potential referent during preview affects the likelihood that it will be fixated when it is temporarily consistent with the input, viz. its name is a cohort of the intended referent.

The second factor is that use of a display with a small number of pictured referents or objects and a limited set of potential actions creates a more restricted environment than language processing in most natural contexts, while at the same time imposing more demands on the participant than most psycholinguistic tasks. In order to address these closed set issues, we will consider two cases: the first from spoken word recognition and the second from reference resolution.

4.1. Spoken word recognition

In the Allopenna et al. paradigm, the potential response set on each trial is limited to four pictured items. If participants adopted a task-specific verification strategy, such as implicitly naming the pictures, then the unfolding input might be evaluated against these activated names, effectively bypassing the usual activation process, and leading to distorted results. Even if participants do not adopt such a strategy, the visual world methodology might be limited if the effects of the response alternatives mask effects of non-displayed alternatives (e.g., neighborhood effects in the entire lexicon). This would restrict its usefulness for investigating many issues in spoken word recognition, in particular issues about the effects of lexical neighborhoods, i.e., the set of words in the lexicon that are similar to the target word. Here, an analogy might be helpful. Researchers often use lexical priming paradigms to probe for whether an exemplar of a particular class of lexical competitor is active, e.g., cohorts or rhymes. However, these paradigms are not well suited for asking questions about the aggregate effects of the number and frequency of potential competitors. In order to investigate this class of question, researchers have found it more useful to measure response time to a target word, e.g., auditory lexical decision, which more closely approximates a processing load measure.

4.2. Implicit naming

The issue of implicit naming has been addressed most directly by Dahan and Tanenhaus (2005) in a study that varied the amount of preview time, 300 or 1000 ms, for four-picture displays with minimal phonological overlap between the names of the distractors and the target. On a subset of the trials, two of the pictures were visually similar (e.g., a picture of a snake and a coiled rope) and the instruction referred to one of the pictures (e.g., *click on the snake*). The particular pictures chosen as the two referents shared some features associated with a prototypical visual representation of one or both words. For example, the pair *snake–rope* was selected because the picture of a coiled rope shares some features with the visual representation most often associated with the concept of a snake. When selecting pictures, we sought to minimize their visual similarity so that the objects could be easily differentiated. For example, we chose a snake in a non-coiled position and a rope that was coiled. Thus, visual similarity was maximized between the prototypical visual representation of one of the concepts, the referent, and the picture associated with the other concept, the competitor, and minimized between the competitor picture and the picture of the referent concept.

Several aspects of the results provide strong evidence against implicit naming. With longer preview, one would expect increased likelihood of implicit naming. However, preview duration did not affect the magnitude of visual similarity effects (looks to visually similar competitors). Moreover, even in the 1000 ms condition, the magnitude of visual similarity effects was not affected by whether or not the competitor was fixated during preview; the naming hypothesis predicts that effects would be eliminated or weakened with preview because the encoded name of the picture would not match the unfolding target. Finally, similarity effects were larger when the target had a competitor that was chosen to share visual features of its prototype representation compared to when that competitor was the referent. Thus visual similarity effects were due to the fit between the picture and the conceptual representation of the picture, not simply surface visual confusability. This suggests that mapping of the word onto its referent picture is mediated by a visual/conceptual match between the activated lexical form of the target and the picture. This hypothesis is further supported by additional analyses of the effects of fixation to a competitor during preview on the likelihood that it will be re-fixated during the speech input reported by Dahan et al. in this book and evidence that a spoken word triggers looks to potential referents when the participant is engaged in a visual search task to identify the location of a dot when it appears on a random location within a schematic scene (Salverda & Altmann, 2005).

4.3. Sensitivity to hidden competitors

Perhaps the strongest test of the sensitivity of visual world studies comes from studies that look for effects of non-displayed or "hidden competitors". A recent study by Magnuson, Dixon Tanenhaus, and Aslin (in press) illustrates this approach. Magnuson et al. examined the temporal dynamics of neighborhood effects using two different metrics: neighborhood density, a frequency-weighted measure defined by the neighborhood activation model (Luce & Pisoni, 1998), and a frequency-weighted measure of cohort density. The referent was displayed along with three semantically unrelated pictures, with names that had little phonological overlap with the referent (all names were monosyllabic). Crucially, none of the referent's neighbors were either displayed or named throughout the course of the experiment. The results showed clear effects of both cohort and neighborhood density, with cohort density effects dominating early in the recognition process and neighborhood effects emerging relatively late. Proportion-of-fixation curves showing the effects of frequency, cohort density and neighborhood density are presented in Figure 3.

These results demonstrate that the processing neighborhood for a word changes dynamically as the word unfolds. It also establishes the sensitivity of the paradigm to the entire lexicon. To a first approximation then, when competitors are displayed, the paradigm can be used to probe specific representations, however, the aggregate effects of competitors can be observed in the timing of fixations to the target referent.

Magnuson et al.'s results complement Dahan, Magnuson, Tanenhaus, and Hogan's (2001) finding that misleading coarticulatory information delays recognition more when it renders the input temporarily consistent with a (non-displayed) word, compared to when it does not. In addition, simulations using the Allopenna et al. linking hypothesis successfully captured differences between the effects of misleading coarticulatory information with displayed and non-displayed competitors. Whether the non-displayed competitor logic can be extended to higher-level sentence processing remains to be seen.

4.4. Sentence processing

Much trickier issues about the effects of the display come into play in higher-level processing. For example, one could argue that in the Tanenhaus et al. (1995) study displaying an apple on a towel and an apple on a napkin increases the salience of a normally less accessible sense compared to circumstances where the alternative referents are introduced linguistically. One could make a similar argument about the effects of action on the rapidity with which object-based affordances influence ambiguity resolution in studies by Chambers and colleagues (Chambers et al. 2002, 2004). In these studies, the issue of implicit naming seems prima facie to be less plausible. However, one might be concerned about task-specific strategies. For example, in Chambers et al. (2002), participants were confused, as indexed by fixations when they were told to *Pick up the cube. Now put the cube in the can,* and there were two cans. The confusion was reduced or eliminated, however, when the cube would only fit in one of the cans.

Note that one might attribute this to problem solving, and not as Chambers et al. argued to the effects of action and affordance on referential domains. The problem-solving argument would be that participants were able to ignore the competitor as soon as they picked up the cube (and learned that they were supposed to put it in something) because they were simply looking for a possible goal. However, if this were the case, then the same pattern of results would be predicted to occur with the instruction used an indefinite article, e.g., *Pick up the cube. Now put it in a can.* However, participants were now

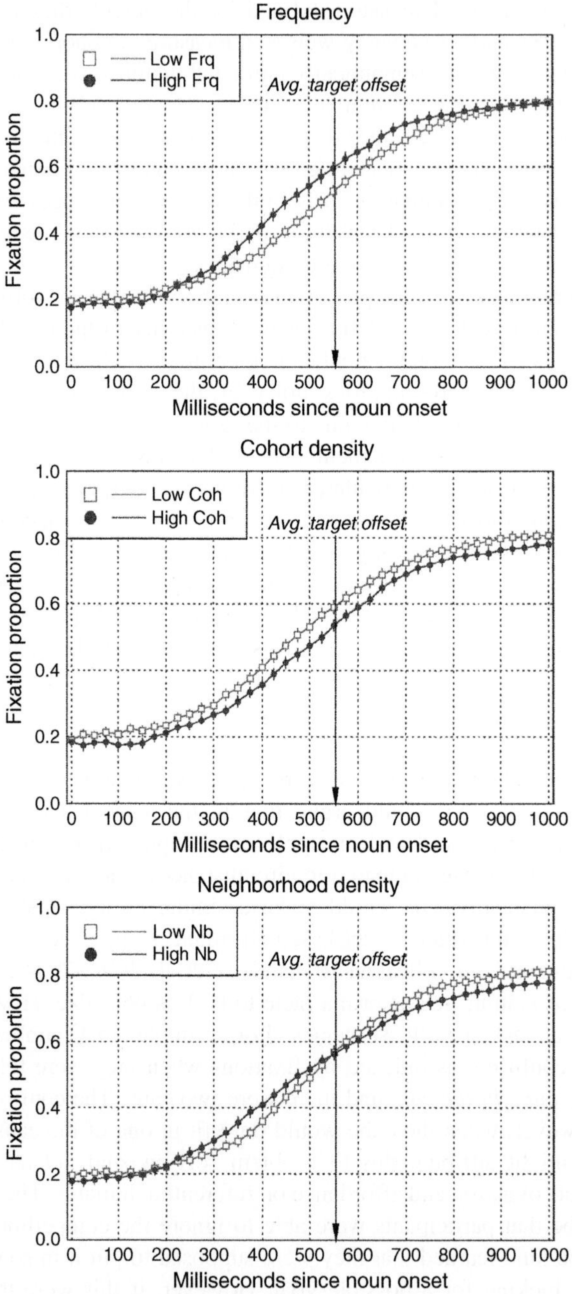

Figure 3. Top to bottom, Time course of frequency effects, cohort density and neighborhood density. From Magnuson, Dixon, Tanenhaus, and Aslin (in press).

confused when the cube would only fit in one of the cans, despite the fact that there was still only one possible action. Thus the referential domain in conjunction with the referential conditions of the article influenced processing even though the article had no effect on the possible action. This strategy of pitting linguistic effects against potential problem-solving strategies that would bypass linguistic processing is one way to evaluate whether or not the task and the display introduce problematic task-specific processing strategies.

Perhaps the most general caution for researchers using the visual world paradigm in both production and comprehension is to be aware that while the visual world displays entities that can be used to infer the representations that the listener is developing, it also serves as a context for the utterance itself. Note that the fact that information in a display affects processing is not itself any more problematic than the observation that reference resolution, e.g., is affected by whether or not potential referents are introduced linguistically in a prior discourse. Both types of contexts can make potential referents salient.

One sometimes encounters the argument that the visual world paradigm can be informative about language processing only if gaze patterns to a potential referent in a display are not affected by the other characteristics of the display. This argument is no more or less valid than the comparable argument that fixations in reading can only inform us about word recognition or reference resolution if fixations to a word are unaffected by the context in which the fixated word occurs. What is crucial, however, is whether the nature of the interactions with the display sheds light on language processing or whether it introduces strategies that mislead or obscure the underlying processes.

Two examples might help illustrate this point. The first is taken from Tanenhaus, Chambers, and Hanna (2004, see also Tanenhaus & Brown-Schmidt, to appear) and illustrates how the visual world paradigm has clarified our understanding of pre-nominal scalar adjectives and definite reference. The second uses another hypothetical example to contrast examples of results that would and results that would not suggest that the visual display was leading to strategies that distort "normal" language comprehension.

Prior to visual world studies, a standard psycholinguistic account of the processing of the sentence, *After putting the pencil below the big apple, James moved the apple onto the towel*, would have gone something like this. When the listener encounters the scalar adjective *big*, interpretation is delayed because a scalar dimension can only be interpreted with respect to the noun it modifies (e.g., compare a big building and a big pencil). As *apple* is heard, lexical access activates the apple concept, a prototypical apple. The apple concept is then modified resulting in a representation of a *big apple*. When *apple* is encountered in the second clause, lexical access again results in activation of a prototypical *apple* concept. Because *apple* was introduced by a definite article, this representation would need to be compared with the memory representation of the *big apple* to decide whether the two co-refer.

Now, consider how real time interpretation proceeds in a context which includes a pencil, two apples, one small prototypical red apple, the other a large misshapen green apple, and a towel, taking into account recent results from visual world studies. At *big*, the listener's attention would be drawn to the larger of the two apples, because a scalar adjective signals a contrast among two or more entities of the same semantic type (Sedivy et al., 1999). Thus *apple* will be immediately interpreted as the misshapen green apple, even though a more prototypical red apple is present in the display. And, when *the apple* is encountered in the second clause, the red apple would be ignored in favor of the large green apple, because that apple has been introduced into the discourse, but not as the most salient entity (Dahan, Tanenhaus, & Chambers, 2002).

The standard account does not make sense when we try to generalize it to processing in the context of concrete referents. In contrast, the account that emerges from visual world research does generalize to processing in the absence of a more specific context. In particular, the scalar *big* would still be interpreted as the member of a contrast set picked out by size; it is just that the contents of the contrast set are not instantiated until the noun, *apple* has been heard. And, any increase in processing difficulty when *the apple* is processed would not be reflecting an inference to establish that the referent is the big apple, but rather the shift in discourse focus from the previously focused entity (the pencil) to a previously mentioned entity (the apple). In this example, then, the display clearly changes processing, but in ways that clarify (but do not distort) the underlying processes.

Now consider the discourse: *The man returned home and greeted his pet dog. It/ The beast/A beast then began to lick/attack him.* Well-understood principles of reference assignment mandate that *it* should refer to the dog, and *a beast* to an animal other than the dog. Now imagine the same discourse in the context of a display containing a man standing in front of an open door to a hut in a jungle village, a dog with a collar, a tiger, and a rabbit. Compared to appropriate control conditions, we would expect a pattern of looks indicating that *it* was interpreted as the dog, and *a beast* as the tiger, regardless of whether the verb was *lick* or *attack*. If, however, *it* were interpreted as the tiger when the verb was *attack*, or if *a beast* were to be interpreted as the dog for *lick*, then we would have a clear case of the display distorting the comprehension process. This conclusion would be merited because these interpretations *would* violate well-understood principles of reference resolution. Now consider the definite noun phrase, *the beast*. This referential expression could either refer to the mentioned entity, the pet dog, or it could introduce another entity. In the discourse-alone condition, a listener or reader would most likely infer that *the beast* refers to the dog, because no other entity has been mentioned. However, in the discourse with display condition, a listener might be more likely to infer that *the beast* refers to the tiger. Here the display changes the interpretation, but it does *not* change the underlying process; the display simply makes accessible a potential unmentioned referent, which is consistent with the felicity conditions for the type of definite reference used in the discourse. Indeed, we would expect the same pattern of interpretation if the tiger had been mentioned in the discourse.[1]

[1] This observation is due to Gerry Altmann. The example presented here is adapted from one presented by Simon Garrod in a presentation at the 2003 Meeting of the CUNY Sentence Processing Conference.

Thus far investigations of the effects of using a display and using a task have not uncovered any evidence that the display or the task is distorting the underlying processes. To the contrary, the results have been encouraging for the logic of the visual world approach. However, it will be crucial in further work to explore the nature of the interactions between the display, the task, and linguistic processing in much greater detail. Moreover, the ability to control and understand the context in which the language is being produced and understood, which is one of the most powerful aspects of the visual world paradigm, depends in large part on developing a better understanding of these interactions.

5. Conclusion

This chapter has provided a general introduction to how psycholinguists are beginning to use eye movements to study spoken language processing. We have reviewed some of the foundational studies, discussed issues of data analysis and interpretation, and discussed issues that arise in comparing eye movement reading studies to visual world studies. We have also discussed some of the issues that arise because the visual world introduces a context for the utterance.

The following chapters each contribute to issues that we have discussed and illustrate the wide range of questions to which the visual world paradigm is now being applied. Dahan, Tanenhaus, and Salverda examine how preview affects the likelihood that a cohort competitor will be looked at as a target word unfolds, contributing to our understanding of the effects of the display on the inferences we can draw about spoken word recognition in visual world studies. Bailey and Ferreira, 2005 show how the visual world paradigm can be used to investigate expectations introduced by disfluency, extending investigations of spoken language processing to the kinds of utterances one frequently encounters in real life, but infrequently encounters in psycholinguistic experiments. Wheeldon, Meyer, and van der Meulen (this book) extend work on eye movements in production by asking whether fixations can shed light on the locus of speech errors in naming. The surprising results have important theoretical and methodological implications for the link between fixations and utterance planning. Finally Knoeferle explores how thematic role assignment is affected by when information in a scene becomes relevant, contributing to our understanding of the interplay between the scene and the unfolding language. She also provides a general framework for understanding these interactions.

Acknowledgments

This work was supported by NIH grants DC005071 and HD 27206. Thanks to Delphine Dahan, Anne Pier Salverda, and Roger Van Gompel for helpful comments.

References

Allopenna, P. D., Magnuson, J. S., & Tanenhaus, M. K. (1998). Tracking the time course of spoken word recognition: Evidence for continuous mapping models. *Journal of Memory and Language, 38*, 419–439.

Altmann, G. T. M., & Kamide, Y. (1999). Incremental interpretation at verbs: Restricting the domain of subsequent reference. *Cognition, 73*, 247–264.

Altmann, G. T. M., & Kamide, Y. (2004). Now you see it, now you don't: Mediating the mapping between language and the visual world. In J. M. Henderson, & Ferreira, F. (Eds.), *The interface of language, vision, and action: Eye movements and the visual world.* New York: Psychology Press.

Altmann, G. T. M., & Steedman, M. J. (1988). Interaction with context during human sentence processing. *Cognition, 30*, 191–238.

Arnold, J. E., Eisenband, J., Brown-Schmidt, S., & Trueswell, J. C. (2000). The rapid use of gender information: Evidence of the time course of pronoun resolution from eyetracking. *Cognition, 76*, B13–B26.

Arnold, J. A., Tanenhaus, M. K. Altmann, R. J., & Fagnano, M. (2004). The old and, theee, uh, new: Disfluency and reference resolution. *Psychological Science, 9*, 578–582.

Bailey, K. G. D., & Ferreira, F. (2003). Disfluencies affect the parsing of garden-path sentences. *Journal of Memory and Language, 49*, 183–200.

Bailey, K. G. D., & Ferreira, F. (2005). The disfluent hairy dog: Can syntactic parsing be affected by non-word disfluencies? In J. Trueswell, & M. K. Tanenhaus (Eds.), *Approaches to studying world-situated language use: Bridging the language-as-product and language-as-action traditions*. Cambridge, MA: MIT Press.

Bock, J. K., Irwin, D. E., & Davidson D. J. (2004). Putting first things first. In J. M. Henderson, & F. Ferreira (Eds.), *The interface of language, vision, and action: Eye movements and the visual world.* New York: Psychology Press.

Bock, J. K., Irwin, D. E., Davidson, D. J., & Levelt, W. J. M. (2003). Minding the clock. *Journal of Memory and Language, 48*, 653–685.

Boland, J. E. (2005). Visual arguments.*Cognition, 95*, 237–274.

Brown-Schmidt, S., Campana, E., & Tanenhaus, M. K. (2005). Real-time reference resolution in a referential communication task. In J. C. Trueswell, & M. K. Tanenhaus (Eds.), *Processing world-situated language: Bridging the language-as-action and language-as-product traditions.* Cambridge, MA: MIT Press.

Brown-Schmidt, S., & Tanenhaus, M. K. (2006). Watching the eyes when talking about size: An investigation of message formulation and utterance planning. *Journal of Memory and Language, 54*, 592–609.

Chambers, C. G., Tanenhaus, M. K., Eberhard, K. M., Filip, H., & Carlson, G. N. (2002). Circumscribing referential domains during real-time language comprehension. *Journal of Memory & Language, 47*, 30–49.

Chambers, C. G., Tanenhaus, M. K., & Magnuson, J. S. (2004). Action-based affordances and syntactic ambiguity resolution. *Journal of Experimental Psychology: Learning, Memory & Cognition, 30*, 687–696.

Clark, H. H. (1992). *Arenas of language use*. Chicago: University of Chicago Press.

Clifton, C., Traxler, M., Mohamed, M. T., Williams, R., Morris, R., & Rayner, K. (2003). The use of thematic role information in parsing: Syntactic processing autonomy revisited. *Journal of Memory and Language, 49*, 317–334.

Cooper, R. M. (1974). The control of eye fixation by the meaning of spoken language: A new methodology for the real-time investigation of speech perception, memory, and language processing. *Cognitive Psychology, 6*, 84–107.

Dahan, D., Magnuson, J. S., & Tanenhaus, M. K. (2001). Time course of frequency effects in spoken-word recognition: Evidence from eye movements. *Cognitive Psychology, 42*, 317–367.

Dahan, D., Magnuson, J. S. Tanenhaus, & Hogan, E. (2001). Subcategorical mismatches and the time course of lexical access: Evidence for lexical competition. *Language and Cognitive Processes, 16*, 507–534.

Dahan, D., & Tanenhaus, M. K. (2004). Continuous mapping from sound to meaning in spoken language comprehension: Evidence from eye movements. *Journal of Experimental Psychology: Learning, Memory and Cognition, 30*, 498–513.

Dahan, D., & Tanenhaus, M. K. (2005). Looking at the rope when looking for the snake: Conceptually mediated eye movements during spoken-word recognition. *Psychological Bulletin & Review, 12*, 455–459.

Dahan, D., Tanenhaus, M. K., & Chambers, C. G. (2002). Accent and reference resolution in spoken language comprehension. *Journal of Memory and Language, 47*, 292–314.

Eberhard, K. M., Spivey-Knowlton, M. J., Sedivy, J. C., & Tanenhaus, M. K. (1995). Eye-movements as a window into spoken language comprehension in natural contexts. *Journal of Psycholinguistic Research, 24*, 409–436.

Ferriera, F. & Bailey, K. G. B., This volume.

Ferreira, F., & Henderson, J. M. (2004). The interface of vision, language, and action. In J. M. Henderson & F. Ferreira (Eds.), *The interface of language, vision, and action: Eye movements and the visual world.* New York: Psychology Press.

Frazier, L., & Rayner, K. (1982). Making and correcting errors during sentence comprehension: Eye movements in the analysis of structurally ambiguous sentences. *Cognitive Psychology, 14*, 178–210.

Griffin, Z. M. (2004a). Why look? In J. M. Henderson, & F. Ferreira (Eds.), *The interface of language, vision, and action: Eye movements and the visual world.* New York: Psychology Press.

Griffin, Z. M. (2004b). The eyes are right when the mouth is wrong. *Psychological Science, 15*, 814–821.

Griffin, Z. M., & Bock, J. K. (2000). What the eyes say about speaking. *Psychological Science, 11*, 274–279.

Hanna, J. E., Tanenhaus, M. K., & Trueswell, J. C. (2003). The effects of common ground and perspective on domains of referential interpretation. *Journal of Memory and Language, 49*, 43–61.

Hayhoe, M., & Ballard, D. (2005). Eye movements in natural behavior. *Trends in Cognitive Sciences 9*, 188–194.

Henderson, J. M., & Ferreira, F. (Eds.). (2004). *The interface of language, vision, and action: Eye movements and the visual world.* New York: Psychology Press.

Huetting, F., & Altmann, G. T. M. (2004). The online processing of ambiguous and unambiguous words in context: Evidence from head mounted eye tracking. In M. Carreiras, & C. Clifton (Eds.), *The on line study of sentence comprehension: Eyetracking, ERP and beyond.* New York, NY: Psychology Press.

Ju, M., & Luce, P. A. (2004). Falling on sensitive ears: Constraints on bilingual lexical activation. *Psychological Science, 15*, 314–318.

Kaiser, E., & Trueswell, J. C. (2004). The role of discourse context in the processing of a flexible word-order language. *Cognition, 94*, 113–147.

Kamide, Y., Altmann, G. T. M., & Haywood, S. L. (2003). Prediction and thematic information in incremental sentence processing: Evidence from anticipatory eye movements. *Journal of Memory and Language, 49*, 133–156.

Keysar, B., Barr, D. J., Balin, J. A., & Brauner, J. S. (2000). Taking perspective in conversation: The role of mutual knowledge in comprehension. *Psychological Science, 11*, 32–38.

Knoeferle, P., Crocker, M. W., Scheepers, C., & Pickering, M. J. (2005). The influence of the immediate visual context on incremental thematic role-assignment: Evidence from eye-movements in depicted events. *Cognition, 95*, 95–127.

Lucas, M. (1999). Context effects in lexical access: A meta-analysis. *Memory and Cognition, 27*, 375–398.

Luce, R. D. (1959). *Individual choice behavior.* New York: Wiley.

Luce, P. A., & Pisoni, D. B. (1998). Recognizing spoken words: The neighborhood activation model. *Ear and Hearing, 19*, 1–36.

Magnuson, J. S. (2005). Moving hand reveals dynamics of thought. *Proceedings of the National Academy of Sciences, 102*, 9995–9996.

Magnuson, J. S., Dixon, J. F., Tanenhaus, M. K., & Aslin, R. N. (in press). Dynamic similarity in spoken word recognition. *Cognitive Science.*

Marslen-Wilson, W. (1987). Functional parallelism in spoken word recognition. *Cognition, 25*, 71–102.

Marslen-Wilson, W. (1990). Activation, competition, and frequency in lexical access. In G. T. M. Altmann (Ed.), *Cognitive models of speech processing. Psycholinguistic and computational perspectives.* Hove, UK: Erlbaum.

Marslen-Wilson, W. (1993). Issues of process and representation in lexical access. In G. T. M. Altmann, & R. Shillcock (Eds.), *Cognitive models of speech processing: The second Sperlonga meeting.* Hove, England UK: Lawrence Erlbaum Associates.

Matin, E., Shao, K., & Boff, K. (1993). Saccadic overhead: information processing time with and without saccades. *Perception & Psychophysics, 53*, 372–380.

McClelland, J. L., & Elman, J. L. (1986). The TRACE model of speech perception. *Cognitive Psychology, 18*, 1–86.

McMurray, B., Aslin, R. N., Tanenhaus, M. K., Spivey, M. J., & Subik, D. (2005). Gradient sensitivity to sub-phonetic variation in voiced-onset time in words and syllables, manuscript submitted for publication.

McMurray, B., Tanenhaus, M. K., & Aslin, R. N. (2002). Gradient effects of within-category phonetic variation on lexical access. *Cognition, 86*, B33–B42.

Meyer, A. S., Sleiderink, A. M., & Levelt, W. J. M. (1998). Viewing and naming objects. *Cognition, 66*, B25–B33.

Novick, J. M., Trueswell, J. C. & Thompson-Schill, S. L. (2005). Cognitive control and parsing: Reexamining the role of Broca's Area in sentence comprehension. *Cognitive, Affective, & Behavioral Neuroscience, 5*, 263–281.

Rayner, K. (1998). Eye movements in reading and information processing: Twenty years of research. *Psychological Bulletin, 124*, 372–422.

Rayner, K., & Duffy, S. A. (1986) Lexical complexity and fixation times in reading: Effects of word frequency, verb complexity, and lexical ambiguity. *Memory & Cognition, 14*, 191–201.

Rossion, B., & Pourtois, G. (2004). Revisiting Snodgrass and Vanderwart's object pictorial set: The role of surface detail in basic-level object recognition. *Perception, 33*, 217–236.

Runner, J. T., Sussman, R. S., & Tanenhaus, M. K. (2003). Assignment of reference to reflexives and pronouns in picture noun phrases: Evidence from eye movements. *Cognition, 81*, B1–13.

Runner, J. T., Sussman, R. S., & Tanenhaus, M. K. (2006). Assigning referents to reflexives and pronouns in picture noun phrases. Experimental tests of Binding Theory. *Cognitive Science, 30*, 1–49.

Salverda, A. P., & Altmann, G. (2005, September). *Cross-talk between language and vision: Interference of visually-cued eye movements by spoken language*. Poster presented at the AMLaP Conference, Ghent, Belgium.

Sedivy, J. C., Tanenhaus, M. K., Chambers, C. G., & Carlson, G. N. (1999). Achieving incremental interpretation through contextual representation: Evidence from the processing of adjectives. *Cognition, 71*, 109–147.

Sereno, S. C., & Rayner, K. (1992). Fast priming during eye fixations in reading. *Journal of Experimental Psychology: Human Perception and Performance, 18*, 173–184.

Simpson, G. B. (1984). Lexical ambiguity and its role in models of word recognition. *Psychological Bulletin, 96*, 316–340.

Snodgrass, J. G., & Yuditsky, T. (1996). Naming times for the Snodgrass and Vanderwart pictures. *Behavioral Research Methods, Instruments & Computers, 28*, 516–536.

Spivey-Knowlton, M. J. (1996). Integration of visual and linguistic information: Human data and model simulations. Ph.D. dissertation, University of Rochester, New York.

Spivey, M., & Geng, J. (2001). Oculomotor mechanisms activated by imagery and memory: Eye movements to absent objects. *Psychological Research, 65*, 235–241.

Spivey, M., Grosjean, M., & Knoblich, G. (2005). Continuous attraction toward phonological competitors. *Proceedings of the National Academy of Sciences, 102*, 10393–10398.

Spivey, M., & Marian, V. (1999). Cross talk between native and second languages: Partial activation of an irrelevant lexicon. *Psychological Science, 10*, 281–284.

Spivey, M. J., Richardson, D. C., & Fitneva, S. A. (2004). Thinking outside the brain: Spatial indices to visual and linguistic information. In J. M. Henderson, & Ferreira, F. (Eds.), *The interface of language, vision, and action: Eye movements and the visual world*. New York: Psychology Press.

Spivey, M. J., Tanenhaus, M. K., Eberhard, K. M., & Sedivy, J. C. (2002). Eye movements and spoken language comprehension: Effects of visual context on syntactic ambiguity resolution. *Cognitive Psychology, 45*, 447–481.

Swinney, D. (1979). Lexical access during sentence comprehension: (Re)consideration of context effects. *Journal of Verbal Learning and Verbal Behavior, 18*, 645–660.

Tanenhaus, M. K. (2004). On-line sentence processing: past, present and, future. In M. Carreiras, & C. Clifton, Jr. (Eds.), *On-line sentence processing: ERPS, eye movements and beyond* (pp. 371–392). New York: Psychology Press.

Tanenhaus, M. K., & Brown-Schmidt, S. (in press). Language processing in the natural world. To appear in Moore, B. C. M., Tyler, L. K., & Marslen-Wilson, W. D. (Eds.), *The perception of speech: from sound to meaning*. Theme issue of Philosophical Transactions of the Royal Society B: Biological Sciences.

Tanenhaus, M. K., Chambers, C. G., & Hanna, J. E. (2004). Referential domains in spoken language comprehension: Using eye movements to bridge the product and action traditions. In J. M. Henderson, & F. Ferreira (Eds.), *The interface of language, vision, and action: Eye movements and the visual world*. New York: Psychology Press.

Tanenhaus, M. K., Leiman, J. M., & Seidenberg, M. S. (1979). Evidence for multiple stages in the processing of ambiguous words in syntactic contexts. *Journal of Verbal Learning and Verbal Behavior, 18*, 427–441.

Tanenhaus, M. K., Spivey-Knowlton, M. J., Eberhard, K. M., & Sedivy, J. E. (1995). Integration of visual and linguistic information in spoken language comprehension. *Science, 268*, 1632–1634.

Tanenhaus, M. K., & Trueswell, J. C. (2005). Using eye movements to bridge the language as action and language as product traditions. In J. C. Trueswell, & M. K. Tanenhaus (Eds.), *Processing world-situated language: Bridging the language-as-action and language-as-product traditions*. Cambridge, MA: MIT Press.

Trueswell, J. C., Sekerina, I., Hill, N., & Logrip, M. (1999). The kindergarten-path effect: Studying on-line sentence processing in young children. *Cognition, 73*, 89–134.

Trueswell, J. C., & Tanenhaus, M. K. (Eds.), (2005). *Processing world-situated language: Bridging the language-as-action and language-as-product traditions*. Cambridge, MA: MIT Press.

Trueswell, J. C., Tanenhaus, M. K., & Garnsey, S. M. (1994). Semantic effects in parsing: Thematic role information in syntactic ambiguity resolution. *Journal of Memory and Language, 33*, 285–318.

Wheeldon, L. R., Meyer, A. S., & van der Muelen, F. F. this volume.

Yee, E., Blumstein, S., & Sedivy, J. C. (2000). *The time course of lexical activation in Broca's aphasia: Evidence from eye movements*. Poster presented at the 13th Annual CUNY Conference on Human Sentence Processing, La Jolla, California.

Chapter 21

THE INFLUENCE OF VISUAL PROCESSING ON PHONETICALLY DRIVEN SACCADES IN THE "VISUAL WORLD" PARADIGM

DELPHINE DAHAN

University of Pennsylvania, USA

MICHAEL K. TANENHAUS and ANNE PIER SALVERDA

University of Rochester, USA

Eye Movements: A Window on Mind and Brain
Edited by R. P. G. van Gompel, M. H. Fischer, W. S. Murray and R. L. Hill

Abstract

We present analyses of a large set of eye–movement data that examine how factors associated with the processing of visual information affect eye movements to displayed pictures during the processing of the referent's name. We found that phonetically driven fixations are affected by display preview, by the ongoing uptake of visual information, and by the position of pictures in the visual display. Importantly, lexical frequency associated with a picture's name affects the likelihood of refixating this picture and the timing of initiating a saccade away from this picture, thus supporting the use of eye movements as a measure of lexical processing.

Eye movements have increasingly become a measure of choice in the study of spoken-language comprehension, providing fine-grained information about how the acoustic signal is mapped onto linguistic representations as speech unfolds. Typically, participants see a small array of pictured objects displayed on a computer screen, hear the name of one of the pictures, usually embedded in an utterance, and then click on the named picture using a computer mouse. Participants' gaze location is monitored. Of interest are the saccadic eye movements observed as the name of the picture unfolds until the appropriate object is selected. Early research revealed that, as the initial sounds of the target picture's name are heard and processed, people are more likely to fixate on an object with a name that matches the initial portion of the spoken word than on an object with a non-matching name. Moreover, fixations to matching pictures are closely time-locked to the input, with signal driven–fixations occurring as quickly as 200 ms after the onset of the word (Allopenna, Magnuson, & Tanenhaus, 1998; also see Cooper, 1974; Tanenhaus, Spivey-Knowlton, Eberhard, & Sedivy, 1995).

Subsequent research has established that eye movements are a powerful tool for investigating the processes by which speech is perceived and interpreted, especially the time course of these processes. Allopenna et al. (1998) showed that the proportion of looks to each picture in the display over time can be closely predicted by the strength of the evidence that the name of the object is being heard. Strength of evidence for each object's name was computed by transforming word activations predicted by a connectionist model of spoken-word recognition, TRACE (McClelland & Elman, 1986), into fixation proportions over time using the Luce choice rule (Luce, 1959) over the set of four word alternatives. Subsequent work has shown that eye movements are extremely sensitive to fine-grained phonetic and acoustic details in the spoken input (Dahan, Magnuson, Tanenhaus, & Hogan, 2001; McMurray, Tanenhaus, & Aslin, 2002; Salverda, Dahan, & McQueen, 2003).

The use of eye movements to visual referents as a measure of lexical processing requires the use of a circumscribed "visual world", which is most often perceptually available to listeners before the target picture's name is heard. This world provides the context within which the input is interpreted because, in most studies, the referent object is present on the display. Furthermore, for eye movements directed to visual referents to reflect processing of the spoken signal, information extracted from each type of stimulus must interact at some level. These aspects raise two interrelated questions: At what level does information extracted from the visual display interact with processing of the spoken input, and does the influence of the display limit the degree to which the results will generalize to less constrained situations?

Here, we lay out three possible ways by which visual and spoken stimuli might interact to constrain gaze locations. One possibility is that previewing the display before the spoken input begins provides a closed set of phonological alternatives against which a phonological representation of the speech input is later evaluated. This view assumes that the phonological forms associated with the pictured objects have been accessed before the spoken input begins, either because listeners implicitly prename the pictures to themselves or because the pictures automatically activate their names. When the spoken signal becomes available, no lexical processing per se is initiated. Instead, participants match the phonological representation of the spoken input with the phonological form associated with each location on the display. The proportion of looks to each picture, then, reflects the goodness of match between the phonological representation of the input and the phonological form associated with the picture, bypassing normal lexical processing (see Henderson & Ferreira, 2004). For instance, participants' fixations to the picture of a candle would reflect the close match between the phonological form */kændl/* associated with the picture's location on the display and the first sounds of the spoken input */kæn/*. Findings from the "visual world" eye-tracking paradigm could not then be generalized to spoken-word recognition in less constrained situations.

Another possibility also assumes an implicit naming of the pictures, but differs from the first one by hypothesizing that the speech signal activates lexical-form representations. Thus, eye movements would reflect the goodness of match between activated lexical representations and phonological forms associated with each picture location. This view differs from the previous one by predicting that the degree of activation of lexical representations should modulate the probability of fixating displayed pictures. The impact of lexical factors, such as frequency and neighborhood density (i.e., the number and frequency of words that are phonologically similar to the spoken word) should be observable.

A third possibility assumes no implicit naming of the pictures. Displayed pictures would be associated with visual, and probably also conceptual, representations. The spoken input would activate lexical-form representations, which in turn activate semantic representations. Eye movements would reflect the goodness of match between activated semantic representations and the visual/conceptual representations associated with each picture location. On this view, then, the effects observed in the paradigm are not mediated by names of the pictures that have been accessed during preview.

Two findings argue against the hypothesis that speech processing in the "visual world" paradigm bypasses lexical processing. First, the probability with which a picture with a name that matches the input is fixated over time varies as a function of the lexical frequency of its name (Dahan, Magnuson, & Tanenhaus, 2001). Second, the time course of looks to the target picture is affected by the degree of match of its name to non-displayed words (Dahan, Magnuson, Tanenhaus, & Hogan, 2001), including the phonological neighborhood density of its name, i.e., the number and frequency of similar-sounding words in the lexicon (Magnuson, 2001; Magnuson, Tanenhaus, Aslin, & Dahan, 2003; Magnuson, Dixon, Tanenhaus, & Aslin, in press). These findings indicate that the set of lexical alternatives considered during the processing of the spoken word extends beyond those associated with the pictures present on the display. Taken together, these two findings demonstrate that eye movements in the "visual world" paradigm reflect the involvement of the lexicon during speech processing.

A recent result from our laboratories cannot be easily accounted for if one assumes that eye movements solely reflect the match between the activated lexical-form representations and preactivated names of the displayed objects, as assumed in the second view just described. Upon hearing a spoken word (e.g., snake), listeners are more likely to temporarily fixate on the picture of an object that shares visual features but no phonological similarity with the referent's name (e.g., a rope) than on the picture of a visually and phonologically unrelated object (e.g., a couch; Dahan & Tanenhaus, 2005; see Huettig & Altmann, 2004, for a similar finding). Moreover, these looks are not delayed compared to looks to pictures with matching names. Thus, the probability of fixating the displayed pictures reflects, at least to some degree, the mapping of lexical–semantic representations onto conceptual and visual representations associated with these pictures.

The findings just reviewed are important for answering questions about potential limitations of the "visual world" paradigm. However, as use of the paradigm grows, it becomes increasingly important to understand how preview and other characteristics of the display influence the nature of the picture/speech interaction. Previous research has always included some preview with the display, although its duration has varied across studies. Moreover, most previous research has not compared trials with fixations to critical pictures during preview and trials with no such fixations (one exception is Dahan and Tanenhaus [2005], who reported a similar visual-similarity effect on trials with and without a fixation to the visual-competitor picture during preview). Lexical factors that have been shown to account for the probability of fixating a critical picture overall may have differential effects when the picture was fixated before the onset of the relevant spoken input and when it was not. Another aspect that has not been examined is the position of pictures in the display. Picture positions have typically been randomized. However, the position of a picture may affect the probability that it will be fixated during preview, or merely attended without being overtly fixated. It is therefore of interest to evaluate the effect of the position of a critical picture on fixation probability. Finally, the speech/picture matching process can be based on representations associated with each picture location established prior to the speech input, whereby later influencing the probability of launching a fixation to a given location, or while the picture is being fixated, whereby affecting the duration of the current fixation. Past research has rarely, if ever, distinguished these dependent variables, let alone evaluated how lexical factors might differentially influence them. Here, we report analyses on a data set (from a study conducted in Dutch for different purposes, see Dahan & Gaskell, in press), where two factors were systematically varied: whether or not listeners were able to preview the display before the target picture's name began, and the lexical frequency of the name of the target and the name of a displayed onset-overlapping competitor (high frequency/low frequency or low frequency/high frequency). By varying lexical frequency and preview time, we are able to examine the nature of the interaction between information extracted from the visual display and the output of phonetic processing.

1. Method

1.1. Participants

Participants were 69 college students from the University of Nijmegen, the Netherlands. Thirty-nine participants took part in the no-preview version of the experiment, and thirty in the preview version.

1.2. Materials

Twenty-eight pairs of picturable Dutch nouns overlapping at onset were selected. One of the nouns had a high frequency of occurrence, and the other, a low frequency (e.g., *koffie* [coffee] and *koffer* [suitcase]). Based on the CELEX database (Baayen, Piepenbrock, & van Rijn, 1993), the high-frequency items had an average log frequency per million of 1.7 ($\sigma = 0.6$), compared to 0.8 for the low-frequency items ($\sigma = 0.5$). In order to form a four-item display, two additional phonologically unrelated picturable nouns were associated with each onset-overlapping pair (e.g., *hond* [dog] and *spiegel* [mirror]). Each noun was matched for frequency with one item of the pair. The high frequency–matched distractors had an average log frequency of 1.7 ($\sigma = 0.5$), and the low frequency–matched distractors, of 0.7 ($\sigma = 0.5$). Finally, 70 filler trials were constructed; 35 were composed of four phonologically and semantically unrelated words; the other 35 trials included two onset-overlapping words, neither of them playing the role of target during the experiment.

Black-and-white line-drawing pictures were assigned to each of the 392 words. A picture-naming task was administered to an independent group of 15 Dutch speakers. Naming responses to each item of an experimental pair were coded to evaluate the identification of the pictured object and the use of its intended name. On average, high- and low-frequency pictures were recognized and labeled as intended and equally so (respectively 95 and 94% correct picture identification, and 88 and 87% correct picture labeling). Spoken stimuli were recorded by a female native speaker of Dutch. The average duration of the experimental target words was 528 ms (538 ms for high-frequency words, 518 ms for low-frequency words).

1.3. Design and procedure

The frequency of the target word and the competitor word was varied for each experimental item pair. On a given trial, the target picture was high-frequency, and its competitor was low-frequency (i.e., the high-frequency condition), or vice versa (i.e., the low-frequency condition). For a given participant, half of the 28 experimental trials were tested in the high-frequency condition, the other half in the low-frequency condition.

Participants were seated at a comfortable distance from a computer screen. Eye movements were monitored with an SMI Eyelink system, sampling at 250 Hz. The head-mounted eye tracker was first fitted onto the participant's head, and a brief calibration procedure was performed. On each trial, a central fixation point appeared on the screen for 500 ms, followed by a blank screen for 600 ms. Then, a 5×5 grid with four pictures, four geometric shapes, and a central cross appeared on the screen (see Figure 1), either concurrently with or 500 ms before the presentation of the referent's name (i.e., the no-preview or preview conditions, respectively). Prior to the experiment, participants were instructed that they would hear a word referring to one of the pictured objects on the screen. Their task was to click on the picture and move it above or below the geometric shape adjacent to it, using the computer mouse. Positions of the pictures were randomized across four fixed positions of the grid. The positions of the geometric shapes were

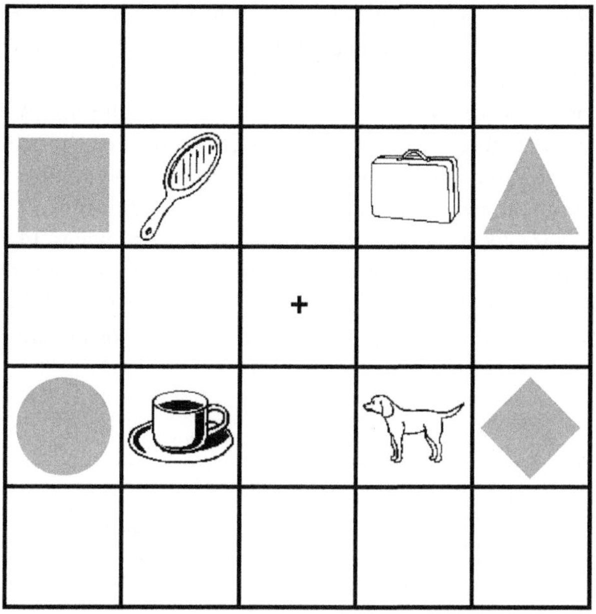

Figure 1. Example of an experimental display (koffie [coffee], koffer [suitcase], hond [dog], spiegel [mirror]).

fixed. The edges of the pictures were approximately 4 cm apart; the distance between the central cross and the closest edge was roughly 3 cm. (One centimeter corresponded to approximately one degree of visual arc.) Participants were under no time pressure to perform the action. After the participant moved the picture, the experimenter pressed a button to initiate the next trial. Every five trials, a central fixation point appeared on the screen, allowing for automatic drift correction.

2. Results and discussion

The data were first parsed into fixations and saccades. Saccade onsets and offsets were automatically detected using the thresholds for motion (0.2), velocity (30/s), and acceleration ($8000/s^2$). Fixation duration corresponded to the time interval between successive saccades. Fixation location was assessed by averaging the *x* and *y* coordinates of the fixation's samples, and by superimposing the fixation location onto the displayed grid and pictures. Fixations that fell within the grid cell containing a picture were hand-coded as fixations to that picture. All other fixations were coded as fixations to the grid, without further distinction. Fixations were coded from the beginning of the trial (i.e., the appearance of the display) until the target picture was fixated and clicked on.

Twenty experimental trials (accounting for 1% of the data) were excluded from the analyses because of poor calibration or track loss (5 trials), failure to fixate on the target object while or before clicking on it (10 trials) or selecting the wrong object (5 trials).

Does preview affect eye-movement behavior? When the spoken word was presented concurrently with the appearance of the display (i.e., the no-preview condition), the mean number of fixations per trial, including the initial fixation, was 4.5 (4.2 in the high-frequency condition and 4.8 in the low-frequency condition). When the pictures were displayed 500 ms before the onset of the spoken word (i.e., the preview condition), 4.8 fixations occurred from spoken-word onset until the end of the trial (4.5 in the high-frequency condition, and 5.2 in the low-frequency condition).

A close look at gaze locations reveals a clear effect of preview. Figure 2 illustrates the distribution of fixations to each of five possible locations, starting from the fixation concurrent with the onset of the spoken word (fixation 0) until the end of the trial. These locations were: (1) the target picture; (2) the competitor picture (i.e., the picture

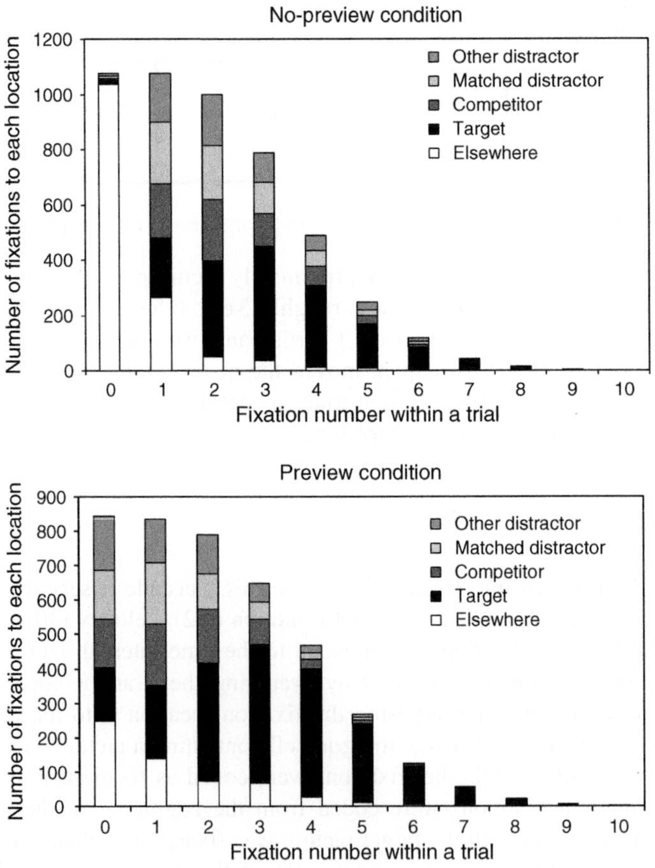

Figure 2. Number of fixations to each location (target, competitor, matched distractor, other distractor, or elsewhere) for each fixation performed during a trial, in the no-preview and preview conditions.

with a name that overlapped with the spoken word at onset); (3) the matched distractor picture (i.e., the distractor matched for frequency with the competitor); (4) the other distractor picture; and (5) elsewhere on the grid. In the no-preview condition, the grid was fixated on the vast majority of all initial fixations (fixation 0). On the next fixation, all four pictures were fixated equally frequently ($\chi^2_{(3)} = 6.15, p > 0.10$). A preference for fixating the target over the other picture locations emerged only on the next fixation (fixation 2, $\chi^2_{(3)} = 72.1, p < 0.0001$). Despite the phonetic overlap between the initial portion of the spoken word and the name of the competitor picture, participants showed no greater tendency to launch a fixation to the competitor than to the matched distractor ($\chi^2_{(1)} = 1.8, p > 0.10$).

The fixation distribution in the preview condition showed a different pattern. On fixation 0 (the fixation concurrent with or immediately following the onset of the spoken word), all four pictures received a roughly equal number of fixations ($\chi^2_{(3)} = 1.3, p > 0.50$). On fixation 1, however, the target picture received more fixations than the other pictures ($\chi^2_{(3)} = 22.6, p < 0.001$). As more fixations were realized, the proportion of fixations to the target increased; importantly, the proportion of fixations to the competitor became larger than that to the matched distractor at fixations 2 and 3 ($\chi^2_{(1)} = 11.3, p < 0.001$ and $\chi^2_{(3)} = 4.2, p < 0.05$, respectively). Thus, only when participants were able to briefly preview the display were more fixations launched to the competitor than to its matched distractor.

In order to test the influence of preview on gaze location throughout the trial, we established, for each trial, whether the competitor picture was fixated or not after the onset of the speech input. In the no-preview condition, the competitor picture was fixated at least once on 526 of the 1072 trials (49%) compared to 516 of the trials (48%) for the matched distractor. In order to compare the probability of fixating the competitor and distractor pictures while preserving the assumption of independence between observations, we restricted our analysis to the trials with a fixation to only one of these two pictures. The number of trials with a fixation to the competitor but not to the distractor was not significantly greater than the number of trials with a fixation to the distractor but not to the competitor (265 vs 255, $\chi^2_{(1)} < 1$). Thus, when participants had not been pre-exposed to the display, they were equally likely to fixate the competitor or the distractor upon hearing the target picture's name. When participants had 500 ms of preview, however, the competitor received more fixations than its matched distractor. The number of trials with (at least) one fixation to the competitor after spoken-word onset, after excluding the trials where the competitor happened to be fixated at spoken-word onset (113 trials), was 343 out of 727 trials (47%). By comparison, the number of trials with at least one fixation to the distractor after spoken-word onset, after excluding the 128 trials with a straddling fixation to the distractor, was 242 out of 712 trials (34%). The number of trials with a fixation to the competitor but not to the distractor was 174; the number of trials with a fixation to the distractor but not to the competitor was 94 ($\chi^2_{(1)} = 23.9, p < 0.0001$).

Thus, while preview had little effect on the total number of fixations per trial, it did influence where these fixations were directed. Without time to preview the display, participants may not have apprehended the display rapidly enough to direct their attention (and thus their gaze) toward pictures that matched the input during the brief portion where the speech signal was ambiguous between target and competitor interpretations. One may argue that the lack of preview did not offer participants the opportunity to prename the pictures. However, as will soon become clear, an examination of *how* preview affected gaze location as the spoken input became available speaks against this interpretation.

Does having fixated a picture during preview affect the likelihood of refixating it later in the trial? To clarify the nature of the information extracted when fixating on a picture during preview, we examined how previewing a picture affected the likelihood of refixating that picture after spoken-word onset. For each of the 727 trials in the preview condition with no fixation to the competitor concurrent with the onset of the spoken word, we determined whether the competitor had been fixated during preview (i.e., before spoken-word onset) and after spoken-word onset (see Table 1). This trial categorization was done separately for high-frequency trials (i.e., where the competitor was of low frequency) and low-frequency trials (i.e., where the competitor was of high frequency). For comparison, trials were similarly categorized according to fixations to the matched distractor before or after spoken-word onset. Overall, having fixated the competitor before the onset of the spoken word did *not* affect the likelihood of fixating it afterwards ($\chi^2_{(1)} < 1$). However, there was a significant interaction when we considered only the low-frequency trials (i.e., when the competitor had a high-frequency name) ($\chi^2_{(1)} = 5.2, p < 0.05$), but not when we considered only the high-frequency trials (i.e., when the competitor had a low-frequency name) ($\chi^2_{(1)} = 1.3, p > 0.20$).[1] Thus, when the competitor picture had a high-frequency name (i.e., in low-frequency trials), participants were more likely to fixate that picture as the spoken input unfolded if they had previously

Table 1
Distribution of trials in the preview condition as a function of the occurrence of a fixation to the competitor or matched-distractor picture before or after spoken-word onset. (Note: The 113 trials with a fixation to the competitor and the 128 trials with a fixation to the distractor concurrently with the onset of the spoken word were excluded.)

			Fixation after word onset?			
			High-frequency trials		Low-frequency trials	
			no	yes	no	yes
Fixation before word onset?	competitor	no	187	149	174	167
		yes	15	7	8	20
	distractor	no	223	110	226	116
		yes	15	9	6	7

[1] A log linear model where the three-way interaction term was omitted yielded a significantly poorer fit to the data than the model that included it ($L^2 = 5.9, p < .05$).

fixated it. In contrast, if the picture had a low-frequency name, having fixated the competitor picture did not affect the likelihood of refixating it. There was no influence of previewing the matched-distractor picture on the probability of fixating it after the onset of the spoken word on either type of trial (High- vs Low-Frequency trials, see Table 1). Thus, regardless of frequency, fixating the distractor picture during preview did not affect the likelihood of refixating it.

The interaction between the frequency of a picture's name and the likelihood of refixating it when the spoken input is consistent with this name suggests that the information extracted during picture preview interacts with the outcome of lexical processing, that is, at a level where lexical-frequency biases operate. This finding argues against a view in which the phonetic input is mapped onto pre-activated picture names, thereby bypassing normal lexical processing of the spoken word. However, a possible objection to this conclusion could be raised if one were to assume that during preview, people may leave a fixated picture before having activated a name for it, and that this is more likely to occur for pictures with low-frequency names than for pictures with high-frequency names because low-frequency words are accessed more slowly than high-frequency words. On this view, as the spoken input becomes available, people would orient their attention to pictures with a pre-activated name that matches the phonetic input, and thus would be more likely to refixate high-frequency pictures, for which the name would more often be available, than low-frequency pictures, for which the name would not be available. Note that this view predicts some (albeit weak) tendency to refixate the picture of a low-frequency competitor, and the data did not support this prediction (in fact, numerically, the effect was in the opposite direction).

Nonetheless, we addressed this objection by examining the duration of fixations to competitor pictures with high- or low-frequency names that occurred after spoken-word onset. In order to neutralize the impact of preview, we restricted this analysis to the no-preview condition. On 96% of the trials, as the pictures appeared, the participant's fixation was on a location of the grid without a picture. Thus, by the time the spoken word began, participants had not fixated any of the pictures. We examined the durations of the fixations to competitor pictures that immediately followed the onset of the spoken word. Of the 1072 fixations, 99 were launched to competitor pictures with high-frequency names and 97 to competitor pictures with low-frequency names. (For inferential statistics purposes, analyses were limited to participants who made such fixations on both high- and low-frequency trials.) Fixations to competitors with high-frequency names were on average 208 ms long, as opposed to 186 ms for fixations to competitors with low-frequency names ($t_{(30)} = 2.2, p < 0.05$). If people were to name the picture they are currently fixating in order to evaluate the match between the speech input and the picture's name, we would have expected to observe longer fixations to competitor pictures with low-frequency names than to pictures with high-frequency name, or no difference between fixations to the two types of pictures if lexical frequency has a negligible effect on fixation duration. Instead, we observed longer fixations to high-frequency competitors than to low-frequency competitors. This finding, together with the influence of a picture name's frequency on the likelihood of refixation reported above, argues against the hypothesis that people prename the pictures during preview or name the currently fixated picture in order to evaluate the match between picture name(s) and the phonetic input.

Longer fixations to competitor pictures with high-frequency names than to competitor pictures with low-frequency names and a greater tendency to refixate high-frequency pictures than low-frequency pictures once spoken input becomes available suggest that the match between lexical representations activated by the spoken input and information extracted from the display (either during preview or during an on-going fixation) is high for high-frequency pictures, but relatively weak for low-frequency pictures. While the results reported so far argue against the first view discussed in the Introduction, they cannot alone distinguish between the other two. Recall that the second view posits that pre-activated picture names are matched to word forms activated by the speech input. The third view, on the other hand, posits no picture prenaming; it assumes that semantic representations associated with word forms activated by the speech input are matched to visual and conceptual representations associated with the pictures' locations. Both views are consistent with the frequency-modulated gaze behavior just reported. However, only the third one can also account for the Dahan and Tanenhaus (2005) results, where listeners tended to fixate a picture that shares visual but little phonological similarity with the concept associated with the target picture's name more than they fixated visually and phonologically unrelated pictures. Thus, the nature of the representation that mediates the mapping between lexical processing and information extracted from picture preview appears to be semantic/conceptual, rather than phonological.

How does the position of the pictures in the display affect fixations? Results reviewed so far have revealed the importance of previewing the display early in the trial. Signal-driven fixations to pictures with names that are temporarily consistent with the spoken input require some exposure with the display. While some information can be extracted parafoveally, fixating a given picture during preview may increase the likelihood of fixating this picture later in the trial. Given the importance of having extracted information on a given location, it would be useful to know whether fixations are directed to particular spatial locations more often than others, both during preview and as the spoken word unfolds. Although picture locations are typically randomized in eye-movement studies to minimize the risk of confounds with location, choosing the spatial arrangement of competitors and distractors might maximize the sensitivity of the paradigm.

Figure 3 summarizes the distribution of fixation locations over the course of a trial, distinguishing fixations to the upper left picture, the upper right picture, the lower left picture, the lower right picture, or elsewhere on the display. A similar pattern emerges for both preview and no-preview conditions. There is a strong tendency to initially fixate on the picture located in the upper left cell, and then to move to another picture by performing either a horizontal saccade (and landing on the upper right picture) or a vertical saccade (and landing on the lower left picture).

A potential implication of the unbalanced distribution of fixations across picture locations is that whether a competitor picture is fixated or not during a trial may strongly depend on its position in the display. To address this question, we determined the position of the competitor picture on trials where it was fixated after spoken-word onset and on

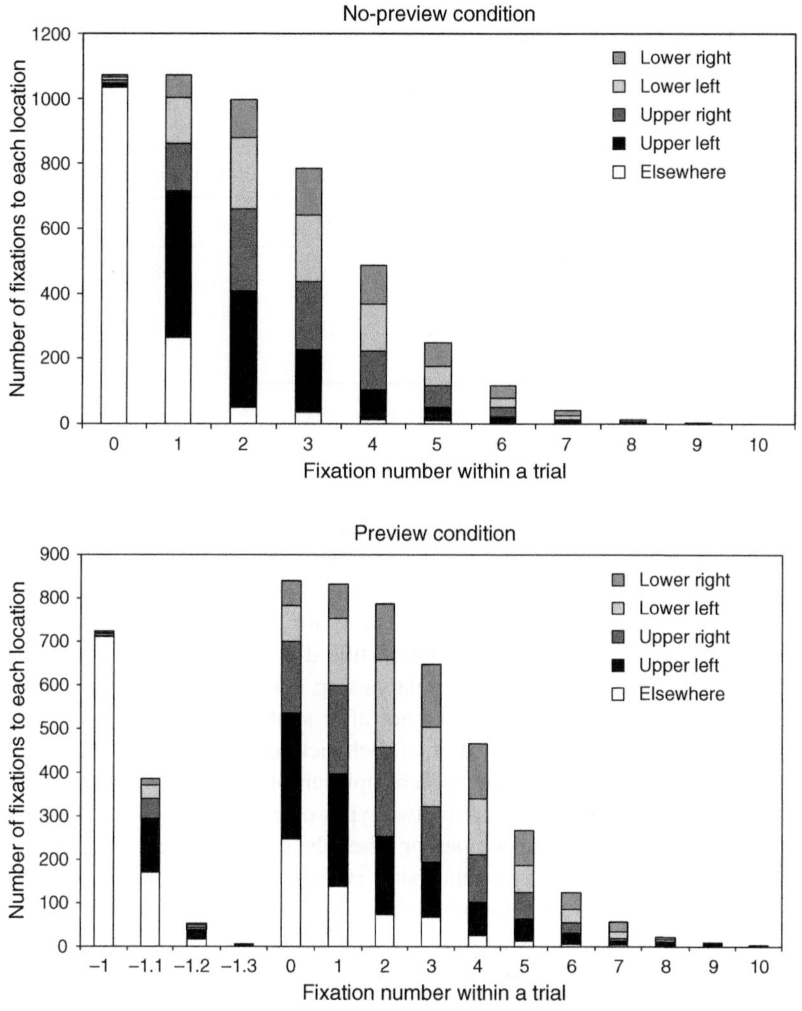

Figure 3. Number of fixations to each location in the display (upper left, upper right, lower left, lower right, or elsewhere) for each fixation performed during a trial, in the no-preview and preview conditions. (Note: The fixation concurrent with or immediately after the onset of the spoken word is labeled as fixation 0. In the 500-ms preview condition, the fixation realized before the onset of the spoken word is labeled fixation −1; when more than one such fixation was realized, they are labeled as fixations −1.1, −1.2, etc. . .).

trials where it was not fixated. We conducted this analysis on the preview condition trials. To neutralize the impact of preview on the likelihood of revisiting the picture, we excluded trials where the competitor had been fixated before or concurrently with the onset of the spoken word. We established the distribution of the remaining 677 trials

Table 2
Distribution of trials in the preview condition, distinguishing whether the competitor or the matched-distractor picture was fixated or not after spoken-word onset, as a function of the position of the picture (upper left, upper right, lower left, lower right). (Note: Trials on which either picture was fixated before or concurrently with the onset of the spoken word were excluded.)

		competitor or distractor position			
		upper left	upper right	lower left	lower right
competitor fixation?	no	71	83	88	119
	yes	90	85	85	56
distractor fixation?	no	105	110	126	108
	yes	43	41	63	79

across all the four picture positions, distinguishing whether the competitor was fixated after spoken-word onset or not. A similar trial classification was established for fixations to the matched distractor (Table 2). Analyses revealed that the distribution of trials across all four possible positions of the competitor picture affected whether the competitor was fixated or not ($\chi^2_{(3)} = 22.1$, $p < 0.0005$). This did not interact with trial frequency (the log linear model omitting the three-way interaction did not significantly differ from the saturated model, $L^2 = 2.5$, $p > 0.40$). A relationship between distractor-picture position and whether the distractor was fixated or not after spoken-word onset was also found ($\chi^2_{(3)} = 10.5$, $p < 0.05$). Thus, the position of each picture on the display affected whether or not that picture was fixated as the speech input unfolded. However, the influence of picture position was different between the two types of pictures. In particular, there were *fewer* trials with a fixation to the competitor when the competitor was in the lower right position than when it was in any of the other positions. By contrast, there were *more* trials with a fixation to the distractor when this distractor was in the lower right position than when it was in any of the other positions. To formally evaluate this interaction, we considered the distribution of trials with a fixation to either the competitor or the distractor (but not to both) across all four possible picture positions. There was a significant interaction between the type of picture fixated and its position ($\chi^2_{(3)} = 16.9$, $p < 0.0005$): When fixated, the competitor tended to be in the upper left or right position, and much less so in the lower right position; by contrast, the fixated distractor tended to be in the lower left or right positions, and less so in the upper left and right positions.

These results suggest the following interpretation. Early in the trial, people's attention is drawn toward the upper positions of the display and some information about the objects displayed there may be acquired even if no fixation is launched to these locations. Upon hearing the spoken word, people are more likely to fixate on these positions if the competitor is located in one of these positions because of the match between the picture and the representation temporally activated by the spoken word. By contrast, if one of these positions is occupied by the distractor, participants' attention is not drawn toward these already explored locations when hearing the spoken word and participants continue to explore the display, thus fixating on the lower positions of the display. This finding is important because it demonstrates that the specific position of a competitor picture will affect whether this picture will be fixated as spoken input becomes available. Experimenters might want to control more systematically (rather than randomize) where critical pictures are located in the display, or perhaps include position as a factor during data analysis.

3. Conclusions

Examining the role of picture preview and display characteristics in the set-up typically used in eye-movement studies of spoken-word recognition revealed a number of interesting findings that shed light on the nature of the representations that mediate phonetically driven fixations. First, some preview is required to observe signal-driven fixations to competitors that are temporarily consistent with the input. When the display is briefly available before the speech input begins, participants are able to extract relevant visual and/or conceptual information associated with the pictured objects, even when they have not fixated on any of the pictures. This information guides their subsequent saccades to relevant pictures. Without preview, initial fixations mainly serve the purpose of extracting visual or conceptual information about the displayed pictures, and thus fail to reflect the temporary activation of lexical candidates triggered by the early portion of the spoken input.

Second, whether or not a picture is fixated during preview influences the likelihood that it will be refixated during speech processing, but only if this picture is consistent with a high-frequency interpretation of the speech input. Furthermore, the first fixation to a picture with a name that matches the unfolding speech input is longer when the picture name is high frequency compared to when it is low frequency. These two new findings provide strong evidence that even in the very circumscribed set of possible referents that the display offers, the speech signal is processed in terms of possible lexical interpretations, rather than directly matched to pre-activated picture names (as assumed by the first view introduced earlier). This extends Dahan et al.'s (2001) results by showing effects of competitor frequency *even* when the competitor picture was fixated during preview, casting doubts on the view that participants engage in covert picture naming.

Third, some spatial locations attract a disproportionate number of fixations, especially early in the trial, and the position of pictures in the display, in conjunction with the speech input, affects whether or not the picture is fixated.

Overall, the contingent analyses reported here shed light on how the display might influence lexical processing and help mitigate concerns about potential task-specific strategies that using displays of pictures might in principle have introduced. The results are encouraging for researchers who want to exploit the properties of the eye-tracking paradigm to explore issues of representation and process in spoken-word recognition.

Acknowledgments

This work was supported by grants from the National Institutes of Health (DC 005071 to Michael K. Tanenhaus) and the National Science Foundation (Human and Social Dynamics 0433567 to Delphine Dahan).

References

Allopenna, P. D., Magnuson, J. S., & Tanenhaus, M. K. (1998). Tracking the time course of spoken word recognition using eye movements: Evidence for continuous mapping models. *Journal of Memory and Language, 38*, 419–439.

Baayen, R. H., Piepenbrock, R., & van Rijn, H. (1993). *The CELEX lexical database (CD-ROM)*. Philadelphia, PA: Linguistic Data Consortium, University of Pennsylvania.

Cooper, R. M. (1974). The control of eye fixation by the meaning of spoken language. A new methodology for the real time investigation of speech perception, memory, and language processing. *Cognitive Psychology, 6*, 84–107.

Dahan, D., & Gaskell, M. G. (in press). The temporal dynamics of ambiguity resolution: evidence from spoken-world recognition. *Journal of Memory and Language.*

Dahan, D., Magnuson, J. S., & Tanenhaus, M. K. (2001). Time course of frequency effects in spoken word recognition: Evidence from eye movements. *Cognitive Psychology, 42*, 317–367.

Dahan, D., Magnuson, J. S., Tanenhaus, M. K., & Hogan, E. M. (2001). Subcategorical mismatches and the time course of lexical access: Evidence for lexical competition. *Language and Cognitive Processes, 16*, 507–534.

Dahan, D., & Tanenhaus, M. K. (2005). Activation of visually-based conceptual representations during spoken-word recognition. *Psychonomic Bulletin and Review, 12*, 453–459.

Henderson, J. M., & Ferreira, F. (2004). Scene perception for psycholinguists. In J. M. Henderson & F. Ferreira (Eds.), *The interface of language, vision, and action: Eye movements and the visual world* (pp. 1–58). New York: Psychology Press.

Huettig, F. & Altmann, G. T. M. (2004). Language-mediated eye movements and the resolution of lexical ambiguity. In M. Carreiras & C. Clifton (Eds.) The on-line study of sentence comprehension: Eye-tracking, ERP, and beyond (pp. 187–207). New york, NY: Psychology Press.

Luce, R. D. (1959). *Individual choice behavior: A theoretical analysis.* New York: Wiley.

Magnuson, J. S. (2001). *The microstructure of spoken word recognition.* Unpublished doctoral dissertation. University of Rochester.

Magnuson, J. S., Dixon, J. A., Tanenhaus, M. K., & Aslin, R. N. (in press). The dynamics of lexical competition during spoken word recognition. *Cognitive Science.*

Magnuson, J. S., Tanenhaus, M. K., Aslin, R. N., & Dahan, D. (2003). The time course of spoken word learning and recognition: Studies with artificial lexicons. *Journal of Experimental Psychology: General, 132*, 202–227.

McClelland, J. L., & Elman, J. L. (1986). The TRACE model of speech perception. *Cognitive Psychology, 18*, 1–86.

McMurray, B., Tanenhaus, M. K., & Aslin, R. N. (2002). Gradient effects of within category phonetic variation on lexical access. *Cognition, 86*, B33–B42.

Salverda, A. P., Dahan, D., & McQueen, J. M. (2003). The role of prosodic boundaries in the resolution of lexical embedding in speech comprehension. *Cognition, 90*, 51–89.

Tanenhaus, M. K., Spivey-Knowlton, M. J., Eberhard, K. M., & Sedivy, J. C. (1995). Integration of visual and linguistic information in spoken language comprehension. *Science, 268*, 1632–1634.

Chapter 22

THE PROCESSING OF FILLED PAUSE DISFLUENCIES IN THE VISUAL WORLD

KARL G. D. BAILEY

Andrews University, USA

FERNANDA FERREIRA

University of Edinburgh, UK

Eye Movements: A Window on Mind and Brain
Edited by R. P. G. van Gompel, M. H. Fischer, W. S. Murray and R. L. Hill

Abstract

One type of spontaneous speech disfluency is the filled pause, in which a filler (e.g. *uh*) interrupts production of an utterance. We report a visual world experiment in which participants' eye movements were monitored while they responded to ambiguous utterances containing filled pauses by manipulating objects placed in front of them. Participants' eye movements and actions suggested that filled pauses informed resolution of the current referential ambiguity, but did not affect the final parse. We suggest that filled pauses may inform the resolution of whatever ambiguity is most salient in a given situation.

The most common type of overt interruption of fluent speech, or disfluency, is the filled pause (Bortfield, Leon, Bloom, Schober, & Brennan, 2001). Speakers produce filled pauses (e.g. *uh* or *um*) for a variety of reasons, such as to discourage interruptions or to gain additional time to plan utterances (Schacter, Christenfeld, Ravina, & Bilous, 1991). While speakers may benefit from producing filled pauses because they gain planning time, listeners may also use the presence of filled pauses to inform language comprehension (Bailey & Ferreira, 2003; Brennan & Schober, 2001; Brennan & Williams, 1995; Clark & Fox Tree, 2002). Thus, given the prevalence of filled pauses, and the use of such pauses by listeners, a complete model of language comprehension should account for how these disfluencies are handled.

In order to construct such a model of disfluency processing, it is necessary to describe and test possible hypotheses about how disfluencies might affect language comprehension. Evidence that supports one such hypothesis, cueing of upcoming structure, comes from a series of experiments involving grammaticality judgments (Bailey & Ferreira, 2003). This hypothesis is built on the observation that filled pauses occur in a particular distribution with respect to syntactic (Clark & Wasow, 1998), semantic (Schacter et al., 1991) or pragmatic (Smith & Clark, 1993) structure. In the case of syntactic structure, filled pauses (and other disfluencies, such as repetitions) are most likely to occur immediately prior to the onset of a complex syntactic constituent (Clark & Wasow, 1998; Ford, 1982; Hawkins, 1971; Shriberg, 1996;). Filled pauses are also likely after the initial word in a complex constituent, especially after function words (Clark & Wasow, 1998). Thus, the cueing hypothesis assumes that listeners might be able to use the presence of a recent filled pause to predict that an ambiguous structure should be resolved in favor of a more complex analysis (Bailey & Ferreira, 2003). In a garden path utterance like [1] below, the filled pause might act as a "good" cue, because it correctly predicts the ultimate structure of the utterance; in [2], the filled pause might be a "bad" cue, because it leads the listener to predict the onset of a new constituent.

[1] While the man hunted the uh uh *deer ran* into the woods.

[2] While the man hunted the *deer* uh uh *ran* into the woods.

Grammaticality judgments supported this cueing hypothesis: [1] was judged grammatical more often than [2], which suggested that [1] is easier to process (Bailey & Ferreira, 2003). However, [1] and [2] confound "good" and "bad" cues with the presence of delay between the ambiguous head noun and the disambiguating verb. This type of delay has led to the same pattern of results in utterances with lexical modifiers (i.e., prenominal adjectives and relative clauses) in place of the disfluencies in [1] and [2] (Ferreira & Henderson, 1991). To avoid this confound, Bailey and Ferreira (2003) tested whether filled pauses that did not introduce delays between temporarily ambiguous head nouns and disambiguating verbs might also affect grammaticality judgments of spoken utterances depending on their location. Disfluencies were placed in two different locations in coordination ambiguity utterances prior to the onset of the temporarily ambiguous head noun. The "good" cue location in [3] below was consistent with the ultimate sentence coordination structure, while the "bad" cue in [4] was consistent with an noun phrase coordination structure (based on the assumption that listeners take a disfluency to be indicative of an upcoming complex constituent).

[3] Sandra bumped into the busboy and the uh uh waiter told her to be careful.

[4] Sandra bumped into the uh uh busboy and the waiter told her to be careful.

Participants were more likely to judge an utterance with a "good" cue disfluency (as in [3]) as grammatical than an utterance with a "bad" cue (as in [4]). This pattern of results was replicated with environmental noises replacing the disfluencies, but, importantly, not with adjectives, suggesting that it is the presence of a non-propositional interruption that is the cue, not the form of that interruption.

However, the results of Bailey and Ferreira (2003), while promising, are based on offline judgments following the end of the utterance. In other words, the grammaticality judgment task makes it possible to see that filled pauses have had an effect consistent with the cueing hypothesis by the time the utterance is finished, but it is not possible to chart the time course of that effect, nor to observe when processing of the disfluency takes place.

A recently rediscovered methodology that allows spoken language comprehension to be monitored on a moment by moment basis is the visual world paradigm (Cooper, 1974; Tanenhaus, Spivey-Knowlton, Eberhard, & Sedivy, 1995). In this paradigm (henceforth, the VWP), participants listen to utterances while viewing a concurrent array of clip art images on a computer screen (e.g. Altmann & Kamide, 1999) or while interacting with a set of objects within reach (e.g. Tanenhaus et al., 1995). The objects or images which make up the constrained visual world and the relationships between them serve as a context for a concurrent referring utterance (Tanenhaus et al., 1995). Inferences about language comprehension are drawn from listeners' eye movement patterns: The eyes are naturally directed to objects that are related to concurrent language processing (Cooper, 1974). In the VWP, utterances can be presented without distortion and it is not necessary to instruct listeners to look at objects which are related to concurrent speech.

Two particular patterns of eye movements have been used to draw inferences about comprehension: anticipatory and confirmatory eye movements. Anticipatory eye movements (Altmann & Kamide, 1999, 2004; Kamide, Altmann, & Haywood, 2003) are saccades launched to objects before they are directly referenced by the utterance. Confirmatory eye movements (e.g. Spivey, Tanenhaus, Eberhard, & Sedivy, 2002) are made in response to a direct reference to an object and can include fixations on possible referents of a constituent (Tanenhaus et al., 1995) or on disconfirmed competitors (Sedivy, Tanenhaus, Chambers, & Carlson, 1999; Kamide et al., 2003). The presence of confirmatory eye movements is most easily seen in the probability of fixating a given object because participants may simply continue to fixate an object that they were already looking at due to an anticipatory eye movement launched prior to direct reference.

The cueing hypothesis would predict that eye movements during a filled pause should reflect a more complex parse of material currently being processed, and that saccades would be launched to objects consistent with that analysis. Confirmatory eye movements during a later ambiguous referring expression would then identify which of a set of possible parses had been selected and the time course of that selection (as the probability of fixating a given object rises and falls).

In this chapter, we will present an experiment that directly tests whether a cueing mechanism can modulate the interpretation of a fully ambiguous utterance in the presence of a fully ambiguous visual world. As described earlier, the position of a disfluency can affect the probability that an utterance is judged grammatical (Bailey & Ferreira, 2003), suggesting that disfluencies may cue the parser to expect a certain structure. The strongest form of the cueing hypothesis, then, predicts that a fully ambiguous utterance will immediately be interpreted differently based solely on the location of a disfluent interruption. We do not find evidence for this strong hypothesis, but do find support for a weaker form, in which the disfluent interval introduced by the filled pause may allow the parser to further process any existing ambiguities. Depending on the demands of the task, final interpretations may or may not be affected by the disfluency cue. Nevertheless, we suggest that filled pauses provide a unique window on sentence processing in general, because they show what ambiguities are relevant at that point in the utterance.

1. Experiment

In order to test whether filled pauses can change the interpretation of an otherwise fully ambiguous utterance, the concurrent visual world must not constrain the interpretation of that utterance. Previous studies using otherwise fully ambiguous prepositional phrase ambiguities (Spivey et al., 2002; Tanenhaus et al., 1995) used visual worlds that constrained the interpretation of utterances such as [5] below. The objects in these displays required participants to arrive at the same semantic representation as in the disambiguated utterance [6], and disallowed the interpretation in [7].

[5] Put the apple on the towel in the box.
[6] Put the apple that's on the towel in the box.
[7] Put the apple on the towel that's in the box.

Two different constrained display types have been used (Spivey et al., 2002). The first, referred to as the one-referent display, contained a target object (e.g. an apple on a towel), a distractor object (e.g. a frog on a mitten), a goal location (e.g. a box), and a distractor location (e.g. a towel). The second, two-referent, display was identical to the first, except that the distractor object matched the target object in part (e.g. an apple on a mitten). Note that in both displays, the only possible action in response to [5] is for the apple that is on the towel to be placed in the empty box because there is no towel inside a box.

In order to modify these displays so that they did not constrain the interpretation of our utterances, we replaced the distractor location described above (a towel by itself) with a modified goal (e.g. a towel in a box as opposed to the unmodified goal, an empty box; see Figure 1). Thus, in the one-referent display, it would be possible to place an apple that is on a towel into an empty box (a modified theme interpretation) in response to [5] or an apple onto a towel that is in a box (a modified goal interpretation). In the case of the two-referent display, of course, the display still constrains the interpretation of the utterance because of the presence of a second apple. The modified goal interpretation in [7] is possible, but only if the listener violates syntactic or discourse constraints and

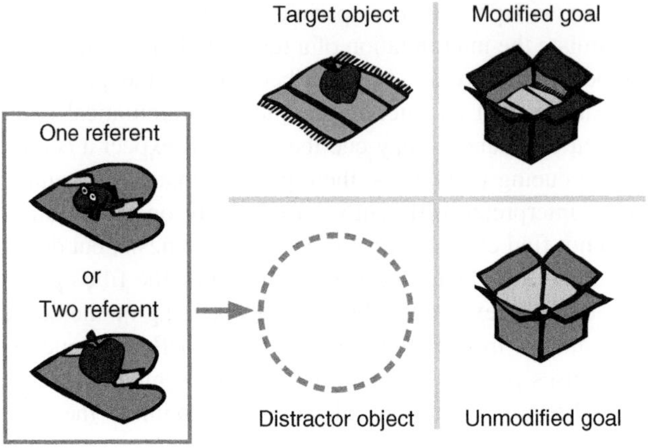

Figure 1. Fully ambiguous displays used in experiment. Only the distractor object differed between the one referent and the two referent displays.

uses the phrase "on the towel" twice: once to identify which apple to pick up (the apple that's on the towel) and once to identify where to place the apple (the towel that's in the box). These interpretations are unlicensed because a single constituent cannot play more than one role in a sentence; nevertheless, we have observed that participants occasionally behave as if that is the interpretation that they have obtained, perhaps because they have engaged in "good-enough processing" (Ferreira, Bailey, & Ferraro, 2002).

According to the strong form of the cueing hypothesis, disfluencies placed before one of the two possible modified noun phrases in [5] should bias the parser to prefer the corresponding structure. Specifically, a filled pause placed as shown in [8] is predicted to yield a modified theme interpretation similar to [6], whereas a filled pause placed as in [9] is predicted to yield modified goal interpretations similar to [7].

[8] Put the uh uh apple on the towel in the box.

[9] Put the apple on the uh uh towel in the box.

For the strong form of the cueing hypothesis to be supported, these interpretations should be seen both immediately (in saccades to appropriate goal objects during filled pauses), as the utterance unfolds (in fixations during the ambiguous noun *towel*), and in the overall interpretation of the utterance (the participants' actions).

1.1. Materials and methods

1.1.1. Participants

Sixteen participants from the Michigan State University community participated in this experiment in exchange for credit in an introductory psychology course or payment ($7.00). All participants were native speakers of English, and had normal hearing and normal or corrected to normal vision.

1.1.2. Materials

Twenty-four prepositional phrase ambiguity utterances were constructed for this experiment using nouns from a set of 12 possible target objects and 12 possible goal objects. Utterances were recorded and digitized at 10 kHz using the Computerized Speech Laboratory (Kay Elemetrics), and then converted to wav format. Each utterance was recorded in two ways: once as an utterance with two disfluencies, as in [10] below, and once as a fluent utterance with two instances of *that's*, as in [11].

[10] Put the uh uh apple on the uh uh towel in the box.

[11] Put the apple that's on the towel that's in the box.

Utterances like [8] and [9] were created from [10], and like [6] and [7] were created from [11] by excising the appropriate disfluency or word. Participants are relatively insensitive to the removal of disfluencies from utterances (Brennan & Schober, 2001; Fox Tree, 1995, 2001; Fox Tree & Schrock, 1999;) and thus this procedure was used to control the prosody of the various utterances. The removal of a single disfluency or word from an utterance did not result in utterances that participants found odd or strange. In the experiment, each participant heard only one version of any given utterance.

Forty-eight filler utterances were also recorded and grouped with the 24 critical utterances into trials of 3 utterances each. A further 72 utterances were recorded to create 24 trials composed of only fillers. The types and proportions of syntactic structures used in the filler utterances and the interleaving of filler and critical trials were identical to those used in Spivey et al. (2002). Filled pauses occurred on half of filler trials and were placed at a variety of different locations within the sentences.

Displays consisted of a 2×2 grid (see Figure 1), and objects were set up according to the description provided by Spivey et al. (2002), so that depending on the height and posture of a given participant, the center of each object (or set of objects) was separated by 10–15° of visual angle from the center of each of its adjacent neighbors (note that previous studies did not report the angular distance between objects). In experimental trials, the possible theme referents (the target and distractor objects) were always on the left, and were each placed equally in both the proximal and distal positions across trials. The possible goal referents (modified and unmodified) were always on the right, and likewise were each placed equally in both the proximal and distal positions across trials. The possible theme and goal referents for filler utterances were equally likely to occur in any of the four positions, and any object on the table could be referenced as a target object in filler utterances. In all, 48 displays were created, one for each set of three utterances. Of the 24 critical displays seen by any participant, 12 were two-referent displays and 12 were one-referent displays. A new random ordering of trials adhering to the interleaving requirements was created for every fourth participant in this experiment.

1.1.3. Apparatus

The eyetracker used in this experiment was an ISCAN model ETL-500 head-mounted eyetracker (ISCAN Incorporated) with eye and scene cameras located on a visor.

Participants were able to view 103° of visual angle horizontally and 64° vertically. No part of the object display was occluded at any time by the visor. Eye position was sampled at 30 Hz and merged with scene video data. Eye position in this merged video was later hand-coded relative to Regions of Interest (henceforth, ROIs) frame by frame, starting with the onset of each critical utterance and ending with the movement of an object to a new location.

1.1.4. Procedure

After a participant was introduced to the objects and apparatus, and had provided informed consent, the eyetracker was placed on the participant's head and adjusted. Depending on the height of each participant, participants either stood or were seated at a table. Participants' eye positions were calibrated to the scene by recording pupil and corneal reflection locations while they looked at nine predetermined targets. The sentence comprehension task was introduced to the participant via three practice utterances involving the movement of a single object from one location to another. The practice utterances did not contain any lexical or syntactic ambiguities.

Immediately before beginning a trial, the experimenter set up the appropriate objects in front of the participant, which allowed 20–30 s of view time prior to the onset of the first utterance in the trial (as in Spivey et al., 2002 and Trueswell, Sekerina, Hill, & Logrip, 1999). Participants were instructed to respond as quickly as possible prior to practice trials, but were not reminded thereafter. In addition, no "Look at the center cross" instruction was given prior to the start of each trial (cf. Spivey et al., 2002), as pilot studies indicated that participants tended to perseverate in fixating the center cross when this instruction was given.

Design. The four utterance types (theme and goal disfluencies, and theme and goal modifiers) were combined with the two displays (one and two referent) to create eight unique conditions for this experiment. Three trials in each condition were presented to each participant, for a total of twenty-four critical trials. Each display occurred in each condition an equal number of times over the course of the experiment.

1.2. Results and discussion

The analysis of eye tracking data presented here differs somewhat from previous studies using this version of the VWP. These studies (Spivey et al., 2002; Tanenhaus et al., 1995; Trueswell et al., 1999) calculated probabilities of fixating particular objects at each sampling interval during arbitrary time segments that did not take into account variations in word length across individual utterances. Probabilities in this study, on the other hand, were calculated separately for each ROI and each word in each utterance and then averaged (see Altmann & Kamide, 2004, who described this procedure). These probabilities were then arcsine-transformed (Winer, 1971) and submitted to a 2 (cue location: theme or goal) by 2 (cue type: disfluency or modifier) by 2 (number of possible theme referents: one or two) ANOVA. In addition, behavioral responses to disfluent instructions in the current experiment were classified as either modified goal (e.g. towel in box) directed or unmodified goal (e.g. empty box), and were submitted to a 2 (number of referents) by 2 (location of disfluency) ANOVA. Unambiguous controls were not included in the behavioral analysis as participants moved an object to the appropriate goal on over 90% of trials.

Participants were more likely to move a target object to the unmodified goal ($F_{1,15} = 23.6$, $p < 0.001$) in the two-referent display (70.8% of trials with a theme disfluency; 64.6% with a goal disfluency) than in the one-referent display (37.5% of trials with a theme disfluency; 35.4% with a goal disfluency). The location of the disfluency had no effect ($F < 1$) on participants' actions and there was no significant interaction between the number of referents and disfluency location ($F < 1$). The effect of number of referents on the final interpretation of the utterance is not surprising, as the two-referent display should have constrained the interpretation of the utterance (due to the presence of two apples), while the one-referent display should not have. However, the lack of effect of disfluency location on the final interpretation of the utterances, even in the one-referent display, is evidence against the strong form of the cueing hypothesis, and suggests that disfluencies were not interpreted as strong predictors of the syntactic parse.

Eye movement patterns, on the other hand, did support a form of the cueing hypothesis. Figure 2 shows graphs representing the probability of fixating and launching a saccade to each ROI for each condition in the experiment. Gray polygons represent the probability of fixation on, and black lines the corresponding probability of launching a saccade to that ROI for each word. Each point on the polygons and line graphs corresponds to a single word in each utterance. Content words, disambiguating function words, and disfluencies are indicated above the fixation polygons. The eight conditions form rows, while the four ROIs form columns.

An effect of number of referents is presented in Figure 2; the different display types elicited different patterns of fixation, especially on the distractor and the modified goal. Consistent with previous studies (Spivey et al., 2002; Tanenhaus et al., 1995; Trueswell et al., 1999; the incorrect goal in previous studies corresponds to our modified goal), there is a significant increase in the probability of fixation on the modified goal in the one-referent display relative to the two-referent display during the word *towel* ($F1_{1,15} = 36.2$, $p < 0.001$; $F2_{1,23} = 30.6$, $p < 0.001$). This difference was found for all utterance types, including theme modifiers, which should rule out the modified goal as a possible referent of *towel* because of the preceding *that's*. Main effects of cue location ($F1_{1,15} = 6.29$, $p < 0.03$; $F2_{1,23} = 5.12$, $p < 0.04$), and cue type ($F1_{1,15} = 7.64$, $p < 0.02$; $F2_{1,23} = 7.22$, $p < 0.02$) were also present. The effect of cue type was due to an increased proportion of looks to the modified goal in the disfluency conditions, which would be expected if the language comprehension system treated those utterances as more ambiguous than either modifier condition. A significant interaction between number of referents and cue location ($F1_{1,15} = 6.75$, $p < 0.02$; $F2_{1,23} = 7.99$, $p < 0.02$) was present, but interactions between cue type and number of referents ($F < 1$), between cue type and location ($F1_{1,15} = 3.54$, $p > 0.05$; $F2_{1,23} = 3.61$, $p > 0.05$), and between all three variables ($F < 1$) were nonsignificant. This pattern (Figure 3) is consistent with the prediction that

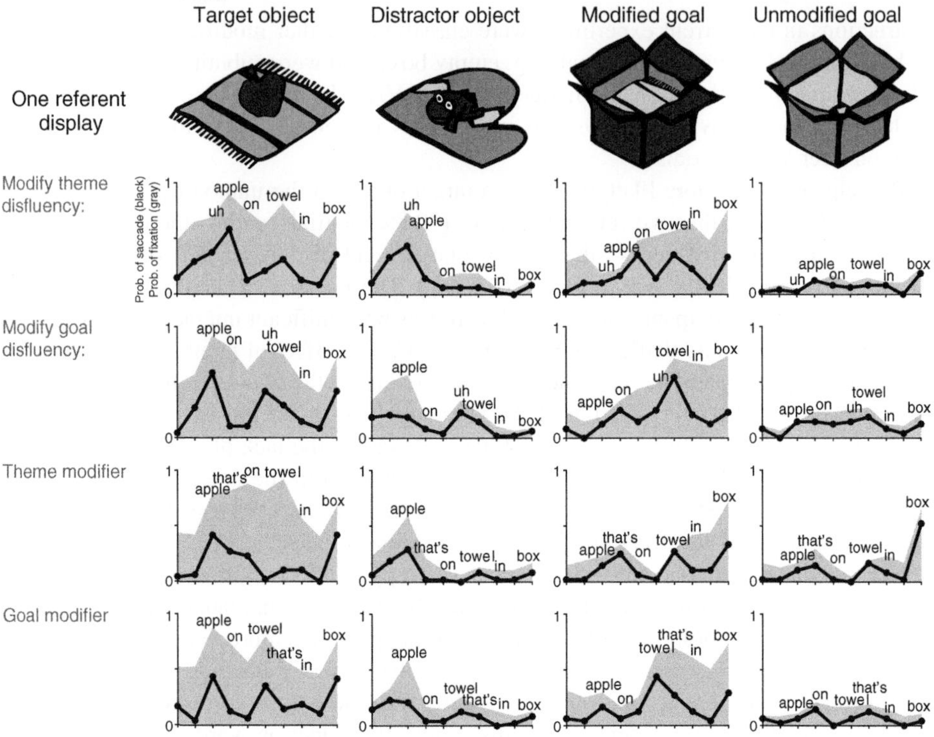

Figure 2. Probability of fixation on or saccade launch to regions of interest for each word in each utterance and display condition in Experiment 1. Gray polygons represent the probability of fixation; black line represent probability of saccade. The locations of content words, disambiguating function words, and disfluencies are indicated above the fixation graphs. The y-axis of each graph represents probability (0–1) and the x-axis the course of the utterance (one word per tick).

theme disfluencies and modifiers should elicit fewer looks to the modified goal (being consistent instead with a modified theme) than the corresponding modify goal utterances, but only in the one-referent display, where the identity of the theme has already been ascertained and the eye movement system is not engaged in deciding between the target and distractor objects. However, separate analyses of the disfluent conditions found only an effect of number of referents ($F1_{1,15} = 29.8$, $p < 0.001$; $F2_{1,23} = 24.9$, $p < 0.001$). The effect of cue location ($F < 1$) and the interaction between number of referents and cue location ($F1_{1,15} = 2.69$, $p > 0.1$; $F2_{1,23} = 2.29$, $p > 0.1$) were not significant, suggesting that the modifier conditions were carrying the overall interaction between the number of referents and cue location. This would suggest that disfluencies were not interpreted by the parser in the same ways as modifiers.

Similar patterns (Figure 4) are also present in the saccade data to the modified goal during *towel*, consistent with confirmatory saccades as the source of fixation patterns.

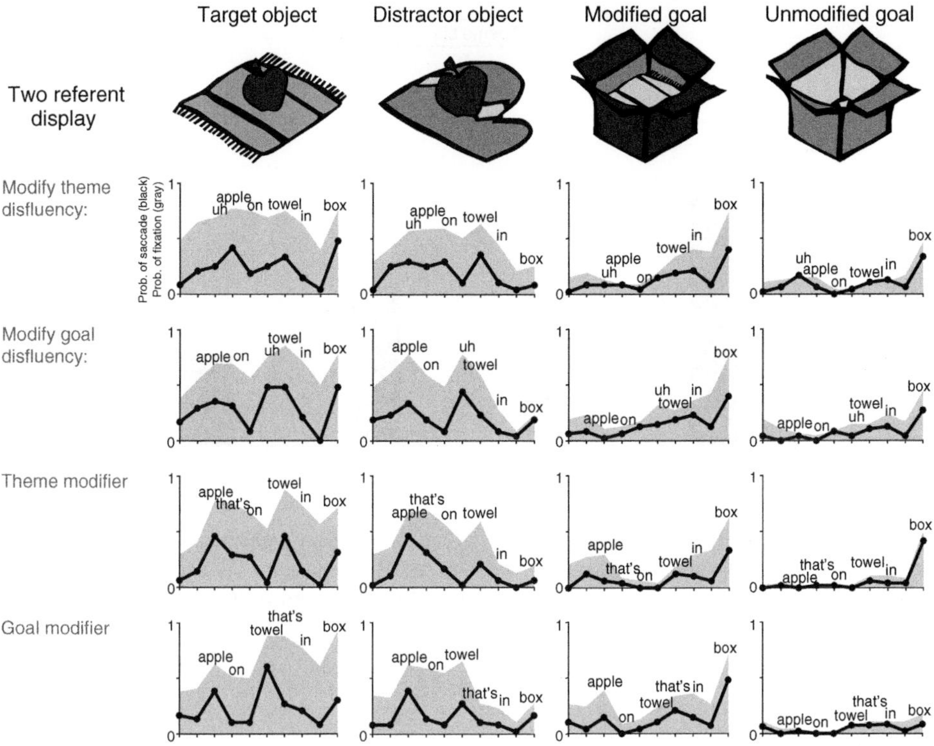

Figure 2. (*continued*)

Main effects of number of theme referents ($F1_{1,15} = 39.6$, $p < 0.001$; $F2_{1,23} = 36.8$, $p < 0.001$), cue location ($F1_{1,15} = 4.68$, $p < 0.05$; $F2_{1,23} = 2.63$, $p > 0.1$), and cue type ($F1_{1,15} = 5.03$, $p < 0.05$; $F2_{1,23} = 3.76$, $p > 0.05$) are again present (although the latter two effects are significant only by participants), as well as a marginal interaction between number of referents and cue location ($F1_{1,15} = 3.18$, $p = 0.095$; $F2_{1,23} = 4.17$, $p = 0.053$). All other interactions were nonsignificant ($F < 1$). Anticipatory saccades, however, may also have contributed to the probability of fixating the modified goal during *towel*, as saccades were also launched to this object during the word *on*, suggesting that these saccades may have been launched based on the expected arguments of the verb *put* (which requires both a theme and a goal when used imperatively; cf. Altmann & Kamide, 1999, 2004).

Separate analyses of disfluent conditions again revealed a main effect of number of referents ($F1_{1,15} = 14.4$, $p < 0.01$; $F2_{1,23} = 15.4$, $p < 0.01$), but only a marginal effect of cue location by participants ($F1_{1,15} = 3.23$, $p < 0.1$; $F2_{1,23} = 1.56$, $p > 0.1$) was present. The interaction between number of referents and cue location was nonsignificant ($F1 = 1.23$; $F2 = 1.87$). The marginal effect of cue location tentatively suggests that disfluencies may have some immediate effect on the parser; however, it is clear that the display itself had a much greater impact on eye-movement patterns.

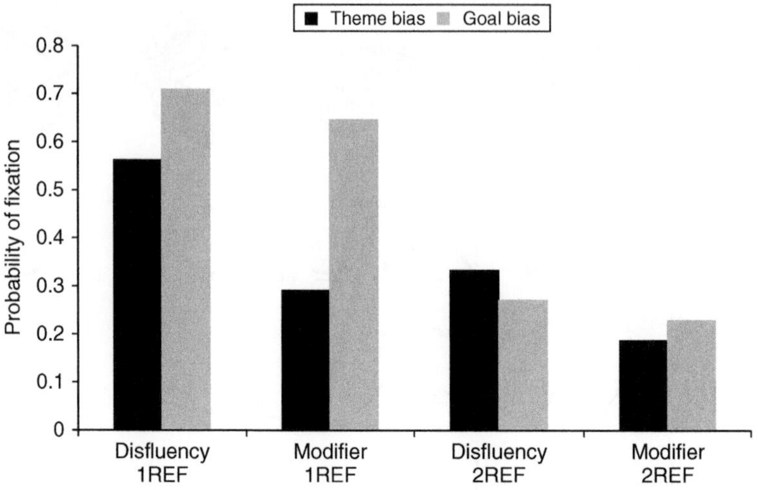

Figure 3. Probability of fixation on the modified goal (e.g. towel in a box) during the word *towel* for each of the eight utterance and display conditions. 1REF and 2REF refer to the number of possible theme referents in the display; theme bias and goal bias refer to the locations of the disfluencies; disfluency and modifier refer to cue types.

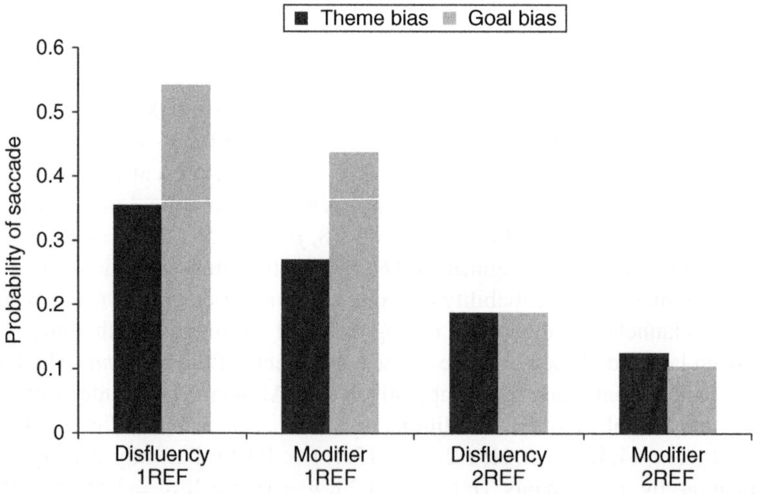

Figure 4. Probability of saccade launch to the modified goal (e.g. towel in a box) during the word *towel* for each of the eight utterance and display conditions. 1REF and 2REF refer to the number of possible theme referents in the display; theme bias and goal bias refer to the locations of the disfluencies; disfluency and modifier refer to cue types.

Disfluencies, then, do not affect the final interpretation of utterances in this study, contrary to what was suggested by previous experiments employing grammaticality judgment tasks (Bailey & Ferreira, 2003), nor do they appear to be strongly biasing the parser during the utterance. The strong form of the cueing hypothesis must therefore be rejected in favor of a hypothesis that can account for both current and previous results. A possible modification might suggest that further processing of the most salient current ambiguities (whether lexical, syntactic, referential, or discourse related) may occur during the disfluency, or that less salient ambiguities may become more salient. As a result, participants' eye movements may reflect the processing of possible resolutions of current ambiguities, and, in cases where the discourse context does not constrain the parse, the final parse may be affected (e.g. Bailey & Ferreira, 2003).

Evidence that ambiguities (not necessarily syntactic) are processed during disfluencies can be seen in the probability of saccade launch to each of the four objects during the filled pauses in the theme and goal disfluency conditions (Figure 5). The probability of launching a saccade (as opposed to probability of fixation) is sensitive to changes in visual attention (and by inference, cognitive operations) during a disfluency (Altmann & Kamide, 2004). As expected, participants are more likely to launch a saccade to the modified goal during the goal disfluency than the theme disfluency, regardless of condition, as indicated by a main effect of cue location (marginal by items; $F1_{1,15} = 5.70$, $p < 0.05$; $F2_{1,23} = 3.21$, $p = 0.086$), a nonsignificant main effect of number of referents, and a nonsignificant interaction between number of referents and cue location. This pattern is consistent with the fact that the goal disfluency occurs later in the utterance, often after the theme has been unambiguously identified.

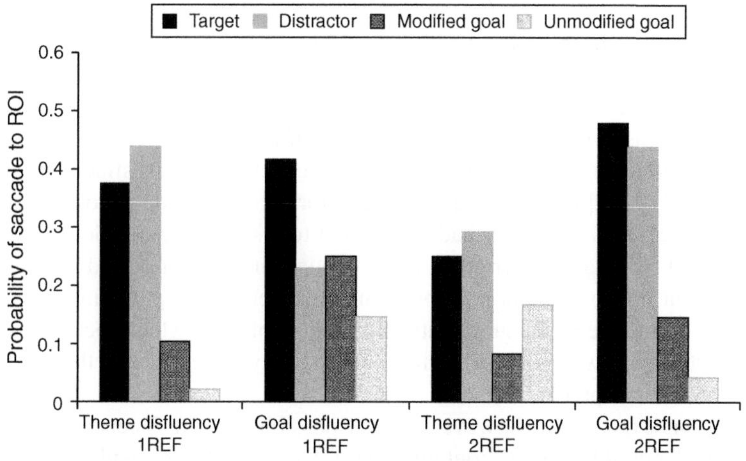

Figure 5. Probability of saccade launch to each of the four regions of interest for each of the disfluent conditions. 1REF and 2REF refer to the number of possible theme referents in the display; theme disfluency and goal disfluency refer to the locations of the disfluencies.

The pattern of results for the distractor object is more complex: Only the interaction between number of referents and cue location is significant ($F1_{1,15} = 5.31$, $p < 0.05$, $F2_{1,23} = 5.89$, $p < 0.03$).

Looks to the distractor are more likely in the theme-disfluency one-referent display condition and the goal-disfluency two-referent display condition. These results are key because they indicate why the display exerts such a powerful influence on parsing in the structurally ambiguous disfluency conditions. When a disfluency occurs later in the utterance (in the goal disfluency conditions), the distractor is still a possible candidate theme in the two-referent condition (leading to many looks to the distractor and few to either goal), as opposed to the one-referent condition (leading to fewer looks to the distractor and more looks to either goal in anticipation of the speaker identifying the goal). Moreover, because the theme disfluency occurs after the word *the*, participants could expect immediate resolution of the referential ambiguity in the one-referent condition, while immediate resolution is not expected until the goal disfluency in the two-referent condition. The interaction between number of referents and cue location thus shows that the language comprehension system is sensitive to immediately upcoming ambiguity resolution.

This ambiguity resolution hypothesis, moreover, suggests a mechanism by which we can account not only for the results in this experiment but also for previous syntactic cueing results (Bailey & Ferreira, 2003). Recall that in the current experiment, the number of possible continuations for any phrase that began with *the* was limited by co-present objects (*the* preceded the disfluency in all of the critical items in this study, as well as in the Bailey & Ferreira, 2003, study). Thus, it was easy for the listener to predict the actual object that would be referenced. However, in Bailey and Ferreira's (2003) study, no context was present. As a result, the number of possible lexical continuations at a disfluency was very large (limited only by the context introduced by the initial utterance fragment). On the other hand, the number of possible syntactic continuations was relatively small. Thus, the parser could have used the disfluent delay to consider less preferred structures, rather than possible lexical items. The grammaticality judgment task may have been sensitive to the occasions on which the parser identified the ultimately correct (less preferred) structure during the disfluency. "Bad" disfluencies may have occurred too early (i.e. [4]) to provide enough information to deduce possible structures, or so late (e.g. [2]) that the parser had committed to a single parse. "Good" disfluencies (i.e. [1] and [3]) may have occurred just late enough that less preferred structures could be identified, but not so late that multiple structures were no longer being considered. In essence, then, the nature of the grammaticality judgment task (which focuses participants on syntactic structure, with relatively little context) may have affected the way in which disfluencies were interpreted.

In the current experiment, however, it was possible for the listener to pick out the complete set of possible lexical continuations for any disfluency and interpret the disfluency as referential uncertainty. The final parse may therefore have been driven by the biases of the parser and the constraints of the display only (i.e. cue location did not lead to differences in commitments), leading to a null effect of disfluency location.

The processing done by the language comprehension system during a disfluency, then, may amount to identifying concurrent ambiguities, but the type of ambiguity that receives further processing depends on the partial parse and discourse context, and on the number of alternative continuations to be considered. Moreover, a particular type of ambiguity may be more salient than others in certain experimental settings (e.g. referential ambiguity in the VWP). This suggests that the language comprehension system uses the delay in propositional input and the distributional cues provided by disfluencies in very flexible ways that fit the comprehension goals of the listener.

Finally, these results suggest that filled pauses may provide an opportunity for studying the relative saliency of a variety of ambiguities during processing in different experimental paradigms, in that they provide a natural interruption of propositional input during which otherwise obscured ambiguity resolution processes may continue to run and thus be more easily observed. In fact, identifying the processes at work in the VWP may be especially important, as a model of cognition and language processing in this paradigm is necessary to ground and guide further study. Additional research is also needed to examine the processes that occur during filled pauses, to test the delay hypothesis described in this chapter, and to further understand the relationship between eye-movement patterns and language comprehension processes in the VWP.

References

Altmann, G. T. M., & Kamide, Y. (1999). Incremental interpretation at verbs: Restricting the domain of subsequent reference. *Cognition, 73*, 247–264.

Altmann, G. T. M., & Kamide, Y. (2004). Now you see it, now you don't: Mediating the mapping between language and the visual world. In J. M. Henderson & F. Ferreira (Eds.), *The interface of language, vision, and action: Eye movements and the visual world.* New York: Psychology Press.

Bailey, K. G. D., & Ferreira, F. (2003). Disfluencies affect the parsing of garden-path sentences. *Journal of Memory and Language, 49*, 183–200.

Bortfield, H., Leon, S., Bloom, J., Schober, M., & Brennan, S. (2001). Disfluency rates in conversation: Effects of age, relationship, topic, role, and gender. *Language and Speech, 44*, 123–147.

Brennan, S. E., & Schober, M. F. (2001). How listeners compensate for disfluencies in spontaneous speech. *Journal of Memory and Language, 44*, 274–296.

Brennan, S. E., & Williams, M. (1995). The feeling of another's knowing: Prosody and filled pauses as cues to listeners about the metacognitive states of speakers. *Journal of Memory and Language, 34*, 383–398.

Clark, H. H., & Fox Tree, J. E. (2002). Using uh and um in spontaneous speaking. *Cognition, 84*, 73–111.

Clark, H. H., & Wasow, T. (1998). Repeating words in spontaneous speech. *Cognitive Psychology, 37*, 201–242.

Cooper, R. M. (1974). The control of eye fixation by the meaning of spoken language: a new methodology for the real-time investigation of speech perception, memory, and language processing. *Cognitive Psychology, 6*(1), 84–107.

Ferreira, F., Bailey, K. G. D., & Ferraro, V. (2002). Good-enough representations in language comprehension. *Current Directions in Psychological Science, 11*, 11–15.

Ferreira, F., & Henderson, J. M. (1991) Recovery from misanalyses of garden-path sentences. *Journal of Memory and Language, 31*, 725–745.

Ford, M. (1982). Sentence planning units: Implications for the speaker's representation of meaningful relations underlying sentences. In J. Bresnan (Ed.), *The mental representation of grammatical relations* (pp. 798–827). MIT Press, Cambridge, MA.

Fox Tree, J. E. (1995) The effects of false starts and repetitions on the processing of subsequent words in spontaneous speech. *Journal of Memory and Language, 34*, 709–738.

Fox Tree, J. E. (2001). Listeners' uses of um and uh in speech comprehension. *Memory & Cognition, 29*(2), 320–326.

Fox Tree, J. E., & Schrock, J. C. (1999). Discourse markers in spontaneous speech: Oh what a difference an oh makes. *Journal of Memory and Language, 40*, 280–295.

Hawkins, P. R. (1971). The syntactic location of hesitation pauses. *Language and Speech, 14*, 277–288.

Kamide, Y., Altmann, G. T. M., & Haywood, S. L. (2003). The time-course of prediction in incremental sentence processing: Evidence from anticipatory eye movements. *Journal of Memory and Language, 49*, 113–156.

Schacter, S., Christenfeld, N., Ravina, B., & Bilous, F. (1991). Speech disfluency and the structure of knowledge. *Journal of Personality and Social Psychology, 60*, 362–367.

Sedivy, J. C., Tanenhaus, M. K., Chambers, C., & Carlson, G. N. (1999). Achieving incremental semantic interpretation through contextual representation. *Cognition, 71*, 109–147.

Shriberg, E. E. (1996). *Disfluencies in SWITCHBOARD*. Proc. International Conference on Spoken Language Processing, Addendum (pp. 11–14). Philadelphia, PA.

Smith, V. L., & Clark, H. H. (1993). On the course of answering questions. *Journal of Memory and Language, 32*, 25–38.

Spivey, M. J., Tanenhaus, M. K., Eberhard, K. M., & Sedivy, J. C. (2002) Effects of visual context in the resolution of temporary syntactic ambiguities in spoken language comprehension. *Cognitive Psychology, 45*, 447–481.

Tanenhaus, M. K., Spivey-Knowlton, M. J. Eberhard, K. M., & Sedivy, J. C. (1995). Integration of visual and linguistic information in spoken language comprehension. *Science, 268*, 1632–1634.

Trueswell, J. C., Sekerina, I., Hill, N. M., & Logrip, M. L. (1999). The kindergarten-path effect: studying on-line sentence processing in young children. *Cognition, 73*, 89–134.

Winer, B. J. (1971). *Statistical principles in experimental design* (2nd ed.). New York: McGraw-Hill.

Chapter 23

SPEECH-TO-GAZE ALIGNMENT IN ANTICIPATION ERRORS

LINDA R. WHEELDON, ANTJE S. MEYER and FEMKE VAN DER MEULEN
University of Birmingham, UK

Eye Movements: A Window on Mind and Brain
Edited by R. P. G. van Gompel, M. H. Fischer, W. S. Murray and R. L. Hill

Abstract

When speakers correctly name several objects, they typically fixate upon all objects in the order of mention and look at each object until about 150 ms before the onset of its name. The duration of gazes to objects that are to be named depends, among other things, on the ease of name retrieval (e.g., the gazes are longer when name agreement is low than when it is high, Griffin, 2001) and the ease of word form encoding (e.g., gazes are longer for objects with long names than with short names, Meyer et al., 2003). This suggests that speakers plan the object names sequentially and do not plan far ahead. We examined whether the same gaze pattern was found when speakers made anticipation errors, or whether such errors were associated with atypically short inspection times of the incorrectly named object or skipping of these objects. The gaze patterns preceding errors were indistinguishable from those preceding correct utterances. The methodological and theoretical implications of this finding are discussed.

It is generally agreed that we generate spoken utterances incrementally: We plan the first part of our utterance before we begin to speak and plan the rest while we are talking (Kempen & Hoenkamp, 1987; Levelt, 1989). Incremental language production allows speakers to use both their own time and the listener's time in the most efficient way, but it is taxing for two reasons: First, speakers must co-ordinate speech planning and speech output in time, and second, they must distribute their processing resources over several concurrent activities. How speakers meet these challenges is the main question we have been addressing in an ongoing research programme. In many of the experiments in this project, we have asked speakers to name sets of pictured objects in utterances such as "*kite doll tap sock whale globe*" (Figure 1). We recorded their eye movements and carried out detailed analyses of the relationship between the speakers' eye movements and their speech output. In these experiments we deliberately minimised grammatical encoding requirements in order to focus on one key component of speech production – lexical retrieval – and its co-ordination with visual information uptake and speech output.

These studies have generated a remarkably consistent picture of the relationship between eye gaze and speech. Speakers usually fixate upon each object they name in the order of mention. Their eyes run slightly ahead of the overt speech, with the saccade from

Figure 1. An example of a display used in Experiments 1 and 2. On the screen, each object covered 3° of visual angle in its longest dimension. The objects were arranged on a virtual oval with a width of 20° and a height of 16°.

one object to the next occurring about 150–200 ms before the onset of the first object's name. The gaze duration for an object (defined as the time between the onset of the first fixation and the offset of the last fixation on the object) depends on the time speakers require to identify the object, select its name and retrieve the corresponding morphological and phonological form (Meyer & Lethaus, 2004; Meyer, Roelofs & Levelt, 2003; Meyer, Sleiderink & Levelt, 1998; Meyer & van der Meulen, 2000; see also Griffin, 2001, 2004a; Griffin & Bock, 2000). In other words, speakers only initiate the shift of gaze to a new object after they have planned the name of the current object to the level of phonological form.

It is perhaps not too surprising to find that speakers usually look at the objects they name; in many cases this is simply necessary to identify the objects. In addition, looking at the objects in the order of mention may support the conceptual ordering and linearisation processes that must be carried out to talk about a spatial arrangement in a sequence of words (e.g., Griffin, 2004a). The more surprising observation is that the speakers' gaze durations usually exceed the time required to identify the objects. All current models of word planning assume that lexical access is based on conceptual, not visual, representations (e.g., Levelt, 1999; Levelt, Roelofs & Meyer, 1999; Rapp & Goldrick, 2000). Therefore one might expect that little would be gained from looking at the objects for longer than necessary to identify them. Different accounts for the speakers' prolonged gazes have been put forward in the literature, which may not be mutually exclusive (for further discussion see Griffin, 2004a; Meyer & Lethaus, 2004). One hypothesis is that gazes to the referent objects play a supporting role in lexical activation (Griffin, 2004a; Meyer, van der Meulen & Brooks, 2004; but see Bock, Irwin, Davidson & Levelt, 2003).

A related hypothesis we are currently exploring is that speakers fixate upon an object they are about to name until they have completed all encoding processes for that object that are not automatic, i.e., all processes that require processing capacity. A commonly held view in the speech production literature is that speakers need processing resources to plan the content of their utterances (e.g., Goldman-Eisler, 1968) but that lexical access is an automatic process (e.g., Levelt, 1989). However, recent experimental studies suggest that lexical retrieval processes might also require processing capacity. Ferreira and Pashler (2002) reported a series of dual-task experiments in which participants had to name target pictures and simultaneously categorise tones as high, medium or low in pitch. Both the picture-naming latencies and the tone-discrimination latencies were affected by the ease of semantic and morphological encoding of the picture names. This suggests that the response to the tone was only prepared after semantic and morphological encoding had been completed, and that these processes are capacity demanding. Recent dual-task experiments carried out in our own lab (Cook & Meyer, submitted) suggest that the generation of the phonological code of words also requires processing capacity. In addition, processing resources are necessary for self-monitoring processes, which compare the phonological representation of an utterance to the conceptual input (Hartsuiker & Kolk, 2001; Postma, 2000; Wheeldon & Morgan, 2002). Thus the only processes that are likely to be automatic appear to be phonetic encoding and articulatory planning processes.

The late shift of eye gaze, occurring after the completion of the phonological encoding processes, is compatible with the hypothesis that speakers focus on each object they name until they have completed the capacity-demanding processes.

If speakers planned their utterances strictly sequentially, dealing with one word at a time, as the speakers' eye movements in the multiple-object naming task suggest, they should never commit anticipatory errors, such as (1) and (2) below (from Fromkin, 1973), in which words or parts of words appear earlier than they should. However, such errors occur quite regularly, and in most error corpora stemming from healthy adult speakers by far outnumber perseverations such as (3) (from Fromkin, 1973).

(1) "a leading list" (instead of "a reading list", p. 243)

(2) "mine gold" (instead of "gold mine", p. 256)

(3) "a phonological fool" (instead of "a phonological rule", p. 244).

The only plausible way of accounting for word and sound anticipations is to assume that speakers do *not* always plan utterances strictly sequentially, but that at least occasionally several words are activated simultaneously. Current models of sentence planning see anticipatory errors as indicative of a forward looking system in which several words of an utterance are being processed in parallel, though usually with differing priorities (e.g., Dell, Burger & Svec, 1997).

There are different ways of reconciling the evidence gained from analyses of speech errors and from eye tracking studies. First, speakers may plan their utterances further ahead when they produce connected speech than when they name individual objects. In the latter case, there is little to gain from advance planning because the names of the objects can be selected independently of each other (e.g., Levelt & Meyer, 2000; Smith & Wheeldon, 1999). Though it may be the case that speakers use different planning strategies for sentences and lists of object names, anticipatory errors are not confined to sentence production, but occur as well when speakers recite lists of words (Dell, Reed, Adams & Meyer, 2000; Shattuck-Hufnagel, 1992) or, as will be shown below, name sets of objects.

Second, in a multiple-object naming task, the object a person is fixating upon may not be the only one whose name is currently activated. Speakers may, for instance, process an extrafoveal object in parallel with the foveated object, as recent results by Morgan and Meyer (2005) suggest. They asked participants to name triplets of objects. During the eye movement from the first to the second object, the object initially shown in the second position (the interloper) was replaced by a new object (the target). The relationship between interloper and target was varied. The gaze duration for the target was shorter when interloper and target were identical or had homophonous names (e.g., animal bat/baseball bat) than when they were unrelated (fork/baseball bat). This demonstrates that the participants had processed the interloper, which they only viewed extrafoveally, and built upon this information when processing the target. One account of this preview effect is that speakers inspect the objects they name sequentially but process them in parallel: A speaker looking at one object and planning its name gives processing priority to that object but simultaneously allocates some processing resources to the next object he/she will have to name. Consequently, that object can be recognised and its name can become activated prior to fixation. On this account, anticipatory errors can arise when the names of extrafoveal objects become highly activated too early and interfere with the selection of the name of the foveated object. More generally, errors can arise in a naming task because in addition to the name of the object that a speaker intends to name next, other object names can become highly activated and compete with the target name to be selected and produced (see also Morsella & Miozzo, 2002; Navarrete & Costa, 2005).

Finally, the tight co-ordination of eye gaze and speech may be a typical feature of correct utterances (which were the only utterances included in our earlier analyses), whereas a different type of co-ordination may be seen in anticipatory errors. Perhaps anticipatory errors arise because speakers sometimes fail to inspect the objects in the order in which they should mention them or because they get ahead of themselves and initiate the shift of visual attention and processing resources to a new object before the name of the present object has become sufficiently activated to be selected. In other words, errors may not always be associated with the premature automatic activation of object names, as discussed before, but may be caused by ill-timed voluntary shifts of eye gaze and attention. The goal of the present study was to discriminate between these two options by comparing the gaze patterns occurring when speakers named sets of objects correctly and when they made anticipation errors.

Our study is similar to a study by Griffin (2004b), who analysed 41 ordering and selection errors from several picture-naming experiments conducted with young and older adults. Griffin also tested the hypothesis that errors would be accompanied by atypical gaze patterns, in particular abnormally short gazes to incorrectly named objects. She found no evidence in support of this hypothesis. We used a larger error corpus than Griffin, which was more homogeneous, as all errors stemmed from two closely related experiments carried out with undergraduate students. In addition, we included only anticipation errors. In our experiments, the speakers named the same sets of objects on up to 20 trials, which allowed us to compare the gaze pattern observed on error trials to the pattern observed when the same speaker named the same set of objects correctly. This had not been possible in Griffin's study.

1. Method

Materials. The errors were collected from two experiments that required participants to name sets of six objects each presented in a circle on a computer screen, as shown in Figure 1. Forty-eight line drawings of objects had been selected from a picture gallery available at the University of Birmingham. Twenty-four objects had monosyllabic names and the others had disyllabic names, which were mono-morphemic and stressed on the first syllable. From this pool of items, we constructed four displays of six objects each with monosyllabic names and four displays of six objects each with disyllabic names. The names of the objects within a set were semantically and phonologically unrelated (see Appendix).

Apparatus. The experiment was controlled by the software package NESU provided by the Max Planck Institute for Psycholinguistics, Nijmegen. The pictures were presented on a Samtron 95 Plus 19-inch screen. Eye movements were monitored using an SMI Eyelink-1 eye tracking system. Throughout the experiment, the *x*- and *y*-coordinates of the participant's point of gaze for the right eye were estimated every 4 ms. The positions and durations of fixations were computed using software provided by SMI. Speech was recorded using a Sony ECM-MS907 microphone and a Sony TCD-D8 DAT recorder.

Participants. Twenty-four undergraduate students of the University of Birmingham participated in each experiment. They were native speakers of English and had normal or corrected-to-normal vision and received payment or course credits for participation.

Procedure. The participants were tested individually, seated in a sound-attenuated booth. At the beginning of the experiment, they received a booklet showing the experimental pictures and the expected names. They studied the pictures and their names and then completed a practice block in which they named the objects shown individually. Any naming errors were corrected by the experimenter. Then the headband of the eye tracking system was placed on the participant's head and the system was calibrated.

In Experiment 1, speakers named the objects in clockwise order, starting at the 12 o'clock position. A fixation point was presented in this position at the beginning of each trial for 700 ms. Speakers were told that on the first presentation of the display, they should aim to name the objects accurately, and on the following seven presentations they should try to name the same objects as quickly as possible. The experimenter terminated the presentation of the objects as soon as the participant had completed the name of the sixth object. The inter-trial interval was 1250 ms. Each display was presented on eight successive trials. Twelve participants named the objects with monosyllabic names, and twelve named those with disyllabic names. After all displays had been tested once, there was a short break, after which all displays were presented and named again.

In Experiment 2, each participant named each monosyllabic and each disyllabic sequence on twenty successive trials. Twelve participants did so at a fast speech rate (3.5 words/second for monosyllabic items and 2.8 words/second for disyllabic items) and twelve at a slower rate (2.3 words/second and 1.8 words/second for monosyllabic and disyllabic items respectively). This allowed us to separate the effects of speech rate and repetition of the materials. In a training block, participants were first extensively trained to produce practice sequences at these rates along with the beat of a metronome and then learned to maintain the rates after the metronome had been switched off. During the main experiment, speakers were instructed to maintain the speed they had been trained to use and they heard a tone at the moment when they should have completed the last word of a sequence. Each of the eight displays was presented on twenty successive trials. The first two trials using each display were considered practice trials and were not included in the analyses. The monosyllabic and disyllabic items were tested in separate blocks, with six participants within each speed condition beginning with the monosyllabic and six participants with the disyllabic items.

2. Results and discussion

One goal of the experiments was to study how the speakers' speech-to-gaze alignment would change when they used different speech rates and after they had extensively practised producing the object names. The relevant findings are reported in full elsewhere (Meyer, van der Meulen & Wheeldon, in prep). Very briefly, we found that the participants usually fixated upon each of the six objects. The likelihood of speakers naming objects without fixating upon them first was less than 10% and did not change much with practice or when different speech rates were used. The participants' eyes ran slightly ahead of their overt speech, i.e., they usually initiated the saccade from one object to the next shortly before they began to say the name of the first object. In Experiment 1, the participants' speech rate increased slightly over the repetitions of the materials, as one would expect. Simultaneously, the gaze durations for the objects decreased, as did the time between the end of the gaze to an object and the onset of its name (the eye–speech lag). In other words, with practice, the co-ordination of eye gaze and speech became tighter. In Experiment 2, the participants maintained a fairly constant fast or slower speech rate across the repetitions of the materials. Their gaze durations for the objects remained stable as well, but the eye–speech lag decreased slightly over the repetitions. All in all, the temporal co-ordination between the speakers' eye gaze and their speech was not greatly affected by practice or the requirement to use different speech rates.

The present study focuses on the participants' speech errors. All utterances were transcribed and the errors were coded following Dell (1986). The error rates were similar for the monosyllabic and the disyllabic sequences (7.6 and 7.3% respectively in Experiment 1 and 7.7 and 11.8% respectively in Experiment 2); therefore the data for these stimuli are collapsed in the analyses that follow.

Table 1 shows the rates of the main error types in each experiment, as well as the percentage of each error type in the total corpus. Thirty-five per cent of the errors were hesitations or

Table 1
The percentages of correct responses and of the different error types in Experiments 1 and 2 are shown along with the percentage of occurrences of each error type in the error corpus. Numbers are collapsed for monosyllabic and disyllabic sequences

	Experiment 1		Experiment 2		% of error corpus	
Response type	Frequency	%	Frequency	%	Frequency	%
Correct responses	1421	92.5	6236	90.2		
Hesitations and pauses	28	1.8	249	3.6	277	35.0
Non-contextual errors	33	2.1	149	2.2	182	23.0
Perseverations	9	0.6	9	0.1	18	2.2
Exchanges	6	0.4	18	0.3	24	3.0
Anticipations	39	2.5	208	3.0	247	31.2
Others	0	0.0	43	0.6	43	5.4
Total errors	115	7.5	676	9.8	791	

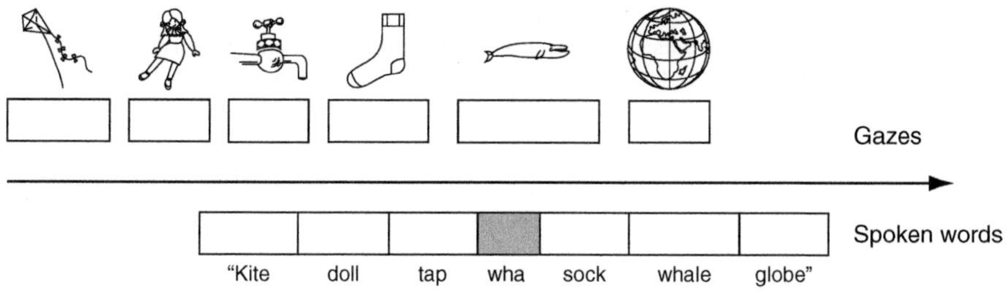

Figure 2. An example of the speech-to-gaze-alignment during the production of an anticipatory error. The order and duration of the gazes to each object are shown as well as the order and duration of the spoken object names. The speech error section is shaded grey. As can be seen, the speech error is not reflected in the gaze pattern.

pauses, and 23% were non-contextual errors, i.e. speakers produced object names that were not part of the current sequence. Among the ordering errors, anticipations were far more frequent than perseverations. This is consistent with the patterns reported in other speech error studies (e.g., Dell, Burger & Svec, 1997). In 86% of the anticipation errors, the object name that was initiated too early was the name of the next object in the sequence. A typical gaze pattern on a trial on which an anticipation error occurred is illustrated in Figure 2.

Our main question was whether the speakers' speech-to-gaze alignment in sequences including an anticipatory error would be different from the alignment in correct utterances. To answer this question, we carried out detailed analyses of a subset of anticipation errors, fulfilling the following criteria: (a) The trial including the error was preceded and followed by a trial in which the same utterance was produced correctly. This criterion was used because the speech-to-gaze alignment on error trials was to be compared to the alignment on trials in which the same objects were named correctly by the same speaker at a similar speech rate and after a similar number of repetitions of the materials. Using this criterion meant that we could not consider errors arising on the first or last repetition of a sequence. (b) The eye movements were recorded for all three trials (i.e. there were no technical errors) and the speaker fixated upon at least three of the six objects on each trial. This criterion was adopted because it makes little sense to say that an object was skipped when the other objects of the sequence were not looked at either. We call the object that was not named correctly on the error trial the target.

There were 104 triplets of trials that met both criteria. As Table 2 shows, on most of these error trials the participants uttered the onset consonant or the first few segments of a later object name and then interrupted and corrected themselves.[1] These errors accounted for 71% of the errors analysed. In almost all of the remaining errors (28% of all errors) the participants uttered the entire name of an upcoming object. Participants stopped and corrected themselves on only 30% of these whole word anticipation trials. On the remaining trials, the target name was simply omitted, i.e. speakers fluently produced five instead of six object names.

[1] The complete error corpus can be obtained from the authors.

Table 2
A break-down of the corpus of 104 anticipatory errors entered for analysis by error type and distance between the target and the anticipated word. In the examples, the anticipated speech is shown in bold and the source of the anticipation is underlined

		Distance from target		
Error type	Examples	Next word	Next word +1	Next word +2
Part of word anticipation				
Single phoneme	*Owl mask web* ***s****-corn sword brush*	16	3	0
Two phonemes	*Lamp coin rope* ***st****-bat straw pie*	33	5	1
Three or more phonemes	*Penguin* ***car****-ladder whistle carrot*	14	1	1
Whole word anticipation	***Whistle eh*** *ladder whistle carrot*	25	2	2
Phoneme anticipation	*Cor****d****-corn sword*	1	0	0

One hypothesis to be examined was that the target object was less likely to be looked at on error trials than on the preceding and following trials. This hypothesis was not borne out: There were two error trials on which the target object was not fixated upon before the corresponding utterance was initiated, compared to three preceding and five following trials on which the corresponding object was not looked at before it was named. In other words, on the majority of the trials, the speakers looked at the target object before naming it – correctly or incorrectly. Interestingly, on all 20 error trials on which the target name was omitted, all six objects were inspected in the order in which they should have been mentioned. In other words, the participants looked at the target object, as they did on correct trials, but failed to mention it.

The second hypothesis was that the target object would be inspected for a shorter time before the onset of an error than before the onset of a correct name. The mean pre-speech gaze durations (the time intervals between the onset of the gaze to the target object and the onset of the corresponding expression) are shown in Table 3. We included only those cases where the target was fixated upon on the error trial, the preceding and the following trial (94 triplets of trials). On seven trials the target was only inspected after the onset of the target name. This happened most often on the trial following the error (four cases). For the remaining trials, we computed the pre-speech gaze duration. On most of these trials, the saccade to the following object occurred before the onset of the target name. Therefore, the pre-speech gaze duration was defined as the time interval between the onset of the first fixation to the target and the end of the last fixation. On 70 trials, the saccade to the next object followed the onset of the target name. In these cases the pre-speech gaze duration was terminated by the onset of the correct or incorrect target name. Contrary to the hypothesis, the mean pre-speech gaze duration did not differ significantly between error trials (312 ms) and the preceding and following trials with (means: 321 and 312 ms respectively, $F < 1$).

Table 3

Mean gaze durations, standard deviations (in ms) and number of observations (n) are shown for the error trials and the correct trials that preceded and followed them. The total gaze duration for the error picture is broken down into gaze duration preceding speech onset and gaze duration following speech onset. The lag between the end of the last fixation on the target and the onset of its name is also shown. Note that the total speech duration is not the sum of pre- and post-speech durations as they are calculated over different subsets of errors

Dependent measure	Preceding trial			Error trial			Following trial		
	Mean	SD	(n)	Mean	SD	(n)	Mean	SD	(n)
Total gaze duration	334	139	(94)	409	237	(94)	325	136	(94)
Pre-speech gaze duration	321	132	(92)	312	151	(93)	312	127	(90)
Post-speech onset gaze duration	87	53	(14)	256	258	(35)	52	71	(21)
Eye–speech lag	104	122	(92)	−19	258	(93)	95	129	(90)

The total gaze duration, however, differed significantly between the three trial types, with longer gaze durations being observed to objects named incorrectly than to objects named correctly, $F(2,272) = 6.2$, $p < 0.01$. This result was due to differences in the participants' speech-to-gaze alignment after the onset of the referring expressions. On 83% of all trials, the saccade to the next object occurred before the onset of the target name. Therefore, the participants were normally looking at the next object when they initiated the correct or incorrect target name. However, occasionally, the order of events was reversed, and the participants were still looking at the target when they initiated its name. This was more likely to be the case on error trials (34%) than on the trials preceding and following them (17%). Considering only the trials on which participants were still looking at the target object after the onset of its name, we found that this post-speech gaze duration was longer on error trials than on correct trials, $t(68) = 4.28$, $p < 0.01$. In addition, speakers were more likely to inspect the target object again, after having fixated another object, when they had made an error than when the utterance had been correct: There were 16 regressions to the target on error trials, compared to a total of three on preceding and following trials. These differences in the gaze patterns between correct and error trials most likely arose because participants detected and corrected their errors. Griffin (2004b) reported similar findings and more extensive analyses of eye movements during error detection and repair.

3. Conclusions

Our experiments demonstrated that speakers make anticipation errors when they name sequences of objects as they do in normal conversational speech. Indeed, most of the ordering errors we observed involved the early intrusion of part of an upcoming picture name in the sequence. Since most of the errors were corrected by the speakers, we assume that they selected the correct concept, but that the *name* of the next object was activated and selected prematurely. Our error data can be explained within standard frameworks, which propose that the early activation of lexical or segmental representations leads to competition with target representations and occasionally results in the selection of incorrect words or sounds (e.g., Dell, Burger & Svec, 1997). Our aim was to determine whether this early activation was in any part due to early voluntary shifts of visual attention and eye gaze to objects whose names were produced too early. If so, we should observe atypical gaze patterns on error trials compared to correct trials. However, similar to Griffin (2004b), our data show that the gaze pattern leading up to anticipation errors is indistinguishable from the pattern leading to correct utterances. Anticipation errors were *not* accompanied by skipping of objects or by early saccades away from the target object.

Where then did the early activation of object names come from? In our experiments, the objects were sized and arranged such that a speaker fixating upon one object could easily identify the neighbouring objects. As discussed in the Introduction, it is possible that speakers allocated their attention in a distributed way across the object they were about to name and the following object (e.g., Cave & Bichot, 1999; Morgan & Meyer, 2005). Most of their attentional resources would be allocated to the object to be named first with the remainder allocated to the following object. Occasionally the representations pertaining to the next object could become highly activated and would be selected instead of those pertaining to the target. The fact that the participants named the same objects on many successive trials may have facilitated such extrafoveal processing of upcoming objects. In addition, the participants may have built up a working memory representation of the name sequences, which may also have contributed to the early activation of object names.

A methodological implication of our research and the research carried out by Griffin (2004b) is that speech errors, and, more generally, the content of a speaker's utterance cannot be predicted on the basis of the gaze pattern. Errors occurred on trials in which the gaze pattern was entirely regular. Gaze patterns therefore do not provide us with the whole picture regarding the availability of words within the speech production system. They are not a reliable predictor of utterance content.

A theoretical implication of our results is that at least in multiple-object naming tasks, anticipatory errors are not linked in any obvious way to the timing of voluntary shifts of eye gaze and the speakers' focus of attention. These errors do not occur as a result of a particular planning strategy. However, one conclusion that should *not* be drawn from these findings is that the speakers' eye movements and their speech planning are unrelated. We found here, as in earlier studies, that on most trials, the speakers looked at all objects they named, in the order in which they *should* name them. Moreover, in earlier research we showed that the gaze durations for objects that were to be named depended on the total time speakers required to identify them and to plan the utterances about them to the level of phonological form (e.g., Levelt & Meyer, 2000; Meyer & van der Meulen, 2000; Meyer, Roelofs & Levett, 2003). We have proposed that the reason why speakers look at the objects until they have completed all of these processes is that these processes require processing resources. In other words, speakers initiate a saccade to a new object as soon as they have completed all capacity-demanding processes for the present object. If this view is correct, eye monitoring can be used to track the time course of the speaker's allocation of processing resources: They demonstrate when and for how long one object is the focus of the speaker's attention and when the focus of attention is moved to a new object.

Acknowledgements

The research was supported by Economic and Social Research Council Grant R000239659 to the first and second authors.

Appendix: Materials of Experiments 1 and 2

A.1. Monosyllabic object names

lamp	coin	rope	bat	straw	pie
pin	toe	spoon	leaf	bow	rat
owl	mask	web	corn	sword	brush
kite	doll	tap	sock	whale	globe

A.2. Disyllabic object names

lemon	toilet	spider	pencil	coffin	basket
saddle	bucket	penguin	ladder	whistle	carrot
barrel	wardrobe	monkey	statue	rabbit	garlic
sausage	dragon	robot	tortoise	candle	orange

References

Bock, K., Irwin, D. E., Davidson, D. J., & Levelt, W. J. M. (2003). Minding the clock. *Journal of Memory and Language, 48*, 653–685.

Cook, A. E., & Meyer, A. S. (submitted). Automatic and capacity demanding processes in word production.

Cave, K. R., & Bichot, N. P. (1999). Visuospatial attention beyond a spotlight model. *Psychonomic Bulletin & Review, 6*, 204–223.

Dell, G. S. (1986). A spreading-activation theory of retrieval in sentence production. *Psychological Review, 93*, 283–321.

Dell, G. S., Burger, L. K., & Svec, W. R. (1997). Language production and serial order: A functional analysis and a model. *Psychological Review, 104*, 123–147.

Dell, G. S., Reed, K. D., Adams, D. R., & Meyer, A. S. (2000). Speech errors, phonotactic constraints, and implicit learning: A study of the role of experience in language production. *Journal of Experimental Psychology: Learning, Memory, and Cognition, 26*, 1355–1367.

Ferreira, V. S., & Pashler, H. (2002). Central bottleneck influences on the processing stages of word production. *Journal of Experimental Psychology: Learning, Memory, and Cognition, 28*, 1187–1199.

Fromkin, V. A. (1973). *Speech errors as linguistic evidence*. The Hague: Mouton.

Goldman-Eisler, F. (1968). *Psycholinguistics: Experiments in spontaneous speech*. New York: Academic Press.

Griffin, Z. M. (2001). Gaze durations during speech reflect word selection and phonological encoding. *Cognition, 82*, B1–B14.

Griffin, Z. M. (2004a). Why look? Reasons for eye movements related to language production. In J. M. Henderson & F. Ferreira (Eds.), *The interface of language vision and action: Eye movements and the visual world* (pp. 213–247). New York: Psychology Press.

Griffin Z. M. (2004b). The eyes are right when the mouth is wrong. *Psychological Science, 15*, 814–821.

Griffin, Z. M., & Bock, K. (2000). What the eyes says about speaking. *Psychological Science, 11*, 274–279.

Hartsuiker, R. J., & Kolk, H. H. J. (2001). Error monitoring in speech production: A computational test of the perceptual loop theory. *Cognitive Psychology, 42*, 113–157.

Kempen, G., & Hoenkamp, E. (1987). An incremental procedural grammar for sentence formulation. *Cognitive Science, 11*, 201–258.

Levelt, W. J. M. (1989). *Speaking: From intention to articulation*. Cambridge: MIT Press.

Levelt, W. J. M. (1999). Models of word production. *Trends in Cognitive Sciences, 3*, 223–232.

Levelt, W. J. M., & Meyer, A. S. (2000). Word for word: Multiple lexical access in speech production. *European Journal of Cognitive Psychology, 12*, 433–452.

Levelt, W. J. M., Roelofs, A. & Meyer, A. S. (1999). A theory of lexical access in language production. *Behavioral and Brain Sciences, 22*, 1–38.

Meyer, A. S., & Lethaus, F., (2004). The use of eye tracking in studies of sentence generation. In J. M. Henderson & F. Ferreira (Eds.), *The interface of language vision and action: Eye movements and the visual world* (pp. 191–211). New York: Psychology Press.

Meyer, A. S., Roelofs, A., & Levelt, W. J. M. (2003). Word length effects in picture naming: The role of a response criterion. *Journal of Memory and Language, 47*, 131–147.

Meyer, A. S., Sleiderink, A. M., & Levelt, W. J. M. (1998). Viewing and naming objects: Eye movements during noun phrase production. *Cognition, 66*, B25–B33.

Meyer, A. S., & van der Meulen, F. F. (2000). Phonological priming of picture viewing and picture naming. *Psychonomic Bulletin & Review, 7*, 314–319.

Meyer, A. S., van der Meulen, F. F., & Brooks, A. (2004). Eye movements during speech planning: Speaking about present and remembered objects. *Visual Cognition, 11*, 553–576.

Meyer, A. S., van der Meulen, F., & Wheeldon, L. R. (in prep). Effects of speed and practice on speech to gaze alignment.

Morgan, J. L., & Meyer, A. S. (2005). Processing of extrafoveal objects during multiple object naming. *Journal of Experimental Psychology: Learning, Memory, & Cognition, 31*, 428–442.

Morsella, E., & Miozzo, M. (2002). Evidence for a cascade model of lexical access in speech production. *Journal of Experimental Psychology: Learning, Memory, & Cognition, 28*, 555–563.

Navarrete, E., & Costa, A. (2005). Phonological activation of ignored pictures: Further evidence for a cascade model of lexical access. *Journal of Memory and Language, 53*, 359–377.

Postma, A. (2000). Detection of errors during speech production: A review of speech monitoring models. *Cognition, 77*, 97–131.

Rapp, B., & Goldrick, M. (2000). Discreteness and interactivity in spoken word production. *Psychological Review, 107*, 460–499.

Shattuck-Hufnagel, S. (1992). The role of word structure in segmental serial ordering. *Cognition, 42*, 213–259.

Smith, M., & Wheeldon, L. (1999). High level processing scope in spoken sentence production. *Cognition, 73*, 205–246.

Wheeldon, L. R., & Morgan, J. L. (2002). Phoneme monitoring in internal and external Speech. *Language and Cognitive Processes, 17*, 503–535.

Chapter 24

COMPARING THE TIME COURSE OF PROCESSING INITIALLY AMBIGUOUS AND UNAMBIGUOUS GERMAN SVO/OVS SENTENCES IN DEPICTED EVENTS

PIA KNOEFERLE

Saarland University, Germany

Eye Movements: A Window on Mind and Brain
Edited by R. P. G. van Gompel, M. H. Fischer, W. S. Murray and R. L. Hill

Abstract

The Coordinated Interplay Account by Knoeferle and Crocker (accepted) predicts that the time course with which a depicted event influences thematic role assignment depends on when that depicted event is identified as relevant by the utterance. We monitored eye movements in a scene during the comprehension of German utterances that differed with respect to when they identified a relevant depicted event for thematic role assignment and structuring of the utterance. Findings confirmed the coordinated interplay account: Gaze patterns revealed differences in the time course with which a depicted event triggered thematic role assignment depending on whether that event was identified as relevant for comprehension early or late by the utterance.

The monitoring of eye movements in scenes containing objects has revealed that the type of visual referential context rapidly influences the initial structuring and interpretation of a concurrently presented utterance (Tanenhaus, Spivey-Knowlton, Eberhard, & Sedivy, 1995). Gaze patterns have further shown that people established reference to an object more rapidly when the scene contained a contrasting object of the same kind than when it did not (Sedivy, Tanenhaus, Chambers, & Carlson, 1999). In more recent research, scenes depicted objects (e.g., a motorbike) and characters (e.g., a man). Semantic information provided by a verb (*ride*) combined with properties of that verb's thematic agent (*man*) enabled the rapid anticipation of the verb's theme (Kamide, Altmann, & Haywood, 2003).

Studies by Knoeferle, Crocker, Scheepers, and Pickering (2005) have extended this work by examining the influence of richer visual environments that contained explicitly depicted agent–action–patient events in addition to characters and objects. For initially structurally ambiguous German sentences, eye movements revealed the rapid, verb-mediated effects of the depicted events on the resolution of structural and thematic role ambiguity.

While there is a growing body of experimental research that examines the effects of increasingly rich, multi-modal settings on online utterance comprehension, findings from this research have not yet been fully and explicitly integrated into theories of online language comprehension (see Knoeferle, 2005). Existing theories of comprehension rather describe the workings of one cognitive system – the language system (e.g. Crocker, 1996; Frazier & Clifton, 1996; MacDonald, Pearlmutter, & Seidenberg, 1994; Townsend & Bever, 2001; Trueswell & Tanenhaus, 1994). Such 'non-situated' comprehension theories account for how various kinds of linguistic and world knowledge are integrated incrementally during comprehension. Knoeferle et al. (2005), however, have argued that to account for the influence of depicted events on thematic role assignment and structural disambiguation, it is necessary to adopt a framework that relates the language system to the perceptual systems, and that offers a suitably rich inventory of mental representations for objects and events in the environment.

General cognitive frameworks (e.g., Anderson et al., 2004; Cooper & Shallice, 1995; Newell, 1990), theories of the language system (e.g., Jackendoff, 2002), and approaches to embodied comprehension (e.g. Barsalou, 1999; Bergen, Chang, & Narayan, 2004; Chambers, Tanenhaus, & Magnuson, 2004; Zwaan, 2004) represent an important step in this direction. Many of these approaches provide frameworks and mental representations for the integration of information from the language comprehension and visual perception systems. Some embodied frameworks, moreover, offer an inventory of comprehension steps (e.g., the activation of a word) in situated settings (e.g., Zwaan, 2004).

These existing approaches do not yet, however, explicitly characterize the time course of situated sentence comprehension and its relation to the time course of visual processes. In brief, an integral part – a processing account – of a theory on situated sentence comprehension is still missing in current theory development.

The first step in characterizing the online interaction of language comprehension, and attention in the scene has been made by Tanenhaus et al. (1995) (see Cooper, 1974). Their research showed the close time-lock between understanding of a word and inspection of appropriate referents in a scene (see Roy & Mukherjee, 2005, for relevant modelling research). Altmann and Kamide (1999) further investigated the time lock between comprehension and attention, and demonstrated that attention to an object can even precede its mention.

What the fixation patterns in these studies do not, however, permit us to determine is whether such a close temporal coordination also holds between the point in time when a relevant object/event is identified by the utterance, and the point in time when that object/event influences comprehension. Findings by Knoeferle et al. (2005) add insights regarding this question: Importantly, gaze patterns revealed that the effects of depicted events were tightly coordinated with when the verb identified these events as relevant for comprehension.

In German, a case-marked article can determine the grammatical function and thematic role of the noun phrase it modifies. Both subject(NOM)–verb–object(ACC) (SVO) and object(ACC)–verb–subject(NOM) (OVS) ordering are possible. The studies by Knoeferle et al. (2005) investigated comprehension of initially structurally ambiguous SVO/OVS sentences when neither case-marking nor other cues in the utterance determined the correct syntactic and thematic relations prior to the sentence-final accusative (SVO) or nominative (OVS) case-marked noun phrase.

For early disambiguation, listeners had to rely on depicted event scenes that showed a princess washing a pirate, while a fencer painted that princess. Listeners heard *Die Prinzessin wäscht/malt den Pirat/der Fechter.* ('The princess (amb.) washes/paints the pirate (ACC)/the fencer (NOM)'). The verb in the utterance identified either the washing or the painting action as relevant for comprehension. Events either determined the princess as agent and the pirate as patient (washing, SVO), or the princess as patient and the fencer as agent (painting, OVS). Eye movements for the SVO compared with the OVS condition did not differ prior to the verb. After the verb had identified the relevant depicted event, and before the second noun phrase resolved the structural and thematic role ambiguity, more anticipatory inspection of the patient (the pirate) for SVO than OVS, and more eye movements to the agent (the fencer) for OVS than SVO revealed which thematic role people had assigned to the role-ambiguous character (the princess). The observed gaze patterns reflected rapid thematic role assignment and structural disambiguation through verb-mediated depicted events.

Based on these and prior findings, Knoeferle (2005) and Knoeferle and Crocker (accepted) explicitly characterized the temporal relationship between sentence comprehension and attention in a related scene as a 'Coordinated Interplay Account' (CIA) of situated utterance comprehension. The CIA identifies two fundamental steps in situated utterance comprehension. First, comprehension of the unfolding utterance guides attention, establishing reference to objects and events (Tanenhaus et al., 1995), and anticipating likely referents (see Altmann & Kamide, 1999). Once the utterance has identified the most likely object or event, and attention has shifted to it, the attended scene information then rapidly influences utterance comprehension (for further details, see Knoeferle and Crocker (accepted)).

If the coordinated interplay outline is correct, then the time course with which a depicted event affects structural disambiguation and thematic role assignment depends on when that depicted event is identified as relevant for comprehension by the utterance. For earlier identification of a relevant depicted event, we would expect to find an earlier influence of that event on comprehension than for cases when identification of a relevant event takes place comparatively later.

The present experiment directly tests the temporal-coordination prediction of the CIA. We examine thematic role assignment and the structuring of an utterance for two types of German sentences that differ with respect to when the utterance identifies relevant role relations in the scene. For initially ambiguous German SVO/OVS sentences, the first noun phrase is ambiguous regarding its grammatical function and thematic role. Scenes were role-ambiguous (i.e., the first-named character was the agent of one, and at the same time the patient of another event). Only when the verb identified one of the two events as relevant for comprehension did it become clear whether the role-ambiguous character had an agent or patient role. The earliest point at which a thematic role can be assigned to the first noun phrase in this scenario is shortly after the verb has identified a relevant depicted event and its associated role relations. We would expect gaze patterns to reveal the influence of depicted events post-verbally for ambiguous sentences if predictions of the coordinated interplay outline are valid (see Knoeferle et al., 2005).

In contrast, for unambiguous sentences, nominative and accusative case-marking on the determiner of the first noun phrase for SVO and OVS sentences respectively should permit identification of the relevant depicted role relations earlier, shortly after the first noun phrase: Object case-marking on the determiner of the first noun phrase can be combined with the event scene, showing the referent of the first noun phrase as the patient of one depicted event. For unambiguous subject-initial sentences, the subject case on the determiner of the first noun phrase can be used to identify its referent as an agent. If listeners rely on this early identification of relevant thematic relations in the scene, gaze patterns should indicate disambiguation tightly temporally coordinated one region earlier than for the ambiguous sentences (on the verb).

1. Experiment

1.1. Method

1.1.1. Participants

There were 32 participants. All were native speakers of German, and had normal or corrected-to-normal vision and hearing.

1.1.2. Materials and design

Figure 1 shows a complete image set for one item. Figures 1a and 1b were presented with initially structurally ambiguous canonical (SVO) and non-canonical (OVS) sentences (see Table 1, sentences 1a and 1b). In addition we included unambiguous sentences that were

Figure 1. Example image set for sentences in Table 1.

Table 1
Example item sentence set for the images in Figure 1

Image	Condition	Sentences
Fig. 1a	SVO ambiguous 1a	Die Frau Orange tritt in diesem Moment den Sir Zwiebel. The Ms Orange (ambiguous) kicks currently the Sir Zwiebel (object). 'Ms Orange kicks currently Sir Onion.'
Fig. 1a	OVS ambiguous 1b	Die Frau Orange schlägt in diesem Moment der Sir Apfel. The Ms Orange (ambiguous) hits currently the Sir Apple (subject). 'Sir Apple hits currently Ms Orange'.
Fig. 1a	SVO unambiguous 2a	Der Herr Orange tritt in diesem Moment den Sir Zwiebel. The Mr Orange (subject) kicks currently the Sir Zwiebel (object). 'Mr Orange kicks currently Sir Onion.'
Fig. 1a	OVS unambiguous 2b	Den Herrn Orange schlägt in diesem Moment der Sir Apfel. The Mr Orange (object) hits currently the Sir Apple (subject). 'Sir Apple hits currently Mr Orange.'
Fig. 1b	SVO ambiguous 1a	Die Frau Orange schlägt in diesem Moment den Sir Apfel. The Ms Orange (ambiguous) hits currently the Sir Apple (object). 'Ms Orange hits currently Sir Apple.'
Fig. 1b	OVS ambiguous 1b	Die Frau Orange tritt in diesem Moment der Sir Zwiebel. The Ms Orange (ambiguous) kicks currently the Sir Onion (subject). 'Sir Onion kicks currently Ms Orange'.
Fig. 1b	SVO unambiguous 2a	Der Herr Orange schlägt in diesem Moment den Sir Apfel. The Mr Orange (subject) hits currently the Sir Apple (object). 'Mr Orange hits currently Sir Apple.'
Fig. 1b	OVS unambiguous 2b	Den Herrn Orange tritt in diesem Moment der Sir Zwiebel. The Mr Orange (object) kicks currently the Sir Onion (subject). 'Sir Onion kicks currently Mr Orange'.

presented with the same images (2a and 2b in Table 1, Figure 1a). The unambiguous versions were created by replacing the ambiguous feminine noun phrase (*Die Frau Orange* (amb.), 'the Ms Orange') of sentences 1a and 1b with an unambiguous subject (NOM) or object (ACC) case-marked masculine noun phrase (*Der Herr Orange* (NOM), 'the Mr Orange (subj)'; *Den Herrn Orange* (ACC), 'The Mr Orange (obj)').[1]

Each image was presented in two versions for counter-balancing reasons, resulting in two images and eight sentences for an item (Table 1, Figure 1). Stereotypicality and plausibility biases were absent in our materials and could not enable disambiguation. The corresponding words for the conditions of an item were matched for length and frequency of lemmas (Baayen, Piepenbrock, & Gulikers, 1995). To exclude the influence of intonational cues on early scene-based disambiguation for ambiguous sentences (e.g., 1a and 1b), the ambiguous sentences were cross-spliced up to the second noun phrase for half of the items. To give an example: For the ambiguous OVS sentence 1b (Figure 1a), the beginning of the SVO sentence 1a (Fig. 1b) was spliced in before the second noun phrase. For ambiguous SVO sentences (e.g., 1a, Fig. 1a), the beginning of the OVS sentence 1b (Fig. 1b) was spliced in prior to the second noun phrase. Crossing ambiguity (ambiguous, unambiguous) with sentence type (SVO, OVS) created four conditions (ambiguous SVO/OVS, and unambiguous SVO/OVS).

For the initially structurally ambiguous sentences *utterance*-based ambiguity resolution could only occur through case-marking on the sentence-final noun phrase. Concurrently presented depicted events, however, showed who-does-what-to-whom, and offered role information for earlier thematic role assignment and structural disambiguation. The verb differed between the SVO condition (*tritt*, 'kicks', sentence 1a) and the OVS condition (*schlägt*, 'hits', 1b, Table 1, Fig. 1a). While the role-ambiguous character (the orange) was the agent of a kicking-event for SVO sentences (orange-kicking-onion), it was the patient of the apple-hitting event (apple-hitting-orange) in OVS sentences (Fig. 1a). If people can use the depicted events rapidly after the verb identified them as relevant, then we would expect more eye movements to the patient for SVO than OVS sentences, and more looks to the agent for OVS than SVO sentences during the post-verbal adverb (see Knoeferle et al., 2005).

For the unambiguous conditions, in contrast to the ambiguous ones, subject and object case-marking on the determiner of the first noun phrase contributed towards identifying the referent of the first noun phrase as either the agent of the kicking-event for SVO (Table 1, 2a, Fig. 1a), or as the patient of the hitting-event for OVS sentences (Table 1, 2b, Fig. 1a). While unambiguous subject case-marking on the determiner of the first noun phrase does not entirely disambiguate the thematic role of that noun phrase (sentences may be passive), this ambiguity is resolved as soon as the main verb appears in second position. As the verb unfolds, its information about relevant events can be combined with the information about thematic roles provided by case-marking. As a result, depicted

[1] We chose fruit and vegetable characters for all stimuli in the present study with the future goal of conducting the experiment with young children. Keeping scenes simple would minimize changes in the materials between the adult and child studies.

events were available earlier for thematic role assignment. If identification of depicted events and their effects upon thematic interpretation are tightly temporally coordinated, we would expect the same gaze patterns, but one region earlier than for ambiguous sentences (i.e., on the verb rather than post-verbally).

There were 24 experimental items. Each participant saw an equal number of SVO and OVS items, and an equal number of initially ambiguous and unambiguous sentences in an individually randomized list. There were 48 filler items. Eight started with an adverbial phrase and images showed two characters with only one performing an action; eight started with an unambiguously case-marked noun phrase, and images had four characters, with two performing an action; eight started with an unambiguously case-marked noun phrase and scenes showed no events; eight described an event that involved one character; sixteen started with a noun phrase coordination, and scenes showed five characters. The fillers ensured that sentences did not always begin with a noun phrase; that there was not always a verb in second position; that there was not always a depicted event. With respect to word order, 32 filler items had a subject–object constituent order and agent–patient relations; eight were dative-initial, and eight were passive sentences. One list contained 24 experimental and 48 filler items. Item trials were separated by at least one filler.

1.1.3. Procedure

An SMI EyeLink I head-mounted eye tracker monitored participants' eye movements with a sampling rate of 250 Hz. Images were presented on a 21-inch multi-scan colour monitor at a resolution of 1024 × 768 pixels concurrently with spoken sentences. Prior to the experiment, the experimenter instructed participants to try to understand both sentences and depicted scenes. There was no other task. Next, the camera was set up to track the dominant eye of participants. Calibration and validation of the eye tracker was performed manually using a nine-point fixation until both were successful. These procedures were always repeated after approximately half of the trials; if necessary they were performed more often. For each trial, the image appeared 1000 ms before utterance onset. The first experimental item for each participant was preceded by three filler items. Between the individual trials, participants fixated a centrally located fixation dot on the screen. This allowed the eye-tracking software to perform drift correction if necessary. The experiment lasted approximately 30 min.

1.1.4. Analysis

The eye-tracker software recorded the XY coordinates of participants' fixations. The coordinates were converted into distinct codes for the characters and background of each image so that participants' fixations were mapped onto the objects of an image (the background, the orange, the onion, the apple, and the distractor objects for Fig. 1a). Characters were coded depending on their event role for the analyses. For the sentences describing Fig. 1a, for instance, the orange would be coded as 'ambiguous' (acting and being acted upon), the apple as 'agent', the onion as 'patient', and the two pot plants as 'distractors'.

Consecutive fixations within one object region (i.e., before a saccade to another region occurred) were added together and counted as one *inspection*. Blinks and out-of-range fixations were added to previous fixations; contiguous fixations of less than 80 ms were pooled and incorporated into larger fixations. We report proportion of inspections and inferential analyses of the number of inspections for individual time regions (Figs 2–5). We rely on this attentional measure since previous studies have shown that it reflects online comprehension processes (e.g., Altmann & Kamide, 1999; Sedivy et al., 1999; Tanenhaus et al., 1995).

The data presented for the individual time regions (Figs 2–5) are based on exact computations of these regions for each individual trial. Word onsets and offsets were marked for the first noun phrase, the verb, the adverb, and the second noun phrase in each item speech file. We computed the proportion of cases per sentence condition for which inspections started within a time region ('NP1', 'VERB', 'ADV', 'NP2'). For the inferential analysis of inspection counts during a time region we used hierarchical log-linear models. These combine characteristics of a standard cross-tabulation chi-square test with those of ANOVA. Log-linear models neither rely upon parametric assumptions concerning the dependent variable (e.g., homogeneity of variance) nor require linear independence of factor levels, and are thus adequate for count variables (Howell, 2002). Inspection counts for a time region were subjected to analyses with the factors *target character* (patient, agent), *sentence type* (SVO, OVS), and either *participants* ($N = 32$) or *items* ($N = 24$). We report effects for the analysis with participants as $LR\chi^2(subj)$ and for the analysis including items as a factor as $LR\chi^2(item)$.

1.2. Results

We describe gaze patterns and report inferential analyses for the four analysis regions (NP1, VERB, ADV, NP2) in the ambiguous (graphs marked 'a') and unambiguous conditions (graphs marked 'b') (see Figs 2–5). For Figs 2–5, 'ambiguous' refers to the role-ambiguous character (the orange in Fig. 1a), 'patient' to the patient of the action performed by the role-ambiguous character (the onion); 'agent' refers to the other agent (the apple), and 'distr' to the distractor objects (the two pot plants) (Fig. 1a).

1.2.1. NP1 region

Figure 2 presents gaze patterns for the NP1 region. People mostly inspected the ambiguous character that is referred to by the first noun phrase. There were no differences in gaze patterns between the initially ambiguous and unambiguous conditions (see Figs 2a and b). Furthermore, there were equally many looks to the patient and agent characters for both ambiguous and unambiguous conditions.

Log-linear analyses for the ambiguous (all $LR\chi^2 < 1$, Fig. 2a) and unambiguous conditions (all $LR\chi^2(subj) < 1$, $ps > 0.2$ by items, Fig. 2b) showed no significant main

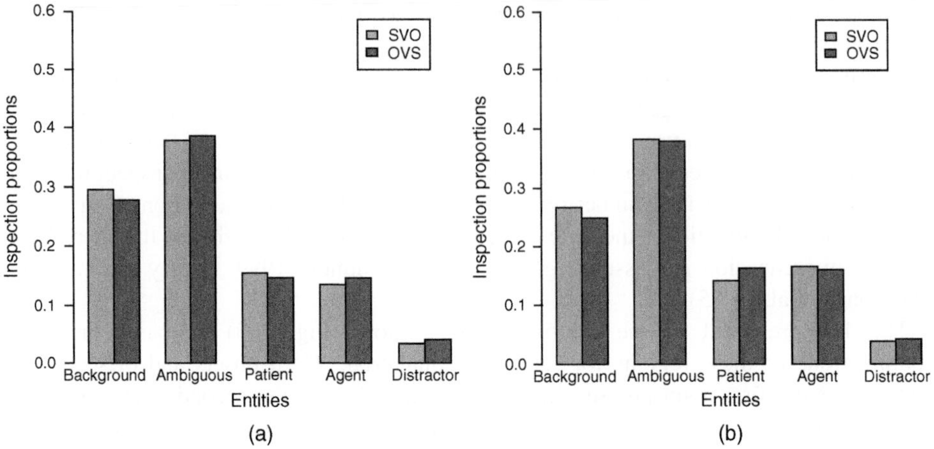

Figure 2. Proportions of inspections to characters on NP1 for the ambiguous (a) and unambiguous (b) conditions.

effects or interactions for sentence type (SVO, OVS) and target character (agent, patient) during the first noun phrase.

1.2.2. Verb region

Gaze patterns for the verb region are presented in Fig. 3. During the verb region, inspections to the first-mentioned ambiguous character (orange) have decreased. Eye-movement patterns to the two target characters (patient, agent) in the ambiguous and unambiguous conditions started to diverge. In the ambiguous conditions, gaze patterns did not yet indicate thematic role assignment and structural disambiguation. Rather, people began to direct their attention more to the patient of the action performed by the ambiguous character than to the other agent (Fig. 3a). This replicates findings from Knoeferle et al. (2005) and Knoeferle and Crocker (accepted), and could be due to visual factors (the first-named ambiguous character is oriented towards the patient), or linguistic expectation of the patient triggered by a main clause preference (e.g., Bever, 1970). The agent was inspected equally often for ambiguous SVO and OVS sentences (Fig. 3a). Log-linear analyses for the ambiguous conditions during the verb region confirmed that the main effect of target character (patient, agent) was significant ($LR\chi^2(subj) = 46.51$, df $= 1$, $p < 0.0001$; $LR\chi^2(item) = 46.51$, df $= 1$, $p < 0.0001$). There was no significant effect of sentence type ($LR\chi^2 < 1$), and no interaction of sentence type and target character ($ps > 0.1$). The difference between inspections to the background/ambiguous character for OVS compared with SVO sentences in the ambiguous conditions was not significant ($ps > 0.2$).

For the unambiguous conditions people also inspected the patient more often than the agent during the verb. At the same time, however, gaze patterns reflected early thematic role assignment and structuring of the utterance during the verb (one region earlier than

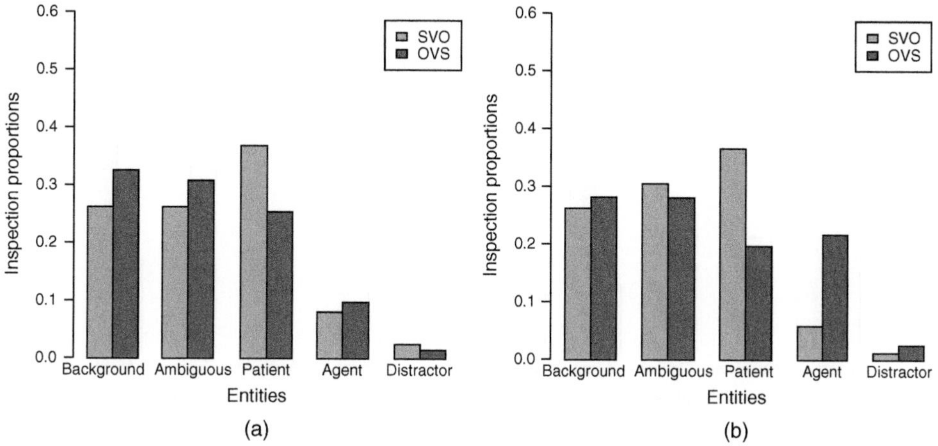

Figure 3. Proportions of inspections to characters on the verb for the ambiguous (a) and unambiguous (b) conditions.

for the initially ambiguous conditions of this study; see also Knoeferle et al. (2005)). People clearly began to inspect the patient more often for unambiguous SVO than OVS sentences, and the agent more often for unambiguous OVS than for SVO sentences during the verb (Fig. 3b). We suggest this is due to the combined use of unambiguous case-marking on the first noun phrase identifying the ambiguous character as either agent (SVO) or patient (OVS) and the relevant depicted event. As soon as case-marking on the determiner of the first noun phrase assigns an agent (SVO) or a patient role (OVS) to the ambiguous character (the orange), and once the verb identifies the relevant depicted event, people anticipate the relevant patient (SVO)/agent (OVS) associated with that event. Log-linear analyses for the verb region corroborated the observed gaze patterns by revealing a main effect of target character (agent, patient) ($LR\chi^2(subj) = 14.89$, df $= 1$, $p < 0.001$; $LR\chi^2(item) = 14.89$, df $= 1$, $p < 0.001$), and a significant interaction of sentence type (SVO, OVS) with target character (agent, patient) ($LR\chi^2(subj) = 31.30$, df $= 1$, $p < 0.0001$; $LR\chi^2(item) = 21.47$, df $= 1$, $p < 0.0001$). The three-way interaction between target character (patient, agent), sentence type (SVO, OVS), and ambiguity (ambiguous, unambiguous) was only significant by items ($LR\chi^2(item) = 8.74$, df $= 1$, $p < 0.01$) ($p > 0.4$ by participants).

1.2.3. Adverb region

Figure 4 shows the eye-movement patterns during the adverb region. In the ambiguous conditions gaze patterns now resemble those of the unambiguous conditions: People looked more often to the patient for SVO than for OVS sentences, and to the agent for OVS compared with SVO sentences, suggesting that at this point in time (post-verbally), the depicted events had enabled thematic role assignment for initially ambiguous sentences

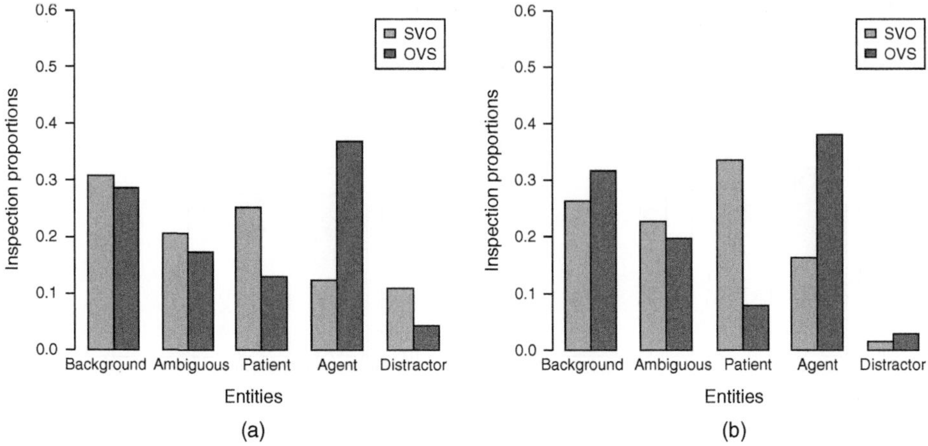

Figure 4. Proportions of inspections to characters on the adverb for the ambiguous (a) and unambiguous (b) conditions.

too (Fig. 4a). Log-linear analyses for the ambiguous conditions corroborated this view by revealing a significant interaction between sentence type (SVO, OVS) and target character (agent, patient) ($LR\chi^2(subj) = 26.85$, df $= 1$, $p < 0.0001$; $LR\chi^2(item) = 20.18$, df $= 1$, $p < 0.0001$). Contrasts confirmed more looks to the patient for SVO than OVS conditions ($LR\chi^2(subj) = 9.59$, df $= 1$, $p < 0.01$; $LR\chi^2(item) = 6.11$, df $= 1$, $p = 0.02$). People inspected the agent more often for OVS than for SVO sentences ($LR\chi^2(subj) = 29.86$, df $= 1$, $p < 0.0001$; $LR\chi^2(item) = 23.35$, df $= 1$, $p < 0.0001$). Differences in gaze proportions to the distractors for OVS compared with SVO conditions were not significant (p's > 0.5).

For the unambiguous conditions, gaze patterns continued to reflect disambiguation and thematic role assignment during the adverb region (Fig. 4b). Log-linear analyses of the adverb region for the unambiguous conditions (Fig. 4b) showed a significant interaction of sentence type (SVO, OVS) and target character (patient, agent) ($LR\chi^2(subj) = 30.67$, df $= 1$, $p < 0.0001$; $LR\chi^2(item) = 44.92$, df $= 1$, $p < 0.0001$). People inspected the patient more often for SVO compared with OVS sentences ($LR\chi^2(subj) = 32.48$, df $= 1$, $p < 0.0001$; $LR\chi^2(item) = 33.76$, df $= 1$, $p < 0.0001$). Contrasts for more looks to the agent in OVS compared with SVO sentences were not significant (p's > 0.06). Differences between OVS and SVO sentence types in inspections to the background were also not significant.

1.2.4. NP2 region

Figure 5 presents inspections to characters for the NP2 region. During the second noun phrase (Fig. 5a), the disambiguation gaze patterns continued for the ambiguous and unambiguous conditions. For the ambiguous conditions, analyses confirmed that the interaction between sentence type (SVO, OVS) and target character (patient, agent) was significant ($LR\chi^2(subj) = 36.45$, df $= 1$, $p < 0.0001$; $LR\chi^2(item) = 35.23$, df $= 1$, $p < 0.0001$).

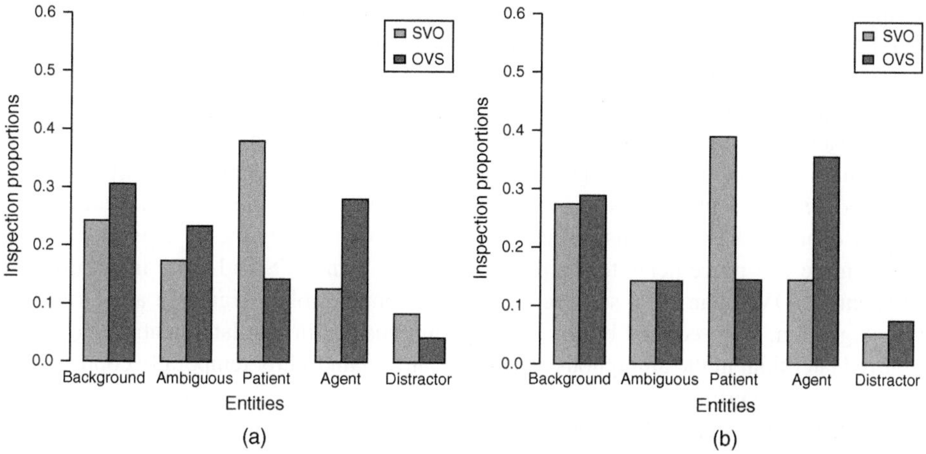

Figure 5. Proportions of inspections to characters on NP2 for the ambiguous (a) and unambiguous (b) conditions.

Contrasts confirmed a significantly higher proportion of inspections to the patient for SVO than for OVS sentences ($LR\chi^2(subj) = 37.86$, df $= 1$, $p < 0.0001$). Contrasts for looks to the agent were not significant (p's > 0.1). For the unambiguous conditions, analyses for the second noun phrase region (Fig. 5b) revealed a significant interaction of sentence type (SVO, OVS) and target character (patient, agent) ($LR\chi^2(subj) = 28.65$, df $= 1$, $p < 0.0001$; $LR\chi^2(item) = 28.82$, df $= 1$, $p < 0.0001$). People inspected the agent more often for OVS than for SVO conditions ($LR\chi^2(subj) = 17.92$, df $= 1$, $p < 0.0001$; $LR\chi^2(item) = 18.58$, df $= 1$, $p < 0.0001$), and the patient more often for SVO compared with OVS sentences ($LR\chi^2(subj) = 22.77$, df $= 1$, $p < 0.0001$; $LR\chi^2(item) = 24.51$, df $= 1$, $p < 0.0001$). Differences of OVS and SVO conditions in inspections to the background/ambiguous character were not significant ($ps > 0.2$).

2. General discussion

The key finding was that the point in time when the utterance identified a relevant depicted event was closely temporally coordinated with the point in time when that depicted event triggered thematic role assignment and structuring of the utterance. As predicted by the Coordinated Interply Account (CIA), an early mediation of the relevant depicted event through unambiguous case-marking on the determiner of the first noun phrase triggered an earlier thematic role assignment than a later, verb-based mediation of the relevant depicted event in initially ambiguous sentences.

For the unambiguous conditions, case-marking on the sentence-initial noun phrase allowed listeners to identify the noun-phrase referent as either the agent of a depicted kicking action, or as the patient of a hitting action performed by another agent (Fig. 1a).

During the ensuing verb, people's anticipatory eye movements to the patient and agent of the relevant depicted event revealed thematic role assignment for SVO and OVS sentences respectively. As predicted by the CIA, these gaze patterns were observed shortly after the first noun phrase, and thus tightly temporally coordinated with the case-marked mediation of a relevant agent–action–patient or a patient–action–agent event.

For initially structurally ambiguous sentences, the relevant depicted event was only identified *after* the sentence-initial ambiguous noun phrase, once the verb had mediated the relevant depicted action and its associated role relations. During the ensuing post-verbal region, more eye movements to the patient for SVO than OVS and more inspections to the agent for OVS than SVO sentences revealed thematic role assignment and structural disambiguation. As predicted by the CIA, disambiguation in this late-identification case occurred later than in the ambiguous conditions, and again tightly temporally coordinated, shortly after the verb.

Despite the close temporal coordination, eye movements revealed thematic role assignment with a certain delay (one region after the utterance identified relevant role relations in the scene). The recording of event-related potentials, however, has revealed the effects of depicted events on structural disambiguation of initially structurally ambiguous utterances during the verb itself (one region earlier than for the ambiguous conditions of this study) (Knoeferle, Habets, Crocker, & Münte, in preparation). These findings suggest an even tighter coordination of comprehension and the influence of depicted events at the neural than attentional level.

The coordinated interplay account builds on and is compatible with prior research. Gaze patterns in studies by Tanenhaus et al. (1995) showed that utterance interpretation rapidly directs attention in the scene (see also Cooper, 1974), and that the type of visual referential context (one possible referent, two possible referents) influences the incremental structuring of the utterance. The rapidity of scene effects was confirmed by the fact that eye movements differed between the two visual context conditions from the onset of the utterance (see also Spivey, Tanenhaus, Eberhard, & Sedivy, 2002).

What the fixation patterns in the studies by Tanenhaus et al. (1995) do not, however, permit us to determine is whether scene information influenced structuring and interpretation of the utterance in a manner closely time-locked to, or independently of, when the utterance identified that scene information as relevant. On one interpretation, comprehension of the utterance (i.e., *the apple*) directed attention in the scene, and this triggered the construction and use of the appropriate referential context (one referent, two referents). A second interpretation is that people acquired the referential context temporally independently of – perhaps even prior to – hearing the utterance, and then accessed that context much as they would access a prior discourse context (e.g., Altmann & Steedman, 1988). The fact that eye-movement patterns differed from the start of the utterance between the two contexts in Tanenhaus et al. (1995) renders it impossible to decide between these two interpretations.

While not directly aimed at investigating the *utterance*-mediated influence of the scene, findings by Sedivy et al. (1999) provide a clearer picture than gaze patterns in Tanenhaus et al. of when, in relation to its identification as relevant by the utterance, scene information influences comprehension. Studies by Sedivy et al. demonstrated that the time course of establishing reference to objects in a scene depended on whether there was referential contrast between two same-category objects (two glasses) or not. In the referential contrast condition, the scene contained two tall objects (a glass and a pitcher), a small glass, and a key. In the no-referential-contrast condition, the scene contained two tall objects (a glass and a pitcher), a key, and a file folder, however, no contrasting object of the category 'glass'. Gaze patterns between the two context types did not differ while participants heard *Pick up the*. Only after people had heard *Pick up the tall*, did they look more quickly at the target referent (the tall glass) than at the other tall object (a pitcher) when the visual context displayed a contrasting object of the same category as the target (a small glass) than when it did not. These results suggest that scene information influences comprehension tightly temporally coordinated with its identification as relevant by the utterance.

The CIA is further compatible with findings by Kamide, Scheepers, and Altmann (2003). Participants in their studies inspected images showing a hare, a cabbage, a fox and a distractor object while hearing *Der Hase frisst gleich den Kohl* ('The hare (NOM/subj) eats soon the cabbage (ACC/object)') or *Den Hasen frisst gleich der Fuchs* ('The hare (ACC/object) eats soon the fox (NOM/subject)'). Anticipatory eye movements to the cabbage and fox during the *post-verbal* region indicated that the nominative (subject) and accusative (object) case-marking on the article of the first noun phrase together with world knowledge about 'what is likely to eat what' extracted at the verb allowed anticipation of the correct post-verbal referent.

Unlike the ambiguous conditions in our study, however, Kamide et al. (2003) observed no effects of unambiguous case-marking (subject, object) on the anticipation of target characters (hare, fox) for the verb region itself. We suggest this difference results from the kinds of information that scenes provided in these two studies (see Henderson & Ferreira, 2004, for discussion on different scene types). In the study by Kamide et al. (2003), scenes showed no explicitly depicted events. In such scenes, an unambiguous subject or object case-marked sentence-initial noun phrase leaves who-is-doing-what-to-whom underspecified: The hare could either be the patient of a future fox-eating action in a passive sentence or the agent of an eating-cabbage action; alternatively, no action between hare and fox might occur.

In contrast, for the present study, when people heard the unambiguously object case-marked noun phrase *Den Herrn Orange* (obj) ('The Mr Orange (obj)'), scenes proffered an explicitly depicted action (hitting) of which Mr Orange was the patient, and of which another character was the agent (Mr Apple, Fig. 1a). There is probably some underspecification in such depicted event scenes too. A depicted hitting-event could be identified as relevant by various verbs, among them *hit*, *bash*, or *beat*. What is no longer underspecified in explicitly depicted agent–action–patient events, however, is who-is-doing-what-to-whom. Object case-marking combined with a depicted hitting-action enabled people to identify the orange as the patient shortly after the first noun phrase, as revealed by anticipatory eye movements to the agent of the hitting event. A subject case-marked sentence-initial noun phrase together with an explicit agent–action–patient event enabled thematic role assignment of an agent role to the orange during the verb.

While the above discussion stresses the role of the utterance in identifying those objects/events that are most informative for comprehension, we do not exclude a more independent use of information from a scene. In increasingly complex settings, however, it will be difficult to inspect all objects in a scene. As a result, the strong guiding role that the CIA accords to the utterance may become even more important for directing attention to informative objects and events. Moreover, once the utterance has identified scene information as relevant, other objects and events that are related to that scene information can also inform comprehension.

Findings from the present experiment clearly confirmed the Coordinated Interplay Account prediction that there is a tight temporal coordination between when the utterance identifies a relevant depicted event, and when that scene event influences thematic role assignment and structuring of the utterance (see also Mayberry, Crocker, & Knoeferle, 2005, for relevant modelling research). Discussion of prior research showed that the CIA is further compatible with, and even predicts, the temporal coordination between utterance comprehension, attention in the scene, and the influence of scene information on comprehension for a range of existing studies.

Taken together the present experimental findings and theoretical discussion represent the first step towards an online processing account of situated sentence comprehension.

Acknowledgements

This research was funded by a PhD scholarship to Pia Knoeferle awarded by the German research foundation (DFG). We would like to thank Matthew W. Crocker for his continued support. Further thanks go to Nicole Kühn for help with data acquisition and analysis.

References

Altmann, G. T. M., & Kamide, Y. (1999). Incremental interpretation at verbs: restricting the domain of subsequent reference. *Cognition, 73*, 247–264.

Altmann, G. T. M., & Steedman, M. (1988). Interaction with context during human sentence processing. *Cognition, 30*, 191–238.

Anderson, J. R., Bothell, D., Byrne, M. D., Douglass, S., Lebiere, C., & Qin, Y. (2004). An integrated theory of the mind. *Psychological Review, 111*, 1036–1060.

Baayen, R. H., Piepenbrock, R., & Gulikers, L. (1995). *The celex lexical database (cd rom).* University of Pennsylvania, Philadelphia, PA: Linguistic Data Consortium.

Barsalou, L. W. (1999). Language comprehension: archival memory or prepared for situated action? *Discourse Processes, 28*, 61–80.

Bergen, B., Chang, N., & Narayan, S. (2004). Simulated action in an embodied construction grammar. In *Proceedings of the 26th Annual Conference of the Cognitive Science Society.* Mahwah, NJ: Lawrence Erlbaum.

Bever, T. (1970). The cognitive basis for linguistic structures. In J. R. Hayes (Ed.), *Cognition and the development of language* (pp. 279–362). New York: Wiley.

Chambers, C. G., Tanenhaus, M. K., & Magnuson, J. S. (2004). Actions and affordances in syntactic ambiguity resolution. *JEP: Learning, Memory, and Cognition, 30*, 687–696.

Cooper, R. (1974). The control of eye fixation by the meaning of spoken language. *Cognitive Psychology, 6*, 84–107.

Cooper, R., & Shallice, T. (1995). Soar and the case for unified theories of cognition. *Cognition, 55*, 115–145.

Crocker, M. W. (1996). *Computational psycholinguistics: An interdisciplinary approach to the study of language*. Dordrecht: Kluwer.

Frazier, L., & Clifton, C. (1996). *Construal*. Cambridge, MA: MIT Press.

Henderson, J. M., & Ferreira, F. (2004). Scene perception for psycholinguists. In J. M. Henderson & F. Ferreira (Eds.), *The interface of language, vision, and action: Eye movements and the visual world.* New York: Psychology Press.

Howell, D. C. (2002). *Statistical methods for psychology* (5th ed.). Pacific Grove: Duxbury.

Jackendoff, R. (2002). *Foundations of language: Brain, meaning, grammar, evolution*. Oxford, UK: Oxford University Press.

Kamide, Y., Altmann, G. T. M., & Haywood, S. (2003). The time course of prediction in incremental sentence processing: evidence from anticipatory eye movements. *Journal of Memory and Language, 49*, 133–156.

Kamide, Y., Scheepers, C., & Altmann, G. T. M. (2003). Integration of syntactic and semantic information in predictive processing: cross-linguistic evidence from German and English. *Journal of Psycholinguistic Research, 32*, 37–55.

Knoeferle, P. (2005). *The role of visual scenes in spoken language comprehension: Evidence from eye tracking*. Published Doctoral Dissertation, Saarland University, http://scidok.sulb.unisaarland.de/volltexte/2005/438/.

Knoeferle, P., & Crocker, M. W. (2006). The coordinated interplay of scene, utterance, and world knowledge: evidence from eye tracking. *Cognitive Science, 30*, 481–529.

Knoeferle, P., Crocker, M. W., Scheepers, C., & Pickering, M. J. (2005). The influence of the immediate visual context on incremental thematic role assignment: Evidence from eye movements in depicted events. *Cognition, 95*, 95–127.

Knoeferle, P., Habets, B., Crocker, M. W., & Muente, T. F. (submitted). Visual scenes trigger immediate syntactic reanalysis: evidence from ERPs during situated spoken comprehension.

MacDonald, M. C., Pearlmutter, N. J., & Seidenberg, M. S. (1994). The lexical nature of syntactic ambiguity resolution. *Psychological Review, 101*, 676–703.

Mayberry, M., Crocker, M. W., & Knoeferle, P. (2005). A connectionist model of sentence comprehension in visual words. *Proceedings of the 27th Annual Conference of the Cognitive Science Society* (pp. 1437–1442). Mahwah, NJ: Lawrence Erlbaum.

Newell, A. (1990). *Unified theories of cognition*. Cambridge, MA: Harvard University Press.

Roy, D., & Mukherjee, N. (2005). Towards situated speech understanding: visual context priming of language models. *Computer Speech and Language, 19*, 227–248.

Sedivy, J. C., Tanenhaus, M. K., Chambers, C. G., & Carlson, G. N. (1999). Achieving incremental semantic interpretation through contextual representation. *Cognition, 71*, 109–148.

Spivey, M. J., Tanenhaus, M. K., Eberhard, K. M., & Sedivy, J. C. (2002). Eye movements and spoken language comprehension: effects of visual context on syntactic ambiguity resolution. *Cognitive Psychology, 45*, 447–481.

Tanenhaus, M. K., Spivey-Knowlton, M. J., Eberhard, K., & Sedivy, J. C. (1995). Integration of visual and linguistic information in spoken language comprehension. *Science, 268*, 632–634.

Townsend, D. J., & Bever, T. G. (2001). *Sentence comprehension: the integration of habits and rules*. Cambridge, MA: MIT Press.

Trueswell, J. C., & Tanenhaus, M. K. (1994). Toward a lexicalist framework for constraint-based syntactic ambiguity resolution. In C. Clifton, L. Frazier, & K. Rayner (Eds.), *Perspectives on sentence processing* (pp. 155–179). Hillsdale, NJ: Lawrence Erlbaum Associates.

Zwaan, R. A. (2004). The immersed experiencer: Towards an embodied theory of language comprehension. In B. H. Ross (Ed.), *The psychology of learning and motivation* (Vol. 44, pp. 35–62). New York: Academic Press.

PART 7

EYE MOVEMENTS AS A METHOD FOR INVESTIGATING ATTENTION AND SCENE PERCEPTION

Edited by

MARTIN H. FISCHER

Chapter 25

VISUAL SALIENCY DOES NOT ACCOUNT FOR EYE MOVEMENTS DURING VISUAL SEARCH IN REAL-WORLD SCENES

JOHN M. HENDERSON and JAMES R. BROCKMOLE

University of Edinburgh

MONICA S. CASTELHANO

University of Massachusetts

MICHAEL MACK

Vanderbilt University

Eye Movements: A Window on Mind and Brain

Edited by R. P. G. van Gompel, M. H. Fischer, W. S. Murray and R. L. Hill

Abstract

We tested the hypothesis that fixation locations during scene viewing are primarily determined by visual salience. Eye movements were collected from participants who viewed photographs of real-world scenes during an active search task. Visual salience as determined by a popular computational model did not predict region-to-region saccades or saccade sequences any better than did a random model. Consistent with other reports in the literature, intensity, contrast, and edge density differed at fixated scene regions compared to regions that were not fixated, but these fixated regions also differ in rated semantic informativeness. Therefore, any observed correlations between fixation locations and image statistics cannot be unambiguously attributed to these image statistics. We conclude that visual saliency does not account for eye movements during active search. The existing evidence is consistent with the hypothesis that cognitive factors play the dominant role in active gaze control.

During real-world scene perception, we move our eyes about three times each second via very rapid eye movements (*saccades*) to reorient the high-resolving power of the fovea. Pattern information is acquired only during periods of relative gaze stability (*fixations*) due to a combination of central suppression and visual masking (Matin, 1974; Thiele, Henning, Buishik, & Hoffman, 2002; Volkman, 1986). *Gaze control* is the process of directing the eyes through a scene in real time in the service of ongoing perceptual, cognitive, and behavioral activity (Henderson, 2003; Henderson & Hollingworth, 1998, 1999).

There are at least three reasons that the study of gaze control is important in real-world scene perception (Henderson, 2003; Henderson & Ferreira, 2004a). First, human vision is active, in the sense that fixation is directed toward task-relevant information as it is needed for ongoing visual and cognitive computations. Although this point seems obvious to eye movement researchers, it is often overlooked in the visual perception and visual cognition literatures. For example, much of the research on real-world scene perception has used tachistoscopic display methods in which eye movements are not possible (though see Underwood, this part; Gareze & Findlay, this part). While understanding what is initially apprehended from a scene is an important theoretical topic, it is not the whole story; vision naturally unfolds over time and multiple fixations. Any complete theory of visual cognition, therefore, requires understanding how ongoing visual and cognitive processes control the direction of the eyes in real time, and how vision and cognition are affected by where the eyes are pointed at any given moment in time.

Second, eye movements provide a window into the operation of selective attention. Indeed, although internal (covert) attention and overt eye movements can be dissociated (Posner & Cohen, 1984), the strong natural relationship between covert and overt attention has recently led some investigators to suggest that studying covert visual attention independently of overt attention is misguided (Findlay, 2004; Findlay & Gilchrist, 2003). For example, as Findlay and Gilchrist (2003) have noted, much of the research in the visual search literature has proceeded as though viewers steadfastly maintain fixation during search, allocating attention only via an internal mechanism. However, visual search is virtually always accompanied by saccadic eye movements (e.g., see the chapters by Hooge, Vlaskamp, & Over, this part; Shen & Reingold, this part). In fact, studies of visual search that employ eye tracking often result in different conclusions than do studies that assume the eyes remain still. As a case in point, eye movement records reveal a much richer role for memory in the selection of information for viewing (e.g. McCarley, Wang, Kramer, Irwin, & Peterson, 2003; Peterson, Kramer, Wang, Irwin, & McCarley, 2001) than research that uses more traditional measures such as reaction time (e.g. Horowitz & Wolfe, 1998). To obtain a complete understanding of the role of memory and attention in visual cognition, it is necessary to understand eye movements.

Third, because gaze is typically directed at the current focus of analysis (see Irwin, 2004, for some caveats), eye movements provide an unobtrusive, sensitive, real-time behavioral index of ongoing visual and cognitive processing. This fact has led to enormous insights into perceptual and linguistic processing in reading (Liversedge & Findlay, 2000; Rayner, 1998; Sereno & Rayner, 2003), but eye movements are only now becoming a similarly important tool in the study of visual cognition generally and scene perception in particular.

1. Fixation placement during scene viewing

A fundamental goal in the study of gaze control during scene viewing is to understand the factors that determine where fixation will be placed. Two general hypotheses have been advanced to explain fixation locations in scenes. According to what we will call the *visual saliency hypothesis*, fixation sites are selected based on image properties generated in a bottom-up manner from the current scene. On this hypothesis, gaze control is, to a large degree, a reaction to the visual properties of the stimulus confronting the viewer. In contrast, according to what we will call the *cognitive control hypothesis*, fixation sites are selected based on the needs of the cognitive system in relation to the current task. On this hypothesis, eye movements are primarily controlled by task goals interacting with a semantic interpretation of the scene and memory for similar viewing episodes (Hayhoe & Ballard, 2005; Henderson & Ferreira, 2004a). On the cognitive control hypothesis, the visual stimulus is, of course, still relevant: The eyes are typically directed to objects and features rather than to uniform scene areas (Henderson & Hollingworth, 1999); however, the relevance of a particular object or feature in the stimulus is determined by cognitive information-gathering needs rather than inherent visual salience.

The visual saliency hypothesis has generated a good deal of interest over the past several years, and in many ways has become the dominant view in the computational vision literature. This hypothesis has received primary support from two lines of investigation. First, computational models have been developed that use known properties of the visual system to generate a *saliency map* or landscape of visual salience across an image (Itti & Koch, 2000, 2001; Koch & Ullman, 1985). In these models, the visual properties present in an image give rise to a 2D map that explicitly marks regions that are different from their surround on image dimensions such as color, intensity, contrast, and edge orientation (Itti & Koch, 2000; Koch & Ullman, 1985; Parkhurst, Law, & Niebur, 2002; Torralba, 2003), contour junctions, termination of edges, stereo disparity, and shading (Koch & Ullman, 1985), and dynamic factors such as motion (Koch & Ullman, 1985; Rosenholtz, 1999). The maps are generated for each image dimension over multiple spatial scales and are then combined to create a single saliency map. Regions that are uniform along some image dimension are considered uninformative, whereas those that differ from neighboring regions across spatial scales are taken to be potentially informative and worthy of fixation. The visual saliency map approach serves an important heuristic function in the study of gaze control because it provides an explicit model that generates precise quantitative predictions about fixation locations and their sequences, and these predictions have been found to correlate with observed human fixations under some conditions (e.g., Parkhurst et al., 2002).

Second, using a *scene statistics* approach, local scene patches surrounding fixation points have been analyzed to determine whether fixated regions differ in some image properties from regions that are not fixated. For example, high spatial frequency content and edge density have been found to be somewhat greater at fixated than non-fixated locations (Mannan, Ruddock, & Wooding, 1996, 1997b). Furthermore, local contrast (the standard deviation of intensity in a patch) is higher and two-point intensity correlation (intensity of the fixated point and nearby points) is lower for fixated scene patches than control patches (Krieger, Rentschler, Hauske, Schill, & Zetzsche, 2000; Parkhurst & Neibur, 2003; Reinagel & Zador, 1999).

Modulating the evidence supporting the visual saliency hypothesis, recent evidence suggests that fixation sites are tied less strongly to saliency when meaningful scenes are viewed during active viewing tasks (Land & Hayhoe, 2001; Turano, Geruschat, & Baker 2003). According to one hypothesis, the modulation of visual salience by knowledge-driven control may increase over time within a scene-viewing episode as more knowledge is acquired about the identities and meanings of previously fixated objects and their relationships to each other and to the scene (Henderson, Weeks, & Hollingworth, 1999). However, even the very first saccade in a scene can often take the eyes in the likely direction of a search target, whether or not the target is present, presumably because the global scene gist and spatial layout acquired from the first fixation provide important information about where a particular object is likely to be found (Antes, 1974; Brockmole & Henderson, 2006b; Castelhano & Henderson, 2003; Henderson et al., 1999; Mackworth & Morandi, 1967).

Henderson and Ferreira (2004a) sorted the knowledge available to the human gaze control system into several general categories. Information about a specific scene can be learned over the short term from the current perceptual encounter (*short-term episodic scene knowledge*) and over the longer term across multiple encounters (*long-term episodic scene knowledge*). Short-term knowledge underlies a viewer's tendency to refixate areas of the current scene that are semantically interesting or informative (Buswell, 1935; Henderson et al., 1999; Loftus & Mackworth, 1978; Yarbus, 1967), enables the prioritization of newly appearing or disappearing objects from a scene (Brockmole & Henderson, 2005a, 2005b), and ensures that objects are fixated when needed during motor interaction with the environment (Land & Hayhoe, 2001). Long-term episodic knowledge involves information about a particular scene acquired and retained over time. Recent evidence suggests that good memory for the visual detail of fixated regions of a viewed scene is preserved over relatively long periods of time (Castelhano & Henderson, 2005; Henderson & Hollingworth, 2003; Hollingworth, 2004; Hollingworth & Henderson, 2002; Williams, Henderson, & Zacks, 2005; for review see Henderson & Castelhano, 2005). The contextual cueing phenomenon shows that perceptual learning of complex visual images can take place relatively rapidly over multiple encounters (Chun & Jiang, 1998), and this effect has been shown to influence eye movements (Peterson & Kramer, 2001). We have recently found that this same type of learning can take place even more rapidly for real-world scenes (Brockmole & Henderson, 2006a). Furthermore, we have shown that these learned representations can facilitate eye movements during search in real-world scenes (Brockmole & Henderson, 2006b). Another interesting example of the influence of episodic scene knowledge on gaze control is the finding that viewers will often fixate an empty scene region when that region previously contained a task-relevant object (Altmann, 2004; Richardson & Spivey, 2000).

A second source of information that can guide gaze is *scene schema knowledge*, the generic semantic and spatial knowledge about a particular category of scene (Biederman, Mezzanotte, & Rabinowitz, 1982; Friedman, 1979; Mandler & Johnson, 1977). Schema knowledge includes information about the objects likely to be found in a specific type of scene (e.g., bedrooms contain beds) and spatial regularities associated with a scene category (e.g., pillows are typically found on beds), as well as generic world knowledge about scenes (e.g., beds do not float in the air). Scene identity can be apprehended and a scene schema retrieved very rapidly (Potter, 1976; Schyns & Oliva, 1994), and schema knowledge can then be used to limit initial fixations to scene areas likely to contain an object relevant to the current task (Henderson et al., 1999).

A third source of information important in gaze control is *task-related knowledge* (Buswell, 1935; Yarbus, 1967). Task-related knowledge can involve a general *gaze control policy* or strategy relevant to a given task, such as periodically fixating the reflection in the rear-view mirror while driving, and moment-to-moment control decisions based on ongoing perceptual and cognitive needs. Gaze control differs during complex and well-learned activities such as reading (Rayner, 1998), tea and sandwich making (Land & Hayhoe, 2001), and driving (Land & Lee, 1994). The distribution of fixations over a given scene changes depending on whether a viewer is searching for an object or trying to memorize that scene (Henderson et al., 1999). Gaze is also strongly influenced by moment-to-moment cognitive processes related to spoken language comprehension and production (Tanenhaus, Spivey-Knowlton, Eberhard, & Sedivy, 1995; see Henderson & Ferreira, 2004b).

2. Present study

As reviewed above, there is abundant evidence that fixation placement during scene viewing is strongly affected by cognitive factors. Most proponents of the visual saliency hypothesis acknowledge that cognitive factors play a role in gaze control, but they tend to focus on the adequacy of a saliency-based approach to account for much of the data on fixation placement (e.g., Parkhurst et al., 2002).

In the present study, we investigated further the degree to which fixation location is related to image properties during scene viewing. First, we collected eye movement data from participants who viewed full-color photographs of real-world outdoor scenes while engaged in a visual search task in which they counted the number of people who appeared in each scene. We then analyzed the fixation data in three ways to investigate the adequacy of the visual saliency hypothesis. First, we compared the fixation data against the predictions generated from an established visual saliency model. Second, we conducted an image statistics analysis to determine whether image properties differed at fixated and non-fixated locations. Third, we tested whether any observed correlations between fixation locations and image statistics might be due to the meaning of the fixated locations. Our conclusion is that the evidence supporting the visual saliency hypothesis is weak, and that the existing evidence is consistent with the hypothesis that cognitive factors play the dominant role in gaze control.

2.1. Method

The eye movements of 8 Michigan State University undergraduates were monitored as they viewed 36 full-color photographs of real-world outdoor scenes displayed on a computer monitor (see, e.g., Figure 1). The photographs were shown at a resolution of 800×600 pixels and subtended 16 deg. horizontally by 12 deg. vertically at a viewing distance of 113 cm. Eye position was sampled at a rate of 1000 Hz from a Fourward Technologies Generation 5.5 Dual Purkinje Image Eyetracker, and raw eye-tracking data were parsed into fixations and saccades using velocity and distance criteria

Figure 1. Top left: Original scene. Top middle: Model-determined salient regions in the scene. Top right: Fixation locations from all participants. Bottom: Scene with salient regions and participant fixations overlaid. Red dots show participant fixations within a salient region. Red tails mark saccade paths that originated in a non-salient region. Green dots denote participant fixations outside of the salient regions. (*See Color Plate 3.*)

(Henderson et al., 1999). The subject's head was held steady with an anchored bite-bar made of dental impression compound. Prior to the first trial, subjects completed a procedure to calibrate the output of the eyetracker against spatial position on the display screen. This procedure was repeated regularly throughout the experiment. Observers were instructed to count the number of people in each photograph. Each participant saw all 36 scenes in a different random order. Each photograph contained between 0 and 6 people and was presented until the participant responded or for 10 s. maximum. Across all search photographs, accuracy on the counting task was 82%, with greater accuracy for scenes with fewer targets present. Accuracy was below 100% because some targets were well hidden and difficult to find in the scenes.

3. Analysis 1: Comparing saliency model predictions to human fixations

A benchmark for the visual saliency hypothesis is the saliency map model of Itti and Koch (2000, 2001). This model produces explicit predictions about where viewers should fixate in complex images. The Itti and Koch model has been shown to predict human fixations reasonably well under some conditions (e.g., Parkhurst et al., 2002), though Turano et al. (2003) demonstrated that the correlations between the model and human fixations were eliminated when the viewing task was active. However, one could argue that this latter result was a consequence of the dynamic interaction between a moving viewer and the real world, a situation for which the model was not specifically developed. In Analysis 1, we examined the degree to which the Itti and Koch saliency map model is able to predict fixation locations in static scenes (the situation for which it was developed) during an active visual search task.

3.1. Do Human Fixations Fall on the Most Visually Salient Regions?

In a first analysis we compared the number of saccadic entries into, and the number of discrete fixations in, the scene regions that the saliency map model specified as most salient. For this analysis, the Itti and Koch bottom-up saliency model posted on the website [http://ilab.usc.edu/toolkit/downloads.shtml] on 16 May 2005, was used to determine the visually salient regions in each of our test scenes. The model "viewed" the scenes for 10 s each (the same amount of time given to the participants) with a foveal radius of 1°. While viewing each scene, the model generated a cumulative saliency map showing the scene regions that it found most salient over the 10 s of viewing. Any region in these cumulative saliency maps that held a value greater than zero was defined as a salient region in the eye movement analysis. To better understand the relationship between the regions the Itti and Koch model found salient and the regions participants fixated while viewing the scenes, two measures of the participants' eye movements were examined, *Salient Region Entries* and *Salient Region Fixations*. "Salient Region Entries" was defined as the proportion of all participant-generated saccades that started outside of a given salient region and landed in that region. This measure captures the degree to which the eyes tended to

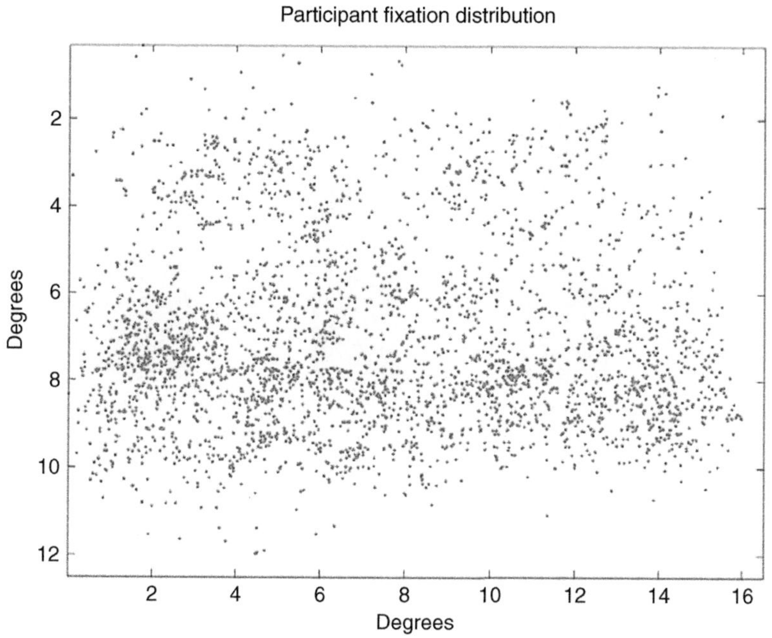

Figure 2. Spatial distribution of all participant-generated fixations across all scenes. The figure shows an overall bias for fixations to be placed along a lower central horizontal band.

saccade to salient regions. "Salient Region Fixations" was defined as the proportion of all participant-generated fixations that fell in a given salient region. This measure reflects all fixations in a salient region regardless of whether the fixation was due to a saccade from beyond that region or within that region. As control contrasts, random fixations of equal number to the participants' were generated by two random models. The first model (*pure random*) simply sampled from all possible fixation locations. To control the participants' bias to fixate in the lower central regions of the scenes (see Figure 2), a second control contrast (*biased random*) used randomly generated fixations based on the probability distribution of fixation locations from the participants' eye movement behavior across all the scenes.

Salient Region Entries. The first analysis was carried out by taking the proportion of saccades that entered a salient region (see Figure 1 for an example). All saccades from the participant trials and an equal number of saccades from the pure- and biased-random models were included in the analysis. A higher proportion of salient region entries means that a greater number of saccades were made into salient regions. If the saliency map model is able to identify regions that capture attention better than chance, the proportion of salient region entries made by participants should be higher than the proportion of salient region entries made by the random models. If the proportion of salient region entries does not differ between participants and the random models, participants are no

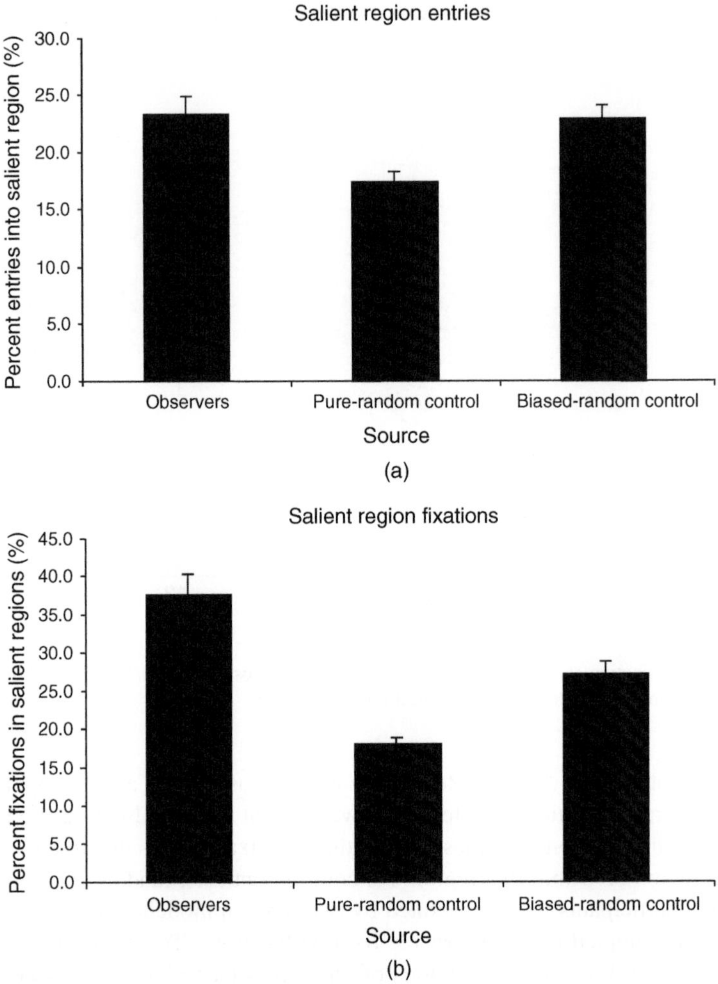

Figure 3. (a) Mean proportion (with standard error) of all participant-generated saccades that moved from outside to inside salient regions, compared with those generated by a pure random model and a biased random model. (b) Mean proportion (with standard error) of all participant-generated fixations that fell within salient regions, compared with those generated by a pure random model and a biased random model.

more likely to saccade to regions identified by the model than they are by chance. As can be seen in Figure 3a, although the saliency map model predicted entries better than did a pure random model, $t(34) = 3.64$, $p < 0.001$, it did not predict entries better than a biased-random model that took into account participants' general tendency to fixate the lower and central regions of all scenes, $t(34) < 1$. Contrary to the visual saliency hypothesis, participants' tendency to move their eyes to specific scene regions was not accounted for by the saliency model.

Salient Region Fixations. The second analysis was carried out by taking the proportion of all fixations that landed in a salient region (see Figure 1 for an example). All fixations from the participant trials and an equal number of fixations from the two random models were included in the analysis. Once again, if participants were to have a higher proportion of salient region fixations than the random model, this would suggest that the model is finding regions that capture attention better than chance. On the other hand, if the proportion of salient region fixations does not differ between participants and the random models, this would suggest that the model is finding regions that are no more likely to be fixated than by chance. Observers fixated salient regions identified by the model more often predicted by the pure-random model, $t(34) = 6.81$, $p < 0.001$, and the biased-random model, $t(34) = 4.02$, $p < 0.001$, indicating that the saliency model predicted the number of fixations in salient regions more accurately than models based on chance (see Figure 3b).

Summary. The Salient Region Entries analysis demonstrates that viewers were no more likely to saccade to a salient scene region (as identified by the saliency map model) than they were by chance. On the other hand, the Salient Region Fixations analysis shows that viewers fixated salient regions more often than would be expected by chance. Together, these data suggest that although the eyes are not specifically attracted to salient regions, they do tend to stick to them once there. The latter result might be taken as at least partial support for the saliency control hypothesis. However, because this hypothesis is supposed to account for the movement of the eyes through a scene rather than the tendency to dwell in a given region, the support is weak. Furthermore, as detailed below, the latter result is also consistent with the possibility that saliency is correlated with "object-ness", and that viewers tend to gaze at objects.

3.2. Do Human Fixations Correspond with Model-Generated Fixation Predictions?

In addition to generating a map of salient scene regions, the saliency model also produces a set of fixations. Therefore, a second way to test the ability of the model to predict human fixations is to compare directly the human- and model-generated fixation locations. We quantified the distance between these fixation locations in two ways, one based on a similarity metric devised by Mannan, Ruddock, & Wooding (1995) and a second using the one that we developed as an extension of this metric.

Mannan, Ruddock, & Wooding (1995) Similarity Metric. The fixation location similarity metric introduced by Mannan et al. (1995) compares the spatial proximity of fixations derived from two unique fixation sets (e.g. model generated and observer generated). The location similarity metric compares the linear distance from one set of fixation locations to the closest fixation in the other set, and vice versa. A high score indicates high similarity. As a control, we also computed the same similarity metric for all pairwise comparisons among participants. If the saliency map model is able to predict the locations of human fixations, its similarity to human observers should be comparable to the similarity of one human viewer to another.

The index of similarity (I_s) introduced by Mannan et al. is based on the squared distances between corresponding fixations in two gaze patterns (D_m and D_{mr}) and is defined in the following manner:

$$I_s = 100\left[1 - \frac{D_m}{D_{mr}}\right], \tag{1}$$

with

$$D_m{}^2 = \frac{n_1 \sum_{j=1}^{n_2} d_{2j}^2 + n_2 \sum_{i=1}^{n1} d_{1i}^2}{2n_1 n_2 (w^2 + h^2)}, \tag{2}$$

where n_1 and n_2 are the number of fixations in the two gaze patterns, d_{1i} is the distance between the ith fixation in the first gaze pattern and its nearest neighbor fixation in the second gaze pattern, d_{2j} is the distance between the jth fixation in the second gaze pattern and its nearest neighbor fixation in the first gaze pattern, and w and h are the width and height of the image of the scene. The calculation of D_{mr} is the same as D_m but with randomly generated gaze patterns of the same size being compared. Similar to a correlation, identical gaze patterns produce an I_s score of 100, random gaze patterns produce an I_s score of 0, and systematically different gaze patterns generate a negative score (Mannan et al., 1995). For our analysis, we examined the first seven fixations each participant produced when viewing each scene and compared them against the first seven fixations produced by the saliency model.

Figure 4a shows the mean similarity score I_s for each participant against all other participants (left bar) and all participants against the model (right bar). As can be seen in the figure, the participants' fixations were significantly less similar to those generated by the saliency model than they were to each other, $t(35) = 7.87, p < 0.001$.

A Unique Assignment Variant of the Mannan et al. (1995) Metric. A potential concern with the Mannan et al. (1995) similarity metric is that it does not take into account the overall spatial variability in the distribution of fixations over an image. For example, if all of the fixations in Set 1 are clustered in one small region of a scene, and there is at least one fixation in that same region in comparison Set 2, all the Set 1 fixations will be compared against that single Set 2 fixation. Another way to compute similarity in the same spirit as the Mannan et al. method that corrects for this issue is to require that each fixation in each set be assigned to a unique fixation in the other set. A metric can then be computed based on the distance of each point in Set 1 to its assigned point in Set 2. Intuitively, this unique-assignment metric better takes into account the overall spatial distributions of fixations. (Unlike the Mannan et al. analysis, this method requires that there be an identical number of fixations in each set.) In our unique-assignment analysis, all possible assignments of each fixation in Set 1 to a unique fixation in Set 2 were examined to find the single assignment that produced the smallest average deviation. This assignment was then used to compute the similarity metric, which is the squared deviation of each fixation point in Set 1 to its mate in Set 2.

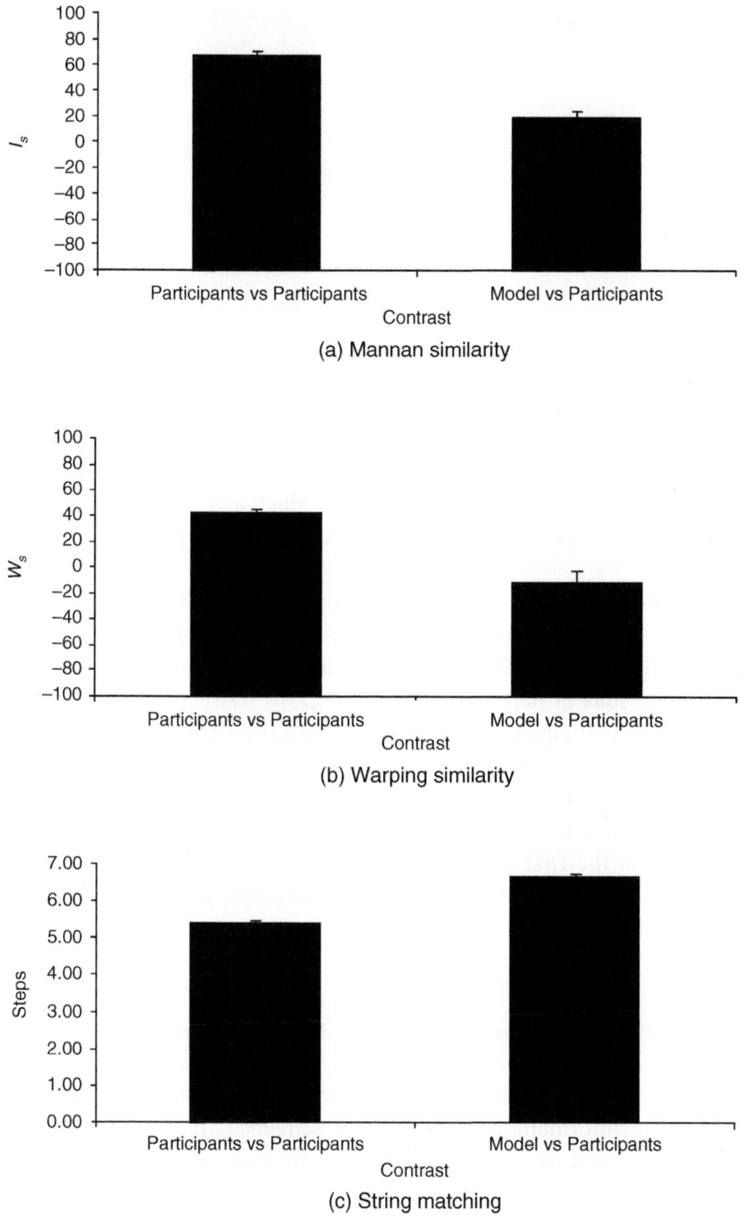

Figure 4. Similarity of participant fixation locations to model-generated locations (left bars) and to each other (right bars) for the Mannan et al. Index of Similarity (a), our Unique-Assignment "Warping" similarity index (b), and the Levenshtein Distance metric for sequence similarity (c).

More precisely, unique-assignment distance (W_s) between two gaze patterns (D_m and D_{mr}) was defined as:

$$W_s = 100\left[1 - \frac{D_w}{D_{wr}}\right], \tag{3}$$

with

$$D_w = \frac{1}{n}\sum_{j=1}^{n} p_j^2, \tag{4}$$

where n is the number of fixations in the gaze patterns, p_j is the distance between the jth unique pair of one fixation from the first set and one fixation from the second set. The calculation of D_{wr} is the same as D_w except that randomly generated gaze patterns of the same size are compared. Identical gaze patterns produce a W_s score of 100, random gaze patterns produce a W_s score of 0, and systematically different patterns generate negative scores.

Again, as a contrast, we also computed the unique-assignment similarity metric for all participants against all other participants. If the saliency model is able to predict human fixations, its similarity to human observers should be comparable to the similarity for all pairwise comparisons of participants. As above, we restricted the analysis to the first seven fixations each participant produced when viewing each scene and the first seven fixations produced by the saliency model.

Figure 4b shows the mean similarity score W_s for each participant against all other participants (left bar) and for all participants against the model (right bar). As can be seen in the figure, as with the first similarity metric, the fixations generated by the saliency model were significantly less similar to those of the participants than were those of the participants to each other, $t(35) = 5.27$, $p < 0.001$.

3.3. Are Human Fixation Sequences Predicted by Model Fixation Sequences?

Both the original Mannan et al. (1995) similarity metric and our unique-assignment variant of it ignore information about fixation sequence. In the case of the Mannan et al. (1995) metric, there is no requirement that fixations be assigned in a one-to-one correspondence across sets, and in the unique-assignment variant, the correspondence is based purely on spatial proximity and so does not take into account the temporal order in which the fixations were produced. It could be that the saliency model does a better job of predicting fixation order (scan pattern) than it does the exact locations of fixations. To investigate this possibility, we computed the Levenshtein Distance, a similarity metric specifically designed to capture sequence. The analysis uses a set of basic transformations to determine the minimum number of steps (character insertion, deletion, and substitution) that would be required to transform one character string into another. This general method is used in a variety of situations including protein sequencing in genetics (Levenshtein, 1966; Sankhoff & Kruskal, 1983). To conduct the analysis, we divided each scene into a grid of 48 regions of about 2° by 2° each. This division allowed some noise in fixation location so that minor deviations from the model would not disadvantage it. Each of the 48 regions was assigned a unique symbol. Each fixation was coded with the symbol assigned to the region in which it fell. We again analyzed the first seven fixations, so each fixation sequence produced a 7-character string. The similarity metric between two strings was the number of steps required to transform one string into another. Identical strings generated a value of 0, and the maximum value was 7. As in the first two analyses, we computed the similarity of each subject's fixation sequence for each scene to the sequence generated for that scene by the model. Again, as a control, we also computed the string metric for all participants against all other participants for each scene. If the saliency model is able to predict human fixations, its similarity to human observers should be comparable to the similarity of the human participants to each other. Figure 4c shows the mean distance score for each participant against all other participants (left bar) and for all participants against the model (right bar). The fixation sequences generated by the saliency model were significantly less similar to those of the participants than were those of the participants to each other, $t(35) = 10.2$, $p < 0.001$.

3.4. Saliency Map Model Comparison Summary

In a first set of analyses, we tested the ability of an implemented saliency map model to predict human fixation locations during an active viewing task. Overall, the results suggested that the model does not do a particularly good job. Human fixations did not land in regions of a scene that the model considered to be visually salient, and the similarity of the participants' fixations to each other was much greater than the similarity of the participants' fixations to model-generated fixations. Of course, the ability of a given model to predict human performance is a function both of those aspects of the model that are theory-inspired and other incidental modeling decisions required for the implementation. One could argue that the spirit of the model is correct, but not the implementation. Similarly, one might argue that the implementation is correct, but not the specific parameter choices. However, it is important to remember that this version of the model has been reported to predict human fixation locations reasonably well under other conditions (Parkhurst et al., 2002; Parkhurst & Neibur, 2003, 2004). The model seems to do a particularly good job with meaningless patterns (such as fractals) and in relatively unstructured viewing tasks. In this context, the present results can be taken to suggest that whereas visual salience (as instantiated in the Itti and Koch saliency map model) does a reasonable job of accounting for fixation locations under some circumstances, it does a poor job when the viewing task involves active search and the image is a real-world scene.

4. Analysis 2: Measuring local image statistics at fixated locations

Several studies have demonstrated that the image properties of fixated regions tend to differ in systematic ways from regions that are not fixated (Krieger et al., 2000; Mannan, et al., 1995, 1996; Mannan, Ruddock, & Wooding, 1997a; Parkhurst & Niebur, 2003; Reinagel & Zador, 1999). Specifically, fixated scene regions tend to be lower in intensity but higher in edge density and local contrast, and are more likely to contain third-order spatial relationships such as T-junctions and curves, than non-fixated regions. These results have been taken to suggest that such regions act as "targets" for fixations. Do these results generalize to an active visual task with real-world scenes?

4.1. Scene Statistics Method

In the present study, we measured the local image statistics associated with the fixations generated by our viewers, and compared those values to the values associated with randomly selected scene locations (see Parkhurst et al., 2002). For each scene image, ensembles of *image patches* were created. These patches had a radius of 1° of visual angle, approximating the spatial extent of foveal vision. Three different types of ensembles were created. In the *subject ensemble*, patches were defined by the subject-selected fixation positions within each image. That is, the center of each patch was defined by the (x, y) coordinates of each fixation. Thus, the subject ensemble was completely constrained by subject behavior. In the *random ensemble*, patches were centered on randomly selected positions within each scene. Thus, the patches in the random ensemble were completely unconstrained and every point in the image was equally likely to be selected. In the *shuffled ensemble*, patches were derived by "shuffling" subject-selected fixation locations from one image onto a different, randomly selected image. Like the biased random control condition in Analysis 1, this shuffled ensemble was used to account for the participants' bias to fixate more centrally in an image (see Parkhurst & Niebur, 2003).

For each ensemble, several measures of local image statistics were calculated. Analyses then focused on evaluating the similarity of the image statistics within each type of ensemble. Image statistics within the subject ensemble are characteristic of those image properties that are fixated. Since it is a random sampling, image statistics within the random ensemble are characteristic of the image properties in the scenes overall. Image statistics within the shuffled ensemble are characteristic of the image properties in those scene regions that tend to be fixated across scenes, such as the lower scene center (see Figure 2). The degree to which the scene statistics of the subject ensembles differ from the random and shuffled ensembles indicates the extent to which fixation location is correlated with particular image statistics.

Three common measures of local image statistics were examined: intensity, contrast, and edge density. These image statistics characterize different properties of image luminance. The luminance of each scene was extracted by converting the scene's RGB values to the CIE $L^* a^* b^*$ colorspace (Oliva & Schyns, 2000) which separates the luminance information of an image into a distinct dimension (L^*). The chromatic information in the a^* and b^* dimensions was discarded, and analyses focused on the values in the L^* dimension. Intensity was defined as the average luminance value of the pixels within a patch (see Mannan et al., 1995). Greater intensity is associated with higher luminance, or a higher degree of subjectively perceived brightness. Local contrast was defined as the standard deviation of luminance within a patch (see Parkhurst & Niebur, 2003; Reinagel & Zador, 1999). Local contrast, then, is a measure of how much the luminance values of pixels within a patch vary from each other. More uniform patches have less contrast. Edge density was defined as the proportion of edge pixels within an image patch. Edge pixels were found by filtering the scenes with a Sobel operator that responds to contours in scenes represented by steep gradients in luminance (see Mannan et al., 1995, 1996). Greater edge density is associated with image patches containing a greater number of contours.

Results. Representative patches from the subject, shuffled, and random ensembles are depicted in Figure 5. Quantitative analyses of the local image statistics available in patches from each ensemble are summarized in Figure 6. For all analyses, the local image

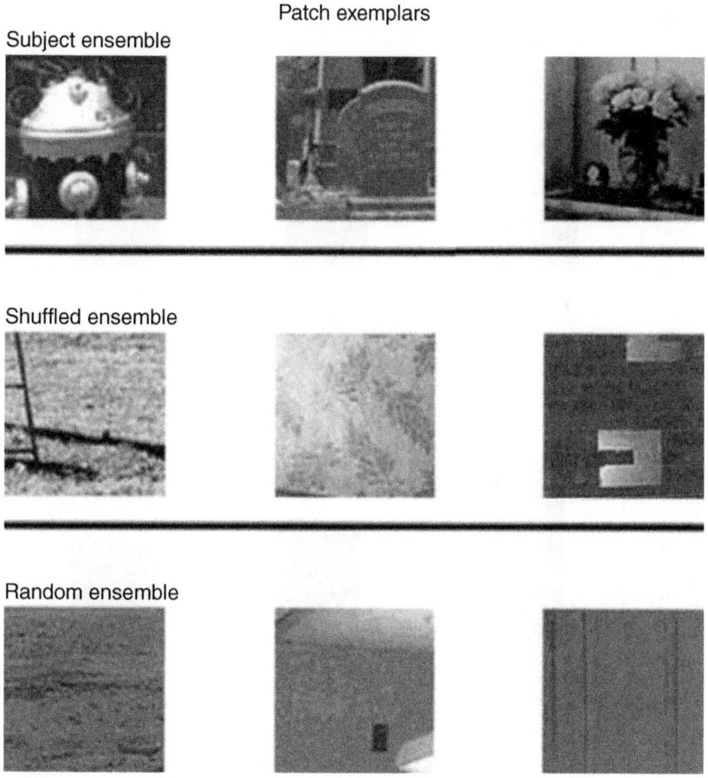

Figure 5. Representative patches from the subject ensembles (determined by participant fixation locations) and for the shuffled and random ensembles used as control conditions. Originals were presented in color.

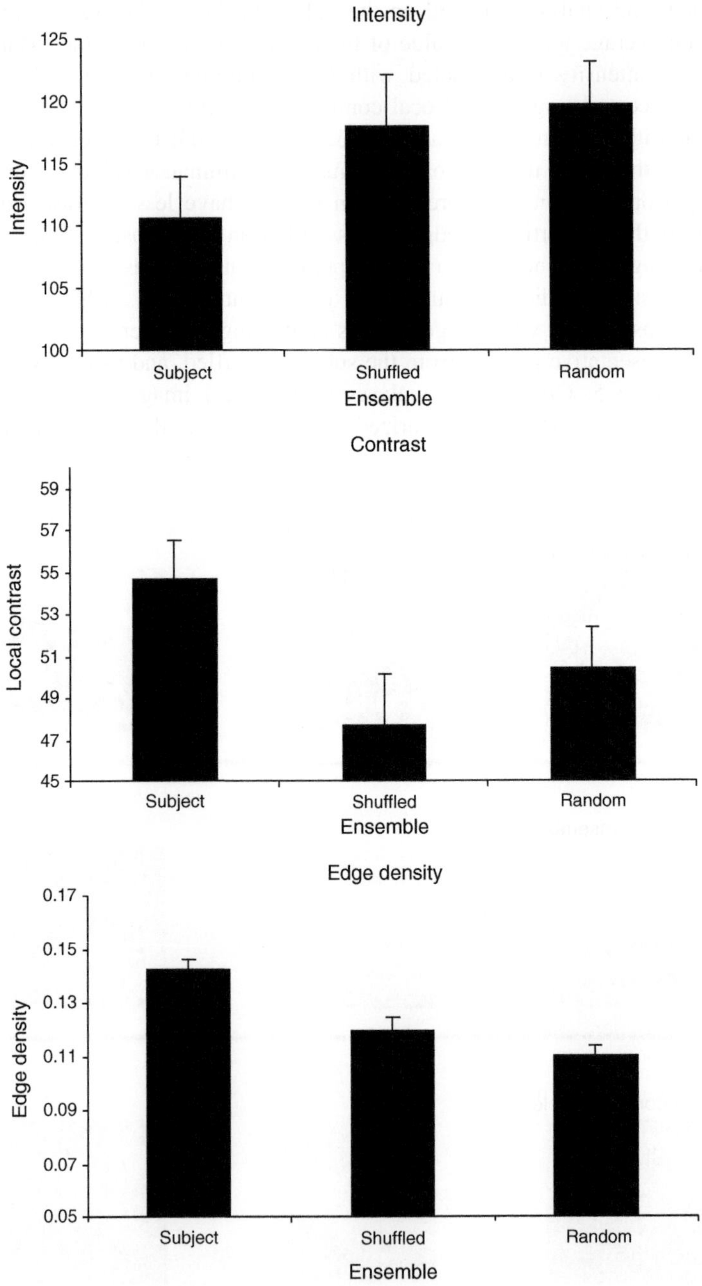

Figure 6. Mean intensity, contrast and edge density (with standard errors) for the subject, shuffled, and random ensembles.

statistics observed at fixation (the subject ensemble) were tested against the random and shuffled ensembles using one-sample t-tests.

Consistent with the prior findings in the literature cited earlier, patches derived from the subject ensembles were reliably different from those from the shuffled and random ensembles for all three local image statistics. Intensity within the subject ensemble patches was 6% lower than that within the shuffled patches ($t(284) = -3.88$, $p < 0.001$), and 8% lower than that in the random patches ($t(284) = -6.55$, $p < 0.001$). Local contrast within the subject ensemble patches was 14% higher than that within the shuffled patches ($t(284) = 6.72$, $p < 0.001$) and 9% higher than that in the random patches ($t(284) = 7.59$, $p < 0.001$). Edge density within the subject ensemble patches was 19% higher than that within the shuffled patches ($t(284) = 8.03$, $p < 0.001$), and 29% higher than that in the random patches ($t(284) = 15.5$, $p < 0.001$).

Summary. Replicating prior results, observers fixated regions that were lower in intensity and higher in local contrast and edge density than either control regions selected randomly or based on fixations from another image. On the face of it, these data could be taken to suggest that regions marked by differences in local image properties compared to the remainder of the scene act as "targets" for fixation, irrespective of the semantic nature of the information contained in those regions (Parkhurst et al., 2002; Parkhurst & Neibur, 2003). However, because these analyses only establish a correlation between fixation locations and image properties, it is also possible that the relationship is due to other factors. In the following section we explore the hypothesis that region meaning is such a factor. Specifically, we measured the semantic informativeness of the subject, shuffled, and random ensembles to determine whether meaning was also correlated with fixation location.

5. Analysis 3: Are fixated scene regions more semantically informative?

The purpose of this analysis was to determine whether fixated regions that have been shown to differ from non-fixated regions in intensity, contrast, and edge density, also differ in semantic informativeness. To investigate this question, an independent group of observers rated the degree to which patches from each ensemble were semantically informative (Antes, 1974; Mackworth & Morandi, 1967).

One hundred patches were selected from each of the subject, shuffled, and random ensembles generated from the scene statistics analysis reported above. These patches met two constraints. First, patches had to be representative of their ensemble supersets (subject, shuffled, random) in terms of local image statistics (as determined above) so that the reliable statistical differences observed would be preserved. Second, a minimum distance of 2° of visual angle was established between the center points of any two patches originating from the same scene so that selected patches could not spatially overlap. Within these constraints, patches were chosen randomly. Patches from all scenes and subjects were represented in the final subset used in Analysis 3.

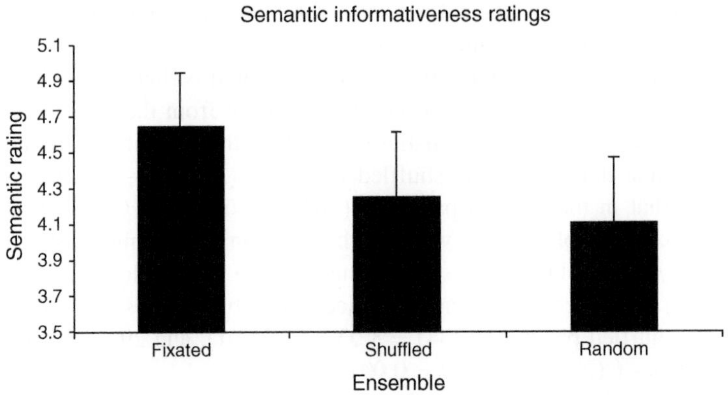

Figure 7. Mean semantic informativeness ratings (with standard errors) for the subject, shuffled, and random ensembles.

Seven Michigan State University undergraduates viewed all 300 selected patches on a computer monitor. Stimuli for presentation were created by placing each patch in the center of a uniform gray background that subtended 16° horizontally and 12° vertically. Each individual patch subtended 2° horizontally and vertically. Presentation order of patches was randomly determined. Using a 7-point Likert-type scale, observers were instructed to rate how well they thought they could determine the overall content of the scene from the small view they were shown.

Results. Mean ratings for each patch type are illustrated in Figure 7. Patches from the subject, shuffled, and random ensembles received mean ratings of 4.65, 4.25, and 4.11, respectively. A one-way repeated-measures ANOVA demonstrated a reliable effect of patch type, $F(2, 12) = 19.4. p < 0.001$, with all pairwise comparisons reliable. Critically, patches from the subject ensemble were judged to be more informative of scene identity than those from the shuffled and random ensembles. This analysis demonstrates that observers in the eyetracking experiment fixated scene regions that were more likely to provide meaningful information about the scene. These results challenge the hypothesis that local scene statistics and semantic informativeness are independent.

Summary. Image statistics of areas selected for eye fixation within scenes differ in systematic ways from areas that are not fixated. A possible interpretation of these results is that fixation position can be accounted for by low-level image statistics (Krieger et al., 2000; Mannan et al., 1995, 1996, 1997a; Parkhust & Niebur, 2003; Reinagel & Zador, 1999). The present results, however, call this interpretation into question. We conclude that examining the relationship between image statistics and fixation location without also measuring the semantic content of fixated regions can provide a partial or even misleading characterization of bottom-up influences on gaze control. Though it is possible that image properties in a scene directly influence gaze control, the results from scene statistics analyses cannot be taken as strong support for it.

6. General discussion

Gaze control during scene perception is critical for timely acquisition of task-relevant visual information. In this study, we tested the hypothesis that the selection of locations for fixation during complex scene viewing is primarily driven by visual salience derived from a bottom-up analysis of image properties. To test this visual saliency hypothesis, we collected eye movement data from participants who viewed full-color photographs of real-world scenes during an active visual search task. We then analyzed the eye movement data in a variety of ways to test the visual saliency hypothesis. In an initial set of analyses, we compared the fixation data against the predictions generated from what is arguably the standard among computational visual saliency models. We found that visual salience, as instantiated by the model, did a poor job of predicting either fixation locations or sequences. In a second set of analyses, we examined whether image properties differ at fixated and non-fixated locations. Consistent with other reports in the literature, we found clear differences in intensity, contrast, and edge density at fixated scene regions compared to regions that were not fixated. However, in a third analysis, we showed that fixated regions also differ in rated meaning compared to regions not fixated. Therefore, any observed correlations between fixation locations and image statistics could be due to the informativeness of fixated locations rather than to differences in the image statistics themselves. Our conclusion is that the evidence supporting the visual saliency hypothesis is weak, and that the existing evidence is consistent with the hypothesis that cognitive factors play the dominant role in gaze control.

6.1. Visual saliency or cognitive control?

To what extent is there strong evidence that gaze is primarily controlled by visual salience? As we have shown, the main sources of evidence, correlation of fixation positions with model-determined visual saliency, and differences in scene statistics at fixated and non-fixated locations, are both problematic. In the case of correlations with saliency model output, there is very little evidence that for active viewing tasks in the real world, existing saliency models do a good job of predicting fixation locations (Turano et al., 2003). In the present study, we showed that an existing saliency model also does a poor job of predicting fixation locations during active visual search in static images.

In the case of analyses showing that image statistics differ at fixated and non-fixated locations, our results suggest that previously reported effects may just as well be due to differences in region meaning as to differences in the image statistics themselves. This confound is probably unavoidable: meaningful objects differ from scene background in their image properties. The important conclusion is that showing differences in image properties at fixated and non-fixated regions cannot be used as unambiguous support for the hypothesis that those properties are driving eye movements. It is at least as likely that the meaning of an image region is responsible for the fact that it was fixated.

The few prior attempts to investigate a causal link between image statistics and fixation point selection have produced mixed results. Mannan and colleagues (1995, 1996) demonstrated that fixation locations in normal and low-pass filtered scenes are similar over the first 1.5 s of viewing. Because the objects in many of the low-pass filtered scenes were not identifiable, these data suggest that human gaze control does not initially select fixation sites based on object identity information. However, Einhauser and König (2003) demonstrated that perceptually detectable modifications to contrast have no effect on fixation point selection, suggesting that contrast does not contribute causally to fixation point selection, though this study has been criticized on methodological grounds (Parkhurst & Niebur, 2004). Nevertheless, Einhauser and König (2003) concluded that top-down, rather than bottom-up, factors determined attentional allocation in natural scenes.

Given that local image statistics associated with semantically informative regions such as objects undoubtedly differ systematically from those of backgrounds, and given our demonstration that such relationships exist for fixated scene regions, the results obtained by investigations linking local image statistics and gaze are entirely consistent with the conclusion that cognitive factors guide eye movements through scenes.

6.2. The special case of sudden onsets

Are there any conditions in which stimulus-based factors have priority over cognitive factors in controlling fixation location during scene viewing? We know of only one definitive case: The top-down direction of the eyes can be disrupted by the abrupt appearance of a new but task-irrelevant object, a phenomenon called *oculomotor capture* (Irwin, Colcombe, Kramer, & Hahn, 2000; Theeuwes, Kramer, Hahn, & Irwin, 1998). We have recently found that during real-world scene viewing, the transient motion signal that accompanies an abruptly appearing new object attracts attention and gaze quickly and reliably: up to 60% of fixations immediately following the onset are located on the new object (Brockmole & Henderson, 2005a, 2005b). This effect is just as strong when the new object is unexpected as when the observer's goal is to search for and identify new objects, suggesting that the allocation of attention to transient onsets is automatic. Thus, visually salient scene regions marked by low-level transient motion signals introduced by sudden changes to a scene can influence gaze in a manner divorced from cognitive control.

6.3. How should stimulus-based and knowledge-based information be combined?

The fact that gaze control draws on stored knowledge implies that image properties about potential fixation targets must somehow be combined with top-down constraints. How is this accomplished? One approach is to construct the initial stimulus-based saliency map taking relevant knowledge (e.g., visual properties of a search target) into account from the outset (Findlay & Walker, 1999; Rao, Zelinsky, Hayhoe, & Ballard, 2002).

A second approach is to combine independently computed stimulus-based and knowledge-based saliency maps so that only salient locations within knowledge-relevant regions are considered for fixation. For example, Oliva, Torralba, Castelhano, & Henderson (2003) filtered an image-based saliency map using a separate knowledge-based map of scene regions likely to contain a specific target. A yet more radical suggestion would be to move away from the concept of an image-based saliency map altogether, and to place primary emphasis on knowledge-based control. For example, in what we might call a *Full Cognitive Control* model, objects would be selected for fixation primarily based on the types of knowledge discussed in the Introduction, such as episodic and schema-based scene knowledge. The visual image in this type of model would still need to be parsed to provide potential saccade targets, but unlike the assumption of the salience hypothesis, the objects and regions would not be ranked according to their inherent visual saliency, but rather would be ranked based on criteria generated from the cognitive knowledge base. For example, if I am searching for the time of the day, and I know I have a clock on the wall in my office, I would rank non-uniform regions in the known location on the wall highly as a saccade target. In this view, visual saliency itself would play no direct role in saccade location selection. Image properties would only be directly relevant to the extent that they support processes needed to segregate potential targets from background. (And of course they would be necessary as input for determining that I'm in my office, how I'm oriented in my office, where the wall is, and so on.) We call this idea that inherent visual saliency plays no role the *Flat Landscape Assumption* and contrast it with the differentially peaked visual saliency landscapes assumed by the saliency hypothesis.

7. Conclusion

What drives eye movements through real-world scenes during active viewing tasks? Despite the recent popularity of the visual saliency hypothesis as an explanation of gaze control, the evidence supporting it is relatively weak. Cognitive factors are a critical and likely dominant determinant of fixation locations in the active viewing of scenes.

Acknowledgements

This research was supported by the National Science Foundation (BCS-0094433 and IGERT ECS-9874541) and the Army Research Office (W911NF-04-1-0078) awarded to John M. Henderson. Affiliations of co-authors are now as follows: James Brockmole, University of Edinburgh; Monica Castelhano, University of Massachusetts at Amherst; Michael Mack, Vanderbilt University. We thank John Findlay and the volume editors for comments on an earlier draft of this chapter. Please address correspondence to John M. Henderson, Psychology, 7 George Square, University of Edinburgh, EH8 9JZ, United Kingdom, or to John. M. Henderson@ed.ac.uk.

References

Altmann, G. T. M. (2004). Language-mediated eye movements in the absence of a visual world: the "blank screen paradigm". *Cognition, 93*, B79–B87.

Antes, J. R. (1974). The time course of picture viewing. *Journal of Experimental Psychology, 103*, 62–70.

Biederman, I., Mezzanotte, R. J., & Rabinowitz, J. C. (1982). Scene Perception: detecting and judging objects undergoing relational violations. *Cognitive Psychology, 14,* 143–177.

Brockmole, J. R., & Henderson, J. M (2005a). Prioritization of new objects in real-world scenes: Evidence from eye movements. *Journal of Experimental Psychology: Human Perception and Performance, 31*, 857–868.

Brockmole, J. R., & Henderson, J. M. (2005b). Object appearance, disappearance, and attention prioritization in real-world scenes. *Psychonomic Bulletin & Review, 12*, 1061–1067.

Brockmole, J. R., & Henderson, J. M. (2006a). Using real-world scenes as contextual cues for search. *Visual Cognition, 13*, 99–108.

Brockmole, J. R., & Henderson, J. M. (2006b). Recognition and attention guidance during contextual cueing in real-world scenes: Evidence from eye movements. *Quarterly Journal of Experimental Psychology, 59*, 1177–1187.

Buswell, G. T. (1935). *How people look at pictures*. Chicago: University of Chicago Press.

Castelhano, M. S., & Henderson, J. M. (2003). *Flashing scenes and moving windows: An effect of initial scene gist on eye movements.* Presented at the Annual Meeting of the Vision Sciences Society, Sarasota, Florida.

Castelhano, M. S., & Henderson, J. M., (2005). Incidental visual memory for objects in scenes. *Visual Cognition, 12*, 1017–1040.

Chun, M. M., & Jiang, Y. (1998). Contextual cueing: Implicit learning and memory of visual context guides spatial attention. *Cognitive Psychology, 36*, 28–71.

Einhauser, W., & König, P. (2003). Does luminance-contrast contribute to a saliency map for overt visual attention? *European Journal of Neuroscience, 17*, 1089–1097.

Findlay, J. M. (2004). Eye scanning and visual search. In J. M. Henderson & F. Ferreira (Eds.), *The interface of language, vision, and action: eye movements and the visual world.* New York: Psychology Press.

Findlay, J. M., & Gilchrist, I. D. (2003). *Active vision: The psychology of looking and seeing*. Oxford, England: Oxford University Press.

Findlay, J. M., & Walker, R. (1999). A model of saccadic eye movement generation based on parallel processing and competitive inhibition. *Behavioral and Brain Sciences, 22*, 661–721.

Friedman, A. (1979). Framing pictures: The role of knowledge in automatized encoding and memory for gist. *Journal of Experimental Psychology: General, 108*, 316–355.

Gareze, & Findlay (this volume).

Hayhoe, M. M., & Ballard, D. H. (2005). Eye movements in natural behavior. *Trends in Cognitive Sciences, 9*, 188–194.

Henderson, J. M. (2003). Human gaze control in real-world scene perception. *Trends in Cognitive Sciences, 7*, 498–504.

Henderson, J. M., & Castelhano, M. S. (2005). Eye movements and visual memory for scenes. In G. Underwood (Ed.), *Cognitive processes in eye guidance* (pp. 213–235). New York: Oxford University Press.

Henderson. J. M., & Ferreira, F. (2004a). Scene perception for psycholinguists. In J. M. Henderson, & F. Ferreira (Eds.), *The interface of language, vision, and action: eye movements and the visual world.* New York: Psychology Press.

Henderson. J. M., & Ferreira, F. (2004b) (Eds.), *The interface of language, vision, and action: eye movements and the visual world.* New York: Psychology Press.

Henderson, J. M., & Hollingworth, A. (1998). Eye movements during scene viewing: An overview. In G Underwood (Ed.), *Eye guidance while reading and while watching dynamic scenes.* (pp. 269–293). Oxford: Elsevier.

Henderson, J. M., & Hollingworth, A. (1999). High-level scene perception. *Annual Review of Psychology, 50*, 243–271.

Henderson, J. M., & Hollingworth, A. (2003). Eye movements and visual memory: Detecting changes to saccade targets in scenes. *Perception & Psychophysics, 65*, 58–71.

Henderson, J. M., Weeks, P. A. Jr., & Hollingworth, A. (1999). Effects of semantic consistency on eye movements during scene viewing. *Journal of Experimental Psychology: Human Perception and Performance, 25*, 210–228.

Hollingworth, A. (2004). Constructing visual representations of natural scenes: The roles of short- and long-term visual memory. *Journal of Experimental Psychology: Human Perception and Performance, 30*, 519–537.

Hollingworth, A., & Henderson, J. M. (2002). Accurate visual memory for previously attended objects in natural scenes. *Journal of Experimental Psychology: Human Perception and Performance, 28*, 113–136.

Hooge, Vlaskamp, & Over (this section).

Horowitz, T. S., & Wolfe, J. M. (1998). Visual search has no memory. *Nature, 394*, 575–577.

Irwin, 2004. Fixation Location and Fixation Duration as Indices of Cognitive Processing. In J. M. Henderson, & F. Ferreira (Eds.), *The interface of language, vision, and action: eye movements and the visual world*. New York: Psychology Press.

Irwin, D. E., Colcombe, A. M., Kramer, A. F., & Hahn, S. (2000). Attentional and oculomotor capture by onset, luminance and color singletons. *Vision Research, 40*, 1443–1458.

Itti., L., & Koch, C. (2000). A saliency-based search mechanism for overt and covert shifts of visual attention. *Vision Research, 40*, 1489–1506.

Itti, L., & Koch, C. (2001). Computational modeling of visual attention. *Nature Reviews: Neuroscience, 2*, 194–203.

Koch, C., & Ullman, S. (1985). Shifts in selective visual attention: towards the underlying neural circuitry. *Human Neurobiology, 4*, 219–227.

Krieger, G., Rentschler, I., Hauske, G., Schill, K., & Zetzsche, C. (2000). Object and scene analysis by saccadic eye-movements: an investigation with higher-order statistics. *Spatial Vision, 13*, 201–214.

Land, M. F., & Hayhoe, M. (2001). In what ways do eye movements contribute to everyday activities? *Vision Research, 41*, 3559–3565.

Land, M. F., & Lee, D. N. (1994). Where we look when we steer. *Nature, 369*, 742–744.

Levenshtein, V. (1966). Binary codes capable of correcting deletions, insertions and reversals. *Soviet Physice – Doklady, 10*, 707–710.

Liversedge, S. P., & Findlay, J. M. (2000). Saccadic eye movements and cognition. *Trends in Cognitive Sciences, 4*, 6–14.

Loftus, G. R., & Mackworth, N. H. (1978). Cognitive determinants of fixation location during picture viewing. *Journal of Experimental Psychology: Human Perception and Performance, 4*, 565–572.

Mackworth, N. H., & Morandi, A. J. (1967). The gaze selects informative details within pictures. *Perception & Psychophysics, 2*, 547–552.

Mandler, J. M., & Johnson, N. S. (1977). Some of the thousand words a picture is worth. *Journal of Experimental Psychology: Human Learning and Memory, 2*, 529–540.

Mannan, S. K., Ruddock, K. H., & Wooding, D. S. (1995). Automatic control of saccadic eye movements made in visual inspection of briefly presented 2-D images. *Spatial Vision, 9*, 363–386.

Mannan, S. K., Ruddock, K. H., & Wooding, D. S. (1996). The relationship between the locations of spatial features and those of fixations made during visual examination of briefly presented images. *Spatial Vision, 10*, 165–188.

Mannan, S. K., Ruddock, K. H., & Wooding, D. S. (1997a). Fixation sequences made during visual examination of briefly presented 2D images. *Spatial Vision, 11*, 157–178.

Mannan, S. K., Ruddock, K. H., & Wooding, D. S. (1997b). Fixation patterns made during brief examination of two-dimensional images. *Perception, 26*, 1059–1072.

Matin, E. (1974). Saccadic suppression: A review and an analysis. *Psychological Bulletin, 81*, 899–917.

McCarley, J. S., Wang, R. F., Kramer, A. F., Irwin, D. E., & Peterson, M. S. (2003). How much memory does oculomotor search have? *Psychological Science, 14*, 422–426.

Oliva A., & Schyns, P. G. (2000). Diagnostic colors mediate scene recognition.*Cognitive Psychology, 41*, 176–210.

Oliva, A., & Torralba, A., Castelhano, M. S., & Henderson, J. M. (2003). Top-down control of visual attention in object detection. *IEEE Proceedings of the International Conference on Image Processing*.

Parkhurst, D. J., & Niebur, E. (2003). Scene content selected by active vision. *Spatial Vision, 6*, 125–154.

Parkhurst, D. J., & Niebur, E. (2004). Texture contrast attracts overt visual attention in natural scenes. *European Journal of Neuroscience, 19*, 783–789.
Parkhurst, D., Law, K., & Niebur, E. (2002). Modeling the role of salience in the allocation of overt visual attention. *Vision Research, 42*, 107–123.
Peterson, M. S., & Kramer, A. F. (2001). Attentional guidance of the eyes by contextual information and abrupt onsets. *Perception & Psychophysics, 63*, 1239–1249.
Peterson, M. S., Kramer, A. F., Wang, R. F., Irwin, D. E., & McCarley, J. S. (2001). Visual search has memory. *Psychological Science, 12*, 287–292.
Posner, M. I., & Cohen Y. (1984). Components of visual orienting. In H. Bouma & D. Bouwhis (Eds.). *Attention and Performance X*. Hillsdale: Erlbaum.
Potter, M. C. (1976). Short-term conceptual memory for pictures. *Journal of Experimental Psychology: Learning Memory, and Cognition, 2*, 509–522.
Rao, R. P. N., Zelinsky, G. J., Hayhoe, M. M., & Ballard, D. H. (2002). Eye movements in iconic visual search. *Vision Research, 421*, 1447–1463.
Rayner, K. (1998). Eye movements in reading and information processing: 20 years of research. *Psychological Bulletin, 124*, 372–422.
Reinagel, P., & Zador, A. M. (1999). Natural scene statistics at the centre of gaze. *Network: Computer and Neural Systems, 10*, 1–10.
Richardson, D. C., & Spivey, M. J. (2000). Representation, space and Hollywood Squares: Looking at things that aren't there anymore. *Cognition, 76*, 269–295.
Rosenholtz, R. (1999). A simple saliency model predicts a number of motion popout phenomena. *Vision Research, 39*, 3157–3163.
Sankhoff, D., & J. B. Kruskal (Eds.), (1983). *Time warps, string edits, and macromolecules: The theory and practice of sequence comparison*. Reading, MA: Addison-Wesley Publishing Company, Inc.
Schyns, P., & Oliva, A. (1994). From blobs to boundary edges: Evidence for time- and spatial-scale-dependent scene recognition. *Psychological Science, 5*, 195–200.
Sereno, S. C., & Rayner, K. (2003). Measuring word recognition in reading: eye movements and event-related potentials. *Trends in Cognitive Sciences, 7*, 489–493.
Shen, & Reingold (this section).
Tanenhaus, M. K., Spivey-Knowlton, M. J., Eberhard, K. M., & Sedivy, J. E. (1995). Integration of visual and linguistic information in spoken language comprehension. *Science, 268*, 632–634.
Theeurves, J., Kramer, A. F., Hahn, S., & Irwin, D. E. (1998). Our eyes do not always go where we want them to go: Capture of the eyes by new objects. *Psychological Science, 9*, 379–385.
Thiele, A., Henning, M., Buischik, K., & Hoffman, P. (2002). Neural mechanisms of saccadic suppression. *Science, 295*, 2460–2462.
Torralba, A. (2003). Modeling global scene factors in attention. *Journal of the Optical Society of America, 20*.
Turano, K. A., Geruschat, D. R., & Baker, F. H. (2003). Oculomotor strategies for the direction of gaze tested with a real-world activity. *Vision Research, 43*, 333–346.
Underwood (this volume).
Volkmann, F. C. (1986). Human visual suppression. *Vision Research, 26*, 1401–1416.
Williams, C. C., Henderson, J. M., & Zacks, R. T. (2005). Incidental visual memory for targets and distractors in visual search. *Perception & Psychophysics, 67*, 816–827.
Yarbus, A. L. (1967). *Eye movements and vision*. New York: Plenum Press.

Chapter 26

CONGRUENCY, SALIENCY AND GIST IN THE INSPECTION OF OBJECTS IN NATURAL SCENES

GEOFFREY UNDERWOOD, LOUISE HUMPHREYS, and ELEANOR CROSS

University of Nottingham, UK

Eye Movements: A Window on Mind and Brain
Edited by R. P. G. van Gompel, M. H. Fischer, W. S. Murray and R. L. Hill

Abstract

Early studies of the inspection of scenes suggested that eye fixations are attracted to objects that are incongruent with the gist of a picture, whereas more recent studies have questioned this conclusion. The two experiments presented here continue to investigate the potency of incongruent objects in attracting eye fixations during the inspection of pictures of real-world scenes. Pictures sometimes contained an object that violated the gist, such as a cow grazing on a ski slope, and the question asked was whether fixations were attracted to these objects earlier than when they appeared in congruent contexts. In both experiments earlier fixation of incongruent objects was recorded, suggesting a role for peripheral vision in the early comprehension of the gist of a scene and in the detection of anomalies.

Are fixations attracted to objects that violate the gist of a scene? Are viewers more likely to look at a wrench on an office desk than at a stapler? Studies of the inspection of scenes have suggested that eye fixations are attracted to objects that are incongruent with the gist of a picture, but more recently results have suggested that incongruous objects have no special status. The present experiments attempt to resolve this debate with new data.

Mackworth and Morandi (1967) recorded eye movements as their participants viewed colour photographs, and found that more eye fixations were made to regions of the scene that were rated as being more informative by a group of independent judges. This effect was present during the first 2 s of inspection as well as later in the sequence long after the gist of the scene would have been acquired. This suggests that objects are analysed in terms of their meaning very early in the inspection of scenes. Loftus and Mackworth (1978) confirmed the effects of top-down processes on the early inspection of scenes. Their viewers were presented with line drawings of recognisable scenes (e.g. a farmyard) in which one particular object was anomalous (an octopus in the farmyard, for example) or congruent (a tractor). Viewers were more likely to look earlier and for longer durations at anomalous objects than at congruent objects placed in the same location, suggesting that the semantic content of pictures influences early inspection and is used to direct attention to these objects. Subsequent research using line drawings of scenes has since questioned whether or not covert attention and overt gaze shifts preferentially seek out objects that have a low semantic plausibility in the scene (De Graef, Christiaens, and d'Ydewalle, 1990; Friedman, 1979; Henderson, Weeks, & Hollingworth, 1999). In a more recent study of the attention given to incongruent objects in photographs of real-world scenes, however, we have reported their effectiveness in attracting early eye fixations (Underwood & Foulsham, 2006). There is considerable inconsistency in the appearance of this congruency effect.

De Graef et al. (1990) proposed that effects of contextual violations are only apparent during later stages of scene inspection, contrary to the notion of schema-driven object perception effective within a single fixation. Henderson et al. (1999) also recorded eye movements during the inspection of line drawings of familiar scenes and similarly found no early effects of incongruous objects. In their first experiment viewers were free to inspect the scene in preparation for a memory test, and in the second they searched for a target object in order to make a present/absent decision. Scenes sometimes contained incongruous objects (for example, a microscope in a bar room), and sometimes the object appeared in a congruous scene (the microscope was in a laboratory setting). There was no evidence of the early fixation of incongruent objects, suggesting that the early inspection of scenes is determined by visual factors as opposed to semantic factors.

This pattern of results prompted Henderson et al. (1999) to suggest a "saliency map" model of eye guidance in scene perception, in which the initial analysis of the image identifies the low-level visual features such as colour and brightness variations. Eye movements are initially guided to the areas of greatest conspicuity in this account, rather than to anomalies of the gist because the meaning of the picture is not identified at this stage. The gist can only be identified after the initial analysis of the visual characteristics of the image. Henderson, Brockmole, Castelhano, and Mack (this volume) have recently revised this account, placing less emphasis on visual saliency. The present experiments exclude the possibility that attention is attracted to incongruous objects because they have high visual conspicuity, by controlling for the saliency of the objects edited into the pictures.

An issue of concern here is the lack of uniformity of stimuli in terms of visual complexity employed across past researchers to investigate the effect of congruency on object perception. The line drawings used by De Graef et al. (1990) and Henderson et al. (1999) were created by tracing and somewhat simplifying the contours of photographs of real-world settings to produce what were described to be "reasonable approximations to the level of complexity found in natural scenes" (Henderson et al., 1999, p. 212). Due to the stylisation and abstraction of these drawings, however, it was often extremely difficult to recognise the target object. This problem is acknowledged by De Graef et al. (1990), who declared that many objects were excluded from analysis due to difficulties with identification. To what extent does a microscope in a bar room, or indeed a cocktail glass in a laboratory (as used by Henderson et al., 1999), contribute to our overall understanding of the gist of the scene? It seems intuitively reasonable to predict that strange and out of place objects, such as an octopus in a farmyard or a tractor on the sea bed, should be fixated relatively early during scene inspection, once the gist has been recognised and when a gist-violating object starts to present difficulties in the resolution of the meaning of the picture.

Our first experiment addressed the question of whether the semantic content of a scene attracts attention, by recording any differences in the initial eye fixations on objects that are placed in expected or unexpected contexts. Whereas Loftus and Mackworth (1978) found that incongruous objects attracted early fixations – an indication of attentional capture following the detection of a scene inconsistency – other studies have found no effect of incongruency (De Graef et al., 1990; Henderson et al., 1999). All three studies used line drawings, but in the Loftus and Mackworth experiment the object of interest may have been drawn with no features overlapping with the background. The published example suggests this, but in contrast De Graef et al. (1990) and Henderson et al. (1999) drew their objects against cluttered backgrounds. The possibility here is that the effect reported by Loftus and Mackworth depends upon visual conspicuity, and that incongruity will be effective only if a conspicuous object stands out from its background. We have previously confirmed the potency of visual conspicuity in picture inspection (Underwood & Foulsham, 2006; Underwood, Foulsham, van Loon, Humphreys, & Bloyce, 2006), and so it is necessary to eliminate low-level visual saliency as a potential confound. When viewers inspect pictures in preparation for a memory test, their early fixations are drawn to regions that are visually distinctive, although in our search tasks and in the search task reported by Henderson et al. (this volume) the effects of saliency are minimal. They are not minimal in free-inspection tasks, however. To control for effects of visual saliency we screened all stimuli for conspicuity differences between objects in the pictures because the task set for the viewers was to inspect in preparation for a memory test. Itti and Koch (2000) have developed software for the determination of the saliency of regions on pictures according to variations in colour, brightness, and orientation, and this program was used to match the saliency values of objects as they appeared in their different contexts.

1. Experiment 1: Incongruous objects in pictures

This experiment investigated whether the semantic content of a scene determines inspection. It investigated whether attention is drawn to objects that are semantically congruent or incongruent to the rest of the scene. The objects from the first experiment were placed in congruent or incongruent backgrounds. Eye movements were recorded during the inspection of these scenes. By comparing the congruent and incongruent conditions, it could be established whether attention is drawn to objects that are appropriate to the scene or inappropriate to the scene. A neutral condition was also included to obtain baseline values for inspection of the same objects in the absence of contextual cues.

1.1. Method

Participants. The volunteers in this experiment were 18 undergraduates from the University of Nottingham and all had normal or corrected-to-normal vision.
Stimuli and apparatus. Three different sets of 60 pictures were created and each figure contained one object of interest. Each set consisted of an equal number of congruent, incongruent, and neutral pictures, defined according to whether the object would normally be found in that location. The 20 objects in each of the three conditions were equally often from indoor and outdoor locations. Examples are shown in Figures 1a (congruent), 1b (incongruent), and 1c (neutral).

The pictures were created by editing the objects into photographs of background scenes using Adobe Photoshop software. The objects were matched so that each indoor object was paired with an outdoor object with similar physical features (for example, a vacuum cleaner and a lawn mower, as in Figure 1a). The congruent pictures were created by placing each object onto its appropriate background (for example, a vacuum cleaner was placed in a hall, as in Figure 1a); the incongruent pictures were created by replacing the objects from the congruent pictures with its matched object (in this example, the vacuum cleaner would be replaced by the lawn mower, as in Figure 1b). The neutral pictures were created by editing each object into one of the neutral backgrounds such as scenes consisting of pictures of brick walls that could be indoors or outdoors (Figure 1c). Objects of interest were always edited into pictures away from the centre, where the initial fixation was to be made. An additional four pictures were created for practice trials. The three sets of pictures contained each background scene and each object of interest, in three different combinations. Objects were permutated across conditions to create the three sets of pictures for use with three groups of participants. This ensured that each object would be seen by each participant, and that over the course of the experiment each object was seen equally often against each of the three backgrounds.

Figure 1a. An example of a congruent indoor picture (top panel) and a congruent outdoor picture (bottom panel). (*See Color Plate 4.*)

Each photograph was processed using the Itti and Koch (2000) software for the determination of the peaks of low-level visual saliency. This program identifies changes in brightness, colour, and orientation and creates a saliency map that is a representation of locations that are visually conspicuous. The rank order of saliency peaks on each picture was used to compare each object as it appeared in the congruous, incongruous, and neutral contexts. A rank of 1 was allocated to the most salient object in the scene. The mean rankings for objects were: congruous context 3.65 (SD = 2.08), incongruous context 3.55 (SD = 1.92), and neutral context 1.05 (SD = 0.22). The congruous and incongruous rankings did not differ ($t < 1$). All but one of the objects in the neutral context was

Figure 1b. An example of an incongruent indoor picture (top panel) and an incongruent outdoor picture (bottom panel). (*See Color Plate 5.*)

identified as the visually most salient object. As a consequence of this inhomogeneity of variance, no statistical comparison was possible between the neutral condition and the other two conditions. It was concluded that the objects in congruous and incongruous contexts had similar saliency rankings, and that these were lower than the ranks of the same objects placed in neutral scenes.

The pictures were presented on a personal computer with a standard colour monitor, which was approximately 60 cm from the seated participant. Stimuli were presented using the E-prime experimental control software. Eye movements were recorded using an SMI iView X RED system with a remote camera for recording eye position every 20 ms.

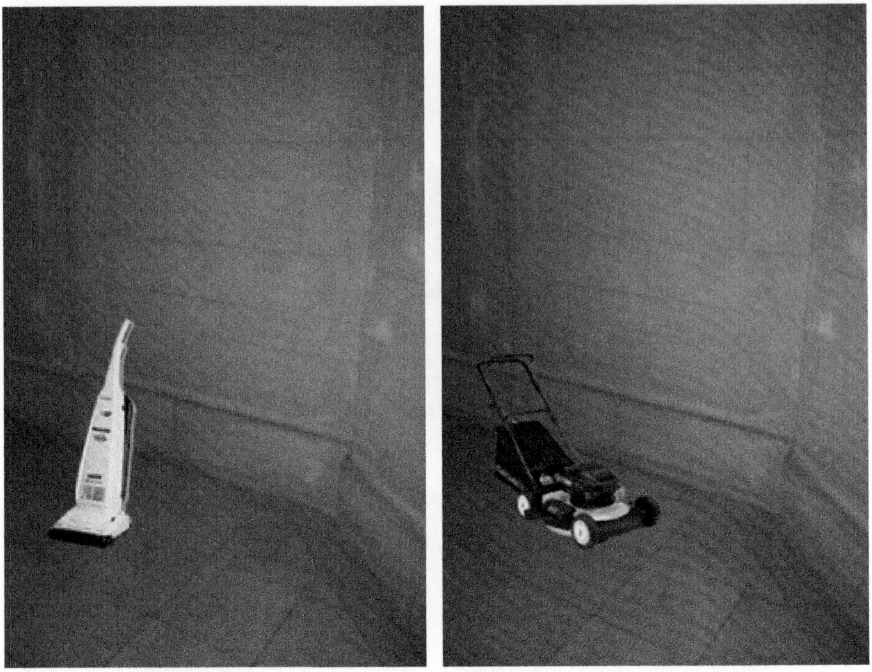

Figure 1c. An example of a neutral picture containing an indoor object (left panel) and a neutral picture containing an outdoor object (right panel). (*See Color Plate 6.*)

Head movements were restricted by requiring participants to rest their head on a chin rest. The spatial accuracy of the system was 0.5 degrees.

Design and procedure. A two-factor within-groups design was used; the factors were object type (indoor and outdoor) and background (congruent, incongruent and neutral). Each participant saw one of the sets of 60 pictures (thus there were three different groups of participants). This design ensured that participants did not see the same object twice during the course of testing, but ensured that each object appeared in each of the three contexts over the course of the whole experiment. Each participant saw each object and each background scene just once during the experiment, but different participants saw the objects and scenes in different combinations.

Participants were first calibrated with the eye-tracker. Calibration involved fixating a central marker, which then appeared at 8 points within the sides of the display area once eye movements had stabilised. Before each picture was presented a fixation cross appeared in the centre of the screen for 1000 ms and each picture was then presented for 5000 ms. After the practice trials were presented, a text display was shown giving instructions for the memory test. In the memory test another picture was presented that had or had not been seen previously. Participants had to decide whether this picture had been previously presented. The picture was displayed until the participant made a response that also terminated the display. At the end of each block of 12 pictures, the same memory test instructions as those given in the practice trial were presented, and a test picture was then presented. The test picture either had been presented in that block or was a new picture. Eye movements were recorded during the presentation of all of the pictures, but not during the memory test.

1.2. Results

Measures of the fixation of the object of interest in each picture are the focus here, and this object will be referred to as the target object. The mean number of fixations before target fixation, and the mean duration of the first gaze on the target were calculated according to the type of background and the type of object (see Table 1). These measures give an indication of the delay in first fixating the object of interest and the amount of attention given to it during the first inspection.

Mean number of fixations before the first fixation of the target. The fixation count prior to target fixation was defined as the number of fixations from the onset of the display (including the initial central fixation) up to the point of first fixation upon the target object, providing a measure of the degree to which the semantic anomaly in a peripheral region of the scene could guide early foveal fixation. The two factors entered into a 2×3 ANOVA were background, which had three levels (congruent, incongruent, and neutral), and object, which had two levels (indoor and outdoor). There was a main effect of background on the number of fixations before fixating on the target object ($F_{2,34} = 130.91$, MSe $= 0.012$, $p < 0.001$). Pairwise comparisons showed that there was a reliable difference between the congruent and incongruent contexts ($p < 0.05$) with target fixation occurring after fewer fixations when the background was incongruent.

Table 1
Means (and standard deviations) of the fixation measures taken in Experiment 1

Context		Mean number of fixations prior to target fixation	Mean first gaze duration on target (ms)
Congruent	Indoor	3.83 (1.44)	963 (472)
	Outdoor	2.91 (0.9)	1156 (451)
Incongruent	Indoor	2.48 (0.63)	1425 (617)
	Outdoor	3.07 (0.70)	1346 (577)
Neutral	Indoor	1.39 (0.37)	2190 (736)
	Outdoor	1.31 (0.47)	2420 (779)

There was earlier fixation of a target object in a neutral context relative to the target in an incongruent and a congruent context (both contrasts at $p < 0.001$). There was no main effect of indoor/outdoor object type ($F_{1,17} = 1.321$), but this factor did interact with background congruency ($F_{2,34} = 5.951$, MSe = 0.016, $p < 0.01$). Pairwise comparisons showed that for pictures containing indoor objects, there were reliable differences between those with congruent and incongruent backgrounds ($p < 0.001$) with fixation occurring earliest when the background was incongruent. There was earlier fixation of objects in neutral contexts in comparison with both the incongruent and congruent backgrounds (both contrasts at $p < 0.001$). For pictures containing outdoor objects, there were significant differences between pictures with incongruent and congruent backgrounds in comparison with objects in neutral contexts (both contrasts at $p < 0.001$). However, there was no significant difference between pictures with congruent and incongruent backgrounds, unlike the comparisons between pictures containing indoor objects.

Duration of the first gaze on the target object. The duration of initial gaze on a target was taken as the sum of the time spent fixating the target object from when the eyes initially landed on the object until the eyes first left that object. A 2 × 3 ANOVA indicated that there was a main effect of background congruency ($F_{2,34} = 89.02$, MSe = 166575, $p < 0.001$). Pairwise comparisons showed that there were longer first gazes on incongruent rather than congruent ($p < 0.01$); and that they were longer on objects in neutral contexts in comparison with incongruent and congruent contexts (both contrasts at $p < 0.001$). There was a main effect of object type, with longer gazes upon outdoor objects ($F_{1,17} = 7.70$, MSe = 51400, $p < 0.05$), and there was an interaction between background and object ($F_{2,34} = 4.50$, MSe = 60116, $p < 0.05$). Pairwise comparisons indicated that for indoor objects all three context conditions differed from each other, with longer gazes on incongruent than on congruent objects ($p < 0.01$), and on neutral objects than on both congruent and incongruent objects (both at $p < 0.001$). For outdoor objects, congruent and incongruent object gaze durations did not differ, but first gazes on objects in neutral scenes were again longer than gazes on indoor or outdoor scenes (both at $p < 0.001$).

1.3. Discussion

When the background was incongruent with the target object, target fixation occurred after fewer fixations than it was congruent, and the gaze duration on the object was longer. When the background was neutral (as opposed to congruent or incongruent) the target object was fixated earlier and attracted longer initial gazes. When comparing the congruent and incongruent conditions the largest differences occurred with objects taken from an indoor scene. When a scene contained an indoor object such as a vacuum cleaner, fixation of the target object occurred sooner and the first inspection was longer when the background was incongruent rather than when it was congruent. With an outdoor object such as a lawnmower, the effect of congruency was mainly upon gaze duration. When an object was placed in a neutral scene the target object was looked at for longer, and was looked at sooner than objects placed in either a congruous or an incongruous background.

However, this result is not surprising since the target object was the only object to inspect in the neutral condition; the target was the only object in the scene.

The results established that fixations were more attracted to an incongruent object within a scene than to a congruent object when placed against a similar background, confirming Loftus and Mackworth's (1978) findings with line drawings and Underwood and Foulsham's (2006) findings with photographs. The congruency effect with rich real-world scenes suggests that cognitive factors are important in governing the early inspection of scenes. It implies a role of scene-specific schemas in governing attention. These schemas generate expectations about what the scene will contain. When we view a scene we are influenced by these expectations and are drawn to objects that are unexpected.

This experiment established that indoor objects were fixated sooner when the background was semantically incongruent with the object than when it was congruent. However, this effect did not occur for outdoor objects placed in indoor scenes. This is curious and we have no explanation of this inconsistency other than to suggest that plausibility may be responsible. It could be argued that it is more surprising to see an indoor object in an outdoor scene than it is to see an outdoor object in an indoor scene. Although an outdoor object placed in an indoor scene may be semantically incongruent with that scene, its presence may not be necessarily unexpected because outdoor objects are sometimes stored indoors. Therefore, it may be more appropriate to describe this as an effect of expectancy or plausibility rather than congruency. The expectancy effect would imply that we are attracted to objects that are not expected to be in a scene. If the outdoor objects could be equally expected in both the congruent and incongruent conditions then we might expect no congruency effect. The second experiment investigates this possible variation in the plausibility of incongruent objects. In addition to fixations on congruent and incongruent objects we introduced bizarre combinations of objects and scenes, to give further emphasis to the implausibility of an object appearing in that specific context.

2. Experiment 2: Bizarre objects in pictures

Natural photographs of scenes were employed in Experiment 1 to investigate the influence of object-scene congruency upon eye guidance. These pictures were paired so that in one background an object appeared either as congruent or as incongruent in relation to the gist of the scene. In this experiment a third condition was introduced in order to create a neutral baseline as an indication of the amount and duration of fixation received by each object independently of distracters and the acquisition of scene gist. In agreement with the findings of Loftus and Mackworth (1978), it was found that viewers located the incongruent object with fewer prior fixations than their congruent counterparts. Also, once fixated, incongruous objects received longer total gaze duration than congruous objects. It was concluded that one is attracted to objects that violate the gist of a scene, and that such eye movements are programmed within the first fixation or two, after the retrieval of the appropriate scene schema. Experiment 1 found inconsistencies in the effects of congruency on the early fixation of indoor and outdoor scenes that may be attributed to differences in the plausibility of an object appearing in the scene. In Experiment 2 we take the plausibility of objects to an extreme, with the introduction of an experimental condition in which the object of interest in a particular context is anomalous and can be best described as bizarre.

2.1. Method

Participants. Twenty-one University of Nottingham undergraduates took part in this experiment. None had taken part in the first experiment, and all had normal or corrected-to-normal vision.

Stimuli and apparatus. Thirty digital images of recognisable natural scenes were used as stimuli. These were displayed on a standard colour monitor at a distance of 60 cm from the seated participant. These images were digitally edited using Adobe Photoshop software so that a target object appeared in three differing scenes. The size and spatial location of the objects were kept constant. For each scene there were three corresponding pictures, providing the basis for the three conditions: plausible, implausible, and bizarre. Each picture contained a sufficient number of distractor items to ensure that the target object had no significance for the participants as a target object in this experiment. Scenes were then grouped so that the contextually consistent object from a given scene, when placed in the same spatial location in its grouped scene, became either implausible or bizarre. For example, a racing car appeared as the consistent target object in a picture of a racetrack, and subsequently appeared as the inconsistent target object on a farmyard, and as the bizarre target object in a picture of a harbour. Figure 2 shows a different example of objects imposed on a scene to create plausible, implausible, and bizarre combinations. The rotation of target objects across conditions was again intended to control for the visual saliency and attractiveness of the individual objects themselves. Each target object was pasted in a way that size and support relations were not violated.

The mean visual saliency values for each picture were again computed using software provided by Itti and Koch (2000). This was performed to avoid any possible correlation between semantic saliency and visual saliency of target objects within different settings. The mean ranks for visual saliency for plausible, implausible, and bizarre pictures were 3.07 ($SD = 2.26$), 2.80 ($SD = 2.12$), and 2.77 ($SD = 1.96$) respectively. A one-factor ANOVA revealed no significant difference in visual saliency across the levels of plausibility ($F < 1$). This analysis shows that target objects placed in a bizarre setting were no more visually salient within that background (in terms of variations in orientation, intensity and colour) than when placed in a plausible or implausible scene.

Participants' eye movements and keyboard responses to each display were recorded using an SMI EyeLink tracker which has a spatial accuracy of 0.5 degrees. Eye position was recorded every 4 ms. The eye-tracker was head-mounted, and a chin-rest was also used to minimise head movement and to ensure a constant-viewing distance of stimuli.

Design and procedure. Each participant viewed 30 pictures, with 10 pictures from each scene group (plausible, implausible, and bizarre). As with Experiment 1, the three

Figure 2a. A sample picture as used in the plausible condition with the skier wearing the pink jacket as the target object. The fixations of one participant are superimposed on the stimulus in this example, with lines indicating the movements between fixations that were themselves shown as circles. Duration of fixation is indicated by the size of the circles with larger circles indicating longer fixation durations than smaller ones. (*See Color Plate 7.*)

Figure 2b. A sample picture with fixation patterns of one viewer as seen in the implausible condition with a snowman appearing as the target object. (*See Color Plate 8.*)

Figure 2c. A sample picture with fixation patterns of one viewer as seen in the bizarre condition with the cow as the target object. (*See Color Plate 9.*)

pictures within a scene group were rotated across participants so that over sets of three participants all were viewed exactly once.

The instructions to participants explicitly stated that the experiment involved a recognition memory test and that their eye movements would be monitored whilst they viewed pictures that they would later be asked to discriminate against comparable new ones. The recognition memory test was administered only during practice. Although we are interested only in the first few fixations, we found that in Experiment 1, with a fixed display period, participants sometimes did not complete their scanning of the picture, and so this was changed in Experiment 2, to give full opportunity to inspect every object shown. Each test picture remained on display until the participant pressed a key on the computer keyboard. Each participant saw ten pictures from each condition, and did not see the same scene more than once. Pictures were presented in a randomised order.

2.2. Results

The means of the measures of interest (number of fixation prior to inspection of the target object and duration of the first fixation) are shown in Table 2.

Mean number of fixations before the first fixation of the target. A one-factor ANOVA revealed a reliable main effect of semantic consistency upon the number of prior fixations ($F_{2,40} = 29.23$, $\text{MSe} = 0.24$, $p < 0.001$). Pairwise comparisons showed that each condition was significantly different to each other (all contrasts at $p < 0.01$). Objects in the bizarre condition were inspected earlier than those in the implausible condition, which in turn were inspected earlier than objects in the plausible contexts.

Table 2
Means (and standard deviations) of the fixation measures taken in Experiment 2

Context	Mean number of fixations prior to target fixation	Mean first gaze duration on target (ms)
Plausible	2.68 (0.75)	759 (219)
Implausible	2.02 (0.51)	841 (274)
Bizarre	1.53 (0.31)	913 (290)

Duration of the first gaze on the target object. A one-factor ANOVA indicated a main effect of context on the duration of initial gaze on the target ($F_{2,40} = 3.84$, MSe = 32577, $p < 0.05$), and pairwise comparisons found that the only reliable difference was that there were longer gazes on objects in bizarre contexts than upon those that were in plausible settings ($p < 0.05$).

2.3. Discussion

The results show that when inspecting a picture in preparation for a short memory test, fewer saccades were made prior to inspection of a target object that is semantically inconsistent with the gist of the scene, and that this effect was enhanced with extremely implausible objects. This result confirms and extends the effect reported in our first experiment here. Target objects in the bizarre condition were preceded by fewer prior fixations than targets in the implausible or plausible condition, and similarly, target objects in the implausible condition were preceded by fewer prior fixations than targets in the plausible condition. The data also show that once fixated, the duration of the first gaze on the target object is significantly longer in the bizarre condition than in the plausible condition. If the duration of the first gaze at object is a reflection of comprehension, then these results indicate that objects that are of great semantic inconsistency are more difficult to process than their plausible counterparts. This is consistent with studies showing that objects are more difficult to detect or name if they are incongruous or that violate the logic of a scene (e.g., Biederman, Mezzanotte, & Rabinowitz, 1982; Davenport & Potter, 2004).

An important finding from these experiments is that fixation patterns in scenes were sensitive to the extent of the implausibility of the object. There were differences between bizarre and implausibly placed objects and this difference may help explain an inconsistency in the literature. Incongruous objects have not always attracted early fixations (e.g. De Graef et al., 1990; Henderson et al., 1999), in contrast with the results of the two experiments here, and in contrast with Loftus and Mackworth's (1978) study as well as other recent experiments that have used photographs of natural scenes

(Underwood & Foulsham, 2006). Possibly, the discrepancy in results found in earlier studies may be explained on the basis of the varying levels of the semantic plausibility of target stimuli employed. It may be that the greater the semantic implausibility of the target object, the more likely it will be to find an effect of semantic inconsistency on object perception. Perhaps the octopus in Loftus and Mackworth's (1978) farmyard was more bizarre than the cocktail glass in Henderson, Weeks, and Hollingworth's (1999) laboratory scene, although we cannot be sure that these examples were typical of the stimuli used in their experiments. Furthermore, this pattern of results cannot be explained by the premise that objects in the bizarre condition were simply more visually salient against a particular background than object–scene pairings in the plausible condition because there was no difference in visual saliency ranks across the three levels of plausibility in our second experiment.

The finding that incongruous objects (and especially those in the bizarre condition) are fixated sooner than non-informative objects can be argued to demonstrate fast analysis of scene semantics, and that the initial fixations on the scene can be determined in part by the detection of a violation of the gist. These findings support the data presented by Loftus and Mackworth (1978) who argued the occurrence of at least three stages of picture viewing. First, the gist of the scene must be determined, which is now known to occur very rapidly within a single fixation (Biederman et al., 1982; Davenport & Potter, 2004; De Graef, 2005; Underwood, 2005). Secondly, objects in peripheral vision must be at least partially identified on the basis of their physical characteristics. Lastly, the viewer must determine the probability of encountering the object within a given scene once the gist has been identified. In this model, fixations will be directed to objects with low *a priori* expectations of being in the scene. Longer durations of the initial gaze upon an inconsistent target object are taken to be due to the process of linking the object to the existing schema for any given scene. They suggest that gist acquisition is analogous to the activation of a schema, and that subsequent fixations are drawn to objects as a means of verifying their association with that schema. The additional time spent gazing at an informative object may reflect the amount of time it takes to add the new object to the many instances of the scene represented within the viewer's personal schema. Alternatively, the longer duration of initial fixation may be due to the viewer actively searching for the appropriate schema from which the inconsistent object belongs as a means of attempting to understand why that object might have been placed within the incongruent scene. Once an incongruent object has been detected, it may take longer to resolve the conflict between the schemas activated by the object and the scene. Hollingworth and Henderson's (2000) 'Attentional Attraction' hypothesis is a potential explanation of the earlier fixation of an inconsistent object as well as longer gaze durations. Here, covert attention is drawn to an object when there is difficulty reconciling the identity of the object with the scene schema. The role of attention may be to make sure that a perceptual mistake has not been made ("is that really a cow on a ski slope?") or to check for additional information that could help to reconcile the conceptual discrepancy.

On the basis of the present data we can conclude that the top-down cognitive mechanisms involved in recognising the gist of a picture and identifying regions of potential semantic inconsistency are used to guide fixations. In order to initiate an eye movement that results in the early fixation of the inconsistent object, the recognition of the scene schema must be completed very early during picture inspection.

References

Biederman, I., Mezzanotte, R. J., & Rabinowitz, J. C. (1982). Scene perception: Detecting and judging objects undergoing relational violations. *Cognitive Psychology, 14*, 143–177.

Davenport, J. L., & Potter, M. C. (2004). Scene consistency in object and background perception. *Psychological Science, 15*, 559–564.

De Graef, P. (2005). Semantic effects on object selection in real-world scene perception. In G. Underwood (Ed.), *Cognitive processes in eye guidance* (pp. 189–211). Oxford: Oxford University Press.

De Graef, P., Christiaens, D., & d'Ydewalle, G. (1990). Perceptual effects of scene context on object identification. *Psychological Research, 52*, 317–329.

Friedman, A. (1979). Framing pictures: The role of knowledge in automatized encoding and memory for gist. *Journal of Experimental Psychology: General, 108*, 316–355.

Henderson, J. M., Brockmole, J. R., Castelhano, M. S., & Mack, M. (2006). Visual saliency does not account for eye movements during visual search in real-world scenes. (This volume.)

Henderson, J. M., Weeks, P. A., & Hollingworth, A. (1999). The effects of semantic consistency on eye movements during scene viewing. *Journal of Experimental Psychology: Human Perception and Performance, 25*, 210–228.

Hollingworth, A., & Henderson, J. M. (2000). Semantic informativeness mediates the detection of changes in natural scenes. *Visual Cognition, 7*, 213–235.

Itti, L., & Koch, C. (2000). A saliency-based search mechanism for overt and covert shifts of visual attention. *Visual Research, 40*, 1489–1506.

Loftus, G. R. (1972). Eye fixations and recognition memory for pictures. *Cognitive Psychology, 3*, 525–551.

Loftus, G. R., & Mackworth, N. H. (1978). Cognitive determinants of fixation location during picture viewing. *Journal of Experimental Psychology: Human Perception and Performance, 4*, 565–572.

Mackworth, N. H., & Morandi, A. J. (1967). The gaze selects informative details within pictures. *Perception and Psychophysics, 2*, 547–552.

Underwood, G. (2005). Eye fixations on pictures of natural scenes: Getting the gist and identifying the components. In G. Underwood (Ed.), *Cognitive processes in eye guidance.* (pp. 163–187). Oxford: Oxford University Press.

Underwood, G., & Foulsham, T. (2006). Visual saliency and semantic incongruency influence eye movements when inspecting pictures. *Quarterly Journal of Experimental Psychology, 18*, 1931–1949.

Underwood, G., Foulsham, T., van Loon, E., Humphreys, L., & Bloyce, J. (2006). Eye movements during scene inspection: A test of the saliency map hypothesis. *European Journal of Cognitive Psychology, 59*, 321–342.

Chapter 27

SACCADIC SEARCH: ON THE DURATION OF A FIXATION

IGNACE TH. C. HOOGE, BJÖRN N. S. VLASKAMP and EELCO A. B. OVER
Utrecht University, The Netherlands

Eye Movements: A Window on Mind and Brain
Edited by R. P. G. van Gompel, M. H. Fischer, W. S. Murray and R. L. Hill

Abstract

Is it the fixated stimulus element or the fixation history that determines fixation duration? We measured 93 922 fixations to investigate how fixation times are adjusted to the demands of a visual search task.

Subjects had to search for an O among C's. The C's could have a large gap (0.220°) or a small gap (0.044°). We varied the proportions of both types of C's in the displays. The main results are that: (1) fixation time depended on the element fixated (small gap C's were 40 ms longer fixated than large gap C's), (2) fixation time on large gap C's decreased with increasing proportion of large gap C's in the display and (3) fixation time on large gap C's depended on the gap size of the previously fixated element.

We conclude that fixation time depends both on the fixation history (expected analysis time) and on the current fixation element (actual analysis time). The contribution of both components on fixation time may depend on the task and the amount of useful information in the displays. Pre-programming of fixation times appears to be conservative such that extension of fixation time occurs in the next fixation whereas shortening of fixation time appears to be delayed.

To inspect a visual scene, observers usually make saccades to direct the fovea to interesting parts of the scene. Between saccades the eyes fixate and visual analysis of the scene may take place. The visual system needs time to analyse the retinal image. This time may vary from several tens of milliseconds to hundreds of milliseconds. It is obvious that fixation times have to be adjusted to these analysis times to make search efficient. If fixation is too short, the retinal image will be overwritten by the next retinal image before it is processed. If fixation is too long, observers spoil time doing nothing.

Fixation times have been a research topic for years (for a good review, see Rayner, 1998) and many authors have shown that longer *mean* fixation times are found in more difficult tasks (Cornelissen, Bruin, & Kooijman, 2005; Hooge & Erkelens, 1996, 1998; Jacobs, 1986; Näsänen, Ojanpää, & Kojo, 2001; Rayner & Fisher, 1987; Vlaskamp, Over, & Hooge, 2005), suggesting that fixations are set to longer durations when the search task requires it. However, this is not a strict one-to-one relation, one may find long and short individual fixation times in a search task with homogenous stimulus material (e.g., one type of non-targets). In addition, it is found in a variety of tasks that the width of the distribution of fixation times is broad and scales with the mean (Harris, Hainline, Abramov, Lemerise, & Camenzuli, 1988).

Usually, individual fixations last from 50 to 700 ms. According to Viviani (1990) at least three processes may take place during fixation: (1) saccade programming, (2) analysis of the foveal image and (3) selection of a new saccade target.

It takes about 150 ms to program a saccade to a target appearing randomly in time and space (Becker & Jürgens, 1979). As stated above, individual saccades may have preceding fixations that are shorter than 150 ms, because multiple saccades may be programmed in parallel (McPeek, Skavenski, & Nakayama, 2000). If two saccades are programmed shortly after each other, the fixation time between these saccades becomes short.

The analysis of the foveal image and the selection of the saccade target also consume time and these processes may occur in parallel with saccade programming (Viviani, 1990; Hooge & Erkelens, 1996). Search is effective if fixations are long enough to allow foveal analysis and saccade target selection to take place. However, it is not clear how saccade programming and these two visual processes are synchronized. Also questionable is how precisely they are synchronized. Hooge and Erkelens (1996) report the occurrence of many return saccades. From this it is hypothesized that they bring back the eyes to a location that was fixated too briefly for completed visual analysis. The majority of the fixations, however, are long enough. How does the brain determine the duration of a fixation? At least, the brain should have knowledge (probably in advance) about the time required for visual analysis to program saccades in such way that search is effective.

An interesting discussion concerns whether fixation times are pre-programmed by the use of the expected analysis time determined during previous fixations or whether the foveal analysis is monitored (process-monitoring) and fixation times are adjusted after visual analysis is complete (Greene & Rayner, 2001; Hooge and Erkelens, 1996, 1998; Rayner, 1978; Rayner & Pollatsek, 1981; Vaughan, 1982; Vaughan & Graefe, 1977).

The main evidence for process-monitoring comes from onset-delay experiments in reading research. In an onset-delay paradigm, certain words disappear at fixation onset and reappear after a certain delay. Mean fixation times on the delayed words are extended by as much as the delay when each block of trials has fixed delay (Rayner & Pollatsek, 1981). With variable delays the extension of the fixation time with short delays is shorter than the onset delay, indicating that there is some sort of pre-programming going on. Rayner and Pollatsek (1981) therefore conclude that control of fixation duration in reading is in accordance with a mixed control model (both pre-programming and process-monitoring). However, for visual search in a stimulus onset-delay paradigm, Vaughan (1982) concludes that the effects for onset delays are in agreement with only process-monitoring.

The main evidence for pre-programming comes from several experiments (Hooge & Erkelens, 1996, 1998; Vaughan & Graefe, 1977). First, Hooge and Erkelens (1996) report the occurrence of many return saccades. Subjects were asked to find an O among C's. They were instructed to respond by continued fixation of the O when found. Hooge and Erkelens (1996) report that after fixation of the O, the subjects often made saccades away from the O to the next C, after which they return to the O. Based on this finding they conclude that before completed analysis of the O, a saccade is programmed that cannot be stopped anymore. To investigate this phenomenon in detail, Hooge and Erkelens (1998) designed the direction-coded search task. In a hexagonal arrangement of C's with one O (the target), subjects had to make saccades in the direction of the gap in the C. The stimulus was designed in such way that if subjects continued making saccades in the direction of the gap of the fixated C, they ended on the O. To be able to make a correctly directed saccade, visual analysis of the fixated C must be complete before the next saccade is programmed. This was often not the case, as the percentage of incorrectly directed saccades ranged from 20 to 35.

Recently, Greene and Rayner (2001) performed a direction-coded search task as well. They concluded from search in denser displays than used in Hooge and Erkelens (1998) that their results "provided evidence for a process-monitoring model of visual search". First, they compared fixation times from their direction-coded condition with fixation times from their uncoded condition. As in Hooge and Erkelens (1998), they found longer fixation times in the direction-coded condition. We do not agree that this is evidence for process-monitoring. In the direction-coded displays the visual task differs from the visual task in the uncoded displays, not only the nature of the element (target or not) but also the direction has to be encoded by the visual system. Pre-programming models would also predict longer fixation times for a more difficult visual task as long as there are enough fixations to build up an estimate of the required fixation time. Secondly, Greene and Rayner (2001) did not see an effect on fixation duration of blocked vs mixed presentation of the coded and uncoded displays (indication for process monitoring). However, Greene and Rayner (2001) cannot rule out the possibility that pre-programming may work at relatively short time scales (for example the time scale of one trial). It even may be possible that observers use the information that there are two types of trials and only recognition of an (un)coded display may trigger the relevant fixation time settings.

In this study we investigated the dynamics of adjustment of fixation duration at the short term. To do so, we designed search displays that contain two kinds of distracters. One distracter resembled the target more than the other. This means that the display

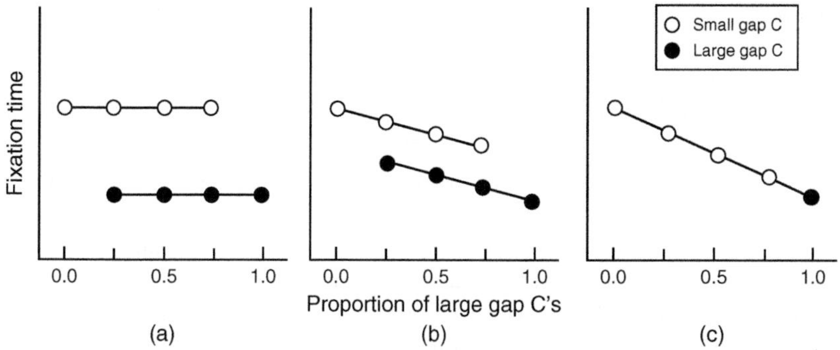

Figure 1. Fixation times predicted by the strict process-monitoring model (a), mixed control model (b) and the strict pre-programming model (c). In the strict process monitoring model the visual analysis time of the element fixated determines the fixation time. In the strict pre-programming model, fixation time is pre-programmed for each fixation and adjusted by the estimated difficulty of the search task. In panel c the two lines overlap completely.

contained difficult (longer analysis time) and easy (shorter analysis time) non-targets. In five conditions we varied the proportion of easy and difficult stimulus elements (1.00/0.00, 0.75/0.25, 0.50/0.50, 0.25/0.75, 0.00/1.00). The two classic models for control of fixation duration yield different predictions for the fixation duration. Process-monitoring predicts that fixation duration depends only on the difficulty of the element fixated. Thus, this model predicts that fixation of easy non-targets is shorter than the fixation of difficult non-targets irrespective of the proportion of easy non-targets (Figure 1a). In contrast, pre-programming models predict fixation durations that depend on the proportions of easy and difficult non-targets (Figure 1b and c). In pre-programming models an estimate for the difficulty of the search task (the analysis time required) is used. Depending on the time used to build up this estimate, fixation times of easy and difficult elements depend on the direct fixation history and thus on the proportions of the two types of non-targets. Extreme pre-programming models would produce data as shown in Figure 1c. Here fixation time does not depend on the analysis time of an individual stimulus element; fixation times are completely determined by the proportion of easy and difficult non-targets.

1. Methods

1.1. Subjects

Six male subjects (BV, CP, EO, GJ, MN and TB; aged 22–27) participated in this experiment. EO and BV are co-authors on this paper and the others were naïve concerning the goals of the experiment.

1.2. Stimuli

The stimuli consisted of 49 elements placed on a hexagonal grid (30.8° × 21.1°) in a 7 × 7 configuration (Figure 2). Each stimulus contained one target (O) and 48 C's (randomly chosen from four orientations). We decided the target to be an O among C's because this combination is known as a serial search task (Treisman & Gormican, 1988). However, we do not believe there is a clear distinction between serial and parallel search (Duncan & Humphreys, 1989); for this experiment a difficult search task is required to make sure that the subjects make saccades to find the target.

We varied the gap in the C to evoke fixations with longer durations and fixations with shorter durations. In analogy to reaction times (Duncan & Humphreys, 1989) increasing target/non-target dissimilarity (in the present experiment larger gaps) usually decreases fixation time (Hooge & Erkelens, 1996). The C could have either a large gap (10 pixels, 0.220°) or a small gap (2 pixels, 0.044°). The proportion of large and small gap C's per display was varied in 5 blocked conditions of 100 trials (1.00/0.00, 0.75/0.25, 0.50/0.50, 0.25/0.75, 0.00/1.00). Thus in one condition all the C's had small gaps, in three conditions the stimulus contained both types of C's and finally in one condition all the C's had large gaps.

To prevent subjects from analyzing elements in the vicinity of the fixated element, individual elements were separated by 4.7°. Moreover each element was surrounded by

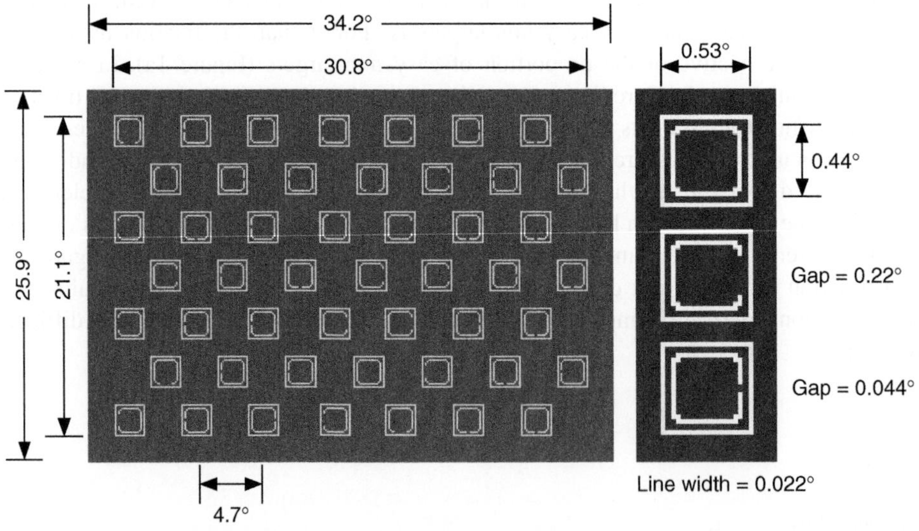

Figure 2. Cartoon of the stimulus and its dimensions. The real stimulus was different from this schematic drawing (e.g., the object size is 0.53° while the centre to centre object distance is about nine times larger (4.7° pixels)). This search display is taken from the 0.50/0.50 condition, it contains 24 large gap C's, 24 small gap C's and one O (the target).

a rectangle. This rectangle is supposed to act as a mask. The stimulus elements measured 24×24 pixels (this resembles $0.53° \times 0.53°$ at 0.64 m). The distance between the elements was 216 pixels (4.7°). Because we used a flat CRT-screen, we report also measures in pixels because at larger viewing angles ($> 10°$) distances on the screen do not scale linearly with viewing angle.

1.3. Set up

Subjects sat in front of a LaCie Blue Electron III 22′ Screen (1600×1200 pixels at 85 Hz) at a distance of 0.64 m. Stimuli were generated by an Apple PowerMac G4 dual 450 using a Matlab program. This Matlab program was based on routines from the Psychophysics Toolbox (Brainard, 1997; Pelli, 1997) and EyeLink Toolbox (Cornelissen, Peters, & Palmer, 2002).

Movements of the left eye were measured at 250 Hz with the EyeLink. The EyeLink is good enough for measuring fixations in a search task (Van der Geest & Frens, 2002; Smeets & Hooge, 2003). Head movements were prevented by the use of a 2-axis bite board. Data were stored on disk and were analysed off-line by a self-written Matlab program.

1.4. Procedure and task

Each of the five experimental conditions started with a calibration (9 dots standard EyeLink calibration). After successful calibration, the subject performed 100 trials. A trial started with the presentation of a fixation marker in the center of the screen. Fixation of this marker was used for on-line drift correction. Then the subjects had to push the space bar to start the presentation of the search display. Subjects were asked to find the O (the target). When found they had to push the right arrow key on the computer keyboard. There were no practice trials.

1.5. Data analysis

The velocity signal of eye movements was searched for peak velocities above 20°/s. Each peak (in the velocity signal) was considered a potential indicator of the presence of a saccade. The exact onset of the saccade was determined by going backward in time to the point where the absolute velocity signal dropped below the average velocity plus two standard deviations during the stable fixation period before the saccade. The exact offset of the saccade was determined by going forward in time to the point where the absolute velocity signal dropped below the average velocity plus two standard deviations during the stable fixation period after the saccade. This method was adopted from Van der Steen and Bruno (1995). This procedure was followed by rejection/acceptance based on a minimum saccade duration of 12 ms and a minimum amplitude of 1.5°. When a saccade was removed, fixation time before and after this saccade and the duration of the saccade were added together.

2. Results

2.1. Search time and number of saccades

In this experiment, we measured 93 922 saccades in 6 subjects. Each search display contained one target (O) and subjects had to search until the O was found. The percentage of being correct was 100 for all subjects in all conditions. Therefore, in the present experiment, search time is a good estimator for the difficulty of the search task.

One of the underlying assumptions in this experiment is that a small gap C is harder to distinguish from the O than a large gap C. Figure 3 shows mean search times for six subjects. As expected search times are longer for the condition with only small gap C's than for the condition having only large gap C's. Search time decreases from 9.4 to 5.8 s with proportion of large gap C's $[F(4, 20) = 25.087,\ p < 0.00001]$. As expected, the present analysis clearly shows that small gap C's are harder to distinguish from the target than large gap C's. This result is in agreement with results from visual search experiments in which target distracter dissimilarity was manipulated and only reaction times were measured (Duncan & Humphreys, 1989).

Figure 4 shows the number of fixations against the proportion of large gap C's. The number of fixations in displays with easy non-targets is smaller than the number of fixations in displays with difficult non-targets. Number of fixations decreases from 34 to 26 with increasing proportion of large gap C's $[F(4, 20) = 9.910, p < 0.0001]$. However, the relative slopes are smaller than those of the search times, which predicts that mean fixation time should decrease with larger proportion of easy non-targets.

As expected, the number of fixations on large gap C's increases with the proportion of large gap C's and number of fixations on small gap C's decreases with the proportions of large gap C's.

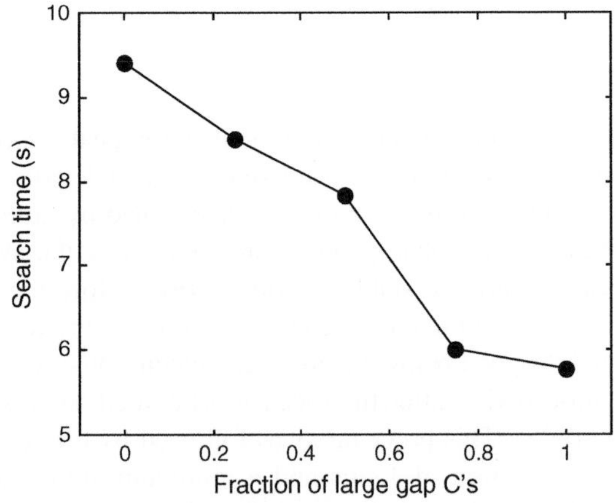

Figure 3. Mean search time vs proportion of large gap C's.

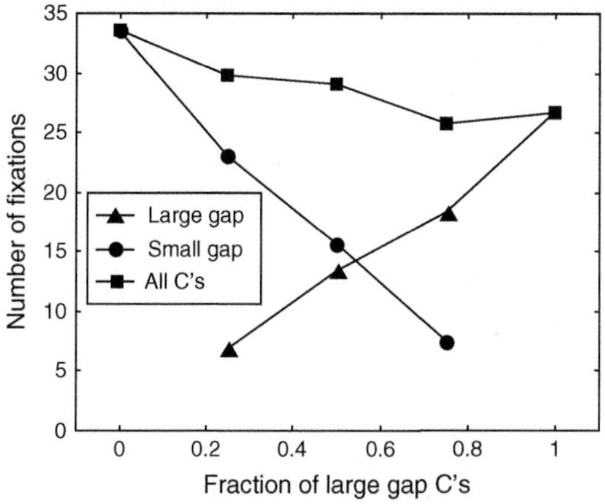

Figure 4. Mean number of fixations vs proportion of large gap C's. Squares denote mean number of fixations. Circles denote mean number of fixations on small gap C's (0.044°). Triangles denote mean number of fixations on large gap C's (0.220°).

2.2. *Fixation time*

As stated in the introduction, both the process-monitoring and pre-programming models yield different predictions with respect to fixation time. Figure 5 shows fixation times on large and small gap C's separately. Fixations on small gap C's are 40 ms longer than fixations on large gap C's [$t(5) = 7.37$, $p = 0.00035$]. Fixation times on small gap C's

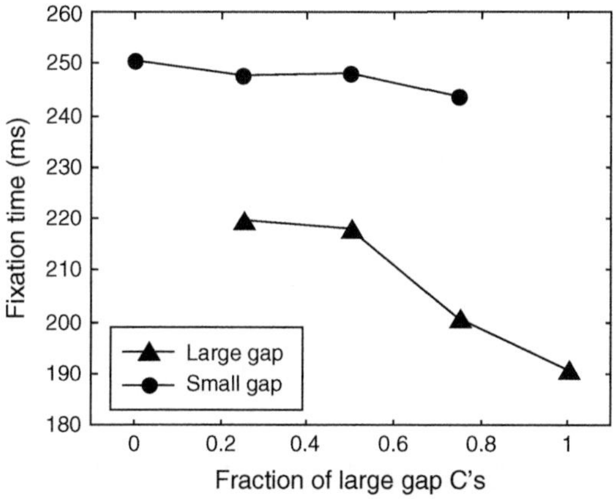

Figure 5. Fixation times for large and small gap C's as function of proportion of large gap C's. Circles denote mean number of fixations on small gap C's (0.044°). Triangles denote mean number of fixations on large gap C's (0.220°).

do not depend on proportion of large gap C's $[F(3, 15) = 0.575, p = 0.64]$. Fixation times on large gap C's decrease from 220 to 192 ms with proportion of large gap C's $[F(3, 15) = 24.22, \; p < 0.00001]$.

Following the rationale of Figure 1, Figure 5 clearly shows that process-monitoring plays a role during search, but we cannot conclude that fixation times are exclusively controlled as in a process-monitoring model. Figure 5 also shows that fixation time on large gap C's depends on the proportion of difficult non-targets (compare to Figure 1a), which is an indication that pre-programming is involved. This will be discussed later.

2.3. *Relation between subsequent fixations*

As discussed in the introduction, in strict process-monitoring models, the analysis time of the currently fixated stimulus element determines the fixation time on that element. In contrast, in pre-programming models an estimate of the analysis time (based on earlier fixations) is used to pre-program fixation time. If control of fixation time has aspects of pre-programming, we expect the direct fixation history to influence succeeding fixation times. To find out whether previously fixated elements influence the fixation time on currently fixated elements, we plotted mean fixation duration on small gap C's after fixation on large gap C's and after small gap C's separately (Figure 6). We did the same for fixation times on large gap C's.

In Figure 5, it is shown that difficult elements (small gap C's) were fixated longer than easy elements. We see the same in Figure 6. But here we see a remarkable interaction $[F(5, 1) = 9.9, p = 0.0254]$: If a fixation on a large gap C is preceded by fixation of a C with small gap, fixation is 18 ms $[t(5) = 6.49, p = 0.00065]$ longer than when it is preceded by the fixation of a large gap C. In case of fixation of a small gap C, the type of fixated C that preceded does not make a difference $[t(5) = 0.588, p = 0.709]$.

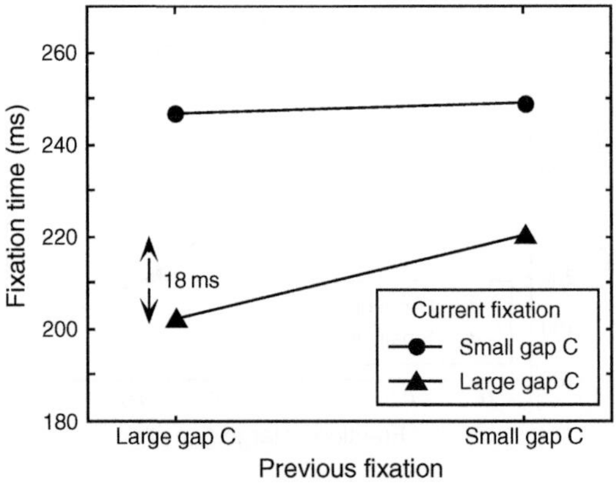

Figure 6. Fixation time on both large and small gap C's vs type of previously fixated element. Circles denote fixations on small gap C's; triangles denote fixations on large gap C's.

In summary, when the observer is confronted with an easier element than during the previous fixation, fixation duration does not change; if the task is more difficult than the previous one, fixation is extended.

3. Discussion

We questioned whether fixation duration is controlled as in a pre-programming or a process-monitoring model. We engaged subjects in a search task that consisted of one target and 48 non-targets. We used easy and difficult non-targets that were present in different proportions in the displays. We validated these stimuli by checking the search times (see Figure 3). For distinguishing the O from the large gap C's lesser time was required than for distinguishing it from the small gap C's.

We compared our data to two extreme models. According to the process-monitoring model, the fixation time should only depend on the element fixated (i.e., longer fixation times on difficult elements). In the second model, fixation time is pre-programmed and the information required to do so is gathered during previous fixations. In our fixation time data, we found evidence for both pre-programming and process-monitoring.

1) Process-monitoring: In mixed displays, fixations on difficult elements were 20–40 ms longer than fixations on easy elements (see Figure 5). This is clear evidence that the foveated stimulus affects fixation time.
2) Pre-programming: (a) Fixation time depended on the proportion of difficult elements in the display (see Figure 5). Easy elements were fixated shorter when the displays contained many large gap C's. (b) We also found that easy elements were fixated longer (up to 18 ms) after fixation on a difficult element than after fixation on an easy element. This is a clear demonstration that fixation history plays a role in the control of fixation duration. The data also suggest that control of fixation duration behaves in a rather conservative way. Extending fixation occurs immediately (difficult after easy leads to extension of fixation time), whereas shortening does not occur immediately (easy after difficult does not lead to shortening of fixation time).

3.1. A new pre-programming strategy

We have a rather speculative alternative explanation of the present data. In this explanation the majority of the fixations is pre-programmed, but process-monitoring still occurs. This explanation was inspired by the new observation of rapid fixation duration increase and slow fixation duration decrease. The former may be caused by a strategy in which saccade programming starts after a fixed amount of time in parallel with visual analysis of the fixated element (McPeek et al., 2000). We will refer to this time as saccade start time (SST). SST is set in such way that analysis of a difficult non-target is possible before the eyes leave the difficult non-target (150 ms after starting saccade programming, Becker & Jürgens, 1979). When the observer fixates an easy non-target it should be efficient to shorten that fixation. However, we assume that the observer finds out that he is fixating an easy non-target after SST. Thus the observer is too late to advance saccade programming. For the next fixation, a shorter SST is set. This explains slow fixation duration decrease. How does this explanation account for rapid fixation duration increase? When the easy non-target fixation is followed by a difficult non-target fixation, the visual system may find out that the object is not an easy non-target. Again we assume that this occurs after the new shorter SST. The saccade that is still being programmed is cancelled and a new saccade is programmed. In this way the fixation of a difficult non-target is extended. Based on the present results we prefer the first explanation instead of this alternative explanation. However, new experiments should be designed to explore this alternative model, because it is in agreement with existing data of, for example, Rayner and Pollatsek (1981) who showed that immediate extension of fixation time is possible.

3.2. Adjustment of fixation duration

Are fixations adjusted to demands of the visual task in this experiment? This is a difficult question because beforehand we do not know the demands of the search task (i.e., distribution of visual analysis times). However, we can compare performance on easy and difficult C's. If the adjustment of fixation duration is similar for easy and difficult C's (e.g., start programming a saccade when there is a certain level of certainty about the nature of the element fixated (Rayner & Pollatsek, 1981)), we expect inspection of easy and difficult C's to be equally efficient. This means, for example, that the relative number of fixations for the easy and difficult C's has to be equal. The relative number of fixations was computed by dividing the number of fixations on easy or difficult C's by their respective numbers. Figure 7 shows that the relative number of fixations on easy C's is lower than the relative number of fixations on difficult C's (The difference is 0.150;

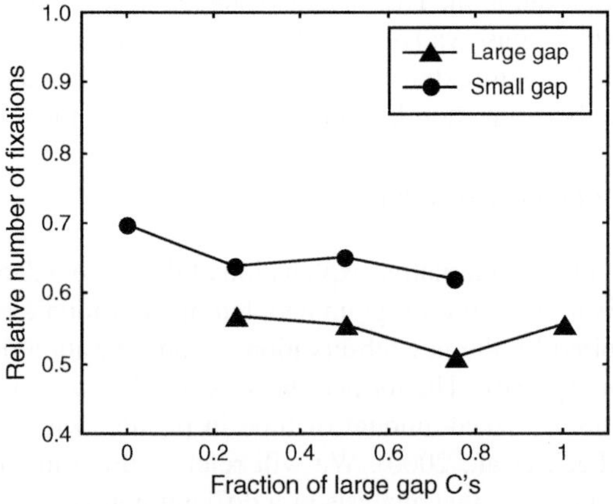

Figure 7. Relative number of fixations. Circles denote fixations on small gap C's; triangles denote fixations on large gap C's. The relative number of fixations is computed by dividing the number of fixations (at large or small gap C's) by the number of elements (with large or small gap).

$t(5) = 4.20$, $p = 0.0043$). This can be interpreted as follows: Search of easy elements was more efficient than search of difficult elements (despite the longer fixation time on difficult elements). This may indicate that fixation times on easy C's are differently adjusted than fixation times on difficult C's, which we see as an argument against process-monitoring. For, what would be the purpose of process-monitoring if its results do not appear in the efficiency of search? However, we also have three alternative explanations.

1. Despite the masks around the stimulus elements, we cannot exclude the possibility that observers have some preview of neighbour elements. If this occurs, it may be possible that in a mixed display difficult C's are more likely to be fixated than easy C's because they resemble the target more. This may explain the higher relative number of fixations on difficult C's.
2. Another explanation is a possible floor effect for fixation time. If the observers had been able to fixate shorter on easy C's they would have done that (at the expense of a higher number of fixations on easy C's). They probably could not search faster on easy C's.
3. Fixation of easy C's is longer when difficult C's are present (see Figure 5). This longer fixation time may cause preview benefits because time is left to analyse neighbours. The effect of preview is described above.

3.3. *The role of the return saccade*

Is accurate adjustment of fixation duration required for successful search? In our opinion, perfect adjustment of fixation time to the demands of the immediate visual task is not necessary because observers always have the possibility to make a return saccade. They can make a return saccade if they have information that the previous fixation was too short to analyse the previously fixated object. At the cost of one additional fixation they may fixate for too short a duration. In search and viewing tasks, observers make many return saccades (Hooge, Over, van Wezel, & Frens, 2005). In fact it would be a smart strategy to decrease fixation time during search until a return saccade is needed to re-inspect a specific part of the scene. The occurrence of a return saccade then may act as a signal to increase fixation time. At the moment this issue is investigated in our lab.

3.4. *Interpreting reaction times*

The majority of the visual search experiments are done with reaction times only. This experiment clearly shows that stimulus properties (such as target/non-target dissimilarity) are not sufficient to explain search times in detail. Figure 3 shows that search time drops for higher fractions of easy non-targets. This drop can be understood if fixation times are taken into account (see Figure 5) because as a consequence of the conservative fixation behaviour, easy non-targets are fixated longer in the presence of difficult non-targets than when presented without the presence of difficult non-targets. The drop in Figure 3 is mainly due to this effect. There is another factor that influenced search times.

In the present experiment efficiency of search is higher for easy elements than for difficult elements (Figure 7). This is a more subtle effect for which eye movement measures are also required. Oculomotor control is a factor that may influence search times especially in tasks that more resemble realistic search tasks in which more than one saccade is required to find the target. Visual search investigators should measure eye movements more often because nowadays eye trackers (such as the EyeLink 2) are cheaper, better and much easier to handle than in the past.

4. Conclusion

In a search task with two kinds of non-targets, we found the fixation time to be dependent on both the immediate stimulus and the fixation history. This is interpreted as evidence for both process-monitoring (Greene & Rayner, 2001) and pre-programming (Hooge & Erkelens, 1998). The contribution of both components on fixation time may depend on the task and the amount of useful information (such as the possibility to preview, homogeneity of the stimulus elements) in the displays.

The process-monitoring component was found to set fixation time differently for easy and difficult stimulus elements. Despite the longer fixation times on difficult elements, search of difficult elements was less efficient, than search of easy elements. The pre-programming component of fixation duration control was found to be conservative such that fixations following fixations on difficult elements were longer than those following fixations on easy elements.

Acknowledgements

We thank Jos van der Geest for providing us with his excellent Matlab routine for computing eye velocity, Ryota Kanai for useful software advice for our eye movement setup and Harold Greene and an anonymous reviewer for helpful comments.

References

Becker, W., & Jürgens, R. (1979). An analysis of the saccadic system by means of double step stimuli. *Vision Research, 19*, 967–983.

Brainard, D. H. (1997). The psychophysics toolbox. *Spatial Vision, 10*, 433–436.

Cornelissen, F. W., Bruin, K. J., & Kooijman, A. C. (2005) The Influence of artificial Scotomas on eye movements during visual search. *Optometry and Vision Science, 82*(1), 27–35.

Cornelissen, F. W., Peters. E., & Palmer, J. (2002). The Eyelink toolbox: Eye tracking with MATLAB and the psychophysics toolbox. *Behavior Research Methods, Instruments and Computers, 34*, 613–617.

Duncan, J., & Humphreys, G. W. (1989). Visual search and stimulus similarity. *Psychological Review, 96*(3), 433–458.

Greene, H. H., & Rayner, K. (2001). Eye movement control in direction coded search. *Perception, 30*, 147–157.

Harris C. M., Hainline L., Abramov I., Lemerise E., & Camenzuli C., 1988. The distribution of fixation durations in infants and naive adults. *Vision Research,. 28*(3), 419–432.

Hooge I. Th. C. & Erkelens, C. J. (1996). Control of fixation duration in a simple visual search task. *Perception and Psychophysics, 58*(7), 969–976.

Hooge I. Th. C., & Erkelens, C. J. (1998). Adjustment of fixation duration in visual search. *Vision Research, 38*(9), 1295–1302.

Hooge, I. Th. C., Over, E. A. B., van Wezel, R. J. A., & Frens, M. A. (2005) Inhibition of return is not a foraging facilitator in saccadic search and free viewing. *Vision Research, 45*, 1901–1908.

Jacobs, A. M. (1986). Eye-movement control in visual search: How direct is visual span control? *Perception and Psychophysics, 39*, 47–58.

McPeek, R. M., Skavenski, A. A., & Nakayama, K. (2000). Concurrent processing of saccades in visual search. *Vision Research, 40*, 2499–2516.

Näsänen, R., Ojanpää, H., & Kojo, I. (2001). Effect of stimulus contrast on performance and eye movements in visual search. *Vision Research, 41*(14), 1817–1824.

Pelli, D. G. (1997). The VideoToolbox software for visual psychophysics: Transforming numbers into movies. *Spatial Vision, 10*, 437–442.

Rayner, K. (1978). Eye movements in reading and information processing. *Psychological Bulletin, 85*, 618–660.

Rayner, K. (1998). Eye movements in reading and information processing: 20 years of research. *Psychological Bulletin, 124*(4), 372–422.

Rayner, K., & Fisher, D. L. (1987). Letter processing during eye fixations in visual search. *Perception and Psychophysics, 42*, 87–100.

Rayner, K., & Pollatsek, A. (1981). Eye movement control during reading: Evidence for direct control. *Quarterly Journal of Experimental Psychology, 33A*, 351–373.

Smeets, J. B. J., & Hooge I. Th. C. (2003) Nature of variability in saccades. *Journal of Neurophysiology, 90*, 12–20.

Treisman, A., & Gormican, S., (1988). Feature analysis in early vision: Evidence from search asymmetries. *Psychological Review, 95*, 15–48.

Van der Geest, J. N., & Frens, M. A. (2002). Recording eye movements with video-oculography and scleral search coils: a direct comparison of two methods. *Journal of Neuroscience Methods, 114*, 185–195.

Van der Steen, J., & Bruno, P. (1995). Unequal amplitude saccades produced by aniseikonic patterns: Effects of viewing distance. *Vision Research, 35*, 3459–3471.

Vaughan, J. (1982). Control of fixation duration in visual search and memeory search: Another look. *Journal of Experimental Psychology: Human Perception and Performance, 8*, 709–723.

Vaughan, J., & Graefe, T. M. (1977). Delay of stimulus presentation after the saccade in visual search. *Perception and Psychophysics, 22*, 201–205.

Viviani, P. (1990). Eye movements in visual search: Cognitive, perceptual and motor control aspects. *Reviews of Oculomotor Research, 4*, 353–393.

Vlaskamp, B. N. S., Over, E. A. B., & Hooge, I. Th. C., (2005). Saccadic search performance: The effect of element spacing. *Experimental Brain Research, 3*, 1–14.

Chapter 28

EFFECTS OF CONTEXT AND INSTRUCTION ON THE GUIDANCE OF EYE MOVEMENTS DURING A CONJUNCTIVE VISUAL SEARCH TASK

JIYE SHEN, AVA ELAHIPANAH, and EYAL M. REINGOLD

University of Toronto, Mississauga, Canada

Eye Movements: A Window on Mind and Brain

Edited by R. P. G. van Gompel, M. H. Fischer, W. S. Murray and R. L. Hill

Abstract

The current study explored the effects of contextual and instructional manipulations on visual search behavior in a conjunctive search task with a distractor-ratio manipulation. Participants' eye movements were monitored when they performed the task. Results from the present investigation demonstrated that block context modulates the temporal dynamics of visual search by shortening fixation durations, reducing the number of fixations, enhancing task-relevant subset selection, and as a result, yielding faster manual responses in those trials that were congruent with the block context than in incongruent trials. The current study also suggests that the presence of top-down influences associated with an instructional manipulation operates over and above any influences of context.

The current study explored the effects of contextual and instructional manipulations on visual search behavior in a conjunctive search task with a distractor-ratio manipulation. In a typical conjunctive visual search task, each trial contains an equal number of distractors of each type; the total number of items within a search display (display size) is manipulated and search efficiency is examined by measuring the change in response time and/or error rate as a function of display size (see Treisman, 1988; Wolfe, 1998 for a review). However, several recent studies have shown that visual search performance is sensitive to the ratio between the types of distractors, even when the total number of items in a display remains constant (e.g., Bacon & Egeth, 1997; Egeth, Virzi, & Garbart, 1984; Kaptein, Theeuwes, & van der Heijden, 1995; Poisson & Wilkinson, 1992; Shen, Reingold, & Pomplun, 2000, 2003; Sobel & Cave, 2002; Zohary, & Hochstein, 1989).

1. Subset-selective processing and distractor-ratio effect

The original feature-integration theory of visual search (Treisman & Gelade, 1980; Treisman, 1988) proposes the existence of preattentive feature maps, one for each stimulus dimension (such as color, shape, orientation, etc.). Information from parallel preattentive processes could only mediate performance in a visual search task if the target was defined by the presence of a unique feature (i.e., feature search), such as searching for a green *X* among red and blue *X*s. However, if the target is defined by a specific combination of features (i.e., conjunction search), such as searching for a green *X* among red *X*s and green *O*s, attention is necessary to locally combine the information from the corresponding feature maps. As a result, participants have to inspect the search display in a serial item-by-item fashion until target detection or exhaustive search.

In contrast to the arguments by the original feature-integration theory, subsequent studies have demonstrated that participants could selectively limit their search to only a subset of distractors to improve search efficiency (e.g., Carter, 1982; Egeth et al., 1984; Friedman-Hill & Wolfe, 1995; Kaptein et al., 1995). For example, Egeth et al. (1984) had participants search for a red *O* (target) among red *N*s (same-color distractors) and black *O*s (same-shape distractors). They varied the number of distractors of one type (i.e., same-color or same-shape) while keeping the number of distractors of the other type constant. Participants were instructed to try to restrict their search to the subset of items that were kept constant in number. Egeth et al. found that search times were independent of the number of items in the uninstructed subset, regardless of whether participants were instructed to attend to the color subset or the shape subset. Similarly, studies that have examined patterns of eye movements during visual search have provided strong evidence of bias in the distribution of saccadic endpoints across different types of distractors. Stimulus dimensions such as color, shape, contrast polarity, and size have been shown to guide the search process (e.g., Bichot & Schall, 1998; Findlay, 1997; Findlay & Gilchrist, 1998; Hooge & Erkelens, 1999; Motter & Belky, 1998; Pomplun, Reingold, & Shen, 2001, 2003; Scialfa & Joffe, 1998; Shen et al., 2000; Shen, Reingold, Pomplun, & D. E. Williams, 2003; D. E. Williams & Reingold, 2001).

Evidence for subset-selective processing also comes from studies using the *distractor-ratio manipulation*. In a color × orientation conjunction search task, Zohary and Hochstein (1989) asked participants to decide whether a red horizontal bar was present among an array of red vertical (same-color distractors) and green horizontal (same-orientation distractors) bars. The search display was presented very briefly (50 ms) and then, after a variable interval (stimulus onset asynchrony, SOA), masked. One critical manipulation in this study was the ratio between the two types of distractors (same-color vs same-orientation) presented in a given array. The number of same-color distractors ranged from 0 to 64, in increments of 4, while the total number of items was held constant at 64. Zohary and Hochstein found that the SOA required to reach a 70% correct response rate was a quadratic function of the number of distractors sharing color with the search target. Specifically, detection was relatively easy for displays with extreme distractor ratios (i.e., either the same-color or same-orientation distractors were rare) but relatively difficult for displays in which the two types of distractors were equally represented. In addition, these investigators found that the performance curve was not completely symmetrical (it took a shorter SOA for participants to reach a certain level of accuracy for displays with few same-color distractors than for displays with a comparable number of same-orientation distractors). The finding that visual-search efficiency in a conjunctive search task depends on the relative frequency of the two types of distractors is consistent with the notion of subset-selective processing and has been referred to as the *distractor-ratio effect* (Bacon & Egeth, 1997).

In addition to the accuracy across SOA measure (Zohary & Hochstein, 1989), the distractor-ratio effect has also been observed in studies measuring response times (Bacon & Egeth, 1997; Poisson & Wilkinson, 1992; Sobel & Cave, 2002). Shen et al. (2000) further examined participants' patterns of eye movements, the spatial distribution of saccadic endpoints in particular, during the search process. They employed a color × shape conjunction search task and systematically manipulated the ratio between the same-color and same-shape distractors in a display. They found a quadratic change in search performance measures such as manual response time, number of fixations per trial, and initial saccadic latency as a function of distractor ratio. Search performance was worse when the ratio between the same-color and same-shape distractors approximated 1:1 and gradually improved as the ratio deviated from 1:1, with performance being best at extreme distractor ratios (i.e., very few distractors of one type). More importantly, Shen et al. demonstrated that when there were very few same-color distractors, participants' saccadic endpoints were biased towards the color dimension whereas when there were very few same-shape distractors, saccades were biased towards the shape dimension. Results from that study suggest that in a distractor-ratio paradigm, participants take advantage of the display information and flexibly switch between different subsets of distractors on a trial-by-trial basis.

2. Bottom-up and top-down processing

Visual attention is currently thought to be controlled by two distinct mechanisms: one is the top-down or goal-directed control, in which the deployment of attention is determined by the observer's knowledge or intentional state, and the other is bottom-up or stimulus-driven control, in which attention is driven by certain aspects of the stimulus, irrespective of the observer's current goals or intent (Egeth & Yantis, 1997; Wolfe, Butcher, Lee, & Hyle, 2003; Yantis, 1998). Various attention and visual search studies have proposed the involvement of both the top-down and bottom-up processes in determining attentional allocation and visual search performance (e.g., Bacon & Egeth, 1994, 1997; Bravo & Nakayama, 1992; Cave & Wolfe, 1990; Hillstrom, 2000; Irwin, Colcombe, Kramer, & Hahn, 2000; Wolfe, 1994).

The finding that participants were able to select and search through the smaller subset of distractors, even when different levels of distractor ratios were randomly mixed in the same block of trials suggests that bottom-up processing driven by the saliency of display items (Koch & Ullman, 1985) plays a role in mediating the distractor-ratio effect (Wolfe, 1994). This has been supported by studies manipulating stimulus discriminability directly. For example, Shen et al. (2003) varied stimulus discriminability along the shape dimension in a color × shape conjunction task (*X* vs *O* was used in the high-discriminability condition whereas in the low-discriminability condition, the shapes were *X* vs *K* for half of the participants and *O* vs *Q* for the other half). These investigators found that, in the high-discriminability condition, participants searched through different subsets of distractors on the basis of distractor ratio (i.e., color subset for displays with very few same-color distractors and shape subset for displays with very few same-shape distractors). In marked contrast, in the low-discriminability condition, saccades were consistently biased towards the color dimension, irrespective of the distractor-ratio manipulation. This suggests that participants searched through the more informative, but not necessarily smaller, subset of distractors. Sobel and Cave (2002) similarly found that in a color × orientation conjunction task, when the difference between orientations was large, participants were able to search through the smaller subset of distractors whereas when orientation differences were small, participants tended to search through the subset of distractors sharing the target color.

The influence of top-down controls in the distractor-ratio effect has been examined in a recent study by Bacon and Egeth (1997), which provided participants with instructions designed to induce particular search strategies (see also Egeth et al., 1984; Kaptein et al., 1995; Sobel & Cave, 2002). With a color × orientation conjunction search task, these investigators manipulated the frequency of trials with very few same-color distractors and trials with very few same-orientation distractors. Participants were informed of which distractor type would be less frequent on most trials and were asked to restrict their search to a subset of distractors (e.g., attend to red items or attend to vertical items). To demonstrate the role of top-down processing in the distractor-ratio effect, these investigators contrasted two experimental conditions, one in which the instruction to the participants was congruent with the optimal search strategy for a given trial (i.e., top-down and bottom-up processes working in concert) with another condition in which the instruction was incongruent with the optimal search strategy (i.e., top-down and bottom-up processes working in opposition). Bacon and Egeth found that search performance was poorer when the optimal search strategy and instruction mismatched than when they matched, and thus demonstrated the role of top-down processing in the distractor-ratio paradigm.

The current study extended Bacon and Egeth (1997) by further investigating the influences of top-down controls on search performance in a distractor-ratio paradigm. With a color × shape conjunction search task, search displays which contained either very few same-color distractors (henceforth color-search displays) or very few same-shape distractors (henceforth shape-search displays) were created. In each block of trials, the frequency of these two types of search displays was manipulated, creating two types of block context. Block context refers to the fact that, on average, the previously experienced trials will have had few distractors with the relevant feature. In half of the blocks, 80% of the trials were color-search displays and 20% of the trials were shape-search displays. In these blocks, color-search displays were congruent with the block context (henceforth, color-congruent trials) whereas the shape-search displays were incongruent with the block context (henceforth, shape-incongruent trials). In the other half of the blocks, 80% of the trials were shape-search displays and 20% of the trials were color-search displays. Thus, in these blocks, color-search displays were incongruent with the block context (henceforth, color-incongruent trials) whereas the shape-search displays were congruent with the block context (henceforth, shape-congruent trials). For both color-search displays and shape-search displays, performance on the congruent trials and incongruent trials was contrasted. A finding of better search performance in the congruent trials than in the incongruent trials (i.e., a congruency effect) would reflect the influence of block context (e.g., Bichot & Schall, 1999; Chun & Jiang, 1999; Irwin, Colcombe, Kramer, & Hahn, 2000; Hillstrom, 2000; see Chun, 2000 for a review).

Furthermore, two groups of participants were included in this experiment. In one condition, similar to Bacon and Egeth (1997), Egeth et al. (1984), and Kaptein et al. (1995), participants were informed of the composition of trials and provided explicit instructions regarding the optimal search strategies (i.e., attend to the color or attend to the shape). For another group of participants, exactly the same search trials were employed except that no explicit instructions concerning search strategies were provided. A comparison across the two conditions permitted a separate evaluation of the influences on search performance due to instruction vs the influences due to block context.

The second goal of the current study is to examine *how* the instructional and contextual manipulations, if observed, influence the spatiotemporal dynamics of visual search behavior. The above-mentioned studies that have examined the top-down controls (Bacon & Egeth, 1997; Egeth et al., 1984; Kaptein et al., 1995; Sobel & Cave, 2002) focused on global search performance measures such as response time and error rate only. Therefore, it remains unknown exactly what aspects of search behavior are changed owing to the top-down controls and whether there are qualitative and quantitative differences in search behavior between the two forms (contextual vs instructional) of top-down controls. For example, could the contextual or instructional congruency effect in manual response time be due to less time spent, upon the presentation of the search display, to segment the search display and decide upon which subset to search through? Could it be that the availability of top-down information makes the rejection of distractor items easier? Or could it be due to a higher bias in the selection of the relevant subset of distractors in the congruent trials, which leads to fewer fixations on the uninformative subset of distractors? To explore these possibilities, the current study also examined participants' eye movements during the search process.

3. Method

3.1. Participants

Twenty-four undergraduate students, with normal or corrected-to-normal visual acuity and normal color vision, participated in three one-hour sessions. They were paid $30 for their participation. A between-subject design was adopted, with half of the participants performing in the instruction condition and the other half in the no-instruction condition. Although this design is less powerful than the within-subject design due to individual differences within each group, it permitted a separate evaluation of the influences on search performance due to instruction vs the influences due to block context without contamination.

3.2. Apparatus

The experiment was run in a lighted room and the luminance of the walls was approximately $30\,cd/m^2$. The eyetracker employed in the current study was the SR Research Ltd. EyeLink I system. This system has high spatial resolution (0.005°) and a sampling rate of 250 Hz (4-ms temporal resolution). Stimulus displays were presented on two monitors, one for the participant (a 19-inch Samsung SyncMaster 900P monitor with a refresh rate of 120 Hz and a screen resolution of 800×600 pixels) and the other for the experimenter. The experimenter monitor was used to give feedback in real-time about the participant's computed gaze position. This allowed the experimenter to evaluate system accuracy and to initiate a recalibration if necessary. In general, the average error in the computation of gaze position was less than 0.5°. Participant made a response by pressing one of two buttons on a response box connected to the EyeLink I system.

3.3. Stimuli and design

Similar to Shen et al. (2000, 2003), display items were created by combining features on two stimulus dimensions: color (red vs green) and shape (*X* vs *O*). All display items were presented in a $15.5° \times 15.5°$ field at a viewing distance of 91 cm. Each individual item subtended 0.8° both vertically and horizontally; the minimum distance between neighboring items was 2.0°. Participants were asked to search for the target item, a green *X*, among distractors that had either the same color (green *O*s, same-color distractors) or the same shape (red *X*s, same-shape distractors) as the target. An equal number of target-present and target-absent trials were used. In each trial, the search display contained either very few color-search displays (see Figure 1a for an example) or very few shape-search displays (see Figure 1b for an example). For both types of search displays, the number of items belonging to the smaller subset was 12 on average (8, 12, or 16) while the total number of items within a display was fixed at 48. The CIE coordinates for the colors were (.582, .350) for red and (.313, .545) for green.

One critical manipulation of the current study was the frequency of the color-search and shape-search displays in a block of trials. In half of the blocks, 80% of the trials presented color-search displays and 20% of the trials presented same-shape distractors, while the opposite distribution of trial types occurred in the other half of blocks. The type of search displays that occurred more frequently within a block of trials was labeled as "congruent" and the infrequent type as "incongruent". Color-congruent and shape-congruent blocks were tested in alternate blocks. The current experiment also manipulated the instructions given to the participants. In one condition (henceforth, instruction condition), participants were provided additional instructions regarding the composition of search trials in a block and the search strategies that worked for the congruent trials (e.g., in a color-congruent block, subjects were given the following instructions: "In this block, most of the displays contain fewer green *O*s than red *X*s. Therefore, you can maximize your search efficiency by paying attention to the green items"). In the other condition (henceforth, no-instruction condition), no such instructions were provided to the participants.

A four-factor factorial design was implemented in the current study, with target presence (present vs absent), display type (color-search vs shape-search), trial congruency (congruent vs incongruent) as within-subject factors and instructional manipulation (instruction vs no instruction) as the between-subject factor. Participants performed a total of 2,160 trials across three individual sessions, which amounted to 432 congruent trials and 108

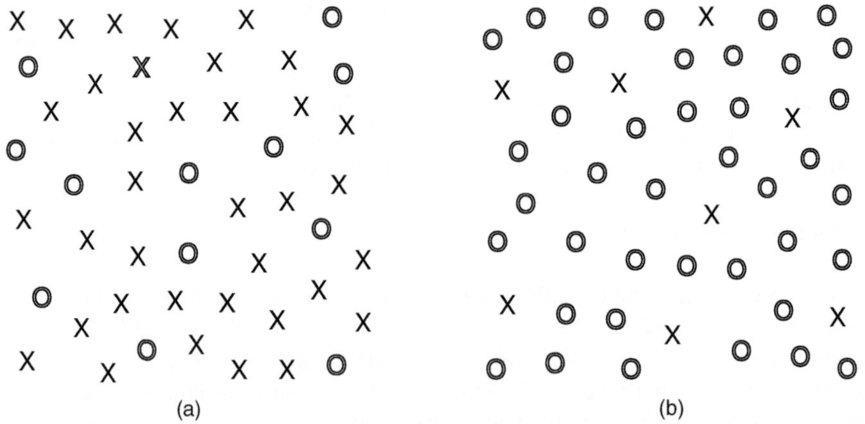

Figure 1. Sample search displays used in the current study (target was a green *X* and the distractors were green *O*s and red *X*s). Red items shown in black; green items shown in white. (a) The panel illustrates a "color-search" display (target present). (b) The panel shows an example of "shape-search" display (target absent).

incongruent trials for each combination of target presence and display type. At the beginning of each session, participants received two practice blocks of 16 trials.

3.4. Procedure

Participants were informed of the identities of the search target and distractor items before the experiment started. They were asked to look for the target item, and indicate whether it was in the display or not by pressing an appropriate button as quickly and accurately as possible.

A 9-point calibration procedure was performed at the beginning of the experiment, followed by a 9-point calibration accuracy test. Calibration was repeated if any point was in error by more than 1° or if the average error for all points was greater than 0.5°. Each trial started with a drift correction in the gaze position. Participants were instructed to fixate on a black dot in the center of the computer screen and then press a start button to initiate a trial. The trial terminated if participants pressed one of the response buttons or if no response was made within 20 s. The time between display onset and the participant's response was recorded as the response time.

4. Results and discussion

Because of an erroneous response, 4.9% (3.3% in target-absent trials and 6.5% in target-present trials) of trials in the instruction condition and 2.0% (0.8% in target-absent trials and 3.2% in target-present trials) of trials in the no-instruction condition were removed. In addition, those trials with a saccade or a blink overlapping the onset of the search display or with an excessively long or short response time (more than 3.0 standard deviations above or below the mean for each cell of the design) were excluded from further analysis. These exclusions accounted for 2.8 and 2.5% of all trials respectively in the instruction condition, and 1.9 and 2.9% of all trials respectively in the no-instruction condition.

The current study examined *how* the instructional and contextual manipulations influence the spatiotemporal dynamics of visual search behavior by analyzing response time and eye movement measures such as number of fixations per trial, latency to move, fixation duration, and saccadic bias. Both response time and fixation number are global search performance measures on how efficiently participants can determine target presence in an array. The initial saccadic latency and fixation duration provide fine-grained temporal information on the search process. These analyses were to extend previous investigation on the patterns of eye movements in visual search tasks (e.g., Binello, Mannan, & Ruddock, 1995; D. E. Williams, Reingold, Moscovitch, & Behrmann, 1997; Zelinsky & Sheinberg, 1997). All these measures were analyzed with a repeated-measures ANOVA, with target presence (2: target present vs target absent), display type (2: color-search vs shape-search), and trial congruency (2: congruent vs incongruent) as within-subject factors and instructional manipulation (2: instruction vs no instruction) as the between-subject factor. In addition, the bias in the distribution of saccadic endpoints (e.g., Findlay, 1997;

Findlay & Gilchrist, 1998; Hooge & Erkelens, 1999; Motter & Belky, 1998; Pomplun, et al., 2001, 2003; Scialfa & Joffe, 1998; Shen et al., 2000, 2003; D. E. Williams & Reingold, 2001) was examined to reveal strategies used by participants in performing conjunctive visual searches.

4.1. Response time

Table 1 shows the average response time as a function of target presence, display type, and trial congruency in both the instruction condition and the no-instruction condition. The repeated-measures ANOVA revealed a significant main effect of target presence, $F(1, 22) = 107.63$, Mse $= 136863.50$, $p < 0.001$, indicating that response time was shorter in target-present trials than in target-absent trials. The instruction condition and the no-instruction condition did not differ in overall response time, $F < 1$. Consistent with our previous studies (Shen et al., 2000, 2003), search was more efficient for the color-search displays than for the shape-search displays, $F(1, 22) = 20.48$, Mse $=$ 19292.14, $p < 0.001$, and this difference was more pronounced in the target-absent trials than in the target-present trials, as indicated by a significant display type $\times$ target presence interaction, $F(1, 22) = 23.60$, Mse $= 17851.35$, $p < 0.001$.

More importantly, the repeated-measures ANOVA revealed a significant effect of trial congruency, $F(1, 22) = 33.86$, Mse $= 21973.47$, $p < 0.001$, indicating that response time was shorter in those trials that were either consistent with the instruction or the block

Table 1
Response time (in ms), number of fixations per trial, initial saccadic latency (in ms), and fixation duration (in ms) as a function of target presence, display type, trial congruency, and instructional manipulation (CT = congruent trial; IT = incongruent trial).

	Target-absent Trials				Target-present Trials			
	Color search		Shape search		Color search		Shape search	
	CT	IT	CT	IT	CT	IT	CT	IT
Response time								
Instruction	1197	1416	1369	1646	835	915	838	915
No instruction	1282	1346	1410	1553	794	856	780	855
Number of fixations								
Instruction	4.28	5.02	4.83	5.82	2.56	2.86	2.52	2.76
No instruction	4.68	4.89	5.14	5.66	2.36	2.67	2.32	2.52
Initial saccadic latency								
Instruction	224	225	232	232	223	224	225	230
No instruction	224	225	229	232	219	217	219	222
Fixation duration								
Instruction	196	205	202	206	205	205	208	209
No instruction	195	199	199	201	212	203	210	218

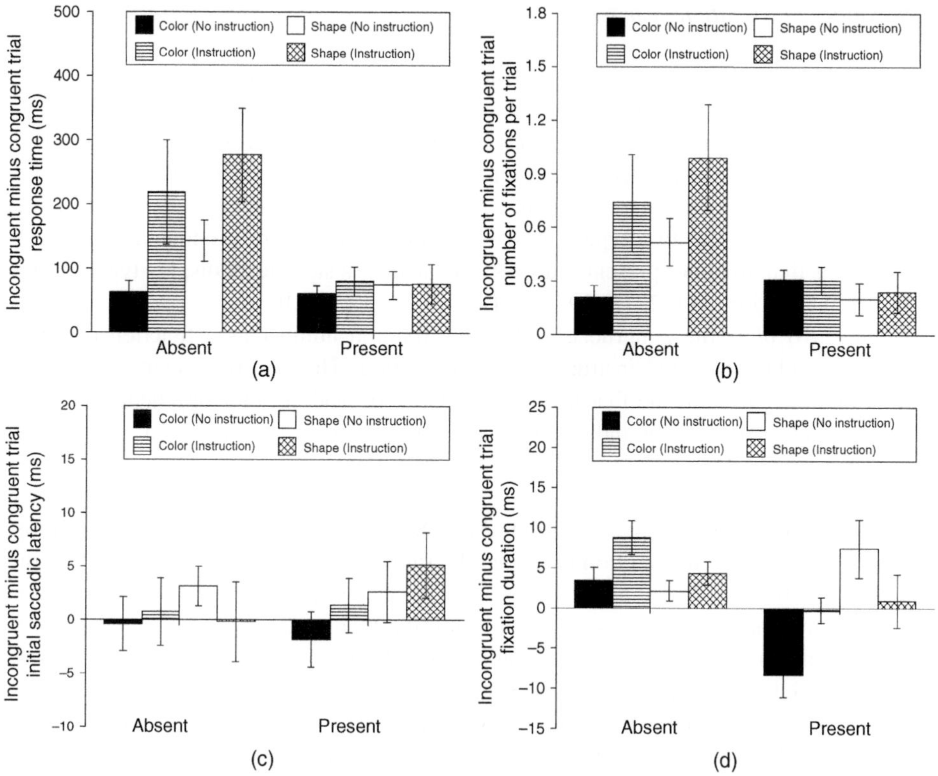

Figure 2. Congruency effect (the difference between the congruent trials and in congruent trials) for response time (a), number of fixations per trial (b), initial saccadic latency (c), and fixation duration (d) as a function of target presence, display type, and instructional manipulation.

context (congruent trials) than in those incongruent trials. Figure 2a plots the amount of congruency effect (i.e., the value of the congruent trials subtracted from that of the incongruent trials) as a function of target presence and display type, in both the instruction condition and the no-instruction condition. The overall congruency effect was greater in target-absent trials than in target-present trials, as indicated by a significant target presence × trial congruency interaction, $F(1, 22) = 11.04$, $\mathrm{Mse} = 11423.61$, $p < 0.001$. The figure also shows that in target-present trials the instruction condition produced the same amount of congruency effect as did the no-instruction condition. However, in target-absent trials, the congruency effect was much stronger in the instruction condition than in the no-instruction condition. These findings were confirmed by a significant interaction of instructional manipulation × target presence × trial congruency, $F(1, 22) = 4.73$, $\mathrm{Mse} = 11423.61$, $p < 0.05$. In addition, the congruency effect for the color-search displays was slightly larger in the shape-search displays, $F(1, 22) = 1.95$, $\mathrm{Mse} = 8328.31$, $p = 0.176$.

4.2. Number of fixations per trial

Table 1 shows that the instruction and no-instruction conditions did not differ in the average number of fixations per trial, $F < 1$. Similar to the response-time data, more fixations were made in the target-absent trials than in the target-present trials, $F(1, 22) = 153.49$, $\text{Mse} = 1.91$, $p < 0.001$. In addition, more fixations were made in the shape-search displays than in the color-search displays, $F(1, 22) = 16.66$, $\text{Mse} = 0.23$, $p < 0.001$; the difference between the two types of search displays was more pronounced in target-absent trials than in target-present trials, as indicated by a significant display type × target presence interaction, $F(1, 22) = 29.75$, $\text{Mse} = 0.21$, $p < 0.001$.

Figure 2(b) plots the congruency effect for fixation number as a function of target presence, display type, and instructional manipulation. The positive congruency values shown in the figure indicate that fewer fixations were made in the congruent trials than in the incongruent trials, as indicated by a significant main effect of trial congruency, $F(1, 22) = 32.81$, $\text{Mse} = 0.28$, $p < 0.001$. Overall, this congruency effect was stronger in target-absent trials than in target-present trials, with a significant interaction between trial congruency and target presence, $F(1, 22) = 8.73$, $\text{Mse} = 0.17$, $p < 0.01$. In addition, although the instruction condition produced the same amount of congruency effect as did the no-instruction condition in target-present trials, the congruency effect in target-absent trials was much stronger in the instruction condition than in the no-instruction condition. This was indicated by a significant interaction of instructional manipulation × target presence × trial congruency, $F(1, 22) = 4.22$, $\text{Mse} = 0.17$, $p < 0.05$. The figure also shows that in target-absent trials, the congruency effect was stronger for shape-search displays than for color-search displays, whereas there was no difference in the amount of congruency effect between the two types of search displays in target-present trials. This was confirmed by a significant target presence × display type × trial congruency interaction, $F(1, 22) = 6.16$, $\text{Mse} = 0.07$, $p < 0.05$.

Results from the current and previous parts reveal a high degree of similarity between response time and number of fixation data. This is not surprising given that both measures reflected global measures of search efficiency and that response time in a visual search task can be accounted for by overt shifts of attention (Findlay & Gilchrist, 1998; Scialfa & Joffe, 1998; D. E. Williams, et al., 1997; Zelinsky & Sheinberg, 1997).

4.3. Initial saccadic latency

Initial saccadic latency is defined as the interval between the onset of a search display and the detection of the first eye movement; this measure may be indicative of the ease of segmenting a search display and identifying an optimal search strategy. Initial saccadic latency was analyzed as a function of target presence, display type, and trial congruency (see Table 1). As in Shen et al. (2000, 2003), the initial saccadic latency was shorter in the color-search displays than in the shape-search displays, $F(1, 22) = 12.48$, $\text{Mse} = 95.51$, $p < 0.01$, and this effect was more pronounced in the target-absent trials than in the target-present trials, as indicated by a significant target presence × display type interaction, $F(1, 22) = 5.78$, $\text{Mse} = 29.26$, $p < 0.05$.

Figure 2(c) plots the amount of congruency effect for the initial saccadic latency. It is clear from the figure that the instructional and contextual manipulation did not have a substantial influence on the initial saccadic latency: There was neither a significant main effect for instruction manipulation, trial congruency, nor a significant interaction between instruction manipulation or trial congruency with other factors, all $F\text{s} < 1.76$, $p\text{s} > 0.199$. This suggests that the execution of first saccades is more influenced by the low-level display characteristics (color-search displays vs shape-search displays). Instead of remaining fixated in the center of the screen to gather initial target information or to check whether the current search display is consistent with optimal search strategy, participants tended to start the search right away (Zelinsky & Sheinberg, 1997). This suggests that top-down controls of visual attention modify the search behavior at a later stage.

4.4. Fixation duration

The repeated-measures ANOVA revealed that the instructional manipulation did not influence the overall fixation duration, $F < 1$. It is clear from Table 1 that average fixation duration was longer in the target-present trials than in the target-absent trials, $F(1, 22) = 10.04$, $\text{Mse} = 326.26$, $p < 0.001$. Consistent with earlier studies (e.g., Galpin & Underwood, in press; Pomplun, Sicheschmidt, Wagner, Clermont, Rickheit, & Ritter, 2001), the longer fixations in the target-present trials were linked to target detection. There was a significant main effect of display type, $F(1, 22) = 18.36$, $\text{Mse} = 44.99$, $p < 0.001$, indicating that fixation duration was longer in the shape-search displays than in the color-search displays. The longer fixation duration, together with more fixations made, may account for the relatively inefficient performance for the shape-search displays compared to the color-search displays.

The repeated-measures ANOVA also revealed a congruency effect for fixation duration, $F(1, 22) = 6.25$, $\text{Mse} = 40.73$, $p < 0.05$, with shorter fixation durations in the congruent trials than in the incongruent trials. There was also a significant instruction manipulation × target presence × display type × trial congruency interaction, $F(1, 22) = 8.13$, $\text{Mse} = 47.38$, $p < 0.05$. Figure 2(d) shows the amount of congruency effect for each cell of the design (target presence × display type × instructional manipulation). As can be seen from the figure, the fixation-duration congruency effect was consistently found in the target-absent trials, regardless of the display type or instructional manipulation. In target-present trials, however, a reliable congruency effect was observed only in the shape-search displays in the no-instruction condition. The figure also shows a reversed congruency effect for the color-search displays in the no-instruction condition. This unexpected finding was probably due to a surprisingly large reversal among three participants in that condition. The effects of congruency on fixation duration suggest that the availability of accurate top-down information makes the rejection of distractor items easier and thus speeds up the search process.

4.5. Saccadic selectivity

Was the distribution of saccadic endpoints influenced by the trial congruency manipulation? If the guidance of eye movements were solely driven by the salience of display items, the same level of saccadic selectivity should be observed across the congruent and incongruent trials, given that exactly the same search displays were used (i.e., with the same bottom-up activation). To examine this, bias in saccadic distribution for each type of search display was calculated – the baseline probability of fixation 25% (i.e., the number of items belonging to the smaller subset, 12 on average, out of a fixed display size of 48 items) was subtracted from the proportion of fixations on the same-color distractors for the color-search displays and from the proportion of fixations on the same-shape distractors for the shape-search displays. As pointed out by Zelinsky (1996; see also D. E. Williams & Reingold, 2001; Shen et al., 2000, 2003), results from target-absent trials can be interpreted more clearly than those from target-present trials where the presence of the target item may influence search behavior. As a result, only target-absent trials were included in the current analysis. Figure 3 plots saccadic bias as a function of display type and trial congruency in both the instruction condition and the no-instruction condition.

A repeated-measures ANOVA was conducted on saccadic bias, with display type (2: color-search vs shape-search) and trial congruency (2: congruent vs incongruent) as within-subject factors and instructional manipulation (2: instruction vs no instruction) as the between-subject factor. Overall, saccadic bias did not differ between the instruction condition and the no-instruction condition, $F(1, 22) = 2.22$, $\text{Mse} = 101.50$, $p = 0.15$. Consistent with the findings from Shen et al. (2000, 2003), saccadic bias was greater in

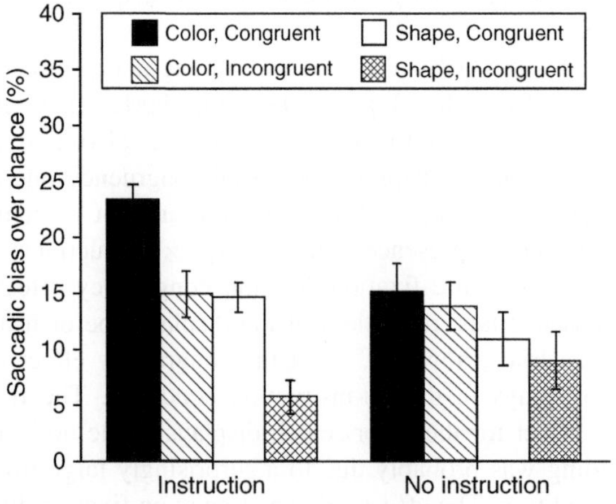

Figure 3. Congruency effect for saccadic bias as a function of display tupe and instructional manipulation (see text for the details on saccadic bias calculation).

the color-search displays, $F(1, 22) = 10.56$, Mse $= 193.32$, $p < 0.001$, implicating color dominance in conjunction search tasks (see also Luria & Strauss, 1975; Motter & Belky, 1998; Pomplun et al., 2001, 2003; L. G. Williams, 1966; D. E. Williams & Reingold, 2001).

The repeated-measures ANOVA revealed a significant main effect of trial congruency, $F(1, 22) = 25.96$, Mse $= 45.35$, $p < 0.001$, indicating stronger saccadic bias in the congruent trials than in the incongruent trials. That is, participants were more likely to direct saccades towards the same-color distractors for color-search displays and towards the same-shape distractors for shape-search displays in those trials that were consistent with the block context or instructional manipulation. This indicates that saccadic selectivity was influenced not only by the bottom-up factors (salience of the display items) but also by the top-down controls (instruction on search strategies or the contextual information). In addition, Figure 3 also shows that the congruency effect was stronger in the instruction condition (with a difference of 8.7% in the overall saccadic bias between the congruent and incongruent trials, $t(11) = 5.03$, $p < 0.001$) than in the no-instruction condition (with a difference of 1.6% in the overall saccadic bias between the congruent and incongruent trials, $t(11) = 2.85$, $p < 0.05$). This was verified by a significant instructional manipulation and trial congruency interaction, $F(1, 22) = 10.19$, Mse $= 45.35$, $p < 0.001$.

5. General discussion

The current study examined the top-down and bottom-up controls of visual attention in a search task with distractor-ratio manipulation. Although exactly the same search displays were used, search performance differed when the top-down information works in concert or against the bottom-up information. Participants' manual response time was shorter in those trials that were congruent with the search instruction or block context than in the incongruent trials. This suggests that visual guidance in a distractor-ratio experiment is not solely based on the information gathered from the current trial, but also modified by the information induced by the instructional manipulation and the block context within which search displays are presented (see also Chun & Jiang, 1999; Sobel & Cave, 2002; see Chun, 2000, for a review). Thus, the current finding supplements our earlier studies on the distractor-ratio effect, which highlighted the bottom-up controls of attention in a search task (Shen et al. 2000, 2003; see also Sobel & Cave, 2002).

The current study differed from Bacon and Egeth (1997) in several important aspects. First of all, Bacon and Egeth (1997) did not separate the effects of instructional and contextual manipulations. In their study, participants were informed of which distractor type would be less frequent on most trials and were asked to restrict their search to a subset of distractors (e.g., attend to red items or attend to vertical items). In contrast, in the present study, a congruency effect was observed both with and without explicit instructions. This finding shows that top-down controls of visual attention come in various forms (explicit instruction vs implicit knowledge; Wolfe et al., 2003). Similar to previous demonstrations of implicit top-down guidance by stimulus features (Wolfe et al., 2003), location context cueing (Chun & Jiang, 1999), knowledge on the probability distribution of certain display types in a group of trials can also induce a bias in deployment of attention during the search process. Future studies may further examine how the guidance of attention and saccadic bias can be modified on a trial-by-trial basis (see Maljkovic & Nakayama, 1994) instead of based on block statistics reported here. The current study also revealed a stronger congruency effect in the instruction condition than in the uninstruction condition. This interaction clearly demonstrates that the presence of top-down influences associated with instructional manipulation operates over and above any influences of implicit block context.

Second, the current study also revealed the spatiotemporal dynamics of visual search behavior by examining the patterns of eye movements accompanying the search process. We found that the congruent trials yielded fewer fixations and shorter fixation durations than did the incongruent trials, whereas the initial saccadic latency did not differ between these two types of trials. In addition, saccadic bias towards the same-color distractors in the color-search displays and towards the same-shape distractors in the shape-search displays was stronger in the congruent trials than in the incongruent trials.

Findings from these oculomotor measures have implications for the study of visual attention. For example, the guided-search model (Cave & Wolfe, 1990; Wolfe, 1994) argues that in a search task, a preattentive parallel process guides the subsequent serial shift of attention through display items. In our earlier studies on the distractor-ratio effect (Shen et al., 2001, 2003), we reported a quadratic change in initial saccadic latency as a function of distractor ratio, with a longer initial saccadic latency in those displays with an approximately 1:1 distractor ratio. It may be speculated that the stronger activation peak associated with extreme distractor ratios results in faster initial saccades. Alternatively, it is possible that the time required for extracting an activation map vary with distractor ratios. In either case, this suggests that the initial saccadic latency is strongly influenced by the bottom-up characteristics of search displays. The current finding of no difference in the initial saccadic latency between the congruent and incongruent trials indicates that participant's initial saccade targeting was not influenced by the informativeness of the top-down information. However, this does not necessarily mean that providing extra instruction or contextual manipulations does not help the initial segmentation of search displays. On the contrary, other eye movement measures indicate that top-down controls modify the search behavior in a later stage of search. The availability of accurate top-down information makes the rejection of distractor items easier, yielding a shorter fixation duration. It also leads to a higher bias in the selection of the relevant subset of distractors. As a result, fewer fixations are necessitated to make a search decision than in those trials in which misleading information is provided. Such a dynamic picture is hard to capture by a study that focuses on manual response only. The temporal dynamics revealed in the current study are useful for the development of the guided search theory (Cave & Wolfe, 1990; Wolfe, 1994) with respect to the interaction between the preattentive stage and the serial stage of processing and the time course of this interaction.

Finally, examination of eye movement patterns reveals that the top-down controls of visual attention (instruction and contextual manipulation) do not completely override the bottom-up contributions. In both the congruent and incongruent trials, saccades were still biased towards the smaller subset of distractors; such a bias was stronger in those trials that were consistent with the block context or instructional manipulations. Thus on inconsistent trials, even with explicit instructions, participants did not stick to the search strategies slavishly by searching entirely through a much larger subset of distractors. Instead, the top-down controls of visual attention were applied flexibly, weighting towards the stimulus dimension that is beneficial to an efficient search. This suggests that human visual search behavior is adaptive and modified flexibly to accommodate the changes to the environment and current task demand (see also Pomplun et al., 2001).

Acknowledgements

Preparation of this manuscript was supported by a grant to Eyal M. Reingold from the Natural Science and Engineering Research Council of Canada (NSERC). We wish to thank Martin Fischer and Adam Galpin for their helpful comments on an earlier draft of the current manuscript.

References

Bacon, W. F., & Egeth, H. E. (1994). Overriding stimulus-driven attentional capture. *Perception & Psychophysics, 55*, 485–496.

Bacon, W. F., & Egeth, H. E. (1997). Goal-directed guidance of attention: Evidence from conjunctive visual search. *Journal of Experimental Psychology: Human Perception & Performance, 23*, 948–961.

Bichot, N. P., & Schall, J. D. (1998). Saccade target selection in macaque during feature and conjunction visual search. *Visual Neuroscience, 16*, 81–89.

Binello, A., Mannan, S., & Ruddock, K. H. (1995). The characteristics of eye movements made during visual search with multi-element stimuli. *Spatial Vision, 9*, 343–362.

Bravo, M. J., & Nakayama, K. (1992). The role of attention in different visual-search tasks. *Perception & Psychophysics, 51*, 465–472.

Carter, R. C. (1982). Visual search with color. *Journal of Experimental Psychology: Human Perception & Performance, 8*, 127–136.

Cave, K. R., & Wolfe, J. M. (1990). Modeling the role of parallel processing in visual search. *Cognitive Psychology, 22*, 225–271.

Chun, M. M. (2000). Contextual cueing of visual attention. *Trends in Cognitive Sciences, 4*, 170–178.

Chun, M. M., & Jiang, Y. (1999). Top-down attentional guidance based on implicit learning of visual covariation. *Psychological Science, 10*, 360–365.

Egeth, H. E., Virzi, R. A., & Garbart, H. (1984). Searching for conjunctively defined targets. *Journal of Experimental Psychology: Human Perception & Performance, 10*, 32–39.

Egeth, H. E., & Yantis, S. (1997). Visual attention: Control, representation, and time course. *Annual Review of Psychology, 48*, 269–197.

Findlay, J. M. (1997). Saccade target selection during visual search. *Vision Research, 37*, 617–631.

Findlay, J. M., & Gilchrist, I. D. (1998). Eye guidance and visual search. In G. Underwood (Ed.), *Eye guidance in reading, driving and scene perception* (pp. 295–312). Oxford: Elservier.

Friedman-Hill, S., & Wolfe, J. M. (1995). Second-order parallel processing: visual search for the odd item in a subset. *Journal of Experimental Psychology: Human Perception & Performance, 21*, 531–551.

Galpin, A. J., & Underwood, G. (2005). Eye movements during search and detection in comparative visual search. *Perception & Psychophysics, 67*, 1313–1331.

Hillstrom, A. P. (2000). Repetition effects in visual search. *Perception & Psychophysics, 62*, 800–817.

Hooge, I. T., & Erkelens, C. J. (1999). Peripheral vision and oculomotor control during visual search. *Vision Research, 39*, 1567–1575.

Irwin, D. E., Colcombe, A. M., Kramer, A. F., & Hahn, S. (2000). Attentional and oculomotor capture by onset, luminance and color singletons. *Vision Research, 40*, 1443–1458.

Kaptein, N. A., Theeuwes, J., & van der Heijden, A. H. C. (1995). Search for a conjunctively defined target can be selectively limited to a color-defined subset of elements. *Journal of Experimental Psychology: Human Perception & Performance, 21*, 1053–1069.

Koch, C., & Ullman, S. (1985). Shifts in selective attention: Towards the underlying neural circuitry. *Human Neurobiology, 4*, 219–227.

Luria, S. M., & Strauss, M. S. (1975). Eye movements during search for coded and uncoded targets. *Perception & Psychophysics, 17*, 303–308.

Maljkovic, V., & Nakayama, K. (1994). Priming of pop-out: I. Role of features. *Memory & Cognition, 22*, 657–672.

Motter, B. C., & Belky, E. J. (1998). The guidance of eye movements during active visual search. *Vision Research, 38*, 1805–1815.

Poisson, M. E., & Wilkinson, F. (1992). Distractor ratio and grouping processes in visual conjunction search. *Perception, 21*, 21–38.

Pomplun, M., Reingold, E. M., & Shen, J. (2001). Peripheral and parafoveal cueing and masking effects on saccadic selectivity. *Vision Research, 41*, 2757–2769.

Pomplun, M., Reingold, E. M., & Shen, J. (2003). Area activation: A computational model of saccadic selectivity in visual search. *Cognitive Science, 27*, 299–312.

Pomplun, M., Sichelschmidt, L., Wagner, K., Clermont, T., Rickheitb, G., & Ritter, H. (2001). Comparative visual search: a difference that makes a difference. *Cognitive Science, 25*, 3–36.

Scialfa, C. T., & Joffe, K. (1998). Response times and eye movements in feature and conjunction search as a function of target eccentricity. *Perception & Psychophysics, 60*, 1067–1082.

Shen, J., Reingold, E. M., & Pomplun, M. (2000). Distractor ratio influences patterns of eye movements during visual search. *Perception, 29*, 241–250.

Shen, J., Reingold, E. M., & Pomplun, M. (2003). Guidance of eye movements during conjunctive visual search: The distractor-ratio effect. *Canadian Journal of Experimental Psychology, 57*, 76–96.

Shen, J., Reingold, E. M., Pomplun, M., & Williams, D. E. (2003). Saccadic selectivity during visual search: The influence of central processing difficulty. In J. Hyönä, R. Radach, & H. Deubel (Eds.), *The mind's eyes: Cognitive and applied aspects of eye movement research* (pp. 65–88). Amsterdam: Elsevier Science Publishers.

Sobel, K. V., & Cave, K. R. (2002). Roles of salience and strategy in conjunction search. *Journal of Experimental Psychology: Human Perception & Performance, 28*, 1055–1070.

Treisman, A. (1988). Features and objects: The fourteenth Bartlett memorial lecture. *The Quarterly Journal of Experimental Psychology, 40A*, 201–237.

Treisman, A., & Gelade, G. (1980). A feature integration theory of attention. *Cognitive Psychology, 12*, 97–136.

Williams, L. G. (1966). The effect of target specification on objects fixated during visual search. *Perception & Psychophysics, 1*, 315–318.

Williams, D. E., & Reingold, E. M. (2001). Preattentive guidance of eye movements during triple conjunction search tasks: The effects of feature discriminability and saccadic amplitude. *Psychonomic Bulletin & Review 8*, 476–488.

Williams, D. E., Reingold, E. M., Moscovitch, M., & Behrmann, M. (1997). Patterns of eye movements during parallel and serial visual search tasks. *Canadian Journal of Experimental Psychology, 51*, 151–164.

Wolfe, J. M. (1994). Guided search 2.0: A revised model of visual search. *Psychonomic Bulletin & Review, 1*, 202–238.

Wolfe, J. M. (1998). Visual search. In H. Pashler (Ed.), *Attention* (pp. 13–71). London: Psychology Press.

Wolfe, J. M., Butcher, S. J., Lee, C., & Hyle, M. (2003). Changing your mind: On the contributions of top-down and bottom-up guidance in visual search for feature singletons. *Journal of Experimental Psychology: Human Perception & Performance, 29*, 483–502.

Yantis, S. (1998). Control of visual attention. In H. Pashler (Ed.), *Attention* (pp. 223–256). London: Psychology Press.

Zelinsky, G. J. (1996). Using eye saccades to assess the selectivity of search movements. *Vision Research, 36*, 2177–2187.

Zelinsky, G. J., & Sheinberg, D. L. (1997). Eye movements during parallel-serial visual search. *Journal of Experimental Psychology: Human Perception & Performance, 23*, 244–262.

Zohary, E., & Hochstein, S. (1989). How serial is serial processing in vision? *Perception, 18*, 191–200.

Chapter 29

ABSENCE OF SCENE CONTEXT EFFECTS IN OBJECT DETECTION AND EYE GAZE CAPTURE

LYNN GAREZE and JOHN M. FINDLAY

University of Durham, UK

Eye Movements: A Window on Mind and Brain

Edited by R. P. G. van Gompel, M. H. Fischer, W. S. Murray and R. L. Hill

Abstract

Research into scene context effects has claimed that the semantic relationship between a scene and an object can affect initial visual processing. Our research contributes to this debate, considering different scene types and methodologies. Using simple line drawings, complex naturalistic photographs and line drawings derived from them, we investigated the detectability of semantically consistent and inconsistent objects in scenes. We failed to find reliable evidence of a consistency effect on scene processing within a single fixation or on subsequent eye movements prior to target fixation.

Since the pioneering studies of Biederman, Mezzanote, & Rabinowitz, et al. (1982), it has been accepted that a single brief glance at a visual scene provides 'gist' information that allows identification of the type of scene being viewed and the spatial layout of the major surfaces (Sanocki, 2003; Schyns & Oliva, 1994). When viewing is prolonged, this gist is supplemented with more detailed information as the eye directs the fovea to successive locations within the scene. An important and unresolved question concerns what information can be extracted from regions that are not foveally fixated. Such information has the potential for supporting the process of supplementation both directly and by contributing to the guidance of the eye scan.

In this chapter we concentrate on the extraction of semantic information from extrafoveal vision. To investigate semantic processing, the semantic relationship between an object and the scene background in which it is located can be manipulated. An object is semantically related to the scene context when its identity is compatible with the scene's 'gist'. For example, consider two objects sharing similar visual features, such as an apple and a ball. An apple is semantically compatible, or *consistent*, with a fruit market context, but a ball in the same location would be *inconsistent* with the scene's meaning. By comparing performance on a given task between objects categorised as semantically consistent and inconsistent with the scene, it is possible to determine whether the object's semantic meaning is affecting the performance measure.

Friedman (1979) recorded eye scans when viewing scenes that might contain inconsistent objects. Objects inconsistent with the scene context elicited significantly longer fixation durations than did consistent objects. Thus semantic inconsistency exerts an immediate effect when an object is viewed foveally. This effect is robust and usually extends to facilitation for inconsistent objects in tasks involving memory and recall, attributable to increased foveal processing (e.g. Pezdek, Whetstone, Reynolds, Askari, & Dougherty, 1989; Lampinen, Copeland, & Neuschatz, 2001). However, the effects of a semantically inconsistent object viewed in extrafoveal vision, prior to direct fixation, are less clear.

Loftus and Mackworth (1978), in the well-known 'octopus in a farmyard' study, reported that, as well as being fixated for longer, inconsistent objects in scenes were more likely to be fixated early in the viewing process than their consistent controls, with saccades directed to such objects from over 7° away as early as the second fixation on the scene. These findings suggested that semantic inconsistency can be detected very rapidly from extrafoveal processing, over 7° from fixation. However, these findings have not been reliably replicated in several subsequent studies (e.g. De Graef, Christiaens, & d'Ydewalle, 1990; Henderson, Weeks, & Hollingworth, 1999) and a number of explanations have been offered for this discrepancy in results (Gareze, 2003).

In an alternative approach, brief scene presentations have been used to determine whether semantic inconsistency can be detected rapidly within a single fixation. Biederman et al. (1982) reported a significant consistency advantage. Participants were faster and more accurate in responding that a consistent object (named by an object label at the start of the trial) was presented at a location specified by a probe after the 150 ms scene presentation. However, Hollingworth and Henderson (1998) proposed that their facilitation for consistent targets could result from response bias. When a replication with adequate controls was implemented, a significant advantage was found for inconsistent objects over consistent objects, a facilitatory effect later replicated (Hollingworth & Henderson, 1999). This inconsistency advantage has also been reported in a change blindness paradigm (Hollingworth & Henderson, 2000) and using an attentional probe paradigm (Gordon, 2004). However, independent researchers have struggled to identify a similar effect when using different experimental stimuli (e.g. Davenport & Potter, 2004) and the process mediating any inconsistent object facilitation remains poorly defined.

The experiments reported here address two concerns that occur with many of the existing studies. The first concern is that the distance between the target object and the participant's fixation position was not controlled, or was not systematically manipulated. For example, in the recent study of Gordon (2004), objects of mean size around 2° were presented at eccentricities (in one condition) of 2.6° (range 0.8–4.7°). For the closest eccentricity, the material is effectively foveal, so this study failed to draw a distinction between foveal, parafoveal and extrafoveal target presentations. The second concern is that studies often repeat trials with the same background scene, sometimes with a different critical object in the same location thus introducing a possible confound with implicit memory effects.

To address these issues, we used a paradigm adapted from Hollingworth and Henderson's work (1998, 1999) in which a brief scene presentation containing a consistent or inconsistent target object was followed by a two-alternative forced-choice display in which the target and a distractor (both consistent or both inconsistent) were presented. Results are presented from this paradigm (Experiments 1–4) and also from a free scene viewing paradigm (Experiment 5) with the same material, comparing eye movements in line drawings and photographs of scenes, containing consistent and inconsistent target objects. By investigating consistency effects during brief scene presentations while manipulating fixation position relative to the target, we aim to determine whether consistency effects occur in foveal, parafoveal or extrafoveal vision. It is possible that consistency might interact with fixation position, accounting for the conflicting data evident in previous studies.

From the free scene viewing task, we can investigate whether semantic inconsistency can be detected immediately upon first fixation on a scene, or whether the effects develop only during scene viewing which allows continued inspection of the image. Henderson and Hollingworth's (1998) review of eye movements during scene viewing indicated that longer fixation durations were found when viewing colour photos than when viewing black and white line drawings. Our comparison will additionally allow us to investigate whether a similar difference occurs between simple line drawings and complex grey-scale photographs.

1. Experiments 1–4: General method

We present four experiments, which make use of the same design but using different images as experimental stimuli in each case. Our intention was to compare the detectability and recognisability of individual target objects when manipulating their consistency with the global scene context and their visual eccentricity, across different image sets. One hundred Durham University students, with normal or corrected-to-normal vision, were recruited for each of the four experiments, with most participants taking part in only one of the four experiments.

As elaborated elsewhere (Gareze, 2003) the 800×600 pixel images were displayed on 15″ monitors. At a viewing distance of 60 cm, the images subtended a visual angle of approximately 19° vertically and 26° horizontally. In each experiment, four target objects could be located within each image, two consistent and two inconsistent with the scene context. As far as possible, the target objects were matched for size and location within the scene.

Figure 1 describes a trial sequence. A stationary cross was presented for 1000 ms, to direct fixation to a specific region of the display, which varied in each trial. Participants were instructed to fixate this cross and were informed that the cross would identify the location of the target in some trials. This location was manipulated experimentally. The coordinates indicated by the fixation cross were selected to either correspond directly with the target or identify a location approximately 3°, 6°, 9° or 12° from the target when presented at a viewing distance of 60 cm (positions 0–4 respectively). For any particular image, each of these eccentric locations was in the same direction relative to the target.

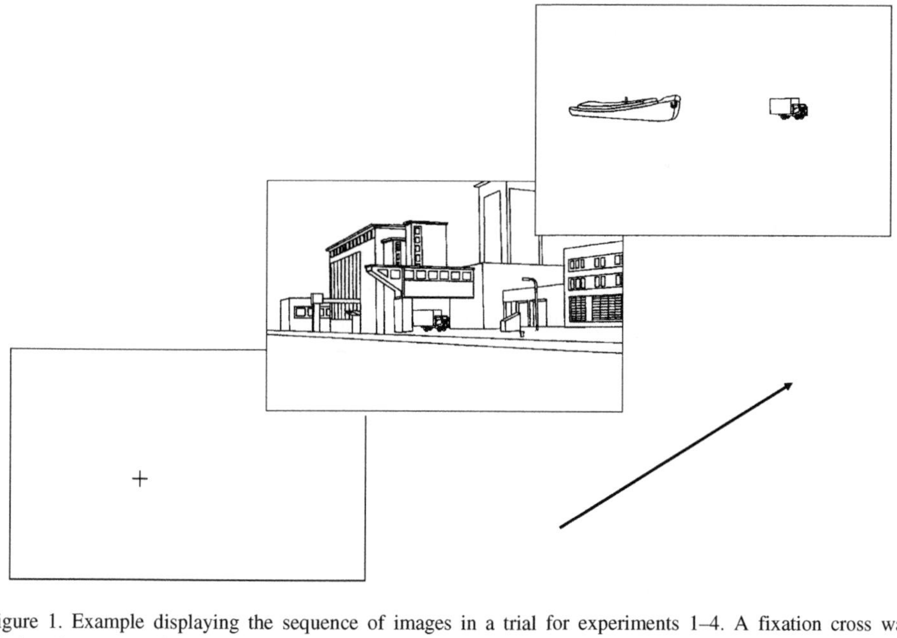

Figure 1. Example displaying the sequence of images in a trial for experiments 1–4. A fixation cross was displayed at a pre-selected location, relative to the subsequent target object, for 1000 ms. A scene image was presented briefly for 120 ms. Finally, a display presented two alternative objects, either both consistent or inconsistent, for selection. In this example, the correct response would be pressing the right-hand button.

This direction was chosen at random with the constraint that all locations had to fit within the dimensions of the image.

With participants fixating the selected point, the scene image was displayed for 120 ms, chosen to prevent any eye movements being executed. A response display was immediately presented which contained two objects, either both semantically consistent or both inconsistent with the scene just presented. One of these objects was always the target and participants were required to make a manual response to indicate which of the two they thought had been presented in the scene. This display remained visible for up to 5000 ms or until a response was recorded, if sooner. An inter-trial interval of 1000 ms followed before the start of the next trial. Practice trials were used to familiarise participants with the procedure. The scene backgrounds used in the practice trials were not used in the experimental trials.

We investigated the effects of scene-target consistency (2 levels: consistent or inconsistent) and fixation location relative to the target (5 levels: 0–4 as above) on response accuracy. Each image could be viewed with any of the five possible fixation positions, which was counterbalanced across participants. As repeated presentations of the same scene backgrounds and response displays might influence responses, each participant viewed only two examples of each scene background, once with a consistent target and once with an inconsistent one. We compared response accuracy in these conditions across different stimuli types in Experiments 1–4; when the images were (1) line drawings drawn from the Leuven library, which will be referred to as simple line drawings, (2) inverted displays of the same set of line drawings, (3) grey-scale photographs of household scenes and (4) line drawings of these photographs (Figure 2).

2. Experiment 1

In Experiment 1, the images used were derived from the Leuven line drawing library. This library consists of a set of line drawing bitmap files of scenes together with similar individual object bitmaps. Objects identified as consistent with a specific scene context could be embedded seamlessly within it so as to occlude the relevant region of scene background.

Inconsistent displays were created by embedding an object consistent with a different scene and manually ensuring appropriate occlusion of the background. As far as possible, object size and location were matched with the consistent object replaced, although the consistency manipulation was the primary consideration. This manipulation was confirmed by a questionnaire study in which naïve participants identified each target object and scene background and rated the likelihood of finding the target in that location.

2.1. Results

Initial analyses showed little evidence of an overall consistency effect (Figure 3a). As expected, accuracy was significantly higher when participants were directed to the precise

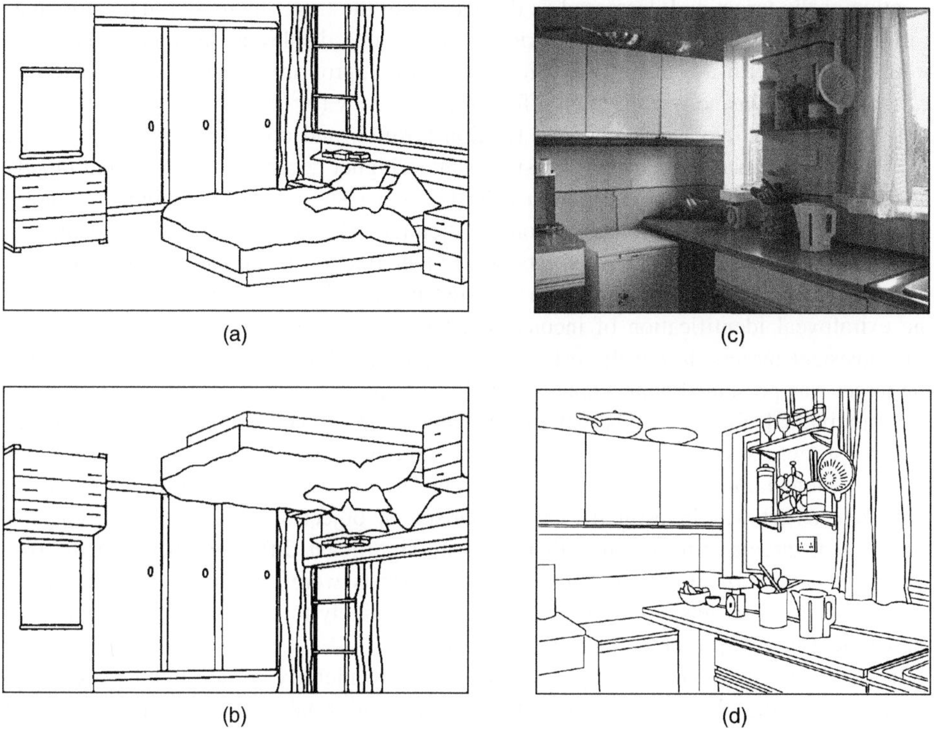

Figure 2. Example images used in (a) Experiment 1 – simple line drawings, (b) Experiment 2 – inverted line drawings, (c) Experiment 3 – grey-scale photographs and (d) Experiment 4 – line drawings of photographs.

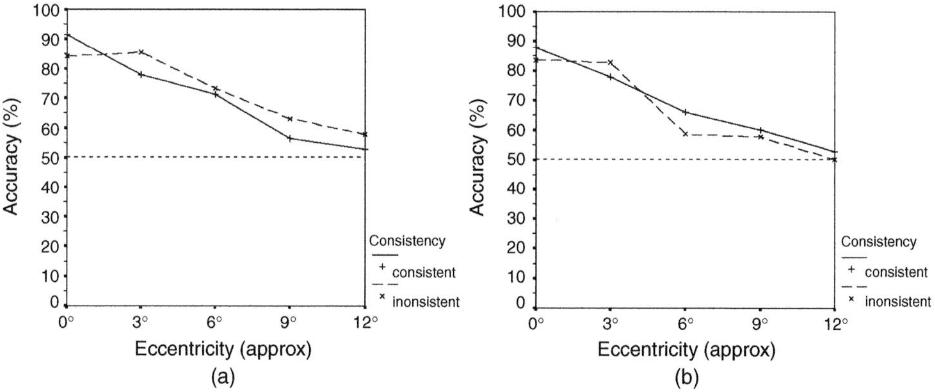

Figure 3. Results of experiments using the Leuven library line drawings (a) Experiment 1 – Simple line drawings from the Leuven library. (b) Experiment 2 – Inverted simple line drawings from the Leuven library. Graphs show the change in accuracy by eccentricity (at 60 cm) and target object consistency. Chance level of 50% is indicated.

location of the target and decreased as distance from the target increased, indicating that participants were fixating the cross prior to each scene display as instructed. The data were analysed with a logistic regression procedure appropriate for the binary outcome variable. There was no overall main effect of consistency ($X^2(1) < 1, p = 0.59$) but there was a significant interaction between fixation position and consistency ($X^2(1) = 4.21$, $p = 0.040$). This occurred because of significantly higher accuracy for consistent targets presented at fixation ($X^2(1) = 5.48, p = 0.019$), but also significantly higher accuracy for inconsistent targets presented 3° from fixation ($X^2(1) = 4.40, p = 0.036$) and slightly higher accuracy for inconsistent targets at all other fixation positions, although none of these comparisons was significant when tested individually. This finding suggests that the extrafoveal identification of inconsistent targets was facilitated above performance for consistent targets during the brief presentation of a line drawing scene, even when the target was presented approximately 12° from fixation.

As individual objects in this experiment were not presented in both a consistent and an inconsistent background, the possibility that an advantage might have been caused by an inadvertent failure to match the inconsistent target's features to that of the matched consistent target was investigated. A comparison of object sizes across consistent and inconsistent targets indicated that there was no significant difference between the two groups ($t(42)<1, p = 0.42$). In fact, consistent targets were on average slightly larger than inconsistent targets, so a size difference was unlikely to result in the advantage for inconsistent targets presented in extrafoveal vision.

It was noted that some participants were unable to identify certain targets even with extended viewing. It was confirmed by a further informal investigation using naïve observers that the target objects were not always identified correctly when presented within a scene, as they were in the experiment. Thus, it is difficult to be certain that the apparent advantage for inconsistent targets at extrafoveal locations can really be attributed to the processing of semantic information.

The presence of an inconsistent object advantage when using images which naïve observers struggled to identify raised the issue of whether the effect produced was generated by the processing of semantic information at all. Visual differences between consistent and inconsistent targets in scenes could have been introduced in the process of creating the images, as inconsistent targets required some manipulation before they could appear compatible with the scene background. To investigate this possibility, Experiment 2 replicated Experiment 1, but displayed inverted images. In every other way, the experiments were identical.

3. Experiment 2

Each image presentation was inverted, to interfere with semantic processing, altering the task to one of matching visual features between the brief inverted scene display and the inverted response display. We hypothesised that if a consistency effect persisted, with improved accuracy for inconsistent targets at extrafoveal locations, this effect could be attributed to visual differences rather than semantic ones. If the inconsistent object advantage were extinguished, this result would suggest that the inversion of the images successfully interfered with semantic processing and abolished the effect.

3.1. Results

Accuracy in this experiment was slightly lower than that in Experiment 1 (Figure 3b), with performance at the furthest fixation position not significantly better than chance for either consistent or inconsistent targets. The broadly similar identification probabilities in the two experiments suggest that orientation invariant features, perhaps of a visual rather than a semantic nature, were mainly used for the task. The absence of a consistency effect at any eccentricity in Experiment 2 indicates that there was no effect of semantic relationship between the target and the scene background when the images were inverted. Thus the presence of an inconsistent object advantage in the peripheral positions in Experiment 1 supports the suggestion of a contribution from semantic processing.

The data from the two experiments were directly compared using a binary logistic regression analysis in which image orientation was added as a variable. For extrafoveal locations, there was a significant main effect of orientation ($X^2(1) = 12.4, p < 0.001$), with higher accuracy for upright images (Experiment 1) than for inverted images (Experiment 2). A significant interaction between orientation and consistency ($X^2(1) = 5.97, p = 0.015$) further suggested that orientation had a greater effect on performance for inconsistent targets than for consistent targets. An additional analysis confirmed this relationship by investigating whether orientation had a significant main effect on consistent and inconsistent trials separately. Image orientation did not significantly affect performance for consistent trials ($X^2(1) < 1, p = 0.84$), but a significant effect was found for inconsistent trials ($X^2(1) = 12.0, p = 0.001$), indicating that performance on inconsistent targets at extrafoveal locations was significantly reduced by inverting the experimental images. These results support the hypothesis that some processing of semantic information might have occurred during the experimental process, as inverting the images reduced accuracy.

As noted previously, a cause for concern with the materials was that it was not possible to display the same object in both a consistent and an inconsistent setting and that participants sometimes struggled to identify the targets and the background scenes. Therefore, the extent to which we can attribute a significant consistency effect to semantic processing is compromised. To overcome this, a set of photographic images was produced in which familiar target objects were manipulated so that the same object could be displayed in both a consistent and an inconsistent background.

4. Experiment 3

In Experiment 3, we investigated whether a consistency effect could be produced using more naturalistic (and identifiable) photographic stimuli. The stimuli were created by identifying familiar backgrounds (household scenes) and objects (household items) to serve as targets. Consistent and inconsistent targets were placed in the same location in a scene and each consistent target also served as an inconsistent target in a different scene background. Therefore, consistent and inconsistent targets at the same location were quite closely matched for size and shape and each object was used as a target twice in different scene backgrounds (Figure 4). Only one instance of each scene was presented to each participant.

4.1. Results

The use of photographic images resulted in improved performance (Figure 5a), with accuracy above 90% for both consistent and inconsistent targets presented at fixation, and performance appeared to plateau above chance level between positions 3 and 4 (9° and 12°), at approximately 66%. However, the change in visual stimuli also completely eradicated any evidence of a consistency effect. Even when the target was presented at fixation, consistent and inconsistent targets were identified with the same accuracy.

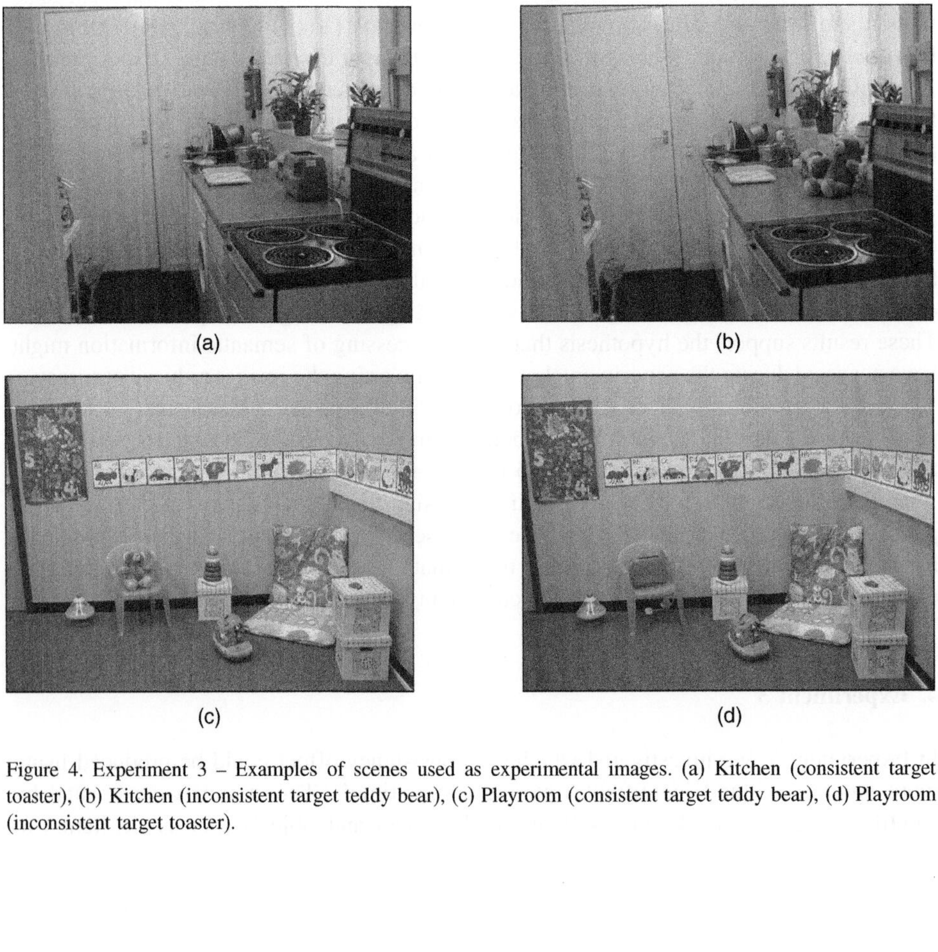

Figure 4. Experiment 3 – Examples of scenes used as experimental images. (a) Kitchen (consistent target toaster), (b) Kitchen (inconsistent target teddy bear), (c) Playroom (consistent target teddy bear), (d) Playroom (inconsistent target toaster).

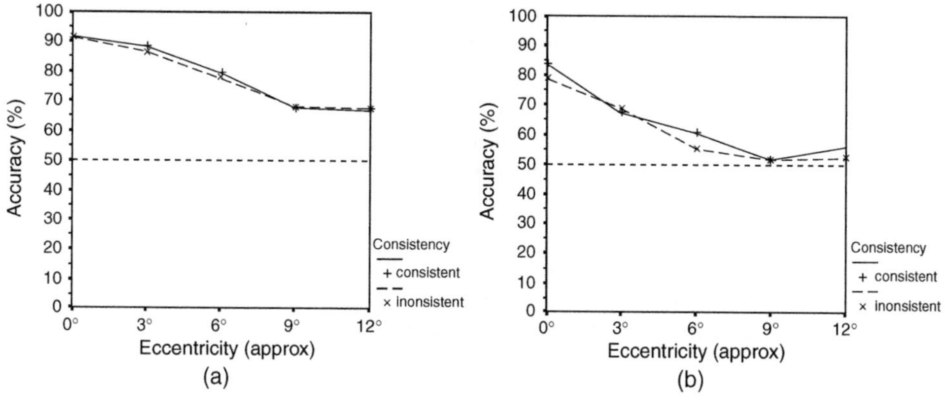

Figure 5. Results of experiments using naturalistic scene stimuli. (a) Experiment 3 – Grey-scale photographs. (b) Experiment 4 – Line drawings derived from grey-scale photographs. Graphs show the change in accuracy by eccentricity (at 60 cm) and target object consistency. Chance level of 50% is indicated.

As each object served as a consistent and an inconsistent target, we analysed the data with an items analysis using a matched pairs *t*-test where accuracy for each object was compared across conditions (although the object size was not necessarily constant). There was no difference in performance at all between an object located in a consistent and an inconsistent background ($t(31)<1$, $p=0.94$). We also compared performance for the size-matched consistent and inconsistent targets, which appeared at the same location. A matched pairs *t*-test again found no significant difference in accuracy ($t(31)<1$, $p=0.95$), indicating that there was no effect of consistency between consistent and inconsistent targets in a scene.

5. Experiment 4

In order to explain the failure to elicit a consistency effect with the photographic stimuli, we investigated whether the nature of the stimuli or their composition could be responsible. In Experiment 4, the photographic scenes were converted to line drawings which maintained a reasonable level of detail, to enable the scenes and the targets to be readily identified. The line drawings were closely matched to the photographs but they were not identical. Again, participants were required to identify which of two line drawings of household objects had appeared in the brief scene presentation.

5.1. Results

In Experiments 1–3, participants with accuracy below 60% were replaced. However, the lack of available participants made this impossible in Experiment 4, and so accuracy was significantly lower in this experiment than in the previous ones. It should be noted that the decrease in accuracy was not solely caused by the inability to replace poorly performing participants, as only 5 participants were replaced in Experiment 3, compared to 31 participants eligible for replacement in Experiment 4.

Even when the target was presented at fixation, accuracy was only 81%, falling below 52% at 9° from fixation. This decrease in accuracy indicated that the increased visual information in the photographic images facilitated performance on this task, rather than hindered it (Figure 5b). There was no evidence of a consistency effect at any position. As in Experiment 3, we compared accuracy for a target according to whether it appeared in a consistent or inconsistent scene and found no evidence of a consistency effect ($t(31) < 1$, $p = 0.38$). Similarly, comparing performance on a consistent and an inconsistent target located in the same scene also failed to reveal any consistency effects ($t(31) = 1.05$, $p = 0.30$).

We compared performance between photographs and line drawings of photographs (Experiments 3 and 4) and found that accuracy decreased equally for consistent and inconsistent targets at all positions when line drawings were displayed, with differences up to 20% between the two conditions. Although a significant main effect of image type was found as expected ($X^2(1) = 90.2$, $p < 0.001$), confirming that accuracy was significantly poorer in Experiment 4 than Experiment 3, there was no evidence of a main effect or interaction involving consistency. These data support the conclusion that the consistency manipulation in these experiments did not produce any evidence of differential processing of consistent and inconsistent targets.

6. Discussion of Experiments 1–4

This series of experiments has enabled us to examine the effect of consistency in a psychophysical identification task. In three of the four experiments, a small advantage for consistent objects when viewed foveally was found, although this effect was absent in Experiment 3 (photographic stimuli). Experiment 3 resulted in the highest overall identification performance (over 90% accuracy with direct fixation) and it is possible that global context has less of an effect when the targets are easily identifiable. In all the experiments, accuracy declined systematically with eccentricity as expected and in no case did the results suggest that the detection superiority for consistent targets was maintained in extrafoveal vision.

However, at extrafoveal locations, the findings were less clear. The only significant difference between performance for consistent and inconsistent targets occurred when presenting upright line drawings obtained from the Leuven library. In Experiment 1, we found a significant advantage for inconsistent targets presented extrafoveally. This advantage was extinguished when the images were inverted. This manipulation would have certainly interfered with the identification of the individual target objects as well as the global scene and was hypothesised to inhibit the processing of semantic information.

Two considerations might cast doubt on this conclusion of an identification advantage for inconsistent stimuli. First, separate testing by questionnaire (Gareze, 2003) showed that identification of the targets used was not always possible even with unlimited viewing. Second, each object was not used both in a consistent and in an inconsistent background and although the object characteristics were balanced as closely as possible, it is possible that subtle visual differences occurred between the two sets. These considerations led to the design of Experiment 3, in which photographic scenes were used with each target object appearing both in a consistent and in an inconsistent background.

In Experiment 3, no differences at all were found between the identification percentages for consistent and inconsistent targets. When the same material was presented in line drawing form (Experiment 4), identification accuracy was reduced. However, this did not result in any significant effects of consistency, although a small non-significant consistency advantage occurred at the foveal position.

To determine whether physical properties of the targets may have influenced consistency effects, we analysed the data considering target object size. Approximate area of each target was calculated by noting the size of the smallest box which could contain it and, within each image set, they were classified as being small, medium or large. In Experiments 1 and 2, small objects were contained within a pixel area of 7000 square pixels or less ($<7°$ square approx.), medium-sized targets within 17 000 square pixels (7°–16° square approx.) and large targets were greater than 17 000 square pixels ($>16°$ square approx.). In both experiments, the same 18 targets were classified as 'small', 16 as 'medium' and 10 were 'large'.

Targets within the photographic stimuli used in Experiments 3 were smaller overall and 24 targets were classified as 'small', less than 4000 square pixels ($<4°$ square), 23 were medium sized, between 4000 and 8000 square pixels (4–8° square), and 17 were considered 'large', greater than 8000 square pixels ($>8°$ square). As converting the photographs to line drawings for Experiment 4 altered the displays slightly, the targets were categorised independently according to the same criteria, with 20 small targets, 23 medium-sized targets and 21 large targets. Each target was presented at each possible eccentricity.

In Experiment 1, large objects produced a significant consistency effect ($X^2(1) = 8.85$, $p = 0.003$), with higher accuracy for inconsistent targets than consistent targets at extrafoveal locations. Medium-sized targets only indicated a trend in this direction and small targets showed no effect of consistency, possibly because of identification difficulties at extrafoveal locations from a brief presentation. However, for all target object sizes, accuracy for consistent objects presented at fixation was slightly, but not significantly, higher than for inconsistent objects. Experiment 2 produced no evidence of a consistency effect mediated by object size, supporting the assumption that semantic information was not obtained from these inverted images.

In Experiment 3, objects of medium size produced an advantage for consistent targets at all fixation positions ($X^2(1) = 7.24$, $p = 0.007$) and conversely, large targets indicated an advantage for inconsistent targets across all fixation positions ($X^2(1) = 14.6$, $p < 0.001$). In Experiment 4, a significant advantage for consistent targets was found for medium-sized objects ($X^2(1) = 4.76$, $p = 0.029$). Accuracy fell to chance level at position 3 (9°) and no difference was found between consistent and inconsistent targets at this position or beyond. This finding suggests that the advantage for medium-sized consistent objects was mediated by their detectability, as the consistency effect was only evident when accuracy was above chance. Unlike Experiment 3, no corresponding effect was found for large objects, with no difference between consistent and inconsistent targets at any fixation position. This result suggests that the conversion from photographs to line drawings may have interfered with the identification of large targets, particularly inconsistent targets, even though accuracy remained high at all fixation positions.

These findings suggest that the exhibition of consistency effects was closely linked to the detectability of the target objects, as consistency effects were limited to conditions in which the target would be more easily identified. Significant consistency effects were not found for small target objects in any experiment, or for medium-sized targets presented beyond 6° from fixation in Experiment 4. The most reliable difference was an advantage for consistent targets over inconsistent targets presented directly at fixation, which was present in three of the four experiments but only significant in Experiment 1. This pattern was also found within the target size analysis. Where visible differences occurred at fixation, the advantage was for consistent targets, with the single exception of large targets in Experiment 3 where no similar effect was found within the whole dataset. This advantage for consistent targets at fixation suggests that a compatible scene context can facilitate accurate responses in this detection task when the target is directly foveated.

Unlike the more universal consistent object advantage at fixation, other significant consistency effects, such as the inconsistent object advantage in extrafoveal vision found in Experiment 1, were apparent only under certain conditions, suggesting that they may have been influenced by other factors unrelated to semantic consistency. The existence of both a consistent object advantage and an inconsistent object advantage for different target sizes within the same data set in Experiment 3 may shed light on why reliable consistency effects have proved difficult to elicit in previous work. Although the number of objects included in each size category may be too small to provide reliable evidence of consistency effects, these data suggest that object size may influence the exhibition of such effects and is worthy of further investigation when considering the often conflicting data in this field.

7. Experiment 5

The results of Experiment 1 suggested that semantically inconsistent objects might be identified more readily than consistent objects from a brief scene presentation, although this result was not replicated in Experiment 3. Experiment 5 investigates whether the effects of semantic inconsistency appear in free viewing. As discussed in the Introduction, following the pioneering work of Loftus and Mackworth (1978), a number of studies have analysed eye scan records during naturalistic scene viewing tasks to investigate whether semantic information can be extracted extrafoveally. Many studies have failed to replicate the original finding but more recent careful work (e.g. Hollingworth & Henderson, 2000) has reopened the question. Hence, we carried out a study in which participants were shown extended (7 s) presentations of the simple line drawing scenes and the grey-scale photographs (used in Experiments 1 and 3 respectively) and their eye movements were recorded for the duration. These data were investigated for evidence that the semantic inconsistency between inconsistent targets and their scene background could be detected prior to direct fixation, compared to scenes containing only consistent objects.

If the inconsistent targets in the Leuven image set were more salient, visually or semantically, than the consistent targets and this difference could be detected extrafoveally, then we would expect the eye movement data to reflect this in measures of saccade behaviour prior to target fixation. The effect of the (more naturalistic) consistency manipulation in photographs on eye movement behaviour was also investigated in this way.

7.1. Method

Twenty-four participants were recruited from Durham University and all had normal, uncorrected vision. A subset of the experimental images used in Experiments 1 and 3 was displayed to each participant in separate blocks, which were counterbalanced across participants. All images were used but the number of times a participant viewed a given scene background was controlled as in the previous experiments and each background was displayed once only. The scenes were presented centrally and subtended approximately $16^{\circ} \times 12^{\circ}$ at a viewing distance of 85 cm. Participants were instructed to view the scenes naturally and that no memory test would follow.

Their eye movements were recorded for the 7 s display duration using a Fourward Technologies Dual Purkinje Generation 5.5 eye tracker. The resolution of the eye tracker was 10 min. of arc. and the sampling rate was every millisecond. The movements of the right eye were monitored but viewing was binocular. Head movements were restrained with a chin rest and two forehead rests. The accuracy of the record was checked every four trials and recalibration occurred when necessary. The eye movement data were analysed offline by a semi-automated procedure. A computer algorithm detected the saccades using a velocity criterion and each record was inspected individually.

7.2. Results

Table 1 summarises the effects of consistency on eye movement behaviour in simple line drawings and photographs. For both image types, a significant effect of consistency was found on measures following direct fixation of the object, such as the first fixation duration (the duration of the first fixation on the target) the first pass fixation duration (the sum of the first and any consecutive fixations on the target before moving the eyes away) and the total fixation duration (the sum of all fixations on the target). As hypothesised, in line drawings, inconsistent targets were fixated for significantly longer than consistent targets ($t(34.1) = -2.76$, $p = 0.008$), although whether this was due to difficulty in reconciling the semantic inconsistency between the object and the scene or difficulty in identifying the line drawing object is still undetermined. In photographs, a similar effect

Table 1
Experiment – Free scene viewing of simple line drawings and grey-scale photographs. Summary of results for consistent and inconsistent line drawings and photographs. Measures show the mean value across the 24 subjects

	Line drawings		Photographs	
Measure	Consistent	Inconsistent	Consistent	Inconsistent
Probability of target fixation (%)	91.7	93.2	89.6	85.4
Number of saccades prior to fixation	4.5	5.2	5.2	5.2
Arrival time (ms)	1309	1613	1856	1803
Saccade amplitude to the target (°)	3.7	3.8	3.3	3.7
First fixation duration (ms)	383	550**	380	433
First pass fixation duration (ms)	573	718	431	549*
Total fixation duration (ms)	1020	1244*	775	1010**

* $p < .05$
** $p < .01$

was found. Although the difference in first fixation time did not reach statistical significance ($t(46) = -1.56$, $p = 0.13$), there was a significant effect of consistency on the first pass ($t(46) = -2.58$, $p = 0.013$) and on the total fixation durations ($t(46) = -3.40$, $p = 0.001$), with longer fixations on inconsistent targets. Upon first fixation, the photographs of inconsistent household objects did not elicit significantly longer fixations than those of consistent household objects, perhaps because they were not sufficiently inconsistent with the scene context or unrecognisable within the background.

To investigate the possibility of a consistency effect prior to target fixation, we considered the evidence that inconsistent targets were fixated earlier than consistent targets. However, there was no evidence that the eyes were directed towards inconsistent targets any sooner than consistent targets in either image type. Although there was a slight difference in the time taken to fixate the target and the number of saccades executed prior to target fixation in line drawings, this difference was in the opposite direction to the hypothesis that inconsistent targets would be fixated sooner.

Saccade amplitudes towards the targets were comparable between consistent and inconsistent targets, in both line drawings and photos, indicating that the objects were selected as saccade targets from approximately the same level of extrafoveal processing. The mean saccade size was approximately 3.75° for saccades to both consistent and inconsistent line drawing targets. This value is considerably less than the eccentricity at which many of the targets were presented in Experiment 1 and which produced evidence of facilitated performance for inconsistent targets compared to consistent targets. As the number of fixations executed prior to target fixation was also comparable across consistency conditions and across image types, we can conclude that targets were selected for fixation from equivalent levels of extrafoveal processing regardless of whether the scenes were simple line drawings or complex grey-scale photographs.

For the photographs, we also compared the behaviour for each target in two different scene contexts and two matched targets in the same scene background. This counter-balancing of targets allowed us to investigate the possibility of consistency effects more closely. However, again, there was no evidence of a consistency effect prior to target fixation, with the only significant differences being in fixation measures. This lack of evidence towards increased visual salience of inconsistent targets indicates that the inconsistent advantage in brief presentations for line drawings (Experiment 1) was not manifest in Experiment 5. Whatever process facilitated object detection in the scene regions containing inconsistent targets in brief presentations failed to influence the eye movement pattern during natural scene viewing.

Targets in the photographs were fixated later (by an average of 369 ms) than those in line drawings, although the number of saccades executed prior to target fixation was approximately the same. This discrepancy suggests that fixations on distractors (which would all be consistent) in photographs were longer than in line drawings (although this is not true of fixations on targets). Total fixation durations on targets in line drawings were considerably longer than in photos (over 200 ms) but this finding can be explained by the relative differences in scene composition between line drawings and photographs. Fewer items to explore in line drawings would result in more refixations on targets than in photographic images, producing longer total fixation times for targets in line drawing images than for those in complex photos containing many distractors to explore.

To investigate whether fixation durations on distractors were longer in photographs than in line drawings, we divided the 7000 ms presentation time into 1000 ms time bins and allocated individual fixation durations to the time bin in which the fixation started. We excluded fixations directed at targets (as consistency effects were seen) and also the final fixation in each trial, as this fixation was terminated artificially by the disappearance of the image. Therefore, considerably fewer fixations were allocated to the final bin than the preceding ones.

Figure 6 displays the mean fixation duration (including 95% confidence limits) across the time course of the trial, for line drawings and photographs. In both cases, the mean fixation duration increases from the shortest fixations, commencing within the first 1000 ms, to a maximum value for fixations beginning 3000–5000 ms into the trial. As the trial elapses beyond this point, fixation durations appear to shorten. This trend for shorter fixations at the start of the trial has been noted previously (as reviewed in Findlay and Gilchrist, 2003). Using computer-rendered images of room interiors, Unema, Pannasch, Joos, & Velichkovsky (2005) found that an asymptotic fixation duration was reached after approximately 3.4 s during a 20 s trial, after which fixation duration levelled off, while our data indicate fixation duration decreasing towards the end of the trial, perhaps motivated by the imminent disappearance of the display.

Within each time bin, the mean fixation duration was longer for photographs than for line drawings, by an average difference of 33 ms. This stable effect accounts in part

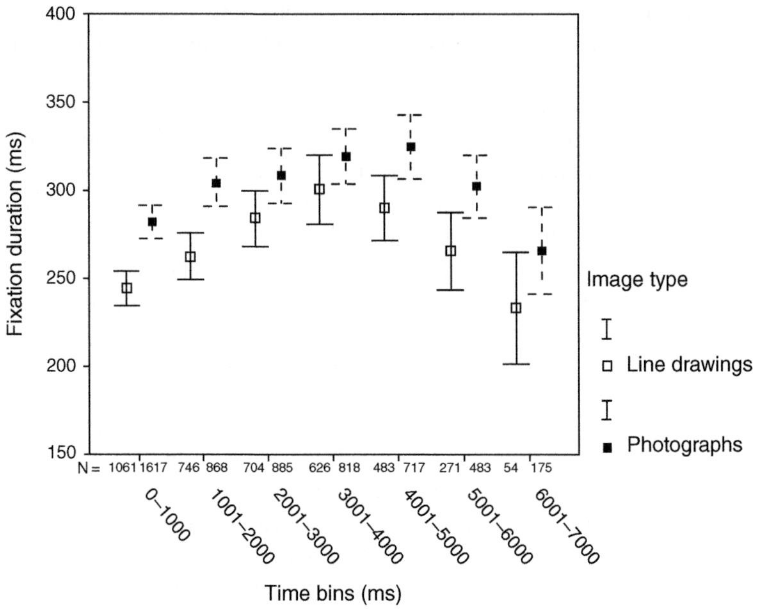

Figure 6. Experiment 5 – Free scene viewing of simple line drawings from the Leuven library and grey-scale photographs. Mean fixation duration by time for line drawings and photographs. Error bars display 95% confidence limits around the mean. N = number of fixations within time bin.

for the discrepancy discussed above, where the number of saccades executed prior to target fixation was approximately the same for line drawings and photographs but the time at which the target was fixated was later for photographs. The fixations investigated in this analysis consisted of those directed at regions other than the target, indicating that non-target-directed fixations were longer when viewing photographs than line drawings, although this effect was not seen on direct target fixations. The increased fixation durations for fixations on photographs, compared to line drawings, are compatible with previous reports (Henderson & Hollingworth, 1998). Although the photographs used in this experiment were grey-scale, a similar effect was seen which could reflect additional foveal processing of the more complex and detailed visual image.

8. General discussion

These findings taken as a whole suggest that the semantic inconsistency manipulation generated in the creation of these stimuli, whether subtle and naturalistic as in the photographic stimuli or stronger and less natural as in the Leuven line drawings, is not detected from extrafoveal vision during casual scene viewing. This is not to claim that it is impossible to identify semantic information about a target which has not yet been fixated, but simply that it does not occur habitually when viewing a scene with no explicit instructions.

In Experiment 1, performance for inconsistent targets was facilitated during a brief presentation of a line drawing image. Although this facilitation was extinguished with the inversion of the images (Experiment 2), we cannot conclude that the facilitation was generated by the immediate detection of an inconsistent object in the scene because independent, naïve participants often failed to correctly identify the targets in question, even during unlimited scene viewing. If the effect was not mediated by the successful identification of the inconsistent targets, then the facilitation cannot be due to the perceived discrepancy between the target's identity and the scene context.

This concern motivated the design of the more recognisable and naturalistic photographs used in Experiment 3, which failed to produce reliable evidence of consistency effects. The line drawings derived from them also failed to elicit any consistency effects (Experiment 4). Although the manipulation of consistency with these stimuli was less strong than with the line drawings used in Experiment 1, as all items could be found in a house and therefore were unusually, rather than impossibly, located, participants reliably rated the inconsistent targets as more unlikely than their matched consistent controls (Gareze, 2003). Therefore, the absence of a consistency effect could be due to an insufficiently strong consistency manipulation, indicating that the effect is unlikely to occur under naturalistic circumstances, or due to the improved recognisability of inconsistent targets, rendering them less salient in extrafoveal vision.

Previous work by Hollingworth and Henderson (1998, 1999, 2000) and Gordon (2004) directly contradicts the results of these experiments. In order to reconcile the two, it is necessary to consider the differences between these experiments, which may modulate such a discrepancy. To begin with, one of the major concerns the current experiments aim to address is the repeated display of the same or similar scenes, often up to eight times each. In these experiments, the number of times a participant viewed each scene background was strictly limited to prevent any learning effects. It was hypothesised that as inconsistent objects are better represented in memory, it would be plausible for an inconsistent target viewed in one background to influence performance on a subsequent repeated presentation. As the appropriate detection of scene gist is a crucial component of any evidenced consistency effects, the continued presentation of the same scene would be worth avoiding.

The importance of fixation position relative to the target has also been underestimated in previous work. Most experiments have involved a central fixation position with targets appearing at differing eccentricities within a scene, which does not consider the possibility that semantic information may be processed differently, if at all, in foveal and extrafoveal vision. Gordon (2004) presented targets at two eccentricities (near and far) but these varied across scenes. In one experiment, the mean near eccentricity was 2.9° but the range of eccentricities (0.8°–5.3°) overlapped with that of far eccentricities (3.4°–8.8°), which had a mean of 6.7°. Additionally, although object size was recorded, it was not manipulated systematically. Targets of approximately 2° square in size were presented at the above eccentricities, making many of the displays foveal or near foveal in presentation. In this way, it was impossible to distinguish between possible foveal and extrafoveal consistency effects.

Similarly, although object size was recorded, there was no analysis of whether object size affected performance on the task or any evident consistency effects. The results of the current series of experiments suggest that other object variables, such as object size, may modulate the expression of consistency effects. Significant and sometimes opposing consistency effects were found with data sets which showed no overall effect of the consistency manipulation, suggesting that the closer analysis of object-specific visual features may be valuable in the study of consistency effects.

A further concern relating to previous work which we have attempted to address is the nature of the visual stimuli used in these experiments. While many experiments have used line drawings derived from real-world scenes, few have used photographs of actual scenes. Our attempt to develop a set of naturalistic visual scenes resulted in a less extreme consistency manipulation than that found in the more frequently used stimuli. However, the failure to elicit consistency effects with these images calls into question the applicability of these effects to real-world viewing. Although some significant effects were found for certain target object sizes, the specificity of the conditions in which they occurred argues against the common occurrence of these effects during everyday scene viewing.

The investigation of eye movement behaviour also failed to provide evidence of differential extrafoveal processing for consistent and inconsistent targets. The only significant differences were found in fixation measures, confirming that inconsistent objects are fixated for longer than consistent ones (e.g. Henderson et al., 1999; De Graef et al., 1990). Even the line drawings, which exhibited an inconsistent object advantage in extrafoveal vision in Experiment 1, failed to produce any evidence of the earlier fixation of inconsistent targets. Despite previous findings of a reliable advantage for inconsistent objects presented extrafoveally in brief presentations paradigms (e.g. Gordon, 2004; Hollingworth & Henderson, 1998, 1999, 2000), no similar effect was found using these stimuli, suggesting that the effect may be difficult to elicit under more natural viewing conditions.

Acknowledgements

We are grateful to Peter De Graef for permission to use these scenes from the set available at ftp://michotte.psy.kuleuven.ac.be/pub/line_drw/ and for suggesting the manipulation in Experiment 2. We are also grateful to Lora Findlay for creating the stimuli used in Experiment 4 using Adobe Photoshop.

References

Biederman, I., Mezzanote, R. J., & Rabinowitz, J. C. (1982). Scene perception: Detecting and judging objects undergoing relational violations. *Cognitive Psychology, 14*, 143–177.

Davenport J. L., & Potter M. C. (2004). Scene consistency in object and background perception. *Psychological Science, 15*, 559–564.

De Graef, P., Christiaens, D., & d'Ydewalle, G. (1990). Perceptual effects of scene context on object identification. *Psychological Research, 52*, 317–329.

Findlay, J. M., & Gilchrist, I. D. (2003). *Active vision: The psychology of looking and seeing*. Oxford: Oxford University Press.

Friedman, A. (1979). Framing pictures: The role of knowledge in automatized encoding and memory for gist. *Journal of Experimental Psychology: General, 108*, 316–355.

Gareze, L. (2003). The role of foveal and extrafoveal vision in the processing of scene semantics. PhD Thesis, University of Durham.

Gordon, R. D. (2004). Attentional allocation during the perception of scenes. *Journal of Experimental Psychology: Human Perception and Performance, 30*, 760–777.

Henderson, J. M., & Hollingworth, A. (1998). Eye movements during scene viewing: an overview. In G. Underwood (Ed.), *Eye guidance in reading and scene perception.* (pp. 269–293). Amsterdam, North-Holland: Elsevier.

Henderson, J. M., Weeks, P. A., Jr., & Hollingworth, A. (1999). The effects of semantic consistency on eye movements during complex scene viewing. *Journal of Experimental Psychology: Human Perception and Performance, 25*, 210–228.

Hollingworth, A., & Henderson, J. M. (1998). Does consistent scene context facilitate object perception? *Journal of Experimental Psychology: General, 127*, 398–415.

Hollingworth, A., & Henderson, J. M. (1999). Object identification is isolated from scene semantic constraint: Evidence from object type and token discrimination. *Acta Psychologica, 102*, 319–343.

Hollingworth, A., & Henderson, J. M. (2000). Semantic informativeness mediates the detection of changes in scenes. *Visual Cognition, 7*, 213–235.

Lampinen, J. M., Copeland, S. M., & Neuschatz, J. S. (2001). Recollections of things schematic: Room schemas revisited. *Journal of Experimental Psychology: Learning, Memory and Cognition, 27*, 1211–1222.

Loftus, G. R., & Mackworth, N. H. (1978). Cognitive determinants of fixation location during picture viewing. *Journal of Experimental Psychology: Human Perception and Performance, 4*, 565–572.

Pezdek, K., Whetstone, T., Reynolds, K., Askari, N., & Dougherty, T. (1989). Memory for real-world scenes: The role of consistency with schema expectation. *Journal of Experimental Psychology: Learning, Memory and Cognition, 15*, 587–595.

Sanocki, T. (2003). Representation and perception of scenic layout. *Cognitive Psychology, 47*, 43–86.

Schyns, P. G., & Oliva, A. (1994). From blobs to boundary edges: evidence for time and spatial scale dependent scene recognition. *Psychological Science, 5*, 195–200.

Unema, P. J. A., Pannasch, S., Joos, M., & Velichkovsky, B. M. (2005). Time course of information processing during scene perception: The relationship between saccade amplitude and fixation duration. *Visual Cognition, 12*, 473–494.

PART 8

EYE MOVEMENTS IN NATURAL ENVIRONMENTS

Edited by

ROGER P. G. VAN GOMPEL

Chapter 30

LEARNING WHERE TO LOOK

MARY M. HAYHOE

University of Rochester, USA

JASON DROLL

University of California, USA

NEIL MENNIE

Univeristy of Nottingham, UK

Eye Movements: A Window on Mind and Brain
Edited by R. P. G. van Gompel, M. H. Fischer, W. S. Murray and R. L. Hill

Abstract

How do the limitations of attention and working memory constrain acquisition of information in the context of natural behavior? Overt fixations carry much information about current attentional state, and are a revealing indicator of this process. Fixation patterns in natural behavior are largely determined by the momentary task. The implication of this is that fixation patterns are a learnt behavior. We review several recent findings that reveal some aspects of this learning. In particular, subjects learn the structure and dynamic properties of the world in order to fixate critical regions at the right time. They also learn how to allocate attention and gaze to satisfy competing demands in an optimal fashion, and are sensitive to changes in those demands. Understanding exactly how tasks exert their control on gaze is a critical issue for future research.

A central feature of human cognition is the strict limitation on the ability to acquire visual information from the environment, set by limitations in attention. Related to this are the limits in retaining this information, set by the capacity of working memory. We are far from understanding how the organization of the brain leads to these limitations. We also have little understanding of how they influence the way that visual perception operates in the natural world, in the service of everyday visually guided behavior. Consideration of how the limited processing capacity of cognition influences acquisition of visual information leads us to the problem of how such acquisition is controlled. It is not really possible to address the question of precisely what information is selected from the image, and when it is selected, in the context of traditional experimental paradigms, where the trial structure is designed to measure a particular visual operation over repeated instances, each of short duration. In natural behavior, on the other hand, observers control what information is selected from the image and when it is selected. By observing natural behavior, knowledge of the task structure often allows quite well constrained inferences about the underlying visual computations, on a time scale of a few hundred milliseconds.

1. Eye movements and task structure

How can we study the acquisition of information in the natural world? Although incomplete, eye movements are an overt manifestation of the momentary deployment of attention in a scene. Covert attentional processes, of course, mean that other information is processed as well, but overt fixations carry a tremendous amount of information about current attentional state, and provide an entrée to studying the problem (Findlay & Gilchrist, 2003). Investigation of visual performance in natural tasks is now much more feasible, given the technical developments in monitoring eye, head, and hand movements in unconstrained observers, as well as the development of complex virtual environments. This allows some degree of experimental control while allowing relatively natural behavior. In natural behavior, the task structure is evident, and this allows the role of individual fixations to be fairly easily interpreted, because the task provides an external referent for the internal computations. In contrast, when subjects simply passively view images, the experimenter often has little control of, and no access to, what the observer is doing. When viewing pictures, observers may be engaged in object recognition, remembering object locations and identity, or performing some other visual operation. Immersion in a real scene probably calls for different kinds of visual computations, because observers may be interacting with the objects in the scene. When viewing images of scenes, some regularities in fixation patterns can be explained by image properties such as contrast or chromatic salience. However, these factors usually account for only a modest proportion of the variance (Itti & Koch, 2001; Mannan, Ruddock & Wooding, 1997; Parkhurst, Law, & Neibur, 2002).

Over the past ten years, a substantial amount of evidence has accumulated about deployment of gaze during ongoing natural behavior. In extended visuomotor tasks such as driving, walking, sports, playing a piano, hand-washing, and making tea or sandwiches, the central finding is that fixations are tightly linked to the performance of the task (Hayhoe, Shrivastrava, Mruczek & Pelz, 2003; Land & Furneaux, 1997; Land & Lee, 1994; Land, Mennie, & Rusted, 1999; Patla & Vickers, 1997; Pelz & Canosa, 2001; Turano, Geruschat, & Baker, 2003). Subjects exhibit regular, often quite stereotyped fixation sequences as they step through the task. Very few irrelevant areas are fixated. Figure 1 shows an example of the clustering of fixations on task-specific regions when a subject makes a sandwich. This is hard to capture in a still image, but can be clearly appreciated in video sequences such as those in Hayhoe et al. (2003). A feature of the relationship of the fixations to the task is that they are tightly linked, in time, to the actions (Land et al., 1999; Hayhoe et al., 2003). The temporal linkage has been demonstrated clearly by Johansson, Westling, Bäckström, & Flanagan (2001), who measured fixation locations and hand path while a subject picked up a bar and maneuvered the tip past an obstacle, to contact a switch. Fixations were made at critical points such as the tip of the obstacle while the bar was moved around it, and then on the switch once the bar had cleared the obstacle. Gaze arrived at the critical point just before the action, and departed just as the action was accomplished. This is illustrated in Figure 2.

This aspect of natural behavior, where observers acquire the specific information they need just at the point it is required in the task, was called a "just-in-time" strategy (Ballard, Hayhoe, & Pelz, 1995). In their experiment, subjects copied a pattern of colored blocks (the Model) using pieces in a Resource area, which they picked up and placed in

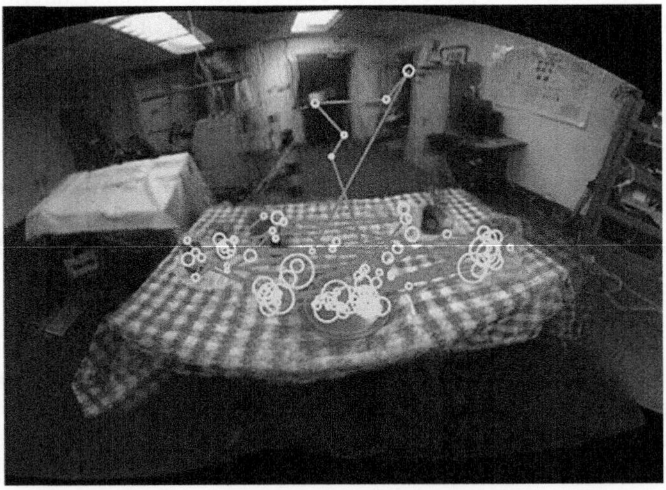

Figure 1. Fixations made by an observer while making a peanut butter and jelly sandwich, indicated by yellow circles. Images were taken from a camera mounted on the head, and a composite image mosaic was formed by integrating over different head positions using a method described in Rothkopf and Pelz (2004) et al. (The reconstructed panorama shows artifacts because the translational motion of the subject was not taken into account.) Fixations are shown as yellow circles, with a diameter proportional to fixation duration. The red lines indicate the saccades. Note that almost all fixations fall on task relevant objects. (*See Color Plate 10.*)

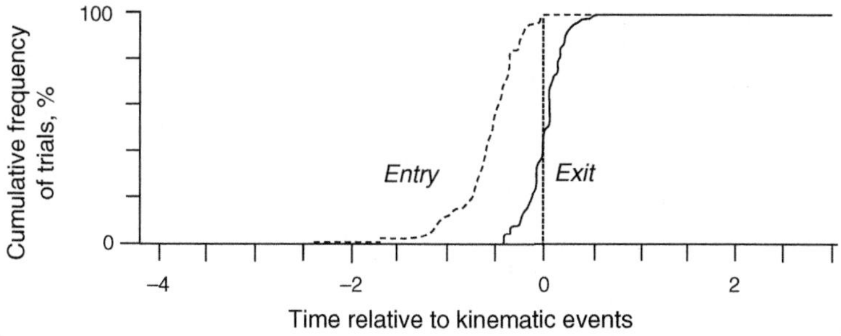

Figure 2. Time gaze arrives at and departs from a point near the obstacle, relative to the time at which the bar is closest to the obstacle. Adapted from Johansson et al. (2001).

the area where the copy was being made. When subjects copied a particular block, they typically fixated a block in the Model, then looked at a block of the same color in the Resource while they picked it up, then looked back at the Model block, presumably to get information about location for placement, and then finally to the copy area where the block was placed in the appropriate location. Thus subjects appeared not to memorize the relatively simple model patterns, but simply to fixate individual blocks to get the information they need at that moment.

2. Learning where to look

Implicit in much of the research on natural tasks is the finding that the observed pattern of eye movements is a consequence of learning at several levels (Land & Furneaux, 1997; Land, 2004; Chapman & Underwood, 1998). For example, in tea making and sandwich making, observers must have learnt what objects in the scene are relevant, and how to locate them in visual search, since almost no fixations fall on irrelevant objects. In driving, Shinoda, Hayhoe, & Shrivastava (2001) showed that approximately 45% of fixations fell in the neighborhood of intersections. As a consequence of this, subjects were more likely to notice Stop signs located at intersections as opposed to signs in the middle of a block, suggesting that subjects have learnt that traffic signs are more likely to be located around intersections. At a more detailed level, subjects must learn the optimal location for the information they need. For example, when pouring tea, fixation is located at the tip of the teapot spout (Land et al., 1999). Presumably, flow from the spout is best controlled by fixating this location. Similarly, in walking, observers must learn where and when to look at locations critical for avoiding obstacles while controlling direction and balance (Patla et al., this volume). Subjects must learn not only the locations at which relevant information is to be found, but also the order in which the fixations must be made in order to accomplish the task. Thus, a subject must locate the peanut butter and the bread before picking them up, pick up the knife before spreading, and so on. This means that a complete understanding of fixations in natural behavior will require an understanding of the way tasks are learnt and represented in the brain.

Another way in which learning is critical for deployment of gaze and attention is that observers must learn the dynamic properties of the world in order to distribute gaze and attention where they are needed. When making tea or sandwiches, items remain in stable locations with stable properties, for the most part. In a familiar room, the observer need only update the locations of items that are moved, or monitor items that are changing state (for example, water filling the kettle). In dynamic environments, such as driving, walking, or in sports, more complex properties must be learnt. Evidence for such learning is the fact that saccades are often pro-active; that is, they are made to a location in a scene in advance of an expected event. For example, in Land & MacLeod's investigation of cricket, batsmen anticipated the bounce point of the ball, and more skilled batsmen arrived at the bounce point about 100 ms earlier than less skilled players (Land & McLeod, 2000). The ability to predict where the ball will bounce depends on previous experience of the cricket ball's trajectory. These saccades were always preceded by a fixation on the ball as it left the bowler's hand, showing that batsmen use current sensory data in combination with prior experience of the ball's motion to predict the location of the bounce. This suggests that observers have stored internal models of the dynamic properties of the world that can be used to position gaze in anticipation of a predicted event.

There is considerable evidence for the role of internal models of the body's dynamics in the control of movement (e.g. Wolpert, Miall & Kawato, 1998). Such models predict the internal state of the body as a consequence of a planned movement, and help mitigate the problem of delays in sensory feedback about body posture. Similar delays in processing visual information about events in the world suggest a similar need for models of the *environment,* particularly in dynamic situations. However, the need for internal models of the environment is less well established. Indeed, the body of evidence in the past has suggested the contrary, that observers construct only minimal representations of the world (Ballard, Hayhoe, Pook, & Rao, 1997; O'Regan, 1992; Simons, 2000). To build internal models of the visual environment, observers must be able to accumulate visual information over the time-varying sequence of visual images resulting from eye and body movements. Visual representations that span fixations are typically thought to be very impoverished. It is generally agreed that, following a change in gaze position, observers retain in memory only a small number of items, consistent with the capacity limits of visual working memory, together with information about scene "gist," and other higher level semantic information (Irwin & Andrews, 1996; Hollingworth & Henderson, 2002). However, some kind of internal model of the environment, such as memory for spatial structure, seems necessary to ensure coordinated movement (Chun & Nakayama, 2000; Loomis & Beall, 2004).

Hayhoe, Mennie, Sullivan, and Gorgos (2005) provide further evidence of the existence of sophisticated internal models of the structure of the environment. Such models may be used to predict upcoming events and plan movements in anticipation of those events. In this study, eye, head, and hand movements were recorded while subjects caught balls thrown with a bounce. Three participants stood in a triangular formation, and threw a ball around the circle. Initially, subjects threw a tennis ball around the circle of three participants. Each throw was performed with a single bounce approximately mid-way between the participants. One of the throwers then changed the ball without warning, to one with greater elasticity (bounciness).

Similar to batsmen in cricket, when catching a ball, subjects initially fixated the hands of the thrower, then made a saccade to the anticipated bounce point, and then pursued the ball until it was close to the hands. Average departure time of gaze from the hands of the thrower was 61 ms after the ball left the hands. Gaze then arrived at a point a little above the anticipated bounce location an average of 53 ms before the bounce. (Note that the ASL 501 eyetracker used in these experiments has a real-time delay of approximately 50 ms. This value was used to correct the latency measurements.) Subjects maintained gaze at this location until the ball came into the fovea, and then made a smooth pursuit movement, maintaining gaze on the ball until the catch. Since the minimum time to program a saccadic eye movement is 200–250 ms (in the absence of any kind of anticipation or preparation), the saccade from the hands to the bounce point must be at least partially under way prior to the release of the ball. The landing points of the saccades relative to the actual bounce point clustered within about 5° laterally, and about 15° vertically above the bounce point. Thus subjects appeared to be targeting a region just above the bounce point, rather than the bounce point itself. This presumably facilitates the subsequent tracking movement by allowing time to capture the ball's trajectory after the bounce. The tight lateral clustering of the saccade landing points relative to the bounce point suggested that subjects were using information from the early part of the throw to target the likely location of the bounce.

2.1. Adjusting to the ball's dynamic properties

Ability to pursue the ball depended on experience with the ball's dynamic properties. When the tennis ball was unexpectedly replaced with a bouncier ball, subjects were unable to track the ball, and instead made a series of saccades. Within a few trials, subjects were once again able to accurately pursue the ball. A crude evaluation of pursuit accuracy was made by measuring the proportion of time gaze was close to the ball, in the period between bounce and catch. Improvement in pursuit performance over six trials is shown in Figure 3, which shows the pursuit accuracy improving rapidly over the first three trials, close to the performance level with the tennis ball. The ability to make accurate pursuit movements in this context therefore depends on the knowledge of the dynamic properties of the new ball. The adjustment in performance was quite rapid, and uniform across subjects, suggesting that adjusting to such changes in the environment is an important feature of natural behavior. (The ability to pursue the tennis ball accurately on the first trial presumably reflects either its slower speed, or that its motion is closer to subjects' prior expectations.) The latency of the first saccade from hands to bounce point also changed over the course of a few trials. Arrival time at the bounce point advanced by about 100 ms over the first six trials following the change from tennis to bouncy ball. The earlier arrival of the eye at the bounce point is accompanied by earlier departure from the hands at the point of release. Thus, anticipatory saccades and pursuit movements reveal that acquisition of visual information is planned for a predicted state of the world.

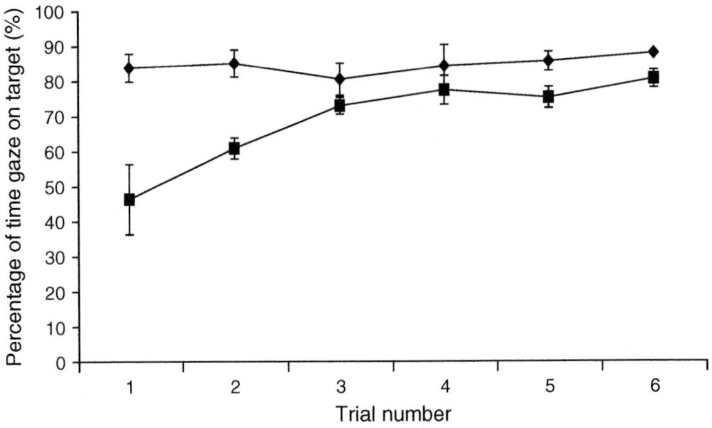

Figure 3. Pursuit accuracy for tennis ball (upper points) and more elastic ball (lower points) as a function of trial number, averaged over five subjects.

Such predictions must be based on a stored memory representation of some kind. The precision of the predictions reveals the quality of the information in the stored memory, or internal model. The spatial and temporal precision of the anticipatory saccades, and the fine-tuning of these movements following a change in the ball's dynamic properties indicate that subjects have an accurate internal model of the ball's spatiotemporal path, and rapidly update this model when errors occur. Rapid adjustment of performance suggests that such prediction is a ubiquitous feature of visually guided behavior.

3. Neural substrate for learning where to look

Control of eye movements, visual acquisition, and memory use by the immediate task entails several different kinds of learning. As mentioned above, observers must learn the structure of tasks such as making tea, they must learn where to look to get the information they need, they must learn the properties of the world, and how those properties change, and they must learn how to allocate attention and fixations in an optimal manner. Recent developments in neurophysiology help us understand how some of this learning might come about. Much research supports a reward-based learning mechanism involving dopamine. Considerable evidence for this comes from experiments by Schultz that show that dopaminergic neurons in the substantia nigra pars compacta in the basal ganglia behave in ways predicted by mathematical models of reinforcement (Montague, Hyman, & Cohen, 2004; Schultz, 2000). This reward system is integral to the generation of saccadic eye movements. Saccade-related areas in the cortex (frontal eye fields, dorso-lateral pre-frontal, and lateral intra-parietal) all converge on the caudate nucleus in the basal ganglia, and the cortical-basal ganglia-superior colliculus circuit appears to regulate the control of fixation and the timing of planned movements. This is achieved by regulation of tonic inhibition exerted by the substantia nigra pars reticulata on the superior colliculus, the mid-brain saccade generator. Such regulation is a critical component of task control of fixations. Hikosaka and colleagues have demonstrated that caudate cell responses reflect both the target of an upcoming saccade and the reward expected after making the movement (Hikosaka, Takikawa, & Kawagoe, 2000; Watanabe, Lauwereyns, & Hikosaka, 2003). Since some kind of sensitivity to reinforcement is necessary for learning, and saccadic eye movements demonstrate such sensitivity, the neural substrate for learning where to look in the context of a task is present in the basal ganglia.

Other areas involved in saccade target selection and generation also exhibit sensitivity to reward. In the lateral intra-parietal area (LIP), the neurons involved in saccadic targeting respond in a graded manner to both the amount of expected reward and the probability of a reward, in the period prior to execution of the response (Dorris & Glimcher, 2004; Glimcher, 2003; Platt & Glimcher, 1999; Sugrue, Corrado, & Newsome, 2004). Sensitivity to both these variables is critical for linking fixation patterns to task demands. Cells in the supplementary eye fields also signal the animal's expectation of reward and monitor the outcome of saccades (Stuphorn, Taylor, & Schall, 2000). Sensitivity to stimulus probability is also revealed in build-up neurons in the intermediate layers of the superior colliculus (Basso & Wurtz, 1998). This parallels psychophysical observations showing that saccade reaction time is similarly influenced by stimulus probability (He & Kowler, 1989).

The neural data showing context-specific responses and the role of reward form a critical substrate for explaining task-directed eye movement patterns. The reward sensitivity of the eye movement circuitry provides a basis for reinforcement learning models that are necessary for understanding how these elemental processes are organized to compose the complex gaze patterns observed in everyday behaviors. Theoretical work such as that by Sprague & Ballard (2003) shows how a graphical agent in a virtual environment can learn to allocate gaze sequentially to areas in the environment important for walking and avoiding obstacles. To choose between ongoing competing tasks such as avoiding obstacles and controlling direction of locomotion, in their model, uncertainty increases (together with an attendant cost) when gaze is withheld from an informative scene location. Fixation is allocated to the task that would have the greatest cost if the relevant information were not updated. They show that such a cost is calculable within the reinforcement learning framework described by Schultz and others. In the context of a complex behavioral sequence, a single eye movement does not, of course, generate a primary reward such as a drop of juice, as in the neurophysiological experiments; but all complex behaviors involve secondary reward of some kind, and the acquisition of information is always a critical step in achieving behavioral goals.

4. Specialized computations during fixations

There is far too much information in visual scenes to process at once. Even at the point of fixation, multiple kinds of information are available. It seems likely that, in the context of natural behavior, the task controls the specific information that is selected within a given fixation. For example, when driving around a bend in the road, drivers fixate the tangent point of the curve (Land & Lee, 1994). Interestingly, the angle of gaze with respect to the body is a measure of the necessary change in the steering angle required for navigating the bend. Thus gaze directly provides the control variable needed for the momentary task. When making a sandwich, subjects will fixate the handle of the knife when picking it up, but the tip of the knife when spreading the peanut butter. In the first case, the subject needs the location and orientation of the handle to guide the pick up action. In the latter case, the tip needs to be fixated to control the spreading action. When first viewing the tabletop scene, subjects make a series of short duration fixations on the relevant objects, such as the peanut butter jar. In this case the fixation on the peanut butter is presumably for the purpose of recognition, and perhaps locating it for future use (Hayhoe et al., 2003). Subsequent fixations will be for guiding the grasping action or for removing the lid. In the absence of ongoing behavior, we are inclined to think that the job of vision is primarily object recognition, but these examples remind us of the complexity of the information one can get while fixating an object, and the variety of operations that vision must perform. This specificity is indicated not only by the ongoing actions and the point in the task, but also by the durations of the fixations, which may vary over a range from less than 100 ms to several seconds (Hayhoe et al., 2003; Pelz et al., 2000). A large component of this variation appears to depend on the particular information required for that point in the task, fixation being terminated when the particular information is acquired (Hayhoe Bensinger, & Ballard, 1998; Henderson, 2003; Pelz et al., 2000). For example, in a task where subjects were required to tap a pre-determined sequence of lights on a table top, fixations in the search phase of the experiment, while subjects are locating the lights, are much shorter than fixations while subjects are tapping, where gaze is used for guiding the hand (Epelboim et al., 1995).

Figure 4 illustrates our conception of the way visual information acquisition is organized by the task (see also Land et al., 1999; Schwartz, Reed, Montgomery, Palmer, & Mayer, 1991). At the most general level, acquisition is driven by a larger cognitive goal, such as making a sandwich. To accomplish this goal, the observer must perform a sequence of micro-tasks, such as grasping the peanut butter, grasping the lid, removing the lid, and so on. A micro-task, such as grasping the peanut butter, involves a fixation on the jar. During that fixation, specific visual information will be acquired, such as the size of the jar, which is necessary in order to plan the grasp. These task-specific computations have been referred to as "visual routines" (Ballard et al., 1997; Hayhoe, 2000; Roelfsema, Lamme, & Spekreijse, 2000; Ullman, 1984). The idea of visual routines is that even quite low-level visual information, such as color, requires specialized computation of some kind, and is not done automatically by the visual system. We hypothesize that the visual computations operating while the subject is guiding a grasp do not necessarily involve computing other information, such as the color of the peanut butter, that are not necessary at that moment. Such information might of course be stored in working memory, or in some longer-term memory representation of the scene.

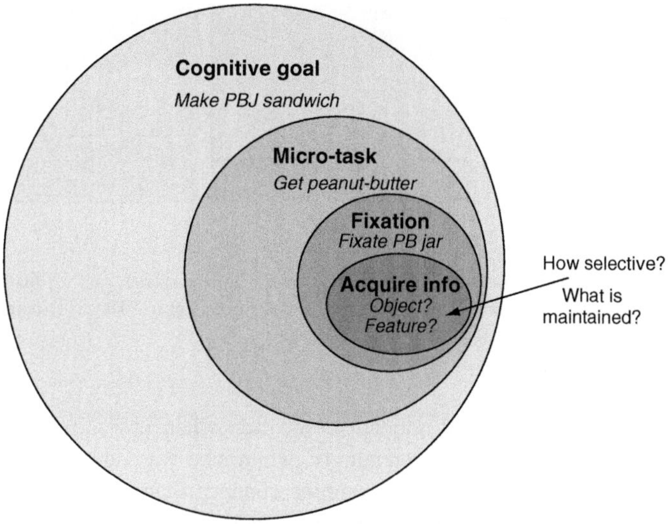

Figure 4. Hierarchy of operations involved in natural tasks such as sandwich making. Within a given fixation, subjects acquire information. Experiments by Droll et al. (2005) investigate how task-specific this information is, and whether it is stored in working memory.

The hypothesis that very specific information is acquired within a fixation is, the strictly, a hypothesis about neural mechanisms. That is, the cortical state (including V1) is different when subjects are involved in different tasks, even when the retinal input is the same. However, it is possible to get supporting evidence psychophysically. Two studies have provided evidence that highly task-specific information is extracted in different fixations (Hayhoe et al., 1998; Triesch, Ballard, Hayhoe, & Sullivan, 2003). Both these studies used the technique of making changes to task relevant information in the scene, and looking for an influence of the changes on task performance, as well as at sensitivity to the changes. In the Triesch et al. experiment, subjects picked up virtual bricks and placed them on one of two virtual "conveyor belts", as illustrated in Figure 5. The bricks were of two different heights, and subjects sorted them onto the conveyor belts according to different rules that varied the point in the task at which the brick height was relevant. In the first case, subjects picked up the bricks according to their location, and size was irrelevant. In the second, the bricks were picked up on the basis of their size, and then all placed on the same conveyor belt, so size was relevant for pickup only. In the third, bricks were placed on different belts depending on their size, so size was relevant for both pickup and placement. On a small proportion of trials, the brick changed size by about 20% while it was being moved to the belt. Although the change in size was very obvious when subjects were attending to it, when they were performing the task, subjects rarely reported seeing the changes, either during the task or upon subsequent questioning. They were most unaware of the change when size was irrelevant, suggesting that information about size might not have been present in the subject's visual representation of the brick

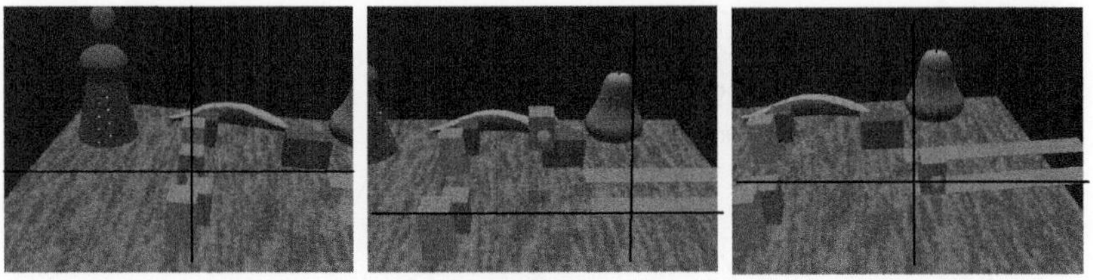

Figure 5. View of the virtual work-space as a subject (a) picks up, (b) carries, and (c) places a brick on a conveyor belt. The dot visible in front of the lifted brick is one of the fingers. The cross-hair shows fixation. Adapted from Triesch et al. (2003). (*See Color Plate 11.*)

in these trials. Size changes were noticed more often when it was relevant for pickup or putdown (conditions 2 and 3). Interestingly, when size was relevant only for pickup (condition 2), subjects were less likely to notice changes than when it was relevant for placement as well (condition 3), suggesting that in the second condition they did not retain a representation of size in working memory when it was no longer needed after they had picked it up. Thus the experiment supported the idea that specific information about the brick's properties, in this case its size, was only perceptually encoded in conditions 2 and 3, when it was needed for pickup, and only retained in working memory in condition 3, when it was needed for putdown as well. In each case, the subject is "attending" to the brick as he/she fixates it, picks it up, moves, and places it. This experiment reveals that a general undifferentiated concept of attention is too crude to reveal the subtleties of the ongoing visual computations. Another finding in the experiment was that subjects were often unaware of the change in size even when they were directly fixating the brick when the change occurred. This happened on some trials when subjects tracked the brick while they moved it across the workspace, so that they were fixating the brick at the point when it changed. (Note, this differs from the fixation pattern shown in Figure 5, where the subject saccades directly to the conveyor belt after pickup.) This suggests that subjects may not represent particular stimulus features such as size when they are performing other computations such as guiding the arm. This appears to be a form of "inattentional blindness", described extensively by Mack & Rock (1998). The pervasive nature of inattentional blindness in perception is taken advantage of by magicians. The strength of this phenomenon in the context of magical tricks is described by Tatler and Kuhn in this volume.

4.1. Task-specific representations and Object Files

The idea that task relevance guides top-down selection of even simple feature information contrasts with the idea that visual information is represented and maintained in the form of object files (Kahneman, Treisman, & Gibbs, 1992). This theory posits that when attention is directed to an object in a scene, a temporary representation called an "object file" is created and held in visual short-term memory, and about three or four object files and their spatial locations can be held in memory at any one time (Gordon & Irwin, 1996; Irwin & Andrews, 1996). Object file theory is consistent with claims that the units of short-term visual memory are integrated objects, not simple features (Luck & Vogel, 1997; Vogel, Woodman, & Luck, 2001). The Treisch et al. experiment suggests that the concept of object files might not extend to natural behavior, where task demands are dynamic and specific to the immediate needs of the observer. To explore this question further, Droll, Hayhoe, Triesch, & Sullivan (2005) conducted an experiment similar in concept to that of Treisch et al. The goal of the experiment was to identify more precisely the information acquired and held in memory and provide a more definitive test of the hypothesis that the visual information acquired in a fixation may be limited to the specific feature required by the task. In the Triesch et al. (2003) experiment the only feature of the bricks that was used in the task was height. Is it the case that brick features that were not required, such as color and shape, were never encoded? This would mean that fixating an object and attending to it would not necessarily bind the features of the brick into some object representation, or object file.

The basic task used by Droll et al. was to select one brick from an array, and to sort this brick onto one of two conveyor belts. The bricks were defined by several features (color, height, width, and texture), and a pickup cue indicated which feature value was relevant for a particular trial. After picking up the brick, a put-down cue was displayed to guide the sorting decision. The brick was placed on the appropriate conveyor belt, removing the brick from the scene, and initiating a new trial with a new pickup cue and array of bricks. Thus, because the put-down cue was presented after pickup, the put-down decision was separated in time and space from pickup, and the representations of the relevant object feature needed to be maintained until the put-down decision was made. In one condition, subjects performed a task in which only one feature dimension was relevant for both pickup and put-down (e.g. color). In another condition, different features were used for pickup and put-down (e.g. color for pickup, height for put-down). The task sequence is illustrated in Figure 6.

This experiment also used the strategy of changing the bricks on a small proportion of trials, but in this case it was possible to change either the feature that was relevant for pick up or putdown, or one of the other features that was not relevant in that block of trials. Subjects indicated whether they saw a change by placing the brick in a "trash can" (the black square in Figure 6). Subjects were about twice as likely to notice the feature change when that feature was relevant to the task as when it was irrelevant (either for pickup or for put-down, or for both). This supports the hypothesis that subjects preferentially represent the task-specific features of the objects. Objects are not necessarily stored in working memory as bound entities. Thus, understanding what visual operations are occurring during a fixation requires knowledge of the current task.

Memory vs Just-in-time representations: Another important finding in Droll et al.'s experiment was subjects' behavior on trials when a task-relevant feature was changed, but subjects still failed to notice that a brick had changed. On these trials, subjects sorted the bricks onto one of the conveyor belts instead of placing it in the trash. How do the subjects

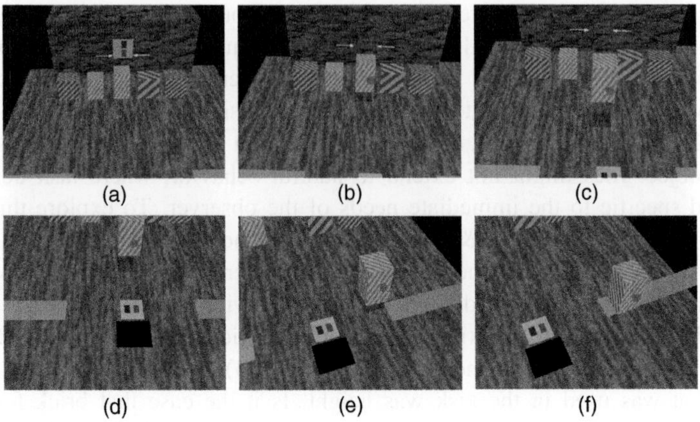

Figure 6. Scene during a single trial of the *One Feature* condition when brick color was task relevant. Fingertips are represented as small red spheres. In a single trial, a subject (a) selects a brick based on the pick-up cue, (b) lifts the brick, (c) brings it towards themselves, (d) decides on which conveyor belt the brick belongs based on a put-down cue, (e) guides the brick to the conveyor belt, (f) sets the brick on the belt where the brick is carried off. In other trials, subjects may have used width, height, or stripes for the pick-up or put-down decision. Adapted from Droll et al. (2005). (*See Color Plate 12.*)

sort the bricks? Figure 7 shows the two possibilities. Either the subject can treat the brick as if it retained its old feature, or else he/she can sort it according to its current feature. Given that the subject invariably fixates the brick either before or during placement, one might expect that the new feature state would be clearly visible, and that the subjects would sort the brick on the basis of its current (changed) feature. Instead, they almost always sorted it by its old feature, despite the fact that they fixated the brick, with its new feature state, for an average of 750 ms after the change. This indicates that subjects are using

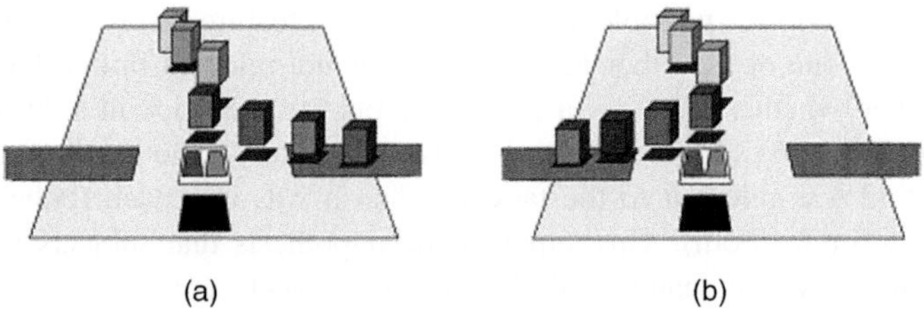

Figure 7. Two possible sorting decisions following a missed feature change. (a) Subjects may sort the brick by the old, pre-change, feature, in which case changes are missed due to a failure to update the new visual information. (b) Subjects may sort also the brick by the new, post-change, feature, in which case changes are missed due to a failure to maintain visual information. When subjects performed blocks of trials using the same feature for the put-down decision, missed changes were most often sorted by the old feature (85%). (*See Color Plate 13.*)

their memory of the brick feature, rather than its actual current state, to make the sorting decision. (Note that the put-down cue is not available until the block has been picked up, so subjects cannot plan the put-down movement immediately after pickup, but must fixate the put-down cue first.) This finding is important because failure to detect changes to items in scenes has traditionally been interpreted as a failure to retain information from the pre-change image in memory (O'Regan, 1992; Rensink, 2000). Although this is undoubtedly the case in many situations, in the current context, sorting by the old feature reveals that subjects do indeed maintain a memory representation of the previous state of the brick, but simply fail to update their representation with the new information following a change. Change blindness as a consequence of failure to update the internal representation is consistent with the suggestion of Henderson & Hollingworth (1999) and Simons & Rensink (2005). It seems likely that in many situations the information in the visual scene is stable. Changes like those in the Droll et al. experiment would be impossible in real scenes. Subjects presumably take advantage of prior experience about the stable properties of normal scenes to accumulate information about scenes in internal representations. Such a strategy makes sense given the limited bandwidth that is set by attention on the accrual of information. As long as the observer has accurate knowledge of the stable properties of scenes, there is no point in constantly updating the information in the image if it takes up attentional resources. It appears that even such low level information as color or size takes computational resources. The fact that subjects were fixating the brick while sorting onto the wrong conveyor belt suggests that attentional resources were devoted to the placement action, rather than to updating information about the brick. This phenomenon may be related to "inattentional blindness," similar to the phenomena mentioned above (Mack & Rock, 1998). Note that the strategy of retaining information in memory in the brick sorting task is in contrast to the "just-in-time" strategy exhibited by observers in the block copying task (Ballard et al., 1995). In one case, observers opt to retain information in memory, and in the other they minimize memory load by fixating the block just at the point when the information is needed. Both are ways that the human cognitive system can deal with attentional and memory limitations. An important question for future research is: What determines which strategy is used?

In Droll et al.'s experiment, subjects treated the block properties as stable. An interesting observation in their experiment was subjects' behavior in subsequent trials after a feature change was successfully detected. Figure 8 shows the total time spent fixating the brick on trials following a noticed change, and also the total duration of the hand movement between lifting and placement on the conveyor belt. Both fixations and hand movements were significantly longer on the trial immediately following a detected change, by as much as 400 ms. This effect fell off sharply over the next few trials. Thus subjects were taking more time to perform the task after they observed the unlikely event where a brick changed its properties. This suggests that they reallocated attentional resources to the brick on the next trial, when events violated their expectations. However, since the event was not repeated, they reverted to the prior strategy. This suggests that internal models of the environment are dynamic, and may often be more strongly influenced by the most

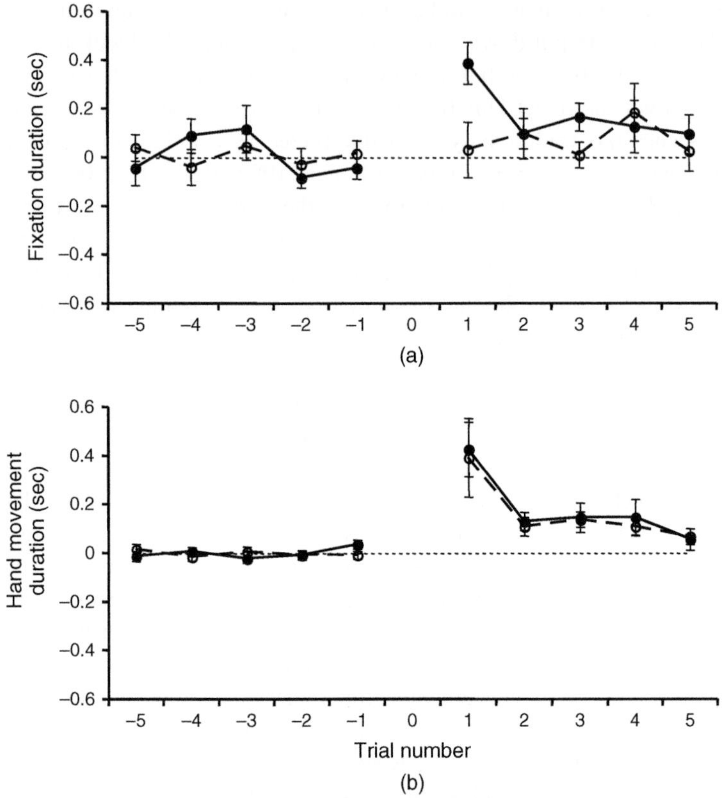

Figure 8. Changes in fixation duration (a) and hand movement duration (b) in trials before and after subjects detected a feature change in a brick during sorting. Negative trial numbers indicate trials before the detected change; positive trial numbers indicate trials following a detected change. Subjects used either the same feature for both pick-up and put-down (dashed line), or different features for each operation (solid line). Error bars represent standard errors across subject mean. Adapted from Droll et al. (2005).

recent events. Recall that similar dynamic updating was required as subjects learned to track a ball with a different elasticity.

5. Summary

The last decade has seen tremendous advances in our knowledge of the way gaze is used in everyday behavior. This chapter attempts to capture some of the insights from this work in order to point toward areas of future exploration. Perhaps the most important insight is that gaze patterns reflect extensive learning at several levels. At the most general level, observers must learn the sequence of operations required to perform tasks. They must learn where to look in a scene to get the information they need for component sub-tasks. They must learn the structure and dynamic properties of the world in order to fixate critical regions at the right time. They must learn how to allocate attention and gaze to satisfy competing demands in an optimal fashion and be sensitive to changes in those demands. There are many questions about the precise way that learning affects gaze patterns. Perhaps the most critical issue is understanding exactly how tasks exert their control on gaze. A growing understanding of the importance of reward in modulating the underlying neural mechanisms and theoretical developments using reinforcement learning models of complex behavior provides us with the tools to understanding how tasks are represented in the brain, and how they control acquisition of information through use of gaze.

Acknowledgements

This work was supported by NIH grants EY 05729 and RR 09283. The authors wish to thank Brian Sullivan, Keith Gorgos and Jennifer Semrau for assistance with the experiments and Keith Parkins for programming support.

References

Ballard, D. H., Hayhoe, M., & Pelz, J. B. (1995). Memory representations in natural tasks. *Journal of Cognitive Neuroscience, 7(1)*, 66–80.

Ballard, D. H., Hayhoe, M. M., Pook, P. K., & Rao, R. P. N. (1997). Deictic codes for the embodiment of cognition. *Behavioral and Brain Sciences, 20*, 723–767.

Basso, M., & Wurtz, R. (1998). Modulation of neuronal activity in superior colliculus by changes in target probability. *Journal of Neuroscience, 18*, 7519–7534.

Chapman, P., & Underwood, G. (1998). Visual search of dynamic scenes: Event types and the role of experience in viewing driving situations. In Underwood, G. (Ed.), *Eye Guidance in Reading and Scene Perception* 369–394). Oxford: Elsevier.

Chun, M., & Nakayama, K. (2000). On the functional role of implicit visual memory for the adaptive deployment of attention across scenes. *Visual Cognition,* 7, 65–82.

Dorris, M. C., & Glimcher, P. W. (2004). Activity in posterior parietal cortex is correlated with the subjective desirability of an action. *Neuron, 44*, 365–378.

Droll, J., Hayhoe, M., Triesch, J., & Sullivan, B. (2005). Task demands control acquisition and maintenance of visual information. *Journal of Experimental Psychology: Human Perception and Performance, 31*, 1416–1438.

Epelboim, J., Steinman, R., Kowler, E., Edwards, M., Pizlo, Z., Erkelens, C., & Collewijn, H. (1995). The function of visual search and memory in sequential looking tasks. *Vision Research, 35*, 3401–3422.

Findlay, J., & Gilchrist, I. (2003). *Active Vision*. Oxford: Oxford University Press.

Glimcher, P. (2003). The neurobiology of visual-saccadic decision making. *Annual Review of Neuroscience, 26*, 133–179.

Gordon, R. D., & Irwin, D. E. (1996). What's in an object file? Evidence from priming studies. *Perception & Psychophysics, 58(8)*, 1260–1277.

Hayhoe, M. (2000). Visual routines: a functional account of vision. *Visual Cognition, 7*, 43–64.

Hayhoe, M., Bensinger, D., & Ballard, D. (1998). Task constraints in visual working memory. *Vision Research, 38*, 125–137.

Hayhoe, M., Mennie, N., Sullivan, B. & Gorgos, K. (2005). *The role of internal models and prediction in catching balls.* Proceedings of AAAI Fall Symposium Series.

Hayhoe, M., Shrivastrava, A., Mruczek, R., & Pelz, J. (2003). Visual memory and motor planning in a natural task. *Journal of Vision, 3*, 49–63.

He, P., & Kowler, E. (1989). The role of location probability in the programming of saccades: Implications for center-of-gravity tendencies. *Vision Research, 29*, 1165–1181.

Henderson, J. (2003). Human gaze control during real-world scene perception. *Trends in Cognitive Science, 7*, 498–504.

Henderson, J. & Hollingworth, A. (1999). The role of fixation position in detecting scene changes across saccades. *Psychological Science, 10*, 438–443.

Hikosaka, O., Takikawa, Y., & Kawagoe, R. (2000). Role of the basal ganglia in the control of purposive saccadic eye movements. *Physiological Review, 80*, 953–978.

Hollingworth, A., & Henderson, J. (2002). Accurate visual memory for previously attended objects in natural scenes. *Journal of Experimental Psychology: Human Perception and Performance, 28*, 113–136.

Irwin, D. E., & Andrews, R. V. (Eds.). (1996). *Integration and accumulation of information across saccadic eye movements.* Cambridge: MIT Press.

Itti, L., & Koch, C. (2001). Computational modeling of visual attention. *Nature Review Of Neuroscience, 2*, 194–203.

Johansson, R., Westling, G., Bäckström, A., & Flanagan, J. R. (2001). Eye-hand coordination in object manipulation. *Journal of Neuroscience, 21*, 6917–6932.

Kahneman, D., Treisman, A., & Gibbs, B. J. (1992). The reviewing of object files: Object-specific integration of information. *Cognitive Psychology, 25*, 175–219.

Land, M. (2004). Eye movements in daily life. In *The Visual Neurosciences* (vol 2), Chalupa, L. & Werner, J. (Eds.), 1357–1368, MIT Press.

Land, M., & Furneaux, S. (1997). The knowledge base of the oculomotor system. *Philosophical Transactions of the Royal Society of London B, 352*, 1231–1239.

Land, M. F., & Lee, D. N. (1994). Where we look when we steer. *Nature* (London), *369*, 742–744.

Land, M. F., & McLeod, P. (2000). From eye movements to actions: how batsmen hit the ball. *Nature Neuroscience, 3*, 1340–1345.

Land, M. F., Mennie, N., & Rusted, J. (1999). Eye movements and the roles of vision in activities of daily living: making a cup of tea. *Perception, 28*, 1311–1328.

Loomis, J., & Beall, A. (2004). Model-based control of perception/action. In Vaina, L. et al. (Eds.), *Optic Flow and Beyond* (pp. 421–441), Netherlands: Kluwer.

Luck, S. J., & Vogel, E. K. (1997). The capacity of visual working memory for features and conjunctions. *Nature, 390* (6657), 279–281.

Mack, A., & Rock, I. (1998). *Inattention blindness.* Cambridge, MA: MIT Press.

Mannan, S. Ruddock, K. H. & Wooding, D. S., (1997). Fixation patterns made during brief examination of two-dimensional images. *Perception, 26*, 1059–1072.

Montague, P. R., Hyman, S. E., Cohen, J. D. (2004). Computational roles for dopamine in behavioral control. *Nature, 431*, 760–767.

O'Regan, J. K. (1992). Solving the 'real' mysteries of visual perception: The world as an outside memory. *Canadian Journal of Psychology, 46*, 461–488.

Parkhurst, D., Law, K., & Niebur, E.(2002). Modeling the role of salience in the allocation of overt visual attention. *Vision Research, 42*, 107–123.

Patla, A., & Vickers, J. (1997). Where do we look as we approach and step over an obstacle in the travel path? *NeuroReport, 8*, 3661–3665.

Patla, A., & Vickers, J. (2003). How far ahead do we look when required to step on specific locations in the travel path during locomotion? *Experimental Brain Research, 148*, 133–138.

Pelz, J. B., & Canosa, R. (2001). Oculomotor Behavior and Perceptual Strategies in Complex Tasks, *Vision Research, 41*, 3587–3596.

Pelz, J. B., Canosa, R., Babcock, J., Kucharczyk, D., Silver, A., & Konno, D. (2000). Portable eyetracking: A study of natural eye movements. *Proceedings of the SPIE:* Vol. 3959. *Human vision and electronic imaging* (pp. 566–583). San Jose, CA: SPIE.

Platt, M. L., & Glimcher, P. W. (1999). Neural correlates of decision variables in parietal cortex. *Nature, 400*, 233–238.

Rensink, R. A. (2000). The dynamic representation of scenes. *Visual Cognition, 7*, 17–42.

Roelfsema, P., Lamme,V., & Spekreijse, H. (2000). The implementation of visual routines. *Vision Research, 40*, 1385–1411.

Rothkopf, C. A., & Pelz, J. B. (2004). Head movement estimation for wearable eye tracker. *Proceedings ACM SIGCHI Eye tracking research & applications symposium* (pp. 123–130). San Antonio, Texas.

Schultz, W. (2000). Multiple reward signals in the brain. *Nature reviews: neuroscience, 1*, 199–207.

Schwartz, M., Reed, E., Montgomery, M., Palmer, C., & Mayer, N. (1991). The quantitative description of action disorganization after brain damage: a case study. *Cognitive Neuropsychology, 8*, 381–414.

Shinoda, H., Hayhoe, M. M., & Shrivastava, A. (2001). Attention in natural environments. *Vision Research, 41*, 3535–3546.

Simons, D. J. (2000). Change blindness and visual memory. *A Special Issues of Visual Cognition*. Hove, UK: Psychology Press.

Simons, D., & Rensink, R. (2005). Change blindness: Past, present, and future. *Trends in Cognitive Science, 9*, 16–20.

Sprague, N., & Ballard, D. (2003). Eye movements for reward maximization. In *Advances in Neural Information Processing Systems*, 16, MIT Press.

Stuphorn, V., Taylor, T., & Schall, J. (2000). Performance monitoring by the supplementary eye field. *Nature, 408*, 857–860.

Sugrue, L. P., Corrado, G., & Newsome, W. (2004). Matching behavior and the encoding of value in parietal cortex. *Science, 304*, 1782–1787.

Triesch, J., Ballard, D., Hayhoe, M., & Sullivan, B. (2003). What you see is what you need. *Journal of Vision, 3*, 86–94.

Turano, K., Geruschat, D., & Baker, F. (2003). Oculomotor strategies for the direction of gaze tested with a real-world activity. *Vision Research, 43*, 333–346.

Ullman, S. (1984).Visual Routines. *Cognition, 18*, 97–157.

Vogel, E. K., Woodman, G. F., & Luck, S. J. (2001). Storage of features, conjunctions and objects in visual working memory. *Journal of Experimental Psychology: Human Perception and Performance, 27*(1), 92–114.

Watanabe, K., Lauwereyns, J., & Hikosaka, O. (2003). Neural correlates of rewarded and unrewarded movements in the primate caudate nucleus. *Journal of Neuroscience, 23*, 10052–10057.

Wolpert, D., Miall, C., & Kawato, M. (1998). Internal models in the cerebellum. *Trends in Cognitive Science, 2*, 338–347.

Chapter 31

OCULOMOTOR BEHAVIOR IN NATURAL AND MAN-MADE ENVIRONMENTS

JEFF B. PELZ

Rochester Institute of Technology, USA

CONSTANTIN ROTHKOPF

University of Rochester, USA

Eye Movements: A Window on Mind and Brain
Edited by R. P. G. van Gompel, M. H. Fischer, W. S. Murray and R. L. Hill

Abstract

Three subjects performed two tasks (*Free-view* and *Walking*) in two environments (*Man-made* and *Wooded*). Fixation duration and saccade size were extracted from eye movements monitored with a wearable eyetracker. Mean, median, and modal fixation durations were shorter during the *Free-view* task in the *Man-made* environment than in the *Wooded* environment. However, despite the difference in characteristics and predictability of the paths between the *Man-made* and the *Wooded* environments (paved walkway and dirt path, respectively), there was not a significant environment difference in the distributions of fixation duration during the *Walking* task. There were no significant differences in the distribution of saccade sizes across any of the conditions. While there was no significant environment effect on fixation duration and saccade-size distributions during the *Walking* task, subjects significantly increased the fraction of time gaze was directed to the path immediately before them from 35% on paved walkways to 62% on dirt paths.

Chapters 2, 3, and 4 in this book demonstrate a rich history of oculomotor research. From the earliest studies, eye-movement research has pursued two parallel goals: research *on* eye movements to understand the oculomotor system, and research *with* eye movements where the movements are used as externally visible markers of attention to probe perception and cognition. Delabarre's (1898) first paper on recording eye movements began, "Many problems suggest themselves to the psychologist whose solution would be greatly furthered by an accurate method of recording the movements of the eye." (Delabarre, 1898, p. 572). The amount of data available in the environment would overwhelm our processing capacity; cortical limitations require that only a small subset of the available information be selected for processing. Eye movements are used to select that subset, so monitoring those movements (which Buswell termed "symptoms of perception") provides a pointer to the selected information and the strategies employed in performing that selection. These strategies are critical to daily perception and action, yet they are rarely consciously selected, so they are not available to self-report.

A given sequence of eye movements is the result of the information available in the environment, a subset of elements in the environment that are critical to ongoing tasks and are within the cortical limitations, and the motivation of an observer. The "visual stimulus" (typically an image of limited extent in laboratory experiments) is only one element in the equation. The results of Yarbus's classical experiments (1961, 1967) demonstrated that the intended "independent variable" (the visual stimulus) can actually be less important than the explicit or implicit task. Land (this book) discusses oculomotor strategies during active behavior where, like in Yarbus's classic work, the *task* can have at least as much influence as does the *stimulus*. This can be cast as the interaction of "bottom-up" (stimulus-driven) vs "top-down" (task-dependent) control. The bottom-up influence of an image has been proposed in which a "saliency map" is calculated based on the image properties (Itti & Koch, 2000, Parkhurst & Niebur, 2003). The goal of such models is to predict the likelihood that image elements will draw fixations, independent of top-down effects. Land (this book) discusses the issue of saliency vs task – top-down vs bottom-up in a number of domains. It is clear from many studies that at least an element of top-down, task-dependent control is necessary to reveal behavior (e.g., Findlay & Walker, 1999).

Oculomotor behavior can be examined at a number of different levels beyond isolated movements of the eye. Perhaps the simplest is to examine the relative frequency of the angular extent of individual saccadic eye movements, and the duration of the intervening fixations. The distributions of fixation duration have been shown to vary between tasks. Perhaps most studied is the overlearned task of reading. Eye movements related to reading are relatively easy to monitor because the subject is stationary, the stimulus is static, and only horizontal eye movements need to be recorded. Silent reading has a mean fixation duration of 225 ms, oral reading has a duration of 275 ms (Rayner, 1998). A wide range of other tasks have mean fixation durations of approximately 300 ms. Henderson & Hollingworth (1998) reported mean fixations of 330 ms for free-viewing images on a small display, the same value reported by Pelz & Canosa (2001) for subjects walking in a hallway. Considerably longer mean fixation durations have been reported for tasks requiring manipulation of objects, such as 500 ms while making a pot of tea (Land, Mennie, & Rusted, 1999) and 450 ms while manipulating small parts to construct a plastic model (Pelz et al., 2000), though in both cases, the mean duration was affected by a small number of very long fixations, some lasting several seconds.

The mean extent of saccades is also task dependent. Rayner (1984) reported a mean saccade size of 1.5° during oral reading, increasing to 2° for silent reading (though measures scale with the font size). While free-viewing images on a small display, Henderson & Hollingworth (1998) reported average saccade sizes of 2.4°. Pelz et al. (2000) reported that the mean scaled with display size, up to an average of 10° for a 50" display subtending about 50°. Task-dependence extends to active tasks as well. Turano and colleagues reported mean sizes ranging from 3.1 to 5.6° for subjects walking down a hallway with the goal of locating a specific doorway (Turano, Geruschat, & Baker, 2003); Pelz & Canosa (2001) reported a mean saccade size of 11° for subjects walking down a hallway without such a specific target. Mean saccade sizes as large as 19° have been reported for complex tasks such as making a pot of tea (Land et al., 1999), though again the mean was affected by a relatively small number of very large gaze changes.

Andrews & Coppola (1999) examined the degree to which fixation duration and saccade size varied by observer and task while viewing five "visual environments". Subjects' eye movements were recorded in a range of conditions; in the drak, viewing repetitive textures, viewing photographs, during visual search, and while reading. They found idiosyncratic differences between individuals that covaried significantly within two groupings; 1) viewing photographs, simple patterns, and in the dark, and 2) visual search and reading. There was no significant covariance between the two groupings.

The distributions of fixation durations and the saccade magnitude represent low-level characteristics of subjects' oculomotor behavior; monitoring the direction of gaze during a task can reveal higher-level, though subconscious, strategies adopted by subjects. Examining eye movements while walking provides an opportunity to examine how vision guides action under variable environmental and task demands. One would expect that the degree to which subjects fixate the surface on which they plan to walk will vary based on its physical characteristics, its predictability, and its visibility. In a study of how patients with retinitis pigmentosa navigate with significant visual field loss, Turano et al. (2001) had normal subjects navigate a hallway wearing a head-mounted display with an integrated eyetracker. The head-mounted display provided the same view to both eyes from a single head-mounted camera. The display's limited field-of-view required that subjects actively acquire information that would normally be available from the periphery. Even with the reduced cues available for moving in the hallway, however, fewer than 25% of normal subjects' fixations were directed toward the floor or boundaries between the floor and walls. Patla & Vickers (2003) studied the gaze behavior of normal subjects as they walked a predetermined path. They measured the gaze of subjects with normal vision as they walked a 10 m path under three conditions; 1) stepping on regularly spaced footprints, 2) stepping on irregularly spaced footprints, and 3) a control condition with no markings on the floor. Under all three conditions the predominant gaze behavior was what Patla and Vickers termed "traveling gaze", in which subjects' gaze moved at the same rate as, and at a fixed distance ahead of, the subjects. The travel gaze accounted for over 60% of the total gaze duration, even when there were no specific targets for foot placement. Given the predictability of the surface, and lack of explicit instructions on foot placement in the control condition, it is surprising that Patla and Vickers' subjects focused on the floor immediately before them 60% of the time.

In the experiment presented here, we extend the investigation of task-dependent oculomotor behavior to consider the effect of environment. Incorporating the environment in the experimental design allows us to probe two issues. We can (1) examine the effect of task and environment on low-level fixation duration and saccade-amplitude metrics, and (2) address Patla and Vicker's (2003) surprising findings regarding gaze direction while walking without explicit directions on foot placement. By monitoring gaze as subjects walk in man-made and natural environments, we seek to understand how the varying physical characteristics and the predictability of the path affect oculomotor behavior. A paved surface should require less active monitoring because the surface is more predictable and because peripheral acuity is sufficient to detect typical variations. Walking on an uneven dirt path, however, is expected to require more active guidance, and therefore more frequent foveation.

1. Methods

Subjects performed two different tasks within two environments. The two Environments we tested were outside of an apartment complex (*Man-made Environment*) and in the dense woods (*Wooded Environment*). Within each *Environment* subjects were required to perform two *Tasks*: *Free-viewing* (standing in one place and looking about the scene) and *Walking*. The *Walking Task* in the *Man-made Environment* was performed on a paved path, and the *Walking Task* in the *Wooded Environment* was performed on a dirt trail.

Three adult volunteers from the student population at the Rochester Institute of Technology participated and received an honorarium for their participation. Subjects had normal or corrected-to-normal vision. The experimental protocol was approved by the Institutional Review Board at the Rochester Institute of Technology. Participants provided their informed consent.

Eye-movement records were collected using a custom-built, wearable eyetracker that allowed subjects' eye movements to be monitored without limiting head or trunk movement. The eyetracker has a scene camera placed directly above the participant's eye. The scene camera has a field-of-view of approximately $40° \times 30°$. Figure 1 shows the system consisting of lightweight headgear and backpack (see Babcock & Pelz, 2004 for a detailed description of the system). The image from the scene camera was used for analysis of subjects' gaze records, and to extract head movements (see below). Rather than perform eyetracking in real-time, the system uses a video multiplexer to record eye and scene images onto a single videotape. The video is later de-multiplexed and processed in a laboratory eyetracking system. Figure 2a shows the raw anamorphic eye and scene images; Figures 2b and 2c show the resultant scene image with overlay cursor indicating gaze

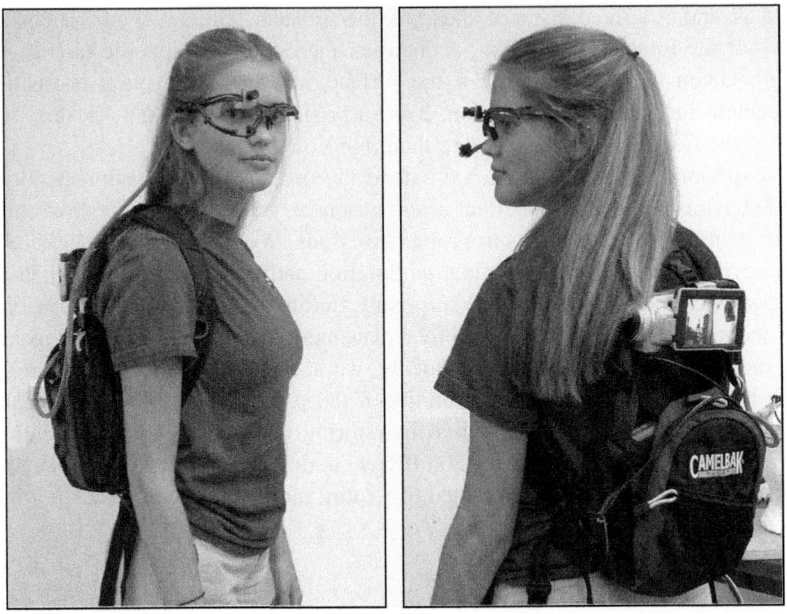

Figure 1. Wearable eyetracker records eye and scene camera images for offline analysis.

(a) (b) (c)

Figure 2. a) Multiplexed scene and eye images b) Wooded Environment c) Man-made Environment.

position for the *Wooded* and *Man-made* environments respectively. A semitransparent eye image is superimposed on the scene image to aid in analysis. The eye image allows blinks and track loss to be identified and be excluded from analysis.

Because gaze position was calculated off-line after completion of the experiments, the field calibration required only that the participant follow a target moved about the scene in front of the subject. The target was moved to cover a range of approximately $\pm 20°$ horizontal and $\pm 10°$ vertical.

The eye and scene images were demultiplexed and fed into a laboratory computer equipped with ISCAN Model PCI-726/PCI-636 processing cards. Calibration was completed by correlating the recorded eye and scene images collected at five positions in the field. The wearable eyetracker provides a video record of gaze superimposed over the scene image and a data stream consisting of 60 Hz horizontal and vertical eye-in-head position. The statistics of saccade amplitudes and duration and inter-saccade periods were obtained with software written to extract saccades from the gaze angle data. Because the eye movements were generated as part of ongoing natural behavior, it was rare for the eye-in-head signal to be stable due to nearly constant linear and/or rotational Vestibular-Ocular-Reflex (VOR) movements. Parsing the eye-movement patterns into "'fixations" and "saccades" was accomplished by parallel analysis using three algorithms with different classification criteria: velocity threshold, hidden Markov model of eye velocity distributions, and an adaptive velocity algorithm (Salvucci & Goldberg, 2000; Sicuranza & Mitra, 2000). Note that in this context "'fixation" represents periods during which a portion of the visual scene is stabilized on the retina, and not necessarily the case where the eye is stable in the orbit. Fixation duration histograms were created by counting fixations within 100 ms wide bins based on the 60-Hz eyetracker data.

Head motion was estimated from the video recording of the scene camera. A sparse-motion field was obtained by tracking specified points in successive video frames. The points were selected according to Tomasi and Kanade's algorithm (1991) and tracked across frames using a pyramidal implementation of the basic feature tracker described by Lucas and Kanade (1981). The egomotion of the scene camera was calculated using methods described in Tian, Tomasi, & Heeger (1996). The estimated rotational head motion was aligned with the eyetracking recording via the timestamp from the video track (see Rothkopf & Pelz, 2004 for a description).

Because the subjects were free to make unconstrained head and trunk movements, a "'fixation" in this context refers to a period during which a given point of regard is stabilized on the retina. In fact it was relatively rare for the eye to actually be still in the orbit. Only when making small eye movements during the *Free-view* task were traditionally defined fixations observed. The majority of the time the head and body were undergoing linear and/or rotational motion resulting in a "base" of linear and rotational vestibular eye movements upon which saccades were superimposed. For the most part, there were no objects in motion within the scene that would illicit smooth-pursuit or optokinetic eye movements (although when a person did come into view s/he was invariably fixated).

Figure 3 shows horizontal and vertical eye-in-head movements for a 5-second *Free-view* segment in the *Man-made* environment. The eye moves through 60°. Except for the 300 ms fixation at ~1700 ms, the eye is in constant motion. The saccades move gaze left and right as head movements and VOR return eye-in-head toward the midline. Figure 4 illustrates the horizontal and vertical head movements over the same period. Figure 5 shows the integrated horizontal and vertical gaze position over the 5-second period as the integrated eye and head movements form a series of gaze fixations and saccadic movements, despite the near-constant motion of eye and head. Comparing Figures 3 and 5 illustrates subjects' ability to make very large gaze changes (a range of 120° is evident in Figure 5), while limiting eye-in-head movements to 60° as seen in Figure 3.

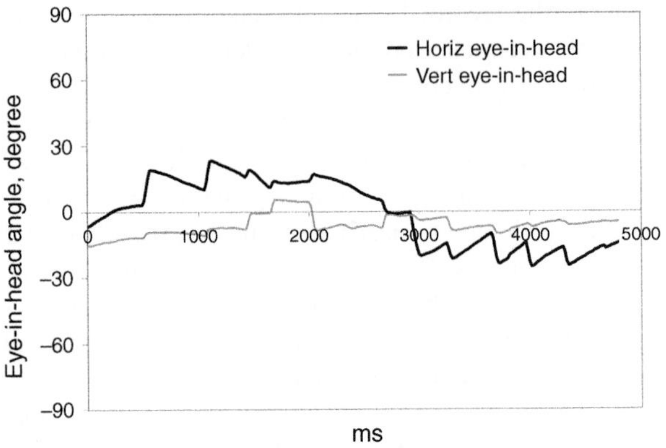

Figure 3. Horizontal and vertical eye-in-head signals for 5-second *Free-view* task.

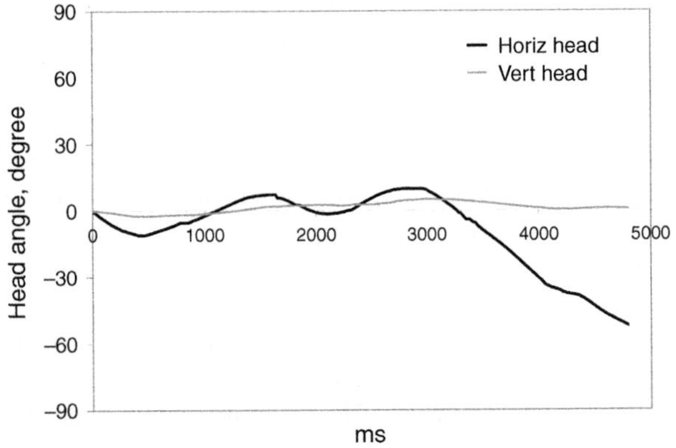

Figure 4. Horizontal and vertical head signals.

In addition to extracting the low-level measures of fixation duration and saccade size, gaze behavior while *Walking* was classified in terms of fixation location. Each gaze period was classified as "Path" if gaze fell directly on the path within approximately 3 m in front of the subject, and as "Away" if the gaze was directed to the side or above the "Path" area. This classification was completed manually with a JVC BR-DV600U VTR controlled by a lab computer with custom software developed for the analysis. Rather than code individual fixations for gaze direction while *Walking* on paved or dirt trails, the video record was coded to identify periods of gaze, defined as contiguous fixations within a specified region. The performance was rated by describing the fraction of the total trial during which gaze was directed to the path so that it could be compared to the results of Turano et al., (2001) and Patla & Vickers (2003).

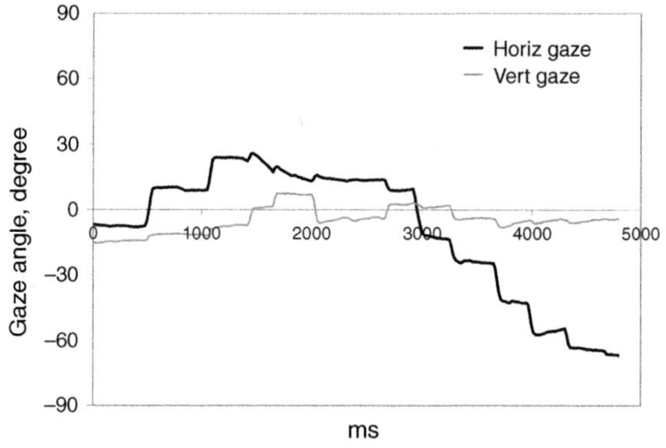

Figure 5. Horizontal and vertical gaze signals for 5-second *Free-view* task.

2. Results

2.1. Fixation duration

Figure 6 shows the distribution of fixation durations for the three subjects during the *Free-view* task in the *Man-made* environment. While the mean fixation duration varied from 250 to 440 ms, the variation was due largely to differences in the frequency of fixations over 500 ms. Median fixation duration varied from 190 to 225 ms, and modal fixation durations for all three subjects were only 150 ms, as seen in Table 1.

The average fixation duration distribution across the three subjects in the *Free-view* task in the *Man-made* environment is seen in Figure 7, along with the distribution for

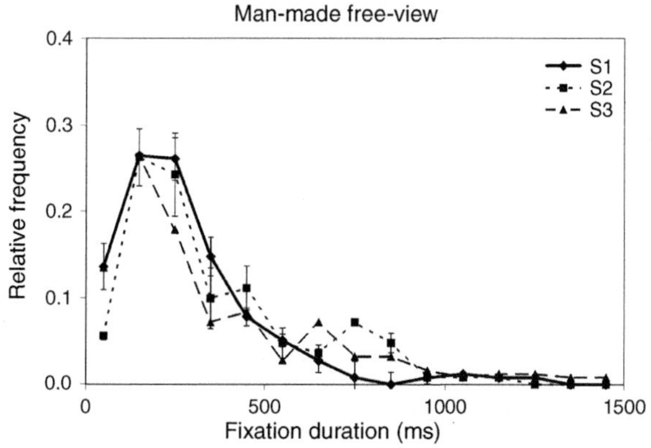

Figure 6. Distributions of fixation durations for three subjects in *Free-view* task in the *Man-made* environment (Error bars represent ±1 SEM).

Table 1
Fixation duration for the *Free-view* task in the *Man-made* environment (ms)

Subject	Mean	Median	Mode
S1	250	190	150
S2	365	225	150
S3	440	210	150

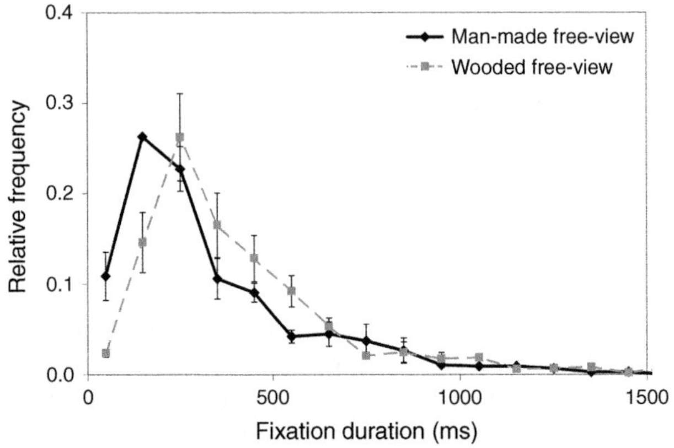

Figure 7. Mean fixation duration for *Free-view* task in *Man-made* (solid line) and *Wooded* (dashed line) environments (Error bars represent ±1 SEM).

the same task in the *Wooded* environment. Table 2 shows the mean, median, and modal fixation durations for the *Free-view* and *Walking* task in both *Man-made* and *Wooded* environments. Note the shift toward longer fixation durations in the *Wooded* environment in the *Free-view* task.

A two-way repeated measures ANOVA (within subjects; Environment × Task) shows a significant Environment (*Man-made* vs *Wooded*) effect in the skewness of the two

Table 2
Mean, median, and modal fixation duration for the *Free-view* and *Walking* task in the *Man-made* and *Wooded* environments (ms) across three subjects

Task	Environment	Mean	Median	Mode
Free-view	Man-made	375	210	150
Free-view	Wooded	440	290	250
Walking	Man-made	305	200	250
Walking	Wooded	305	210	250

distributions ($F(1, 2) = 25.44$; $p < 0.05$), but no main effect of Task ($F(1, 2) = 0.11$, $p = 0.77$), or interaction ($F(1, 2) = 0.15$, $p = 0.74$). The primary difference between the distributions is in the shift away from very short fixations in the *Free-view* task in the *Wooded* Environment. Paired two-sample *t*-tests show significantly fewer fixations in the bins centered at 50 and 150 ms in the *wooded* than in the *Man-made* environments during the *Free-view* task ($p < 0.01$ and $p < 0.05$ respectively).

The log-fixation duration distribution approximates a Gaussian distribution, and highlights the differences in the bounded shorter fixations. As seen in Figure 8, the difference in this log-duration space appears as a narrowing of the distribution in the *Wooded* environment. A two-way repeated measures ANOVA (within subjects; Environment × Task) in this space revealed significant effects in the standard deviation of the log-fixation distributions. There was a significant [Environment × Task] interaction ($F(1, 2) = 26.75$; $p < 0.05$). The nature of the interaction is evident in Figures 9a and 9b; the significant difference in the distributions in the *Man-made* and *Wooded* environments is due to variation in the *Free-view* task, but not while *Walking*. The effect of Environment approached significance ($F(1, 2) = 17.72$, $p = 0.052$), and no main effect of Task was evident ($F(1, 2) = 1.56$, $p = 0.34$).

2.2. Saccade size

The average eye-in-head saccade size was 5.1°, varying from as low as 3.7° for free view in the *Wooded* environment, to as high as 5.8° for *Free-view* in the *Man-made* environment, though no significant differences were found in the distribution of saccade

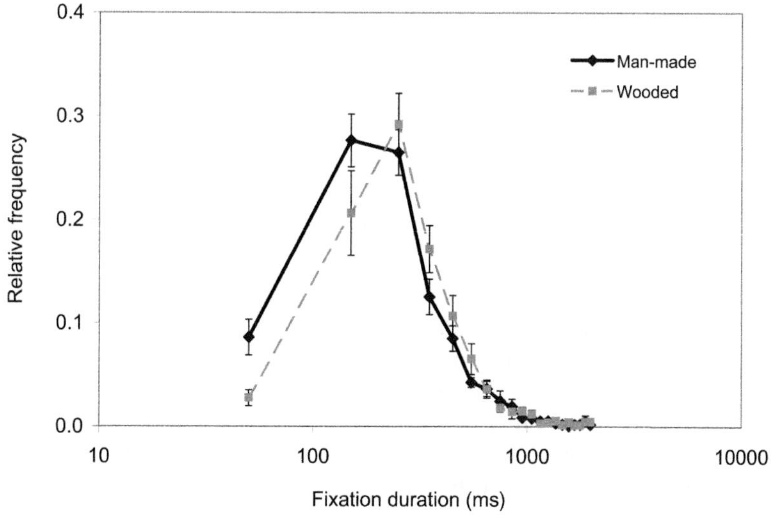

Figure 8. Mean (log-scale) fixation duration in *Man-made* (solid line) and *Wooded* (dashed line) environments (Error bars represent ±1 SEM).

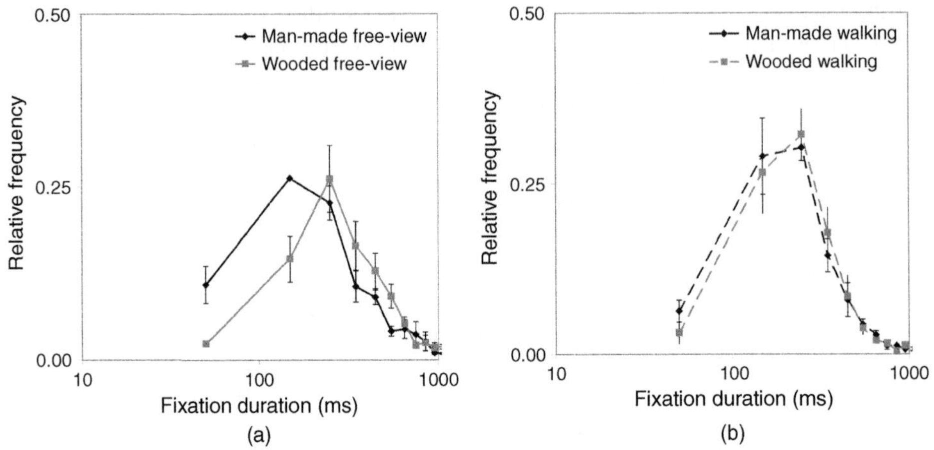

Figure 9. a) Mean (log-scale) fixation duration for *Free-view* task in *Man-made* and *Wooded* environments b) Mean (log-scale) fixation duration for *Walking* task in *Man-made* and *Wooded* environments (Error bars represent ±1 SEM).

sizes across Environment ($F(1, 2) = 0.52$, $p = 0.49$), Task ($F(1, 2) = 0.03$, $p = 0.88$), or any interaction ($F(1, 2) = 0.63$, $p = 0.45$). As is often the case with unconstrained eye movements, the mean saccade size was inflated by a relatively small number of very large gaze changes (e.g., Land et al., 1999; Pelz & Canosa, 2001); the median saccade size across all subjects and conditions never exceeded 3°.

2.3. 'Path' gaze distribution during walking

Figures 9b and 10, and Table 2 all indicate very little difference by (and no significant effect of) Environment on performance during the *Walking* task. This may be surprising, given the variation in predictability in footing between the two environments. While there was little difference in the low-level metrics of fixation duration and saccade size, Environment caused a dramatic difference in high-level gaze behavior, as measured by the fraction of gaze directed to the path immediately in front of the subjects. Recall that gaze during the *Walking* task was manually encoded as "Path" when subjects directed their gaze at the region where they would be taking the next few paces. Figure 11 shows the fraction of gaze coded as "Path" in the *Man-made* and *Wooded* environments. When navigating the uneven dirt path, subjects devoted 55–75% of gaze to the near "Path" region, similar to the Patla and Vickers (2003) result. One subject maintained a high fraction on the 'Path' region while walking on the paved walkway, but the others dropped to 15–25%, similar to that reported by Turano et al. (2001). Paired two-sample t-tests showed a significant increase in 'Path' fixations from 35% on the paved walkway in the *Man-made* environment to 62% on the dirt path in the *Wooded* environment ($t = 2.94$, $p < 0.05$).

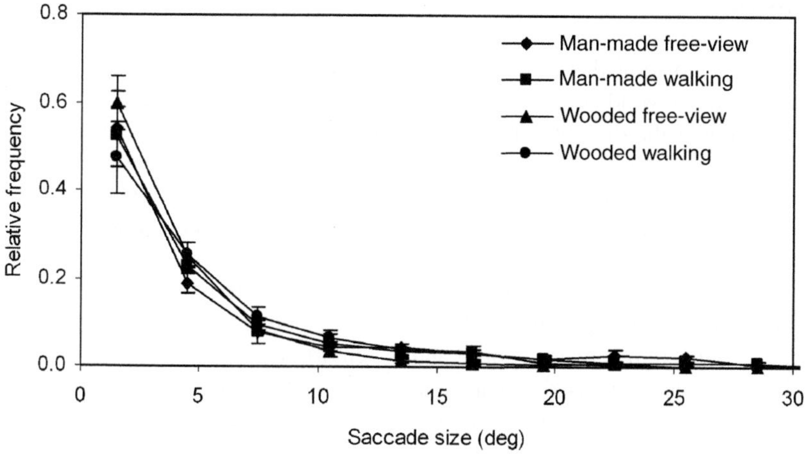

Figure 10. Distribution of saccade size for *Free-view* and *Walking* tasks in in *Man-made* and *Wooded* environments (Error bars represent ±1 SEM).

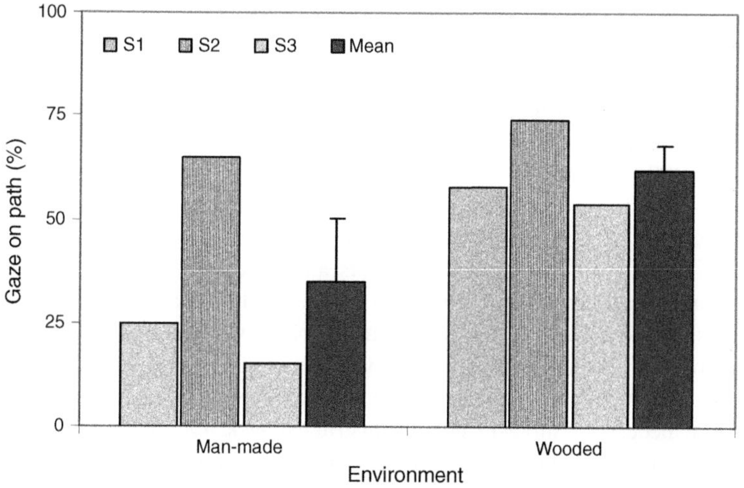

Figure 11. Fraction of trial duration during which gaze is directed toward the 'Path' region, within approximately 3 meters of the subject in *Man-made* and *Wooded* environments. Shaded bars show fraction of gaze on 'Path' region for subjects S1, S2, and S3. Solid bars show the average across subjects in each environment. (Error bars represent 1 SEM).

3. Discussion

Distributions of fixation duration and saccade size are low-level behavioral measures that can provide a window into oculomotor performance at a specific level. In the present experiment the distribution of fixation durations proved to be sensitive to some Environment and Task effects. Fixation durations were different in the two environments. Mean, median, and modal fixation durations were shorter in the *Man-made* environment than in the *Wooded* environment. The differences were due largely to a drop in the frequency of very short fixations (less than 200 ms) in the *Free-view* task in the *Wooded* environment. It is interesting to note that there was little difference between *Man-made* and *Wooded* environments in the *Walking* task, despite the fact that there is a more direct motor connection between subject and the environment while walking. In the *Free-view* task, where vision is used to gather information, but not to support motor actions, the low-level measure of fixation duration was more strongly affected by the type of environment.

Unlike the fixation-duration and saccade-size distributions, analysis of gaze position during the *Walking* task provided a more direct measure of the value of visual information gathered during foveations. Despite significant between-subject variation, there was a dramatic increase in the fraction of gaze directed to the path directly in front of the subject while walking in the *Wooded* environment. The difference between walking in the two environments can be expressed in terms of the 'predictability' of the travel path. When walking on a paved walkway there is an expectation that changes will be slow in coming, and significant changes will be signaled by differences that can be captured with peripheral vision. Walking on a dirt path, however, is not as predictable, and differences might require foveal acuity.

Unlike the results reported by Patla and Vickers (2003), two of the three subjects in the present study spent less than 25% of their time focusing on the 3-meter area directly before them in the man-made environment. These results are closer to those reported by Turano et al. (2001), in which normal subjects attempted to navigate a hallway with a limited field of view. Even in that case gaze was directed away from the immediate path approximately 75% of the time. Patla and Vickers' (2003) result is surprising given the predictability of the regularly spaced footprints and especially in the condition with no specific instructions regarding placement of the feet while walking. It is possible that those subjects focused more on the specific travel path even when there were no explicit instructions to do so because subjects were well aware that the focus of the experiment was gaze placement on the path. In this experiment, as well as Turano et al. (2001), walking was a necessary subtask, not the instructed task. One element that should be considered in 'real-world' tasks is the ample supply of alternate fixation targets. Unlike a laboratory task that may present only a blank wall as an alternative to fixating the direct *Path* region, both *Man-made* and *Wooded* environments presented a huge number of potential fixation targets to compete with the *Path* region.

As we move from lab-based experiments with 2D stimuli toward exploring natural oculomotor behavior in mobile observers, the interactions between environment, task, and top-down motivation will become ever more evident. As we try to understand truly natural oculomotor behavior, it is important that the eyetrackers we use to monitor that behavior interfere as little as possible with natural behavior. Minimizing both the physical constraints and the obtrusiveness of the systems is critical.

References

Andrews, T., & Coppola, D. (1999). Idiosyncratic characteristics of saccadic eye movements when viewing different visual environments, *Vision Research, 39*, 2947–1953.

Babcock, J. S., & Pelz, J. (2004). Building a lightweight eyetracking headgear, *ETRA 2004: Eye Tracking Research and Applications Symposium,* San Antonio, Texas.

Delabarre, E. B. (1898). A method of recording eye-movements," *American Journal of Psychology, 9*, 572–574.

Findlay, J. M. & Walker, R. (1999). A model of saccade generation based on parallel processing and competitive inhibiton. *Behavioral and Brain Science, 22*(4), 661–721.

Henderson, J. M., & Hollingworth, A. (1998). Eye movements during scene viewing: an overview. In G. Underwood (Ed.), *Eye guidance in reading and scene perception* (pp. 269–293). Amsterdam: Elsevier.

Itti. L., & Koch, C. (2000). A saliency-based search mechanism for overt and covert shifts of visual attention. *Vision Research, 40*, 1489–1506.

Land, M. F. (**this volume**). Fixations strategies during active behaviour; a brief history.

Land, M. F., Mennie, N., & Rusted, J. (1999). The roles of vision and eye movements in the control of activities of daily living. *Perception, 28*, 1311–1328.

Lucas, B. D., & Kanade, T. (1981). An iterative image registration technique with an application to stereo vision. *Proceedings of the 7th International Conference on Artifical Intelligence, Aug 24–28*, Vancouver, British Columbia, 674–679.

Parkhurst, D., & Niebur, E. (2003). Scene content selected by active vision. *Spatial Vision, 16*(2), 125–154.

Patla, A., & Vickers, J. (2003). How far ahead do we look when required to step on specific locations in the travel path during locomotion? *Experimental Brain Research, 148*, 133–138.

Pelz, J. B., & Canosa, R. (2001). Oculomotor Behavior and Perceptual Strategies in Complex Tasks, *Vision Research, 41*, 3587–3596.

Pelz, J. B., Canosa, R., Babcock, J., Kucharczyk, D., Silver, A., & Konno, D. (2000). Portable eyetracking: A study of natural eye movements, Proceedings of the SPIE, *Human Vision and Electronic Imaging,* San Jose, CA.

Rayner, K. (1984). Visual search in reading, picture perception and visual search. A tutorial review. In H. Bouma and D. Bouwhuis (Eds.), *Attention and Performance X*. 67–96. Hillsdale, NJ: Lawerance Erbaum.

Rayner, K. (1998). Eye movements in reading and information processing: 20 years of research. *Psychological Bulletin, 124*(3), 372–422.

Rothkopf, C., & Pelz, J. B. (2004). Head movement estimation for wearable eye Tracker, *ETRA 2004: Eye Tracking Research and Applications Symposium,* San Antonio, Texas.

Salvucci, D. D., & Goldberg, J. H. (2000). Identifying fixations and saccades in eye-tracking protocols. *Proceedings of the Eye Tracking Research and Applications Symposium,* ACM Press, New York.

Sicuranza, G., & Mitra, S. (2000). *Nonlinear image processing* (*Communications, Networking and Multimedia*), Academic Press, Orlando, FL.

Tian, T. Y., Tomasi, C., & Heeger, D. J. (1996). Comparison of approaches to egomotion computation. *Proceedings of the IEEE Conference on Computer Vision and Pattern Recognition,* 315–320. San Francisco, CA.

Tomasi, C., & Kanade, T. (1991). Shape from motion from image streams: A factorization method - 3. Detection and Tracking of Point Features, Technical Report CMU-CS-91-132, Carnegie Mellon University.

Turano, K. A., Geruschat, D. R., & Baker, F. H., (2003). Oculomotor strategies for the direction of gaze tested with a real-world activity, *Vision Research, 43*, 333–346.

Turano, K. A., Geruschat, D. R., Baker, F. H., Stahl, J. W., & Shapiro, M. D. (2001). Direction of gaze while walking a simple route: Persons with normal vision and persons with retinitis pigmentosa. *Optometry and Vision Science, 78*, 667–675.

Wade, N. J. (**this volume**). Scanning the seen: Vision and the origins of eye movement research.

Westheimer, G (**this volume**). Eye movement research in the 1950s.

Yarbus, A. L. (1961). Eye movements during examination of complex objects (Russian) *Biofizika, 6.*

Yarbus, A. L. (1967). *Eye movements and vision* (B. Haigh, Trans.). New York: Plenum Press. (from original work published 1955–1955–1961).

Chapter 32

GAZE FIXATION PATTERNS DURING GOAL-DIRECTED LOCOMOTION WHILE NAVIGATING AROUND OBSTACLES AND A NEW ROUTE-SELECTION MODEL

AFTAB E. PATLA, S. SEBASTIAN TOMESCU, MICHAEL GREIG, and ALISON NOVAK
University of Waterloo, Canada

Eye Movements: A Window on Mind and Brain
Edited by R. P. G. van Gompel, M. H. Fischer, W. S. Murray and R. L. Hill

Abstract

Spatial-temporal gaze fixation patterns along with locomotion data as participants selected a safe route around obstacles to reach an exit point were analyzed to determine what information is critical and when it is needed for route selection. The results suggest that routes are not planned a priori but are based on visual information acquired during locomotion. During locomotion, gaze was intermittently fixated at various locations which provided useful information for steering control and collision avoidance. A new route-selection model was developed that more accurately predicted participants' travel paths than two previous models, the on-line control and avoid-a-crowd models. The new model was guided by gaze fixation data and identifies safe corridors while minimizing path deviations from the end-goal to select a route.

Obtaining information about environmental features that are located at a distance is essential to move safely to the intended goal. While other sensory modalities such as the auditory system and the olfactory system can provide information about environmental features that are located at a distance, no modality comes close to providing as accurate and precise information about environmental features, static and dynamic, as vision. It is no wonder that most animal species rely on the visual system to guide them safely through their environment. However, the challenge has always been to identify what information is used and how it is used to regulate locomotor patterns.

One approach is to study gaze behavior: where people look and how gaze patterns change when approaching a target can provide unique insights into the nature of visual information used for planning and controlling limb movements in a cluttered terrain (Patla 2004; Sherk & Fowler, 2000). Our eyes provide us with a detailed visual image through intermittent foveation of specific locations/areas despite non-homogenous retinal acuity (Land, 1999). Saccades and fixations are the dominant eye-movement patterns observed during the performance of a variety of perceptual tasks such as viewing a picture (Yarbus, 1967), simple motor tasks like pointing (Neggers & Bekkering, 2001), approaching and stepping over an obstacle (Patla & Vickers, 1997) and performing complex procedural motor tasks such as tea making (Land, Mennie, & Rusted, 1999). Where the eyes fixate is not random: the location/object being fixated provides essential information for both perception and the control of action (Land, 1999). Therefore gaze patterns can provide very useful information on what is relevant to the task. Furthermore, the temporal relationship between the onset of gaze fixations and body movement changes can distinguish between online visual control (Hollands, Marple-Horvat, Henkes, & Rowan, 1995; Land & Lee, 1994; Neggers & Bekkering, 2001) and feed-forward control (Patla & Vickers, 2003).

Until recently, gaze behavior during locomotion was studied for tasks where the environment and/or instruction specified the action such as stepping over an obstacle (Patla & Vickers, 1997), changing direction of locomotion (Hollands, Patla, & Vickers, 2002), or stepping on specific targets (Hollands et al., 1995; Patla & Vickers, 2003). In these studies, no action choices had to be made. Therefore gaze fixations were guided by the task requirement, be it avoiding a particular obstacle or stepping on a particular target.

When the task only specifies the end goal in a cluttered environment, it is unclear how routes are selected. Specifically, is route selection based on on-line control using visual information about obstacles as they are encountered on the path to the end goal? Or does some preplanning occur in order to determine a route in a cluttered environment in advance? There is no research in the literature that has empirically examined gaze behavior during path selection in a cluttered environment. Hence, our primary objective in the current study is to document spatio-temporal patterns of gaze behavior while individuals walked through a cluttered environment to a goal that was visible from the start. This is the first study to document gaze behavior when the environment and/or instruction do not specify the travel path to be taken. While the description of the travel path that individuals took is useful, by itself it does not shed light on the type of information and/or the strategies used by individuals for path selection. We monitored participants' gaze patterns during this task to find out what they fixated on and therefore what was important for route selection and locomotion (Land & Hayhoe, 2001), and how they used vision to guide their travel path selection. Since major changes in travel path involve steering around obstacle(s), one would expect gaze fixations on objects that force participants to turn to be linked tightly with steering onset (Land & Lee, 1994).

The second objective was to evaluate two models, the on-line control and avoid-a-crowd model (Patla, Tomescu, & Ishac, 2004), which make different predictions regarding the travel path people take and to determine which is most consistent with the observed gaze patterns. Finally, we propose a new model, in part guided by the observed spatio-temporal gaze patterns, that more accurately predicts route selection (Fajen, Beem, & Warrren, 2002; Fajen & Warren, 2003; Patla et al., 2004).

1. Methods and materials

1.1. Participants

Five healthy University of Waterloo students volunteered for the study (2 females, 3 males; Age range: 19–31 years, average 23.6 years). The protocol was approved by the Office of Research Ethics at the University of Waterloo.

1.2. Experimental protocol

Twelve standard traffic pylons were used as obstacles that had to be avoided during walking. Each pylon had a square base ($l = 0.35$ m) on which a cone was mounted ($h = 0.72$ m, $d = 0.28$ m). A 9×13 grid with square cells ($l = 0.35$ m) was marked on the laboratory floor. The entire grid measured 4.55 m $\times$ 3.15 m. The 12 pylon locations on the grid cells were randomly generated with the conditions that (1) no two pylons were placed in adjacent cells (a pylon was never in one of the eight cells around another pylon) and (2) pylons were not placed directly in front of an entrance or exit point. There were three entrance and two exit points to the grid. A schematic diagram of the experimental setup is shown in Figure 1a.

Gaze location was monitored by an Applied Sciences Laboratories (ASL, Bedford, USA) 501 eye-tracker, which is a head-mounted device monitoring the eye and the scene (reflecting what the individual sees) to determine eye gaze relative to the head-mounted optics (Hollands et al., 2002; Land & Hayhoe, 2001; Land et al., 1999; Patla & Vickers, 1997, 2003). The eye-tracking system weighs 8–9 oz, has an accuracy of 0.5°, and a resolution of 0.25°. Video information of the eye and the scene, together with video information collected by a room camera giving a frontal plane perspective of the environment, were combined via two digital mixers (Videonics, model MX-1) and recorded to DVD at 30 Hz. A typical video frame is shown in Figure 1b, with the recording from the eye camera in the top left corner, the room camera in the left panel and the scene camera in the right panel. Eye gaze is shown by the black square on the scene camera (the cursor on one of the pylons in Figure 1b).

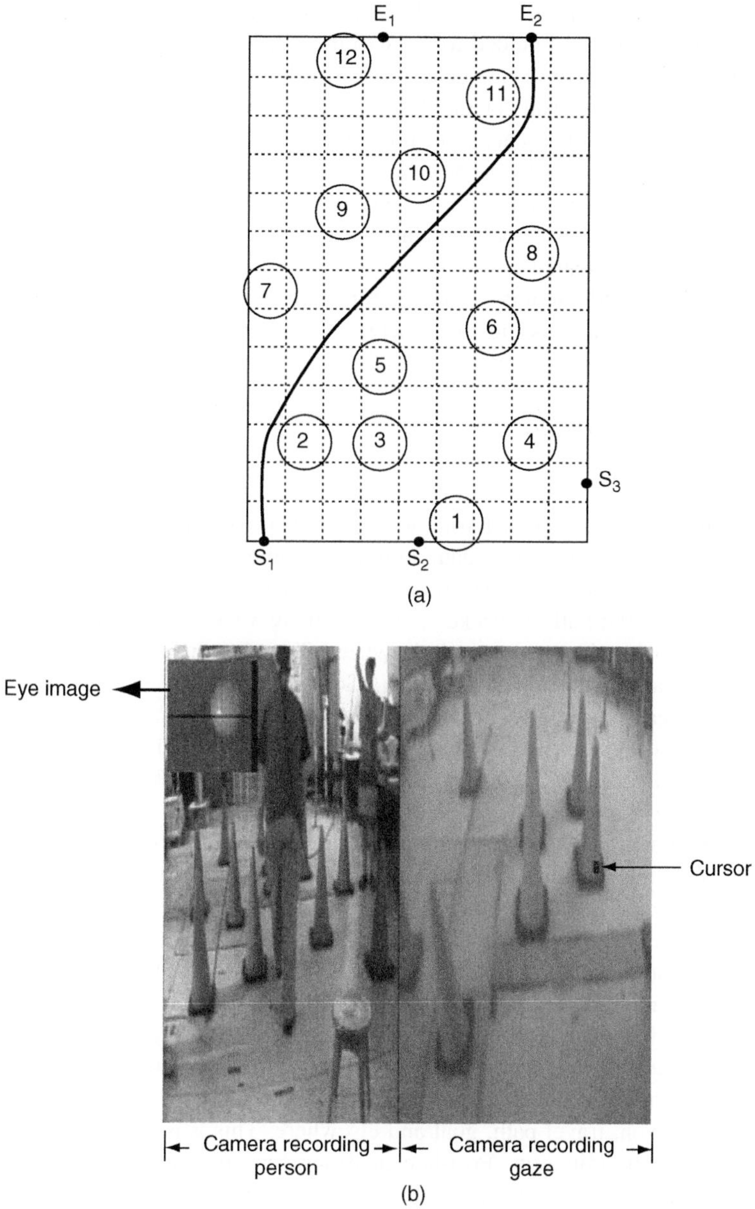

Figure 1. (a) A schematic diagram of the experimental setup with the 12 randomly arranged pylons, shown as circles, the three start positions (S1 – S3) and the two exits or goals (E1 and E2). The solid line indicates the path taken by four out of five participants for the specific pylon arrangement. (b) Left: Video frame from the gaze analysis setup with an image of the eye in the top left hand corner. Right: Image of the person walking around the pylons with the gaze location indicated by the black square on the pylon.

The participants were informed that they had to walk to one of the two end points from their starting point and guided to walk to the starting point without looking at the pylon arrangement. They were instructed to hold a board obstructing their view of the environment ahead and to wait for the "go" signal. The board was removed when they heard the "go" signal. Participants were told to proceed as fast as possible to the goal, which was defined by two vertical posts spaced 35 cm apart. Five obstacle arrangements combined with six combinations of start and end points resulted in 30 trials for each participant. For each obstacle arrangement, the start and end points were randomized. Each trial in effect presented a different obstacle arrangement from the participant's point-of-view, and posed a unique challenge for route selection. We used the same pylon arrangements and start and end positions between participants to evaluate if the selected paths were similar across participants.

1.3. Data analyses and results

In 86% of the trials four out of five participants chose the same route. There were no collisions with any pylons. For each trial, we traced the travel path taken by the participant and then analyzed the gaze fixation data. Frame-by-frame analysis of the combined video data was conducted to identify gaze fixations (Hollands et al., 2002; Land & Hayhoe, 2001; Land et al., 1999; Patla & Vickers, 1997; Patla & Vickers, 2003) on four locations: the travel path, goal, pylon region and elsewhere. Travel path fixations are fixations on the floor which fall on the travel path area. Goal is defined as either fixations on the goal posts, the area between the posts or an area slightly ahead of goal posts. The pylon region included the area defined by the gaze cursor around the pylons. Fixations on other pylons and spatial locations were categorized as elsewhere. A fixation was defined as gaze stabilized on a location for three consecutive frames ($\sim$0.1 s) or longer (Patla & Vickers, 1997). Each fixation was recorded relative to the onset of gait initiation. After the appearance of the gaze cursor in the video data (reflecting the participant's eye gaze) following the removal of the board, foot lift off for the first step identified the time for gait initiation. The gaze fixation data were grouped in two categories, those occurring between the appearance of the gaze cursor and gait initiation, and those occurring between the onset of gait initiation and reaching the goal. Typical spatial-temporal gaze fixation patterns for two trials from two different participants are shown in Figure 2.

A one-way repeated measures ANOVA was carried out on the relative gaze fixation frequency to determine if the fixation frequency was influenced by gaze location which includes pylon region, travel path, goal and elsewhere. This was done for both fixations before and after gait initiation. Post-hoc analyses for this and subsequent ANOVAs involved LSMEANS tests. Significance levels were set at 0.05 for all analyses.

1.3.1. Gaze fixation characteristics prior to gait initiation

Gait was initiated 0.49 s (SD: 0.134) after the appearance of the gaze cursor (which appeared when the board obstructing the participant's view was removed). On average

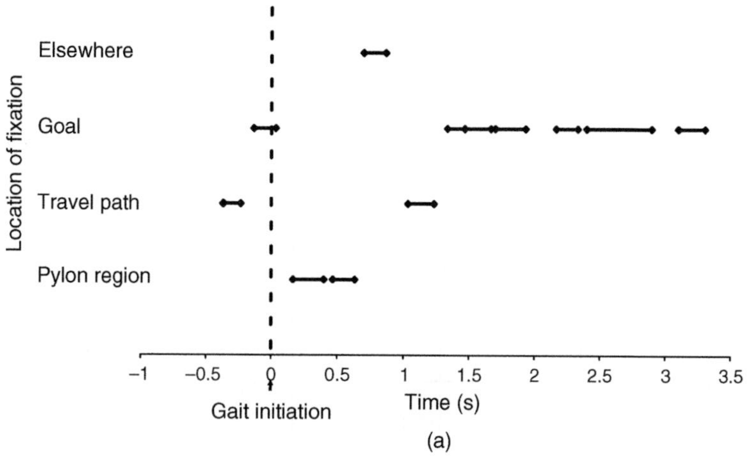

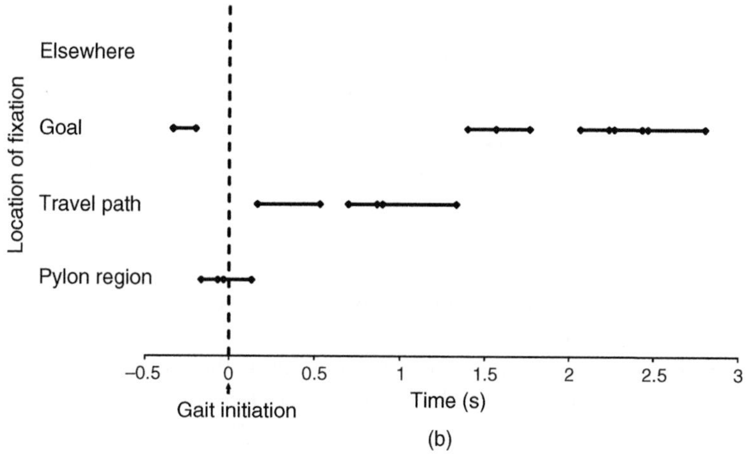

Figure 2. Two examples of spatial-temporal gaze fixation patterns. The dashed vertical line at 0 s represents gait initiation. The four gaze fixation locations were the pylon region, the travel path, the goal and elsewhere. The onset and offset of the gaze fixations is indicated by filled diamonds with the connecting lines representing the duration of the gaze fixations. Times between gaze fixations are taken up by saccades and/or blinks.

one fixation occurred prior to gait initiation; the average duration of this fixation was 305 ms (SD: 51 ms). The relative frequency of gaze fixation on the different locations during this period is summarized in Figure 3. Statistical analyses revealed a significant effect of location on frequency of gaze fixation ($F(3, 12) = 4.70$, $p = 0.02$). Post-hoc analyses showed that the majority of time, participants looked at the goal (41%); they revealed no differences in frequency of gaze fixation between the other locations.

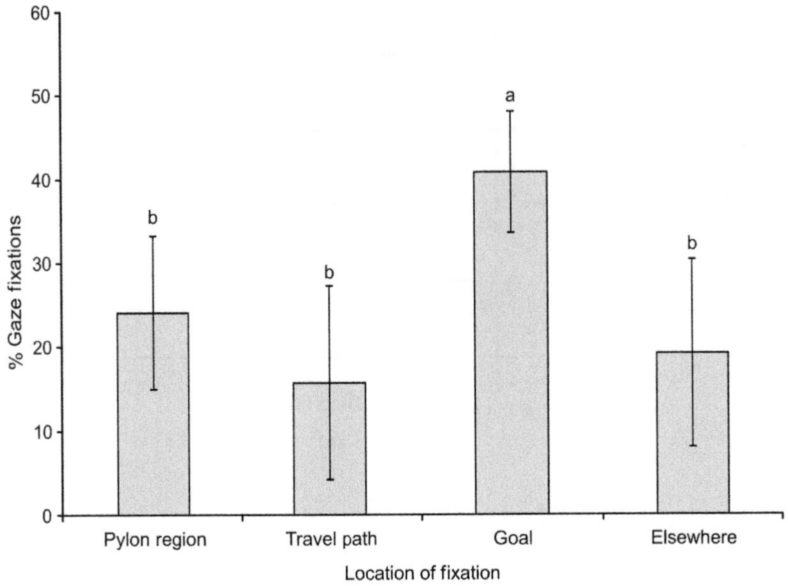

Figure 3. Percentage of gaze fixations (mean +/− 1 SD) on the four locations prior to onset of gait initiation: travel path, goal, elsewhere and pylon region. The letters on the bars show the results from the post-hoc analyses: same letter indicates no significant difference.

1.3.2. Gaze fixation characteristics during task performance

The frequency of gaze fixation on the four locations during the walking trial is summarized in Figure 4. Statistical analysis revealed a significant effect of gaze fixation location on the frequency of gaze fixation ($F(3, 12) = 54.41$, $p < 0.001$). Participants spent more time looking at the goal (64%) than at any of the other locations. The pylon region accounted for 19% of gaze fixations, which was not different from fixations on the travel path (13%). Fixations elsewhere (4%) were significantly different from fixations on the other locations. The pylons fixated on were predominantly (~94%) pylons bordering the travel path.

Next we divided the total travel time between gait initiation and the point in time when participants reached the goal posts into four travel phases of equal duration (1–4). The relative frequencies of gaze fixations on the travel path and pylon region surrounding the travel path were combined into a single category called path/pylon. The pylons in this case were the ones bordering the path. A two-way repeated measure ANOVA with gaze location (goal and path/pylon) and travel phase (phases 1–4) was done on the relative gaze fixation frequencies. The distribution of fixation durations on the goal and path/pylon was calculated in 100 ms increment intervals. Statistical characteristics of these distributions were determined.

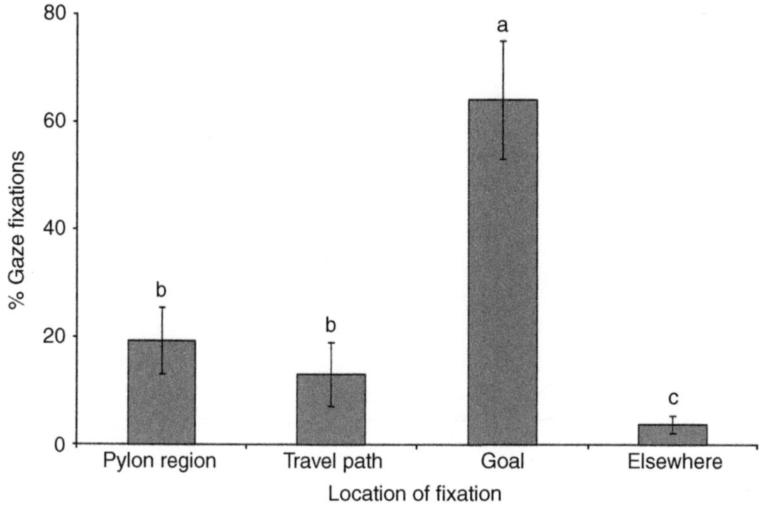

Figure 4. Percentage of gaze fixations (mean +/– 1 SD) on the four locations after the onset of gait initiation: travel path, goal, elsewhere and pylon region. The letters on the bars show the results from the post-hoc analyses: same letter indicates no significant difference.

The frequency of gaze fixation on the goal and path/pylon region varied as a function of travel phase. Statistical analyses revealed a significant interaction between gaze fixation location and travel phase ($F(3, 12) = 6.67$, $p < 0.001$). While the frequency of gaze fixation on the goal increased slightly over time, the frequency of gaze fixation on the path/pylon region decreased as participants got closer to the goal (Figure 5a). The distribution of gaze fixation durations for the two locations, goal and path/pylon, are shown in Figure 5b, which shows that they are very similar. Statistical properties of the two distributions are summarized in Table 1.

1.3.3. Characteristics of gaze fixations on turn pylons

Pylons around which a turn took place were identified as turn pylons. A turn was defined as a change in travel direction and was determined from upper body yaw rotation which precedes a change in foot placement (Patla, Adkin, & Ballard, 1999). The travel area was defined by a 9 (columns) by 13 (rows) grid. The percentage of turn pylons fixated on were determined for turns occurring in rows 1–3, 4–6, 7–9, 10–12. The rows allow us to define four spatial components of the travel path. A one-way repeated measures ANOVA was carried out on the percentage of fixated turn pylons as a function of row grouping. Depending on the direction of the turn, the turn pylon was classified as being either on the inside edge or on the outside edge of the turn. The distribution of the times between the onset of gaze fixation on the turn pylon and the turn was determined for each row grouping. Statistical characteristics of these distributions were determined.

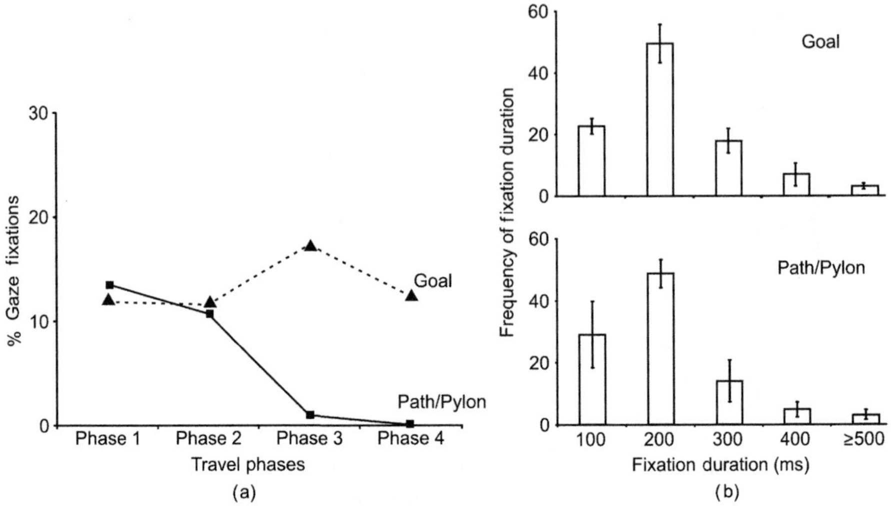

Figure 5. (a) Percentage of gaze fixations on the goal and path/pylon region as a function of travel phase (rows 1–3, 4–6, 7–9, 10–12). (b) Frequency distributions of gaze fixation durations on the goal and path/pylon region in 100 ms increments. The mean +/− 1SD values are shown for the frequency values.

Table 1
Distribution statistics of the gaze fixation durations on the goal and the path/pylon region

	Goal	Path/Pylon
Mean (ms)	218 (14.90)	204 (30.96)
Median (ms)	155 (10.88)	142 (22.75)
Mode (ms)	200	200

The percentage of fixations on pylons along the travel path edges was 17.4% (SD: 6.6): this translates into 1–2 travel path pylons that were fixated on during each walking trial. Half of the travel path pylons were turn pylons (49%). The percentage of turn pylons fixated on was significantly influenced by the four set of rows (rows 1–3; 4–6; 7–9; 10–12) ($F(3, 12) = 13.59$, $p < 0.001$); post hoc analyses revealed that for turns occurring in rows 1–3, the frequency of gaze fixation was lower than in the other rows, which were not statistically different from each other (Figure 6a). For cases where the turn pylon was fixated, the fixation distributions for the time between the onset of gaze fixation and the onset of the turn are shown in Figure 6b. The statistics of these distributions are summarized in Table 2. The turn pylon that participants fixated was predominantly on the inside edge of the turn (80%, SD: 14%).

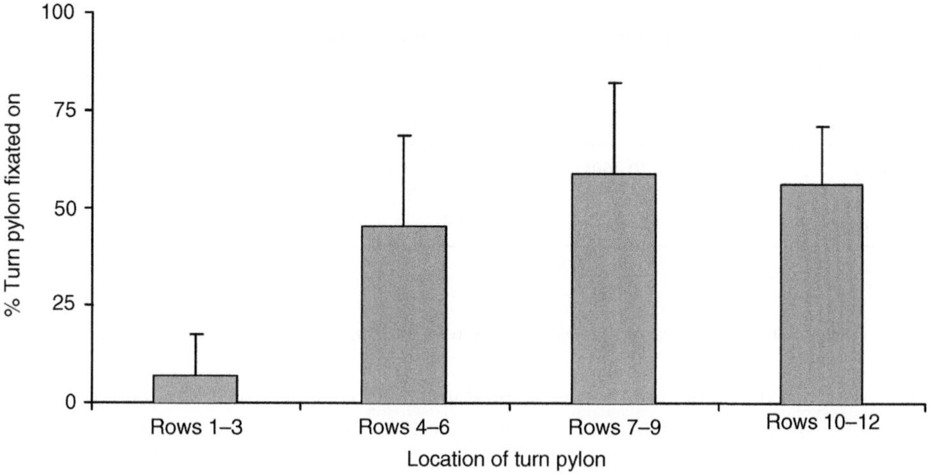

(a)

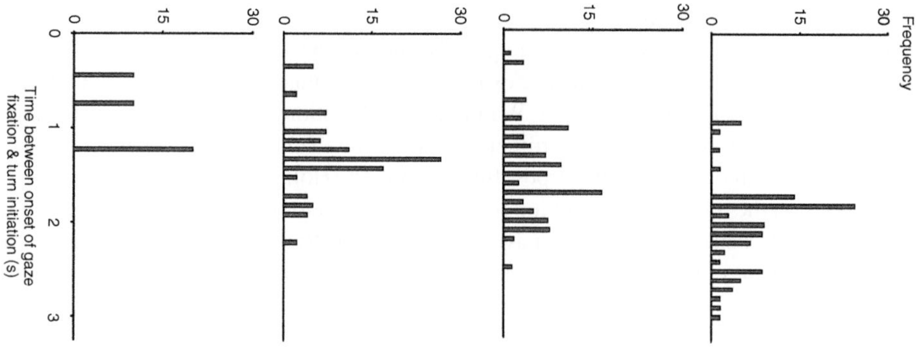

(b)

Figure 6. (a) Percentage of fixated turn pylons in the four travel phases (mean +/ − 1SD). (b) Frequency distribution of the time between the onset of the gaze fixation on the turn pylon and turn initiation (in seconds) for the four travel phases.

1.4. Discussion

Route selection around obstacles during walking is not random, but systematic. The challenge is to discover how people control their body movements and the nature of the visual information used for collision free route selection.

Table 2
Distribution statistics of the time lag between the gaze onset on the turn pylon and the turn initiation for the four travel phases (SD in brackets)

	Rows 1–3	Rows 4–6	Rows 7–9	Rows 10–12
Mean (s)	0.925 (0.460)	1.309 (0.199)	1.517 (0.159)	2.039 (0.115)
Median (s)	1.000 (0.459)	1.341 (0.221)	1.444 (0.204)	1.981 (0.179)
Modal interval (s)	1.2–1.3	1.3–1.4	1.7–1.8	1.8–1.9

1.4.1. Where and when people look reveals what is important for route selection

Gaze fixations during locomotion were predominantly on the goal, travel path and pylon region (~96%); only 4% of gaze fixations were in the elsewhere category. The three locations, goal, travel path and pylon region are all relevant for the task. Similar task specific fixation locations have been observed in a variety of other locomotor tasks (Hollands et al., 2002; Patla & Vickers, 1997, 2003) and other everyday tasks (Land & Hayhoe, 2001; Land et al., 1999); rarely do we see fixations on irrelevant objects or locations. This supports 'top-down' control of gaze behavior (Land & Furneaux, 1997; Shinoda, Hayhoe & Shrivastava, 2001).

Researchers have suggested two strategies for monitoring where one is heading: either monitoring optic flow (Warren, Kay, Zosh, Duchon, & Sahuc, 2001) or perceived target location (Rushton, Harris, Lloyd, & Wann, 1998). Fixations on the goal can be used to track one's heading during goal directed locomotion. This checking of where one is heading is a common feature of purposeful locomotion (Hollands et al., 2002).

Fixations on the travel path and pylon region can be used for path planning, to determine the presence of obstacles for collision avoidance (Patla & Vickers, 1997) and the location where steering may be required (Hollands et al., 2002). What is interesting is how the frequency of gaze fixations on the goal and the path/pylon region changes with time. As participants get closer to the goal, fixations on the travel path or pylon region decreased to almost zero. This could in part be due to the simple fact that the number of pylons that are visible decreases as participants approach the goal. Earlier on in the travel path the number of fixations on the goal and path/pylon region was almost equal. These fixations in the early phase of travel allow for path planning; later on during travel fixations on the goal guide the body to the appropriate exit location. Models of how these intermittent gaze fixations on the goal and path/pylons assist path planning and selection are discussed below.

Gaze fixation duration was on average around 200 ms for both fixations on the goal and on the path/pylon region (Table 1). In a block copying task, Ballard, Hayhoe, Li, and Whitehead (1992) found that the duration of fixation varied depending on the type of information that was being acquired: identifying appropriate color blocks required longer fixation durations than identifying the location of the block in the model. In the task in the current study, the goal and path/pylon region provide spatial information to guide the travel path. This explains why the average gaze fixation duration on the goal and the path/pylon region is similar.

1.4.2. Route planning is based on visual information acquired during locomotion, not on information acquired prior to gait initiation.

In the simpler task of single obstacle avoidance, we have shown that when individuals had approximately 1.5 s to view the environment from a standing posture and complete the task blindfolded, the success rate was about 50% (Patla, 1998; Patla & Greig, 2006). Gait initiation times in the current study clearly suggest that individuals did not spend any appreciable time scanning the environment and planning a route before they started walking: half a second with one gaze fixation on average is probably not enough to plan the whole route in a complex environment. This is consistent with our previous work (Patla et al., 2004). Therefore it must be visual information acquired during locomotion that is used for route selection. The advantage of visual information acquired during locomotion over visual sampling from a standing posture has been shown for the simple task of single obstacle avoidance (Patla, 1998; Patla & Greig, 2006). It is the retinal stimulation as one moves about the environment, termed optic flow, which provides rich information for guidance of whole body movements (Gibson, 1958).

The majority of gaze fixations prior to gait initiation were on the goal. Since there were two possible end goals, a fixation prior to gait initiation allows the individual to identify where he/she should be heading and probably helps in the planning of the direction of the first step. What is surprising is that people did not fixate the goal before gait initiation in every trial. For the two start positions S1 and S2 in Figure 1a, the goals are straight ahead, so a fixation anywhere in the travel area would allow individuals to identify the location of the goal in their upper visual field. Researchers studying mental maze solving tasks have similarly shown that people acquire information from the upper visual field through intermittent gaze fixations at different points along the route (Crowe, Averbeck, Chafee, Anderson, & Georgopoulos, 2000). In contrast, for start position S3, the end goals are in the participant's right peripheral visual field, so the goal at E2 is out of view. While gaze fixations allow people to gather detailed visual information about visual features at specific locations, simple identification of the goal can be achieved in the peripheral visual field. This may explain the low number of fixations on the end goal prior to gait initiation.

1.4.3. Visual information is used in both feed-forward and on-line control of whole body movement

To reach the goal in the current study, modifications in gait direction were primarily required for changing direction. What visual information is needed for changing walking direction? First, participants need to know where steering has to be executed. Second, they need to know the new direction of travel in order to determine the magnitude of change in direction. In this study the location of direction change was normally around a pylon and the difference between the current direction of travel and the goal determined the magnitude of change. Fixations on turn pylons provide the first piece of information while fixations on the goal provide the second piece of information. Surprisingly, only about 50% of the turn pylons were fixated; for turns occurring in the early phases of the trials (rows 1–3) this number was even lower (~10%). Fixations on the goal were much more frequent, as discussed before. This suggests that visual information about the location of the turn is acquired in the peripheral visual field, consistent with previous work on gaze behavior during steering control (Hollands et al., 2002). In Hollands et al.'s study, the location and magnitude of direction change were specified by a clearly marked tape (a turn mat and light cue). The results showed that individuals primarily fixated on where they were currently heading for and where they would be going following the direction change; fixations on the turn mat and the light cue accounted for less than 15% of gaze durations. In the current study the pylons were very salient and could also be easily identified using peripheral vision, which explains why people often did not fixate them. In contrast, Land and Lee (1994) showed consistent gaze fixations on the tangent point of the road curvature during driving. This difference in results may in part be due to the consequence of error if the turn is not initiated at the correct location.

When individuals do fixate on the turn pylons, are they using it to control their steering on-line or are they using it to plan their turn in advance? We need to look at the time lag between the onset of the gaze fixation on the turn pylon and the initiation of the turn in order to address this question. But before that we need to establish what time lag would indicate that vision is used for on-line control. Unlike other gait adaptations, such as step length and width modulation, which can be successfully completed in a single step, steering control requires a minimum of two steps for successful completion (Patla et al., 1989, 1991). Therefore the minimum time lag between the onset of a gaze fixation and the initiation of a turn is approximately 1.2 s. The distribution statistics of this time lag, shown in Table 2, suggests that how vision is used for steering control depends on where the turn occurs. For turns occurring in rows 4–6, the average lag of 1.3 s indicates on-line steering control, whereas for rows 7–12, the time lags are longer, suggesting that vision is used primarily for planning turns ahead. Land and Lee (1994) showed that during driving, the time lag between a gaze fixation and turning the steering wheel was approximately 0.75 s, also suggesting that vision is used for on-line steering control. Similar to what was found by Land and Lee (1994), participants in the current task also predominantly fixated on the inside edge of the turn.

1.4.4. Evaluating models of route selection

We compared the performance of two models in their ability to predict appropriate route selection: the on-line control (Fajen & Warren, 2003) and avoid-a-crowd model (Patla et al., 2004). We also evaluated if the observed gaze patterns in the current study are compatible with these models.

The on-line control model claims that each obstacle in the travel path is avoided by minimizing the deviation from the current walking direction (Fajen & Warren, 2003;

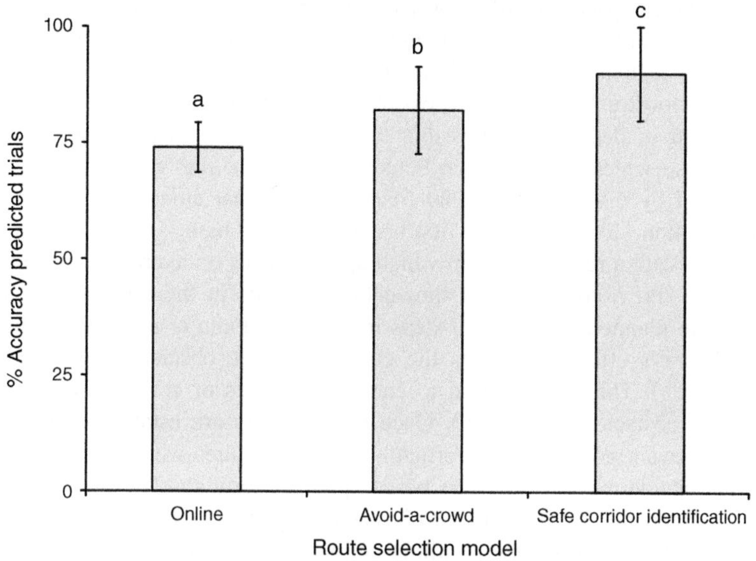

Figure 7. Percentage of trials accurately predicted by the three route selection models discussed in the text. The letters on the bars show the results from the post-hoc analyses: same letter indicates no significant difference.

Shiller, 2003). This model correctly predicts the travel path in 74% of the trials in the current study (Figure 7). While its success rate is impressive, it is the lowest among the models that we compared. Gaze fixation data also suggest that participants did not use the strategy predicted by the online control model. One would have predicted more frequent gaze fixations on the pylons because they determine the changes in locomotor direction involved in the travel path. Participants fixated only on 1–2 pylons (out of a possible 6–8) bordering the travel path.

A more successful model, proposed by Patla et al. (2004), is the 'avoid-a-crowd' strategy. This model predicts that participants circumvent clusters of obstacles and minimize the number of turns rather than weave in and out among obstacles. While this model is better than the on-line control model in predicting travel paths (82% vs 74%; Figure 7), one would have expected more consistent gaze fixations on turn pylons because turn pylons are key features for the cluster identification routine (Patla et al., 2004). In the current study, only approximately 50% of the turn pylons were fixated.

1.4.5. A new route selection model based on gaze fixation results accurately predicts the selected travel path

It appears that avoiding clusters of obstacles and minimizing the number of turns in the travel path is not enough to predict the travel paths chosen by most participants. Gaze fixations on the travel path and the pylons surrounding the path suggest on-line monitoring of the travel area. We therefore propose a new model that involves on-line evaluation of which spaces afford an obstacle-free passage, while minimizing changes in the current travel direction and minimizing the deviation from the end-goal. In addition to the obstacle free area, the model takes into account the width of the corridor in order to ensure that an unobstructed passage is possible without having to modify body orientation.

The first step in determining possible travel paths is to map the areas which are visible and those which are invisible, beginning with the start point on the grid. This is accomplished by dividing the visual field into rays that emanate from the observer and terminate upon falling upon the first obstacle in their path. The rays are spread in a circular pattern with a radius of 2.5 m which approximates on average three-step lengths during walking. Previous work has shown that changes in the direction of the travel path have to be planned at least two steps in advance (Patla et al., 1991). The circular pattern of the rays effectively maps the contours of the obstacles over the specified radius (Figure 8a). This is then used to construct wedges of corridors of unobstructed space between obstacles (Figure 8b). Once the corridors are established the challenge is to extract relevant geometrical information. Four measures were used to quantify the corridors and decide which corridor is best to proceed through. The first measure is the WIDTH of the corridor between the two obstacles (Figure 8c): it should be large enough to allow a person to walk through without collision, with a larger width preferred. The second factor is the AREA of each corridor as mapped by the rays. An example of four areas (A1–A4) is shown in Figure 8b. Once again the larger the obstruction-free area, the more preferred it is. While the area indirectly includes the width of the corridor, it is possible to have a large area with a narrow width and therefore area alone is not enough to choose the best corridor. The next two factors account for directional changes that are required from the current heading, termed the BACKWARD ANGLE (Figure 8d), defined as the angle between the current line of progression and the line connecting the current location to the midpoint of the selected corridor, and the FORWARD ANGLE (Figure 8c), which is the angle between the selected path (defined by the line connecting the current location with the mid-point of the selected corridor) and a straight line to the goal. Small BACKWARD ANGLES are preferred to minimize abrupt sharp turns, while small FORWARD ANGLES minimizes the deviation from the end-goal. All factors are scaled from 0 to 1 using linear functions between the maximum and the minimum distances and angles (see Figure 8e).

To select the optimum corridor the four factors are assigned different weights to arrive at a suitability index for each corridor; the weighting is modified if the exit is within the radius of the rays projecting from the current position (see Figure 8f for the specific weightings used). The weights were determined using the data from a previous study (Patla et al., 2004) and were then fixed. Therefore the model does not involve mere curve fitting of the selected travel paths; the model actually predicts travel paths. The model chooses the corridor with the highest index and the algorithm is repeated with the midpoint of that corridor (indicated by open circles in Figure 9a) becoming the new starting point for the subsequent corridor search. This algorithm thus takes into account local information about the spacing between the obstacles along with global information about the goal's location in order to plan a path segment.

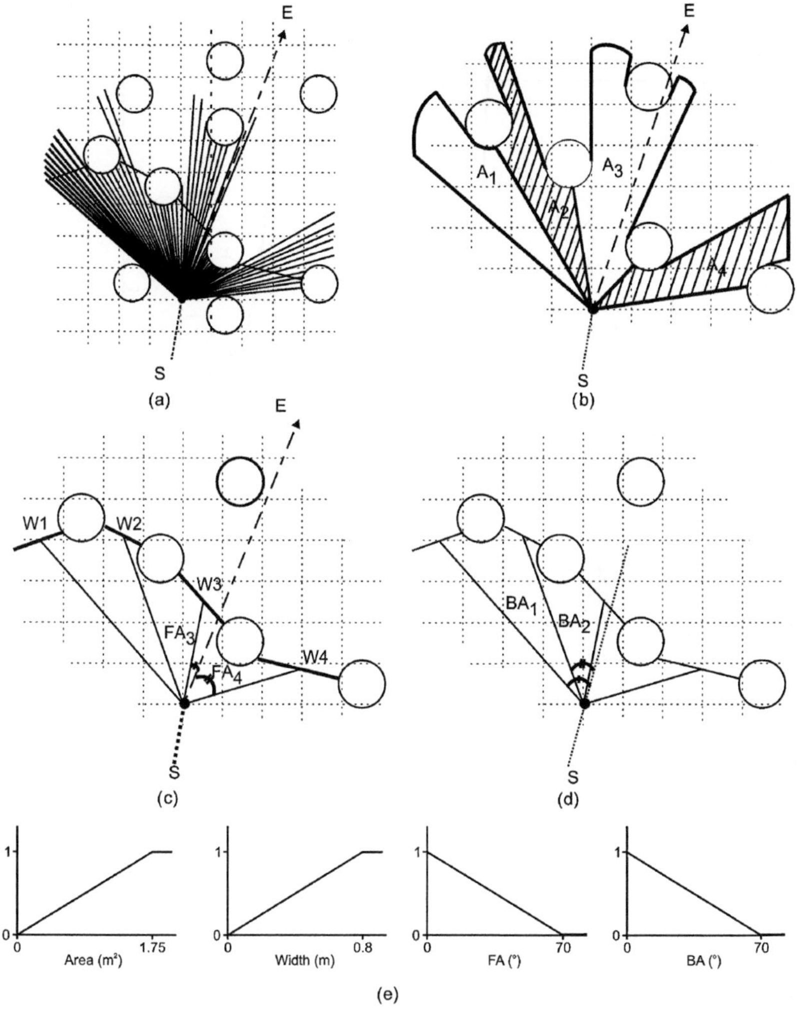

Normal weighting formula

Corridor index (i) = 1.25 * Area(i) +
0.50 * Width(i) +
0.75 * Forward angle(i) +
0.75 * Back angle(i)

Exit within sight weighting formula

Corridor Index (i) = 0.50 * Area(i) +
1.00 * Width(i) +
1.75 * Forward angle(i) +
0.50 * Back angle(i)

(f)

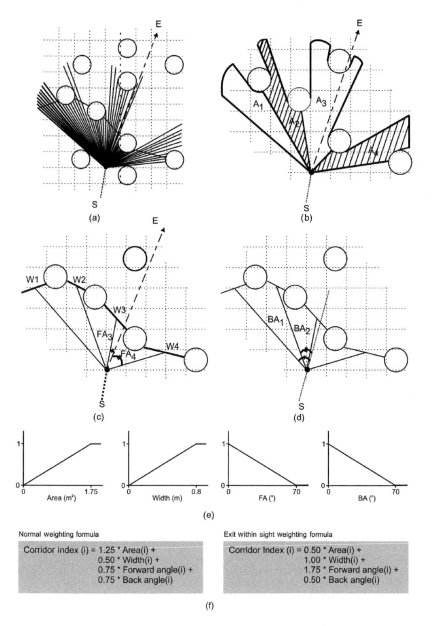

Figure 8. (a) Rays representing visual identification of obstacles and corridors from a given point on the grid. (b) Four possible obstruction-free travel areas are identified (A1 – A4). (c) Corridor width (W1 – W4) and forward angle representing the deviation from the end goal for two corridors (FA3 and FA4, indicated by arcs) influence which corridor is selected. (d) The factor backward angle representing the deviation from the current heading for two corridors (BA1 and BA2, indicated by arcs). (e) Linear weighting for the four factors area, width, forward angle and backward angle. (f) The weighting formulas for determining the suitability index for a corridor. The weightings on the left are used before the exit is in sight; the weightings on the left when the exit is in sight.

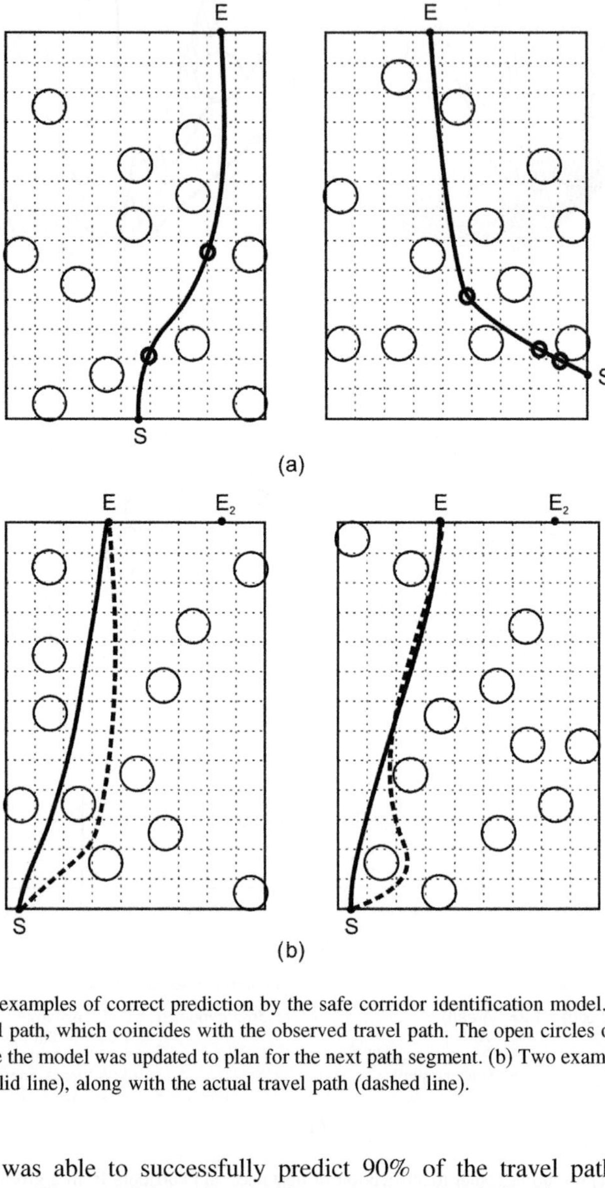

Figure 9. (a) Two examples of correct prediction by the safe corridor identification model. The line represents the predicted travel path, which coincides with the observed travel path. The open circles on the line represent the locations where the model was updated to plan for the next path segment. (b) Two examples of unsuccessful path prediction (solid line), along with the actual travel path (dashed line).

This model was able to successfully predict 90% of the travel paths, significantly higher than the on-line model or the avoid-a-crowd model ($F(2, 8) = 13.13$, $p < 0.001$; Figure 7). Two successfully predicted paths are shown in Figure 9a. The 10% of paths that were not correctly predicted were different between participants. This indicates that it is individual variation and not the generalizability of the model that accounts for less than 100% success. Two examples of unsuccessful predictions are shown in Figure 9b. In both examples, the participant's initial direction of travel was appropriate for the second exit point, but the travel direction was corrected later, leading to a different path. It is possible that in these trials the participants did not fixate on the required exit point and initiated their travel based on incorrect information.

The weights assigned to the four factors that define the safe corridor to be taken clearly highlight that minimizing the number of direction changes alone does not predict the paths taken by individuals. The size of both the obstacle free areas and the corridor widths influence the selected path. In contrast to the avoid-a-crowd model, in this model there is on-line planning of path segments. The number of path segments depends on the environmental features: the more cluttered the environment, the greater the number of path segments (see Figure 9a). The gaze fixation data are consistent with the model: intermittent fixations between the goal and the path/pylon region enable participants to plan the next path segment and/or check if the planned path segment is safe and appropriate.

Conclusion

Gaze fixation data reveal how visual information is used for safe passage around obstacles during goal-directed locomotion. Behavioral data alone do not tell us what information is important and when it is used. While models of route selection are useful to formalize the putative strategies used for travel, accuracy of prediction alone is not enough to validate them. Gaze fixation data can provide support for the models and guide the development of new models. The current study is an initial attempt to relate gaze fixation data to route-selection strategies. The eyes are indeed windows into the visual-motor transformations that are essential for adaptive human locomotion.

Acknowledgements

S. Sebastian Tomescu is, at the time of writing, at Queen's University. Supported by grants from NSERC, Canada and Office of Naval Research, USA. We acknowledge the detailed insight provided by Dr Roger Van Gompel.

References

Ballard, D. H., Hayhoe, M. M., Li, F., & Whitehead, S. D. (1992). Hand-eye coordination during sequential tasks. *Philosophical Transactions of the Royal Society, Series B, 337*, 331–339.

Crowe, D. A., Averbeck, B. B., Chafee, M. V., Anderson, J. H., & Georgopoulos, A. P. (2000). Mental maze solving. *Journal of Cognitive Neuroscience, 12*(*5*), 813–827

Fajen, B. R., & Warren, W. H. (2003). Behavioral dynamics of steering, obstacle avoidance, and route selection. *Journal of Experimental Psychology: Human Perception and Performance, 29*(2), 343–362.

Fajen, B. R., Beem, N., & Warren, W. H. (2002). Route selection emerges from the dynamics of steering and obstacle avoidance. *2nd. Annual Conference of the Vision Sciences Society*, Sarasota, Florida.

Gibson, J. J. (1958). Visually controlled locomotion and visual orientation in animals. *British Journal of Psychology, 49*, 182–189.
Patla, A. E., & Greig, M. A. (2006). Any way you look at it, successful obstacle negotiation needs visually guided on-line foot placement regulation during the approach phase. *Neuroscience Letters, 397*(1–2): 110–114.
Hollands, M., Marple-Horvat, D., Henkes, S., & Rowan, A. K. (1995). Human eye movements during visually guided stepping. *Journal of Motor Behavior, 27*, 155–163.
Hollands, M. A., Patla, A. E., & Vickers, J. N. (2002). "Look where you're going!": Gaze behaviour associated with maintaining and changing the direction of locomotion. *Experimental Brain Research, 143*, 221–230.
Land, M. F., & Lee, D. N. (1994). Where we look when we steer. *Nature, 369*, 42–744.
Land, M. F. (1999). Motion and vision: Why animals move their eyes. *Journal of Comparative Physiology, 195*, 341–352.
Land M. F., & Furneaux, S. (1997). The knowledge base of the oculomotor system. *Philosophical Transactions of Royal Society of London B Biological Sciences, 352(1358)*, 1231–1239.
Land, M. F., Mennie, N., & Rusted, J. (1999). The roles of vision and eye movements in the control of activities of daily living. *Perception, 28*, 1311–1328
Land, M. F., & Hayhoe, M. (2001). In what ways do eye movements contribute to everyday activities? *Vision Research, 41*, 3559–3565
Neggers, S. F. W., & Bekkering, H. (2001). Gaze anchoring to a pointing target is present during the entire pointing movement and is driven by a non-visual signal. *Journal of Neurophysiology, 86*, 961–970.
Patla, A. E., Robinson, C., Samways, M., & Armstrong, C. J. (1989). Visual control of step length during overground locomotion: Task-specific modulation of the locomotor synergy. *Journal of Experimental Psychology: Human Perception & Performance, 15*(3), 603–617.
Patla, A. E., Prentice, S. D., Robinson, C., & Neufeld, J. (1991). Visual control of locomotion: Strategies for changing direction and for going over obstacles. *Journal of Experimental Psychology: Human Perception & Performance, 17*(3), 603–634.
Patla, A. E., & Vickers, J. N. (1997). Where and when do we look as we approach and step over an obstacle in the travel path? *NeuroReport, 8*, 3661–3665.
Patla, A. E. (1998). How is human gait controlled by vision? Ecological Psychology, 10(3-4): 287–302.
Patla, A. E., Adkin, A., & Ballard, T. (1999). On-line steering: Coordination and control of body center of mass, head and body re-orientation. *Experimental Brain Research, 12994*, 629–634.
Patla, A. E., & Vickers, J. N. (2003). How far ahead do we look when required to step on specific locations in the travel path during locomotion. *Experimental Brain Research, 148*, 133–138.
Patla, A. E., Tomescu, S. S., & Ishac, M. (2004). What visual information is used for navigation around obstacles in a cluttered environment? *Canadian Journal of Physiology and Pharmacology, 82*(9), 682–692.
Patla, A. E. (2004). Gaze behaviours during adaptive human locomotion: Insights into the nature of visual information used to regulate locomotion. In L. Vania, S. Rushton, & S. Beardsley (Eds.), *Optic Flow and Beyond* (pp. 383–400). Germany: Kluwer Academic Publishers.
Rushton, S. K., Harris, J. M., Lloyd, M. R., & Wann, J. P. (1998). Guidance of locomotion on foot uses perceived target location rather than optic flow, *Current Biology, 8*, 1191–1194
Sherk, H., & Fowler, G. A. (2000). Optic flow and the visual guidance of locomotion in the cat. In Markus Lappe (Ed.), *Neuronal Processing of Optic Flow* (pp. 141–167). New York, N.Y.: Academic Press.
Shiller, Z. (2003). On-line dynamic obstacle avoidance. 2nd. International Symposium on *Adaptive Motion of Animals and Machines*, Kyoto, Japan, 56.
Shinoda, H., Hayhoe, M. M., & Shrivastava, A. (2001). What controls attention in natural environments? *Vision Research, 41*, 3535–3545
Warren, W. H. Jr., Kay, B. A., Zosh, W. D., Duchon, A. P., & Sahuc, S. (2001). Optic flow is used to control human walking. *Nature Neuroscience, 4*(2), 213–216.
Yarbus, A. L. (1967). *Eye movements and vision*. New York: Plenum Press.

Chapter 33

DON'T LOOK NOW: THE MAGIC OF MISDIRECTION

BENJAMIN W. TATLER
University of Dundee
GUSTAV KUHN
University of Durham

Eye Movements: A Window on Mind and Brain
Edited by R. P. G. van Gompel, M. H. Fischer, W. S. Murray and R. L. Hill

Abstract

How do magicians misdirect their audiences? We recorded eye movements as observers watched a magician perform a trick on a live one-to-one basis. All observers watched the trick twice. Half of the observers were informed in advance that they would be watching a trick; half were not. Observers tended to follow the magician's gaze, particularly in the second half of the trick. Even informed observers were susceptible to the magician's social cues for joint attention, following his gaze during the trick. While knowing that they would be watching a trick was not sufficient for observers to defeat the magician's misdirection, watching the trick a second time was; all observers were able to describe how the magician made a cigarette disappear after viewing the trick a second time. Our findings not only demonstrate an everyday example of inattentional blindness, but also that social cues for joint attention provide the magician with a powerful means of misdirecting his audience successfully.

Picture this: you are at a magic show and the magician announces that he is going to make a donkey appear behind a curtain in the middle of the stage. He walks to the curtain, which is lying on the floor, and raises it above his head. A moment later he drops the curtain to the ground to reveal a real live donkey behind it, to the amazement of you and the rest of the audience. How can the magician have performed such an impressive trick?

Now let us watch the trick again to reveal its secret: when the magician raises the curtain above his head, his glamorous (and probably scantily clad) assistant walks across the front of the stage. As they do this, the magician replaces the curtain on the floor, walks to the side of the stage, collects a donkey and drags it over behind the curtain. The magician raises the curtain again and waits for his moment to drop it to reveal the 'magically appeared' donkey. You are somewhat less impressed. Surely this is not the same trick? Surprisingly, it is exactly as it was performed the first time. How could it be that you missed such an obvious act as the magician walking over to drag a donkey on stage?

Ridiculous though it may seem, this illustrates the way in which many magic tricks are performed: the magician diverts the observers by directing their attention to a distracting act, while at the same time performing what would be an otherwise obvious act (see Lamont & Wiseman, 1999). Magicians have for many years accomplished their misdirection by combining processes that psychologists have learnt about only recently: our tendency to look where others look – social attention – and our inability to spot rather obvious events under certain circumstances – change blindness and inattentional blindness.

The fallibility of our visual sense in detecting unexpected events has become the focus of particular interest in recent years. Observers can fail to notice what would seem otherwise to be a very large change to a complex scene provided that change is accompanied by a brief interruption to viewing (for a review, see Rensink, 2002). These changes can include changing the colour of an object, moving it to another position in the scene or removing it completely. Our surprising inability to detect seemingly obvious changes has become known as change blindness, and can be observed when changes are made during an eye movement (e.g., Grimes, 1996; McConkie & Currie, 1996), a blink (e.g., O'Regan, Deubel, Clark, & Rensink, 2000), or an artificial flicker of the image (Rensink, O'Regan, & Clark, 1995, 1997, 2000).

However, it is not only abrupt changes occurring during periods where the scene is occluded that can go unnoticed by observers. Unexpected events, lasting for several seconds, can occur in full view of an observer and yet not be detected. Professional pilots can fail to notice an aeroplane across the runway in a simulator and go on to land through it (Haines, 1991). Observers watching teams pass a basketball to each other can fail to notice a person in a gorilla suit (Simons & Chabris, 1999) or a woman carrying an umbrella (Neisser, 1967) walk through the midst of the players. Such failure to detect unexpected events that occur in full view has become known as inattentional blindness (see Mack & Rock, 1998), and is thought to arise because attention has been allocated to a particular task and subset of objects in the visual display.

An important concern in studies of change blindness and inattentional blindness has been to relate these phenomena to everyday vision and to consider their implications for how vision operates normally (see Most, Scholl, Clifford & Simons, 2005, for a similar position). Researchers have attempted to devise more realistic situations in which to explore these phenomena (e.g., Levin & Simons, 1997; Simons & Levin, 1998). However, even these situations remain somewhat removed from 'normal' experience. A more familiar situation in which these phenomena can be studied, and one that many of us have experienced, is magic. The misdirection employed by magicians in many of their tricks parallels inattentional blindness paradigms; it occurs seemingly in full view of the observer, yet is not noticed.

For the magician to be successful, the audience must be misdirected when the crucial part of the trick is performed. How does the magician do this? From an early age our attention is strongly influenced by other people. This tendency to attend to locations indicated by others is known as shared (or joint) attention (e.g., Tomasello, 1995, 1999). A particularly strong cue for shared attention is where somebody else is looking; we show a strong and somewhat automatic tendency to follow someone else's gaze (e.g., Driver et al., 1999; Langton, Watt, & Bruce, 2000; Scaife & Bruner, 1975; Triesch, Teuscher, Deák, & Carlson, 2006). Indeed, under certain conditions people automatically imitate another person's gaze (Ricciardelli, Bricolo, Aglioti, & Chelazzi, 2002). Perhaps this is how the magician misdirects us? If social cues strongly influence how we direct our attention, then doing something as simple as looking away from the act that the magician wishes to conceal, might be enough to make us attend to the wrong location and so miss an act that otherwise would have been obvious.

In this chapter we consider the details of how a magician (the second author, GK) achieves his misdirection when performing a simple magic trick. The way in which the trick was performed will now be described.

1. Our magic trick

We developed a trick in which the magician made a cigarette and lighter 'disappear'. The trick was performed 'live' by the magician, in front of the observer, on a one-to-one basis.

Figure 1 shows the typical progress of the trick. Facing the observer, the magician asks if they would mind if he smokes. He looks at and reaches for the cigarette packet and removes a cigarette. While moving the cigarette towards his mouth with one hand (the 'cigarette hand'), he turns to look at the lighter and reaches for this with his other hand (the 'lighter hand'). The cigarette is deliberately placed in the mouth the wrong way around. The magician then brings the lighter up to attempt to light the cigarette. There then follows the first misdirection by the magician. 'Noticing' that the cigarette is the wrong way around, the magician turns his head to the side, removes the cigarette and turns it around to replace it. Throughout this manoeuvre the magician keeps his gaze fixed on the cigarette and the hand manipulating it. While performing this manipulation of the

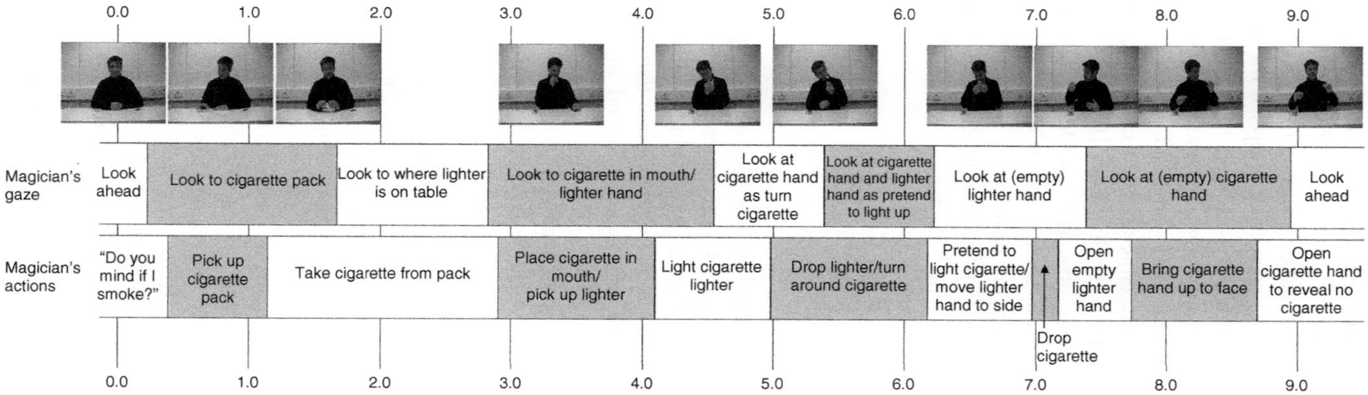

Figure 1. The trick used in the present study. Top row: Images illustrating various points of the trick. Middle row: Where the magician was looking. The start of the magician's gaze event corresponds to when the magician began his head turn towards that target. Bottom row: What the magician was doing. The numbered vertical lines indicate time (in seconds) from the start of the trick. Timings of the magician's actions and gaze are the average of all trials recorded in this study.

cigarette, the magician lowers his other hand, holding the lighter, towards the tabletop and drops the lighter onto his knees. The magician then brings his (now empty) hand back up to his face, while turning his head back towards the observer. He attempts to light the (now correctly positioned) cigarette and makes a show of being surprised that the lighter is no longer in his hand. This surprise is used to accomplish the second misdirection of this trick. Upon 'noticing' that the lighter is no longer in his hand, the magician turns his head to the side, raises his (now empty) lighter hand and opens it while looking at it. At the same time, the magician lowers his other hand, which is holding the cigarette, towards the tabletop. When it is near the tabletop, the magician drops the cigarette onto his knees. When dropped, the cigarette is usually about 10–15 cm above the tabletop and is therefore dropped in full view of the observer, visible for about 140 ms of its drop (calculated from the video records of the trick – see below). Shortly after he opens his empty lighter hand, the magician brings the now empty cigarette hand back up toward his face and turns to look at it, feigning surprise that the cigarette has also 'disappeared'. Finally, the magician turns back to face the observer (a video clip of this trick can be viewed online at http://www.dur.ac.uk/gustav.kuhn/.)

2. The present study

In a previous report, we considered whether the success of the magician's misdirection at the time of the cigarette drop in the above trick was dependent upon an overt misdirection of the eyes or a covert misdirection of attention (Kuhn & Tatler, 2005). In general, observers failed to spot how the trick was performed when they watched it for the first time, but all spotted how it was done when they watched the magician perform the trick for a second time. Detecting the cigarette drop was not dependent upon where the observer was fixating as it dropped; on trials where the drop was spotted, observers were still often fixating the lighter hand (as would be predicted for a successful misdirection). This result implied that it was the deployment of covert visual attention rather than overt gaze that determined whether the observer detected the cigarette drop.

One question raised by the Kuhn and Tatler study is how the magician achieves his misdirection. By considering the relationship between the magician's gaze and the observer's gaze we can evaluate the role of social cues (shared or joint attention) during the trick and whether these form the basis for the magician's success at misdirection. Each participant watched the magician perform the trick twice. This allowed us to consider whether different gaze strategies are employed to observe the trick for the second time. Performing the trick a second time also allows us to consider whether the observer can overcome the normal tendency for joint attention and not attend to the locations intended by the magician. We divided our participants into two groups: half were told before the experiment that they were about to watch a magic trick in which a cigarette and lighter would be made to disappear and that they should try to work out how this was done. The remaining observers did not know that they were about to watch a magic trick. Given that task can have such a profound effect upon where observers fixate when viewing static scenes (e.g., Buswell, 1935; Yarbus, 1967), we expected prior knowledge to have a significant impact upon how observers viewed the performances. We expected uninformed observers to be more prone to the misdirection and thus follow the magician's gaze more closely than informed observers. Our hypothesis for the combined effects of prior knowledge and repetition is that these two factors will interact such that informed observers are more able to ignore the magician's social cues for joint attention and that both groups are less prone to this misdirection on the second performance of the trick.

3. Procedure

Twenty participants (mean age $= 21.65$, $SD = 6.3$) took part in this experiment; half of whom knew in advance that they were about to watch the trick, half of whom did not. Each observer saw the trick performed twice.

Eye movements of the observers were recorded while they watched the magic trick, using Land's custom-built head-mounted eye-tracker (for details, see Land 1993; Land & Lee 1994). We defined seven possible gaze target regions: the magician's face, the cigarette hand, the cigarette packet, the cigarette, the lighter hand, other positions on the magician's body, and other locations in the scene (such as items on the table that were not related to the trick). Consecutive fixations within a single region were summed and treated as a single gaze duration. Data from two of the observers were discarded at this stage due to poor quality of the recorded videos. The direction of the magician's gaze was determined from the participant's eye-movement video record, by observing the magician's head movements throughout the performances of the trick. This measure is crude, but can be used to achieve a reasonably good idea of what the magician is looking at throughout the trick.

We were interested in how both prior knowledge and repetition of the trick influenced viewing strategies. Because we did not want to confound our data with any strategic differences that might arise in association with detecting the cigarette drop, we excluded the data for the two observers who spotted the cigarette drop on the first performance of the trick. The others all spotted it on the second performance. Eight of these remaining observers were previously informed that they would be watching the trick and eight were uninformed.

Given that the trick was performed live to each observer, it is important to ensure that there is a reasonable degree of consistency in the way in which this trick was performed. We divided the trick up into a series of easily identifiable actions carried out by the magician (depicted in Figure 1) and recorded the time at which these happened throughout each performance. We found no interaction between performance number (the two performances of the trick) and the prior knowledge of the observers on the timings of the actions performed by the magician throughout the trick, $F(7, 8) = 0.660$, $p = 0.701$. Similarly, we found no interaction between performance number and the prior knowledge of the observers on the times at which the magician moved his gaze to each of the targets depicted in Figure 1 throughout the trick, $F(7, 8) = 0.436$, $p = 0.866$. Thus, there were no systematic differences in the way that the trick was performed to informed or uninformed observers or on the first and second performances of the trick to each participant. Thus any differences in the strategic deployment of gaze by the observers that arise from prior knowledge or repetition of the trick will not be artefacts of the way in which the trick was performed.

4. A typical observer

Figure 2 shows the data for one of the uninformed observers watching the first performance of the trick. There was a reasonably close association between where the magician was looking and where the observer fixated. The observer in Figure 2 was successfully misdirected: she was watching the cigarette (as was the magician) while the lighter was dropped, and was watching the lighter hand while the cigarette was dropped. We will now consider how observers' gaze behaviour throughout the trick and around the time of these two instances of misdirection is influenced by prior knowledge that they will see the trick, or by seeing the trick for a second time.

5. Misdirection, prior knowledge, and repetition

One measure of the inspection strategies employed by observers is to consider the total amount of time spent gazing at each of the possible gaze targets while watching the trick. Gaze was categorised as having been directed to one of seven possible regions: the magician's face, the cigarette hand, the cigarette packet, the cigarette, the lighter hand, other positions on the magician's body, and other locations in the scene (such as items on the desk that were not related to the trick). A three-way mixed design ANOVA (gaze target region, informed/uninformed, first/second performance) showed a main effect of gaze target, $F(6, 9) = 49.81$, $p < 0.001$. Table 1 summarises the simple pairwise comparisons between the seven possible gaze target regions. No other main effects were significant (all $p > 0.462$). Observers spent most of their time looking at the magician's face and his two hands as might be expected given the nature of the trick.

There was a two-way interaction between gaze target region and the prior knowledge of the participant, $F(6, 9) = 5.74$, $p = 0.010$ (Figure 3a). Simple pairwise comparisons showed that informed participants spent longer looking at the magician's body (other than his hands and face), $p < 0.001$. Given that the magician's gaze is directed primarily at either of his hands or at the observer throughout the trial (see Figure 1), fixations of the magician's body might represent attempts to avoid being 'captured' by the magician's misdirection. Because the magician's body is effectively the centre of the visual scene throughout the performance, fixating here might also provide an optimal location for viewing the trick. However, there were no other differences between informed and uninformed observers (all $p > 0.096$), suggesting that any strategic differences between the two groups of observers were small.

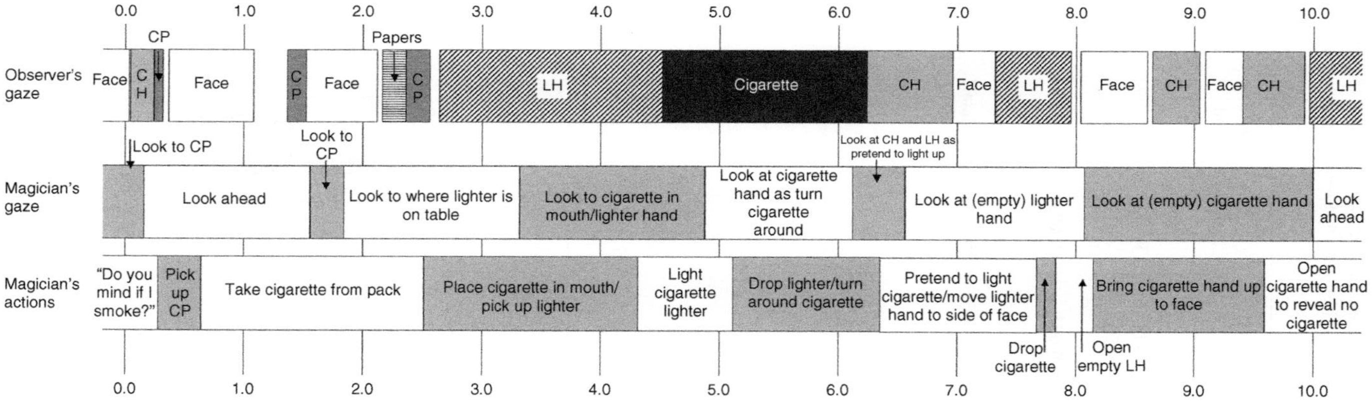

Key to abbreviations:
CH = cigarette hand CP = cigarette packet LH = lighter hand

Figure 2. A typical observer. Top row: Where the observer fixated. Middle row: Where the magician looked. Bottom row: What the magician was doing. The numbered vertical lines indicate time (in seconds) from the start of the trick. This figure demonstrates the similarity between where the magician looks and where the observer looks. For example, after just under two seconds, the magician turned his head towards the lighter on the table. Approximately one second later, as the magician reached for the lighter and picked it up, the observer directed her gaze to the hand that the magician was using to pick up the lighter and both the magician and observer watched this hand as it was brought towards the face.

Table 1
Results of simple pairwise comparisons for the total time spent gazing at each of the seven possible gaze targets when watching the trick. Boldface p-values indicate differences that were significant at the < 0.050 level. Mean total gaze time (along with the standard deviation) is given for each target region on the left of the table

	Face	Cigarette hand	Cigarette packet	Cigarette	Lighter hand	Body	Other
Face $\bar{X} = 2315$, SD = 240.1	–	.346	**<0.001**	**0.004**	0.985	**<0.001**	**<0.001**
Cigarette hand $\bar{X} = 3056$, SD = 154.7	–	–	**<0.001**	**<0.001**	**0.001**	**<0.001**	**<0.001**
Cigarette pack $\bar{X} = 380$, SD = 103.9	–	–	–	0.817	**0.001**	>0.999	>0.999
Cigarette $\bar{X} = 854$, SD = 149.1	–	–	–	–	0.115	0.659	**0.030**
Lighter hand $\bar{X} = 1575$, SD = 150.1	–	–	–	–	–	**0.001**	**<0.001**
Body $\bar{X} = 450$, SD = 72.7	–	–	–	–	–	–	0.183
Other $\bar{X} = 157$, SD = 63.7	–	–	–	–	–	–	–

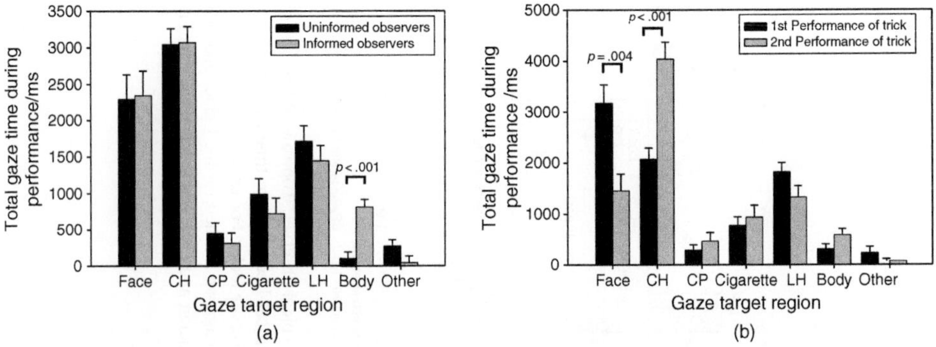

Figure 3. (a) Mean total gaze times in each target region (+1 SE) for the informed and uninformed observers. (b) Mean total gaze times in each target region (+1 SE) for the two performances of the trick.

The only other significant interaction was a two-way interaction between gaze target region and performance number, $F(6, 9) = 3.69$, $p = 0.039$ (Figure 3b). Simple pairwise comparisons showed that observers spent more time fixating the magician's face during the first performance of the trick than during the second performance, $p = 0.004$, and that they spent more time fixating the cigarette hand during the second performance of the trick than during the first performance, $p < 0.001$. No other pairwise comparisons were significant (all $p > 0.118$). Spending more time looking at the cigarette hand and less time looking at the magician's face on the second trial is an entirely plausible strategy for working out how the magician made the cigarette 'disappear'.

Our measure of the overall time spent gazing at a region does not account for how gazes to this region are distributed throughout the performance; for example, the 2.3 s for which observers watch the cigarette hand (Figure 3a) may be a single gaze at a certain point in the trick, or may be a series of short gazes to the cigarette hand distributed throughout the trick. We therefore examined more directly what was fixated at various points in the performance of the trick and whether this differed according to the prior knowledge of the observers or whether they were viewing the trick for the first or second time.

Figure 4 shows plots of the allocation of gaze to each of the seven possible gaze target regions throughout the course of the performance of the trick by all observers. While informed and uninformed observers were largely similar in the way they watched the trick, there were some differences. During the first performance, uninformed observers tended mainly to fixate the cigarette hand only during a period between 2.5 and 4.5 s into the trick and again at the end of the trick. Looks to the cigarette hand by informed observers were less clustered around these two time periods, with more looks to the cigarette hand throughout the middle portions of the trick. This result might suggest a strategic difference in the way that informed and uninformed observers watched the trick, with informed observers showing a tendency to monitor the cigarette over a more extended section of the trick than uninformed observers.

The more striking differences in inspection behaviour arose between the first and second performances of the trick. All observers seemed to adopt a strategy of largely watching the cigarette hand through much of the second performance of the trial. There were also fewer looks to the magician's face throughout the trial.

If we consider the lighter and cigarette drops, it can be seen that the magician's misdirection appears to be effective. Shortly before the lighter was dropped, the magician turned his head to watch the cigarette and cigarette hand as he turned the cigarette around (see also Figure 1). At this time during the first performance of the trick, informed observers tended to be looking at the cigarette hand as expected, but uninformed observers tended to be looking at the magician's face (Figure 4). Although the latter is not entirely as might be expected (or as the magician might hope), both represent an effective misdirection from the lighter hand as the lighter is dropped.

In the case of the cigarette drop all participants tended to follow the magician's gaze during the first performance of the trick and be looking at the lighter hand at this time. Because both groups of observers tended to watch the cigarette hand throughout most of the second performance of the trick, the magician's misdirection was less effective.

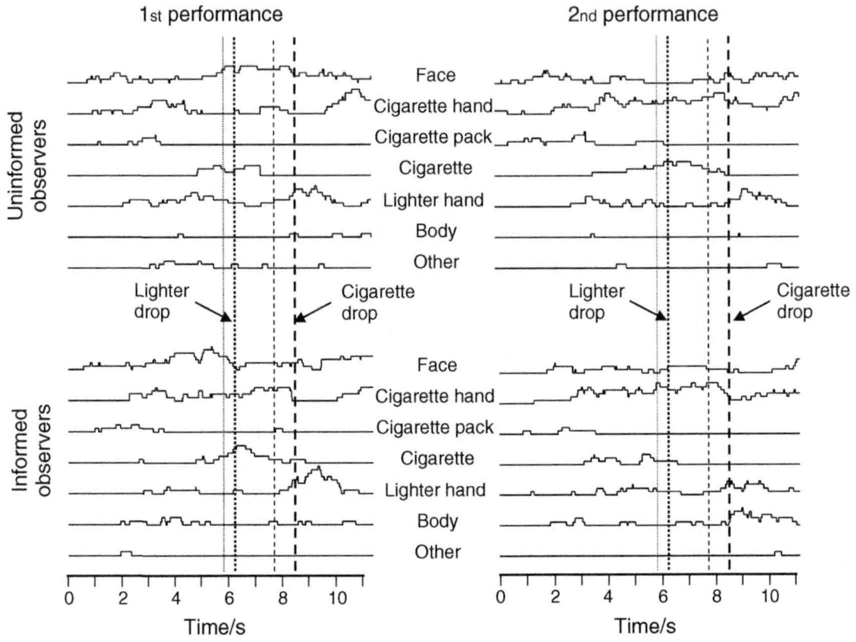

Figure 4. Plots of the total number of observers looking at each of the 7 possible gaze target regions over the course of the performance of the trick. In each plot, the baseline for each target region is 0 (no observers fixated that region at that time) and the maximum was 8 (all observers fixated that region at that time). Plots show these data for both uninformed (upper panels) and informed (lower panels) observers and for both the first (left panels) and second (right panels) performances of the trick. The timings of the lighter and cigarette drops are also indicated by the bold dashed lines. The faint dashed lines indicate when the magician turned his head towards the target of misdirection before the drop (i.e. to the cigarette hand just before the lighter drop and to the lighter hand just before the cigarette drop – see Figure 1).

However, it is interesting to note that when the cigarette was dropped, the observers did again show a tendency to be misdirected to the lighter hand. Here again arises a difference between the two groups of observers. When uninformed observers watched the trick a second time, most fixated the lighter hand around the time of the cigarette drop, whereas informed observers fixated either the lighter hand or the magician's body (in equal number). Fixating the body at this time might represent a 'disengagement' strategy; defeating the magician's misdirection by fixating a 'neutral' location in the scene from which the cigarette drop might be detected. Thus although both groups of observers successfully detected the cigarette drop on the second performance, their viewing strategies at the time of this event appear to be somewhat different.

One interesting observation in the above results is that detecting the cigarette did not appear to be dependent upon fixating on or very near to the dropping cigarette (as was also found in Kuhn & Tatler, 2005); observers detected the drop in the second trial even when not fixating the cigarette at the time it was dropped. Our results are therefore not entirely in accordance with the traditional view that attention and eye movements are necessarily intimately linked (e.g., Peterson, Kramer, & Irwin, 2004). Recent studies have argued that visual encoding of a scene is largely biased by what is selected for fixation (e.g., Hollingworth & Henderson, 2002; Tatler, Gilchrist, & Land, 2005). Hollingworth, Schrock, and Henderson (2001) found a large effect of fixation position in a change detection task, with most detections being accompanied by fixation of the changing object. From this result they suggested that fixation of the changing object was instrumental in detecting the change. Given this, our result that detection of the cigarette drop was not heavily dependent upon fixating on or near it is rather unexpected. Our data argue that, in the case of our magic trick, attention and eye position can be de-coupled, such that the eye may be misdirected but covert attention may not. This intriguing possibility clearly deserves closer examination in the future.

6. Do observers follow the magician's gaze?

Throughout this chapter we have alluded to the role of the magician's gaze in misdirecting the observer. As we mentioned in the introduction, social cues such as where a person is looking can strongly influence an observer's direction of attention; an observer will often look where someone else directs their gaze to (e.g., Ricciardelli et al., 2002). However, this is not always the case: a speaker's gaze at their own gestures does not necessarily result in an eye movement to that gesture by the observer (Gullberg, 2002).

Figure 5 shows observers watching the first performance of the trick. For the first section of the trick, there is a relatively loose coupling between where the magician looks and where the observers look. However, from about the time that the magician turns the cigarette around a much closer association develops. This close association between the magician's gaze and the observer's gaze is consistent with what is known from the social attention literature, where people show a strong tendency to follow another person's gaze (e.g., Driver et al., 1999; Langton et al., 2000; Scaife & Bruner, 1975; Triesch et al., 2006). In current work, our group has been exploring in more detail the issue of social attentional cuing in magic. By varying the availability of social cues for shared attention we have found that the effectiveness of a magic trick varies; the fewer social cues provided by the magician, the less effective the trick (Kuhn & Land, 2006). It is interesting that the close association between the gazes of the magician and the observer is not maintained throughout the trick but only develops for the second half of the performance, when the crucial acts of the trick (the disappearances) occur. What (if any) extra cues for directing the observer's attention are employed by the magician during this half of the trick that reinforce the shared attention are not entirely clear and warrant further investigation.

In order to explore the congruence between the magician's and observer's gaze in more detail, we divided the trick into 10 sections of equal duration and calculated the proportion of time for which the magician and observer were looking at the same gaze target. Because

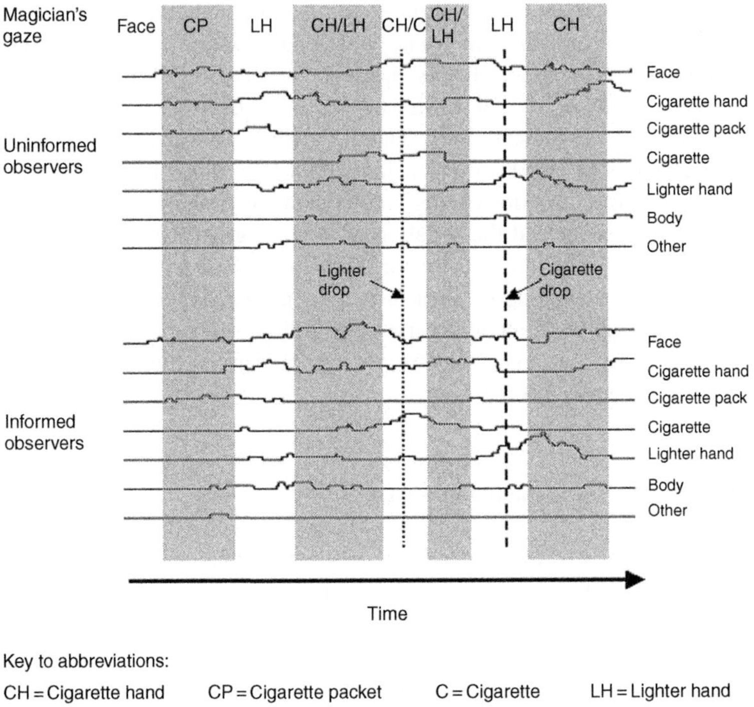

Figure 5. Where observers fixated during the first performance of the trick for uninformed (upper) and informed (lower) observers (plotted using the same conventions as in Figure 4 above). The superimposed white and grey bands indicate where the magician was looking throughout the trial and therefore where the observer might have been expected to fixate if following the magician's gaze. The dashed lines indicate when the lighter and cigarette drops occurred.

the observer's gaze will necessarily lag the magician's even if they are entirely following his gaze, we offset the observer's gaze by 500 ms. A three-way mixed design ANOVA (section in trick, informed/uninformed, first/second performance) showed a main effect of the section within the trial, $F(9, 126) = 21.05,\ p < 0.001$. There was no main effect of prior knowledge, $F(1, 14) = .29,\ p = 0.597$, underlining our previous suggestion that prior knowledge has relatively little influence on visual inspection behaviour when watching the trick. There was a significant two-way interaction between prior knowledge and the section of the trick, $F(9, 126) = 2.41,\ p = 0.015$. Simple pairwise comparisons showed two significant differences between informed and uninformed observers: congruence was higher for informed observers in the fourth section of the trick, $p = 0.009$ (just before the lighter drop), and for uninformed observers in the seventh section, $p = 0.037$ (just before the cigarette drop).

There was a significant three-way interaction between prior knowledge, first or second performance of the trick, and section of the trick, $F(9, 126) = 3.51,\ p = 0.001$ (Figure 6).

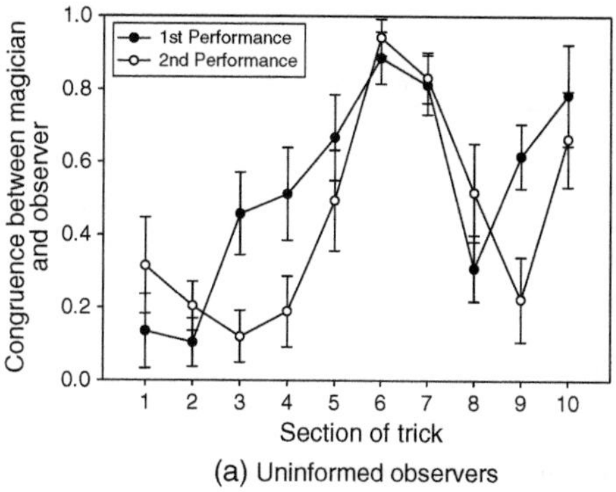

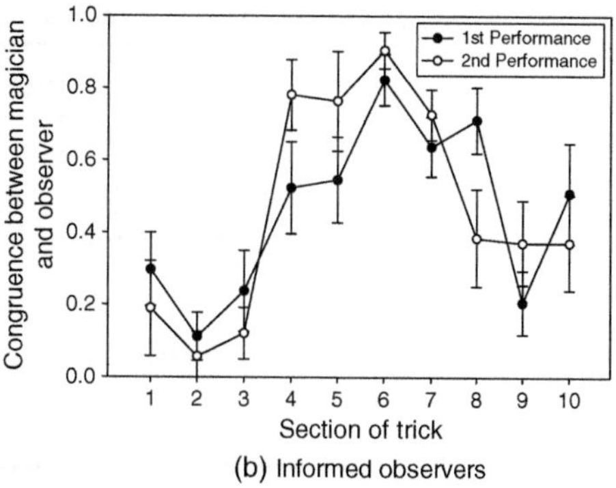

Figure 6. (a) Mean congruence between the magician's gaze and observer's gaze (± 1 SE) for uninformed observers when watching the two performances of the trick for each of ten equal duration sections of the trick. (b) Mean congruence between the magician's and observer's gaze (± 1 SE) for informed observers when watching the two performances of the trick for each of ten equal duration sections of the trick.

Simple pairwise comparisons showed one difference that approached significance between the two performances of the trick for informed observers: congruence was higher in the eighth section for the first performance, $p = 0.053$, perhaps suggesting greater disengagement from the magician's misdirection at this point (around the time of the cigarette drop) during the second performance. For uninformed observers, congruence was higher for the first performance of the trick during the third ($p = 0.033$; roughly when the cigarette is initially placed in mouth), fourth ($p = 0.088$; just before the lighter drop) and ninth ($p = 0.028$; just after the cigarette drop) sections of the trick, but higher for the second performance of the trick during the first section ($p = 0.043$; the initial engagement of the observer by the magician). Thus, uninformed observers appeared to be less likely to be led by the magician's gaze during the first half of the second performance than in the first half of the first performance. While we found differences between informed and uninformed participants and between the two performances of the trick, these were relatively few and for most sections of the trick there was little difference between the observer groups or performances.

The misdirection at the time of the cigarette drop is interesting because (unlike the lighter drop) the observers were generally looking at the cigarette hand until misdirected to the lighter hand. This event can therefore allow further insights into the dynamics of misdirection. Figure 7 shows that, in the first trial, the magician tended to look at the lighter hand about 740 ms before dropping the cigarette from his other hand. On average, observers followed the magician's gaze and looked at the lighter hand 400 ms after the magician. Therefore, they tended to move their eyes to the lighter hand on average 340 ms before the cigarette was dropped. By directing his gaze to the lighter hand some three quarters of a second before dropping the cigarette, the magician appears to allow sufficient time to ensure that the observer is likely to have shifted their gaze to the lighter hand before the cigarette is dropped.

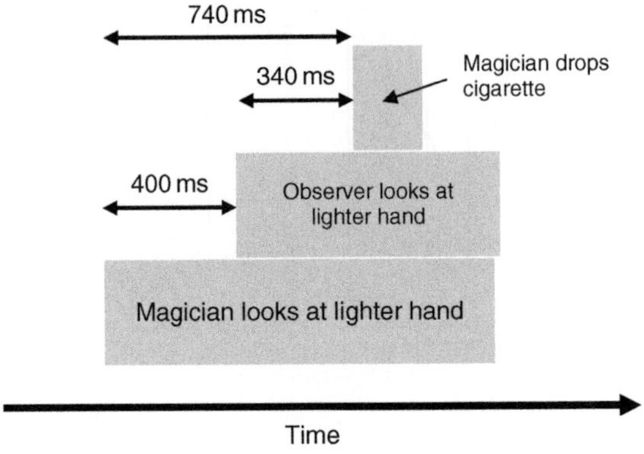

Figure 7. The average timings between when the magician turned his gaze towards the lighter hand to misdirect the observer, when the observer shifted their gaze to the lighter hand, and when the cigarette was dropped.

7. Conclusions

In this chapter we have seen that when watching a magician perform a magic trick live, our capacity and propensity to follow another person's gaze appears to be an important aspect of a magician's success in achieving misdirection.

As we suggested in our previous article (Kuhn & Tatler, 2005), detecting the manner of the 'disappearances' was not dependent upon fixating the object around the time it was dropped. Clearly, this not only argues that detection was achieved using covert visual attention, but also highlights the possibility that overt eye position and covert attention can be de-coupled by observers watching this trick. Our data also raise the interesting possibility that while the oculomotor system may still be misdirected by the magician even for informed observers or during the second performance, the attentional system seems somewhat less prone to this misdirection.

There were large differences in the viewing strategies of observers on the first and second trials; observers appeared to be adopting a strategy of watching the cigarette and cigarette hand throughout the second performance. Despite these differences and an apparent attempt to defeat the magician's misdirection, observers still seemed to be prone to the magician's misdirection, often tending to look to the now empty lighter hand when the cigarette was dropped.

It is surprising how little difference there was between the uninformed and informed observers. Given such different priorities and knowledge for the two groups of observers when viewing the trick, more profoundly different viewing strategies might have been expected. Instructions given to observers before viewing a static scene can have profound effects upon the subsequent viewing strategies (e.g., Buswell, 1935; Yarbus, 1967), yet in this case they appear to have little effect. Informed observers were prone to the magician's misdirection when they watch the trick for the first time. Our finding of only small differences between the strategies of uninformed and informed observers is consistent with our previous study that showed little difference in detection rates for spotting the cigarette drop in the same magic trick (Kuhn & Tatler, 2005). That even the informed observers followed the magician's gaze demonstrates the dominance of social cues for joint attention: even when trying to defeat the magician's misdirection these observers were unable to defeat their propensity for joint attention.

Acknowledgements

We thank Catherine Hughes for her invaluable help in recruiting participants for the experiment. Tom Sgouros kindly provided us with stories of unexpectedly effective stage magic, upon which our anecdote at the start of this chapter is based. We are very grateful to Boris Velichkovsky, Mike Land and Roger Van Gompel for their helpful comments on a previous version of this chapter.

References

Buswell, G. T. (1935). *How people look at pictures: A study of the psychology of perception in art.* Chicago: University of Chicago Press.

Driver, J., Davis, G., Kidd, P., Maxwell, E., Ricciardelli, P., & Baron-Cohen, S. (1999). Gaze perception triggers reflexive visuospatial orienting. *Visual Cognition, 6*, 509–541.

Grimes, J. (1996). On the failure to detect changes in scenes across saccades. In K. Atkins (Ed.), *Perception: Vancouver studies in cognitive science* (Vol. 2, pp. 89–110). New York: Oxford University Press.

Gullberg, M. (2002). Eye movements and gestures in human interaction. In J. Hyönä, R. Radach, & H. Deubel (Eds.), *The mind's eyes: Cognitive and applied aspects of eye movements* (pp. 685–703). Oxford: Elsevier.

Haines, R. F. (1991). A breakdown in simultaneous information processing. In G. Obrecht & L. W. Stark (Eds.), *Presbyopia research: From molecular biology to visual adaptation* (pp. 171–175). New York: Plenum Press.

Hollingworth, A., & Henderson, J. M. (2002). Accurate visual memory for previously attended objects in natural scenes. *Journal of Experimental Psychology-Human Perception and Performance, 28*(1), 113–136.

Hollingworth, A., Schrock, G., & Henderson, J. M. (2001). Change detection in the flicker paradigm: The role of fixation position within the scene. *Memory & Cognition, 29*(2), 296–304.

Kuhn, G., & Land, M. F. (2006). There is more to magic than meets the eye. *Current Biology, 16*(22), R950.

Kuhn, G., & Tatler, B. W. (2005). Magic and fixation: Now you don't see it, now you do. *Perception, 34*(9), 1155–1161.

Lamont, P., & Wiseman, R. (1999). *Magic in theory.* Hartfield: Hermetic Press.

Land, M. F. (1993). *Eye-head coordination during driving.* Paper presented at the IEEE Systems, Man and Cybernetics, Le Touquet.

Land, M. F., & Lee, D. N. (1994). Where we look when we steer. *Nature, 369*(6483), 742–744.

Langton, S. R. H., Watt, R. J., & Bruce, V. (2000). Do the eyes have it? Cues to the direction of social atention. *Trends in Cognitive Sciences, 4*(2), 50–59.

Levin, D. T., & Simons, D. J. (1997). Failure to detect changes to attended objects in motion pictures. *Psychonomic Bulletin & Review, 4*(4), 501–506.

Mack, A., & Rock, I. (1998). *Inattentional blindness.* Cambridge, MA: MIT Press.

McConkie, G. W., & Currie, C. B. (1996). Visual stability across saccades while viewing complex pictures. *Journal of Experimental Psychology-Human Perception and Performance, 22*(3), 563–581.

Most, S. B., Scholl, B. J., Clifford, E. R., & Simons, D. J. (2005). What you see is what you set: Sustained inattentional blindness and the capture of awareness. *Psychological Review, 112*(1), 217–242.

Neisser, U. (1967). *Cognitive psychology*. New York: Appleton Century Crofts.

O'Regan, J. K., Deubel, H., Clark, J. J., & Rensink, R. A. (2000). Picture changes during blinks: Looking without seeing and seeing without looking. *Visual Cognition, 7*(1–3), 191–211.

Peterson, M. S., Kramer, A. F., & Irwin, D. E. (2004). Covert shifts of attention precede involuntary eye movements. *Perception & Psychophysics, 66*, 398–405.

Rensink, R. A. (2002). Change detection. *Annual Review of Psychology, 53*, 245–277.

Rensink, R. A., O'Regan, J. K., & Clark, J. J. (1995). Image flicker is as good as saccades in making large scene changes invisible. *Perception, 24*(suppl.), 26–27.

Rensink, R. A., O'Regan, J. K., & Clark, J. J. (1997). To see or not to see: The need for attention to perceive changes in scenes. *Psychological Science, 8*(5), 368–373.

Rensink, R. A., O'Regan, J. K., & Clark, J. J. (2000). On the failure to detect changes in scenes across brief interruptions. *Visual Cognition, 7*(1–3), 127–145.

Ricciardelli, P., Bricolo, E., Aglioti, S. M., & Chelazzi, L. (2002). My eyes want to look where your eyes are looking: exploring the the tendency to imitate another individual's gaze. *Neuro Report, 13*(17), 2259–2264.

Scaife, M., & Bruner, J. S. (1975). The capacity for joint visual attention in the infant. *Nature, 253*, 265–266.

Simons, D. J., & Chabris, C. F. (1999). Gorillas in our midst: sustained inattentional blindness for dynamic events. *Perception, 28*(9), 1059–1074.

Simons, D. J., & Levin, D. T. (1998). Failure to detect changes to people during a real-world interaction. *Psychonomic Bulletin & Review, 5*(4), 644–649.

Tatler, B. W., Gilchrist, I. D., & Land, M. F. (2005). Visual memory for objects in natural scenes: From fixations to object files. *Quarterly Journal of Experimental Psychology Section A-Human Experimental Psychology, 58*(5), 931–S960.

Tomasello, M. (1995). Joint attention as social cognition. In C. Moore & P. J. Dunham (Eds.), *Joint attention: Its origins and role in development* (pp. 103–130). Hillsdale, NJ: Lawrence Erlbaum Associates.

Tomasello, M. (1999). *The cultural origins of human cognition*. Cambridge, MA: Harvard University Press.

Triesch, J., Teuscher, C., Deák, G., & Carlson, E. (2006). Gaze following: why (not) learn it? *Developmental Science, 9*(2), 125–147.

Yarbus, A. L. (1967). *Eye movements and vision*. New York: Plenum Press.

INDEX

Printed and bound by CPI Group (UK) Ltd, Croydon, CR0 4YY
08/05/2025
01865013-0001